From Abbeville to Yemassee

SCANning South Carolina's Natural History

SCAN: The Second Ten Years
January 1986 - December 1995

Published by the
South Carolina Association of Naturalists
Columbia, South Carolina
1999

Suggested cataloging data:

From Abbeville to Yemassee : SCANning South Carolina's natural history : SCAN, the second ten years, January 1986-December 1995. — Columbia, S.C. : South Carolina Association of Naturalists, 1999.

xiv, 378 p. : ill. ; 28 cm.

Contains field trip reports and species lists.
Bibliography: p. 250-253.
Includes index.

1. South Carolina Association of Naturalists.
2. Natural history — South Carolina.
3. South Carolina — Description.
I. South Carolina Association of Naturalists.

QH105.S6S58 1999
577.09575 — ddc21

ISBN 0-9673974-0-5

Published by:

South Carolina Association of Naturalists (SCAN)
Post Office Box 5651
Columbia, South Carolina 29250-5651

Printed and bound in the United States of America by:

Image Printing, Inc.
Dave Baker, owner
10 Commerce Alley
Camden, South Carolina 29020
Telephone: 803-424-0508

Book Committee:

Virginia E. Winn — Committee chairman, book editor, layout/design artist, digital imaging, photographer
Lynn B. Smith — Writer, photographic editor, photographer, proofreader
Dee Hope — Copy editor, proofreader
Janet C. Ciegler — Graphic artist, writer, proofreader
Kenneth A. Boni — Proofreader

The committee would like to thank the following for their assistance:

Robert Skinder — Technical assistance with indexing
Sara Lee Simons — Technical advice, photographer
Jennifer Moss — Common name index
Randall L. Miller — Scientific name index designer
Michele James — Common name index, ISBN, copyright information
Heidi Hoerman — Indexing coordinator, common name index, cataloging information
Lynn Greenlee — Photographer
Betty Gilbert (deceased) — Photographer
Randy Emmitt — Digital imaging, photographer
Jill Chappelle-Fail — Technical assistance with software
Alex Ciegler — Photographer
Fran Beals — Photographer

Front Cover Photographs:

Milkweed, *Asclepias* sp. (Photo by A. Ciegler); Jan Ciegler, Lynn Smith, Mary Garland Douglass, Gary Sowell, and Patty Shrum take to the water at Savage Bay, Kershaw County, Sandhills region (Photo by L. Greenlee); Wild onion, *Allium* sp. (Photo by A. Ciegler).

Back Cover Photographs:

Immature red-shouldered hawk, *Buteo lineatus* (Photo by R. Emmitt); Water lily, *Nymphaea odorata* (Photo by A. Ciegler); Pandora sphinx moth larva, *Eumorpha pandorus* (Photo by L. Greenlee).

Title Page Photograph:

Sphinx moth larva covered with braconid wasp cocoons, *Ipanteles* sp. (Photo by L. Greenlee).

Dedication Photograph:

Samuel Langford Outen, digitally retouched by Winn & Winn Photographic Restoration. (Photo by V. Winn).

This Book is Dedicated to

Samuel Langford Outen

March 6, 1927 - March 4, 1997
Our Beloved Friend and Benefactor

Acknowledgements

As chairman, I would first like to thank the other members of the book committee for all their assistance. They were Kenneth A. Boni, Janet C. Ciegler, Dee Hope, and Lynn B. Smith. As chairman, I was responsible for the administrative details plus doing the layout for the book. Without the other members, I would not have had a book to lay out.

Lynn and Jan coordinated the individual site write-ups and directions. Lynn did the site write-ups, while Jan wrote the directions and drew all the directional maps. On several occasions Jan or Lynn would actually return to the location to obtain correct current information. Now that's dedication!

Jan also supplied all the up-dated flora/fauna lists. I certainly appreciated her patience with all the needed changes.

Without Lynn I never could have sorted through the many slides that now enhance the book. As we progressed, she would "remember" we had a particular slide for the space under discussion and simply zip right to the perfect photo.

Lynn was also particularly helpful in writing up the "Science & Nature Notes". If they were not already available from one of our newsletters she would write-up whatever I needed.

Dee was the primary proofreader. He must have read the book at least four times. Having an English major on the committee certainly does help, especially when the chairman is known for her creative spelling and typing. Lynn and Jan also helped proofreading. Ken was responsible for the last major proofreading which was certainly a challenge.

During a committee meeting at about the halfway point, we decided to do a major update on the common names. Everyone jumped right in. Dee tackled the flora, Jan and Lynn the fauna, and Ken the fungi. The book was then re-layed out with the new information. I can only say thank you for what I know were some really late nights.

Thanks are also due Heidi Hoerman and her students Randall L. Miller, Michele M. James and Jennifer Moss. Heidi is an instructor in the College of Library and Information Science Department of the University of South Carolina and a SCAN member. As a project for her students in her indexing class, she agreed for them to take on the major challenge of indexing this book, and obtaining the ISBN, copyright, and Library of Congress cataloging information. She wanted her students to have a "real life experience". Believe me, we could not have agreed more. Jill Chappelle-Fail, assistant technology coordinator in the same department, helped the students with software requirements.

Randy designed the scientific name index and wrote the plotting programs. Robert Skinder, librarian at the University of South Carolina, assisted Randy in conceptualizing the design. Heidi, Michele and Jennifer handled the common name index, while Michele also took care of the ISBN and copyright information.

We would like to thank Image Printing and its owner, Dave Baker, in Camden, SC for helping us provide the book in a format that allowed us to include large numbers of photographs for little additional expense. Thanks for allowing us to 'go wild'.

Of course we couldn't go wild without photographs. Those who provided slides are SCAN members Alex Ciegler, Randy Emmitt, Lynn Greenlee, Sara Lee Simons, Lynn Smith, Betty Gilbert (now deceased, sadly) and myself. Choosing among those submitted was difficult. Lynn Smith and I chose on the basis of composition and exposure for reproduction, as many of us had similar photographs of the same subject, a hazard of all attending the same field trips. Occasionally, it was necessary to use a photograph from a different site or season to show the main subject effectively.

I am most grateful to Randy Emmitt for agreeing to digitize all the slides and photographs for inclusion in the book. Randy is a nature photographer living in North Carolina, offering scanning services as well as selling his photos via the web. His work is excellent and I encourage everyone to check out his website at http://www.naturevision.com and enjoy browsing. As a result of his scanning our slides, Randy is now a member of SCAN. Scanning our slides is a project I certainly appreciate and it made my job much easier.

There are two additional SCAN members who need to be thanked. Sara Lee Simons gave me much needed guidance when we first began this project. I had many questions about the first book that she graciously answered. Although the committee decided to use photographs instead of drawings, I would like to thank Anna K. Singley for submitting a cover design as well as allowing us to re-use the illustrations from the first book, had we so chosen. That support was greatly appreciated.

Lastly, I would like to acknowledge our benefactor, Sam Outen. Without his forethought we would not have had the necessary money to produce this book. Here's to you, Sam!

Virginia Winn,
Book Committee Chairman

20 Years of Leadership
1976 - 1995

(Photo by L. Greenlee)

Presidents of SCAN
Charles Lewis, Danny Smith, Jan Ciegler, Lynn Smith, Virginia Winn, Ted Cottingham, Alex Ciegler (center front), Rudy Mancke (our founder and longest running patriarch), Bill Carter, and Lynn Greenlee

The South Carolina Association of Naturalists, or SCAN as we are most often called, was formed in 1976 and is still a viable organization over 20 years later. We are composed of a wide and diverse membership with many varied interests, but most of all we are people who have an amateur or professional interest in the natural history of South Carolina.

SCAN publishes a monthly newsletter announcing the upcoming field trip with all pertinent information. Since one of our main objectives is to share factual natural history information, the newsletter contains a list of the plants (flora) and animals (fauna) we saw on the previous field trip. Other information of interest to naturalists is often included in these newsletters.

SCAN maintains a web site where the public can obtain field trip and membership information. Please visit us at http://members.aol.com/LynnBSmith/scan.

SCAN has at least one outdoor field trip per month between February and December. We always meet on the 4th Saturday of the month, except in November (3rd Saturday) and December (2nd Saturday). Members meet at the field trip site promptly at 10:30 am, unless otherwise stated in the newsletter, and then the fun begins.

SCAN members must be 18 years of age or older. However, because we feel that it is important to get children involved in environmental stewardship, SCAN tries to organize two special trips a year on which the members may bring their children.

SCAN's membership runs on a calendar year from January to December, with yearly dues payable at our only indoor meeting of the year, the annual business meeting, which is always held in January. If you would like to learn more about membership visit our web site or write us at:

South Carolina Association of Naturalists
Post Office Box 5651
Columbia, South Carolina 29250-5651
http://members.aol.com/LynnBSmith/scan

South Carolina Association of Naturalists

Post Office Box 5651, Columbia, South Carolina 29250-5651

Dear Reader,

Although a relatively small state, South Carolina has been blessed with an almost unbelievable variety of natural wealth. And, like so much wealth of a more materialistic nature, these riches were not always obvious to many of our state's inhabitants.

Therefore, over twenty years ago, the South Carolina Association of Naturalists (SCAN) was formed with the purpose of documenting these treasures and, in the process, providing a learning and enriching experience to its members and to the public. Membership was not restricted to those already knowledgeable in the various fields of natural science but was open to all who shared this interest, amateur and professional, neophyte and expert, alike. Their initial efforts were documented in a book covering the first ten years of the group's existence.

The gathering and compilation of this information provided SCAN members with an unexpected perspective on what many regard as the mysticism of life — beauty, mystery, the wonders of revealed simplicity and complexity — an ever shifting and changing kaleidoscope of nature's gifts. And there was the added bonus of fellowship that can only be shared by those pursuing like interests.

But the years continued to pass, another of Nature's inexplicables, and, like any organization, SCAN's membership waxed and waned. Yet progress continued on what many began to regard as the pursuit of Nature's 'Holy Grail'. The current book now documents the second ten years of SCAN's endeavors, yet its members know that, as from cornucopia, the wealth being poured out can never be fully absorbed, only appreciated in its vastness.

It is with considerable pleasure that I applaud the efforts of the small book committee that led to the production of this book. They should be proud of their contribution and of the contributions of the SCAN members who shared in this endeavor. And still the effort goes on, a joyful search and documentation of nature's mysteries. As at its conception, SCAN welcomes the participation of all citizens of South Carolina and, yes, interested citizens of its neighboring states, to its wanderings and wonderings.

Dee Hope
President, SCAN (1999)

Table of Contents

Sites in South Carolina by County

(Photo by V. Winn)

SCAN members never stay together. Woods Dargan, Darlington County, Sandhills region.

(Photo by A. Ciegler)

SCAN members enjoy a day on the Ashepoo River, Colleton County, Atlantic Coast Flatwoods region. Jan Ciegler (center) looks for beetles even when paddling.

Sites in Georgia by County

Sites in North Carolina by County

Alphabetically by Site

Science and Nature Notes

Membership Applications
are located at the end of the book.

Introduction

Welcome to SCAN's second book. We hope you will find the information and photographs we have included to be useful and enjoyable. If you are familiar with the first book, ***South Carolina Association of Naturalists: The First Ten Years***, you will find this book organized in a different manner to make it useful both to members and non-members. The counties, in alphabetical order, are used to organize the descriptions of the sites of SCAN field trips and corresponding flora and fauna lists from the second ten years. As with all the flora and fauna lists SCAN produces, these were created through the contributions of members who were willing to keep a list of what they saw on the trip that day. The expertise of these contributors is mostly "advanced amateur" but a lot of gracious help came from professionals. SCAN members include people with a wide range of interests and areas of expertise; all the expert observations were combined to make one list per trip. It is definitely a case of the sum being greater than the individual parts. While we did our best we admit there may be a few cases of mistaken identity on these pages.

This book may be useful to readers in several ways. SCAN members can use it for reminiscing about trips they remember with fondness. Other readers may well find, from the descriptions, sites that are nearby or of special interest and can use the directions and map to visit the site personally. If you are near a site and in a similar habitat, the flora and fauna list will give a good idea about what you might be seeing. Persons with a particular interest can select sites that have had similar species observed. Remember that the time of year and current weather conditions will strongly affect the actual species observed.

To aid in finding sites containing habitats in which the reader is interested, a map of South Carolina's physiographic regions is also included. These regions are based on communities of flora and fauna found there rather than exact geologic boundaries. The following descriptions start in the mountains and proceed to the sea.

The Blue Ridge is the area of highest altitude exhibiting montane plants and animals. Two South Carolina sites that exemplify these communities are Caesars Head State Park in Greenville County and the Coon Branch and Lower Whitewater Falls trails in Oconee County.

The largest region is the **Southern Piedmont** covering from the immediate foothills of the mountains to the fall line which is the geologic boundary of the coastal plain. This area features soils with a high amount of clay and is easily eroded. Most of this region has been farmed in the past, often with poor conservation practices, increasing the amount of erosion. Rare is the location with forests older than 60 to 70 years. Typical examples of this region are found at Croft State Park in Spartanburg County, Kings Mountain State Park in York County and the various recreation areas in Sumter National Forest that are not in Oconee County. Unique geologic features found scattered along the region are the granite monadnocks and outcrops; Kings Mountain in York County and Glassy Mountain in Pickens County are monadnocks. Forty Acre Rock, part of Flat Creek Heritage Preserve, is a granite outcrop with solution pools on the surface and a very unique set of flora.

The fall line, where the rocks underlying the Piedmont end, coincides with sand dunes that mark the remains of an Eocene-aged ocean edge, thus creating the **Sandhills.** It is rolling and very sandy. Plants in this region must adapt to the lower water availability caused by soils of very low water-holding capacity. Soil temperatures are often very high in summer. Carolina Sandhills National Wildlife Refuge and Cheraw State Park in Chesterfield County and Goodale State Park in Kershaw County provide classic examples of the region.

Just below the fall line and the Sandhills lies the **Southern Coastal Plain**. It is usually gently rolling with adequate moisture for abundant plant growth. It has the basic conditions for good agricultural lands. Poinsett State Park in Sumter County gives a good look at the variety available in this region.

Atlantic Coast Flatwoods encompasses the large low area that acts as the outer coastal plain. Much of the land is poorly drained providing many swamps and the special geologic feature known as the Carolina Bay. Francis Marion National Forest west of US 17 is mostly classic flatwoods. Cartwheel Bay and Lewis Ocean Bay heritage preserves in Horry County have both flatwoods and bay habitats. Woods Bay in Sumter County is a well-preserved deep water Carolina Bay.

The junction of the mainland and the Atlantic Ocean gives the physiographic region known as **Tidewater.** The water level varies daily from the rise and fall of the tides but is also affected by the amount of rainfall in the higher areas of the state. This area is generally located east of US 17 throughout the length of the state. South Carolina is blessed with several state parks that give easy access to beaches, marsh, and maritime forests. Hunting Island in Beaufort County, Edisto Beach in Colleton County, and Huntington Beach in Georgetown County are all very rewarding state parks to visit.

SCAN's philosophy is one of learning by seeing and experiencing. Reading about these sites will get you started, but you will have the most fun and gain the most knowledge by using the information in the book or SCAN's first book to prepare yourself for a pleasant day "in the field". The generosity and foresight of Sara Lee Simons to enable the publishing of the first ten years of data and of Sam Outen to allow the continuation of the data publication is generosity of the highest order.

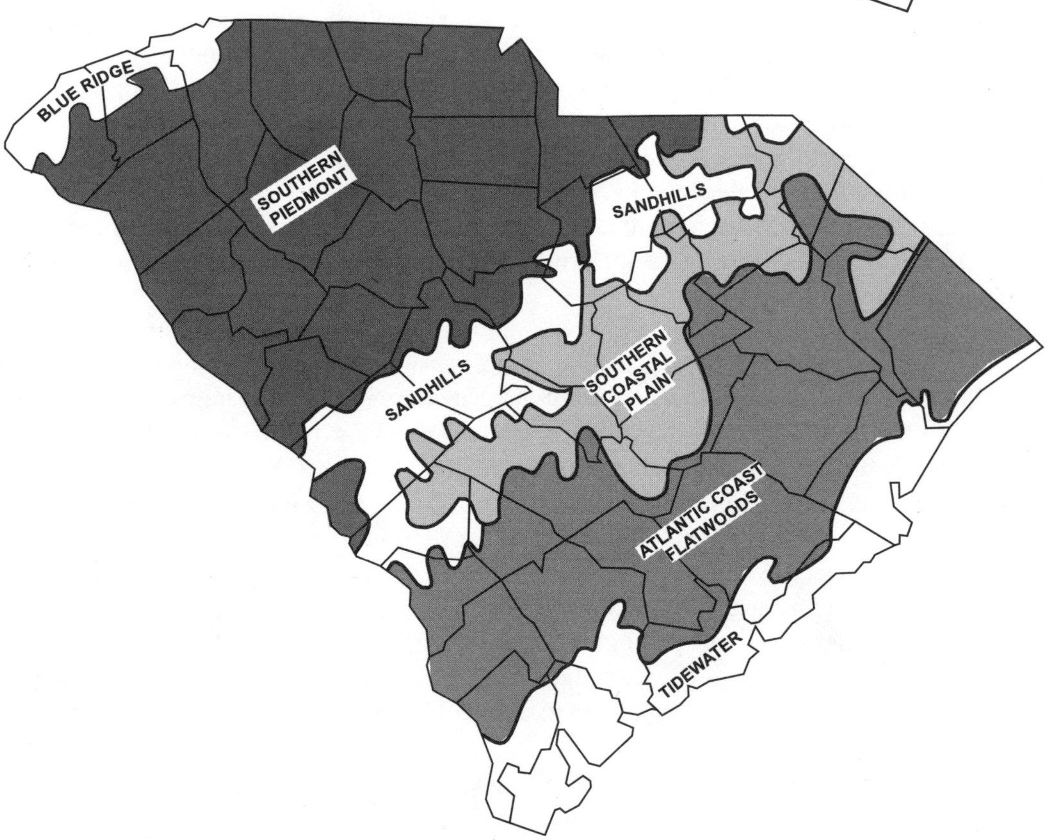

Abbeville County

Parson's Mountain Lake Recreation Area

(Sumter National Forest)
Abbeville County
July 23, 1994

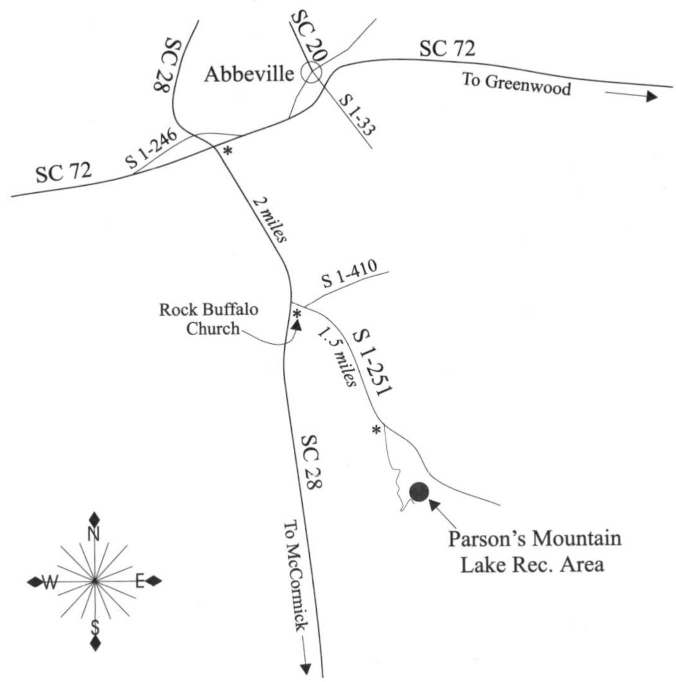

Parson's Mountain is a low monadnock in the western piedmont. It is an area of usually hot humid July weather, but this day it was cloudy and still in the low 80's in the early afternoon. This may account for the listings of a diverse nature. One plant highlight was the American chestnut on one side of the trail immediately opposite a chinquapin. Another American chestnut was loaded with nuts. Mushrooms also thrived in the warm humid weather that preceded the trip.

One of the most exciting aspects of the day was the interesting variety of reptiles and amphibians encountered. While examining a mysterious egg that was deposited on a log, the smallest ringnecked snake observed by anyone in the group was discovered. Under the bark of another log, a female five-lined skink was guarding her eggs. Picking the eggs up in her mouth, she slowly rearranged them in a less exposed position. Female skinks often protect their eggs during incubation. During the course of the day, newly-hatched ground skinks and fence lizards were spotted.

Don't miss the view from the top of the forestry tower if it is open when you visit.

Directions to Parson's Mountain Lake Recreation Area: Southwest of Abbeville, drive south on SC 28 about 2 miles beyond the intersection with SC 72 to Rock Buffalo Church. Turn left at the church onto S 1-251 and go southeast 1.5 miles; turn right at the sign to the Parson's Mountain Lake Recreation Area and drive to the parking lot at the end of the road.

FLORA

Lichen
Pertusaria texana
Pertusaria multipunctoides
Lecanora varia
Phlyctidia ludoviciensis
Phlyctis willeyi
Bulbothrix goebelii
Hypotrachyna livida
Parmelina aurulenta
Parmelina galbina
Parmelina horrescens
Parmelina minarum
Parmotrema hypotropum
Parmotrema michauxianum
Parmotrema perforatum
Xanthoparmelia tasmanica
Flavoparmelia baltimorensis
Flavoparmelia caperata
Canoparmelia caroliniana
Punctelia rudecta
Rimelia reticulata
Usnea strigosa
Anaptychia palmatula
Buellia stellulata
Buellia punctata
Buellia curatellae
Physcia americana

Bracken
Pteridium aquilinum
Southern Lady Fern
Athyrium asplenioides
Christmas Fern
Polystichum acrostichoides
Broad Beech-Fern
Thelypteris hexagonoptera
Ebony Spleenwort
Asplenium platyneuron
Resurrection Fern
Polypodium polypodioides
Loblolly Pine
Pinus taeda
Short-leaf Pine
Pinus echinata
Eastern Red-cedar
Juniperus virginiana
Cane; Switch Cane
Arundinaria gigantea
Spike Grass
Uniola sessiliflora
River Oats
Uniola latifolia
Wild Rye Grass
Elymus virginicus
Johnson Grass
Sorghum halepense
Nut-rush
Scleria oligantha
Fringed Greenbrier
Smilax bona-nox
Sawbrier; Greenbrier
Smilax glauca
Fly-poison
Amianthium muscaetoxicum
Yellow Star-grass
Hypoxis hirsuta
Downy Rattlesnake-plantain
Goodyera pubescens
Crane-fly Orchid
Tipularia discolor
Shagbark Hickory
Carya ovata

Parson's Mountain Lake Recreation Area (continued)
July 23, 1994

Pale Hickory
Carya pallida
Mockernut Hickory
Carya tomentosa
Pignut Hickory
Carya glabra
Tag Alder
Alnus serrulata
Hop-hornbeam
Ostrya virginiana
American Hornbeam; Ironwood
Carpinus caroliniana
Beech
Fagus grandifolia
American Chestnut
Castanea dentata
Allegheny Chinquapin
Castanea pumila
White Oak
Quercus alba
Post Oak
Quercus stellata
Red Oak
Quercus rubra
Black Oak
Quercus velutina
Scarlet Oak
Quercus coccinea
Southern Red Oak; Spanish Oak
Quercus falcata falcata
Black Jack Oak
Quercus marilandica
Willow Oak
Quercus phellos
Winged Elm
Ulmus alata
Red Mulberry
Morus rubra
Virginia Snakeroot
Aristolochia serpentaria
Heartleaf
Hexastylis arifolia
Black Cohosh
Cimicifuga racemosa
Liverleaf
Hepatica americana
Tulip-poplar
Liriodendron tulipifera
Umbrella-tree
Magnolia tripetala
Cucumber-tree
Magnolia acuminata
Dwarf PawPaw
Asimina parviflora
Sweet-shrub; Carolina-allspice
Calycanthus floridus
Sassafras
Sassafras albidum
Bloodroot
Sanguinaria canadensis
Alumroot
Heuchera americana
Sweet-gum
Liquidambar styraciflua
Witch-hazel
Hamamelis virginiana
Blackberry; Brambles
Rubus sp.
Cocklebur
Agrimonia pubescens microcarpa
Hawthorn
Crataegus flabellata
Black Cherry; Wild Cherry
Prunus serotina
Redbud; Judas-tree
Cercis canadensis
Wild Sensitive-plant
Cassia nicitans
Wild Indigo
Baptisia sp.
Leadplant
Amorpha herbacea
Pencil-flower
Stylosanthes biflora
Naked-flower Tick-trefoil
Desmodium nudiflorum
Dollarleaf
Desmodium rotundifolium
Bristly Locust
Robinia hispida
Tephrosia
Tephrosia spicata
Climbing Butterfly-pea
Centrosema virginianum
Flowering Spurge
Euphorbia corollata
Prostrate Spurge
Euphorbia supina
Poison-ivy
Rhus radicans
Poison-oak
Rhus toxicodendron
Winged Sumac
Rhus copallina
Smooth Sumac
Rhus glabra
American Holly
Ilex opaca
Hearts-a'burstin'-with-love
Euonymus americanus
Red Maple
Acer rubrum
Southern Sugar Maple
Acer saccharum floridanum
New-Jersey-tea
Ceanothus americanus
Virginia-creeper
Parthenocissus quinquefolia
Muscadine; Bullace
Vitis rotundifolia
Basswood; Linden
Tilia americana
St.-Andrew's-cross
Hypericum hypericoides
St.-John's-wort
Hypericum mutilum
Bird-foot Violet
Viola pedata
Primrose-leaved Violet
Viola primulifolia
Passionflower; Maypops
Passiflora incarnata
Russian-olive
Elaeagnus umbellata
English Ivy
Hedera helix
Black Snakeroot
Sanicula marilandica
Snakeroot; Sanicle
Sanicula smallii
Queen Anne's-lace; Wild Carrot
Daucus carota
American Carrot
Daucus pusillus
Meadow Parsnip
Thaspium barbinode
Black Tupelo; Blackgum
Nyssa sylvatica sylvatica
Flowering Dogwood
Cornus florida
Spotted Wintergreen
Chimaphila maculata
Pine-sap
Monotropa hypopithys
Azalea
Rhododendron sp.
Sourwood
Oxydendrum arboreum
Trailing-arbutus
Epigaea repens
Sparkleberry
Vaccinium arboreum
Deerberry; Squaw-huckleberry
Vaccinium stamineum
Low-bush Blueberry
Vaccinium vacillans
Whorled Loosestrife
Lysimachia quadrifolia
Persimmon
Diospyros virginiana
Carolina Silverbell
Halesia carolina
Bigleaf Snowbell
Styrax grandifolia
Yellow Jessamine
Gelsemium sempervirens
Polypremum
Polypremum procumbens
Indian-hemp
Apocynum cannabinum
Milkweed
Asclepias sp.
Angle-pod
Matelea sp.
Phlox
Phlox carolina
Skullcap
Scutellaria elliptica
Carpenterweed
Prunella vulgaris
False Foxglove
Aureolaria pectinata
Downy False Foxglove
Aureolaria virginica
Trumpet Vine
Campsis radicans
Beech-drops
Epifagus virginiana
English Plantain
Plantago lanceolata
Button-bush
Cephalanthus occidentalis
Poor-jo (or Po-jo)
Diodia teres
Partridge-berry
Mitchella repens
Bedstraw
Galium circaezans
Black-haw; Nanny-berry
Viburnum prunifolium
Prenanthes
Prenanthes sp.
Rattlesnake-weed
Hieracium venosum
Elephant's-foot
Elephantopus tomentosus
Eupatorium
Eupatorium sp.
Pussy-toes
Antennaria plantaginifolia
Goldenrod
Solidago sp.
Silphium
Silphium dentatum
Black-eyed-Susan
Rudbeckia hirta
Coreopsis
Coreopsis major
Witches'-butter
Tremella mesenterica
Cinnabar-red Chanterelle
Cantharellus cinnabarinus
Fiber-vase
Thelephora sp.
Varnish Shelf
Ganoderma curtisii
Turkey-tail
Trametes versicolor
Violet-toothed Polypore
Trichaptum biformis
Common Scaber-stalk
Leccinum scabrum
Old-man-of-the-woods
Strobilomyces floccopus
Salmon Waxy-cap
Hygrophorus pratensis
Peppery Milky
Lactarius piperatus
Voluminous-latex Milky
Lactarius volemus
Emetic Russula
Russula emetica
Fragile Russula
Russula fragilis
Variable Russula
Russula variata
Leaflike Oyster
Hohenbuehelia petaloides
Rounded Earthstar
Geastrum saccatum

Elegant Stinkhorn
Mutinus elegans
Scrambled-egg Slime
Fuligo septica

FAUNA

Juxtaposed Forest Snail
Triodopsis juxtidens
Great Common Zonite (snail)
Mesomphix vulgatus
Almond Millipede
Sigmoria aberrans
White Micrathena (spider)
Micrathena mitrata
Orchard Spider
Leucauge venusta
Writing Spider
Argiope aurantia
Orb-weaver
Neoscona sp.
Arrowhead Spider
Verrucosa arenata
Scorpion
Centruroides vittatus
Short-winged Green Grasshopper
Dichromorpha viridis
Boll's Sand Grasshopper
Spharagemon bolli
Sand Grasshopper
Spharagemon cristatum
Spur-throated Grasshopper
Melanoplus sp.
Round-headed Katydid
Amblycorypha floridana
Fork-tailed Bush Katydid
Scudderia furcata
Shield-backed Katydid
Atlanticus gibbosus
Field Cricket
Gryllus sp.
Witch Hazel Cone Gall (aphid)
Hormaphis hamamelidis
Huckleberry Oak Gall (wasp)
Callirhytis lustrans
Thread-waisted Wasp
Ammophila sp.
Oak Blister Gall (midge)
Cincticornia pustulata
Fruit Fly, Pomace Fly
Drosophila melanogaster
House Fly
Musca domestica
Eastern Pondhawk
Erythemis simplicicollis
Bar-winged Skimmer
Libellula axilena
Ebony Jewelwing
Calopteryx maculata
Blue-fronted Dancer Damselfly
Argia apicalis
Tiger Beetle
Cicindela unipunctata
Red-bellied Tiger Beetle
Cicindela rufiventris
Pedunculate Ground Beetle
Pasimachus depressus
Woodland Ground Beetle
Evarthrus sp.
Notch-mouthed Ground Beetle
Dicaelus dilitatus
Whirligig Beetle
Dineutus emarginatus
Obscure Rove Beetle
Gyrophaena corruscula
Gyrophaena dissimilis
Atheta pennsylvanica
Atheta flaviola
Aleochara sp.
Passalus; Bess-beetle
Popilius disjunctus
Hister-like Dung Beetle
Ateuchus histeroides
Japanese Beetle
Popillia japonica
Firefly; Lightning Beetle
Photinus pyralis
Tumbling Flower Beetle
Mordella melaena
Sap-feeding Beetle
Pallodes sp.
Pleasing Fungus Beetle
Megalodacne heros
Triplax macra
Shining Leaf Beetle
Lema sayi
Flower Weevil
Aulobaris ibis
Pipevine Swallowtail
Battus philenor
Zebra Swallowtail
Eurytides marcellus
Eastern Tiger Swallowtail
Papilio glaucus
Red Admiral
Vanessa atalanta rubria
Red-spotted Purple (butterfly)
Limenitis arthemis astyanax
Carolina Satyr (butterfly)
Hermeuptychia sosybius
Silver-spotted Skipper
Epargyreus clarus
Bream
Lepomis sp.
Fowler's Toad
Bufo woodhousei
Green Frog; Bronze Frog
Rana clamitans
Yellow-bellied Slider
Chrysemys scripta
Carolina Anole
Anolis carolinensis
Eastern Fence Lizard
Sceloporus undulatus
Five-lined or SE 5-lined Skink
Eumeces sp.
Ground Skink
Scincella lateralis
Black Racer
Coluber constrictor
Ringneck Snake
Diadophis punctatus
Great Egret
Casmerodius albus
Great Blue Heron
Ardea herodias
Turkey Vulture
Cathartes aura
Chimney Swift
Chaetura pelagica
Red-bellied Woodpecker
Melanerpes carolinus
Pileated Woodpecker
Dryocopus pileatus
Acadian Flycatcher
Empidonax virescens
Crow
Corvus sp.
Carolina Chickadee
Parus carolinensis
Carolina Wren
Thryothorus ludovicianus
Red-Eyed Vireo
Vireo olivaceus
Northern Cardinal
Cardinalis cardinalis
Eastern Meadowlark
Sturnella magna
Summer Tanager
Piranga rubra
Beaver (sign)
Castor canadensis
White-tailed Deer
Odocoileus virginianus

Aiken County

Aiken State Park
Aiken County
June 27, 1992

Hot, oppressive June rarely brings out the best in a site. Even so, the varied terrain of Aiken State Park was interesting. The park encompasses a sand ridge, man-made ponds, a floodplain, a bog and the bank of the Edisto River. The nature trail provided a sampling of each. Some of the less common specimens seen were hammocksweet azalea, loblolly-bay, and the comet darner dragonfly.

Directions to Aiken State Park: From Columbia, follow SC 302 through South Congaree and Pelion for 35 miles to Wagener; continue on SC 302 for 9 miles more to Kitchings Mill. Turn left onto S 2-53 and go 2 miles further, crossing the South Fork of the Edisto River. Park entrance is on the right.

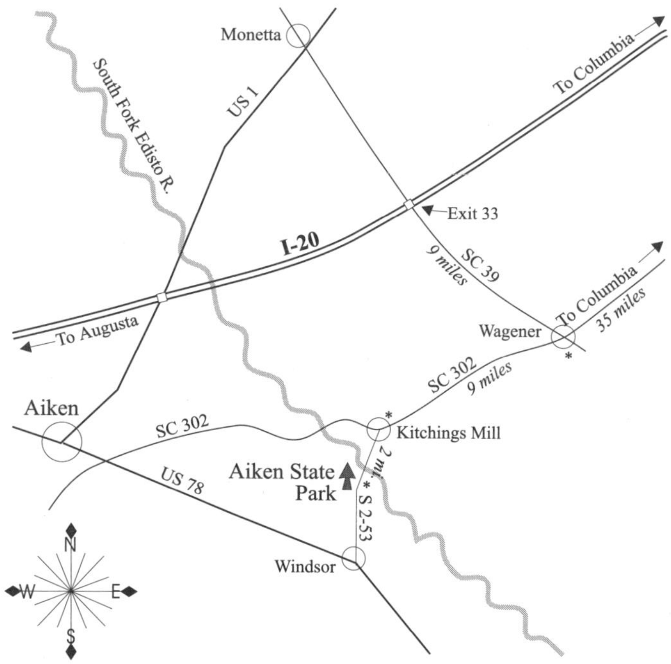

Aiken State Park (continued)
June 27, 1992

FLORA

Lichen
Leptogium cyanescens
Graphis scripta
Graphis mosquitensis
Graphis lineola
Graphis poitaeoides
Graphis sp.
Bacidia schweinitzii
Leptogium austroamericanum
Lecidea russula
Cladina subtenuis subtenuis
Cladina subtenuis cinerascens
Cladonia chlorophaea f. intermedia
Cladonia chlorophaea f. pterygota
Cladonia coniocraea f. phyllostrota
Cladonia coniocraea f. subpellucida
Cladonia cristatella vestita
Cladonia incrassata f. incrassata
Cladonia incrassata f. squamulosa
Cladonia leporina
Cladonia rei
Pertusaria ostiolata
Pertusaria paratuberculifera
Pertusaria trachythallina
Pertusaria subpertusa
Pertusaria tetrathalmia
Pertusaria sinusmexicani
Haematomma puniceum
Lecanora cupressi
Lecanora pallida
Lecanora varia
Ochrolechia africana
Phlyctis sp.
Bulbothrix confoederata
Bulbothrix goebelii
Bulbothrix laevigatula
Hypotrachyna livida
Hypotrachyna pustulifera
Parmelina minarum
Parmeliopsis ambigua
Parmotrema dilitatum
Parmotrema michauxianum
Parmotrema perforatum
Parmotrema rampoddense
Canoparmelia caroliniana
Punctelia rudecta
Ramalina americana
Usnea ceratina
Usnea strigosa
Buellia curatellae
Buellia disciformis
Buellia sp.
Physcia aipolia
Heterodermia obscurata
Pyxine caesiopruinosa
Pyxine caesiopruinosa f. aurea
Normandina pulchella

Cinnamon Fern
Osmunda cinnamomea
Royal Fern
Osmunda regalis spectabilis
Bracken
Pteridium aquilinum
Netted Chain Fern
Woodwardia areolata
Resurrection Fern
Polypodium polypodioides
Long-leaf Pine
Pinus palustris
Loblolly Pine
Pinus taeda
Pond Pine
Pinus serotina
Bald-cypress
Taxodium distichum
Eastern Red-cedar
Juniperus virginiana
Bur-reed
Sparganium americanum
Cane; Switch Cane
Arundinaria gigantea
Dulichium (sedge)
Dulichium arundinaceum
Golden-club
Orontium aquaticum
Arrowleaf Arum
Peltandra virginica
Bogmoss
Mayaca sp.
Spanish-moss
Tillandsia usneoides
Greenbrier; Bullbrier
Smilax rotundifolia
Sawbrier; Greenbrier
Smilax glauca
Bamboo Smilax
Smilax laurifolia
Spoonleaf Yucca
Yucca filamentosa filamentosa
Wax-myrtle; Southern Bayberry
Myrica cerifera
Swamp Bayberry
Myrica heterophylla
Mockernut Hickory
Carya tomentosa
Tag Alder
Alnus serrulata
Post Oak
Quercus stellata
Scrubby Post Oak
Quercus margaretta
Live Oak
Quercus virginiana
Turkey Oak
Quercus laevis
Spanish Oak
Quercus falcata
Water Oak
Quercus nigra
Willow Oak
Quercus phellos
Winged Elm
Ulmus alata
False Nettle
Boehmeria cylindrica
Wire-plant
Stipulicida setacea
Tulip-poplar
Liriodendron tulipifera
Sweetbay
Magnolia virginiana
Redbay
Persea borbonia
Sassafras
Sassafras albidum
Virginia-willow
Itea virginica
Sweet-gum
Liquidambar styraciflua
Witch-alder
Fothergilla gardenii
Blackberry; Brambles
Rubus sp.
Red Chokeberry
Sorbus arbutifolia arbutifolia
Hawthorn
Crataegus sp.
Black Cherry; Wild Cherry
Prunus serotina
Carolina Cherrylaurel
Prunus caroliniana
Sensitive-brier
Schrankia microphylla
Gopherweed
Baptisia perfoliata
Rattlebox
Crotalaria angulata
American Wisteria
Wisteria frutescens
Chinese Wisteria
Wisteria sinensis
Tephrosia
Tephrosia spicata
Butterfly-pea
Clitoria mariana
Tread-softly
Cnidoscolus stimulosus
Poison Sumac
Rhus vernix
Poison-ivy
Rhus radicans
Poison-oak
Rhus toxicodendron
Winged Sumac
Rhus copallina
Titi
Cyrilla racemiflora
American Holly
Ilex opaca
Inkberry
Ilex glabra
Sweet Gallberry
Ilex coriacea
Red Maple
Acer rubrum
Muscadine; Bullace
Vitis rotundifolia
Loblolly-bay
Gordonia lasianthus
St.-Andrew's-cross
Hypericum hypericoides
Black Tupelo; Blackgum
Nyssa sylvatica sylvatica
Swamp Tupelo; Swamp Blackgum
Nyssa sylvatica biflora
Flowering Dogwood
Cornus florida
Sweet Pepperbush
Clethra alnifolia tomentosa
Spotted Wintergreen
Chimaphila maculata

Hammocksweet Azalea
Rhododendron viscosum serrulatum
Mountain-laurel
Kalmia latifolia
Fetterbush
Lyonia lucida
Dog-hobble
Leucothoe axillaris
Coastal Honeybells
Leucothoe racemosa
Sourwood
Oxydendrum arboreum
Huckleberry
Gaylussacia sp.
Sparkleberry
Vaccinium arboreum
Squaw-huckleberry
Vaccinium stamineum stamineum
Highbush Blueberry
Vaccinium corymbosum
Persimmon
Diospyros virginiana
Horse-sugar; Sweetleaf
Symplocos tinctoria
Water Ash; Carolina Ash
Fraxinus caroliniana
Yellow Jessamine
Gelsemium sempervirens
Small Floating-heart
Nymphoides cordata
Butterfly-weed
Asclepias tuberosa
Sandhill Milkweed
Asclepias humistrata
White Morning-glory
Bonamia patens
Cross Vine
Anisostichus capreolata
Catalpa
Catalpa sp.
Button-bush
Cephalanthus occidentalis
Japanese Honeysuckle
Lonicera japonica
Possumhaw Viburnum
Viburnum nudum
American Elderberry
Sambucus canadensis

FAUNA

Walkingstick
Diapheromera femorata
Common Eastern Velvet-ant
Dasymutilla occidentalis occidentalis
Comet Darner
Anax longipes
Eastern Pondhawk
Erythemis simplicicollis
Great Blue Skimmer
Libellula vibrans
Eastern Amberwing (dragonfly)
Perithemis tenera
Violet-masked Glider
Tramea carolina
Ebony Jewelwing
Calopteryx maculata
Variable Dancer (damselfly)
Argia fumipennis
Pearl Crescent
Phyciodes tharos
Common Buckeye (butterfly)
Junonia coenia
Southern Cricket Frog
Acris gryllus
Pine Woods Treefrog
Hyla femoralis
Green Frog; Bronze Frog
Rana clamitans
Eastern Mud Turtle (sign)
Kinosternon subrubrum
Carolina Anole
Anolis carolinensis
Northern Bobwhite
Colinus virginianus
Yellow-billed Cuckoo
Coccyzus americanus
Red-bellied Woodpecker
Melanerpes carolinus
Red-headed Woodpecker
Melanerpes erythrocephalus
Great Crested Flycatcher
Myiarchus crinitus
Eastern Wood-pewee
Contopus virens
Acadian Flycatcher
Empidonax virescens
American Crow
Corvus brachyrhynchos
Tufted Titmouse
Parus bicolor
Wood Thrush
Hylocichla mustelina
White-eyed Vireo
Vireo griseus
Prothonotary Warbler
Protonotaria citrea
Black-and-white Warbler
Mniotilta varia
Pine Warbler
Dendroica pinus
Hooded Warbler
Wilsonia citrina
Yellowthroat
Geothlypis trichas
Northern Cardinal
Cardinalis cardinalis
Rufous-sided Towhee
Pipilo erythrophthalmus
White-tailed Deer
Odocoileus virginianus

(Photo by L. Greenlee)

Mary Garland Douglass and Oscar LaBorde key flora.

Savannah River Bluffs Heritage Preserve
Aiken County
June 23, 1990

Savannah River Bluffs Heritage Preserve provides a wide variety of special features in the realm of natural history. River bluffs on the high side, with the slope leading down to the river, provide a high concentration of uncommon plants. The mild temperatures that day helped contribute to one of the longest flora-fauna lists a single day excursion has produced. A red-tailed hawk was unusually disturbed by human presence. Was it nesting? No nest was found. Ripe blackberries added an unexpectedly sweet snack; a sphinx moth larva covered with braconid wasp cocoons reminded us of the harshness of nature. Unusual plants and much insect activity abounded. The rocky shoals spider-lily was abundant, blooming along the banks and the shoals of the river.

Directions to Savannah River Bluffs Heritage Perserve: From I-20 west of Columbia, take Exit 1 (SC 230, Martintown Road) south toward North Augusta. Go 0.5 miles to a "SMILE" gas station, then turn right on the first paved road, Plantation Road. After 0.2 miles,turn right again on Old Plantation Road. Drive 0.3 miles and park before the sign indicating private property. Walk 0.1 mile along the road through private property to trail sign on the left. Follow the trail to a power line right-of-way, turn left along the power line, then right to the Savannah River Bluffs loop trail.

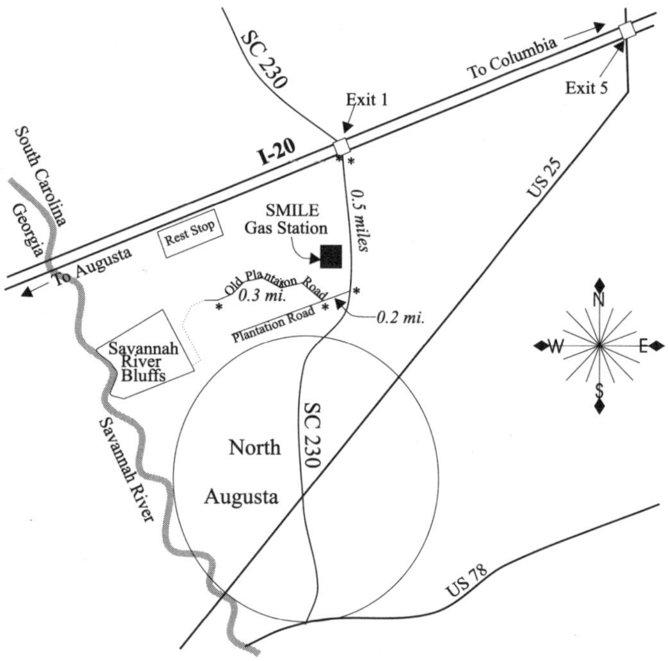

Savannah River Bluffs Heritage Preserve (continued)
June 23, 1990

FLORA

Lichen
Cladina evansii
Cladonia chlorophaea
Cladonia floerkeana
Cladonia mateocyatha
Hypotrachyna livida
Parmelina dissecta
Parmotrema michauxianum
Parmotrema perforatum
Parmotrema robustum
Parmotrema subtinctorium
Parmotrema tinctorum
Parmotrema xanthinum
Pseudoparmelia caroliniana
Xanthoparmelia plittii
Xanthoparmelia subramigera
Canoparmelia crozalsiana
Punctelia rudecta
Tuckermannopsis oakesiana
Physcia millegrana
Physcia sorediosa
Common Grape Fern
Botrychium dissectum
Hairy Lip Fern
Cheilanthes lanosa
Spreading Bladder Fern
Cystopteris protrusa
Christmas Fern
Polystichum acrostichoides
Ebony Spleenwort
Asplenium platyneuron
Resurrection Fern
Polypodium polypodioides
Loblolly Pine
Pinus taeda
Bald-cypress
Taxodium distichum
Eastern Red-cedar
Juniperus virginiana
Pondweed
Potamogeton sp.
Brazilian Elodea
Egeria densa
Cane; Switch Cane
Arundinaria gigantea
Spike Grass
Uniola sessiliflora
River Oats
Uniola latifolia
Quaking Grass
Briza minor
Melic Grass
Melica mutica
Wild Rye Grass
Elymus virginicus
Oat Grass
Danthonia spicata
Needle Grass
Stipa avenacea
Brachyelytrum (grass)
Brachyelytrum erectum
Oplismenus (grass)
Oplismenus setarius
Microstegium (grass)
Microstegium vimineum
Spike-rush
Eleocharis obtusa
Bulrush
Scirpus koilolepis
Nut-rush
Scleria oligantha
Sedge
Carex retroflexa
Carex rosea
Carex cephalophora
Carex vulpinoidea
Carex willdenowii
Carex blanda
Carex amphibola
Carex flaccosperma
Carex oxylepis
Carex debilis
Carex caroliniana
Carex complanata
Carex crinita
Carex intumescens
Dwarf Palmetto
Sabal minor
Arrowleaf Arum
Peltandra virginica
Spanish-moss
Tillandsia usneoides
Dayflower
Commelina erecta
Roseling; Spiderwort
Tradescantia rosea
Spiderwort
Tradescantia hirsuticaulis
Soft Rush
Juncus effusus
Path Rush
Juncus tenuis
Woodrush
Luzula echinata
Fringed Greenbrier
Smilax bona-nox
Sawbrier; Greenbrier
Smilax glauca
False Solomon's Seal
Smilacina racemosa
Bellwort
Uvularia perfoliata
Wild Yam
Dioscorea villosa
Spider-lily
Hymenocallis occidentalis
Yellow Star-grass
Hypoxis hirsuta
False Aloe
Agave virginica
Blue-eyed-grass
Sisyrinchium mucronatum
Southern Blue Flag
Iris virginica
Black Walnut
Juglans nigra
Pignut Hickory
Carya glabra
Tag Alder
Alnus serrulata
River Birch
Betula nigra
White Oak
Quercus alba
Post Oak
Quercus stellata
Red Oak
Quercus rubra
Swamp Red Oak
Quercus shumardii
Black Oak
Quercus velutina
Spanish Oak
Quercus falcata
Water Oak
Quercus nigra
Laurel Oak
Quercus laurifolia
Slippery Elm
Ulmus rubra
American Elm
Ulmus americana
Winged Elm
Ulmus alata
Hackberry
Celtis occidentalis georgiana
Red Mulberry
Morus rubra
Wood-nettle
Laportea canadensis
False Nettle
Boehmeria cylindrica
Virginia Snakeroot
Aristolochia serpentaria
Heartleaf
Hexastylis arifolia
Common Chickweed
Stellaria media
Windflower
Thalictrum thalictroides
Liverleaf
Hepatica americana
Coralbeads
Cocculus carolinus
Southern Magnolia
Magnolia grandiflora
PawPaw
Asimina triloba
Arabis
Arabis laevigata
Virginia-willow
Itea virginica
Mock-orange
Philadelphus inodorus
Alumroot
Heuchera americana
Sweet-gum
Liquidambar styraciflua
Witch-hazel
Hamamelis virginiana
American Sycamore
Platanus occidentalis
Indian Strawberry
Duchesnea indica
Blackberry
Rubus argutus
Redroot; Avens
Geum canadense
Cocklebur
Agrimonia pubescens microcarpa

Hawthorn
Crataegus punctata
Cockspur Hawthorn
Crataegus crus-galli
Littlehip Hawthorn
Crataegus spathulata
Downy Serviceberry
Amelanchier arborea
Black Cherry; Wild Cherry
Prunus serotina
Redbud; Judas-tree
Cercis canadensis
Wild Senna
Cassia marilandica
Yellowwood
Cladrastis lutea
White Wild Indigo
Baptisia alba
Indigo-bush
Amorpha fruticosa
Naked-flower Tick-trefoil
Desmodium nudiflorum
Black Locust
Robinia pseudo-acacia
Tephrosia
Tephrosia spicata
Groundnut
Apios americana
Hog-peanut
Amphicarpa bracteata
Wood-sorrel
Oxalis florida
Wild Geranium
Geranium maculatum
Racemed Milkwort
Polygala polygama
Polygala
Polygala verticillata
Tragia
Tragia urticifolia
Poison-ivy
Rhus radicans
Winged Sumac
Rhus copallina
American Holly
Ilex opaca
Possumhaw Holly
Ilex decidua
Holly
Ilex sp.
Hearts-a'burstin'-with-love
Euonymus americanus
Bladdernut
Staphylea trifolia
Boxelder
Acer negundo
Red Maple
Acer rubrum
Southern Sugar Maple
Acer saccharum floridanum
Chalk Maple
Acer saccharum leucoderme
Painted Buckeye
Aesculus sylvatica
Bottlebrush Buckeye
Aesculus parviflora

Spotted Touch-me-not
Impatiens capensis
Supple-jack; Rattan-vine
Berchemia scandens
Buckthorn
Rhamnus caroliniana
Virginia-creeper
Parthenocissus quinquefolia
Muscadine; Bullace
Vitis rotundifolia
Pepper-vine
Ampelopsis arborea
Basswood; Linden
Tilia caroliniana
St.-Andrew's-cross
Hypericum hypericoides
St.-John's-wort
Hypericum mutilum
St.-John's-wort
Hypericum punctatum
Rockrose, Frostweed
Helianthemum rosmarinifolium
Violet
Viola sp.
Yellow Passionflower
Passiflora lutea
Thorny Elaeagnus
Elaeagnus pungens
Russian-olive
Elaeagnus umbellata
Water-milfoil
Myriophyllum brasiliense
Devil's-walking-stick
Aralia spinosa
Pennywort
Hydrocotyle verticillata
Sweet Chervil
Osmorhiza longistylis
Wild Chervil
Chaerophyllum tainturieri
Trepocarpus
Trepocarpus aethusae
Mock Bishop's-weed
Ptilimnium capillaceum
Flowering Dogwood
Cornus florida
Spotted Wintergreen
Chimaphila maculata
Piedmont Azalea
Rhododendron canescens
Sparkleberry
Vaccinium arboreum
Squaw-huckleberry
Vaccinium stamineum stamineum
Elliott's Blueberry
Vaccinium elliottii
Shooting-star
Dodecatheon meadia
Persimmon
Diospyros virginiana
American Ash; White Ash
Fraxinus americana
White Ash; American Ash
Fraxinus americana biltmoreana
Forestiera
Forestiera ligustrina

Fringe-tree
Chionanthus virginicus
Chinese Privet
Ligustrum sinense
Yellow Jessamine
Gelsemium sempervirens
Indian-pink
Spigelia marilandica
Angle-pod
Matelea gonocarpa
Baby-blue-eyes
Nemophila microcalyx
Heliotrope
Heliotropium amplexicaule
Brazilian Vervain
Verbena braziliensis
Beautyberry
Callicarpa americana
Lop-seed
Phryma leptostachya
Common Skullcap
Scutellaria integrifolia
Skullcap
Scutellaria elliptica
Carpenterweed
Prunella vulgaris
Pennyroyal
Hedeoma pulegioides
Shrubby Savory
Satureja georgiana
Moth Mullein
Verbascum blattaria
Downy False Foxglove
Aureolaria virginica
Cross Vine
Anisostichus capreolata
Trumpet Vine
Campsis radicans
Ruellia
Ruellia caroliniensis
Bracted Plantain
Plantago aristata
Button-bush
Cephalanthus occidentalis
Bluets
Houstonia longifolia
Cleavers
Galium aparine
Sweet Breath-of-spring
Lonicera fragrantissima
Japanese Honeysuckle
Lonicera japonica
Black-haw; Nanny-berry
Viburnum prunifolium
Venus'-looking-glass
Specularia perfoliata
Small Venus'-looking-glass
Specularia biflora
Wahlenbergia
Wahlenbergia marginata
Rattlesnake-weed
Hieracium venosum
Dwarf Dandelion
Krigia virginica
Small's Ragwort
Senecio smallii

Thistle
Carduus sp.
Elephant's-foot
Elephantopus tomentosus
Climbing Hempweed
Mikania scandens
Pussy-toes
Antennaria plantaginifolia
Rabbit-tobacco
Gnaphalium sp.
Daisy Fleabane
Erigeron annuus
Bearsfoot
Polymnia uvedalia
Silphium
Silphium dentatum
Spilanthes
Spilanthes americana repens
Verbesina
Verbesina sp.
Bitterweed
Helenium amarum
Rust
Gymnosporangium sp.

FAUNA

River Horn Snail
Elimia sp.
White-lipped Forest Snail
Mesodon thyroidus
Juxtaposed Forest Snail
Triodopsis juxtidens
River Horn Snail
Goniobasis sp.
Copper Great Zonite (snail)
Mesomphix cupreus
Asiatic Clam
Corbicula fluminea
Asiatic Clam
Corbicula manilensis
Orchard Spider
Leucauge venusta
Green-striped Grasshopper
Chortophaga viridifasciata
Snowy Tree Cricket
Oecanthus fultoni
Leaf-footed Bug
Acanthocephala femorata
Froghopper
Prosapia bicincta
Goldenrod Ball Gall (fly)
Eurosta solidaginis
Dragonslayer
Hagenius brevistylus
Pied Skimmer
Libellula luctuosa
Ebony Jewelwing
Calopteryx maculata
Woodland Ground Beetle
Evarthrus sp.
Whirligig Beetle
Dineutus sp.
Large Rove Beetle
Staphylinus sp.
Passalus; Bess-beetle
Popilius disjunctus

Savannah River Bluffs Heritage Preserve (continued)
June 23, 1990

June Beetle
Phyllophaga sp.
Japanese Beetle
Popillia japonica
Tick Trefoil Leaf-miner
Pachyschelus laevigatus
Eyed Elater
Alaus oculatus
Soldier Beetle
Cantharis sp.
Firefly; Lightning Beetle
Lucidota atra
Firefly; Lightning Beetle
Photinus pyralis
Tumbling Flower Beetle
Mordellistena fuscata
Darkling Beetle
Strongylium sp.
Sap-feeding Beetle
Pallodes sp.
Flower Weevil
Odontocorynus scutellum-album
Odontocorynus pulverulentus
Black Swallowtail
Papilio polyxenes asterius
Common Buckeye (butterfly)
Junonia coenia
Carolina Satyr (butterfly)
Hermeuptychia sosybius
Little Wood-satyr (butterfly)
Megisto cymela
Zabulon Skipper
Poanes zabulon
Polyphemus Moth
Antheraea polyphemus
Luna Moth
Actias luna
Fowler's Toad
Bufo woodhousei
Cope's Gray Treefrog
Hyla chrysoscelis
Carolina Anole
Anolis carolinensis
Copperhead
Agkistrodon contortrix
Red-tailed Hawk
Buteo jamaicensis
Yellow-billed Cuckoo
Coccyzus americanus
Pileated Woodpecker (tracks)
Dryocopus pileatus
Barn Swallow
Hirundo rustica
Blue Jay
Cyanocitta cristata
American Crow
Corvus brachyrhynchos
Tufted Titmouse
Parus bicolor
Carolina Chickadee
Parus carolinensis
Blue-gray Gnatcatcher
Polioptila caerulea
Eastern Bluebird
Sialia sialis
Northern Parula
Parula americana
Northern Cardinal
Cardinalis cardinalis
Rufous-sided Towhee
Pipilo erythrophthalmus
Summer Tanager
Piranga rubra
Eastern Cottontail (scat)
Sylvilagus floridanus
Gray Squirrel
Sciurus carolinensis
Raccoon
Procyon lotor
White-tailed Deer (tracks)
Odocoileus virginianus

Savannah River Site
Aiken & Barnwell Counties
August 27, 1994

The natural areas on the Savannah River Site, a facility of the U.S. Dept. of Energy, are monitored by the Savannah River Ecology Lab. The staff of the Lab conducted our field trip by special arrangement. The trip started at a Carolina bay completely enclosed in a drift fence (nonoperational in August) used to monitor reptiles and amphibians, and some insect populations. Next was a cypress swamp with a floating boardwalk and a nice female cottonmouth at the entrance, which was caught by lab personnel to be measured and recorded and later released back at the boardwalk. Lastly, we visited the live animal collection and learned about ongoing research projects at the lab. SCAN was happy just observing the dragonflies sharing the alligator pool with the 'gators.

Directions to Savannah River Site: From Aiken, take SC 19 south through New Ellenton, cross US 278, and continue until Road 1 forks off to the right. Travel on Road 1 for 3-5 miles. At the yellow light, slow down and go through the guard gate. The first large building on the left is the Savannah River Ecology Lab; take the second entrance to the parking lot for visitors. NOTE: The Savannah River Site cannot be visited without prior arrangement.

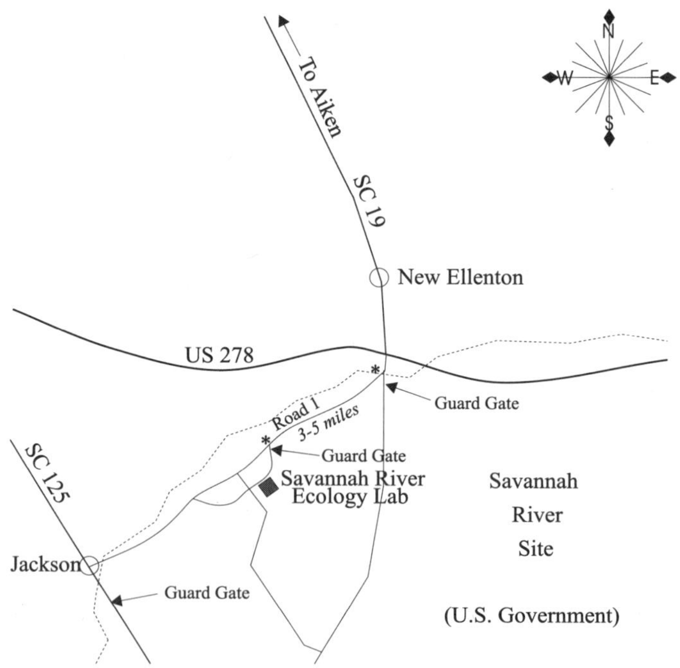

FLORA

Lichen
Graphis scripta
Glyphis cicatricosa
Phaeographis dendritica
Bacidia polychroa
Leptogium austroamericanum
Cladonia sobolescens
Cladonia leporina
Pertusaria copiosa
Pertusaria ostiolata
Pertusaria paratuberculifera
Pertusaria texana
Pertusaria tetrathalmia
Pertusaria sinusmexicani
Maronea constans
Lecanora cupressi
Lecanora pallida
Lecanora varia
Lecanora hybocarpa
Ochrolechia africana
Bulbothrix goebelii

Bulbothrix goebelii f. minor
Bulbothrix laevigatula
Hypotrachyna livida
Hypotrachyna pustulifera
Parmelina aurulenta
Parmotrema hypotropum
Parmotrema michauxianum
Parmotrema perforatum
Parmotrema praesorediosum
Parmotrema rampoddense
Punctelia rudecta
Rimelia reticulata
Usnea strigosa
Buellia curatellae
Buellia disciformis
Physcia aipolia
Physcia alba
Heterodermia albicans
Pyxine caesiopruinosa
Anthracothecium leucostomum
Anthracothecium sp.
Trypethelium mastoideum
Sensitive Fern
Onoclea sensibilis
Loblolly Pine
Pinus taeda
Bald-cypress
Taxodium distichum
Common Cat-tail
Typha latifolia
Creeping Water-plantain
Echinodorus cordifolius
Cane; Switch Cane
Arundinaria gigantea
Smut Grass
Sporobolus poiretii
Baggy-knees Grass
Sacciolepis striata
Microstegium (grass)
Microstegium vimineum
Wool-grass
Scirpus cyperinus
Umbrella-grass
Fuirena squarrosa
Dwarf Palmetto
Sabal minor
Yellow-eyed-grass
Xyris sp.
Spanish-moss
Tillandsia usneoides
Greenbrier; Bullbrier
Smilax rotundifolia
Fringed Greenbrier
Smilax bona-nox
Lizard's-tail
Saururus cernuus
Black Willow
Salix nigra
Tag Alder
Alnus serrulata
White Oak
Quercus alba
Post Oak
Quercus stellata
Water Oak
Quercus nigra
Laurel Oak
Quercus laurifolia
American Elm
Ulmus americana
Winged Elm
Ulmus alata
Water-elm; Planer-tree
Planera aquatica
Sugarberry; Hackberry
Celtis laevigata
Red Mulberry
Morus rubra
False Nettle
Boehmeria cylindrica
Jumpseed
Tovara virginiana
Polygonum
Polygonum sp.
Water-lily
Nymphaea odorata
Water-shield
Brasenia schreberi
Sassafras
Sassafras albidum
Virginia-willow
Itea virginica
Sweet-gum
Liquidambar styraciflua
Blackberry
Rubus argutus
Blackberry; Brambles
Rubus sp.
Chickasaw Plum
Prunus angustifolia
Black Cherry; Wild Cherry
Prunus serotina
Redbud; Judas-tree
Cercis canadensis
Wild Sensitive-plant
Cassia nicitans
Wild Indigo
Baptisia sp.
Sericea
Lespedeza cuneata
Groundnut
Apios americana
Three-seeded Mercury
Acalypha sp.
Poison-ivy
Rhus radicans
Poison-oak
Rhus toxicodendron
Winged Sumac
Rhus copallina
Red Maple
Acer rubrum
Red Buckeye
Aesculus pavia
Spotted Touch-me-not
Impatiens capensis
Virginia-creeper
Parthenocissus quinquefolia
Muscadine; Bullace
Vitis rotundifolia
Grape
Vitis sp.
Pepper-vine
Ampelopsis arborea
Pineweed
Hypericum gentianoides
St.-John's-wort
Hypericum mutilum
Hypericum
Hypericum sp.
Prickly-pear
Opuntia compressa
Water Loosestrife
Decodon verticillatus
Pale Meadow-beauty
Rhexia mariana
Common Meadow-beauty
Rhexia virginica
Ludwigia
Ludwigia suffruticosa
Ludwigia sp.
Marsh Pennywort
Hydrocotyle umbellata
Swamp Tupelo; Swamp Blackgum
Nyssa sylvatica biflora
Water Tupelo
Nyssa aquatica
Flowering Dogwood
Cornus florida
Blueberry
Vaccinium sp.
Persimmon
Diospyros virginiana
Water Ash; Carolina Ash
Fraxinus caroliniana
Ash
Fraxinus sp.
Miterwort
Cynoctotum mitreola
Polypremum
Polypremum procumbens
Swamp Milkweed
Asclepias perennis
Ground-ivy; Ponyfoot
Dichondra carolinensis
Phlox
Phlox sp.
Verbena
Verbena sp.
Beautyberry
Callicarpa americana
Dotted Horsemint
Monarda punctata
Chinese-lanterns
Physalis virginiana virginiana
Monkey-flower
Mimulus alatus
Gerardia
Agalinis sp.
Trumpet Vine
Campsis radicans
Bladderwort
Utricularia biflora
Button-bush
Cephalanthus occidentalis
Poor-jo (or Po-jo)
Diodia teres
Mexican-clover
Richardia scabra
Partridge-berry
Mitchella repens
Japanese Honeysuckle
Lonicera japonica
Cardinal-flower
Lobelia cardinalis
Fireweed
Erechtites hieracifolia
Elephant's-foot
Elephantopus sp.
Blazing-star
Liatris secunda
Dog-fennel
Eupatorium capillifolium
Eupatorium compositifolium
Climbing Hempweed
Mikania scandens
Rabbit-tobacco
Gnaphalium obtusifolium
Goldenrod
Solidago sp.
Black-knot
Apiosporina morbosa

FAUNA

Arrow-shaped Spiny-back
Micrathena sagittata
Orb-weaver
Neoscona arabesca
Green Lynx Spider
Peucetia viridans
Jumping Spider
Phidippus cardinalis
Olive-green Swamp Grasshopper
Paroxya clavuliger
Meadow Katydid
Orchelimum volantum
Short-horned Grasshopper
Metaleptea brevicornis
Autumn Yellow-winged Grasshpr.
Arphia xanthoptera
Boll's Sand Grasshopper
Spharagemon bolli
Spur-throated Grasshopper
Melanoplus sp.
Leather-color Bird Grasshopper
Schistocerca alutacea
Cone-headed Katydid
Neoconocephalus retesus
Backswimmer
Notonecta sp.
Dog-day Cicada
Tibicen canicularis
Argid Sawfly
Argidae Family
Leafy Oak Gall (wasp)
Andricus foliatus
Common Eastern Velvet-ant
Dasymutilla occidentalis occidentalis
Cypress Flower Gall (midge)
Itonida anthici
Common Green Darner
Anax junius
Eastern Pondhawk
Erythemis simplicicollis

Savannah River Site (continued)
August 27, 1994

Blue Dragonlet
Erythrodiplax minuscula
Golden-winged Skimmer
Libellula auripennis
Slaty Skimmer
Libellula incesta
Great Blue Skimmer
Libellula vibrans
Common Whitetail (dragonfly)
Libellula lydia
Blue Dasher
Pachydiplax longipennis
Wandering Glider
Pantala flavescens
Violet-masked Glider
Tramea carolina
Black-mantled Glider
Tramea lacerata
Ebony Jewelwing
Calopteryx maculata
Variable Dancer (damselfly)
Argia fumipennis
Lilypad Forktail (damselfly)
Ischnura kellicotti
Fragile Forktail (damselfly)
Ischnura posita
Colorful Foliage Ground Beetle
Nemotarsus elegans
Crawling Water Beetle
Haliplus triopsus
Peltodytes shermani
Peltodytes sp.
Predacious Diving Beetle
Coptotomus interrogatus
Burrowing Water Beetle
Copius inflatus
Whirligig Beetle
Dineutus serrulatus serrulatus
Water Scavenger Beetle
Tropisternus collaris striolatus
Berosus sp.
Enochrus ochraceus
Marsh Beetle
Scirtes sp.
Soldier Beetle
Chauliognathus marginatus
Blister Beetle
Pyrota lineata
Lizard Beetle
Languria lecontei
Dusky Lady Beetle
Scymnus sp.
Multicolored Asian Lady Beetle
Harmonia axyridis
Oval Leaf Beetle
Metachroma longicollis
Broad-bodied Leaf Beetle
Zymogramma suturalis
Flea Beetle
Disonycha sp.
Systena frontalis
White-fringed Beetle
Graphognathus leucoloma
Graphognathus sp.
Fruit Weevil
Anthonomus tectus
Cloudless Sulphur
Phoebis sennae eubule
Monarch (butterfly)
Danaus plexippus
Silver-spotted Skipper
Epargyreus clarus
Long-tailed Skipper
Urbanus proteus
Least Skipper
Ancyloxypha numitor
Polyphemus Moth
Antheraea polyphemus
Southern Cricket Frog
Acris gryllus
Carolina Anole
Anolis carolinensis
Broadhead Skink
Eumeces laticeps
Six-lined Racerunner
Cnemidophorus sexlineatus
Cottonmouth
Agkistrodon piscivorus
Reddish Egret
Egretta rufescens
Turkey Vulture
Cathartes aura
Red-shouldered Hawk
Buteo lineatus
Wild Turkey
Meleagris gallopavo
American Crow
Corvus brachyrhynchos
Gray Catbird
Dumetella carolinensis
White-eyed Vireo
Vireo griseus
Yellowthroat
Geothlypis trichas
Northern Cardinal
Cardinalis cardinalis
Rufous-sided Towhee
Pipilo erythrophthalmus
Rabbit
Sylvilagus sp.
Wild Pig (sign)
Sus scrofa

Silver Bluff Audubon Preserve
Aiken County
September 25, 1993

Silver Bluff is an Audubon sanctuary noted for its wood storks and its hardwood river bottom. Dan Connelly, preserve manager, gave us background information on management techniques currently in use, and showed us the significant areas to visit. On this day, about 160 wood storks were present as well as many other birds. Ecologically varied sites such as cypress-tupelo river bottomland (dry from drought), managed longleaf pine woods, and meadow-forest edges were visited. The beginning of autumn was much in evidence with beautiful stands of fall wildflowers, including for many of us the first sighting of bur cucumber (*Sicyos angulatus*) in bloom. It was a "grand day".

Directions to Silver Bluff Audubon Preserve: From downtown Aiken, take SC 19 south approximately 2 miles; angle right onto SC 302 (Silver Bluff Road). After 12 miles cross SC 125, where Silver Bluff Road becomes County Road 32. Continue for about 3.5 miles more, crossing the railroad track. After going 0.5 miles past the railroad, there are chain-link fences on both sides of the road; go through the gate on the left.

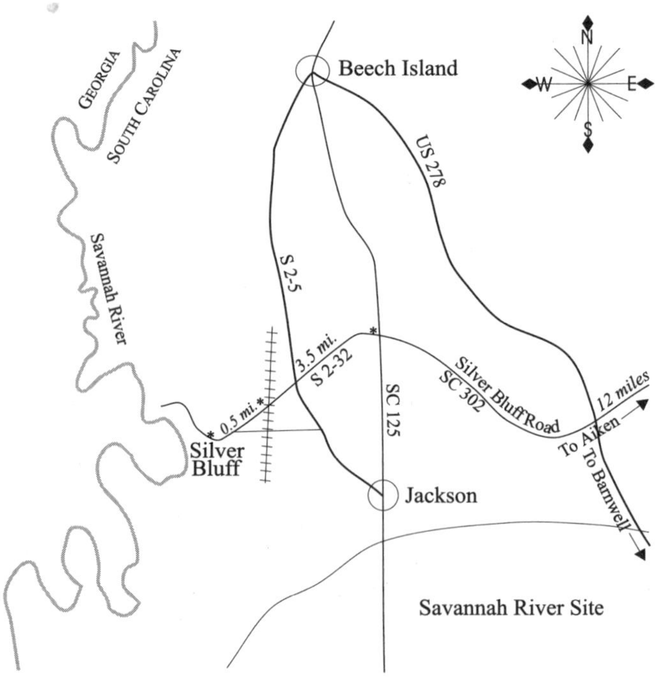

FLORA

Lichen
Cladonia chlorophaea
Cladonia cristatella
Cladonia polycarpoides
Cladonia peziziformis
Haematomma puniceum
Lecanora subfusca
Lecanora cupressi
Lecanora pallida
Lecanora varia
Ochrolechia africana
Bulbothrix goebelii
Hypotrachyna livida
Parmotrema hypotropum
Parmotrema michauxianum
Parmotrema perforatum
Parmotrema subisidiosum
Canoparmelia caroliniana
Punctelia rudecta
Rimelia reticulata
Ramalina americana
Usnea strigosa
Buellia punctata
Heterodermia obscurata
Royal Fern
Osmunda regalis spectabilis
Bracken
Pteridium aquilinum
Southern Lady Fern
Athyrium asplenioides
Sensitive Fern
Onoclea sensibilis
Netted Chain Fern
Woodwardia areolata
Ebony Spleenwort
Asplenium platyneuron
Resurrection Fern
Polypodium polypodioides
Loblolly Pine
Pinus taeda
Bald-cypress
Taxodium distichum
Cane; Switch Cane
Arundinaria gigantea
Spike Grass
Uniola sessiliflora
River Oats
Uniola latifolia
Three Awn Grass
Aristida tuberculosa
Goose Grass
Eleusine indica
Wingstem
Leersia lenticularis
Spanish-moss
Tillandsia usneoides
Dayflower
Commelina virginica
Commelina communis
Asian Spiderwort
Aneilema keisak
Sweet-scented Smilax
Smilax smallii
Sarsaparilla-vine
Smilax pumila
Ladies'-tresses
Spiranthes sp.
Lizard's-tail
Saururus cernuus
Wax-myrtle; Southern Bayberry
Myrica cerifera
Hickory
Carya sp.
American Hornbeam; Ironwood
Carpinus caroliniana
Overcup Oak
Quercus lyrata
Water Oak
Quercus nigra
Red Mulberry
Morus rubra
False Nettle
Boehmeria cylindrica
Clearweed
Pilea pumila
Jumpseed
Tovara virginiana
Jointweed
Polygonella americana
Mexican-tea
Chenopodium ambrosioides
Pokeweed; Pokeberry
Phytolacca americana
Coralbeads
Cocculus carolinus
Tulip-poplar
Liriodendron tulipifera
Sassafras
Sassafras albidum
Spicebush
Lindera benzoin
Virginia-willow
Itea virginica
Climbing-hydrangea
Decumaria barbara
Sweet-gum
Liquidambar styraciflua
American Sycamore
Platanus occidentalis
Hawthorn
Crataegus sp.
Sicklepod
Cassia obtusifolia
Painted-leaf
Euphorbia heterophylla
Poison-ivy
Rhus radicans
Winged Sumac
Rhus copallina
Boxelder
Acer negundo
Red Maple
Acer rubrum
Buckeye
Aesculus sp.
Spotted Touch-me-not
Impatiens capensis
Virginia-creeper
Parthenocissus quinquefolia
Muscadine; Bullace
Vitis rotundifolia
Pepper-vine
Ampelopsis arborea
Basswood; Linden
Tilia caroliniana
St.-Andrew's-cross
Hypericum hypericoides
Devil's-walking-stick
Aralia spinosa
Swamp Tupelo; Swamp Blackgum
Nyssa sylvatica biflora
Water Tupelo
Nyssa aquatica
Spotted Wintergreen
Chimaphila maculata
Persimmon
Diospyros virginiana
Horse-sugar; Sweetleaf
Symplocos tinctoria
Yellow Jessamine
Gelsemium sempervirens
Angle-pod
Matelea sp.
Tie-vine; Jacquemontia
Jacquemontia tamnifolia
Beautyberry
Callicarpa americana
Cross Vine
Anisostichus capreolata
Trumpet Vine
Campsis radicans
Squaw-root
Conopholis americana
Poor-jo (or Po-jo)
Diodia teres
Richardia
Richardia brasiliensis
Partridge-berry
Mitchella repens
Japanese Honeysuckle
Lonicera japonica
American Elderberry
Sambucus canadensis
Bur Cucumber
Sicyos angulatus
Wahlenbergia
Wahlenbergia marginata
Cardinal-flower
Lobelia cardinalis
Common Ragweed
Ambrosia artemisiifolia
Ironweed
Vernonia sp.
Elephant's-foot
Elephantopus tomentosus
Dog-fennel
Eupatorium compositifolium
Mistflower; Ageratum
Eupatorium coelestinum
Climbing Hempweed
Mikania scandens
Marsh-fleabane
Pluchea camphorata
Horseweed; Mare's-tail
Erigeron canadensis canadensis
Frost Aster
Aster pilosus
Goldenrod
Solidago sp.
Spilanthes
Spilanthes americana repens
Spanish-needles
Bidens bipinnata
Velvet Bluespread
Pulcherrichium caeruleum
Turkey-tail
Trametes versicolor
Common Split-gill
Schizophyllum commune
Crowded Parchment
Stereum complicatum
Lemon-yellow Lepiota
Lepiota lutea
Emetic Russula
Russula emetica
Green Brittle-gill
Russula virescens
Granular Puffball
Arachnion album
Puffball
Calvatia utriformis

FAUNA

Writing Spider
Argiope aurantia
Orb-weaver
Neoscona domiciliorum
Brownish-gray Fishing Spider
Dolomedes tenebrosus
Six-spotted Fishing Spider
Dolomedes triton
Green Lynx Spider
Peucetia viridans
Yellow Crab Spider
Misumenoides formosipes
Meadow Grasshopper
Orchelimum sp.
Differential Grasshopper
Melanoplus differentialis
Cave Cricket
Ceuthophilus sp.
Field Cricket
Gryllus assimilis
Brown Stink Bug
Euschistus sp.
Broad-headed Bug
Alydus eurinus
Red Bug
Largus succinctus
Ambush Bug
Phymata fasciata
Water-scorpion
Ranatra americana
Froghopper
Prosapia bicincta
Spittlebug
Philaenus sp.
Verbesina Axillary Flower Gall
Rhopalomyia sp.
Common Green Darner
Anax junius
Golden-winged Skimmer
Libellula auripennis

Silver Bluff Audubon Preserve (continued)
September 25, 1993

Great Blue Skimmer
Libellula vibrans
Eastern Amberwing (dragonfly)
Perithemis tenera
Sparkling Jewelwing Damselfly
Calopteryx dimidiata
Blue-fronted Dancer Damselfly
Argia apicalis
Powdered Dancer (damselfly)
Argia moesta
Woodland Ground Beetle
Pterostichus sp.
Colorful Foliage Ground Beetle
Lebia analis
Crawling Water Beetle
Peltodytes muticus
Whirligig Beetle
Dineutus sp.
Gyrinus borealis
Water Scavenger Beetle
Tropisternus striolatus
Passalus; Bess-beetle
Popilius disjunctus
Dermestid Beetle
Cryptorhopalum picicorne
Black Blister Beetle
Epicauta pennsylvanica
Sap-feeding Beetle
Pocadius sp.
Spotless Lady Beetle
Cycloneda sanguinea
Shining Flower Beetle
Olibrus sp.
Acylomus sp.
Shining Leaf Beetle
Lema sayi
12-spotted Cucumber Beetle
Diabrotica duodecimpunctata
Goldbug
Metriona bicolor
Fungus Weevil
Euparius marmoreus
White-fringed Beetle
Graphognathus leucoloma
Graphognathus sp.
Flower Weevil
Centrinaspis grisescens
Centrinaspis sp.
Minute Seed Weevil
Aleutes tachygonoides
Pipevine Swallowtail
Battus philenor
Spicebush Swallowtail
Papilio troilus
Cloudless Sulphur
Phoebis sennae eubule
Little Yellow (butterfly)
Eurema lisa
Sleepy Orange (butterfly)
Eurema nicippe
Eastern Tailed-blue(butterfly)
Everes comyntas
Zebra (butterfly)
Heliconius charitonius tuckeri
Variegated Fritillary
Euptoieta claudia
Pearl Crescent
Phyciodes tharos
Common Buckeye (butterfly)
Junonia coenia
Red-spotted Purple (butterfly)
Limenitis arthemis astyanax
Hackberry Emperor (butterfly)
Asterocampa celtis
Carolina Satyr (butterfly)
Hermeuptychia sosybius
Banded Tussock Moth
Halysidota tessellaris
American Alligator
Alligator mississippiensis
Yellow-bellied Slider
Chrysemys scripta
Carolina Anole
Anolis carolinensis
Ground Skink
Scincella lateralis
Double-crested Cormorant
Phalacrocorax auritus
Little Blue Heron
Egretta caerulea
Snowy Egret
Egretta thula
Great Egret
Casmerodius albus
Great Blue Heron
Ardea herodias
Wood Stork
Mycteria americana
White Ibis
Eudocimus albus
Canada Goose
Branta canadensis
Wood Duck
Aix sponsa
Killdeer
Charadrius vociferus
Spotted Sandpiper
Actitis macularia
Least Sandpiper
Calidris minutilla
Turkey Vulture
Cathartes aura
Bald Eagle
Haliaeetus leucocephalus
Red-shouldered Hawk
Buteo lineatus
Northern Bobwhite
Colinus virginianus
Rock Dove; Pigeon
Columba livia
Mourning Dove
Zenaida macroura
Barred Owl
Strix varia
Belted Kingfisher
Ceryle alcyon
Red-bellied Woodpecker
Melanerpes carolinus
Red-headed Woodpecker
Melanerpes erythrocephalus
Downy Woodpecker
Picoides pubescens
Pileated Woodpecker
Dryocopus pileatus
Bank Swallow
Riparia riparia
Northern Rough-winged Swallow
Stelgidopteryx serripennis
Barn Swallow
Hirundo rustica
Blue Jay
Cyanocitta cristata
American Crow
Corvus brachyrhynchos
Tufted Titmouse
Parus bicolor
Carolina Wren
Thryothorus ludovicianus
White-eyed Vireo
Vireo griseus
Red-Eyed Vireo
Vireo olivaceus
Pine Warbler
Dendroica pinus
Northern Cardinal
Cardinalis cardinalis
Common Grackle
Quiscalus quiscula
Virginia Opossum (tracks)
Didelphis virginiana
Eastern Fox Squirrel
Sciurus niger
Raccoon
Procyon lotor
White-tailed Deer
Odocoileus virginianus

(Photo by L. Smith)
Nodding ladies-tresses, *Spiranthes cernua*.

(Photo by L. Greenlee)
Common green darner, *Anax junius*.

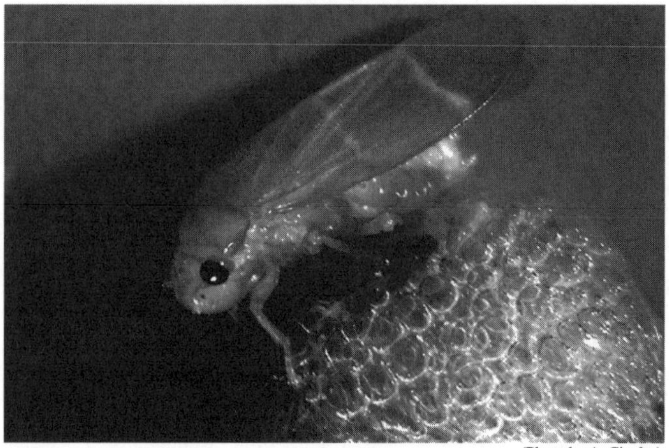

(Photo by A. Ciegler)
Spittlebug, *Philaenus* sp., hatching. This image was photographed using a microscope.

Allendale County

Chert Quarry
Allendale County
March 26, 1988

Allendale County contains some outstanding outcrops of chert (a dense form of quartz) which were popular with early Indians and still are with present-day rock hounds. These outcrops are privately owned, but one owner was gracious enough to allow us to visit, explore, and collect. The open field near the outcrop and the old road down to the river provided lots of living material to examine as well as fossils to study preserved in the attractive rocks. The area is typical river bottomland. One highlight for many SCAN members was the sighting of a swallowtail kite. Thank you to regional director Frank Hill for getting us permission to visit this site.

Privately owned. Not accessible to the public. Map not included.

FLORA

Sensitive Fern
Onoclea sensibilis
Ebony Spleenwort
Asplenium platyneuron
Loblolly Pine
Pinus taeda
Bald-cypress
Taxodium distichum
Common Cat-tail
Typha latifolia
Cane; Switch Cane
Arundinaria gigantea
Uniola Grass
Uniola laxa
River Oats
Uniola latifolia
Dwarf Palmetto
Sabal minor
Spanish-moss
Tillandsia usneoides
False Garlic
Allium bivalve
Wild Yam
Dioscorea villosa
Blue-eyed-grass
Sisyrinchium sp.
Lizard's-tail
Saururus cernuus
Eastern Cottonwood
Populus deltoides
Wax-myrtle; Southern Bayberry
Myrica cerifera
River Birch
Betula nigra
Hop-hornbeam
Ostrya virginiana
American Hornbeam; Ironwood
Carpinus caroliniana
Swamp Chestnut Oak
Quercus michauxii
Water Oak
Quercus nigra
Willow Oak
Quercus phellos
American Elm
Ulmus americana
Winged Elm
Ulmus alata
Heartleaf
Hexastylis arifolia
Common Chickweed
Stellaria media
Mouse-ear Chickweed
Cerastium glomeratum
Pearlwort
Sagina decumbens
Redbay
Persea borbonia
Mouse-ear Cress
Arabidopsis thaliana
Sweet-gum
Liquidambar styraciflua
American Sycamore
Platanus occidentalis
Swamp Haw
Crataegus viridis
Parsley-leaved Hawthorn
Crataegus marshallii
Redbud; Judas-tree
Cercis canadensis
Coral-bean
Erythrina herbacea
Poison-ivy
Rhus radicans
Winged Sumac
Rhus copallina
American Holly
Ilex opaca
Hearts-a'burstin'-with-love
Euonymus americanus
Red Maple
Acer rubrum
Silver Maple
Acer saccharinum
Red Buckeye
Aesculus pavia
Spotted Touch-me-not
Impatiens capensis
Supple-jack; Rattan-vine
Berchemia scandens
Virginia-creeper
Parthenocissus quinquefolia
Muscadine; Bullace
Vitis rotundifolia
Pennywort
Hydrocotyle sp.
Buckthorn Bumelia
Bumelia lycioides
Yellow Jessamine
Gelsemium sempervirens
Ground-ivy; Ponyfoot
Dichondra carolinensis
Phacelia
Phacelia dubia
Common Henbit
Lamium amplexicaule
Woolly Mullein
Verbascum thapsus
Toadflax
Linaria canadensis
Cross Vine
Anisostichus capreolata
Partridge-berry
Mitchella repens
Small Bluets
Houstonia pusilla
Dwarf Dandelion
Krigia virginica
Tree-ear
Auricularia auricula
Gelatinous-pored Polypore
Caloporus dichrous
Crowded Parchment
Stereum complicatum
False Turkey-tail
Stereum ostrea
Oyster Mushroom
Pleurotis sp.
Earthstar Scleroderma
Scleroderma geaster
Pine-oak Rust
Cronartium quercuum

FAUNA

Six-spotted Fishing Spider
Dolomedes triton
Wolf Spider
Pardosa sp.
Carolina Wolf Spider
Lycosa carolinensis
Green Lynx Spider
Peucetia viridans
Predacious Diving Beetle
Laccophilus fasciatus
Bidessus sp.
Water Scavenger Beetle
Hydrochus inaequalis
Passalus; Bess-beetle
Popilius disjunctus
Dung Beetle
Onthophagus orpheus
Aphodiine Dung Beetle
Ataenius sp.
Soldier Beetle
Cantharis bilineatus
Darkling Beetle
Alobates pennsylvanica
False Darkling Beetle
Eustrophinus bicolor
Asclera ruficollis
Pleasing Fungus Beetle
Tritoma thoracica
Flower Weevil
Centrinaspis grisescens
Eastern Tiger Swallowtail
Papilio glaucus
Falcate Orangetip
Anthocharis midea
Eastern Tailed-blue(butterfly)
Everes comyntas
Pearl Crescent
Phyciodes tharos
Red-spotted Purple (butterfly)
Limenitis arthemis astyanax

Chert Quarry (continued)
March 26, 1988

Carolina Satyr (butterfly)
Hermeuptychia sosybius
Silver-spotted Skipper
Epargyreus clarus
Juvenal's Duskywing
Erynnis juvenalis
Common Checkered-skipper
Pyrgus communis
Polyphemus Moth (cocoon)
Antheraea polyphemus
Snapping Turtle
Chelydra serpentina
Carolina Anole
Anolis carolinensis
Ground Skink
Scincella lateralis
Great Blue Heron
Ardea herodias
Turkey Vulture
Cathartes aura
Black Vulture
Coragyps atratus
American Swallow-tailed Kite
Elanoides forficatus
Red-bellied Woodpecker
Melanerpes carolinus
Yellow-bellied Sapsucker
Sphyrapicus varius
Pileated Woodpecker
Dryocopus pileatus
American Crow
Corvus brachyrhynchos
Tufted Titmouse
Parus bicolor
Carolina Chickadee
Parus carolinensis
Carolina Wren
Thryothorus ludovicianus
Ruby-crowned Kinglet
Regulus calendula
Blue-gray Gnatcatcher
Polioptila caerulea
White-eyed Vireo
Vireo griseus
Northern Parula
Parula americana
Yellow-throated Warbler
Dendroica dominica
Pine Warbler
Dendroica pinus
Northern Cardinal
Cardinalis cardinalis
White-throated Sparrow
Zonotrichia albicollis
Swamp Sparrow
Melospiza georgiana
Beaver (gnaw mark)
Castor canadensis
Wild Pig
Sus scrofa
White-tailed Deer (tracks)
Odocoileus virginianus

Anderson County

Beaverdam Creek Waterfowl Management Area and Simpson Agricultural Experiment Station

Anderson County
November 19, 1988

This visit to a popular birding site in the Piedmont of South Carolina taught the SCAN planning committee a lesson. Do not plan a trip within a few miles of either Clemson University or the University of South Carolina football stadium on the day of the Clemson/Carolina football game! Beaverdam Creek Waterfowl Management Area includes marshy areas plus small ponds and open fields, some of which are flooded for duck feeding under the current management, but were not in 1988. The day was very cold and misty, so birds were not active, and were hard to see when they did move. The site clearly had potential for good birding on a better day.

Beaverdam Creek is currently managed by the Department of Natural Resources for waterfowl. Read the signs about access. The area is currently closed to all but hunters from November 1 to February 15 each winter. Contact the DNR office in Clemson at (864) 654-1671 for up-to-date information.

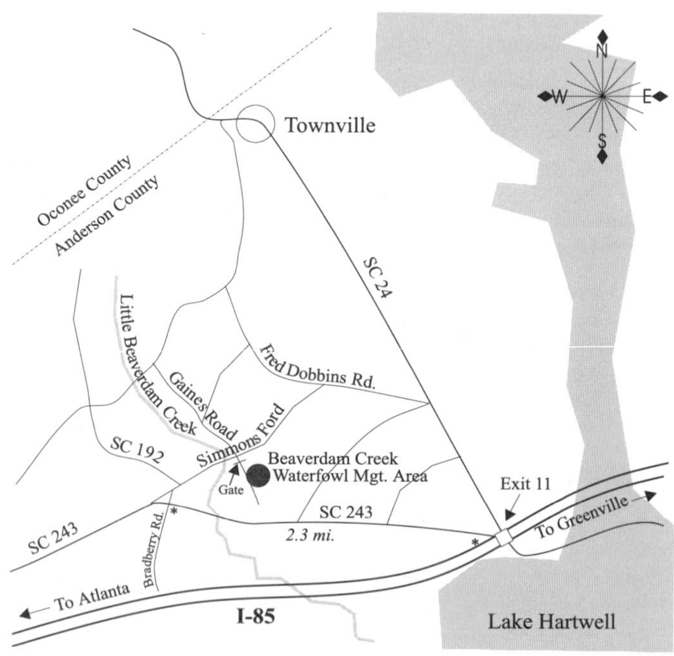

Directions to Beaverdam Creek Waterfowl Management Area: From I-85 west of Greenville, take Exit 11 (SC 24/243). Drive on SC 243 west for about 2.3 miles, and then turn right onto Bradberry Road. Immediately turn right again at the stop sign (Simmons Ford Road). Continue past the solid waste recycling area on left, cross a small creek, then turn right onto Gaines Road. Park along the side of the old road bed. The access road is closed and gated.

The afternoon portion of the day was spent at Simpson Experiment Station which is a typical example of Piedmont farm country, consisting fields, small woodlands, and fence rows. It was disappointing because of the steady medium rain that moved in and continued most of the afternoon. Almost no animals were moving, and because it was so late after the first frost of the fall, little plant material was observable. Between the gaps in the rain, a few birds made an appearance.

The station is owned by Clemson University, College of Agricultural Sciences.

Directions to Simpson Experiment Station: From I-85 west of Greenville, take Exit 21 (US 178) north toward Liberty and Pickens. After about 7 miles, turn left onto County Road 231. Drive to the "T" and turn right. At the 4-way intersection, turn left onto County Road 57 west toward Pendleton. The experiment station is on the left in about 3.2 miles.

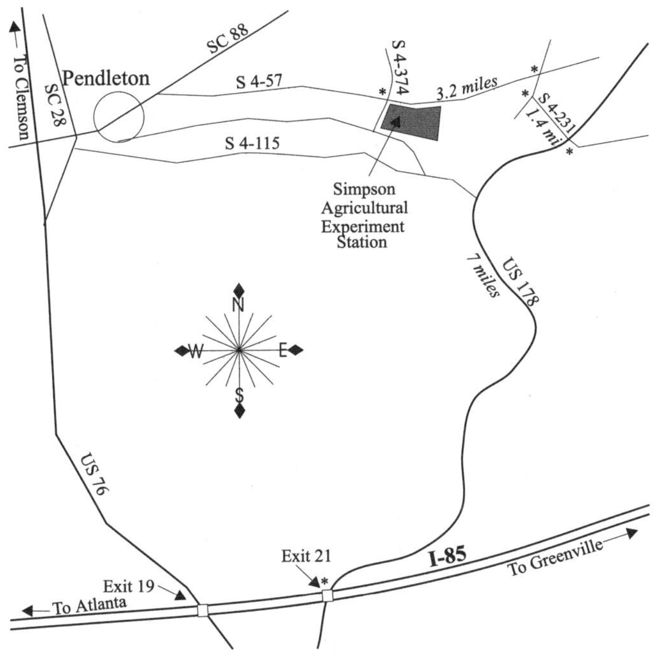

FLORA

Common Grape Fern
Botrychium dissectum
Ebony Spleenwort
Asplenium platyneuron
Loblolly Pine
Pinus taeda
Eastern Red-cedar
Juniperus virginiana
Cane; Switch Cane
Arundinaria gigantea
Giant Reed
Arundo donax
Yellow Foxtail Grass
Setaria glauca
Beard Grass; Plume Grass
Erianthus contortus
Johnson Grass
Sorghum halepense
Fringed Greenbrier
Smilax bona-nox
Sawbrier; Greenbrier
Smilax glauca
Black Willow
Salix nigra
Tag Alder
Alnus serrulata
River Birch
Betula nigra
Hackberry
Celtis occidentalis georgiana
Mistletoe
Phoradendron serotinum
Curly Dock; Sour-grass
Rumex crispus
Pokeweed; Pokeberry
Phytolacca americana
Tulip-poplar
Liriodendron tulipifera
Poor-man's-pepper
Lepidium virginicum
Sweet-gum
Liquidambar styraciflua
Rose
Rosa micrantha
Swamp Rose
Rosa palustris
Black Cherry; Wild Cherry
Prunus serotina
Mimosa; Silk-tree
Albizia julibrissin
Honey Locust
Gleditsia triacanthos
Cranesbill
Geranium carolinianum
Poison-ivy
Rhus radicans
Smooth Sumac
Rhus glabra
Mountain Holly
Ilex ambigua montana
Red Maple
Acer rubrum
Anoda
Anoda cristata
Hibiscus
Hibiscus sp.
Meadow-beauty
Rhexia sp.
Silky Dogwood
Cornus amomum
Persimmon
Diospyros virginiana
Common Henbit
Lamium amplexicaule
Trumpet Vine
Campsis radicans
Button-bush
Cephalanthus occidentalis
Cocklebur
Xanthium strumarium
Common Groundsel
Senecio vulgaris
Rabbit-tobacco
Gnaphalium obtusifolium
Black-knot
Apiosporina morbosa
Alder Cone Fungus
Taphrina alni-incanae

FAUNA

Green Lynx Spider
Peucetia viridans
Carolina Mantid (egg)
Stagmomantis carolina
Chinese Mantid (egg)
Tenodera aridifolia sinensis
Hackberry Gall Psyllid (aphid)
Pachypsylla venusta
Woolly Alder Aphid
Prociphilus tesselatus
Spiny Rose Gall (wasp)
Diplolepis bicolor
Rove Beetle
Tachinus axillaris
Tachinus pallipes
Small Carrion Beetle
Catops sp.
Eastern Tent Caterpillar (egg)
Malacosoma americanum
Upland Chorus Frog (call)
Pseudacris triseriata
Painted Turtle
Chrysemys picta
Great Blue Heron
Ardea herodias
Northern Harrier
Circus cyaneus
Sharp-shinned Hawk
Accipiter striatus
Red-tailed Hawk
Buteo jamaicensis
American Kestrel
Falco sparverius
Belted Kingfisher
Ceryle alcyon
Eastern Phoebe
Sayornis phoebe
Blue Jay
Cyanocitta cristata
American Crow
Corvus brachyrhynchos
Ruby-crowned Kinglet
Regulus calendula
Eastern Bluebird
Sialia sialis
Brown Thrasher
Toxostoma rufum
Yellow-rumped Warbler
Dendroica coronata
Northern Cardinal
Cardinalis cardinalis
Song Sparrow
Melospiza melodia
White-crowned Sparrow
Zonotrichia leucophrys
Eastern Meadowlark
Sturnella magna
Common Grackle
Quiscalus quiscula
Least Shrew
Cryptotis parva
Beaver (dam)
Castor canadensis
White-tailed Deer (remains)
Odocoileus virginianus

Bamberg County

Cathedral Bay Heritage Preserve

Bamberg County

July 23, 1988

Cathedral Bay, also known as Chitty Bay or Pond, is an excellent example of a Carolina Bay, an elliptical or oval basin found throughout the Atlantic Coastal Plain with the long axis of the ellipse oriented northwest-southeast. Soils of the bay interior differ markedly from soils outside the bay, interior soils being high in organic matter and hydric. This particular bay is a striking example of a blackwater cypress bay, with a remote, stately interior. The dominant vegetation is an almost pure stand of pond-cypress, a close relative of bald-cypress. The thick carpet of foot-tall grass adds to the cathedral-like atmosphere. The flora-fauna list does not do the site justice as a line of thunderstorms arrived soon after SCAN members did.

Directions to Cathedral Bay Heritage Preserve: From Columbia, take US 321 south through Gaston and Denmark about 60 miles to Olar. Turn left onto SC 64 toward Ehrhardt and Walterboro. After about 2 miles, stop at the Heritage Preserve sign on the right, just after a field and before tall woods. Cathedral Bay is inside the woods.

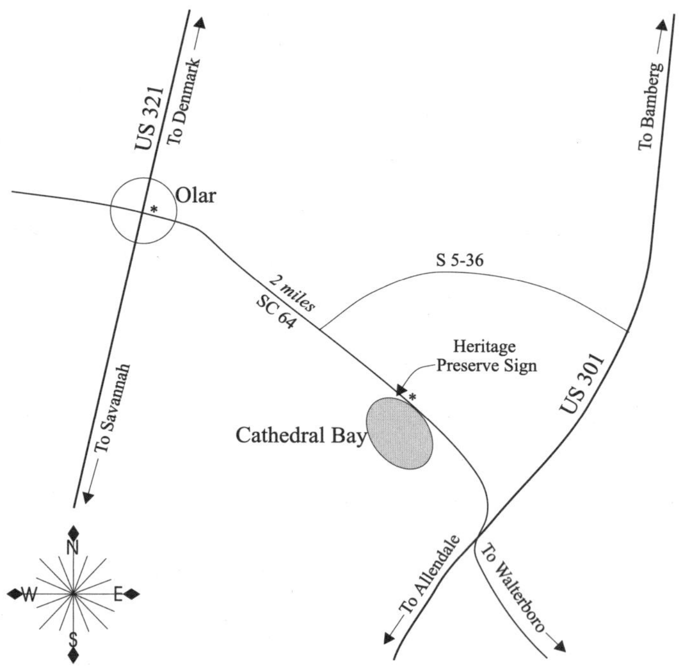

FLORA

Netted Chain Fern
Woodwardia areolata
Ebony Spleenwort
Asplenium platyneuron
Resurrection Fern
Polypodium polypodioides
Loblolly Pine
Pinus taeda
Pond-cypress
Taxodium ascendens
Fringed Greenbrier
Smilax bona-nox
Sawbrier; Greenbrier
Smilax glauca
Bamboo Smilax
Smilax laurifolia
Redroot
Lachnanthes caroliniana
Water Oak
Quercus nigra
Sassafras
Sassafras albidum
Myrtle-leaved Holly
Ilex cassine myrtifolia
Pepper-vine
Ampelopsis arborea
Wild Lettuce
Lactuca sp.
Silphium
Silphium astericus

FAUNA

Crab-shaped Spiny-back
Gasteracantha cancriformis
Orb-weaver
Araneus bicentenarius
Little Yellow (butterfly)
Eurema lisa
Broad-winged Hawk
Buteo platypterus
Red-tailed Hawk
Buteo jamaicensis
Northern Bobwhite
Colinus virginianus
American Crow
Corvus brachyrhynchos
Carolina Wren
Thryothorus ludovicianus
Northern Cardinal
Cardinalis cardinalis

Rivers Bridge State Park
Bamberg County
July 23, 1988

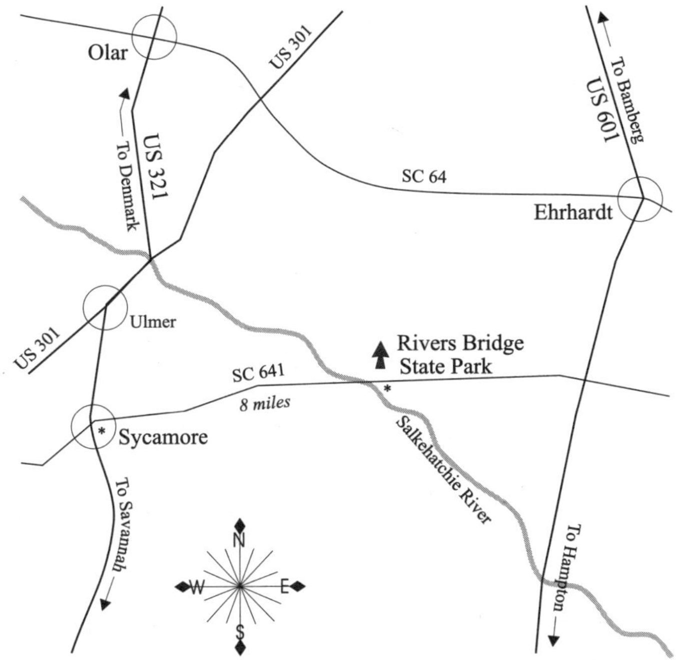

Rivers Bridge was a back-up in case it rained at Cathedral Bay, which it did. After lunch in the picnic shelter the rain stopped, allowing us to explore a pleasant river bottom area. The park superintendent was kind enough to guide his unexpected visitors to the Confederate breastworks built in a futile effort to stop Sherman's march to Columbia. Although they didn't stop Sherman, they did succeed in providing a habitat for an unusually complex riverside community.

Directions to Rivers Bridge State Park: From Columbia, take US 321 south through Gaston and Denmark about 70 miles to Sycamore in Allendale County. Turn left (east) onto SC 641 and drive about 8 miles. Just after crossing the Salkehatchie River and entering Bamberg County, watch for the sign indicating the left turn into Rivers Bridge State Park.

FLORA

Lichen
Parmotrema perforatum
Usnea sp.
Cinnamon Fern
Osmunda cinnamomea
Bracken
Pteridium aquilinum
Southern Lady Fern
Athyrium asplenioides
Sensitive Fern
Onoclea sensibilis
Virginia Chain Fern
Woodwardia virginica
Loblolly Pine
Pinus taeda
Bald-cypress
Taxodium distichum
Beak-rush
Rhynchospora sp.
Golden-club
Orontium aquaticum
Dayflower
Commelina sp.
Roseling; Spiderwort
Tradescantia rosea graminea
Southern Rein Orchid
Habenaria flava
Lizard's-tail
Saururus cernuus
American Hornbeam; Ironwood
Carpinus caroliniana
White Oak
Quercus alba
Swamp Chestnut Oak
Quercus michauxii
Heartleaf
Hexastylis arifolia
Sweet-gum
Liquidambar styraciflua
Hawthorn
Crataegus sp.
Rattlebox
Crotalaria angulata
Pencil-flower
Stylosanthes biflora
Beggar-lice
Desmodium sp.
Tephrosia
Tephrosia sp.
Tread-softly
Cnidoscolus stimulosus
Croton
Croton glandulosus septentrionalis
Flowering Spurge
Euphorbia corollata
Poison-ivy
Rhus radicans
Winged Sumac
Rhus copallina
Boxelder
Acer negundo
Red Maple
Acer rubrum
Violet
Viola sp.
Prickly-pear
Opuntia compressa
Pennywort
Hydrocotyle sp.
Blackgum
Nyssa sylvatica
Sweet Pepperbush
Clethra alnifolia
Yellow Jessamine
Gelsemium sempervirens
Sabatia
Sabatia calycina
Ground-ivy; Ponyfoot
Dichondra carolinensis
Trumpet Vine
Campsis radicans
Dyschoriste
Dyschoriste oblongifolia
Water-willow
Justicia ovata
Partridge-berry
Mitchella repens
Bluets
Houstonia purpurea
Cardinal-flower
Lobelia cardinalis
Rabbit-tobacco
Gnaphalium obtusifolium
Silky Parchment
Stereum striatum
Gemmed Amanita
Amanita gemmata
Zoned-cap Collybia
Crinipellis zonata

FAUNA

Mottled Sand Grasshopper
Spharagemon collare
Spur-throated Grasshopper
Melanoplus sp.
Ebony Jewelwing
Calopteryx maculata
Dingy Ground Beetle
Discoderus sp.
Tick Trefoil Leaf-miner
Pachyschelus laevigatus
Blister Beetle
Epicauta sp.
Convergent Lady Beetle
Hippodamia convergens
Gray-sided Oak Weevil
Pandeleteius hilaris
Black Swallowtail
Papilio polyxenes asterius
Eastern Tiger Swallowtail
Papilio glaucus
Eastern Musk Turtle
Sternotherus odoratus
Cattle Egret
Bulbicus ibis
Yellow-billed Cuckoo
Coccyzus americanus
Ruby-throated Hummingbird
Archilochus colubris
Belted Kingfisher
Ceryle alcyon
Red-bellied Woodpecker
Melanerpes carolinus
Downy Woodpecker
Picoides pubescens
Acadian Flycatcher
Empidonax virescens
Tufted Titmouse
Parus bicolor
Summer Tanager
Piranga rubra

Barnwell County

Barnwell State Park
Barnwell County
November 16, 1991

Barnwell State Park sits on the transition between inner coastal plain and sandhills habitat. It is a mildly rolling terrain that leads to interesting microhabitats. The trail around the main lake in the park provided ladies'-tresses orchids (*Spiranthes*) still in bloom in a seepage area at the base of a rich hardwood slope. A variety of frogs, a banded water snake about to shed its skin, and an abundance of dragonflies and damselflies provided interesting fauna in the midst of the varied flora.

Directions to Barnwell State Park: From Columbia, take US 321 south to Denmark (about 50 miles.) Turn right (west) onto US 78 and travel about 8 miles to Blackville. Turn left (south) on SC 3 toward Barnwell. The park is 2-3 miles on the right.

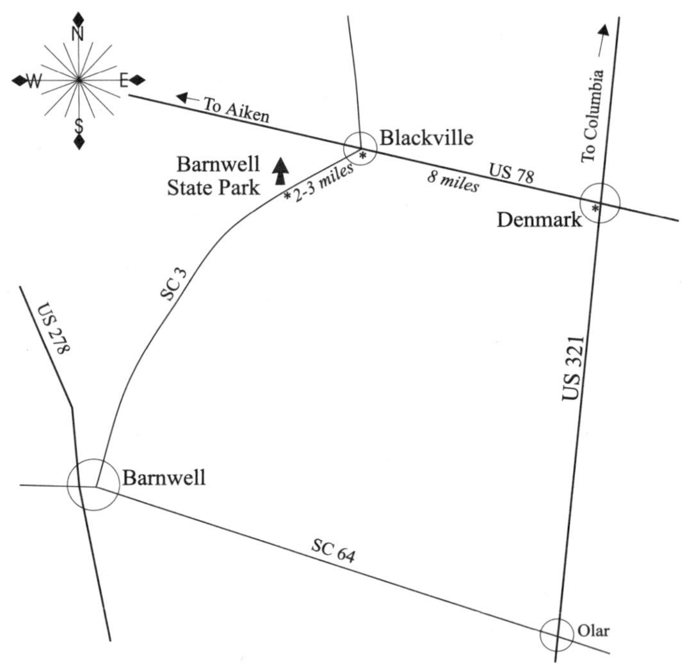

FLORA

Lichen
Leptogium burnetiae
Graphis scripta
Graphis sp.
Gyrostomum scyphuliferum
Coenogonium sp.
Bacidia schweinitzii
Brigantiaea leucoxantha
Lecidea russula
Cladonia polycarpoides
Pertusaria ostiolata
Pertusaria paratuberculifera
Pertusaria subpertusa
Pertusaria texana
Haematomma puniceum
Lecanora cupressi
Lecanora varia
Ochrolechia africana
Bulbothrix goebelii
Hypotrachyna formosana
Hypotrachyna livida
Parmelina horrescens
Parmotrema crinitum
Parmotrema mellissii
Parmotrema rampoddense
Parmotrema reticulatum
Parmotrema subisidiosum
Parmotrema ultralucens
Pseudoparmelia caroliniana
Buellia punctata
Physcia cloantha
Heterodermia albicans
Heterodermia hypoleuca
Heterodermia obscurata
Pyrenula cruenta
Pyrenula sp.
Trypethelium tropicum
Trypethelium virens
Porina rhaphidosperma

Meadow Spikemoss
Selaginella apoda
Southern Grape Fern
Botrychium biternatum
Cinnamon Fern
Osmunda cinnamomea
Royal Fern
Osmunda regalis spectabilis
Bracken
Pteridium aquilinum
Southern Lady Fern
Athyrium asplenioides
Virginia Chain Fern
Woodwardia virginica
Netted Chain Fern
Woodwardia areolata
Ebony Spleenwort
Asplenium platyneuron
Resurrection Fern
Polypodium polypodioides
Loblolly Pine
Pinus taeda
Pond Pine
Pinus serotina
Eastern Red-cedar
Juniperus virginiana
Common Cat-tail
Typha latifolia
Spotted Pondweed
Potamogeton pulcher
Cane; Switch Cane
Arundinaria gigantea
Uniola Grass
Uniola laxa
Rice Cut Grass
Leersia oryzoides
Dulichium (sedge)
Dulichium arundinaceum

Arrowleaf Arum
Peltandra virginica
Jack-in-the-pulpit
Arisaema triphyllum
Duckweed
Lemna sp.
Water-meal
Wolffia papulifera
Bogmat
Wolffiella floridana
Yellow-eyed-grass
Xyris sp.
Spanish-moss
Tillandsia usneoides
Soft Rush
Juncus effusus
Carrion-flower
Smilax ecirrhata
Fringed Greenbrier
Smilax bona-nox

Sawbrier; Greenbrier
Smilax glauca
Red-berried Greenbrier
Smilax walteri
Bamboo Smilax
Smilax laurifolia
Indian Cucumber-root
Medeola virginiana
Fly-poison
Amianthium muscaetoxicum
Wild Yam
Dioscorea villosa
Burmannia
Burmannia biflora
Small Green Wood Orchid
Habenaria clavellata
Nodding Ladies'-tresses
Spiranthes cernua
Crane-fly Orchid
Tipularia discolor

Black Willow
Salix nigra
Wax-myrtle; Southern Bayberry
Myrica cerifera
Swamp Bayberry
Myrica heterophylla
Mockernut Hickory
Carya tomentosa
Pignut Hickory
Carya glabra
Tag Alder
Alnus serrulata
White Oak
Quercus alba
Post Oak
Quercus stellata
Overcup Oak
Quercus lyrata
Live Oak
Quercus virginiana
Black Oak
Quercus velutina
Spanish Oak
Quercus falcata
Black Jack Oak
Quercus marilandica
Water Oak
Quercus nigra
Red Mulberry
Morus rubra
False Nettle
Boehmeria cylindrica
Mistletoe
Phoradendron serotinum
Heartleaf
Hexastylis arifolia
Polygonum
Polygonum sp.
Pokeweed; Pokeberry
Phytolacca americana
Water-lily
Nymphaea odorata
Tulip-poplar
Liriodendron tulipifera
Sweetbay
Magnolia virginiana
Southern Magnolia
Magnolia grandiflora
Redbay
Persea borbonia
Sassafras
Sassafras albidum
Bloodroot
Sanguinaria canadensis
Virginia-willow
Itea virginica
Climbing-hydrangea
Decumaria barbara
Sweet-gum
Liquidambar styraciflua
Witch-hazel
Hamamelis virginiana
Cocklebur
Agrimonia sp.
Red Chokeberry
Sorbus arbutifolia arbutifolia

Shadbush; Serviceberry
Amelanchier sp.
Wild Plum
Prunus americana
Black Cherry; Wild Cherry
Prunus serotina
Carolina Cherrylaurel
Prunus caroliniana
Mimosa; Silk-tree
Albizia julibrissin
American Wisteria
Wisteria frutescens
Chinese Wisteria
Wisteria sinensis
Orange Milkwort
Polygala lutea
Poison Sumac
Rhus vernix
Poison-ivy
Rhus radicans
Winged Sumac
Rhus copallina
Titi
Cyrilla racemiflora
American Holly
Ilex opaca
Inkberry
Ilex glabra
Sweet Gallberry
Ilex coriacea
Red Maple
Acer rubrum
Supple-jack; Rattan-vine
Berchemia scandens
Virginia-creeper
Parthenocissus quinquefolia
Muscadine; Bullace
Vitis rotundifolia
Devil's-shoestring
Sida rhombifolia
St.-Andrew's-cross
Hypericum hypericoides
St.-John's-wort
Hypericum virginicum
St.-John's-wort
Hypericum walteri
Hypericum
Hypericum sp.
Thorny Elaeagnus
Elaeagnus pungens
Water Loosestrife
Decodon verticillatus
Pale Meadow-beauty
Rhexia mariana
Water-willow
Ludwigia decurrens
Ludwigia
Ludwigia sp.
English Ivy
Hedera helix
Devil's-walking-stick
Aralia spinosa
Pennywort
Hydrocotyle sp.
Sanicle; Snakeroot
Sanicula canadensis

Black Tupelo; Blackgum
Nyssa sylvatica sylvatica
Swamp Tupelo; Swamp Blackgum
Nyssa sylvatica biflora
Flowering Dogwood
Cornus florida
Sweet Pepperbush
Clethra alnifolia
Spotted Wintergreen
Chimaphila maculata
Piedmont Azalea
Rhododendron canescens
Azalea
Rhododendron sp.
Fetterbush
Lyonia lucida
Dog-hobble
Leucothoe axillaris axillaris
Coastal Honeybells
Leucothoe racemosa
Dangleberry
Gaylussacia frondosa
Sparkleberry
Vaccinium arboreum
Highbush Blueberry
Vaccinium corymbosum
Blueberry
Vaccinium sp.
Horse-sugar; Sweetleaf
Symplocos tinctoria
Styrax
Styrax sp.
Ash
Fraxinus sp.
Fringe-tree
Chionanthus virginicus
Wild Olive
Osmanthus americana
Chinese Privet
Ligustrum sinense
Yellow Jessamine
Gelsemium sempervirens
Miterwort
Cynoctotum mitreola
Dodder
Cuscuta sp.
Hydrolea
Hydrolea quadrivalvis
Beautyberry
Callicarpa americana
Skullcap
Scutellaria sp.
Blue Water-hyssop
Bacopa caroliniana
Shade Mudflower
Micranthemum umbrosum
Cross Vine
Anisostichus capreolata
Trumpet Vine
Campsis radicans
Button-bush
Cephalanthus occidentalis
Richardia
Richardia brasiliensis
Partridge-berry
Mitchella repens

Possumhaw Viburnum
Viburnum nudum
American Elderberry
Sambucus canadensis
Cardinal-flower
Lobelia cardinalis
Elephant's-foot
Elephantopus sp.
Climbing Hempweed
Mikania scandens
Marsh-fleabane
Pluchea camphorata
Camphorweed
Pluchea sp.
Sea-myrtle; Groundsel-tree
Baccharis halimifolia
Aster
Aster dumosus
Aster sp.
Beggar-ticks
Bidens laevis
Black-knot
Apiosporina morbosa
Tree-ear
Auricularia sp.
Smoky Polypore
Bjerkandera sp.

FAUNA

White-lipped Forest Snail
Triodopsis albolabris
Lesser Mystery Snail
Campeloma decisum
Crab-shaped Spiny-back
Gasteracantha cancriformis
Basilica Spider (egg)
Mecynogea lemniscata
Orb-weaver
Neoscona hentzi
Marbled Orb-weaver
Araneus marmoreus
Arrowhead Spider
Verrucosa arenata
Long-jawed Orb-weaver
Tetragnatha elongata
Nursery-web Spider
Pisaurina mira
Jumping Spider
Metaphidippus sp.
Common Meadow Katydid
Orchelimum vulgare
Cattail Toothpick Grasshopper
Leptysma marginicollis
American Bird Grasshopper
Schistocerca americana
Carolina Mantid
Stagmomantis carolina
Green Stink Bug
Nezara viridula
Periodical Cicada
Magicicada sp.
Spiny Witch Hazel Gall (aphid)
Hamamelistes spinosus
Oak-apple (gall wasp)
Amphibolips inanis

Barnwell State Park (continued)
November 16, 1991

Red Imported Fire Ant
Solenopsis invicta
Bald-faced Hornet
Vespula maculata
Yellowjacket
Vespula sp.
Paper Wasp (nest)
Polistes sp.
Phantom Cranefly
Bittacomorpha clavipes
Woodland Ground Beetle
Pterostichus sp.
False Bombadier Beetle
Galeritula bicolor
Crawling Water Beetle
Peltodytes sexmaculatus
Peltodytes shermani
Predacious Diving Beetle
Hydroporus clypealis
Hydroporus sp.
Coptotomus interrogatus
Whirligig Beetle
Dineutus emarginatus
Dineutus assimilis
Gyrinus sp.
Red Net-winged Beetle
Dictyopterus aurora
Darkling Beetle
Uloma punctulata
Pleasing Fungus Beetle
Tritoma sp.
Cloudless Sulphur
Phoebis sennae eubule
Sleepy Orange (butterfly)
Eurema nicippe
Zebra (butterfly)
Heliconius charitonius tuckeri
Anglewing (butterfly)
Polygonia sp.
American Lady (butterfly)
Vanessa virginiensis
Common Buckeye (butterfly)
Junonia coenia
Confused Eusarca Moth
Eusarca confusaria
Regal Moth
Citheronia regalis
Mosquitofish
Gambusia sp.
Southern Cricket Frog
Acris gryllus
Green Treefrog
Hyla cinerea
Spring Peeper
Hyla crucifer
Squirrel Treefrog
Hyla squirella
Southern Leopard Frog
Rana sphenocephala
Eastern Box Turtle
Terrapene carolina
Carolina Anole
Anolis carolinensis
Ground Skink
Scincella lateralis
Banded Water Snake
Nerodia fasciata
Great Blue Heron
Ardea herodias
American Woodcock
Scolopax minor
Turkey Vulture
Cathartes aura
Red-tailed Hawk
Buteo jamaicensis
Northern Bobwhite
Colinus virginianus
Belted Kingfisher
Ceryle alcyon
Red-bellied Woodpecker
Melanerpes carolinus
Northern Flicker
Colaptes auratus
Pileated Woodpecker
Dryocopus pileatus
Eastern Phoebe
Sayornis phoebe
Carolina Chickadee
Parus carolinensis
Brown Creeper
Certhia americana
Golden-crowned Kinglet
Regulus satrapa
Ruby-crowned Kinglet
Regulus calendula
American Robin
Turdus migratorius
Pine Warbler
Dendroica pinus
Red-winged Blackbird
Agelaius phoeniceus
Gray Squirrel
Sciurus carolinensis
Beaver (gnaw mark)
Castor canadensis

Beaufort County

Hunting Island State Park
Beaufort County
October 28, 1989

Hunting Island State Park is the southernmost park on our coast. It is a classic barrier island with the front beach exposed to the Atlantic Ocean and a broad tidal creek and expansive marsh on the inland side. This visit by SCAN was shortly after Hurricane Hugo hit the state; the beach erosion caused by this and earlier storms was impressive. During this visit, high tide reached the treeline and completely covered the open beach. A long line of toppled trees and other vegetation, along with still standing trees rapidly eroding, marked the high tide line.

October brings lots of butterflies to the coast. Mating monarchs and a close relative, the queen, were seen. The queen and zebra butterflies are rarely seen further north at this time of year. A few mild winters, however, have helped the zebra increase its range north and inland. A rich maritime forest helped to add to the variety of species seen.

Directions to Hunting Island State Park: From I-95 south of Walterboro, take exit 42 (US 21) south past Yemassee toward Beaufort. US 21 will join US ALT-17 for about 18 miles, then joins US 17 (north) for 11 miles. At Gardens Corner, make the right turn toward Beaufort. In Beaufort, follow Business 21 over the bridge to St. Helena Island. Continue on US 21 all the way to Hunting Island. The Visitors' Center is at the second left turn after reaching Hunting Island; follow the signs.

(Photo by L. Greenlee)

Mating monarch butterflies, *Danaus plexippus*.

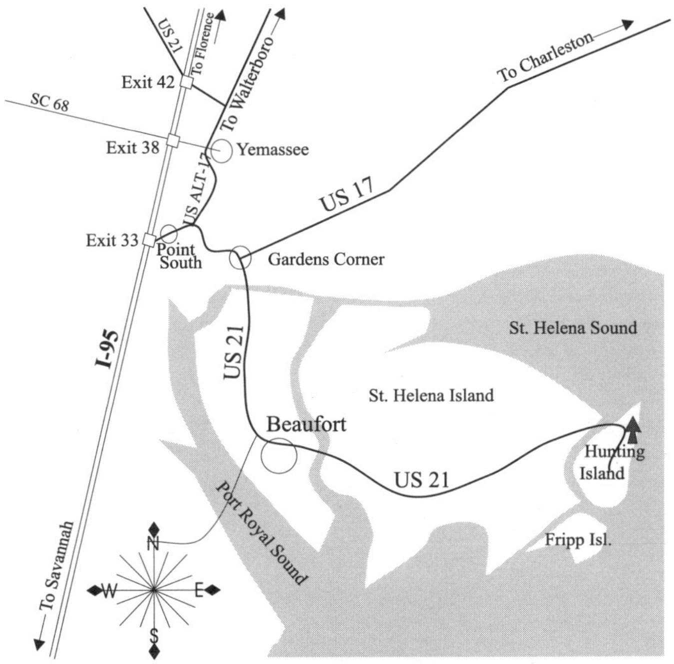

FLORA

Long-leaf Pine
Pinus palustris
Slash Pine
Pinus elliotti
Southern Red-cedar
Juniperus silicicola
Common Cat-tail
Typha latifolia
Purple-top
Tridens flavus
Sand Grass
Triplasis purpurea
Uniola Grass
Uniola laxa
Sea Oats
Uniola paniculata
Coastal Dropseed
Sporobolus virginicus
Crowfoot Grass
Dactyloctenium aegyptium
Finger Grass
Chloris petraea
Cord Grass
Spartina patens
Smooth Cord Grass
Spartina alterniflora
Dune Sandbur
Cenchrus tribuloides
Oplismenus (grass)
Oplismenus setarius
Baggy-knees Grass
Sacciolepis striata
Seaside Panic Grass
Panicum amarum
Switch Grass
Panicum virgatum
Broomsedge
Andropogon virginicus
Indian Grass
Sorghastrum sp.
Fimbristylis (sedge)
Fimbristylis spadicea
Nut-rush
Scleria triglomerata
Saw Palmetto
Serenoa repens
Dwarf Palmetto
Sabal minor
Cabbage Palmetto
Sabal palmetto
Spanish-moss
Tillandsia usneoides
Black Needle Rush
Juncus roemerianus
Fringed Greenbrier
Smilax bona-nox
Spanish-bayonet
Yucca aloifolia
Three Birds Orchid
Triphora trianthophora
Wax-myrtle; Southern Bayberry
Myrica cerifera
Live Oak
Quercus virginiana
False Nettle
Boehmeria cylindrica
Mexican-tea
Chenopodium ambrosioides
Seabeach Orach
Atriplex arenaria
Sea-blight
Suaeda linearis
Saltwort
Batis maritima
Iresine
Iresine rhizomatosa
Redbay
Persea borbonia
Sweet-gum
Liquidambar styraciflua
Dewberry
Rubus trivialis
Bladderpod
Glottidium vesicarium
Wild Indigo
Indigofera caroliniana
Coral-bean
Erythrina herbacea
Wood-sorrel
Oxalis dillenii
Hercules'-club
Zanthoxylum clava-herculis
Silverleaf Croton
Croton punctatus
Three-seeded Mercury
Acalypha sp.
Spurge
Euphorbia ammannioides
Spurge
Euphorbia polygonifolia
Poison-ivy
Rhus radicans
Winged Sumac
Rhus copallina
Yaupon
Ilex vomitoria
Virginia-creeper
Parthenocissus quinquefolia
Muscadine; Bullace
Vitis rotundifolia
Rockrose
Helianthemum corymbosum
Prickly-pear
Opuntia compressa
Devil-joint
Opuntia drummondii
Devil's-walking-stick
Aralia spinosa
Dune Pennywort
Hydrocotyle bonariensis
Sea-lavender
Limonium carolinianum
Limonium nashii
Wild Olive
Osmanthus americana
Milkweed-vine
Cynanchum palustre
Verbena
Verbena scabra
Fog-fruit
Lippia nodiflora
Beautyberry
Callicarpa americana
Blue-curls
Trichostema dichotomum
Wood Sage; Germander
Teucrium canadense
Ground-cherry
Physalis viscosa maritima
Gerardia
Agalinis tenella
Trumpet Vine
Campsis radicans
Richardia
Richardia brasiliensis
Trailing Bluets
Houstonia procumbens
Bedstraw; Purple-bead
Galium hispidulum
Creeping Cucumber
Melothria pendula
Marsh-elder
Iva frutescens
Beach-elder
Iva imbricata
Fireweed
Erechtites hieracifolia
Thistle
Carduus sp.
Dog-fennel
Eupatorium capillifolium
Climbing Hempweed
Mikania scandens
Camphorweed
Pluchea purpurascens
False-willow
Baccharis angustifolia
Groundsel-tree
Baccharis glomeruliflora
Sea-myrtle; Groundsel-tree
Baccharis halimifolia
Salt-marsh Aster
Aster tenuifolius
Heterotheca
Heterotheca nervosa nervosa
Camphorweed
Heterotheca subaxillaris
Sea Ox-eye
Borrichia frutescens
Carbon-balls (fungus)
Daldinia concentrica
Bear's-head Tooth
Hericium coralloides
Artist's Conk
Ganoderma applanatum
Chicken Mushroom
Laetiporus sulphureus
Crowded Parchment
Stereum complicatum
Meadow Mushroom
Agaricus campestris
Citron Amanita
Amanita citrina
White Fiber-head
Inocybe geophylla
Emetic Russula
Russula emetica
Yellow Oyster-mop
Tricholomopsis sulfureoides

FAUNA

Sea-whip
Leptogorgia virgulata
Rubbery Bryozoan
Alcyonidium hauffi
Marsh Periwinkle; Mud Snail
Littorina irrorata

Hunting Island State Park (continued)
October 28, 1989

Knobbed Whelk (shell)
Busycon carica
Sea-whip Barnacle
Conopea galeata
Common Blue Crab
Callinectes sapidus
Ghost Crab
Ocypode quadrata
Fiddler Crab
Uca sp.
Predacious Diving Beetle
Agabus punctatus
Water Scavenger Beetle
Tropisternus lateralis nimbatus
Hydrochara soror
Berosus infuscatus
Palmetto Tortoise Beetle
Hemisphaerota cyanea
Clouded Sulphur
Colias philodice
Cloudless Sulphur
Phoebis sennae eubule
Little Yellow (butterfly)
Eurema lisa
Sleepy Orange (butterfly)
Eurema nicippe
Gray Hairstreak
Strymon melinus
American Snout (butterfly)
Libytheana carinenta
Gulf Fritillary
Agraulis vanillae
Zebra (butterfly)
Heliconius charitonius tuckeri
Phaon Crescent
Phyciodes phaon
Common Buckeye (butterfly)
Junonia coenia
Monarch (butterfly)
Danaus plexippus
Queen (butterfly)
Danaus gilippus
Long-tailed Skipper
Urbanus proteus
Southern Flannel Moth (larva)
Megalopyge opercularis
Polyphemus Moth (cocoon)
Antheraea polyphemus
Pink-spotted Hawk Moth
Agrius cingulatus
Bella Moth
Utetheisa bella
Virginia Ctenucha Moth
Ctenucha virginica
American Alligator
Alligator mississippiensis
Carolina Anole
Anolis carolinensis
Six-lined Racerunner
Cnemidophorus sexlineatus
Glass Lizard
Ophisaurus sp.
Black Racer
Coluber constrictor
Common Loon
Gavia immer
Brown Pelican
Pelecanus occidentalis
Double-crested Cormorant
Phalacrocorax auritus
Little Blue Heron
Egretta caerulea
Great Egret
Casmerodius albus
Great Blue Heron
Ardea herodias
Hooded Merganser
Lophodytes cucullatus
Clapper Rail
Rallus longirostris
Semipalmated Plover
Charadrius semipalmatus
Black-bellied Plover
Pluvialis squatarola
Willet
Catoptrophorus semipalmatus
Dunlin
Calidris alpina
Sanderling
Calidris alba
Semipalmated Sandpiper
Calidris pusilla
Least Sandpiper
Calidris minutilla
Ruddy Turnstone
Arenaria interpres
Laughing Gull
Larus atricilla
Ring-billed Gull
Larus delawarensis
Herring Gull
Larus argentatus
Great Black-backed Gull
Larus marinus
Royal Tern
Sterna maxima
Caspian Tern
Sterna caspia
Black Skimmer
Rhynchops niger
Osprey
Pandion haliaetus
American Kestrel
Falco sparverius
Belted Kingfisher
Ceryle alcyon
Red-bellied Woodpecker
Melanerpes carolinus
Yellow-bellied Sapsucker
Sphyrapicus varius
Pileated Woodpecker
Dryocopus pileatus
Tree Swallow
Tachycineta bicolor
Blue Jay
Cyanocitta cristata
Carolina Chickadee
Parus carolinensis
Brown-headed Nuthatch
Sitta pusilla
Ruby-crowned Kinglet
Regulus calendula
Gray Catbird
Dumetella carolinensis
Northern Mockingbird
Mimus polyglottos
Brown Thrasher
Toxostoma rufum
White-eyed Vireo
Vireo griseus
Yellow-rumped Warbler
Dendroica coronata
Pine Warbler
Dendroica pinus
Northern Cardinal
Cardinalis cardinalis
Red-winged Blackbird
Agelaius phoeniceus
Common Grackle
Quiscalus quiscula
Boat-tailed Grackle
Quiscalus major
Virginia Opossum
Didelphis virginiana
Gray Squirrel
Sciurus carolinensis
Raccoon
Procyon lotor
White-tailed Deer
Odocoileus virginianus

(Photo by A. Ciegler)
Thistle, *Carduus* sp.

Pinckney Island National Wildlife Refuge
Beaufort County
February 22, 1986

Pinckney Island is a large island just inland from Hilton Head. The entire island plus several smaller islands across Mackey's Creek make up the wildlife refuge. Pinckney Island, named for Charles Pinckney, a signer of the U.S. constitution and early owner of the island, makes up the largest portion of the refuge, and the only part readily accessible by automobile. It has tidal marsh, maritime forest, open fields and several freshwater ponds. The bird population, especially during the winter and migration seasons is usually quite diverse. The freshwater ponds attract a lot of mammals, amphibians and reptiles as well. There were few developed trails when SCAN visited in 1986. The addition of many miles of trails since then should make the exploring of this refuge even more rewarding. It is a classic tidewater habitat.

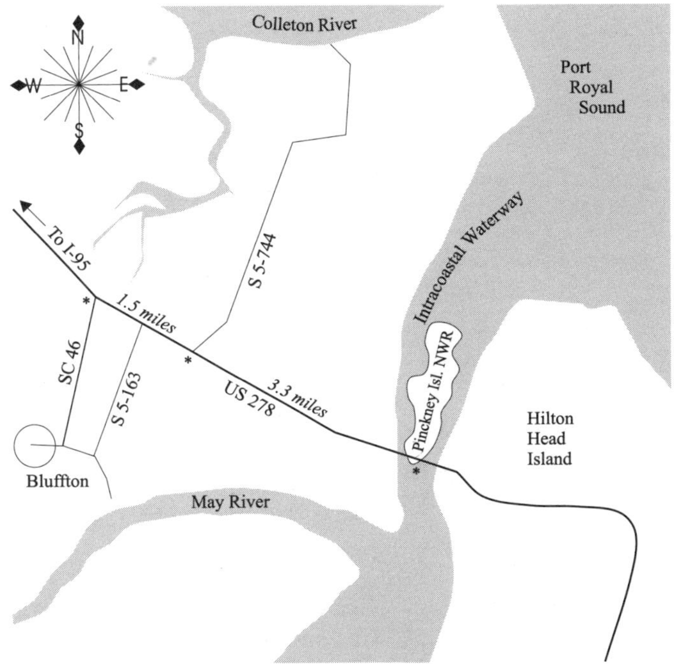

Directions to Pinckney Island National Wildlife Refuge: From I-95 near Ridgeland, take exit 21, US 278, east toward Hilton Head. Follow the signs; US 278 makes many twists and turns. After passing the junction with SC 46 at Bluffton, it is about 4 miles to the bridge over Mackey's Creek. Ahead you will see the higher bridge over Skull Creek and the Intracoastal Waterway. Between the bridges is the entrance on the north side of the highway. Approaching Hilton Head, it is the last left turn before crossing the bridge to Hilton Head. Leaving Hilton Head it is the first right turn after coming over the bridge from Hilton Head.

FLORA

Spleenwort
Asplenium platyneuron
Resurrection Fern
Polypodium polypodioides
Long-leaf Pine
Pinus palustris
Eastern Red-cedar
Juniperus virginiana
Common Cat-tail
Typha latifolia
Cane; Switch-cane
Arundinaria gigantea
Purple Top
Tridens flavus
Broomsedge
Andropogon virginicus
Dwarf Palmetto
Sabal minor
Cabbage Palmetto
Sabal palmetto
Spanish-moss
Tillandsia usneoides
Fringed Greenbriar
Smilax bona-nox
Spanish-bayonet
Yucca aloifolia
Narcissus
Narcissus sp.
Poplar; Cottonwood
Populus sp.
Wax-myrtle; Southern Bayberry
Myrica cerifera
Live Oak
Quercus virginiana
Mistletoe
Phoradendron serotinum
Glasswort
Salicornia bigelovii
Perennial Glasswort
Salicornia virginica
Saltwort
Batis maritima
Chickweed
Stellaria media
Redbay
Persea borbonia
Corydalis
Corydalis micrantha australis
Tansy Mustard
Descurainia pinnata
Sweet-gum
Liquidambar styraciflua
Dewberry
Rubus trivialis
Rattlebox
Crotalaria spectabilis
Bladderpod
Glottidium vesicarium
Wild Geranium
Geranium carolinianum
Hercules' club
Zanthoxylum clava-herculis
Chinaberry
Melia azedarach
Winged Sumac
Rhus copallina
Yaupon
Ilex vomitoria
Red Maple
Acer rubrum
Red Buckeye
Aesculus pavia
St.-Andrews-Cross
Hypericum hypericoides
Violet
Viola sp.
Ludwigia
Ludwigia alternifolia
Devils-walking-stick
Aralia spinosa
Centella
Centella asiatica
Sea-lavender
Limonium sp.
Yellow Jessamine
Gelsemium sempervirens
Polypremem
Polypremem procumbeas
Brazilian Vervain
Verbena brasiliensis
Fog-fruit
Lippia nodiflora
Blue-curls
Trichostema dichotomum
Common Henbit
Lamium amplexicaule
Bacopa
Bacopa monnieri
Woolly Mullein
Verbascum thapsus
Cross Vine
Anisostichus capreolata

Pinckney Island National Wildlife Refuge (continued)
February 22, 1986

Trumpet Vine
Campsis radicans
Trailing Bluets
Houstonia procumbens
False-willow
Baccharis angustifolia
Sea-myrtle; Groundsel-tree
Baccharis halimifolia
Sea Ox-eye
Borrichia frutescens
Shaggy Ink Cap
Coprinus comatus
Common Split-gill
Schizophyllum commune
Beaked Earthstar
Geastrum pectinatum
Mustard-yellow Polypore
Phellinus gilvus
Chocolate Tube Slime
Stemonitis splendens

FAUNA

Elongate Cannibal Snail
Euglandina rosea
Writing Spider (egg sac)
Argiope aurantia
Wolf Spider
Lycosa sp.
Carolina Mantid (egg mass)
Stagmomantis carolina
Potter Wasp (nest)
Eumenes fraterna
Goldenrod Ball Gall Fly
Eurosta solidaginis
Goldenrod Bunch Gall Midge
Rhopalomyia solidaginis
Cloudless Sulphur
Phoebis sennae eubule
Snout Butterfly
Libytheana bachmannii
Southern Flannel Moth
Megalopyga opercularis
Green Treefrog
Hyla cinerea
Carolina Anole
Anolis carolinensis
Ground Skink
Scincella laterale
Pied-billed Grebe
Podilymbus podiceps
Double-crested Cormorant
Phalacrocorax auritus
Great Egret
Casmerodius albus
Bufflehead
Bucephala albeola
Red-breasted Merganser
Mergus serrator
American Coot
Fulica americana
Ring-billed Gull
Larus delawarensis
Turkey Vulture
Cathartes aura
American Kestrel
Falco sparverius
Red-bellied Woodpecker
Melanerpes carolinus
Northern Flicker
Colaptes auratus
Yellow-bellied Sapsucker
Sphyrapicus varius
Eastern Phoebe
Sayornis phoebe
Blue Jay
Cyanocitta cristata
American Crow
Corvus brachyrhynchos
Tufted Titmouse
Parus bicolor
Carolina Chickadee
Parus carolinensis
Carolina Wren
Thryotherus ludovicianus
Marsh Wren
Cistothorus palustris
Ruby-crowned Kinglet
Regulus calendula
Eastern Bluebird
Sialia sialis
Gray Catbird
Dumetella carolinensis
Northern Mockingbird
Mimus polyglottos
Yellow-rumped Warbler
Dendroica coronata
Pine Warbler
Dendroica pinus
Northern Cardinal
Cardinalis cardinalis
Rufous-sided Towhee
Pipilo erythrophthalmus
Savannah Sparrow
Passerculus sandwichensis
White-throated Sparrow
Zonotrichia albicollis
Red-winged Blackbird
Agelaius phoeniceus
Purple Finch
Carpodacus purpureus
Virginia Opossum
Didelphis virginiana
Gray Fox
Urocyon cinereoargenteus
Raccoon
Procyon lotor
River Otter
Lutra canadensis
Bobcat
Felis rufus
White-tailed Deer
Odocoileus virginianus

Spring Island
Beaufort County
February 27, 1993

Spring Island in Beaufort County is surrounded by tidewater, but it includes a pond with lots of bird life, a natural waterfall, and the ruins of the original tabby-built (tabby uses coquina shells) plantation house. Spring Island Development Company hired a naturalist, Bruce Lampright, to select environmentally sensitive areas to be protected on the island in addition to establishing guidelines for individual lot development. Owners can't cut down a tree without Bruce's permission, not even if the owner is rich and famous! One of the highlights of the winter woods on the island is a huge concentration of *Trillium cuneatum* and the opportunity to see fox squirrels and wintering birds. Tree swallows were a bird highlight for this trip. Bruce Lampright's guidance on this trip added greatly to its enjoyment and was highly appreciated.

Privately owned by Spring Island Development Company. Advance permission needed to visit.

Directions to Spring Island: From I-95 between Walterboro and Ridgeland, take exit 28 (Coosawhatchie-Hilton Head) and follow SC 462 east, merging with US 278 after about 10 miles. Follow US 278 for 8 miles more, then turn left onto SC 170 by the Beaufort-Jasper Career Education Center. Follow SC 170 about 3.5 miles to the Callawassie Store/BP Station, and turn right at the store. Go 1.2 miles to the gatehouse at Callawassie Island. Guard at the gate can direct you to the Spring Island Reception Center; advance arrangements must be made.

(Photos by A. Ciegler)

Spring Island has beautiful ponds and marshes.

FLORA

Sea Lettuce
Ulva lactuca
Lichen
Leptogium cyanescens
Graphis scripta
Graphis elegans
Phaeographis dendritica
Phaeographis punctiformis
Bacidia dryina
Bacidia assulata
Leptogium austroamericanum
Scoliciosporum umbrinum
Micarea chlorosticta
Cladonia pityrea
Pertusaria ostiolata
Pertusaria paratuberculifera
Pertusaria subpertusa
Pertusaria texana
Pertusaria sinusmexicani
Pertusaria sp.
Haematomma elatinum
Lecanora cupressi
Lecanora pallida
Lecanora varia
Ochrolechia africana
Bulbothrix confoederata
Bulbothrix goebelii
Hypotrachyna livida
Parmelina horrescens
Parmelina minarum
Parmotrema hypotropum
Parmotrema michauxianum
Parmotrema perforatum
Parmotrema praesorediosum
Parmotrema stuppeum
Parmotrema subisidiosum
Parmotrema subtinctorium
Parmotrema tinctorum
Canoparmelia caroliniana
Punctelia rudecta
Rimelia reticulata
Ramalina willeyi
Usnea strigosa
Buellia punctata
Dirinaria picta
Dirinaria purpurascens
Dirinaria sp.
Physcia aipolia
Physcia millegrana
Heterodermia albicans
Pyxine caesiopruinosa
Pyxine eschweileri
Rinodina sophodes
Physconia enteroxantha
Caloplaca citrina
Pyrenula cruenta
Pyrenula mamillana
Staurothele diffractella
Arthothelium interveniens
Bracken
Pteridium aquilinum
Netted Chain Fern
Woodwardia areolata
Ebony Spleenwort
Asplenium platyneuron
Resurrection Fern
Polypodium polypodioides
Long-leaf Pine
Pinus palustris
Loblolly Pine
Pinus taeda
Spruce Pine
Pinus glabra
Eastern Red-cedar
Juniperus virginiana
Cane; Switch Cane
Arundinaria gigantea

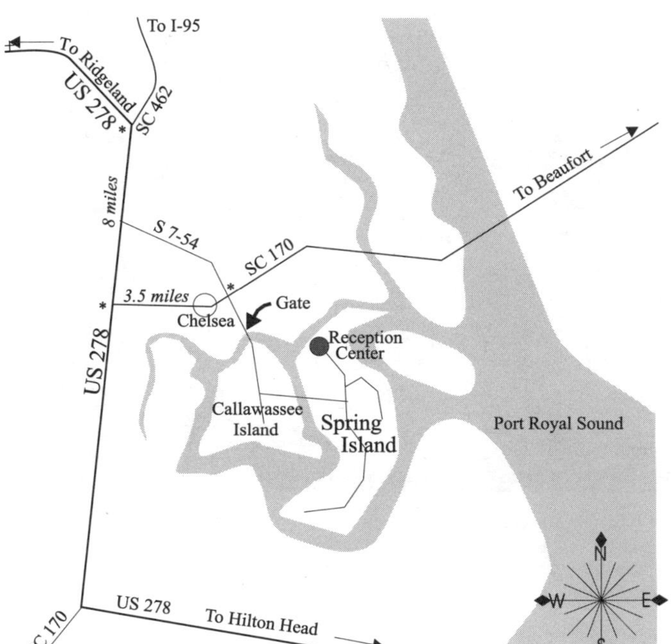

Dropseed
Sporobolus sp.
Cord Grass
Spartina patens
Dune Sandbur
Cenchrus tribuloides
Broomsedge
Andropogon virginicus
Saw Palmetto
Serenoa repens
Dwarf Palmetto
Sabal minor
Cabbage Palmetto
Sabal palmetto
Jack-in-the-pulpit
Arisaema triphyllum
Spanish-moss
Tillandsia usneoides
Black Needle Rush
Juncus roemerianus
Greenbrier; Bullbrier
Smilax rotundifolia
Fringed Greenbrier
Smilax bona-nox
Little-sweet-Betsy
Trillium cuneatum
Spanish-bayonet
Yucca aloifolia
Daylily
Hemerocallis fulva
Field Garlic
Allium vineale
Narcissus
Narcissus tazetta x poeticus
Daffodil
Narcissus pseudo-narcissus
Atamasco-lily
Zephyranthes atamasco
Lizard's-tail
Saururus cernuus
Wax-myrtle; Southern Bayberry
Myrica cerifera
Black Walnut
Juglans nigra
Pecan
Carya illinoensis
Bitternut Hickory
Carya cordiformis
Mockernut Hickory
Carya tomentosa
Pignut Hickory
Carya glabra
Hop-hornbeam
Ostrya virginiana
White Oak
Quercus alba
Post Oak
Quercus stellata
Live Oak
Quercus virginiana
Water Oak
Quercus nigra
Willow Oak
Quercus phellos
Laurel Oak
Quercus laurifolia
Sugarberry; Hackberry
Celtis laevigata
Perennial Glasswort
Salicornia virginica
Tulip-poplar
Liriodendron tulipifera
Southern Magnolia
Magnolia grandiflora
Redbay
Persea borbonia
Sassafras
Sassafras albidum
Bloodroot
Sanguinaria canadensis
Wild Radish
Raphanus raphanistrum
Sweet-gum
Liquidambar styraciflua
Dewberry
Rubus trivialis
Black Cherry; Wild Cherry
Prunus serotina
Carolina Cherrylaurel
Prunus caroliniana
Redbud; Judas-tree
Cercis canadensis
Chinese Wisteria
Wisteria sinensis
Hercules'-club
Zanthoxylum clava-herculis
Chinaberry
Melia azedarach
American Holly
Ilex opaca
Dahoon; Cassena
Ilex cassine cassine
Yaupon
Ilex vomitoria
Red Maple
Acer rubrum
Red Buckeye
Aesculus pavia
Muscadine; Bullace
Vitis rotundifolia
Devil's-walking-stick
Aralia spinosa
Flowering Dogwood
Cornus florida
Spotted Wintergreen
Chimaphila maculata
Sparkleberry
Vaccinium arboreum
Sea-lavender
Limonium carolinianum
Limonium sp.
Horse-sugar; Sweetleaf
Symplocos tinctoria
American Ash; White Ash
Fraxinus americana
Wild Olive
Osmanthus americana
Privet; Ligustrum
Ligustrum japonicum
Chinese Privet
Ligustrum sinense
Yellow Jessamine
Gelsemium sempervirens

Spring Island (continued)
Feburary 27, 1993

Common Henbit
Lamium amplexicaule
Woolly Mullein
Verbascum thapsus
Cross Vine
Anisostichus capreolata
Partridge-berry
Mitchella repens
Trailing Bluets
Houstonia procumbens
Japanese Honeysuckle
Lonicera japonica
American Elderberry
Sambucus canadensis
Elephant's-foot
Elephantopus sp.
Groundsel-tree
Baccharis glomeruliflora
Sea-myrtle; Groundsel-tree
Baccharis halimifolia
Bearsfoot
Polymnia uvedalia
Sea Ox-eye
Borrichia frutescens

FAUNA

Marsh Periwinkle; Mud Snail
Littorina irrorata
Ribbed Mussel
Geukensia demissa
Eastern Oyster (shell)
Crassostrea virginica
Mud Crab
Panopeus sp.
Mud Fiddler Crab
Uca pugnax
Leafy Oak Gall (wasp)
Andricus foliatus
Conical Woolly Oak Gall (wasp)
Andricus perplexus
Small Bullet Oak Gall (wasp)
Andricus maxwelli
Horned Oak Gall (wasp)
Andricus clavigerus
Wool-bearing Gall (wasp)
Andricus lanigera
Stem Gall (wasp)
Plagiotrochus sp.
Conical Oak Gall (wasp)
Callirhytis conifera
Bullet Gall (wasp)
Cynips sp.
Oak Leaf Gall (wasp)
Belenocnema fossoria
Marginal Oak Leaf Gall (wasp)
Dryophanta sp.
Pedunculate Ground Beetle
Ardistomis sp.
Seed-eating Ground Beetle
Amara sp.
Notch-mouthed Ground Beetle
Dicaelus elongatus
Small Carrion Beetle
Catops sp.
Hickory Girdler Beetle (sign)
Oncideres cingulata
Marbled Salamander
Ambystoma opacum
Mole Salamander
Ambystoma talpoideum
Southern Leopard Frog
Rana sphenocephala
Corn Snake
Elaphe guttata
Common Loon
Gavia immer
Pied-billed Grebe
Podilymbus podiceps
Anhinga; Snakebird
Anhinga anhinga
Double-crested Cormorant
Phalacrocorax auritus
Tricolored Heron
Egretta tricolor
Snowy Egret
Egretta thula
Great Egret
Casmerodius albus
Great Blue Heron
Ardea herodias
Bufflehead
Bucephala albeola
Sora
Porzana carolina
Common Moorhen
Gallinula chloropus
Killdeer
Charadrius vociferus
Willet
Catoptrophorus semipalmatus
Greater Yellowlegs
Tringa melanoleuca
Laughing Gull
Larus atricilla
Bonaparte's Gull
Larus philadelphia
Ring-billed Gull
Larus delawarensis
Forster's Tern
Sterna forsteri
Turkey Vulture
Cathartes aura
Black Vulture
Coragyps atratus
Red-tailed Hawk
Buteo jamaicensis
Osprey
Pandion haliaetus
Northern Bobwhite
Colinus virginianus
Wild Turkey
Meleagris gallopavo
Mourning Dove
Zenaida macroura
Belted Kingfisher
Ceryle alcyon
Red-bellied Woodpecker
Melanerpes carolinus
Northern Flicker
Colaptes auratus
Downy Woodpecker
Picoides pubescens
Yellow-bellied Sapsucker
Sphyrapicus varius
Eastern Phoebe
Sayornis phoebe
Tree Swallow
Tachycineta bicolor
Blue Jay
Cyanocitta cristata
American Crow
Corvus brachyrhynchos
Tufted Titmouse
Parus bicolor
Carolina Chickadee
Parus carolinensis
White-breasted Nuthatch
Sitta carolinensis
Carolina Wren
Thryothorus ludovicianus
Ruby-crowned Kinglet
Regulus calendula
Eastern Bluebird
Sialia sialis
Hermit Thrush
Catharus guttatus
American Robin
Turdus migratorius
Northern Mockingbird
Mimus polyglottos
European Starling
Sturnus vulgaris
Yellow-rumped Warbler
Dendroica coronata
Pine Warbler
Dendroica pinus
Palm Warbler
Dendroica palmarum
Northern Cardinal
Cardinalis cardinalis
Song Sparrow
Melospiza melodia
Chipping Sparrow
Spizella passerina
Dark-eyed Junco
Junco hyemalis
White-throated Sparrow
Zonotrichia albicollis
Red-winged Blackbird
Agelaius phoeniceus
Boat-tailed Grackle
Quiscalus major
American Goldfinch
Carduelis tristis
Virginia Opossum (sign)
Didelphis virginiana
Rabbit (scat)
Sylvilagus sp.
Gray Squirrel
Sciurus carolinensis
Eastern Fox Squirrel
Sciurus niger
Raccoon (tracks)
Procyon lotor
River Otter
Lutra canadensis
White-tailed Deer
Odocoileus virginianus

(Photo by L. Greenlee)

Bloodroot, *Sanguinaria canadensis*.

Victoria Bluff Heritage Preserve
Beaufort County
February 22, 1992

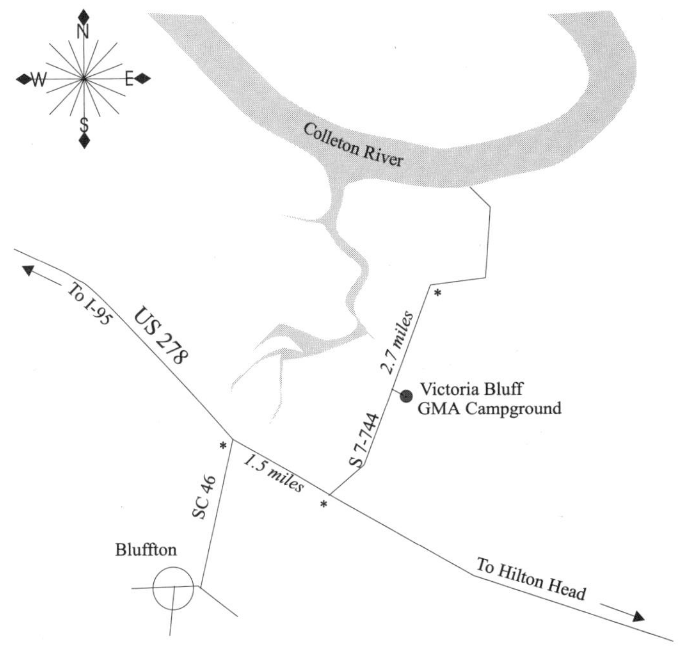

Victoria Bluff Heritage Preserve was purchased to preserve an example of the longleaf pine/saw palmetto community in South Carolina. This community is much more common in Georgia and Florida, but very rare in South Carolina. There are also a number of lime sinks on the property. Among the notable plants seen were whisk fern (very primitive), pondspice, the rare needle palm along with the other three native palm species, and rare species of *Lyonia* and *Vaccinium*. There were several species of frogs, many birds -- including a flock of bobwhites who marched across the road to bid the group farewell -- and other uncommon plants as well.

Directions to Victoria Bluff Heritage Preserve: From I-95 near Ridgeland, take exit 21, US 278, east toward Hilton Head. Follow the signs; US 278 takes many twists and turns. After passing the junction with SC 46 at Bluffton, go 1.5 miles further and turn left onto County Road 744 where the signs point to "Coastal Zone Education Center" and "Waddell Mariculture Center." Go 2.7 miles north to a power line right-of-way where some trails begin, or stop at the campground for other trails.

FLORA

Lichen
Leptogium cyanescens
Graphis scripta
Graphis beaumontii
Graphis afzelii
Graphis sp.
Thelotrema monosporum
Thelotrema sp.
Coenogonium interplexum
Coenogonium moniliforme
Bacidia phacodes
Bacidia incompta
Bacidia suffusa
Bacidia sp.
Leptogium austroamericanum
Cladonia calycantha
Cladonia capitata
Cladonia chlorophaea
Cladonia coniocraea
Cladonia cylindrica
Cladonia didyma
Cladonia furcata
Cladonia ochrochlora
Pertusaria leucostoma
Pertusaria ostiolata
Pertusaria paratuberculifera
Pertusaria propinqua
Pertusaria texana
Pertusaria sinusmexicani
Pertusaria sp.
Haematomma puniceum
Lecanora cupressi
Lecanora pallida
Lecanora varia
Bulbothrix confoederata
Bulbothrix goebelii
Hypotrachyna livida
Parmelina galbina
Parmelina horrescens
Parmeliopsis subambigua
Parmotrema perforatum
Parmotrema rampoddense
Parmotrema reticulatum
Parmotrema robustum
Parmotrema tinctorum
Pseudoparmelia caroliniana
Usnea strigosa
Buellia punctata
Buellia conspirans
Buellia sp.
Dirinaria picta
Physcia americana
Physcia millegrana
Heterodermia albicans
Heterodermia obscurata
Pyxine caesiopruinosa
Phaeophyscia pusilloides
Pyrenula cruenta
Mycoporum pyrenocarpum
Arthothelium taediosum
Arthothelium sp.
Trypethelium mastoideum
Trypethelium virens

Floating Liverwort
Ricciocarpus natans
Riccia fluitans

Whiskfern
Psilotum nudum

Cinnamon Fern
Osmunda cinnamomea

Royal Fern
Osmunda regalis spectabilis

Bracken
Pteridium aquilinum

Virginia Chain Fern
Woodwardia virginica

Netted Chain Fern
Woodwardia areolata

Ebony Spleenwort
Asplenium platyneuron

Resurrection Fern
Polypodium polypodioides

Long-leaf Pine
Pinus palustris

Slash Pine
Pinus elliotti

Loblolly Pine
Pinus taeda

Pond Pine
Pinus serotina

Cane; Switch Cane
Arundinaria gigantea

Spike Grass
Uniola sessiliflora

Oplismenus (grass)
Oplismenus setarius

Nut-rush
Scleria sp.

Sedge
Carex sp.

Saw Palmetto
Serenoa repens

Dwarf Palmetto
Sabal minor

Cabbage Palmetto
Sabal palmetto

Needle Palm
Rhapidophyllum hystrix

Duckweed
Lemna sp.

Spanish-moss
Tillandsia usneoides

Soft Rush
Juncus effusus

Fringed Greenbrier
Smilax bona-nox

Bamboo Smilax
Smilax laurifolia

Sarsaparilla-vine
Smilax pumila

Spoonleaf Yucca
Yucca filamentosa filamentosa

Southern Twayblade
Listera australis

Crane-fly Orchid
Tipularia discolor

Lizard's-tail
Saururus cernuus

Willow
Salix sp.

Wax-myrtle; Southern Bayberry
Myrica cerifera

Mockernut Hickory
Carya tomentosa

Pignut Hickory
Carya glabra

Victoria Bluff Heritage Preserve (continued) February 22, 1992

Live Oak
Quercus virginiana
Southern Red Oak; Spanish Oak
Quercus falcata falcata
Water Oak
Quercus nigra
Laurel Oak
Quercus laurifolia
False Nettle
Boehmeria cylindrica
Sweetbay
Magnolia virginiana
Southern Magnolia
Magnolia grandiflora
Redbay
Persea borbonia
Sassafras
Sassafras albidum
Pondspice
Litsea aestivalis
Sweet-gum
Liquidambar styraciflua
Witch-hazel
Hamamelis virginiana
Poison-ivy
Rhus radicans
American Holly
Ilex opaca
Dahoon; Cassena
Ilex cassine cassine
Yaupon
Ilex vomitoria
Inkberry
Ilex glabra
Sweet Gallberry
Ilex coriacea
Red Maple
Acer rubrum
Southern Sugar Maple
Acer saccharum floridanum
Muscadine; Bullace
Vitis rotundifolia
Loblolly-bay
Gordonia lasianthus
St.-Andrew's-cross
Hypericum hypericoides
Bog White Violet
Viola lanceolata
Devil-joint
Opuntia drummondii
Pennywort
Hydrocotyle sp.
Swamp Tupelo; Swamp Blackgum
Nyssa sylvatica biflora
Sweet Pepperbush
Clethra alnifolia
Indian-pipe
Monotropa uniflora
Staggerbush
Lyonia ferruginea
Fetterbush
Lyonia lucida
Dangleberry
Gaylussacia frondosa
Sparkleberry
Vaccinium arboreum
Black Highbush Blueberry
Vaccinium atrococcum
Blueberry
Vaccinium myrsinites
Persimmon
Diospyros virginiana
Horse-sugar; Sweetleaf
Symplocos tinctoria
Wild Olive
Osmanthus americana
Yellow Jessamine
Gelsemium sempervirens
Cross Vine
Anisostichus capreolata
Button-bush
Cephalanthus occidentalis
Partridge-berry
Mitchella repens
Trailing Bluets
Houstonia procumbens
Vanilla-plant
Trilisa odoratissima
Sea-myrtle; Groundsel-tree
Baccharis halimifolia
Carbon-balls (fungus)
Daldinia sp.
Red Russula
Russula sp.

FAUNA

Basilica Spider
Mecynogea lemniscata
Brownish-gray Fishing Spider
Dolomedes tenebrosus
Nursery-web Spider
Pisaurina mira
Florida Harvester Ant
Pogonomyrmex badius
Florida Carpenter Ant
Camponotus abdominalis
Bald-faced Hornet
Vespula maculata
Woodland Ground Beetle
Odontonyx sp.
Predacious Diving Beetle
Agabus johannis
Water Scavenger Beetle
Hydrobius melaenus
Enochrus ochraceus
Large Rove Beetle
Staphylinus maxillosus
Hister Beetle
Epierus pulicarius
Hister depurator
Platysoma carolinus
Passalus; Bess-beetle
Popilius disjunctus
Earth-boring Dung Beetle
Geotrupes egeriei
Darkling Beetle
Uloma imberbis
Darkling Beetle (larva)
Alobates pennsylvanica
Convergent Lady Beetle
Hippodamia convergens
Cloudless Sulphur
Phoebis sennae eubule
Salt Marsh Moth (cocoon)
Estigmene acrea
Spring Peeper
Hyla crucifer
Pine Woods Treefrog
Hyla femoralis
Little Grass Frog
Limnaoedus ocularis
Brimley's Chorus Frog
Pseudacris brimleyi
Bullfrog
Rana catesbiana
Carolina Anole
Anolis carolinensis
Ground Skink
Scincella lateralis
Common Loon
Gavia immer
Brown Pelican
Pelecanus occidentalis
Double-crested Cormorant
Phalacrocorax auritus
Bonaparte's Gull
Larus philadelphia
Turkey Vulture
Cathartes aura
Red-shouldered Hawk
Buteo lineatus
Osprey
Pandion haliaetus
Northern Bobwhite
Colinus virginianus
Mourning Dove
Zenaida macroura
Red-bellied Woodpecker
Melanerpes carolinus
Northern Flicker
Colaptes auratus
Downy Woodpecker
Picoides pubescens
Yellow-bellied Sapsucker
Sphyrapicus varius
Pileated Woodpecker
Dryocopus pileatus
American Crow
Corvus brachyrhynchos
Tufted Titmouse
Parus bicolor
Carolina Chickadee
Parus carolinensis
Carolina Wren
Thryothorus ludovicianus
Ruby-crowned Kinglet
Regulus calendula
Brown Thrasher
Toxostoma rufum
White-eyed Vireo
Vireo griseus
Solitary Vireo
Vireo solitarius
Black-and-white Warbler
Mniotilta varia
Yellow-rumped Warbler
Dendroica coronata
Yellow-throated Warbler
Dendroica dominica
Pine Warbler
Dendroica pinus
Rufous-sided Towhee
Pipilo erythrophthalmus
Eastern Mole (sign)
Scalopus aquaticus
White-tailed Deer (tracks, remains)
Odocoileus virginianus

(Photo by L. Smith)

Pondspice, *Litsea aestivalis*.

Berkeley County

Francis Marion National Forest

Berkeley County
June 22, 1991

Three different locations, showing three distinctly different habitats, were visited in Francis Marion National Forest. The group saw a diversity of Low Country flora and fauna, some of which are rare or endangered. The first area, known as the Honey Hill Lime Sinks, contained limestone sinks (ponds in limestone depressions.) Here the alkaline soil fields less common plants such as pondspice and southern spicebush. Dragonflies were abundant. The next site was a pine savanna with abundant pitcher plants and other wet area plants. The last site, a pond-cypress savanna, featured several species of native orchids. Not even Hurricane Hugo damage could hide the beauty of these areas.

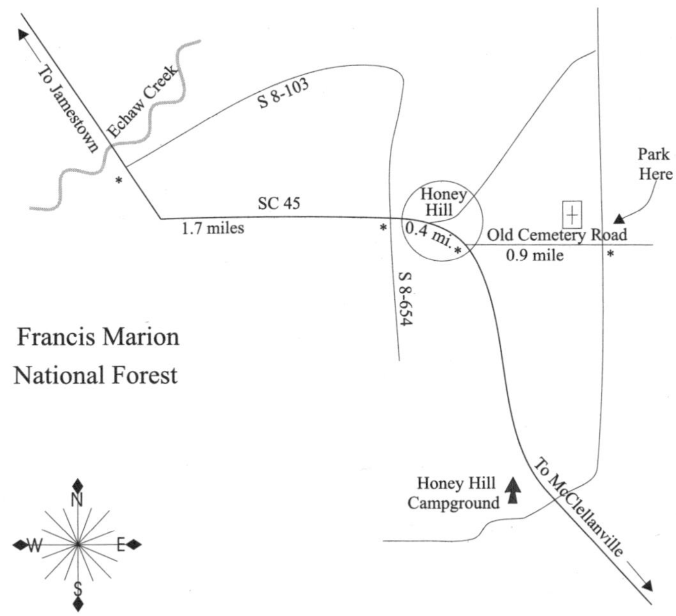

Directions to Francis Marion National Forest: From Moncks Corner, follow US ALT-17 about 36 miles north to Jamestown. Turn right onto SC 45 heading east to Honey Hill, another 10.5 miles. After reaching the crossroads in Honey Hill (SC 8-654), continue on SC 45 another 0.4 mile to Old Cemetery Road, the second dirt road to the left. Fork left here and go approximately 0.9 mile to the old cemetery. Turn left at the cemetery and park on the right. The lime sinks are a short distance east.

FLORA

Lichen
Collema nigrescens
Lobaria sp.
Graphis scripta
Graphis nitida
Graphis mosquitensis
Graphis elegans
Graphis sp.
Glyphis cicatricosa
Phaeographis dendritica
Leptogium austroamericanum
Baeomyces fungoides
Cladina evansii
Cladonia balfourii
Cladonia capitata
Cladonia didyma
Cladonia leporina
Cladonia verticillata
Pertusaria copiosa
Pertusaria ostiolata
Pertusaria paratuberculifera
Pertusaria trachythallina
Pertusaria texana
Haematomma ochrophaeum
Lecanora hageni
Lecanora cupressi
Lecanora varia
Ochrolechia africana
Ochrolechia farinacea
Phlyctis sp.
Bulbothrix confoederata
Bulbothrix goebelii
Hypotrachyna livida
Parmotrema dilitatum
Parmotrema hypotropum
Parmotrema michauxianum
Parmotrema perforatum
Parmotrema rampoddense
Parmotrema subisidiosum
Parmotrema subtinctorium
Parmotrema tinctorum
Parmotrema xanthinum
Pseudoparmelia caroliniana
Ramalina americana
Usnea strigosa
Buellia punctata
Physcia millegrana
Heterodermia albicans
Heterodermia granulifera
Heterodermia hypoleuca
Heterodermia obscurata
Heterodermia casarettiana
Phaeophyscia adiastola
Phaeophyscia ciliata
Phaeophyscia imbricata
Melanotheca cruenta
Trypethelium tropicum
Trypethelium virens

Southern Bog Clubmoss
Lycopodium adpressum
Cinnamon Fern
Osmunda cinnamomea
Bracken
Pteridium aquilinum
Virginia Chain Fern
Woodwardia virginica
Netted Chain Fern
Woodwardia areolata
Resurrection Fern
Polypodium polypodioides
Long-leaf Pine
Pinus palustris
Loblolly Pine
Pinus taeda
Bald-cypress
Taxodium distichum
Pond-cypress
Taxodium ascendens
Arrowhead; Duck-potato
Sagittaria graminea graminea
White-bracted Sedge
Dichromena latifolia
Umbrella-grass
Fuirena squarrosa
Beak-rush
Rhynchospora macrostachya
Yellow-eyed-grass
Xyris sp.
Pipewort; Hatpins
Eriocaulon decangulare
Bog-buttons
Lachnocaulon anceps
Dayflower
Commelina communis
Bamboo Smilax
Smilax laurifolia
False Asphodel
Tofieldia racemosa
Colic-root; Stargrass
Aletris aurea
Redroot
Lachnanthes caroliniana

Francis Marion National Forest (continued)
June 22, 1991

Flag
Iris tridentata
Burmannia
Burmannia biflora
Rose Pogonia
Pogonia ophioglossoides
Spreading-pogonia
Cleistes divaricata
Grass-pink
Calopogon pulchellus
Green Adder's-mouth
Malaxis unifolia
Wax-myrtle; Southern Bayberry
Myrica cerifera
Mockernut Hickory
Carya tomentosa
Allegheny Chinquapin
Castanea pumila
Post Oak
Quercus stellata
Live Oak
Quercus virginiana
Turkey Oak
Quercus laevis
Spanish Oak
Quercus falcata
Black Jack Oak
Quercus marilandica
Water Oak
Quercus nigra
Blue-jack Oak
Quercus incana
Running Oak
Quercus pumila
Polycarpon
Polycarpon tetraphyllum
Sweetbay
Magnolia virginiana
Redbay
Persea borbonia
Pondspice
Litsea aestivalis
Southern Spicebush
Lindera melissaefolium
Trumpet Pitcher-plant
Sarracenia flava
Hooded Pitcher-plant
Sarracenia minor
Sundew
Drosera intermedia
Drosera capillaris
Sweet-gum
Liquidambar styraciflua
Red Chokeberry
Sorbus arbutifolia
Rattlebox
Crotalaria purshii
Crotalaria angulata
Zornia
Zornia bracteata
Goat's-rue
Tephrosia virginiana
Tephrosia
Tephrosia hispidula

Groundnut
Apios americana
Dollar-weed
Rhynchosia reniformis
Milk-pea
Galactia volubilis
Polygala
Polygala incarnata
Orange Milkwort
Polygala lutea
Polygala
Polygala ramosa
Pine-barren Milkwort
Polygala cymosa
Flowering Spurge
Euphorbia corollata
Titi
Cyrilla racemiflora
American Holly
Ilex opaca
Dahoon; Cassena
Ilex cassine cassine
Myrtle-leaved Holly
Ilex cassine myrtifolia
Inkberry
Ilex glabra
Red Maple
Acer rubrum
Muscadine; Bullace
Vitis rotundifolia
Piriqueta
Piriqueta caroliniana
St.-Andrew's-cross
Hypericum hypericoides
St.-Peter's-wort
Hypericum stans
St.-John's-wort
Hypericum fasciculaatum
Rockrose, Frostweed
Helianthemum rosmarinifolium
Smooth Meadow-beauty
Rhexia alifanus
Yellow Meadow-beauty
Rhexia lutea
Pale Meadow-beauty
Rhexia mariana
Slender Seed-box
Ludwigia virgata
Mermaid-weed
Proserpinaca pectinata
Devil's-walking-stick
Aralia spinosa
Intelligence-plant
Centella asiatica
Button Snakeroot
Eryngium yuccifolium
Black Tupelo; Blackgum
Nyssa sylvatica sylvatica
Swamp Tupelo; Swamp Blackgum
Nyssa sylvatica biflora
Sweet Pepperbush
Clethra alnifolia
Dwarf Azalea
Rhododendron atlanticum

Fetterbush
Lyonia lucida
Staggerbush
Lyonia mariana
Dangleberry
Gaylussacia frondosa
Low-bush Blueberry
Vaccinium tenellum
Highbush Blueberry
Vaccinium corymbosum
Creeping Blueberry
Vaccinium crassifolium
Persimmon
Diospyros virginiana
Storax; Snowbell
Styrax americana
Sandhill Milkweed
Asclepias humistrata
Dodder
Cuscuta sp.
Buchnera
Buchnera floridana
Purple Bladderwort
Utricularia purpurea
Lobelia
Lobelia boykinii
Lobelia nuttallii
Lobelia sp.
Marsh-fleabane
Pluchea rosea
Black-root
Pterocaulon pycnostachyum
White-topped Aster
Aster paternus
Aster
Aster squarrosus
Stiff-leaved Aster
Aster linariifolius
Heterotheca
Heterotheca nervosa
Coreopsis
Coreopsis major
Sneeze-weed
Helenium flexuosum
Granular Puffball
Arachnion album
Splash-cups
Cyathus striatus
Wolf's-milk Slime
Lycogala epidendrum

FAUNA

Crab-shaped Spiny-back
Gasteracantha cancriformis
Stink Bug
Mormidea sp.
Green Stink Bug
Nezara viridula
Brown Water-scorpion
Ranatra fusca
Golden-winged Skimmer
Libellula auripennis
Blue Dasher
Pachydiplax longipennis
Violet-masked Glider
Tramea carolina

Glider
Tramea sp.
Tiger Beetle
Cicindela punctulata
Predacious Diving Beetle
Graphoderus liberus
Whirligig Beetle
Dineutus carolinus
Common Tumblebug
Canthon laevis
Flower Scarab Beetle
Trigonopeltastes delta
Trichiotinus piger
Horned Fungus Beetle
Bolitotherus cornutus
False Blister Beetle
Oxacis sp.
Oval Leaf Beetle
Rhabdopterus sp.
Flea Beetle
Altica marevagans
Pinhole Borer Beetle
Platypus flavicornis
Palamedes Swallowtail
Papilio palamedes
Red-banded Hairstreak
Calycopis cecrops
Little Metalmark (butterfly)
Calephelis virginiensis
Common Wood-nymph (butterfly)
Cercyonis pegala pegala
Grapeleaf Skeletonizer Moth
Harrisina americana
Luna Moth
Actias luna
Pitcher-plant Borer Moth (larva)
Papaipema appassionata
Half-Yellow Moth
Tarachidia semiflava
Southern Cricket Frog
Acris gryllus
Cricket Frog
Acris sp.
Pine Woods Treefrog
Hyla femoralis
Little Grass Frog
Limnaoedus ocularis
Bullfrog
Rana catesbiana
Southern Leopard Frog
Rana sphenocephala
Cattle Egret
Bulbicus ibis
Turkey Vulture
Cathartes aura
White-breasted Nuthatch
Sitta carolinensis
Blue-gray Gnatcatcher
Polioptila caerulea
White-eyed Vireo
Vireo griseus
Northern Parula
Parula americana
Summer Tanager
Piranga rubra
White-tailed Deer (bones)
Odocoileus virginianus

Lake Marion (Bird Island Area)
Berkeley County
June 25, 1988

This trip was unique for SCAN, as we spent the day on pontoon boats exploring Lake Marion, its islands and inlets. We even waded along the shallow edges of an island that wading birds use for a large rookery. Near the dam we could see large garfish below us. Osprey had utilized still-standing tree trunks out in the lake for nesting sites, and were feeding their young. In the Red Hill area we explored the shore for exposed fossils and artifacts. Glenn Quattlebaum, our Southern Regional Director, was able to share his knowledge gained by growing up in the area, which lent a special flavor to the trip.

Directions to Lake Marion (Bird Island Area): From Manning, take SC 260 south about 9 miles, passing through Jordan and crossing Wyboo Creek. Before reaching Game Warden Slew, stop at the public boat ramp on the right. There, a rented boat may go to Bird Island.

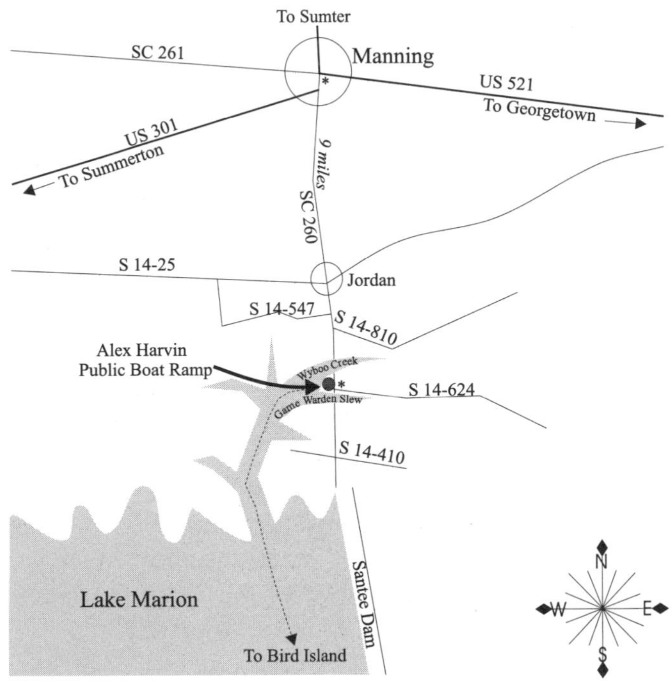

FLORA

Bald-cypress
Taxodium distichum
Water-weed
Elodea sp.
Giant Reed
Arundo donax
Black Willow
Salix nigra
Swamp Cottonwood
Populus heterophylla
Wax-myrtle; Southern Bayberry
Myrica cerifera
Water Hickory
Carya aquatica
Overcup Oak
Quercus lyrata
Laurel Oak
Quercus laurifolia
Water-elm; Planer-tree
Planera aquatica
Alligator-weed
Alternanthera philoxeroides
Pokeweed; Pokeberry
Phytolacca americana
Lotus; Pond-nuts
Nelumbo lutea
Sweet-gum
Liquidambar styraciflua
American Sycamore
Platanus occidentalis
Poison-ivy
Rhus radicans
Boxelder
Acer negundo
Red Maple
Acer rubrum
Muscadine; Bullace
Vitis rotundifolia
Pepper-vine
Ampelopsis arborea
Rose Mallow; Wild Cotton
Hibiscus moscheutos
Passionflower; Maypops
Passiflora incarnata
Water-primrose
Ludwigia uruguayensis
Blackgum
Nyssa sylvatica
Water Tupelo
Nyssa aquatica
Brazilian Vervain
Verbena braziliensis
Trumpet Vine
Campsis radicans
Button-bush
Cephalanthus occidentalis
Spongy-toothed Polypore
Spongipellis pachydon
Tiger Sawgill
Lentinus tigrinus
Bitter Oyster
Panellus stipticus

FAUNA

Dobsonfly (egg)
Corydalis cornutus
Eastern Pondhawk
Erythemis simplicicollis
Eastern Amberwing (dragonfly)
Perithemis tenera
Bluet (damselfly)
Enallagma sp.
Predacious Diving Beetle
Hydrovatus pustulatus
Burrowing Water Beetle
Suphisellus sp.
Whirligig Beetle
Dineutus sp.
Tumbling Flower Beetle
Mordella melaena
Darkling Beetle
Diaperis maculata
Hoplocephala viridipennis
Platydema erythrocera
Platydema sp.
Convergent Lady Beetle
Hippodamia convergens
Oval Leaf Beetle
Paria sellatus
Broad-bodied Leaf Beetle
Chrysomela scripta
Flea Beetle
Disonycha caroliniana
Altica sp.
Cloudless Sulphur
Phoebis sennae eubule
Pearl Crescent
Phyciodes tharos
Common Buckeye (butterfly)
Junonia coenia
Red-spotted Purple (butterfly)
Limenitis arthemis astyanax
Viceroy; Mimic
Limenitis archippus
Fall Webworm Moth
Hyphantria cunea
Largemouth Bass
Micropterus salmoides
Anhinga; Snakebird
Anhinga anhinga
Double-crested Cormorant
Phalacrocorax auritus
Green-backed Heron
Butorides striatus
Little Blue Heron
Egretta caerulea
Cattle Egret
Bulbicus ibis
Great Egret
Casmerodius albus
Great Blue Heron
Ardea herodias
White Ibis
Eudocimus albus
Laughing Gull
Larus atricilla
Turkey Vulture
Cathartes aura
Osprey
Pandion haliaetus
Rock Dove; Pigeon
Columba livia

Lake Marion (continued)
June 25, 1988

Chimney Swift
Chaetura pelagica
Red-headed Woodpecker
Melanerpes erythrocephalus
Eastern Kingbird
Tyrannus tyrannus
Purple Martin
Progne subis
Fish Crow
Corvus ossifragus
Loggerhead Shrike
Lanius ludovicianus
Northern Mockingbird
Mimus polyglottos
Red-winged Blackbird
Agelaius phoeniceus
Raccoon (tracks)
Procyon lotor
White-tailed Deer (bones)
Odocoileus virginianus

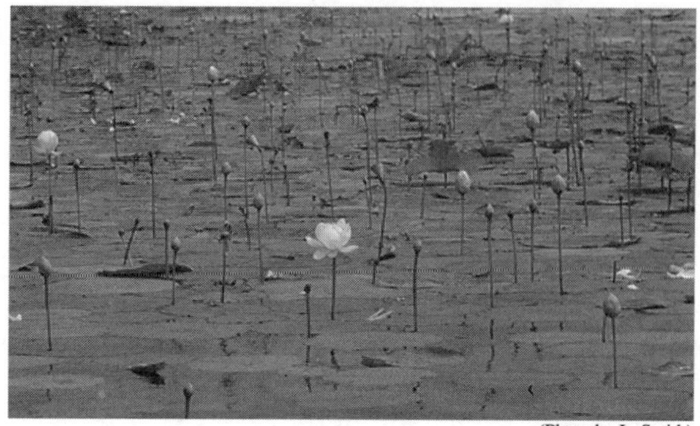
(Photo by L. Smith)
Lotus or pond-nuts, *Nelumbo lutea*.

Old Santee Canal State Park
Berkeley County
December 14, 1991 & May 9, 1992

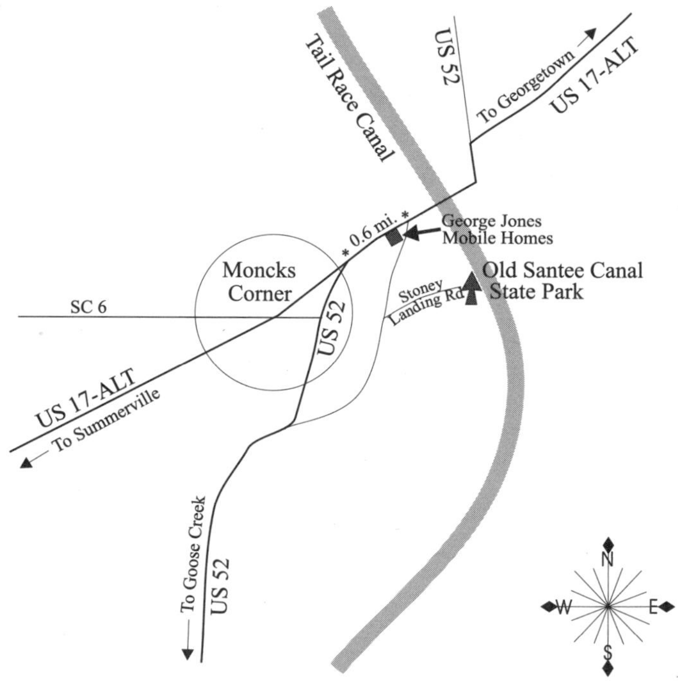

This state park, with the old canal and Low Country vegetation, makes a nice combination of human and natural history. The interpretive center is spectacular with its entrance through a model of one of the old locks complete with water gates. Birds were a significant portion of both visits. Still swollen from a recent meal, a juvenile yellow rat snake revealed a pattern of blocks along its body, quite different from the stripes it will acquire later in life. The spring trip also provided good wildflowers such as *Silene* and *Spigelia*. The viewing of a colony of hairy flower bees tunneling into a limestone embankment (see below) constituted an unusual treat (no one was stung during the close-up photography binge).

Directions to Old Santee Canal State Park: From I-26 near Summerville, take Exit 199 (US 17-ALT) north to the far side of Moncks Corner. Where US 52 merges with US 17-ALT, stay on 17-ALT and go about 0.6 mile further. Just after passing George Jones Mobile Home Sales, turn right; if you cross the bridge over the canal, you missed the turn. Proceed about one mile and look for the sign for the Old Santee Canal State Park. Turn left there and follow Stoney Landing Road to the park.

Science and Nature Notes: Hairy Flower Bees

Hairy flower bees (*Anthophora* sp.) seen at Old Santee Canal State Park, as well as on the land over the waterfall at Peach Tree Rock in Lexington County and along the trail at Eastatoe Creek Heritage Preserve in Pickens County, are representative of a number of related Apidae species that are all large and hairy. They nest in tunnels, usually in clay banks. Although gregarious, forming colonies up to hundreds in size, they are not truly social in that they don't share a single hive. Each bee digs a separate 3/8" hole over which it builds a mud entrance tube. The tube is smoothed on the inside by water lapped up by the bee and carried to the nest. These tunnels extend into the bank leading to an inner cluster of cells which are waterproofed by a thin layer of salivary excretion. Additionally, these inner cells contain a semifluid mass of pollen and nectar which provide food for the young. The larvae remain in the cells through the winter, pupate in the spring, and emerge in early summer.

FLORA

Lichen
Leptogium cyanescens
Graphis scripta
Graphis mosquitensis
Medusulina nitida
Sarcographa sp.
Bilimbia lignaria
Bacidia schweinitzii
Leptogium austroamericanum
Lecidea russula
Letrouitia domingense
Pertusaria ostiolata
Pertusaria paratuberculifera
Pertusaria plittiana
Pertusaria texana
Pertusaria valliculata
Haematomma puniceum
Lecania fuscella
Lecanora miculata
Lecanora allophana argentata
Lecanora cupressi
Lecanora pallida
Lecanora varia
Ochrolechia africana
Bulbothrix confoederata
Bulbothrix goebelii
Bulbothrix laevigatula
Hypotrachyna livida
Parmelina aurulenta
Parmelina dissecta
Parmeliopsis subambigua
Parmotrema cetratum
Parmotrema crinitum
Parmotrema cristiferum
Parmotrema michauxianum
Parmotrema praesorediosum
Parmotrema rampoddense
Parmotrema reticulatum
Parmotrema subtinctorium
Pseudoparmelia caroliniana
Usnea strigosa
Buellia caloosensis
Buellia curtisii
Buellia stillingiana
Buellia sp.
Physcia aipolia
Physcia crispa
Heterodermia albicans
Phaeophyscia ciliata
Anthracothecium thelomorphum
Pyrenula cruenta
Pyrenula mamillana
Hyperphyscia syncolla
Trypethelium mastoideum
Trypethelium tropicum
Trypethelium virens
Trypethelium scorites
Trypethelium floridanum
Porina nucula
Lecanactis sp.
Amphiloma lanuginosum
Royal Fern
Osmunda regalis spectabilis
Christmas Fern
Polystichum acrostichoides
Thelypteris (fern)
Thelypteris kunthii
Sensitive Fern
Onoclea sensibilis
Ebony Spleenwort
Asplenium platyneuron
Resurrection Fern
Polypodium polypodioides
Mosquito Fern
Azolla caroliniana
Bald-cypress
Taxodium distichum
Eastern Red-cedar
Juniperus virginiana
Common Cat-tail
Typha latifolia
Pondweed
Potamogeton sp.
Creeping Water-plantain
Echinodorus cordifolius
Brazilian Elodea
Egeria densa
Cane; Switch Cane
Arundinaria gigantea
River Oats
Uniola latifolia
Wild Rye Grass
Elymus virginicus
Needle Grass
Stipa avenacea
Goose Grass
Eleusine indica
Southern Wild Rice
Zizaniopsis miliacea
Broomsedge
Andropogon virginicus
Wool-grass
Scirpus cyperinus
Dwarf Palmetto
Sabal minor
Golden-club
Orontium aquaticum
Giant Duckweed
Spirodela sp.
Duckweed
Lemna sp.
Bogmat
Wolffiella floridana
Spanish-moss
Tillandsia usneoides
Asian Spiderwort
Aneilema keisak
Pickerelweed
Pontederia cordata
Greenbrier; Bullbrier
Smilax rotundifolia
Fringed Greenbrier
Smilax bona-nox
Sweet-scented Smilax
Smilax smallii
Little-sweet-Betsy
Trillium cuneatum
Mound-lily Yucca
Yucca gloriosa
Cinnamon-vine
Dioscorea batatas
Lizard's-tail
Saururus cernuus
Willow
Salix sp.
Eastern Cottonwood
Populus deltoides
Swamp Cottonwood
Populus heterophylla
Wax-myrtle; Southern Bayberry
Myrica cerifera
Black Walnut
Juglans nigra
Mockernut Hickory
Carya tomentosa
Tag Alder
Alnus serrulata
American Hornbeam; Ironwood
Carpinus caroliniana
Allegheny Chinquapin
Castanea pumila
White Oak
Quercus alba
Swamp Chestnut Oak
Quercus michauxii
Live Oak
Quercus virginiana
Swamp Red Oak
Quercus shumardii
Spanish Oak
Quercus falcata
Water Oak
Quercus nigra
Slippery Elm
Ulmus rubra
American Elm
Ulmus americana
Winged Elm
Ulmus alata
Sugarberry; Hackberry
Celtis laevigata
Red Mulberry
Morus rubra
False Nettle
Boehmeria cylindrica
Clearweed
Pilea pumila
Wild Ginger
Asarum canadense
Swamp Dock
Rumex verticillatus
Polygonum
Polygonum sp.
Fire Pink
Silene virginica
Hornwort
Ceratophyllum demersum
Clematis
Clematis sp.
Thimbleweed
Anemone virginiana
Mayapple
Podophyllum peltatum
Tulip-poplar
Liriodendron tulipifera
Southern Magnolia
Magnolia grandiflora
PawPaw
Asimina triloba
Redbay
Persea borbonia
Sassafras
Sassafras albidum
Spicebush
Lindera benzoin
Bloodroot
Sanguinaria canadensis
Virginia-willow
Itea virginica
Climbing-hydrangea
Decumaria barbara
Witch-hazel
Hamamelis virginiana
American Sycamore
Platanus occidentalis
Indian Strawberry
Duchesnea indica
Five-fingers
Potentilla canadensis
Swamp Rose
Rosa palustris
Chickasaw Plum
Prunus angustifolia
Redbud; Judas-tree
Cercis canadensis
Sour Clover
Melilotus indica
Indigo-bush
Amorpha fruticosa
American Wisteria
Wisteria frutescens
Groundnut
Apios americana
Coral-bean
Erythrina herbacea
Hog-peanut
Amphicarpa bracteata
Wild Geranium
Geranium maculatum
Chinese Tallow-tree
Sapium sebiferum
Poison Sumac
Rhus vernix
Poison-ivy
Rhus radicans
Winged Sumac
Rhus copallina
American Holly
Ilex opaca
Yaupon
Ilex vomitoria
Possumhaw Holly
Ilex decidua
Boxelder
Acer negundo
Red Maple
Acer rubrum
Southern Sugar Maple
Acer saccharum floridanum
Red Buckeye
Aesculus pavia
Supple-jack; Rattan-vine
Berchemia scandens

Old Santee Canal State Park (continued) December 14, 1991, and May 9, 1992

New-Jersey-tea
Ceanothus americanus
Buckthorn
Rhamnus caroliniana
Pepper-vine
Ampelopsis arborea
Basswood; Linden
Tilia heterophylla
Yellow Passionflower
Passiflora lutea
Devil's-walking-stick
Aralia spinosa
Water Pennywort
Hydrocotyle ranunculoides
Yellow-flowered Snakeroot
Sanicula gregaria
Wild Chervil
Chaerophyllum tainturieri
Meadow Parsnip
Thaspium barbinode
Goutweed; Bishop's-weed
Aegopodium podagraria
Water Parsnip
Sium suave
Water Hemlock
Cicuta maculata
Swamp Tupelo; Swamp Blackgum
Nyssa sylvatica biflora
Silky Dogwood
Cornus amomum
Dogwood
Cornus asperifolia
Water Pimpernel
Samolus parviflorus
Buckthorn Bumelia
Bumelia lycioides
Persimmon
Diospyros virginiana
Bigleaf Snowbell
Styrax grandifolia
Ash
Fraxinus sp.
Privet; Ligustrum
Ligustrum japonicum
Chinese Privet
Ligustrum sinense
Indian-pink
Spigelia marilandica
Milkweed
Asclepias sp.
Brazilian Vervain
Verbena braziliensis
Beautyberry
Callicarpa americana
Lop-seed
Phryma leptostachya
Ground-cherry
Physalis sp.
Shade Mudflower
Micranthemum umbrosum
Woolly Mullein
Verbascum thapsus
Figwort
Scrophularia marilandica
False Foxglove
Aureolaria sp.
Cross Vine
Anisostichus capreolata
Trumpet Vine
Campsis radicans
Floating Bladderwort
Utricularia inflata
Elytraria
Elytraria caroliniensis
Button-bush
Cephalanthus occidentalis
Bedstraw; Purple-bead
Galium hispidulum
Japanese Honeysuckle
Lonicera japonica
Coral Honeysuckle
Lonicera sempervirens
Black-haw; Nanny-berry
Viburnum prunifolium
Small-leaf Viburnum
Viburnum obovatum
Arrowwood
Viburnum dentatum
American Elderberry
Sambucus canadensis
Cardinal-flower
Lobelia cardinalis
Wild Lettuce
Lactuca sp.
Common Sow-thistle
Sonchus oleraceus
Butterweed
Senecio glabellus
Mistflower; Ageratum
Eupatorium coelestinum
Climbing Hempweed
Mikania scandens
Sea-myrtle; Groundsel-tree
Baccharis halimifolia
Heart-leaved Aster
Aster cordifolius
Climbing Aster
Aster carolinianus
Frost Aster
Aster pilosus
Aster
Aster dumosus
Bearsfoot
Polymnia uvedalia
Green-and-gold
Chrysogonum virginianum
Eclipta
Eclipta alba
Spilanthes
Spilanthes americana repens
Verbesina
Verbesina occidentalis
Verbesina virginica virginica
Melanthera
Melanthera hastata
Beggar-ticks
Bidens sp.

FAUNA

White-lipped Forest Snail
Mesodon thyroidus
Asiatic Clam
Corbicula sp.
Wheel Bug
Arilus cristatus
Woolly Alder Aphid
Prociphilus tesselatus
Cottony Cushion Scale
Icerya purchasi
Potter Wasp
Eumenes fraternus
Bald-faced Hornet
Vespula maculata
Pipe-organ Wasp
Trypoxylon politum
Green Tiger Beetle
Cicindela sexguttata
Tiger Beetle
Cicindela unipunctata
Crawling Water Beetle
Peltodytes sexmaculatus
Whirligig Beetle
Dineutus emarginatus
Gyrinus analis
Gyrinus lugens
Water Scavenger Beetle
Tropisternus mixtus
Marsh Beetle
Cyphon sp.
Metallic Wood-boring Beetle
Taphrocerus sp.
Soldier Beetle
Cantharis sp.
Firefly; Lightning Beetle
Lucidota atra
Branch Twig Borer Beetle
Dysides obscurus
False Blister Beetle
Xanthochroa sp.
Dusky Lady Beetle
Scymnus terminatus
Spotless Lady Beetle
Cycloneda sanguinea
Flat-faced Longhorn Beetle
Hippopsis lemniscata
Broad-bodied Leaf Beetle
Calligrapha rhoda
Flea Beetle
Disonycha sp.
Altica sp.
Leaf-mining Beetle
Chalepus scapularis
Primitive Weevil
Arrhenodes minutus
Leaf-mining Weevil
Eugnamptus collaris collaris
Billbug; Grain Weevil
Sphenophorus sp.
Cloudless Sulphur
Phoebis sennae eubule
Little Yellow (butterfly)
Eurema lisa
Gulf Fritillary
Agraulis vanillae
Long-tailed Skipper
Urbanus proteus
Yellow-bellied Slider
Chrysemys scripta
Carolina Anole
Anolis carolinensis
Skink
Eumeces sp.
Ground Skink
Scincella lateralis
Rat Snake
Elaphe obsoleta
Eastern Hognose Snake
Heterodon platyrhinos
Little Blue Heron
Egretta caerulea
Great Egret
Casmerodius albus
Great Blue Heron
Ardea herodias
White Ibis
Eudocimus albus
Common Moorhen
Gallinula chloropus
Killdeer
Charadrius vociferus
Turkey Vulture
Cathartes aura
Red-shouldered Hawk
Buteo lineatus
Osprey
Pandion haliaetus
Mourning Dove
Zenaida macroura
Chimney Swift
Chaetura pelagica
Belted Kingfisher
Ceryle alcyon
Red-bellied Woodpecker
Melanerpes carolinus
Downy Woodpecker
Picoides pubescens
Pileated Woodpecker
Dryocopus pileatus
Eastern Kingbird
Tyrannus tyrannus
Great Crested Flycatcher
Myiarchus crinitus
Eastern Phoebe
Sayornis phoebe
Acadian Flycatcher
Empidonax virescens
Blue Jay
Cyanocitta cristata
American Crow
Corvus brachyrhynchos
Fish Crow
Corvus ossifragus
Tufted Titmouse
Parus bicolor
Carolina Wren
Thryothorus ludovicianus

Ruby-crowned Kinglet
Regulus calendula
Blue-gray Gnatcatcher
Polioptila caerulea
American Robin
Turdus migratorius
Loggerhead Shrike
Lanius ludovicianus
Gray Catbird
Dumetella carolinensis
Northern Mockingbird
Mimus polyglottos
Red-Eyed Vireo
Vireo olivaceus
Prothonotary Warbler
Protonotaria citrea
Northern Parula
Parula americana
Black-and-white Warbler
Mniotilta varia
Yellowthroat
Geothlypis trichas
Northern Cardinal
Cardinalis cardinalis
Red-winged Blackbird
Agelaius phoeniceus
Common Grackle
Quiscalus quiscula
Boat-tailed Grackle
Quiscalus major

(Photo by L. Greenlee)

Gulf Fritillary, *Agraulis vanillae,* emergies from its chrysalis.

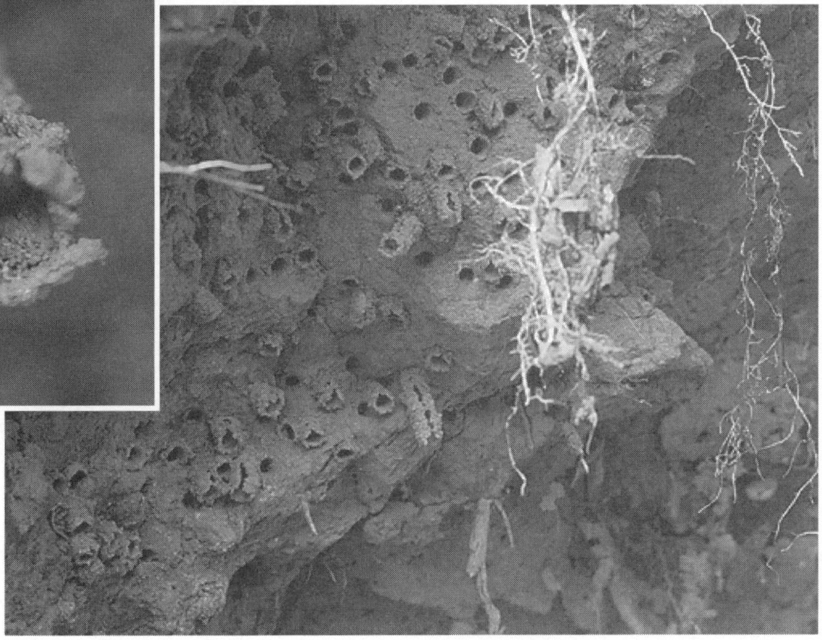

(Photos by L. Smith)

Hairy bees, *Anthophora* sp., build tubes in the banks along trails.

Charleston County

Beaches of Edisto Island
Charleston and Colleton Counties

Edingsville Beach, October 24, 1987
Botany Bay Island, October 22, 1988
Edisto Beach State Park, June 10, 1995

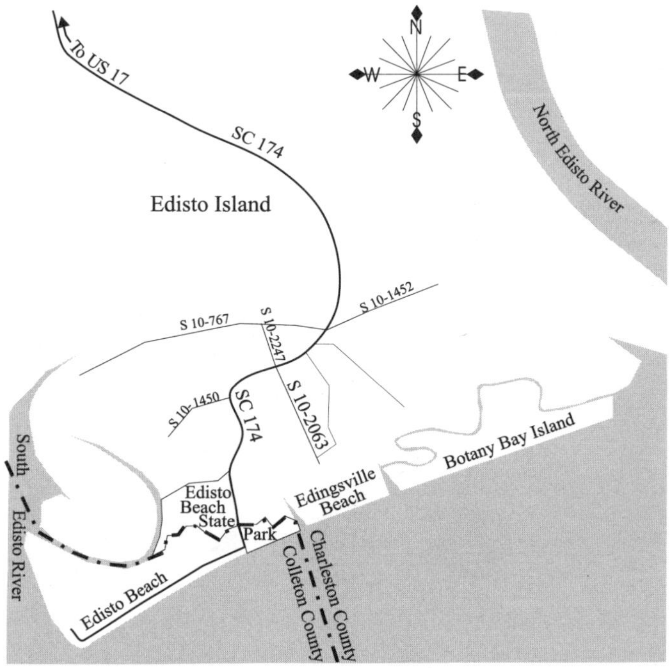

Beachcombing, classic maritime forests, and marshes are the highlights of any visit to this part of the South Carolina coast. The contour of the ocean bottom offshore helps wash up an unusual amount of fossil and shell material onto the beaches in this area. SCAN was fortunate enough to be granted access to Edingsville Beach and Botany Bay Island from the private landowners. We appreciate their helpfulness and kindness in this. The shell and marsh specimens list is enhanced because of their generosity. This area is the coastal edge of the ACE Basin. The Edisto Island part of the state park offers a good look at the marsh and maritime forest areas. It is a good area to explore slowly, seeing all the small parts that make up a natural coastal area.

Directions to the Beaches of Edisto Island: From Walterboro, take SC 64 southeast 14 miles to Jacksonboro. Turn left onto US 17 northeast toward Charleston, and go about 7 miles to SC 174. Turn right (south) onto SC 174, passing through Adams Run, and continue all the way to the Atlantic Ocean, about 25 miles. The Edisto Beach State Park picnic area is on the beach on the left just before the highway turns abruptly to the right. For the village of Edisto Beach, follow SC 174 from the state park to the end of the road. Edingsville Beach can be reached by turning left from SC 174 onto County Road S 10-2063 about 2.5 miles before reaching the state park. The boat landing for Botany Bay Island can be reached by turning left onto the unnumbered dirt road about 0.5 miles after the intersection of SC 174 and County Roads S 10-767/S 10-1452.

FLORA

Lichen
Graphis scripta
Graphis elegans
Glyphis cicatricosa
Phaeographis dendritica
Pertusaria neoscotica
Pertusaria texana
Maronea constans
Haematomma elatinum
Lecanora cupressi
Lecanora pallida rubescens
Lecanora varia
Lecanora sp.
Lecanora sp.
Ochrolechia africana
Bulbothrix goebelii
Bulbothrix goebelii f. minor
Hypotrachyna pustulifera
Parmelina horrescens
Parmelina minarum
Parmotrema dilitatum
Parmotrema hypotropum
Parmotrema michauxianum
Parmotrema perforatum
Parmotrema praesorediosum
Parmotrema tinctorum
Canoparmelia caroliniana
Punctelia rudecta
Ramalina americana
Usnea mutabilis
Buellia punctata
Dirinaria confusa
Dirinaria picta
Physcia millegrana
Schismatomma rediunta

Cinnamon Fern
Osmunda cinnamomea
Royal Fern
Osmunda regalis spectabilis
Bracken
Pteridium aquilinum
Virginia Chain Fern
Woodwardia virginica
Netted Chain Fern
Woodwardia areolata
Ebony Spleenwort
Asplenium platyneuron
Resurrection Fern
Polypodium polypodioides
Loblolly Pine
Pinus taeda
Eastern Red-cedar
Juniperus virginiana
Southern Red-cedar
Juniperus silicicola
Cane; Switch Cane
Arundinaria gigantea

Salt Grass
Distichlis spicata
Sea Oats
Uniola paniculata
Needle Grass
Stipa avenacea
Smut Grass
Sporobolus poiretii
Finger Grass
Chloris petraea
Cord Grass
Spartina patens
Smooth Cord Grass
Spartina alterniflora
Dune Sandbur
Cenchrus tribuloides
Oplismenus (grass)
Oplismenus setarius
Seaside Panic Grass
Panicum amarum
Broomsedge
Andropogon virginicus
Fimbristylis (sedge)
Fimbristylis spadicea
Nut-rush
Scleria triglomerata
Dwarf Palmetto
Sabal minor
Cabbage Palmetto
Sabal palmetto
Coconut
Cocos nucifera
Spanish-moss
Tillandsia usneoides
Black Needle Rush
Juncus roemerianus
Fringed Greenbrier
Smilax bona-nox
Dune Greenbrier
Smilax auriculata
Spanish-bayonet
Yucca aloifolia
Mound-lily Yucca
Yucca gloriosa
Bear-grass
Yucca filamentosa
Blue-eyed-grass
Sisyrinchium mucronatum atlanticum
Wax-myrtle; Southern Bayberry
Myrica cerifera
Coastal Chinquapin
Castanea pumila ashei
White Oak
Quercus alba
Live Oak
Quercus virginiana
Water Oak
Quercus nigra
Laurel Oak
Quercus laurifolia
Hackberry
Celtis occidentalis georgiana
Clearweed
Pilea pumila
Seabeach Orach
Atriplex arenaria
Marsh Orach
Atriplex patula
Glasswort
Salicornia bigelovii
Salicornia europaea
Salicornia sp.
Perennial Glasswort
Salicornia virginica
Sea-blight
Suaeda linearis
Russian-thistle
Salsola kali
Pokeweed; Pokeberry
Phytolacca americana
Purslane
Portulaca oleracea
Southern Magnolia
Magnolia grandiflora
Redbay
Persea borbonia
Sassafras
Sassafras albidum
Sea-rocket
Cakile harperi
Sweet-gum
Liquidambar styraciflua
Black Cherry; Wild Cherry
Prunus serotina
Sensitive-brier
Schrankia microphylla
Black Medic
Medicago lupulina
Beggar-lice
Desmodium sp.
Coral-bean
Erythrina herbacea
Butterfly-pea
Clitoria mariana
Beach-pea; Wild Bean
Strophostyles helvola
Hercules'-club
Zanthoxylum clava-herculis
Silverleaf Croton
Croton punctatus
Chinese Tallow-tree
Sapium sebiferum
Spurge
Euphorbia ammannioides
Euphorbia polygonifolia
Poison-ivy
Rhus radicans
Winged Sumac
Rhus copallina
American Holly
Ilex opaca
Yaupon
Ilex vomitoria
Red Maple
Acer rubrum
Virginia-creeper
Parthenocissus quinquefolia
Muscadine; Bullace
Vitis rotundifolia
St.-Andrew's-cross
Hypericum hypericoides
Tamarisk; Saltcedar
Tamarix gallica
Prickly-pear
Opuntia compressa
Devil's-walking-stick
Aralia spinosa
Dune Pennywort
Hydrocotyle bonariensis
Snakeroot
Sanicula sp.
Flowering Dogwood
Cornus florida
Spotted Wintergreen
Chimaphila maculata
Sea-lavender
Limonium carolinianum
Limonium nashii
Buckthorn Bumelia
Bumelia tenax
Persimmon
Diospyros virginiana
Wild Olive
Osmanthus americana
Yellow Jessamine
Gelsemium sempervirens
Polypremum
Polypremum procumbens
Fog-fruit
Lippia nodiflora
Beautyberry
Callicarpa americana
Horse-nettle
Solanum carolinense
Cross Vine
Anisostichus capreolata
Broadleaf Plantain
Plantago rugelii
Trailing Bluets
Houstonia procumbens
Bedstraw
Galium sp.
Japanese Honeysuckle
Lonicera japonica
Coral Honeysuckle
Lonicera sempervirens
Beach-elder
Iva imbricata
Common Ragweed
Ambrosia artemisiifolia
Common Dandelion
Taraxacum officinale
Elephant's-foot
Elephantopus tomentosus
Elephantopus elatus
Dog-fennel
Eupatorium capillifolium
Marsh-fleabane
Pluchea rosea
Camphorweed
Pluchea purpurascens
Sea-myrtle; Groundsel-tree
Baccharis halimifolia
Salt-marsh Aster
Aster tenuifolius
Salt-marsh Goldenrod
Solidago sempervirens
Goldenrod
Solidago tenuifolia
Sea Ox-eye
Borrichia frutescens
Beggar-ticks
Bidens sp.
Chanterelle
Cantharellus sp.
Chicken Mushroom
Laetiporus sulphureus
Tacky Green Russula
Russula aeruginea
Red Russula
Russula sp.
Earthstar
Geastrum sp.

FAUNA

Sea-whip
Leptogorgia virgulata
Tan Bushy Soft Coral
Plexaura flexuosa
Sea-pork (tunicate)
Aplidium sp.
Brown Rock Urchin
Arbacia punctulata
Five-holed Keyhole Urchin
Mellita quinquiesperforata
Marsh Periwinkle; Mud Snail
Littorina irrorata
Wentletrap (snail)
Epitoneum sp.
Flat Slipper (snail)
Crepidula plana
Southern Moon Snail; Shark-eye
Polinices duplicatus
Baby's-ear (shell)
Sinum perspectivum
Atlantic Oyster-drill
Urosalpinx cinerea
Murex (snail)
Chicoreus sp.
Channeled Whelk
Busycon canaliculatum
Knobbed Whelk
Busycon carica
Lightning Whelk
Busycon contrarium
Florida Horse Conch
Pleuroploca gigantea
Lettered Olive (snail)
Oliva sayana
Olive (snail)
Oliva sp.
Ribbed Mussel
Geukensia demissa
Ponderous Ark
Noetia ponderosa
Winged Pearl Oyster
Pteria colymbus
Stiff Pen (shell)
Atrina rigida
Pen (shell)
Atrina sp.
Eastern Oyster
Crassostrea virginica
Lion's-paw (shell)
Lyropecten nodosus

Beaches of Edisto Island (continued)
October 24, 1987, October 22, 1988, and June 10, 1995

Jingle Shell
Anomia simplex
Cross-hatched Lucine
Divaricella quadrisulcata
Giant Atlantic Cockle
Dinocardium robustum
Common Cockle
Trachycardium auricatum
Atlantic Razor Clam
Ensis directus
Coquina (shell)
Donax sp.
Northern Quahog Clam
Mercenaria mercenaria
Angel-wing (shell)
Cyrtopleura costata
Atlantic Horseshoe Crab
Limulus polyphemus
Tropical Goose Barnacle
Lepas anatifera
Rock Barnacle
Balanus sp.
Hermit Crab
Pagurus sp.
Common Mole Crab
Emerita talpoida
Gulf Calico Crab
Hepatus epheliticus
Mottled Purse Crab
Persephona mediterranea
Spider Crab
Libinia sp.
Speckled Crab
Arenaeus cribrarius
Common Blue Crab
Callinectes sapidus
Lady Crab
Ovalipes ocellatus
Stone Crab
Menippe mercenaria
Purple Marsh Crab
Sesarma reticulatum
Ghost Crab
Ocypode quadrata
Fiddler Crab
Uca sp.
Crab-shaped Spiny-back
Gasteracantha cancriformis
Golden-silk Spider
Nephila clavipes
Daddy-long-legs
Leiobunum sp.
Velvet-ant
Dasymutilla sp.
Red Imported Fire Ant
Solenopsis invicta
Honey Bee
Apis mellifera
Swamp Darner
Epiaeschna heros
Great Blue Skimmer
Libellula vibrans
Wandering Glider
Pantala flavescens
Violet-masked Glider
Tramea carolina
Tiger Beetle
Cicindela dorsalis media
Cicindela sp.
Fiery Searcher; Ground Beetle
Calosoma scrutator
Dingy Ground Beetle
Anisodactylus sp.
Passalus; Bess-beetle
Popilius disjunctus
Tumblebug; Dung Beetle
Glaphyrocanthon viridis
Flower Scarab Beetle
Trigonopeltastes delta
Tick Trefoil Leaf-miner
Pachyschelus laevigatus
Tumbling Flower Beetle
Mordella melaena
Mordella atrata
Mordellistena ambusta
Comb-clawed Beetle
Hymenorus sp.
Sap-feeding Beetle
Pallodes sp.
Spotless Lady Beetle
Cycloneda sanguinea
Hickory Girdler Beetle (tracks)
Oncideres cingulata
Gray-sided Oak Weevil
Pandeleteius hilaris
Black Swallowtail
Papilio polyxenes asterius
Palamedes Swallowtail
Papilio palamedes
Cloudless Sulphur
Phoebis sennae eubule
Sleepy Orange (butterfly)
Eurema nicippe
Gulf Fritillary
Agraulis vanillae
Variegated Fritillary
Euptoieta claudia
Common Buckeye (butterfly)
Junonia coenia
Gemmed Satyr (butterfly)
Cyllopsis gemma
Monarch (butterfly)
Danaus plexippus
Queen (butterfly)
Danaus gilippus
Long-tailed Skipper
Urbanus proteus
Gar Fish (bones)
Lepisosteus sp.
Green Treefrog
Hyla cinerea
Diamondback Terrapin
Malaclemys terrapin
Loggerhead Sea Turtle
Caretta caretta
Carolina Anole
Anolis carolinensis
Six-lined Racerunner
Cnemidophorus sexlineatus
Common Loon
Gavia immer
Brown Pelican
Pelecanus occidentalis
Double-crested Cormorant
Phalacrocorax auritus
Tricolored Heron
Egretta tricolor
Little Blue Heron
Egretta caerulea
Snowy Egret
Egretta thula
Great Egret
Casmerodius albus
Great Blue Heron
Ardea herodias
White Ibis
Eudocimus albus
Clapper Rail
Rallus longirostris
American Avocet
Recurvirostra americana
Semipalmated Plover
Charadrius semipalmatus
Willet
Catoptrophorus semipalmatus
Dunlin
Calidris alpina
Sanderling
Calidris alba
Western Sandpiper
Calidris mauri
Least Sandpiper
Calidris minutilla
Ruddy Turnstone
Arenaria interpres
Franklin's Gull
Larus pipixcan
Laughing Gull
Larus atricilla
Ring-billed Gull
Larus delawarensis
Herring Gull
Larus argentatus
Royal Tern
Sterna maxima
Black Vulture
Coragyps atratus
Sharp-shinned Hawk
Accipiter striatus
Osprey
Pandion haliaetus
American Kestrel
Falco sparverius
Rock Dove; Pigeon
Columba livia
Mourning Dove
Zenaida macroura
Ruby-throated Hummingbird
Archilochus colubris
Belted Kingfisher
Ceryle alcyon
Red-bellied Woodpecker
Melanerpes carolinus
Downy Woodpecker
Picoides pubescens
Hairy Woodpecker
Picoides villosus
Pileated Woodpecker
Dryocopus pileatus
Great Crested Flycatcher
Myiarchus crinitus
Tree Swallow
Tachycineta bicolor
Barn Swallow
Hirundo rustica
Blue Jay
Cyanocitta cristata
American Crow
Corvus brachyrhynchos
Tufted Titmouse
Parus bicolor
Carolina Chickadee
Parus carolinensis
Carolina Wren
Thryothorus ludovicianus
Ruby-crowned Kinglet
Regulus calendula
Blue-gray Gnatcatcher
Polioptila caerulea
American Robin
Turdus migratorius
Northern Mockingbird
Mimus polyglottos
Brown Thrasher
Toxostoma rufum
European Starling
Sturnus vulgaris
White-eyed Vireo
Vireo griseus
Northern Parula
Parula americana
Northern Cardinal
Cardinalis cardinalis
Painted Bunting
Passerina ciris
Rufous-sided Towhee
Pipilo erythrophthalmus
Red-winged Blackbird
Agelaius phoeniceus
Brewer's Blackbird
Euphagus cyanocephalus
Common Grackle
Quiscalus quiscula
Boat-tailed Grackle
Quiscalus major
Summer Tanager
Piranga rubra
Gray Squirrel
Sciurus carolinensis
Raccoon
Procyon lotor
Bottle-nosed Dolphin
Tursiops truncatus
White-tailed Deer (tracks)
Odocoileus virginianus

FOSSILS

Stingray (barb, dental plate)
Myliobatis sp.
Dasyatis sp.
Lemon Shark (tooth)
Negaprion brevirostris
Sand Shark (tooth)
Odontaspis hopei
Sand Tiger Shark (tooth)
Odontaspis taurus
Ground Shark (tooth)
Carcharhinus sp.
Black Drum (tooth)
Pogonias cromis
Pond Turtle (shell fragment)
Emydidae Family

(Photo by L. Greenlee)
Ghost Crab, *Ocypode quadrata.*

Bull Island
(Cape Romain National Wildlife Refuge)
Charleston County
December 9, 1995

Bull Island is the classic barrier island, protecting the marsh and inland areas from the full direct force of the open ocean. The devastation wrought by Hurricane Hugo would have been far worse inland had these islands not dampened some of the force of the huge tides. Most of the island's forest canopy was lost, and will take decades to recover.

As a result, the forest consists of typical post-hurricane maritime plant material. There are some man-made fresh water ponds that support alligators and a large population of wading birds, especially in the winter months. The diversity of plants and animals is enhanced by minimal human intervention. Open sand roads still remain for easy access to the many differing habitats, allowing the visitor to see a lot in a short time. During SCAN's visit, hunters were using part of the refuge, so our choices of areas to visit were limited mostly to the outer beach and main trail across the middle of the island. Even that provided many fascinating things to examine.

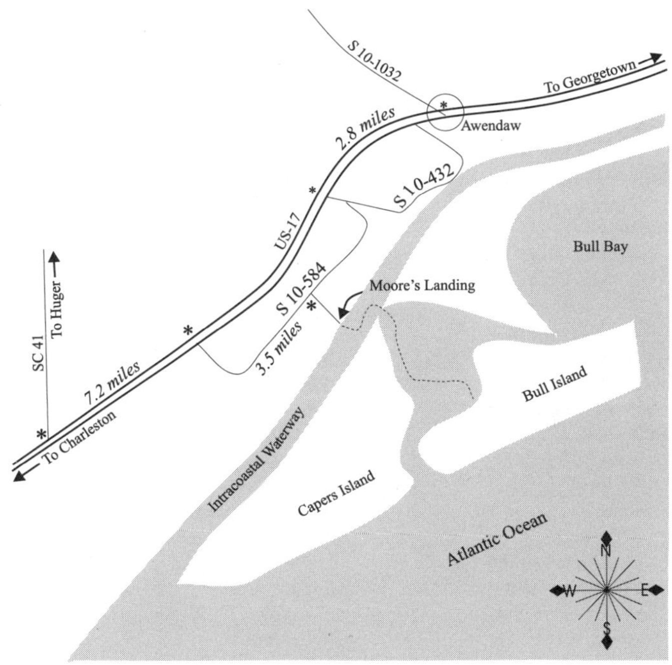

Directions to Bull Island: From Charleston take US-17 north toward Georgetown. After passing through Mount Pleasant and beyond the intersection with I-526, note the junction with SC 41 on the left. Continue north on US-17 for 7.2 miles beyond SC 41; then turn right onto County Road S 10-584. Go 3.5 miles to the Moore's Landing road, turn right, and follow that road to Moore's Landing and the boat dock. (If you miss the turn onto S 10-584 and arrive in Awendaw, turn around, go 2.8 miles to S 10-432, turn left, then take the right fork at the school onto the north end of S 10-584.)

FLORA

Gulfweed
Sargassum fluitans
Lichen
Leptogium cyanescens
Graphis scripta
Graphis elegans
Phaeographis dendritica
Pertusaria tetrathalmia
Haematomma elatinum
Lecanora hypoptoides
Lecanora cupressi
Lecanora pallida rubescens
Bulbothrix goebelii
Parmelina galbina
Parmelina horrescens
Parmotrema dilitatum
Parmotrema hypotropum
Parmotrema michauxianum
Parmotrema perforatum
Parmotrema praesorediosum
Parmotrema tinctorum
Parmotrema ultralucens
Punctelia rudecta
Usnea mutabilis
Usnea strigosa
Buellia dialyta
Buellia curatellae
Dirinaria confusa
Dirinaria picta
Dirinaria purpurascens
Physcia americana
Heterodermia albicans
Heterodermia speciosa
Pyxine caesiopruinosa
Rinodina roboris
Ochrolechia africana
Pyrenula cruenta
Pteridium aquilinum
Resurrection Fern
Polypodium polypodioides
Mosquito Fern
Azolla caroliniana

Bull Island (continued)
December 9, 1995

Loblolly Pine
Pinus taeda
Southern Red-cedar
Juniperus silicicola
Common Cat-tail
Typha latifolia
Cane; Switch Cane
Arundinaria gigantea
Giant Reed
Arundo donax
Sea Oats
Uniola paniculata
Smut Grass
Sporobolus poiretii
Finger Grass
Chloris petraea
Cord Grass
Spartina patens
Smooth Cord Grass
Spartina alterniflora
Sandbur; Sandspurs
Cenchrus sp.
Seaside Panic Grass
Panicum amarum
Broomsedge
Andropogon sp.
White-bracted Sedge
Dichromena colorata
Cabbage Palmetto
Sabal palmetto
Coconut
Cocos nucifera
Duckweed
Lemna sp.
Water-meal
Wolffia sp.
Spanish-moss
Tillandsia usneoides
Black Needle Rush
Juncus roemerianus
Fringed Greenbrier
Smilax bona-nox
Dune Greenbrier
Smilax auriculata
Spanish-bayonet
Yucca aloifolia
Mound-lily Yucca
Yucca gloriosa
Willow
Salix sp.
Wax-myrtle; Southern Bayberry
Myrica cerifera
Live Oak
Quercus virginiana
Laurel Oak
Quercus laurifolia
Hackberry
Celtis occidentalis georgiana
Southern Magnolia
Magnolia grandiflora
Redbay
Persea borbonia
Sassafras
Sassafras albidum

Sea-rocket
Cakile sp.
Dewberry
Rubus trivialis
Blackberry; Brambles
Rubus sp.
Wild Indigo
Baptisia sp.
Rattlebox
Crotalaria sp.
Spotted Medic
Medicago arabica
Sesbania
Sesbania exaltata
Chinese Wisteria
Wisteria sinensis
Coral-bean
Erythrina herbacea
Hercules'-club
Zanthoxylum clava-herculis
Silverleaf Croton
Croton punctatus
Chinese Tallow-tree
Sapium sebiferum
Poison-ivy
Rhus radicans
Poison-oak
Rhus toxicodendron
Winged Sumac
Rhus copallina
American Holly
Ilex opaca
Yaupon
Ilex vomitoria
Virginia-creeper
Parthenocissus quinquefolia
Muscadine; Bullace
Vitis rotundifolia
Pepper-vine
Ampelopsis arborea
Hypericum
Hypericum sp.
Tamarisk; Saltcedar
Tamarix gallica
Rockrose
Helianthemum corymbosum
Prickly-pear
Opuntia sp.
Devil's-walking-stick
Aralia spinosa
Pennywort
Hydrocotyle sp.
Buckthorn Bumelia
Bumelia tenax
Angle-pod
Matelea suberosa
Ground-ivy; Ponyfoot
Dichondra carolinensis
Morning-glory
Ipomoea sp.
Brazilian Vervain
Verbena braziliensis
Fog-fruit
Lippia nodiflora

Beautyberry
Callicarpa americana
Wood Sage; Germander
Teucrium canadense
Toadflax
Linaria canadensis
Cross Vine
Anisostichus capreolata
Trumpet Vine
Campsis radicans
Trailing Bluets
Houstonia procumbens
American Elderberry
Sambucus canadensis
Marsh-elder
Iva frutescens
Beach-elder
Iva imbricata
Pyrrhopappus
Pyrrhopappus carolinianus
Dog-fennel
Eupatorium capillifolium
Mistflower; Ageratum
Eupatorium coelestinum

Climbing Hempweed
Mikania scandens
Rabbit-tobacco
Gnaphalium obtusifolium
False-willow
Baccharis angustifolia
Sea-myrtle; Groundsel-tree
Baccharis halimifolia
Salt-marsh Goldenrod
Solidago sempervirens
Camphorweed
Heterotheca subaxillaris
Sea Ox-eye
Borrichia frutescens

FAUNA

Sea-whip
Leptogorgia virgulata
Deadman's-fingers (sponge)
Briareum asbestinum
Five-holed Keyhole Urchin
Mellita quinquiesperforata

(Photos by A. Ciegler)

Fallen trees sculpt the beaches of Bull Island.

Brown Sea-cucumber
Sclerodactyla briareus
Worm Shell
Vermicularia knorrii
Common Slipper (snail)
Crepidula fornicata
Southern Moon Snail; Shark-eye
Polineces duplicatus
Channeled Whelk
Busycon canaliculatum
Knobbed Whelk
Busycon carica
Lightning Whelk
Busycon contrarium
Whelk (egg case)
Busycon sp.
Lettered Olive (snail)
Oliva sayana
Incongruous Ark
Anadara brasiliana
Ponderous Ark
Noetia ponderosa
Half-naked Pen (shell)
Atrina seminuda
Amber Pen (shell)
Pinna carnea
Eastern Oyster
Crassostrea virginica
Jingle Shell
Anomia simplex
Giant Atlantic Cockle
Dinocardium robustum
Dwarf Surf Clam
Mulinia lateralis
Jack-knife Clam
Tagelus plebeius
Florida Coquina
Donax variabilis
Cross-barred Venus (shell)
Chione cancellata
Disk (shell)
Dosinia discus
Angel-wing (shell)
Cyrtopleura costata
Atlantic Horseshoe Crab
Limulus polyphemus
Rock Barnacle
Balanus sp.
Gulf Calico Crab
Hepatus epheliticus
Spider Crab
Libinia sp.
Common Blue Crab
Callinectes sapidus
Ghost Crab
Ocypode quadrata
Band-winged Grasshopper
Psinidia fenestralis
Red-legged Grasshopper
Melanoplus femurrubrum
Obscure Bird Grasshopper
Schistocerca obscura
Field Cricket
Gryllus sp.
Mole Cricket
Scapteriscus sp.

Antlion; Doodlebug
Myrmeleon sp.
Live-oak Potato Gall (wasp)
Plagiotrochus batatoides
Dung Beetle
Onthophagus hecate
Gulf Fritillary
Agraulis vanillae
Anglewing (butterfly)
Polygonia sp.
Common Buckeye (butterfly)
Junonia coenia
Evergreen Bagworm Moth
Thyridopteryx ephemeraeformis
Clearnose Skate (egg case)
Raja eglanteria
Common Loon
Gavia immer
Horned Grebe
Podiceps auritus
Brown Pelican
Pelecanus occidentalis
Anhinga; Snakebird
Anhinga anhinga
Double-crested Cormorant
Phalacrocorax auritus
Tricolored Heron
Egretta tricolor
Little Blue Heron
Egretta caerulea
Great Blue Heron
Ardea herodias
Bufflehead
Bucephala albeola
Red-breasted Merganser
Mergus serrator
Hooded Merganser
Lophodytes cucullatus
Clapper Rail
Rallus longirostris
Sora
Porzana carolina
Common Moorhen
Gallinula chloropus
American Oystercatcher
Haematopus palliatus
Semipalmated Plover
Charadrius semipalmatus
Killdeer
Charadrius vociferus
Black-bellied Plover
Pluvialis squatarola
Marbled Godwit
Limosa fedoa
Willet
Catoptrophorus semipalmatus
Greater Yellowlegs
Tringa melanoleuca
Red Knot
Calidris canutus
Dunlin
Calidris alpina
Sanderling
Calidris alba
Ruddy Turnstone
Arenaria interpres

Ring-billed Gull
Larus delawarensis
Herring Gull
Larus argentatus
Forster's Tern
Sterna forsteri
Turkey Vulture
Cathartes aura
Northern Harrier
Circus cyaneus
Red-tailed Hawk
Buteo jamaicensis
Osprey (nest)
Pandion haliaetus
American Kestrel
Falco sparverius
Wild Turkey
Meleagris gallopavo
Mourning Dove
Zenaida macroura
Belted Kingfisher
Ceryle alcyon
Red-bellied Woodpecker
Melanerpes carolinus
Red-headed Woodpecker
Melanerpes erythrocephalus
Northern Flicker
Colaptes auratus
Yellow-bellied Sapsucker
Sphyrapicus varius
Eastern Phoebe
Sayornis phoebe
House Wren
Troglodytes aedon
Carolina Wren
Thryothorus ludovicianus
Marsh Wren
Cistothorus palustris
Golden-crowned Kinglet
Regulus satrapa
Ruby-crowned Kinglet
Regulus calendula
Hermit Thrush
Catharus guttatus

Loggerhead Shrike
Lanius ludovicianus
Gray Catbird
Dumetella carolinensis
Northern Mockingbird
Mimus polyglottos
European Starling
Sturnus vulgaris
Yellow-rumped Warbler
Dendroica coronata
Yellowthroat
Geothlypis trichas
Northern Cardinal
Cardinalis cardinalis
Rufous-sided Towhee
Pipilo erythrophthalmus
Song Sparrow
Melospiza melodia
Chipping Sparrow
Spizella passerina
White-throated Sparrow
Zonotrichia albicollis
Swamp Sparrow
Melospiza georgiana
Eastern Meadowlark
Sturnella magna
Red-winged Blackbird
Agelaius phoeniceus
Boat-tailed Grackle
Quiscalus major
Eastern Fox Squirrel (tracks)
Sciurus niger
Red Wolf (tracks)
Canis rufus
Raccoon (tracks)
Procyon lotor
River Otter (sign)
Lutra canadensis
Bobcat; Wildcat
Felis rufus
Bottle-nosed Dolphin
Tursiops truncatus
White-tailed Deer (tracks)
Odocoileus virginianus

(Photo by A. Ciegler)

Red Knot, *Calidris canutus*.

Santee River Delta
Charleston County

The Wedge, March 12, 1988
Washo Reserve, February 25, 1995

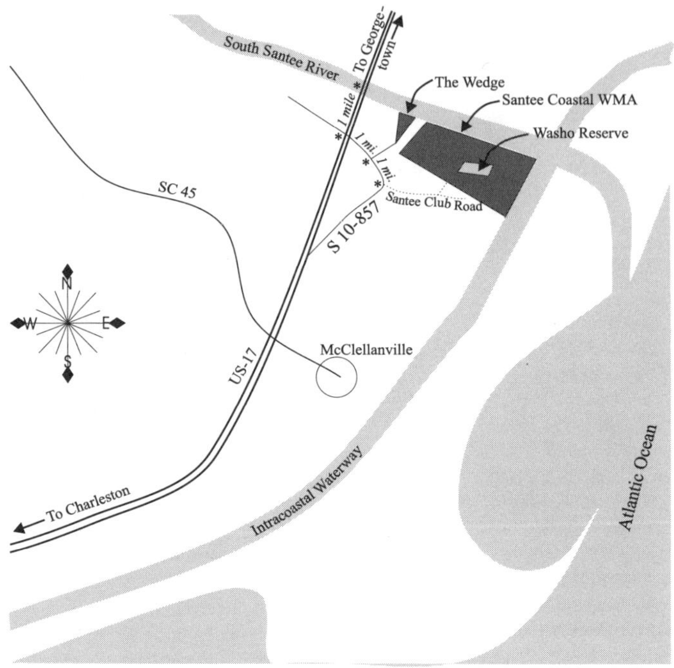

These sites, just a few miles apart on the Santee River delta, feature old rice ponds and both maritime and coastal plain woodlands. In addition, Santee Coastal Reserve has an old freshwater pond known as Washo Reserve. Its function was to provide an emergency supply of freshwater to the people on the coast. Today it has swampy edges and a large colony of nesting osprey in the more open area of the pond. Also, there are large areas of higher, dryer coastal plain forests and old rice ponds with brackish water. Because much of the reserve has not been recently disturbed, less common plants and animals are often observed.

The Wedge, at the time SCAN visited, was a research facility owned and operated by the University of South Carolina for research and teaching in the area of tropical vector-borne diseases. It housed an excellent moth specimen collection collected by the previous plantation owner, Dr. Richard Dominick. Many areas on the plantation had been developed to attract insects for study purposes.It is currently closed to the public.

Washo Reserve is owned by the South Carolina Department of Natural Resources.

Directions to the Santee River Delta: From Georgetown take US-17 south toward Charleston. After about 15 miles, the road crosses the South Santee River and enters Charleston County. Travel 1 mile, then turn left onto County Road S 10-857 (South Santee Road). For The Wedge, drive about 1 mile, then turn left again at the sign to the plantation. The main office is about a mile down the driveway, and the laboratory is through the gate beyond the office. For the Washo Reserve, drive 2 miles from US 17, then bear left on Santee Club Road; bear left again at the next fork in the road. The parking lot is ahead on the right.

FLORA

Lichen
Graphis scripta
Graphis elegans
Glyphis cicatricosa
Gyrostomum scyphuliferum
Medusulina nitida
Coenogonium interplexum
Bacidia clementis
Brigantiaea leucoxantha
Cladonia chlorophaea
Cladonia coniocraea
Cladonia didyma
Cladonia ochrochlora
Cladonia peziziformis
Cladonia subradiata
Pertusaria copiosa
Pertusaria ostiolata
Pertusaria paratuberculifera
Pertusaria subpertusa
Pertusaria texana
Haematomma puniceum
Haematomma elatinum
Haematomma cismonicum
Lecanora cupressi
Lecanora varia
Lecanora hybocarpa
Lecanora caesiorubella subsp. caesiorubella
Ochrolechia africana
Bulbothrix goebelii
Bulbothrix goebelii f. minor
Hypotrachyna livida
Hypotrachyna pustulifera
Parmelina aurulenta
Parmelina minarum
Parmeliopsis subambigua
Parmotrema dilitatum
Parmotrema hypotropum
Parmotrema michauxianum
Parmotrema perforatum
Parmotrema rampoddense
Parmotrema tinctorum
Flavoparmelia caperata
Canoparmelia caroliniana
Punctelia rudecta
Rimelia reticulata
Rimelia subisidiosa
Usnea ceratina
Usnea mutabilis
Usnea rubicunda
Usnea strigosa
Buellia curatellae
Dirinaria aspera
Dirinaria confusa
Dirinaria picta
Dirinaria purpurascens
Heterodermia albicans
Hetrerodermia casarettiana
Heterodermia speciosa
Heterodermia spaltans sp. nov.
Pyxine caesiopruinosa
Pyxine sorediata
Campylothelium amylosporum
Pyrenula cruenta
Eopyrenula leucoplaca
Clathroporina sp.
Calicium abietinum
Arthothelium interveniens
Trypethelium mastoideum
Trypethelium virens
Porina sp.

Peat Moss
Sphagnum sp.

Bracken
Pteridium aquilinum

Sensitive Fern
Onoclea sensibilis

Netted Chain Fern
Woodwardia areolata

Ebony Spleenwort
Asplenium platyneuron

Resurrection Fern
Polypodium polypodioides

Long-leaf Pine
Pinus palustris

Loblolly Pine
Pinus taeda
Bald-cypress
Taxodium distichum
Eastern Red-cedar
Juniperus virginiana
Southern Red-cedar
Juniperus silicicola
Narrow-leaved Cat-tail
Typha angustifolia
Bur-reed
Sparganium americanum
Spotted Pondweed
Potamogeton pulcher
Arrowhead; Duck-potato
Sagittaria sp.
Cane; Switch Cane
Arundinaria gigantea
Spike Grass
Uniola sessiliflora
Wild Rye Grass
Elymus virginicus
Cord Grass
Spartina cynosuroides
Foxtail Grass
Setaria sp.
Gamma Grass
Tripsacum dactyloides
Dwarf Palmetto
Sabal minor
Cabbage Palmetto
Sabal palmetto
Duckweed
Lemna sp.
Bogmat
Wolffiella floridana
Spanish-moss
Tillandsia usneoides
Black Needle Rush
Juncus roemerianus
Rush
Juncus sp.
Fringed Greenbrier
Smilax bona-nox
False Garlic
Allium bivalve
Narcissus
Narcissus sp.
Swamp Willow
Salix caroliniana
Wax-myrtle; Southern Bayberry
Myrica cerifera
White Oak
Quercus alba
Post Oak
Quercus stellata
Live Oak
Quercus virginiana
Black Jack Oak
Quercus marilandica
Water Oak
Quercus nigra
Laurel Oak
Quercus laurifolia
Mistletoe
Phoradendron serotinum

Sorrel; Dock
Rumex sp.
Polygonum
Polygonum sp.
Common Chickweed
Stellaria media
Mouse-ear Chickweed
Cerastium glomeratum
Hornwort
Ceratophyllum demersum
Water-lily
Nymphaea odorata
Sweetbay
Magnolia virginiana
Southern Magnolia
Magnolia grandiflora
Redbay
Persea borbonia
Sassafras
Sassafras albidum
Hairy Bitter Cress
Cardamine hirsuta
Bitter Cress
Cardamine pensylvanica
Virginia-willow
Itea virginica
Sweet-gum
Liquidambar styraciflua
Dewberry
Rubus trivialis
Cherokee Rose
Rosa laevigata
Redbud; Judas-tree
Cercis canadensis
Honey Locust
Gleditsia triacanthos
Sesbania
Sesbania exaltata
Bladderpod
Glottidium vesicarium
American Wisteria
Wisteria frutescens
Cranesbill
Geranium carolinianum
Chinaberry
Melia azedarach
American Holly
Ilex opaca
Yaupon
Ilex vomitoria
Red Maple
Acer rubrum
Muscadine; Bullace
Vitis rotundifolia
St.-Andrew's-cross
Hypericum hypericoides
Primrose-leaved Violet
Viola primulifolia
Violet
Viola sp.
Meadow-beauty
Rhexia sp.
Seed-box
Ludwigia alternifolia
Water Pennywort
Hydrocotyle ranunculoides

Pennywort
Hydrocotyle sp.
Blackgum
Nyssa sylvatica
Water Tupelo
Nyssa aquatica
Flowering Dogwood
Cornus florida
Sweet Pepperbush
Clethra alnifolia
Spotted Wintergreen
Chimaphila maculata
Fetterbush
Lyonia lucida
Blueberry
Vaccinium sp.
Persimmon
Diospyros virginiana
Horse-sugar; Sweetleaf
Symplocos tinctoria
Yellow Jessamine
Gelsemium sempervirens
Periwinkle
Vinca major
Common Henbit
Lamium amplexicaule
Nightshade
Solanum sp.
Corn Speedwell; Veronica
Veronica arvensis
Trumpet Vine
Campsis radicans
Floating Bladderwort
Utricularia inflata
Partridge-berry
Mitchella repens
Small Bluets
Houstonia pusilla
Bedstraw; Purple-bead
Galium hispidulum
American Elderberry
Sambucus canadensis
Facelis
Facelis retusa
Sea-myrtle; Groundsel-tree
Baccharis halimifolia
Carbon-balls (fungus)
Daldinia concentrica
Collybia (mushroom)
Collybia sp.

FAUNA

Cloudless Sulphur
Phoebis sennae eubule
Sleepy Orange (butterfly)
Eurema nicippe
Eastern Tailed-blue(butterfly)
Everes comyntas
Gulf Fritillary
Agraulis vanillae
American Lady (butterfly)
Vanessa virginiensis
Polyphemus Moth
Antheraea polyphemus
Sweetbay Silkmoth
Callosamia securifera

Green Treefrog
Hyla cinerea
American Alligator
Alligator mississippiensis
Carolina Anole
Anolis carolinensis
Ground Skink
Scincella lateralis
Pied-billed Grebe
Podilymbus podiceps
Double-crested Cormorant
Phalacrocorax auritus
Tricolored Heron
Egretta tricolor
Little Blue Heron
Egretta caerulea
Snowy Egret
Egretta thula
Great Egret
Casmerodius albus
Great Blue Heron
Ardea herodias
White Ibis
Eudocimus albus
Mallard
Anas platyrhynchos
Green-winged Teal
Anas crecca
American Wigeon
Anas americana
Northern Pintail
Anas acuta
Northern Shoveler
Anas clypeata
Blue-winged Teal
Anas discors
Ring-necked Duck
Aythya collaris
Greater Scaup
Aythya marila
Lesser Scaup
Aythya affinis
Common Goldeneye
Bucephala clangula
Bufflehead
Bucephala albeola
Red-breasted Merganser
Mergus serrator
Hooded Merganser
Lophodytes cucullatus
Clapper Rail
Rallus longirostris
American Coot
Fulica americana
Killdeer
Charadrius vociferus
Greater Yellowlegs
Tringa melanoleuca
Lesser Yellowlegs
Tringa flavipes
Ring-billed Gull
Larus delawarensis
Royal Tern
Sterna maxima
Turkey Vulture
Cathartes aura

Santee River Delta (continued)
March 12, 1988, and February 25, 1995

Black Vulture
Coragyps atratus
Bald Eagle
Haliaeetus leucocephalus
Northern Harrier
Circus cyaneus
Osprey
Pandion haliaetus
American Kestrel
Falco sparverius
Wild Turkey
Meleagris gallopavo
Mourning Dove
Zenaida macroura
Belted Kingfisher
Ceryle alcyon
Red-bellied Woodpecker
Melanerpes carolinus
Red-headed Woodpecker
Melanerpes erythrocephalus
Northern Flicker
Colaptes auratus
Yellow-bellied Sapsucker
Sphyrapicus varius
Pileated Woodpecker
Dryocopus pileatus
Eastern Phoebe
Sayornis phoebe
Tree Swallow
Tachycineta bicolor
Blue Jay
Cyanocitta cristata
Fish Crow
Corvus ossifragus
Tufted Titmouse
Parus bicolor
Carolina Chickadee
Parus carolinensis
Brown-headed Nuthatch
Sitta pusilla
Ruby-crowned Kinglet
Regulus calendula
Eastern Bluebird
Sialia sialis
Hermit Thrush
Catharus guttatus
American Robin
Turdus migratorius
Northern Mockingbird
Mimus polyglottos
Brown Thrasher
Toxostoma rufum
Cedar Waxwing
Bombycilla cedrorum
Solitary Vireo
Vireo solitarius
Yellow-rumped Warbler
Dendroica coronata
Yellow-throated Warbler
Dendroica dominica
Pine Warbler
Dendroica pinus
Yellowthroat
Geothlypis trichas
Northern Cardinal
Cardinalis cardinalis
Rufous-sided Towhee
Pipilo erythrophthalmus
Song Sparrow
Melospiza melodia
Chipping Sparrow
Spizella passerina
Dark-eyed Junco
Junco hyemalis
White-throated Sparrow
Zonotrichia albicollis
Red-winged Blackbird
Agelaius phoeniceus
Gray Squirrel
Sciurus carolinensis
Eastern Fox Squirrel
Sciurus niger
Beaver (gnaw mark)
Castor canadensis
Raccoon
Procyon lotor
Mink
Mustela vison
River Otter (tracks) (scat)
Lutra canadensis
Wild Pig
Sus scrofa
White-tailed Deer (tracks)
Odocoileus virginianus

(Photo by V. Winn)
Julie Finlayson checks out an old rice-paddy dike for creatures.

(Photos by L. Greenlee)
Polyphemus moth, female (left) and male (above), *Antheraea polyphemus*.

St. Christopher Camp and Conference Center
(Barrier Island Environmental Education Program, Seabrook Island)
Charleston County
October 21 - 23, 1994

St. Christopher Camp is on the remote western end of Seabrook Island, providing a refuge for the natural world on an otherwise highly developed barrier island. We were given the opportunity to explore the beachfront along the mouth of the N. Edisto River. These yielded marine and estuarine life, beach, intertidal zone and dunes habitat, plus over 100 acres of salt marsh, 200 acres of maritime forest containing many sloughs, an extensive freshwater swamp, and 2 freshwater ponds! Many of us gave blood to the cause of the local mosquitoes. Our 'lichen-man' Alex Ciegler discovered a new lichen growing on *Ilex opaca* in the forest which he plans to name *Lecidea christophora* in honor of the camp.

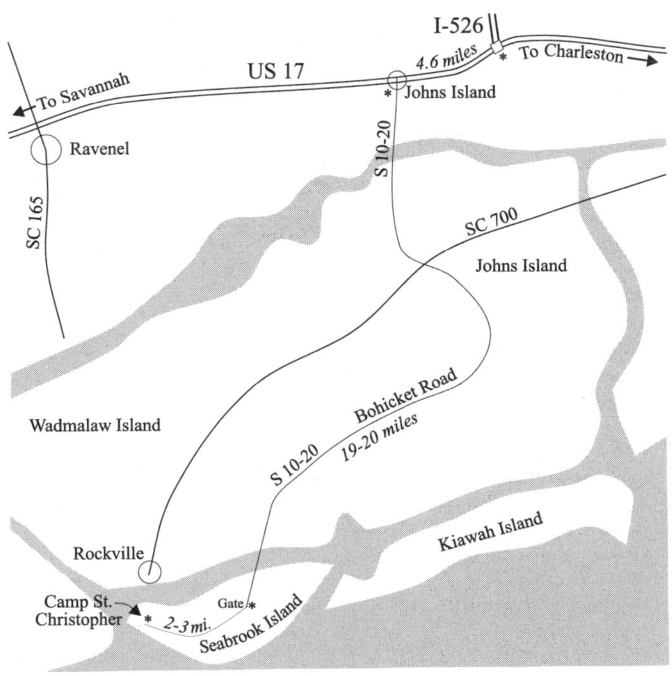

Owned and operated by the South Carolina Diocese of the Episcopal Church. Access by prior arrangement only.

Directions to St. Christopher Camp and Conference Center: From Charleston take US-17 south toward Savannah. Note the intersection with I-526 (Mark Clark Expressway) and continue on US-17 for 4.6 miles past that intersection. Turn left (south) on County Road S 10-20, where signs direct you toward Seabrook and Kiawah Islands. Continue south on S 10-20 down Bohicket Road all the way to Seabrook Island, 19-20 miles. Permission must be obtained in advance to pass through the guarded gate. Travel 2-3 miles, and the entrance to the conference center is on the right. Pass the camp on your left until you see the conference facilities where the offices are located.

Science and Nature Notes: Comb Jellyfishes

In our exploring of the intertidal zone, several comb jellies were caught in the seine. Closely related to jellyfishes, comb jellies look like blobs of transparent jelly. Instead of swimming by contracting a bell, as the jellyfishes do, a comb jelly uses eight rows of ctenes (comblike tufts of microscopic hairs) to propel itself; it swims through the water, engulfing plankton with its huge mouth. Each row of these minute combs is iridescent during the day and luminescent at night. When conditions are right at night, any disturbance of the ocean surface causes these creatures to glow like small light bulbs.

FLORA

Dead Man's Fingers Seaweed
Codium sp.
Sea Lettuce
Ulva sp.
Lichen
Graphis scripta
Graphis leucopepla
Graphis elegans
Lecidea christophera
Letrouitia domingense
Pertusaria ostiolata
Pertusaria paratuberculifera
Pertusaria propinqua
Pertusaria texana
Pertusaria sinusmexicani
Maronea constans
Lecanora cupressi
Lecanora varia
Ochrolechia africana
Bulbothrix goebelii
Bulbothrix goebelii f. minor
Hypotrachyna pustulifera
Parmelina aurulenta
Parmelina minarum
Parmotrema crinitum
Parmotrema hypotropum
Parmotrema michauxianum
Parmotrema perforatum
Parmotrema rampoddense
Parmotrema tinctorum
Parmotrema ultralucens
Canoparmelia caroliniana
Punctelia rudecta
Rimelia reticulata
Ramalina americana
Usnea strigosa
Dirinaria confusa
Dirinaria picta
Dirinaria purpurascens
Physcia americana
Physcia crispa
Physcia millegrana
Physcia arediosa
Heterodermia albicans
Heterodermia spaltans sp. nov.
Campylothelium amylosporum

St. Christopher Camp and Conference Center (continued)
October 21-23, 1994

Pyrenula cruenta
Trypethelium virens
Schismatomma rediunta
Lepraria sp.
Floating Liverwort
Ricciocarpus natans
Bracken
Pteridium aquilinum
Netted Chain Fern
Woodwardia areolata
Ebony Spleenwort
Asplenium platyneuron
Resurrection Fern
Polypodium polypodioides
Loblolly Pine
Pinus taeda
Spruce Pine
Pinus glabra
Southern Red-cedar
Juniperus silicicola
Common Cat-tail
Typha latifolia
Sea Oats
Uniola paniculata
Coastal Dropseed
Sporobolus virginicus
Bermuda Grass
Cynodon dactylon
Cord Grass
Spartina patens
Smooth Cord Grass
Spartina alterniflora
Giant Foxtail Grass
Setaria magna
Dune Sandbur
Cenchrus tribuloides
Oplismenus (grass)
Oplismenus setarius
Seaside Panic Grass
Panicum amarum
Dwarf Palmetto
Sabal minor
Cabbage Palmetto
Sabal palmetto
Spanish-moss
Tillandsia usneoides
Fringed Greenbrier
Smilax bona-nox
Dune Greenbrier
Smilax auriculata
Sarsaparilla-vine
Smilax pumila
Mound-lily Yucca
Yucca gloriosa
Black Willow
Salix nigra
Wax-myrtle; Southern Bayberry
Myrica cerifera
Live Oak
Quercus virginiana
Willow Oak
Quercus phellos
Laurel Oak
Quercus laurifolia
Sugarberry; Hackberry
Celtis laevigata
Red Mulberry
Morus rubra
False Nettle
Boehmeria cylindrica
Water Smartweed
Polygonum densiflorum
Dotted Smartweed
Polygonum punctatum
Climbing Buckwheat; Bindweed
Polygonum convolvulus
Mexican-tea
Chenopodium ambrosioides
Seabeach Orach
Atriplex arenaria
Marsh Orach
Atriplex patula
Perennial Glasswort
Salicornia virginica
Russian-thistle
Salsola kali
Southern Magnolia
Magnolia grandiflora
Redbay
Persea borbonia
Sassafras
Sassafras albidum
Sea-rocket
Cakile harperi
Sweet-gum
Liquidambar styraciflua
Carolina Cherrylaurel
Prunus caroliniana
Rattlebox
Crotalaria spectabilis
Beggar-lice
Desmodium sp.
Sesbania
Sesbania exaltata
Coral-bean
Erythrina herbacea
Beach-pea; Wild Bean
Strophostyles helvola
Hercules'-club
Zanthoxylum clava-herculis
Silverleaf Croton
Croton punctatus
Poison-ivy
Rhus radicans
Winged Sumac
Rhus copallina
American Holly
Ilex opaca
Yaupon
Ilex vomitoria
Red Maple
Acer rubrum
Shellmound-shrub
Sageretia minutiflora
Virginia-creeper
Parthenocissus quinquefolia
Muscadine; Bullace
Vitis rotundifolia
Fox Grape
Vitis labrusca
Pepper-vine
Ampelopsis arborea
Prickly-pear
Opuntia compressa
Devil's-walking-stick
Aralia spinosa
Dune Pennywort
Hydrocotyle bonariensis
Blackgum
Nyssa sylvatica
Flowering Dogwood
Cornus florida
Sparkleberry
Vaccinium arboreum
Buckthorn Bumelia
Bumelia lycioides
Persimmon
Diospyros virginiana
Wild Olive
Osmanthus americana
Yellow Jessamine
Gelsemium sempervirens
Ivy-leaf Morning-glory
Ipomoea hederacea hederacea
Fiddle-leaf Morning-glory
Ipomoea stolonifera
Fog-fruit
Lippia nodiflora
Beautyberry
Callicarpa americana
Wood Sage; Germander
Teucrium canadense
Dotted Horsemint
Monarda punctata
Cross Vine
Anisostichus capreolata
Trumpet Vine
Campsis radicans
Partridge-berry
Mitchella repens
Bedstraw; Purple-bead
Galium hispidulum
American Elderberry
Sambucus canadensis
Marsh-elder
Iva frutescens
Beach-elder
Iva imbricata
Common Ragweed
Ambrosia artemisiifolia
Cocklebur
Xanthium strumarium glabratum
Elephant's-foot
Elephantopus tomentosus
Dog-fennel
Eupatorium capillifolium
Mistflower; Ageratum
Eupatorium coelestinum
Climbing Hempweed
Mikania scandens
Sea-myrtle; Groundsel-tree
Baccharis halimifolia
Salt-marsh Aster
Aster tenuifolius
Aster
Aster sp.
Goldenrod
Solidago sp.
Camphorweed
Heterotheca subaxillaris
Sea Ox-eye
Borrichia frutescens
Wing-stem
Verbesina virginica
Chicken Mushroom
Laetiporus sulphureus
Elegant Polypore
Polyporus varius

(Photo by S. Simons)
Chicken mushroom, *Laetiporus sulphureus*.

(Photo by L. Greenlee)

Mud fiddler crab, *Uca pugnax*.

Crowded Parchment
Stereum complicatum
Fly Agaric
Amanita muscaria
Destroying-angel
Amanita virosa
Deadly Galerina
Galerina autumnalis
Corrugated-cap Milky
Lactarius corrugis
Family Collybia
Clitocybula familia
Little Brown Collybia
Collybia alkalivirens
Common Mycena
Mycena galericulata
Magnolia-cone Mushroom
Strobilurus conigenoides
Rounded Earthstar
Geastrum saccatum

FAUNA

Sea-whip
Leptogorgia virgulata
Deadman's-fingers (sponge)
Briareum asbestinum
Hermit Crab Anemone
Calliactis tricolor
Northern Cup Coral
Astrangia danae
Star Coral
Favia fragum
Beroe's Comb-jelly
Beroe ovata
Five-holed Keyhole Urchin
Mellita quinquiesperforata
Striped Sea-cucumber
Thyonella gemmata
Rubbery Bryozoan
Alcyonidium hauffi
Variable Crust Bryozoan
Schizoporella errata
Cayenne Keyhole Limpet (shell)
Diodoro cayenensis
Marsh Periwinkle; Mud Snail
Littorina irrorata
Common Slipper (snail) (shell)
Crepidula fornicata
Flat Slipper (snail) (shell)
Crepidula plana
Southern Moon Snail; Shark-eye
Polineces duplicatus
Baby's-ear (shell) (shell)
Sinum perspectivum
Thick-lipped Oyster-drill
Eupleura caudata
Channeled Whelk
Busycon canaliculatum
Knobbed Whelk
Busycon carica
Lightning Whelk
Busycon contrarium
Banded Tulip (snail) (shell)
Fasciolaria hunteria
Lettered Olive (snail)
Oliva sayana
Augur Snail
Terebra sp.
Incongruous Ark
Anadara brasiliana
Blood Ark
Anadara ovalis
Transverse Ark
Anadara transversa
Ponderous Ark (shell)
Noetia ponderosa
Pen (shell)
Atrina sp.
Eastern Oyster
Crassostrea virginica
Cat's-paw; Kitten's-paw (shell)
Plicatula gibbosa
Jingle Shell
Anomia simplex
Giant Atlantic Cockle (shell)
Dinocardium robustum
Atlantic Razor Clam (shell)
Ensis directus
Coquina (shell)
Donax sp.
Disk (shell)
Dosinia discus
Northern Quahog Clam (shell)
Mercenaria mercenaria
Angel-wing (shell)
Cyrtopleura costata
Roofing Worm; Shingle Tubeworm
Owenia fusiformis
Atlantic Horseshoe Crab (shell)
Limulus polyphemus
Sea-whip Barnacle
Conopea galeata
Ivory Barnacle
Balanus eburneus
Rock Barnacle
Balanus sp.
Saltmarsh Beachhopper
Orchestia grillus
Ghost Shrimp (scat)
Callianassa sp.
Green-striped Hermit Crab
Clibanarius vittatus
Flat-clawed Hermit Crab
Pagurus pollicaris
Gulf Calico Crab (shell)
Hepatus epheliticus
Mottled Purse Crab
Persephona mediterranea
Speckled Crab
Arenaeus cribrarius
Blue Crab
Callinectes similis
Lady Crab (shell)
Ovalipes ocellatus
Ghost Crab
Ocypode quadrata
Mud Fiddler Crab
Uca pugnax
Sand Fiddler Crab
Uca pugilator
Orchard Spider
Leucauge venusta
Leather-color Bird Grasshopper
Schistocerca alutacea
Palmetto-bug
Periplaneta fuliginosa
Leaf-footed Yucca Bug
Leptoglossus sp.
Water-scorpion
Ranatra sp.
Fire Ant (nest)
Solenopsis sp.
Black Salt-marsh Mosquito
Aedes taeniorhyncus
Fall Woodlands Mosquito
Aedes infirmatus
Golden Salt-marsh Mosquito
Aedes sollicitans
Unmarked Slender Mosquito
Culisetia inornata
Common Green Darner
Anax junius
Eastern Pondhawk
Erythemis simplicicollis
Blue Dasher
Pachydiplax longipennis
Wandering Glider
Pantala flavescens
Eastern Amberwing (dragonfly)
Perithemis tenera
Glider
Tramea sp.
Roseate Skimmer (dragonfly)
Orthemis ferruginea
Tiger Beetle
Cicindela sp.
Crawling Water Beetle
Peltodytes dietrichi
Water Scavenger Beetle
Tropisternus quadristriatus quadristriatus
Smartweed Flea Beetle
Disonycha pensylvanica
Cloudless Sulphur
Phoebis sennae eubule
Little Yellow (butterfly)
Eurema lisa
Gulf Fritillary
Agraulis vanillae
Zebra (butterfly)
Heliconius charitonius tuckeri
Common Buckeye (butterfly)
Junonia coenia
Monarch (butterfly)
Danaus plexippus
Long-tailed Skipper
Urbanus proteus
Atlantic Silverside Fish
Menidia menidia
Skate (egg case)
Raja sp.
Puffer Fish; Globefish
Chilomycterus sp.
Mackerel
Scomber sp.
Green Treefrog
Hyla cinerea
Loggerhead Sea Turtle (egg)
Caretta caretta
Carolina Anole
Anolis carolinensis
Ground Skink
Scincella lateralis
Common Loon
Gavia immer
Brown Pelican
Pelecanus occidentalis
Double-crested Cormorant
Phalacrocorax auritus
Great Egret
Casmerodius albus
Great Blue Heron
Ardea herodias
Black-bellied Plover
Pluvialis squatarola
Willet
Catoptrophorus semipalmatus
Sanderling
Calidris alba
Laughing Gull
Larus atricilla

St. Christopher Camp and Conference Center (continued)
October 21-23, 1994

Ring-billed Gull
Larus delawarensis
Royal Tern
Sterna maxima
Caspian Tern
Sterna caspia
Black Skimmer
Rhynchops niger
Turkey Vulture
Cathartes aura
Red-shouldered Hawk
Buteo lineatus
Red-tailed Hawk
Buteo jamaicensis
Osprey
Pandion haliaetus
Peregrine Falcon
Falco peregrinus
Mourning Dove
Zenaida macroura
Eastern Screech-owl
Otus asio
Belted Kingfisher
Ceryle alcyon
Red-bellied Woodpecker
Melanerpes carolinus
Pileated Woodpecker
Dryocopus pileatus
Eastern Phoebe
Sayornis phoebe
Tree Swallow
Tachycineta bicolor
Purple Martin
Progne subis
Blue Jay
Cyanocitta cristata
American Crow
Corvus brachyrhynchos
House Wren
Troglodytes aedon
Carolina Wren
Thryothorus ludovicianus
Ruby-crowned Kinglet
Regulus calendula
Gray Catbird
Dumetella carolinensis
Northern Mockingbird
Mimus polyglottos
Brown Thrasher
Toxostoma rufum
Red-Eyed Vireo
Vireo olivaceus
Black-throated Blue Warbler
Dendroica caerulescens
Northern Cardinal
Cardinalis cardinalis
Boat-tailed Grackle
Quiscalus major
Summer Tanager
Piranga rubra
Gray Squirrel
Sciurus carolinensis
Bottle-nosed Dolphin
Tursiops truncatus
White-tailed Deer (tracks)
Odocoileus virginianus

FOSSILS

Scallop
Pecten sp.

(Photo by B. Gilbert)
Glasswort, *Salicornia* sp.

Cherokee County

Old Cherokee Ironworks and Silver Mine Ridge
Cherokee County
February 24, 1990

Old Cherokee Ironworks, a pre-Civil War iron smelting company, is an interesting ruins, with natural and cultural history to see. Over the years, the remnants of canals, the kiln, and the commissary have been overgrown with vines and undergrowth. Gravel, made of slag from the furnaces, was used for fill on roads and can still be found. Bricks revealed thumbprints, evidence of their hand craftsmanship.

In the remains of a canal, a large maggot was found that proved to be a rat-tailed maggot. The adult fly is a bee mimic, yellow and black, that feeds on nectar and pollen. The female lays sticky, elongated white eggs in crevices near stagnant water or in manure. The aquatic larva breathes through very long snorkel-like tubes at the tip of the abdomen.

After lunch, we drove to a slightly different area to walk a trail along Doolittle Creek which led to a view of the old entrance to a now defunct silver mine. Chorus frogs serenaded us at both sites. When we returned to our cars, we found a reporter from a local newspaper waiting to find out what so many people found interesting in his part of the county. We hope he went back later and did a story on the historical aspects of the area. SCAN appreciated Dr. Bobby Moss, retired from Lander College, for guiding us to these sites and providing the necessary historical background to appreciate what we were seeing.

Both sites are owned by the Gaffney Historical Society.

Directions to Old Cherokee Ironworks and Silver Mine Ridge: From Gaffney, take US 29 east toward Blacksburg. Turn right (south) onto SC 329, and go 1.0 mile to County Road S 11-50 (Ford Road). Turn left and go approximately 0.5 mile; park along the road. The entrance to the Ironworks is on the left and is hard to spot. From the Ironworks to the Silver Mine, go back to SC 329, turn right (north), and turn right again onto US 29. Cross the Broad River on a steel bridge, and go 1.9 miles past the bridge. Turn right at the sign for Cherokee Falls (S 11-30) and drive approximately 1.5-2.0 miles. As you go around a left hand curve, there will be a pull-off on the left side of the road; park there and walk in.

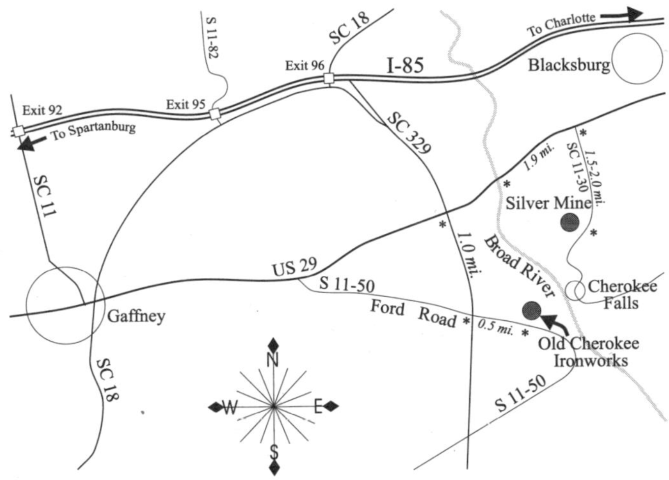

FLORA

Lichen
Cladonia cristatella
Hypotrachyna livida
Parmelia sulcata
Parmotrema hypotropum
Parmotrema perforatum
Pseudoparmelia sp.
Ramalina americana
Usnea strigosa
Usnea trichodea
Usnea sp.
Lepraria sp.

Apple Moss
Bartramia pomiformis

Running-pine
Lycopodium flabelliforme

Christmas Fern
Polystichum acrostichoides

Ebony Spleenwort
Asplenium platyneuron

Loblolly Pine
Pinus taeda

Virginia Pine; Scrub Pine
Pinus virginiana

Eastern Red-cedar
Juniperus virginiana

Cane; Switch Cane
Arundinaria gigantea

Woodrush
Luzula echinata

Dog's-tooth-violet
Erythronium americanum

Crane-fly Orchid
Tipularia discolor

Puttyroot; Adam-and-Eve
Aplectrum hyemale

Tag Alder
Alnus serrulata

River Birch
Betula nigra

American Hornbeam; Ironwood
Carpinus caroliniana

Beech
Fagus grandifolia

Winged Elm
Ulmus alata

Wild Ginger
Hexastylis minor

Tulip-poplar
Liriodendron tulipifera

PawPaw
Asimina triloba

Sweet-shrub; Carolina-allspice
Calycanthus floridus

Bloodroot
Sanguinaria canadensis

Stonecrop
Sedum ternatum

Sweet-gum
Liquidambar styraciflua

Witch-hazel
Hamamelis virginiana

American Sycamore
Platanus occidentalis

Hawthorn
Crataegus sp.

Downy Serviceberry
Amelanchier arborea

Black Cherry; Wild Cherry
Prunus serotina

Poison-ivy
Rhus radicans

Winged Sumac
Rhus copallina

American Holly
Ilex opaca

Hearts-a'burstin'-with-love
Euonymus americanus

Boxelder
Acer negundo

Red Maple
Acer rubrum

Summer Grape
Vitis aestivalis

Halberd-leaved Violet
Viola hastata

Blackgum
Nyssa sylvatica

Flowering Dogwood
Cornus florida

Spotted Wintergreen
Chimaphila maculata

Mountain-laurel
Kalmia latifolia

Dog-hobble
Leucothoe axillaris

Sourwood
Oxydendrum arboreum

Sparkleberry
Vaccinium arboreum

Galax; Galaxy; Coltsfoot
Galax aphylla

Yellow Jessamine
Gelsemium sempervirens

Cross Vine
Anisostichus capreolata

Partridge-berry
Mitchella repens

Viburnum
Viburnum sp.

Pussy-toes
Antennaria solitaria

Green-and-gold
Chrysogonum virginianum australe

Black-knot
Apiosporina morbosa

FAUNA

Asiatic Clam
Corbicula manilensis

Bowl-and-doily Spider
Frontinella pyramitela

Red Velvet Mite
Trombidium sp.

Carolina Mantid (egg)
Stagmomantis carolina

Blackberry Knot Gall (wasp)
Diastrophus nebulosus

Pine Needle Gall (midge)
Janetiella coloradensis

Rat-tailed Maggot
Eristalis tenax

Woodland Ground Beetle
Piesmus mondelus

Shining Leaf Beetle
Lema sp.

Spring Azure (butterfly)
Celastrina ladon

Mourning Cloak
Nymphalis antiopa

Polyphemus Moth (cocoon)
Antheraea polyphemus

Upland Chorus Frog (call)
Pseudacris triseriata

Turkey Vulture
Cathartes aura

American Kestrel
Falco sparverius

Yellow-bellied Sapsucker
Sphyrapicus varius

American Crow
Corvus brachyrhynchos

Golden Mouse
Ochrotomys nuttalli

Raccoon (tracks)
Procyon lotor

White-tailed Deer (scat)
Odocoileus virginianus

(Photo by V. Winn)

Christmas Fern, *Polystichum acrostichoides.*

Chester County

Landsford Canal State Park
Chester County
October 25, 1986, and July 25, 1992

Landsford Canal State Park offers both cultural and natural history lessons, for it encompasses the remnants of an old canal and the surrounding low Piedmont woodlands.

The canal was built in the early 1800's to divert barges around the rocky shoals of this part of the Catawba River. Over 1,000 artisans labored to form locks and bridges of stone, their handiwork sophisticated even by today's standards. Incredibly, the whole project was abandoned soon after completion in 1823. Today, the locks provide excellent habitat for skinks and other reptiles, including one very large copperhead sunning on our President's day-pack on the 1992 trip. It was carefully shooed away to less people-occupied areas!

Diverse habitats include the river, its floodplain, bluffs, and numerous tributaries with some rushing water. Accordingly, a great variety of uncommonly seen plants were noted from the trail. Thanks to Hurricane Hugo and some ice storms, the canopy was much more open in 1992 than on SCAN's earlier visits to the park. In addition, much more of the trail and the canal have been restored. The spider lilies were in bloom along the rocky shoals on the river in July. Many insects utilize the combination of open areas and forest edges, and birds abound in the diverse habitats.

In the years following SCAN's last visit, a pair of bald eagles set up housekeeping.

Directions to Landsford Canal State Park: From I-77 between Columbia and Rock Hill, take Exit 65 and go east on SC 9 to Richburg. As you enter town, turn left onto SC 223. Go about 7 miles and SC 223 will merge into US 21 heading north. Just after that junction, take the first right turn onto S 12-330 and go about 2 miles to the intersection with S 12-327. Continuing straight ahead will lead you to the lower entrance. To reach the main entrance and picnic ground, turn left (north) on S 12-327, go 1.3 miles, then make a right turn at the sign to the park.

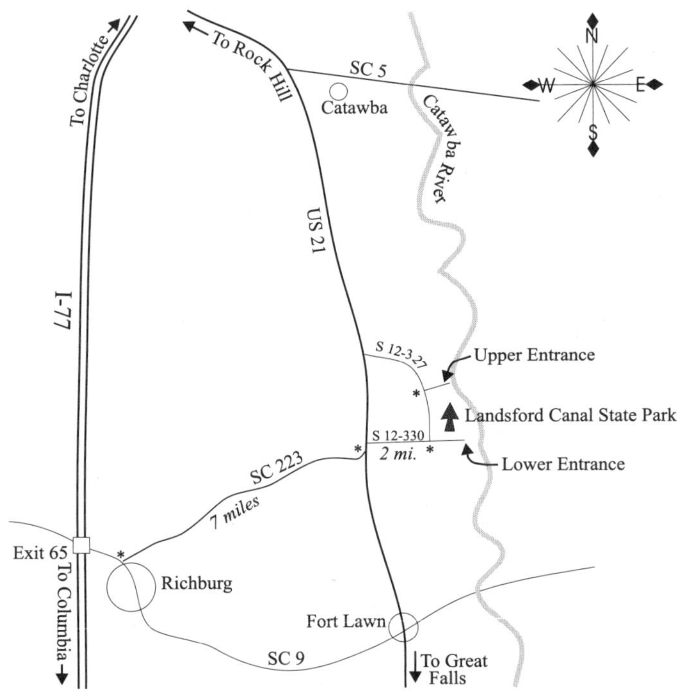

FLORA

Lichen

Collema subflaccidum
Leptogium cyanescens
Graphis mosquitensis
Bacidia polychroa
Bacidia laurocerasi
Bacidia sp.
Leptogium austroamericanum
Cladonia cristatella
Cladonia polycarpoides
Cladonia peziziformis
Pertusaria ostiolata
Pertusaria paratuberculifera
Pertusaria texana
Pertusaria caroliniana sp. nov.
Lecanora cupressi
Lecanora pallida
Lecanora varia
Lecanora allophana
Ochrolechia africana
Hypotrachyna livida
Parmotrema michauxianum
Parmotrema perforatum
Parmotrema xanthinum
Flavoparmelia baltimorensis
Flavoparmelia caperata
Canoparmelia caroliniana
Punctelia rudecta
Rimelia reticulata
Rimelia cetrata
Ramalina americana
Usnea strigosa
Usnea sp.
Buellia disciformis
Physcia americana
Physcia pseudospeciosa
Physcia stellaris
Heterodermia albicans
Heterodermia obscurata
Heterodermia speciosa
Pyxine caesiopruinosa
Pyxine sorediata

Phaeophyscia rubropulchra
Phaeophyscia sciastra
Caloplaca flavorubescens
Arthonia rubella
Horsetail
Equisetum hyemale affine
Grape Fern; Moonwort
Botrychium sp.
Bracken
Pteridium aquilinum
Ebony Spleenwort
Asplenium platyneuron
Pine
Pinus sp.
Eastern Red-cedar
Juniperus virginiana
Water-plantain
Alisma subcordatum
Arrowhead; Duck-potato
Sagittaria latifolia
Cane; Switch Cane
Arundinaria gigantea
River Oats
Uniola latifolia
Paspalum (grass)
Paspalum sp.
Sedge
Carex sp.
Jack-in-the-pulpit
Arisaema triphyllum
Dayflower
Commelina virginica
Asian Spiderwort
Aneilema keisak
Pickerelweed
Pontederia cordata
Carrion-flower
Smilax herbacea
Fringed Greenbrier
Smilax bona-nox
Sawbrier; Greenbrier
Smilax glauca
Greenbrier; Catbrier
Smilax sp.
False Solomon's Seal
Smilacina racemosa
Bear-grass
Yucca filamentosa
Spider-lily
Hymenocallis occidentalis
Crane-fly Orchid
Tipularia discolor
Lizard's-tail
Saururus cernuus
Swamp Cottonwood
Populus heterophylla
Black Walnut
Juglans nigra
Shagbark Hickory
Carya ovata
Mockernut Hickory
Carya tomentosa
Tag Alder
Alnus serrulata
River Birch
Betula nigra
American Hornbeam; Ironwood
Carpinus caroliniana
Beech
Fagus grandifolia
White Oak
Quercus alba
Post Oak
Quercus stellata
Swamp Red Oak
Quercus shumardii
Scarlet Oak
Quercus coccinea
Spanish Oak
Quercus falcata
Water Oak
Quercus nigra
Willow Oak
Quercus phellos
Slippery Elm
Ulmus rubra
American Elm
Ulmus americana
Winged Elm
Ulmus alata
Sugarberry; Hackberry
Celtis laevigata
Hackberry
Celtis occidentalis georgiana
Red Mulberry
Morus rubra
False Nettle
Boehmeria cylindrica
Clearweed
Pilea pumila
Pinkweed
Polygonum pensylvanicum
Nodding Smartweed
Polygonum lapathifolium
Mexican-tea
Chenopodium ambrosioides
Pokeweed; Pokeberry
Phytolacca americana
Wild Columbine
Aquilegia canadensis
Black Cohosh
Cimicifuga racemosa
Coralbeads
Cocculus carolinus
Tulip-poplar
Liriodendron tulipifera
Sweet-shrub; Carolina-allspice
Calycanthus floridus
Sassafras
Sassafras albidum
Spicebush
Lindera benzoin
Stonecrop
Sedum ternatum
Climbing-hydrangea
Decumaria barbara
Sweet-gum
Liquidambar styraciflua
American Sycamore
Platanus occidentalis
Indian Strawberry
Duchesnea indica
Blackberry; Brambles
Rubus sp.
Black Cherry; Wild Cherry
Prunus serotina
Mimosa; Silk-tree
Albizia julibrissin
Redbud; Judas-tree
Cercis canadensis
Honey Locust
Gleditsia triacanthos
Leadplant; Indigo-bush
Amorpha sp.
Beggar-lice
Desmodium sp.
Black Locust
Robinia pseudo-acacia
Hog-peanut
Amphicarpa bracteata
Tree-of-heaven
Ailanthus altissima
Tarweed
Acalypha rhomboidea
Prostrate Spurge
Euphorbia supina
Poison-ivy
Rhus radicans
Winged Sumac
Rhus copallina
Smooth Sumac
Rhus glabra
American Holly
Ilex opaca
Boxelder
Acer negundo
Red Maple
Acer rubrum
Painted Buckeye
Aesculus sylvatica
Buckeye
Aesculus sp.
Virginia-creeper
Parthenocissus quinquefolia
Muscadine; Bullace
Vitis rotundifolia
Grape
Vitis sp.
Pepper-vine
Ampelopsis cordata
Hypericum
Hypericum nudiflorum
Green Violet
Hybanthus concolor
Passionflower; Maypops
Passiflora incarnata
Yellow Passionflower
Passiflora lutea
Russian-olive
Elaeagnus umbellata
Water-willow
Ludwigia decurrens
Ludwigia
Ludwigia sp.
Queen Anne's-lace; Wild Carrot
Daucus carota
Angelica
Angelica venenosa
Blackgum
Nyssa sylvatica
Flowering Dogwood
Cornus florida
Highbush Blueberry
Vaccinium corymbosum
Blueberry
Vaccinium sp.
Water Pimpernel
Samolus parviflorus
Persimmon
Diospyros virginiana
Horse-sugar; Sweetleaf
Symplocos tinctoria
Carolina Silverbell
Halesia carolina
American Ash; White Ash
Fraxinus americana
Chinese Privet
Ligustrum sinense
Yellow Jessamine
Gelsemium sempervirens
Sabatia
Sabatia sp.
Angle-pod
Matelea sp.
Morning-glory
Ipomoea sp.
Beautyberry
Callicarpa americana
Lop-seed
Phryma leptostachya
Woolly Mullein
Verbascum thapsus
Cross Vine
Anisostichus capreolata
Trumpet Vine
Campsis radicans
Catalpa; Catawba-tree
Catalpa speciosa
Squaw-root
Conopholis americana
Ruellia
Ruellia caroliniensis
Button-bush
Cephalanthus occidentalis
Japanese Honeysuckle
Lonicera japonica
Black-haw; Nanny-berry
Viburnum prunifolium
American Elderberry
Sambucus canadensis
Cardinal-flower
Lobelia cardinalis
Ironweed
Vernonia sp.
Elephant's-foot
Elephantopus carolinianus
Elephantopus sp.
Dog-fennel
Eupatorium capillifolium
Rabbit-tobacco
Gnaphalium obtusifolium
Aster
Aster sp.

Landsford Canal State Park (continued)
October 25, 1986, and July 25, 1992

FAUNA

Asiatic Clam
Corbicula fluminea
Almond Millipede
Sigmoria aberrans
Spiny-backed Spider
Micrathena gracilis
Arrow-shaped Spiny-back
Micrathena sagittata
Orchard Spider
Leucauge venusta
Writing Spider
Argiope aurantia
Orb-weaver
Neoscona domiciliorum
Arrowhead Spider
Verrucosa arenata
Wolf Spider
Lycosa sp.
Crab Spider
Xysticus sp.
Daddy-long-legs
Leiobunum sp.
Lone Star Tick
Amblyomma americanum
Spur-throated Grasshopper
Melanoplus sp.
Carolina Mantid
Stagmomantis carolina
Wheel Bug
Arilus cristatus
Poplar Petioleball Aphid
Pemphigus populitransversus
Black-shouldered Spinyleg
Dromogomphus spinosus
Little Corporal Dragonfly
Ladona deplanata
White-tail Dragonfly
Plathemis lydia
Ebony Jewelwing
Calopteryx maculata
Variable Dancer (damselfly)
Argia fumipennis
Blue-ringed Dancer (damselfly)
Argia sedula
Bluet (damselfly)
Enallagma sp.
Rambur's Forktail (damselfly)
Ischnura ramburi
Green Tiger Beetle
Cicindela sexguttata
Red-bellied Tiger Beetle
Cicindela rufiventris
June Beetle
Phyllophaga sp.
Japanese Beetle
Popillia japonica
Grape Beetle
Pelidnota sp.
Green June Beetle
Cotinus nitida
Tick Trefoil Leaf-miner
Pachyschelus laevigatus
Spotted Lady Beetle
Coleomegilla maculata
Shining Leaf Beetle
Lema sayi
Oval Leaf Beetle
Colaspis sp.
Flea Beetle
Oedionychis sp.
Crepidodera nitens
Systena elongata
Broad-nosed Weevil
Nemocestes sp.
Eastern Tiger Swallowtail
Papilio glaucus
Spicebush Swallowtail
Papilio troilus
American Snout (butterfly)
Libytheana carinenta
Pearl Crescent
Phyciodes tharos
American Lady (butterfly)
Vanessa virginiensis
Red-spotted Purple (butterfly)
Limenitis arthemis astyanax
Hackberry Emperor (butterfly)
Asterocampa celtis
Carolina Satyr (butterfly)
Hermeuptychia sosybius
Hoary Edge (skipper)
Achalarus lyciades
Ailanthus Webworm Moth
Atteva punctella
Clymene Moth
Haploa clymene
Largemouth Bass
Micropterus salmoides
Fowler's Toad
Bufo woodhousei
Eastern Box Turtle
Terrapene carolina
Eastern Fence Lizard
Sceloporus undulatus
Five-lined Skink
Eumeces fasciatus
Skink
Eumeces sp.
Copperhead
Agkistrodon contortrix
Great Egret
Casmerodius albus
Great Blue Heron
Ardea herodias
Turkey Vulture
Cathartes aura
Northern Harrier
Circus cyaneus
Osprey
Pandion haliaetus
Wild Turkey
Meleagris gallopavo
Yellow-billed Cuckoo
Coccyzus americanus
Belted Kingfisher
Ceryle alcyon
Red-bellied Woodpecker
Melanerpes carolinus
American Crow
Corvus brachyrhynchos
Fish Crow
Corvus ossifragus
Carolina Chickadee
Parus carolinensis
Carolina Wren
Thryothorus ludovicianus
American Robin
Turdus migratorius
Red-Eyed Vireo
Vireo olivaceus
Northern Cardinal
Cardinalis cardinalis
Rufous-sided Towhee
Pipilo erythrophthalmus
Red-winged Blackbird
Agelaius phoeniceus
Summer Tanager
Piranga rubra
White-tailed Deer
Odocoileus virginianus

(Photo by L. Smith)
Wheel bug, *Arilus cristatus*.

Woods Ferry Recreation Area
(Sumter National Forest)
Chester County
July 24, 1993

This recreation area is established around the remains of a ferry landing on the Chester County side of the Broad River, a large, broad Piedmont river. The river floodplain and its bluff have been incorporated into a picnic area, campground, and nature trail. Although none of the area is unique, it is an excellent example of this kind of habitat. Despite being very hot and dry, the day yielded a wide variety of plant and animal sightings. Among the more rewarding experiences: hearing the call of a wild turkey, the "ka, ka, ka" of a yellow-billed cuckoo, and seeing the splash of bright pink sabatia flowers among the wilted leaves of the majority of trailside plants. The heat did allow insects to be active, leading to a number of very large wheel bugs, beautiful damselflies, bright leaf-eating beetles, and a variety of grasshoppers.

(Photo by L. Greenlee)
Slaty skimmer, *Libellula incesta*.

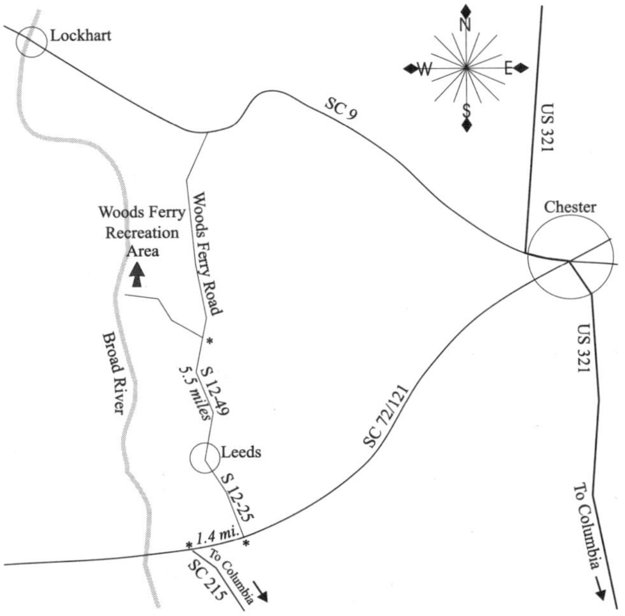

Directions to Woods Ferry Recreation Area: From Columbia take SC 215 until it intersects with SC 72 and 121 in Chester County, about 50 miles. Turn right toward Chester on SC 72/121 and go 1.4 miles. Turn left (north) onto Woods Ferry Road at the sign for Woods Ferry Recreation Area. Travel 5.5 miles, following the signs through Leeds, and turn left. The recreation area is 3.5 miles farther.

FLORA

Lichen
Collema subflaccidum
Leptogium cyanescens
Peltigera canina
Peltigera polydactyla
Graphina sophisticascens
Graphis scripta
Phaeographis dendritica
Bacidia schweinitzii
Bacidia suffusa
Bacidia sp.
Leptogium austroamericanum
Baeomyces fungoides
Cladonia chlorophaea
Cladonia chlorophaea costata
Cladonia chlorophaea f. pterygota
Cladonia sobolescens
Cladonia coniocraea
Cladonia didyma
Cladonia macilenta theiophila
Cladonia polycarpia
Cladonia polycarpoides
Cladonia polycarpoides epiphylla
Cladonia peziziformis
Cladonia rappii
Pycnothelia papillaria
Pertusaria amara
Pertusaria ostiolata
Pertusaria paratuberculifera
Pertusaria subpertusa
Pertusaria texana
Maronea constans
Lecanora cupressi
Lecanora varia
Ochrolechia africana
Bulbothrix goebelii
Bulbothrix goebelii f. minor
Candelaria concolor
Hypotrachyna livida
Parmelina aurulenta
Parmelina galbina
Parmelina horrescens
Parmelina minarum
Parmotrema hypotropum
Parmotrema michauxianum
Parmotrema perforatum
Flavoparmelia caperata
Canoparmelia caroliniana
Canoparmelia crozalsiana
Punctelia rudecta
Melanelia halei
Rimelia reticulata
Rimelia cetrata
Rimelia subisidiosa
Ramalina americana
Usnea mutabilis
Buellia curatellae
Physcia americana
Physcia millegrana
Heterodermia obscurata
Pyxine caesiopruinosa
Pyxine sorediata
Phaeophyscia ciliata
Phaeophyscia rubropulchra
Caloplaca cinnabarina
Caloplaca ferruginea
Opegrapha astraea
Lepraria membranacea
Lepraria finkii

Running-pine
Lycopodium flabelliforme
Grape Fern; Moonwort
Botrychium sp.
Christmas Fern
Polystichum acrostichoides
Ebony Spleenwort
Asplenium platyneuron
Loblolly Pine
Pinus taeda
Short-leaf Pine
Pinus echinata
Eastern Red-cedar
Juniperus virginiana
Common Cat-tail
Typha latifolia
Cane; Switch Cane
Arundinaria gigantea
River Oats
Uniola latifolia
Paspalum (grass)
Paspalum sp.
Panic Grass
Panicum sp.
Microstegium (grass)
Microstegium vimineum
Dayflower
Commelina sp.
Greenbrier; Bullbrier
Smilax rotundifolia
Fringed Greenbrier
Smilax bona-nox
Sawbrier; Greenbrier
Smilax glauca
Indian Cucumber-root
Medeola virginiana
Solomon's-seal
Polygonatum biflorum
Wild Yam
Dioscorea sp.
Downy Rattlesnake-plantain
Goodyera pubescens
Black Willow
Salix nigra
Eastern Cottonwood
Populus deltoides
Black Walnut
Juglans nigra
Shagbark Hickory
Carya ovata
Pignut Hickory
Carya glabra
Tag Alder
Alnus serrulata
River Birch
Betula nigra
American Hornbeam; Ironwood
Carpinus caroliniana
Beech
Fagus grandifolia
White Oak
Quercus alba
Post Oak
Quercus stellata
Red Oak
Quercus rubra
Southern Red Oak; Spanish Oak
Quercus falcata falcata
Water Oak
Quercus nigra
Willow Oak
Quercus phellos
Slippery Elm
Ulmus rubra
American Elm
Ulmus americana
Winged Elm
Ulmus alata
Sugarberry; Hackberry
Celtis laevigata
Red Mulberry
Morus rubra
False Nettle
Boehmeria cylindrica
Heartleaf
Hexastylis arifolia
Meadow-rue
Thalictrum revolutum
Liverleaf
Hepatica americana
Tulip-poplar
Liriodendron tulipifera
Sweet-shrub; Carolina-allspice
Calycanthus floridus
Spicebush
Lindera benzoin
Climbing-hydrangea
Decumaria barbara
Mock-orange
Philadelphus inodorus
Sweet-gum
Liquidambar styraciflua
American Sycamore
Platanus occidentalis
Blackberry
Rubus argutus
Black Cherry; Wild Cherry
Prunus serotina
Mimosa; Silk-tree
Albizia julibrissin
Redbud; Judas-tree
Cercis canadensis
Wild Sensitive-plant
Cassia nicitans
Honey Locust
Gleditsia triacanthos
Pencil-flower
Stylosanthes biflora

Woods Ferry Recreation Area (continued)
July 24, 1993

Sericea
Lespedeza cuneata
Rhynchosia
Rhynchosia sp.
Cranesbill
Geranium carolinianum
Tragia
Tragia sp.
Poison-ivy
Rhus radicans
Poison-oak
Rhus toxicodendron
Winged Sumac
Rhus copallina
American Holly
Ilex opaca
Possumhaw Holly
Ilex decidua
Hearts-a'burstin'-with-love
Euonymus americanus
Bladdernut
Staphylea trifolia
Boxelder
Acer negundo
Red Maple
Acer rubrum
Southern Sugar Maple
Acer saccharum floridanum
Red Buckeye
Aesculus pavia
Spotted Touch-me-not
Impatiens capensis
New-Jersey-tea
Ceanothus americanus
Virginia-creeper
Parthenocissus quinquefolia
Muscadine; Bullace
Vitis rotundifolia
Grape
Vitis sp.
Rose Mallow; Wild Cotton
Hibiscus moscheutos
St.-Andrew's-cross
Hypericum hypericoides
Hypericum
Hypericum sp.
Ludwigia
Ludwigia sp.
Snakeroot
Sanicula sp.
Angelica
Angelica venenosa
Black Tupelo; Blackgum
Nyssa sylvatica sylvatica
Swamp Tupelo; Swamp Blackgum
Nyssa sylvatica biflora
Flowering Dogwood
Cornus florida
Silky Dogwood
Cornus amomum
Spotted Wintergreen
Chimaphila maculata
Sourwood
Oxydendrum arboreum
Sparkleberry
Vaccinium arboreum
Blueberry
Vaccinium sp.
Persimmon
Diospyros virginiana
Carolina Silverbell
Halesia carolina
Bigleaf Snowbell
Styrax grandifolia
Ash
Fraxinus sp.
Chinese Privet
Ligustrum sinense
Yellow Jessamine
Gelsemium sempervirens
Polypremum
Polypremum procumbens
Sabatia
Sabatia sp.
Indian-hemp
Apocynum cannabinum
Milkweed
Asclepias quadrifolia
Angle-pod
Matelea gonocarpa
Glecoma
Glecoma hederacea
Carpenterweed
Prunella vulgaris
Lyre-leaved Sage
Salvia lyrata
Mountain Mint
Pycnanthemum sp.
False Foxglove
Aureolaria sp.
Cross Vine
Anisostichus capreolata
Trumpet Vine
Campsis radicans
Catalpa
Catalpa sp.
Ruellia
Ruellia caroliniensis
English Plantain
Plantago lanceolata
Button-bush
Cephalanthus occidentalis
Partridge-berry
Mitchella repens
Bedstraw
Galium sp.
Japanese Honeysuckle
Lonicera japonica
Black-haw; Nanny-berry
Viburnum prunifolium
American Elderberry
Sambucus canadensis
Wild Lettuce
Lactuca sp.
Pale Indian-plantain
Cacalia atriplicifolia
Elephant's-foot
Elephantopus tomentosus
Eupatorium
Eupatorium sp.
Rabbit-tobacco
Gnaphalium obtusifolium
Silphium
Silphium sp.
Verbesina
Verbesina sp.
Coreopsis
Coreopsis major
Black-knot
Apiosporina morbosa

FAUNA

Asiatic Clam
Corbicula sp.
Blackgum Capsule Gall Mite
Eriophyes nyssae
Wild Cherry Pouch Gall Mite
Eriophyes padi
Globular Elm Leaf Gall Mite
Eriophyes ulmi
Green-striped Grasshopper
Chortophaga viridifasciata
Spur-throated Grasshopper
Melanoplus sp.
Montezuma Katydid
Montezumina modesta
Wheel Bug
Arilus cristatus
Hackberry Blister Gall (aphid)
Pachypsylla vesiculum
Cottonwood Petiole Gall(aphid)
Pemphigus populicaulis
Large Conical Hickory Gall
Phylloxera notabilis
Blackberry Knot Gall (wasp)
Diastrophus nebulosus
Blackberry Branch Gall (wasp)
Diplolepis rubicola
Oak Bullet Gall (wasp)
Disholcaspis globulus
Water Oak Midvein Gall (wasp)
Callirhytis sp.
Common Eastern Velvet-ant
Dasymutilla occidentalis occidentalis
Dogwood Club Gall (midge)
Mycodiplosis clavula
Slaty Skimmer
Libellula incesta
Great Blue Skimmer
Libellula vibrans
Blue Dasher
Pachydiplax longipennis
White-tail Dragonfly
Plathemis lydia
Ebony Jewelwing
Calopteryx maculata
Blue-fronted Dancer Damselfly
Argia apicalis
Powdered Dancer (damselfly)
Argia moesta
Woodland Ground Beetle
Pterostichus chalcites
Agonum sp.
Notch-mouthed Ground Beetle
Dicaelus dilitatus
Crawling Water Beetle
Haliplus fasciatus
Peltodytes sexmaculatus
Peltodytes shermani
Peltodytes sp.
Predacious Diving Beetle
Laccophilus fasciatus
Hydroporus niger
Coptotomus interrogatus
Whirligig Beetle
Dineutus robertsi
Water Scavenger Beetle
Hydrochus excavatus
Tropisternus quadristriatus quadristriatus
Stag Beetle
Pseudolucanus placidus
Passalus; Bess-beetle
Popilius disjunctus
Contractile Scarab Beetle
Cloeotus aphodioides
Japanese Beetle
Popillia japonica
Spotted Lady Beetle
Coleomegilla maculata
Pleasing Fungus Beetle
Megalodacne fasciata
Warty Leaf Beetle
Chlamisus gibbosus
Oval Leaf Beetle
Rhabdopterus picipes
Larger Elm Leaf Beetle
Monocesta coryli
E. Spotted Cucumber Beetle
Diabrotica undecimpunctata howardi
Tortoise Beetle
Deloyola guttata
Hibiscus Seed Weevil
Althaeus hibisci
Flower Weevil
Baris aerea
Cocklebur Weevil
Rhodobaenus tredecimpunctatus
Eastern Tiger Swallowtail
Papilio glaucus
Spicebush Swallowtail
Papilio troilus
Red Admiral
Vanessa atalanta rubria
Red-spotted Purple (butterfly)
Limenitis arthemis astyanax
Carolina Satyr (butterfly)
Hermeuptychia sosybius
Silver-spotted Skipper
Epargyreus clarus
Pine Woods Treefrog
Hyla femoralis
Green Frog; Bronze Frog
Rana clamitans
Skink
Eumeces sp.
Green-backed Heron
Butorides striatus

Black Vulture
Coragyps atratus
Wild Turkey
Meleagris gallopavo
Yellow-billed Cuckoo
Coccyzus americanus
Red-bellied Woodpecker
Melanerpes carolinus
Red-headed Woodpecker
Melanerpes erythrocephalus
Northern Flicker
Colaptes auratus
Pileated Woodpecker
Dryocopus pileatus
Eastern Phoebe
Sayornis phoebe
American Crow
Corvus brachyrhynchos
Fish Crow
Corvus ossifragus
Tufted Titmouse
Parus bicolor
Carolina Chickadee
Parus carolinensis
Blue-gray Gnatcatcher
Polioptila caerulea
Prairie Warbler
Dendroica discolor
Northern Cardinal
Cardinalis cardinalis
Rufous-sided Towhee
Pipilo erythrophthalmus
Field Sparrow
Spizella pusilla
Chipping Sparrow
Spizella passerina
Summer Tanager
Piranga rubra

Chesterfield County

Science and Nature Notes: The Sandhills

The sandhills provide the closest thing to a true desert South Carolina has. The sandy soil naturally retains little moisture. In the area around Carolina Sandhills National Wildlife Refuge, Sandhills State Forest, and Cheraw State Park, the xeric conditions were worsened by poor farming practices in the 1800's and early 1900's as humus and minerals were used up or leached away. Plants that have survived in this habitat have adapted in special ways. They are smaller than their relatives in wetter areas, and are more widely spaced, necessitated by the competition for water and nutrients. Species that adapt most easily often have waxy coatings on their leaves, or the ability, especially in oaks, of turning their leaves from horizontal to vertical when the sun is most intense. This reduces overheating and dehydration from sun above, and heat reflected from the white sand below. Certain grasshopper species also develop extra heavy waxy compounds in their "skins" when they live in the driest, sunniest parts of these areas.

The drawing below is a vertical section through our coastal plain. Sections taken from New Jersey to Mississippi would be similar. The formations are all tilted down and each becomes thicker as it approaches the ocean. The Earth's surface transects these sedimentary strata at a low angle. Hence, although the Tuscaloosa at the sandhills is only several hundred feet thick, its surface exposure will be 5 - 10 miles.; its thickness at Charleston may be several thousand feet. Although dinosaurs were upon the land at this time, fossils are not likely to be found in this area. The high energy environment of deposition and the porous nature of the substrate preclude preservation of animal material. However, petrified wood may be found in many places.

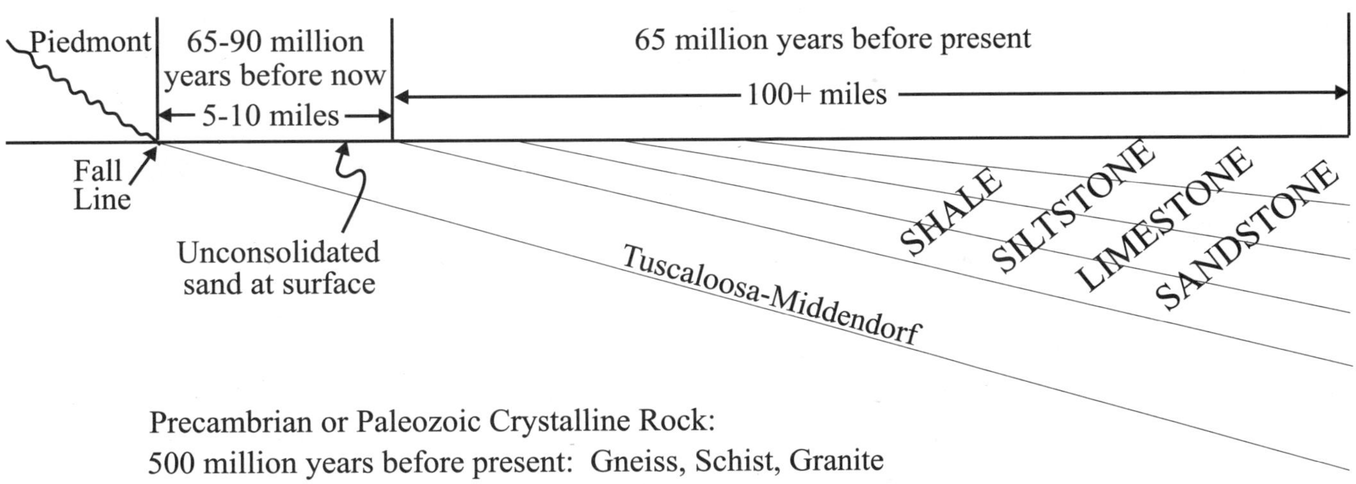

Woods Ferry Recreation Area, Chester County
Carolina Sandhills NWR and Sandhills State Forest, Chesterfield County

Carolina Sandhills National Wildlife Refuge, September 26, 1992 and Sandhills State Forest, September 24, 1994
Chesterfield County

Carolina Sandhills National Wildlife Refuge is 45,348 acres of land, including about 30 small man-made lakes and ponds, that lies along the Fall Line separating the Piedmont plateau from the Atlantic Coastal Plain. Most of the refuge is covered by longleaf pine forest interspersed with scrub oak. The soils are composed mainly of rolling beds of deep sand. Fields and clearings, plus the lakes and ponds, provide many types of habitats available for wildlife. These support a wide diversity of plant and animal species and serve as a refuge for such endangered and/or threatened species as the red-cockaded woodpecker, pine-barrens tree frog, and southern bald eagle. This refuge supports a larger population of red-cockaded woodpecker than all other refuges combined. Unlike other woodpeckers, it nests in cavities of living, but older, longleaf pines. As is often the case in early fall in the sandhills, flora, especially wildflowers both in the sandy areas and in the bogs near the lakes, provided the most rewarding finds of the day.

Directions to Carolina Sandhills National Wildlife Refuge: From Camden, take US 1 northeast 27 miles through Bethune and McBee. After crossing SC 151 in McBee, continue about 4 miles further and turn left into the Carolina Sandhills Refuge Headquarters (pick up refuge maps there). Follow the Wildlife Drive until reaching the sign to Martins's Lake; turn right and go to the parking lot near the lake. The Lake Bee picnic area may be reached by continuing along the Wildlife Drive, or by forking left onto SC 145 just outside of McBee.

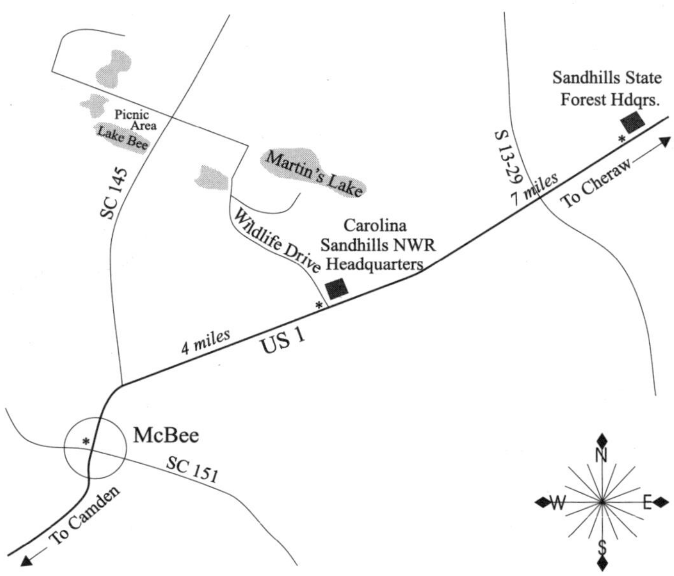

Sandhills State Forest is 46,000 acres, mostly in Chesterfield County. Approximately one-fifth of the forest floor is burned each year to provide a 5-year burn cycle to increase and maintain a longleaf pine habitat. This, along with matching forest management practices, is restoring the natural diversity of the native plant cover and enhances the red-cockaded woodpecker population as well. There is an active program of providing man-made cavities for the woodpeckers, additionally increasing the numbers. On this visit, several small unique habitats such as creek bottoms and seepage areas provided a number of unusual sightings. Despite a morning of hard rain, forester Darryl Jones successfully converted several of us who mistakenly considered the area devoid of interest from 55mph!

Directions to Sandhills State Forest: From Camden, take US 1 northeast 27 miles through Bethune and McBee; continue on US 1 about 11 miles further, passing the Sandhills National Wildlife Refuge Headquarters and crossing S 13-29. (Do not confuse the two headquarters.) The Sandhills State Forest Headquarters is on the left about 1.5 miles after crossing S 13-29. If you come to Patrick and SC 102, turn around and go back about 3 miles.

FLORA

Lichen

Lecidea nylanderi
Lecidea sp.
Cladina evansii
Cladonia apodocarpa
Cladonia beaumontii
Cladonia calycantha
Cladonia capitata
Cladonia cariosa
Cladonia chlorophaea
Cladonia coniocraea
Cladonia cristatella
Cladonia cristatella cristatella
Cladonia cristatella pleurocarpa
Cladonia didyma
Cladonia incrassata
Cladonia leporina
Cladonia leporina ramosa
Cladonia leporina prolifera
Cladonia leporina squamulosa
Cladonia polycarpoides
Cladonia ravenelii
Cladonia strepsilis
Cladonia subsetacea
Cladonia uncialis
Pycnothelia papillaria
Pertusaria amara
Pertusaria globularis
Pertusaria pustulata
Pertusaria propinqua
Pertusaria sp.
Maronea constans
Lecanora hypoptoides
Lecanora miculata
Lecanora rugosa
Lecanora cupressi
Lecanora varia
Lecanora saligna
Ochrolechia africana
Candelaria concolor
Cetraria sp.
Hypotrachyna livida
Parmelina aurulenta
Parmelina galbina
Parmelina horrescens
Parmelina minarum

Parmotrema haitiense
Parmotrema hypotropum
Parmotrema mellissii
Parmotrema michauxianum
Parmotrema perforatum
Parmotrema subtinctorium
Xanthoparmelia sp.
Flavoparmelia caperata
Punctelia appalachensis
Punctelia rudecta
Rimelia reticulata
Tuckermannopsis viridis
Ramalina americana
Usnea rubicunda
Usnea strigosa
Buellia parasema
Buellia punctata
Buellia disciformis
Physcia aipolia
Heterodermia speciosa
Heterodermia casarettiana
Physconia enteroxantha
Caloplaca aurantiaca
Trypethelium tropicum
Trypethelium scorites
Trypethelium sp.
Southern Bog Clubmoss
Lycopodium adpressum
Cinnamon Fern
Osmunda cinnamomea
Royal Fern
Osmunda regalis spectabilis
Bracken
Pteridium aquilinum
Virginia Chain Fern
Woodwardia virginica
Netted Chain Fern
Woodwardia areolata
Resurrection Fern
Polypodium polypodioides
Long-leaf Pine
Pinus palustris
Slash Pine
Pinus elliotti
Pond Pine
Pinus serotina
Short-leaf Pine
Pinus echinata
Virginia Pine; Scrub Pine
Pinus virginiana
Sand Pine
Pinus clausa
Atlantic White-cedar
Chamaecyparis thyoides
Eastern Red-cedar
Juniperus virginiana
Arrowhead; Duck-potato
Sagittaria sp.
Cane; Switch Cane
Arundinaria gigantea
Sand Grass
Triplasis americana
Love Grass
Eragrostis sp.
Orchard Grass
Dactylis glomerata
Wiregrass
Aristida stricta
Aristida sp.
Dropseed
Sporobolus sp.
Grass
Calamovilfa brevipilis
Toothache Grass
Ctenium aromaticum
Beard Grass
Gymnopogon brevifolius
Anthaenantia (grass)
Anthaenantia villosa
Panic Grass
Panicum verrucosum
Little Bluestem
Andropogon scoparius
Broomsedge
Andropogon sp.
Indian Grass
Sorghastrum sp.
Dulichium (sedge)
Dulichium arundinaceum
Umbrella-grass
Fuirena squarrosa
Beak-rush
Rhynchospora chalarocephala
Sedge
Carex glaucescens
Yellow-eyed-grass
Xyris sp.
Pipewort; Hatpins
Eriocaulon sp.
Bog-buttons
Lachnocaulon anceps
Bamboo Smilax
Smilax laurifolia
False Asphodel
Tofieldia racemosa
White-featherling
Tofieldia glabra
Zigadenus
Zigadenus glaberrimus
Burmannia
Burmannia capitata
Nodding Ladies'-tresses
Spiranthes cernua
Swamp Bayberry
Myrica heterophylla
Mockernut Hickory
Carya tomentosa
Tag Alder
Alnus serrulata
Post Oak
Quercus stellata
Scrubby Post Oak
Quercus margaretta
Overcup Oak
Quercus lyrata
Turkey Oak
Quercus laevis
Southern Red Oak; Spanish Oak
Quercus falcata falcata
Black Jack Oak
Quercus marilandica
Blue-jack Oak
Quercus incana
Laurel Oak
Quercus laurifolia
Heartleaf
Hexastylis sp.
Jointweed
Polygonella polygama
Cottonweed
Froelichia floridana
Wire-plant
Stipulicida setacea
Sandwort; Longroot
Arenaria caroliniana
Water-lily
Nymphaea odorata
Water-shield
Brasenia schreberi
Tulip-poplar
Liriodendron tulipifera
Sweetbay
Magnolia virginiana
Redbay
Persea borbonia
Sassafras
Sassafras albidum
Trumpet Pitcher-plant
Sarracenia flava
Sweet Pitcher-plant
Sarracenia rubra
Hooded Pitcher-plant
Sarracenia minor
Frog's-britches
Sarracenia purpurea
Sundew
Drosera intermedia
Drosera capillaris
Virginia-willow
Itea virginica
Sweet-gum
Liquidambar styraciflua
Witch-alder
Fothergilla gardenii
Blackberry; Brambles
Rubus sp.
Red Chokeberry
Sorbus arbutifolia
Black Cherry; Wild Cherry
Prunus serotina
Sensitive-brier
Schrankia microphylla
Wild Indigo
Baptisia sp.
Common Lupine
Lupinus diffusus
Leadplant
Amorpha herbacea
Creeping Bush-clover
Lespedeza repens
Goat's-rue
Tephrosia virginiana
Drum-heads
Polygala cruciata
Polygala
Polygala incarnata
Polygala ramosa
Orange Milkwort
Polygala lutea
Tread-softly
Cnidoscolus stimulosus
Carolina Ipecac
Euphorbia ipecacuanhae
Spurge
Euphorbia curtisii
Euphorbia sp.
Poison Sumac
Rhus vernix
Poison-oak
Rhus toxicodendron
Winged Sumac
Rhus copallina
Titi
Cyrilla racemiflora
American Holly
Ilex opaca
Inkberry
Ilex glabra
Sweet Gallberry
Ilex coriacea
Red Maple
Acer rubrum
Virginia-creeper
Parthenocissus quinquefolia
Muscadine; Bullace
Vitis rotundifolia
St.-Andrew's-cross
Hypericum hypericoides
St.-Peter's-wort
Hypericum stans
Pineweed
Hypericum gentianoides
Hypericum
Hypericum sp.
Passionflower; Maypops
Passiflora incarnata
Prickly-pear
Opuntia compressa
Smooth Meadow-beauty
Rhexia alifanus
Pale Meadow-beauty
Rhexia mariana
Common Meadow-beauty
Rhexia virginica
Rhexia virginica virginica
Intelligence-plant
Centella asiatica
Button Snakeroot
Eryngium yuccifolium
Snakeroot
Eryngium integrifolium
Oxypolis
Oxypolis ternata
Black Tupelo; Blackgum
Nyssa sylvatica sylvatica
Swamp Tupelo; Swamp Blackgum
Nyssa sylvatica biflora
Flowering Dogwood
Cornus florida
Sweet Pepperbush
Clethra alnifolia
Spotted Wintergreen
Chimaphila maculata
Maleberry
Lyonia ligustrina

Carolina Sandhills NWR and Sandhills State Forest (continued) September 26, 1992, and September 24, 1994

Fetterbush
Lyonia lucida
Staggerbush
Lyonia mariana
Coastal Honeybells
Leucothoe racemosa
Sourwood
Oxydendrum arboreum
Trailing-arbutus
Epigaea repens
Dangleberry
Gaylussacia frondosa
Sparkleberry
Vaccinium arboreum
Highbush Blueberry
Vaccinium corymbosum
Creeping Blueberry
Vaccinium crassifolium
Pyxie-moss
Pyxidanthera barbulata
Persimmon
Diospyros virginiana
Horse-sugar; Sweetleaf
Symplocos tinctoria
Yellow Jessamine
Gelsemium sempervirens
Polypremum
Polypremum procumbens
Pine-barren Gentian
Gentiana autumnalis
Blue-star
Amsonia ciliata
Seymeria
Seymeria cassioides
Purple Gerardia
Agalinis purpurea
Gerardia
Agalinis sp.
Bladderwort
Utricularia sp.
Poor-jo (or Po-jo)
Diodia teres
Mexican-clover
Richardia scabra
Sherardia
Sherardia arvensis
Possumhaw Viburnum
Viburnum nudum
Wahlenbergia
Wahlenbergia marginata
Lobelia
Lobelia nuttallii
Lobelia sp.
Thistle
Carduus sp.
Ironweed
Vernonia acaulis
Elephant's-foot
Elephantopus tomentosus
Elephantopus sp.
Blazing-star
Liatris graminifolia
Deer's-tongue
Trilisa paniculata
Carphephorus
Carphephorus bellidifolius
Carphephorus sp.
Joe-pye-weed
Eupatorium fistulosum
Dog-fennel
Eupatorium capillifolium
Eupatorium compositifolium
Eupatorium
Eupatorium pilosum
False Hoarhound
Eupatorium rotundifolium
White Snakeroot
Eupatorium rugosum
Aster
Aster squarrosus
Aster sp.
Stiff-leaved Aster
Aster linariifolius
Slim Goldenrod
Solidago stricta
Gray Goldenrod
Solidago nemoralis
Sweet Goldenrod
Solidago odora
Heterotheca
Heterotheca graminifolia
Heterotheca mariana
Heterotheca gossypina
Rosinweed
Silphium compositum
Sunflower
Helianthus angustifolius
Coreopsis
Coreopsis gladiata
Barbara's-buttons
Marshallia graminifolia
Common Fiber-vase
Thelophora terrestris

FAUNA

Crayfish
Cambarus sp.
Crab-shaped Spiny-back
Gasteracantha cancriformis
Arrow-shaped Spiny-back
Micrathena sagittata
Basilica Spider
Mecynogea lemniscata
Orb-weaver
Neoscona domiciliorum
Orb-weaver
Neoscona hentzi
Long-jawed Orb-weaver
Tetragnatha laboriosa
Carolina Wolf Spider
Lycosa carolinensis
Rabid Wolf Spider
Lycosa rabida
Meadow Grasshopper
Orchelimum sp.
Grasshopper
Orphulella pelidna
Slant-faced Grasshopper
Syrbula admirabilis
Grasshopper
Mermiria intertexta
Boll's Sand Grasshopper
Spharagemon bolli
Spur-throated Grasshopper
Melanoplus sp.
Leather-color Bird Grasshopper
Schistocerca alutacea
Texas Bush Katydid
Scudderia texensis
Thick-legged Camel Cricket
Ceuthophilus crassifemoris
Froghopper
Prosapia bicincta
Bumblebee
Bombus sp.
House Fly
Musca domestica
Amanda's Pennant (dragonfly)
Celithemis amanda
Banded Pennant (dragonfly)
Celithemis fasciata
Eastern Pondhawk
Erythemis simplicicollis
Blue Dragonlet
Erythrodiplax minuscula
Carolina Spreadwing Damselfly
Lestes vidua
Seepage Dancer (damselfly)
Argia bipunctulata
Atlantic Bluet (damselfly)
Enallagma doubledayi
Citrine Forktail (damselfly)
Ischnura hastata
Tiger Beetle
Cicindela tranquebarica
Woodland Ground Beetle
Agonum sp.
Oval Ground Beetle
Oodes amaroides
Predacious Diving Beetle
Hydroporus sp.
Burrowing Water Beetle
Hydrocanthus iricolor
Whirligig Beetle
Dineutus emarginatus
Dineutus carolinus
Dineutus horni
Large Flat-headed Pine Borer
Chalcophora virginiensis
Soldier Beetle
Chauliognathus pennsylvanicus
Black Blister Beetle
Epicauta pennsylvanica
Spotless Lady Beetle
Cycloneda sanguinea
Flower Longhorn Beetle
Typocerus sp.
12-spotted Cucumber Beetle
Diabrotica duodecimpunctata
Skeletonizing Leaf Beetle
Luperodes sp.
Pea & Bean Weevil
Meibomeus musculus
Fruit Weevil
Anthonomus sp.
Cloudless Sulphur
Phoebis sennae eubule
Little Yellow (butterfly)
Eurema lisa
Eastern Tailed-blue(butterfly)
Everes comyntas
Pearl Crescent
Phyciodes tharos
Monarch (butterfly)
Danaus plexippus
Southern Flannel Moth
Megalopyge opercularis
Southern Toad
Bufo terrestris
Southern Cricket Frog
Acris gryllus
Cricket Frog
Acris sp.
Yellow-bellied Slider
Chrysemys scripta
Carolina Anole
Anolis carolinensis
Eastern Fence Lizard
Sceloporus undulatus
Black Racer
Coluber constrictor
Banded Water Snake
Nerodia fasciata
Pied-billed Grebe
Podilymbus podiceps
Great Egret
Casmerodius albus
Great Blue Heron
Ardea herodias
Canada Goose
Branta canadensis
Wood Duck
Aix sponsa
Osprey
Pandion haliaetus
American Kestrel
Falco sparverius
Mourning Dove
Zenaida macroura
Chimney Swift
Chaetura pelagica
Belted Kingfisher
Ceryle alcyon
Red-bellied Woodpecker
Melanerpes carolinus
Northern Flicker
Colaptes auratus
Red-cockaded Woodpecker (sign)
Picoides borealis
Pileated Woodpecker
Dryocopus pileatus

Blue Jay
Cyanocitta cristata
American Crow
Corvus brachyrhynchos
Brown-headed Nuthatch
Sitta pusilla
Carolina Wren
Thryothorus ludovicianus
Eastern Bluebird
Sialia sialis
American Robin
Turdus migratorius
Gray Catbird
Dumetella carolinensis
Northern Mockingbird
Mimus polyglottos
Cedar Waxwing
Bombycilla cedrorum
Pine Warbler
Dendroica pinus
Northern Cardinal
Cardinalis cardinalis
Rufous-sided Towhee
Pipilo erythrophthalmus
Chipping Sparrow
Spizella passerina
White-tailed Deer (tracks)
Odocoileus virginianus

(Photo by L. Greenlee)
Florence High and Mary Garland Douglass busy at work.

Cheraw Fish Hatchery, Cheraw State Park and Sugarloaf Mountain (Sandhills State Forest)
Chesterfield County
March 24, 1990

Although this list covers three sites, the habitats are similar, with some exceptions. This trip was led by Dr. George Sawyer of Coker College and Kalmia Gardens in Hartsville, who gave both a natural history and cultural history interpretation of the area. A little-known area of the state park provided a rich sandhills area featuring *Hudsonia ericoides*, the only known location of the plant in the Carolinas. Sugarloaf Mountain, in contrast, is a high cone-shaped hill made up of layers of sandstone and clay. Once characteristic of the entire area, these layers were as high if not higher than the present "mountain's" peak. At one time several streams ran through the area carrying iron oxide. A natural cement, it combined with the sand as it flowed downstream to form sandstone. As this iron-rich sandstone gradually reached the layer of clay, it formed a layer on top, protecting it from the more rapid erosion of the surrounding area. The result is present-day Sugarloaf Mountain, which rises some 150 feet above the surrounding coastal plain. Several uncommon plants are found on Sugarloaf and its lower neighbor, Horse Shoe Mountain, such as pyxie moss and creeping blueberry.

Directions to Cheraw Fish Hatchery, Cheraw State Park and Sugarloaf Mountain: From Camden, take US 1 northeast 41 miles through McBee and Patrick; continue on US 1 for 7 additional miles past Patrick to the Cheraw Fish Hatchery on the right. Cheraw State Park may be entered by walking through the woods from the Fish Hatchery, or through the main entrance on US 52 just 0.8 miles south of the intersection with US 1. For Sugarloaf Mountain, return 4.8 miles south of Patrick (or 9 miles north of McBee) and turn north on S 13-29. Go about a mile, angle right at the fork onto a dirt road, and then take the next right fork again. You will pass a lake and a camping area on the right, and the "mountain" is on the left.

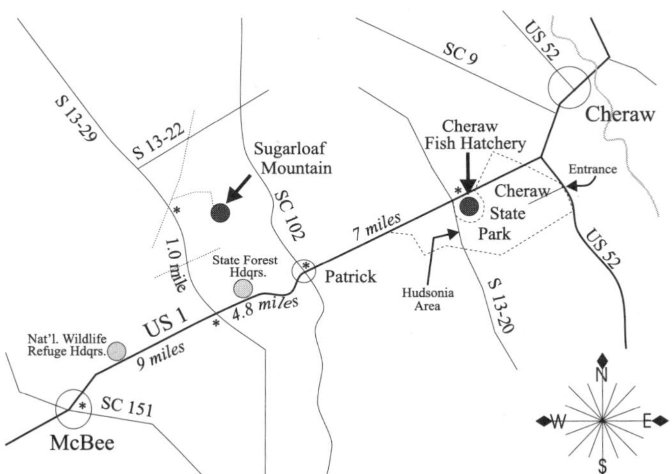

FLORA

Lichen
Cladina evansii
Cladonia caroliniana
Cladonia coniocraea
Cladonia cristatella
Cladonia gracilis
Cladonia leporina
Hypotrachyna livida
Parmotrema perforatum
Ramalina americana
Usnea trichodea
Lepraria sp.
Southern Bog Clubmoss
Lycopodium adpressum
Sand Spikemoss
Selaginella arenicola
Bracken
Pteridium aquilinum
Bradley's Spleenwort
Asplenium bradleyi
Long-leaf Pine
Pinus palustris
Pond Pine
Pinus serotina
Virginia Pine; Scrub Pine
Pinus virginiana
Foxtail Grass
Alopecurus carolinianus
Golden-club
Orontium aquaticum
Bogmoss
Mayaca aubletii
Yellow-eyed-grass
Xyris sp.

Cheraw Fish Hatchery, Cheraw State Park and Sugarloaf Mountain (continued)
March 24, 1990

Bamboo Smilax
Smilax laurifolia
Redroot
Lachnanthes caroliniana
Dwarf Iris
Iris verna
Turkey Oak
Quercus laevis
Jointweed
Polygonella polygama
Water-lily
Nymphaea odorata
Water-shield
Brasenia schreberi
Mouse-tail
Myosurus minimus
Redbay
Persea borbonia
Sassafras
Sassafras albidum
Teesdalia
Teesdalia nudicaulis
Sundew
Drosera intermedia
Red Chokeberry
Sorbus arbutifolia
Water Starwort
Callitriche deflexa
Titi
Cyrilla racemiflora
American Holly
Ilex opaca
Inkberry
Ilex glabra
Red Maple
Acer rubrum
Hudsonia
Hudsonia ericoides
Bird-foot Violet
Viola pedata
Primrose-leaved Violet
Viola primulifolia
Water-milfoil
Myriophyllum sp.
Intelligence-plant
Centella asiatica
Sweet Pepperbush
Clethra alnifolia
Sand-myrtle
Leiophyllum buxifolium
Mountain-laurel
Kalmia latifolia
Staggerbush
Lyonia mariana
Trailing-arbutus
Epigaea repens
Sparkleberry
Vaccinium arboreum
Black Highbush Blueberry
Vaccinium atrococcum
Creeping Blueberry
Vaccinium crassifolium
Pyxie-moss
Pyxidanthera barbulata barbulata
Pyxidanthera barbulata brevifolia
Persimmon
Diospyros virginiana
Horse-sugar; Sweetleaf
Symplocos tinctoria
Yellow Jessamine
Gelsemium sempervirens
Wiry Bladderwort
Utricularia subulata
Floating Bladderwort
Utricularia inflata
Small Bluets
Houstonia pusilla
Facelis
Facelis retusa
Sea-myrtle; Groundsel-tree
Baccharis halimifolia
Woody Goldenrod
Solidago pauciflosculosa

FAUNA

Green Lacewing
Chrysopa sp.
Mantled Baskettail (dragonfly)
Epitheca semiaquea
Pedunculate Ground Beetle
Pasimachus sp.
Woodland Ground Beetle
Agonum sp.
Dingy Ground Beetle
Stenolophus ochropezus
Crawling Water Beetle
Haliplus triopsus
Whirligig Beetle
Dineutus emarginatus
Water Scavenger Beetle
Tropisternus collaris striolatus
Tropisternus blatchleyi blatchleyi
Dung Beetle
Onthophagus sp.
Earth-boring Dung Beetle
Geotrupes blackburnii blackburnii
Garden Beetle
Serica sp.
Rhinoceros Beetle
Strategus sp.
Large Flat-headed Pine Borer
Chalcophora virginiensis
Leaf Buprestid Beetle
Brachys sp.
Soldier Beetle
Silis latilobus
Firefly; Lightning Beetle
Ellychnia sp.
Lady Beetle
Coleomegilla fuscilabris
Twice-stabbed Lady Beetle
Chilocorus stigma
Flower Longhorn Beetle
Rhagium inquisitor
Hickory Girdler Beetle (tracks)
Oncideres cingulata
Shining Leaf Beetle
Lema sp.
Skeletonizing Leaf Beetle
Galerucella nymphaeae
Zebra Swallowtail
Eurytides marcellus
Eastern Tiger Swallowtail
Papilio glaucus
Spicebush Swallowtail
Papilio troilus
Red-banded Hairstreak
Calycopis cecrops
Henry's Elfin (butterfly)
Callophrys henrici
Spring Azure (butterfly)
Celastrina ladon
Luna Moth
Actias luna
Virginian Tiger Moth
Spilosoma virginica
Northern Cricket Frog
Acris crepitans
Southern Cricket Frog
Acris gryllus
Yellow-bellied Slider
Chrysemys scripta
Eastern Fence Lizard
Sceloporus undulatus
Ground Skink
Scincella lateralis
Red-tailed Hawk
Buteo jamaicensis
Brown-headed Nuthatch
Sitta pusilla
Pine Warbler
Dendroica pinus

Clarendon County

Bennetts Bay Heritage Preserve
Clarendon County
April 8, 1989

Bennetts Bay, also shown on older maps as Junkyard Bay, is a classic Carolina Bay with water in the middle surrounded by a dense border of shrubs. A sand ridge is outside that dense border, and it provided one of the more exciting finds of the day, running oak (*Quercus pumila*). Chinquapin (*Castanea pumila*) was also seen. After exploring the drier rim and waiting for the day to warm, most people donned wading gear and got "up close and personal" with the aquatic portion of the bay. That made for a memorable trip for those who risked getting wet. Children were invited on this trip, but most of these "children" were over twenty-one.

Directions to Bennetts Bay Heritage Preserve: From the intersection of US 521 and SC 261 in the center of Manning, take US 521 east toward Greeleyville and Andrews for 1.0 mile. Just before the store, turn right (south) onto County Road S 14-48. Follow this road about 5.2 miles; then turn left onto County Road S 14-211. Go about 3 miles, passing County Roads S 14-262 and S 14-340, and turn left onto a unnamed dirt road shortly before a brick house on the left. Just after the tobacco barn on the right, turn right into a lane, park near the Nature Conservancy sign, and walk the rest of the way to Bennett's Bay.

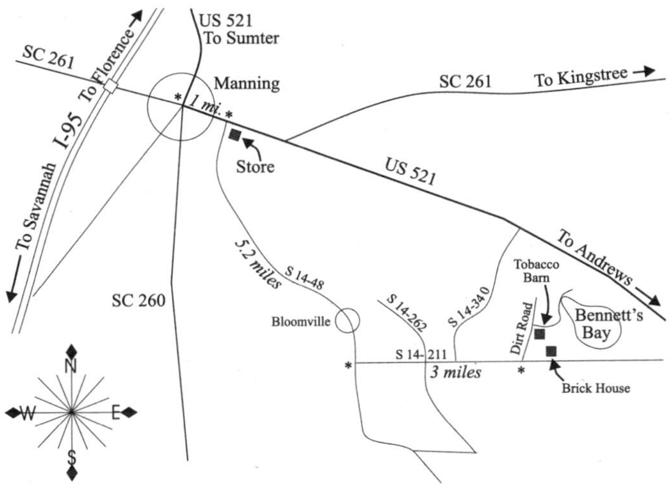

FLORA

Fruticose Lichen
Cladonia sp.
Peat Moss
Sphagnum sp.
Field Horsetail
Equisetum arvense
Cinnamon Fern
Osmunda cinnamomea
Bracken
Pteridium aquilinum
Loblolly Pine
Pinus taeda
Cane; Switch Cane
Arundinaria gigantea
Yellow-eyed-grass
Xyris sp.
Bamboo Smilax
Smilax laurifolia
Bear-grass
Yucca filamentosa
Dwarf Iris
Iris verna
Wax-myrtle; Southern Bayberry
Myrica cerifera
Mockernut Hickory
Carya tomentosa
River Birch
Betula nigra
Allegheny Chinquapin
Castanea pumila
Post Oak
Quercus stellata
Turkey Oak
Quercus laevis
Spanish Oak
Quercus falcata
Black Jack Oak
Quercus marilandica
Water Oak
Quercus nigra
Running Oak
Quercus pumila
Mistletoe
Phoradendron serotinum
Sweetbay
Magnolia virginiana
Redbay
Persea borbonia
Sassafras
Sassafras albidum
Trumpet Pitcher-plant
Sarracenia flava
Sundew
Drosera leucantha
Virginia-willow
Itea virginica
Sweet-gum
Liquidambar styraciflua
Red Chokeberry
Sorbus arbutifolia
Shadbush
Amelanchier canadensis
Black Cherry; Wild Cherry
Prunus serotina
American Wisteria
Wisteria frutescens
Poison-ivy
Rhus radicans
Poison-oak
Rhus toxicodendron
American Holly
Ilex opaca
Inkberry
Ilex glabra
Sweet Gallberry
Ilex coriacea
Red Maple
Acer rubrum
Muscadine; Bullace
Vitis rotundifolia
Loblolly-bay
Gordonia lasianthus
Bog White Violet
Viola lanceolata
Prickly-pear
Opuntia compressa
Devil's-walking-stick
Aralia spinosa
Intelligence-plant
Centella asiatica
Flowering Dogwood
Cornus florida
Sweet Pepperbush
Clethra alnifolia
Pinxter-flower; Pinxter-bloom
Rhododendron nudiflorum
Fetterbush
Lyonia lucida
Dog-hobble
Leucothoe axillaris
Dangleberry
Gaylussacia frondosa
Sheep-laurel; Lamb-kill
Kalmia angustifolia
Horse-sugar; Sweetleaf
Symplocos tinctoria
Fringe-tree
Chionanthus virginicus
Dodder
Cuscuta sp.
Phlox
Phlox nivalis
Lyre-leaved Sage
Salvia lyrata
Toadflax
Linaria canadensis
Possumhaw Viburnum
Viburnum nudum
Dwarf Dandelion
Krigia virginica
Common Split-gill
Schizophyllum commune

FAUNA

Green Tiger Beetle
Cicindela sexguttata
Water Scavenger Beetle
Helocombus bifidus
Rove Beetle
Anthobium sp.
Quedius sp.
Tachinus axillaris
Tachinus pallipes
Oxypoda sp.
Dung Beetle
Onthophagus hecate
Aphodiine Dung Beetle
Aphodius sp.
Ataenius cognatus
Firefly; Lightning Beetle
Ellychnia sp.
Soft-winged Flower Beetle
Attalus scincetus
Attalus circumscriptus
Seven-spotted Lady Beetle
Coccinella septempunctata
Round-necked Longhorn Beetle
Molorchus sp.
Flower Weevil
Centrinaspis perscillus
Eastern Tiger Swallowtail
Papilio glaucus
Spicebush Swallowtail
Papilio troilus
Eastern Tailed-blue(butterfly)
Everes comyntas
Common Buckeye (butterfly)
Junonia coenia
Carolina Satyr (butterfly)
Hermeuptychia sosybius
Rough Green Snake
Opheodrys aestivus
Cattle Egret
Bulbicus ibis
Turkey Vulture
Cathartes aura
Northern Bobwhite
Colinus virginianus
American Crow
Corvus brachyrhynchos
Tufted Titmouse
Parus bicolor
Carolina Chickadee
Parus carolinensis
Brown-headed Nuthatch
Sitta pusilla
Carolina Wren
Thryothorus ludovicianus
Eastern Bluebird
Sialia sialis
Pine Warbler
Dendroica pinus
Northern Cardinal
Cardinalis cardinalis
Rufous-sided Towhee
Pipilo erythrophthalmus

Santee National Wildlife Refuge
Clarendon County
February 25, 1989

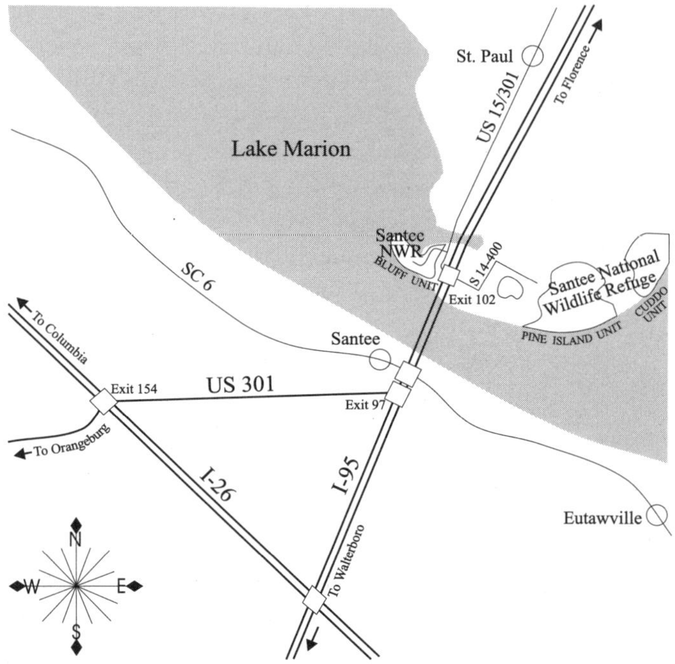

Santee National Wildlife Refuge has been a popular destination for SCAN during winter months. Birds are usually the highlight of such a visit, but they are only one of the many interesting things to be seen. The refuge is made up of separate units separated by several miles, offering a wide variety of habitats. At the visitor's center you can learn about what you are likely to be seeing during your visit and obtain a map of the more remote units. Trails close to the visitor's center are laid out to provide an example of most habitats represented on the refuge. Small towers and benches are provided where watching is likely to be fruitful. Bald eagles are frequently seen in the area. The wooded areas have typical coastal plain plant material and animals. The bay of Lake Marion that comes close to the visitor's area is usually full of waterfowl during the winter season. Other units have ponds full of waterfowl in season and alligators most of the year. Butterflies abound in the open fields.

Directions to Santee National Wildlife Refuge: From Columbia, take I-26 east 45 miles to Exit 154. Turn north on US 301 toward Santee, and follow that for 15 miles where it merges with I-95 heading north toward Summerton and Florence. As soon as it crosses Lake Marion (5 miles), take Exit 102 to US 15/301 heading north toward St. Paul and Manning. US 15/301 bends under the interstate and makes a sharp right turn. After this watch carefully for the sign on the left to Santee National Wildlife Refuge. Turn in there; the headquarters is on the left in about half a mile.

FLORA

Netted Chain Fern
Woodwardia areolata
Long-leaf Pine
Pinus palustris
Bald-cypress
Taxodium distichum
Common Cat-tail
Typha latifolia
Wool-grass
Scirpus cyperinus
Spanish-moss
Tillandsia usneoides
Greenbrier; Catbrier
Smilax sp.
Black Willow
Salix nigra
Wax-myrtle; Southern Bayberry
Myrica cerifera
Water Oak
Quercus nigra
Heartleaf
Hexastylis arifolia
Lotus; Pond-nuts
Nelumbo lutea
Sassafras
Sassafras albidum
Wild Radish
Raphanus raphanistrum
Bitter Cress
Cardamine pensylvanica
Sweet-gum
Liquidambar styraciflua
Sesbania
Sesbania exaltata
Bladderpod
Glottidium vesicarium
Cranesbill
Geranium carolinianum
Chinaberry
Melia azedarach
Poison-ivy
Rhus radicans
Red Maple
Acer rubrum
Silver Maple
Acer saccharinum
Water-milfoil
Myriophyllum brasiliense
Flowering Dogwood
Cornus florida
Yellow Jessamine
Gelsemium sempervirens
Common Henbit
Lamium amplexicaule
Veronica
Veronica sp.
Japanese Honeysuckle
Lonicera japonica

FAUNA

Basilica Spider (egg)
Mecynogea lemniscata
Goldenrod Ball Gall (fly)
Eurosta solidaginis
Pied-billed Grebe
Podilymbus podiceps
Anhinga; Snakebird
Anhinga anhinga
Double-crested Cormorant
Phalacrocorax auritus
Great Egret
Casmerodius albus
Great Blue Heron
Ardea herodias
Tundra Swan
Cygnus columbianus
Canada Goose
Branta canadensis
Mallard
Anas platyrhynchos
American Black Duck
Anas rubripes
Green-winged Teal
Anas crecca
American Wigeon
Anas americana
Ring-necked Duck
Aythya collaris
Lesser Scaup
Aythya affinis
Bufflehead
Bucephala albeola
Hooded Merganser
Lophodytes cucullatus
American Coot
Fulica americana
Killdeer
Charadrius vociferus
American Woodcock
Scolopax minor
Caspian Tern
Sterna caspia
Bald Eagle
Haliaeetus leucocephalus
Northern Harrier
Circus cyaneus
Red-tailed Hawk
Buteo jamaicensis
Mourning Dove
Zenaida macroura

Pileated Woodpecker
Dryocopus pileatus
Eastern Phoebe
Sayornis phoebe
Blue Jay
Cyanocitta cristata
Tufted Titmouse
Parus bicolor
Carolina Chickadee
Parus carolinensis
Carolina Wren
Thryothorus ludovicianus
Ruby-crowned Kinglet
Regulus calendula
Eastern Bluebird
Sialia sialis
American Robin
Turdus migratorius
Northern Mockingbird
Mimus polyglottos
Cedar Waxwing
Bombycilla cedrorum
Black-and-white Warbler
Mniotilta varia
Yellow-rumped Warbler
Dendroica coronata
Rufous-sided Towhee
Pipilo erythrophthalmus
Red-winged Blackbird
Agelaius phoeniceus
White-tailed Deer (tracks)
Odocoileus virginianus

(Photo by R. Emmitt)

Great blue heron, *Ardea herodias*.

Colleton County

Ashepoo River (Canoe Trip)
(ACE Basin)
Colleton County
September 9, 1995

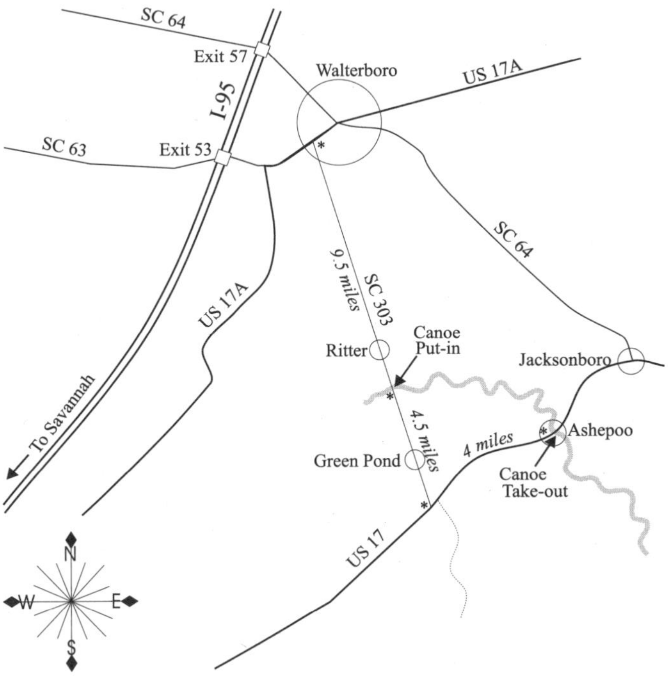

The 5.5 mile section of the Ashepoo River SCAN explored is easily canoed, and is the best way to see the river. This section widens from a relatively narrow blackwater creek to a brackish, tide-influenced river, to a salt marsh. Often there are large mats of floating plants, although on the day SCAN canoed this section, heavy rain had washed them away. Water ash, sweet gum, red maple and swamp dogwood were common. In an area near old rice fields, some wild rice was found flowering along the banks.

Insect larvae were interesting finds on this day. An unusual moth caterpillar was found on a deciduous holly. About an inch and a half long, it was stocky and appeared triangular in cross section, wide at the bottom and narrowing to a ridge on top. It was covered with long gray hairs with two small spots of white hairs; the upper ridge was decorated with a line of tawny hairs. It looked most like a puss caterpillar or a flannel moth. Also found was a colorful black swallowtail larva feeding on water parsnip. He demonstrated his erectile organs on cue; they looked like a pair of yellow horns which appeared behind his head, emitted a formic-acid-like substance, and then disappeared.

Directions to Ashepoo River (to canoe): Southbound on I-95, take Exit 57, turning onto SC 64 toward Walterboro. Proceed through Walterboro, staying to your right, until you come to the junction with SC 303 on your left. Take SC 303 south, for 9.5 miles to the put-in site. You will pass through Ritter. To get to the take-out site continue on SC 303 south, passing through Green Pond, to the junction with US 17. Turn left (north toward Charleston), and go 4 miles. Shortly before reaching the Ashepoo River bridge, turn right onto a private dirt road and follow that to the river bank. Car-shuttle back to the put-in on SC 303 between Ritter and Green Pond.

Ashepoo River (continued)
September 9, 1995

FLORA

Lichen
Leptogium cyanescens
Graphis scripta
Phaeographis dendritica
Lecidea sp.
Pertusaria trachythallina
Pertusaria subpertusa
Maronea constans
Haematomma puniceum
Haematomma elatinum
Haematomma cismonicum
Lecanora cupressi
Lecanora varia
Tephromela atra
Bulbothrix goebelii
Bulbothrix goebelii f. minor
Hypotrachyna livida
Parmelina aurulenta
Parmelina horrescens
Parmelina minarum
Parmotrema crinitum f. ultralucens
Parmotrema mellissii
Parmotrema michauxianum
Parmotrema perforatum
Parmotrema praesorediosum
Parmotrema robustum
Flavoparmelia caperata
Canoparmelia caroliniana
Canoparmelia texana
Punctelia rudecta
Rimelia reticulata
Rimelia subisidiosa
Ramalina americana
Usnea mutabilis
Usnea strigosa
Buellia punctata
Buellia curatellae
Heterodermia albicans
Pyxine sorediata
Physconia enteroxantha
Anthracothecium pyrenuloides
Pyrenula laevigata
Trypethelium virens
Trypethelium floridanum
Royal Fern
Osmunda regalis spectabilis
Southern Lady Fern
Athyrium asplenioides
Virginia Chain Fern
Woodwardia virginica
Netted Chain Fern
Woodwardia areolata
Resurrection Fern
Polypodium polypodioides
Loblolly Pine
Pinus taeda
Bald-cypress
Taxodium distichum
Eastern Red-cedar
Juniperus virginiana
Common Cat-tail
Typha latifolia
Arrowhead; Duck-potato
Sagittaria falcata
Frog's-bit
Limnobium spongia
Cane; Switch Cane
Arundinaria gigantea
Cord Grass
Spartina cynosuroides
Wild Rice
Zizania aquatica
Saw-grass
Cladium jamaicense
Dwarf Palmetto
Sabal minor
Golden-club
Orontium aquaticum
Duckweed
Lemna sp.
Spanish-moss
Tillandsia usneoides
Water-hyacinth
Eichhornia crassipes
Pickerelweed
Pontederia cordata
Wax-myrtle; Southern Bayberry
Myrica cerifera
Water Hickory
Carya aquatica
River Birch
Betula nigra
American Hornbeam; Ironwood
Carpinus caroliniana
Swamp Chestnut Oak
Quercus michauxii
Laurel Oak
Quercus laurifolia
Sugarberry; Hackberry
Celtis laevigata
False Nettle
Boehmeria cylindrica
Mistletoe
Phoradendron serotinum
Tearthumb; Scratchgrass
Polygonum arifolium
Alligator-weed
Alternanthera philoxeroides
Water-lily
Nymphaea odorata
Fanwort
Cabomba caroliniana
Southern Magnolia
Magnolia grandiflora
Redbay
Persea borbonia
Virginia-willow
Itea virginica
Sweet-gum
Liquidambar styraciflua
Swamp Rose
Rosa palustris
Indigo-bush
Amorpha fruticosa
Groundnut
Apios americana
Chinese Tallow-tree
Sapium sebiferum
Winged Sumac
Rhus copallina
American Holly
Ilex opaca
Winterberry; Black-alder
Ilex verticillata
Red Maple
Acer rubrum
Muscadine; Bullace
Vitis rotundifolia
Grape
Vitis sp.
Seashore Mallow
Kosteletskya virginica
Water-milfoil
Myriophyllum pinnatum
Water Parsnip
Sium suave
Swamp Tupelo; Swamp Blackgum
Nyssa sylvatica biflora
Swamp Dogwood; Stiff Cornel
Cornus stricta
Blueberry
Vaccinium sp.
Persimmon
Diospyros virginiana
Water Ash; Carolina Ash
Fraxinus caroliniana
Dodder
Cuscuta sp.
Lavender Marsh Morning-glory
Ipomoea sagittata
Trumpet Vine
Campsis radicans
Button-bush
Cephalanthus occidentalis
Japanese Honeysuckle
Lonicera japonica
Possumhaw Viburnum
Viburnum nudum
Arrowwood
Viburnum dentatum
Cardinal-flower
Lobelia cardinalis
Climbing Hempweed
Mikania scandens
Sea-myrtle; Groundsel-tree
Baccharis halimifolia
Climbing Aster
Aster carolinianus

FAUNA

Lubber Grasshopper
Brachystola magna
Red Imported Fire Ant
Solenopsis invicta
Bald-faced Hornet
Vespa maculata
Paper Wasp
Polistes sp.
River Cruiser (dragonfly)
Macromia sp.
Four-spotted Pennant(dragonfly)
Brachymesia gravida
Eastern Pondhawk
Erythemis simplicicollis
Golden-winged Skimmer
Libellula auripennis
Blue Dasher
Pachydiplax longipennis
Eastern Amberwing (dragonfly)
Perithemis tenera
Burrowing Water Beetle
Copius inflatus
Whirligig Beetle
Gyrinus lugens
Spotted Lady Beetle
Coleomegilla maculata
Skeletonizing Leaf Beetle
Galerucella notulata
Flea Beetle
Systena elongata
Alligatorweed Flea Beetle
Agasicles hygrophila
Black Swallowtail
Papilio polyxenes asterius
Bullfrog
Rana catesbiana
Pig Frog
Rana grylio
American Alligator
Alligator mississippiensis
Carolina Anole
Anolis carolinensis
Banded Water Snake
Nerodia fasciata
Great Egret
Casmerodius albus
White Ibis
Eudocimus albus
Wood Duck
Aix sponsa
Black Vulture
Coragyps atratus
American Kestrel
Falco sparverius
Northern Flicker
Colaptes auratus
Downy Woodpecker
Picoides pubescens
Pileated Woodpecker
Dryocopus pileatus
Acadian Flycatcher
Empidonax virescens
Tree Swallow
Tachycineta bicolor
Blue Jay
Cyanocitta cristata
American Crow
Corvus brachyrhynchos
Tufted Titmouse
Parus bicolor
Carolina Chickadee
Parus carolinensis

White-breasted Nuthatch
Sitta carolinensis
Carolina Wren
Thryothorus ludovicianus
Blue-gray Gnatcatcher
Polioptila caerulea
Northern Mockingbird
Mimus polyglottos
White-eyed Vireo
Vireo griseus
Northern Parula
Parula americana
Pine Warbler
Dendroica pinus
Northern Cardinal
Cardinalis cardinalis
Red-winged Blackbird
Agelaius phoeniceus
Common Grackle
Quiscalus quiscula
Boat-tailed Grackle
Quiscalus major

(Photo by A. Ciegler)
Sam Outen (left) learns to paddle on his first canoe trip.

Bear Island Wildlife Management Area
(ACE Basin)
Colleton County
February 28, 1987 & February 27, 1988

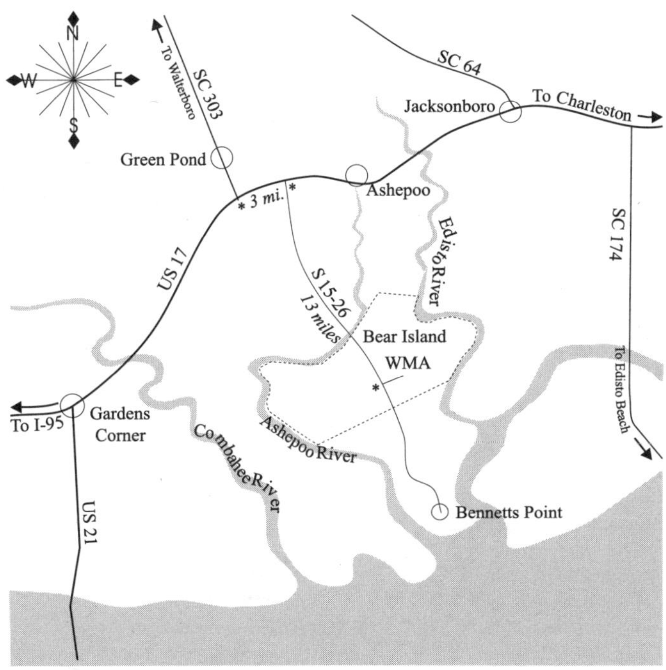

The Bear Island area of what is now the ACE Basin Project was formerly Low Country plantations. This means wooded areas, open fields, and old rice ponds adjacent to tidal creeks. This diverse set of habitats naturally attracts an exceptionally large number of wildlife species. The most obvious of these are the birds, with wintering waterfowl, nesting bald eagles, and migrating song birds heading the list. Mammals are less obvious, but deer, otters, and fox squirrels are commonly seen. Outer coastal plain plant life also abounds. January and February often provide the highest concentration of birds, which is why SCAN chose these dates. Rain can often reduce the visibility and the state suffered a large amount of it on February 28, 1987. Rescheduling in 1988 gave a gorgeous day with lots of birds seen. Eagles put on quite a show on both trips. The 1988 trip gave members the opportunity to spot eaglets in the nest (using binoculars), with the adults soaring overhead against a gorgeous blue sky. A sight (and site) to remember!

Directions to Bear Island Wildlife Management Area: From Walterboro, take SC 303 south through Green Pond to the intersection with US 17; turn left toward Charleston. Drive about 3 miles and then turn right (south) onto S 15-26, crossing over the railroad track. Pass Airy Hall Plantation, cross the Ashepoo River, and continue to the Bear Island WMA, about 13 miles. The entrance and headquarters buildings are on the left.

FLORA

Bracken
Pteridium aquilinum
Mosquito Fern
Azolla caroliniana
Long-leaf Pine
Pinus palustris
Loblolly Pine
Pinus taeda
Ditch-grass
Ruppia maritima
Cane; Switch Cane
Arundinaria gigantea
Cord Grass
Spartina patens
Spartina cynosuroides
Smooth Cord Grass
Spartina alterniflora
Broomsedge
Andropogon virginicus
Andropogon sp.
Bulrush
Scirpus robustus
Cabbage Palmetto
Sabal palmetto
Spanish-moss
Tillandsia usneoides
Black Needle Rush
Juncus roemerianus
Wax-myrtle; Southern Bayberry
Myrica cerifera
Post Oak
Quercus stellata
Live Oak
Quercus virginiana
Cherrybark Oak
Quercus falcata pagodaefolia
Black Jack Oak
Quercus marilandica

Bear Island Wildlife Management Area (continued)
February 28, 1987, and February 27, 1988

Water Oak
Quercus nigra
Sawtooth Oak
Quercus acutissima
Common Chickweed
Stellaria media
Sassafras
Sassafras albidum
Sweet-gum
Liquidambar styraciflua
Sesbania
Sesbania exaltata
Bladderpod
Glottidium vesicarium
Coral-bean
Erythrina herbacea
Cranesbill
Geranium carolinianum
Red Maple
Acer rubrum
Johnny-jump-up
Viola rafinesquii
Marsh-elder
Iva frutescens
Sea-myrtle; Groundsel-tree
Baccharis halimifolia
Violet-toothed Polypore
Trichaptum biformis

FAUNA

Basilica Spider (egg)
Mecynogea lemniscata
Fire Ant
Solenopsis sp.
Palmetto Tortoise Beetle
Hemisphaerota cyanea
Cloudless Sulphur
Phoebis sennae eubule
Sleepy Orange (butterfly)
Eurema nicippe
American Alligator
Alligator mississippiensis
Yellow-bellied Slider
Chrysemys scripta
Pied-billed Grebe
Podilymbus podiceps
Anhinga; Snakebird
Anhinga anhinga
Double-crested Cormorant
Phalacrocorax auritus
American Bittern
Botaurus lentiginosus
Black-crowned Night-Heron
Nyctocorax nyctocorax
Tricolored Heron
Egretta tricolor
Little Blue Heron
Egretta caerulea
Snowy Egret
Egretta thula
Great Egret
Casmerodius albus
Great Blue Heron
Ardea herodias
Glossy Ibis
Plegadis falcinellus
White Ibis
Eudocimus albus
Canada Goose
Branta canadensis
Mottled Duck
Anas fulvigula
Green-winged Teal
Anas crecca
American Wigeon
Anas americana
Northern Pintail
Anas acuta
Northern Shoveler
Anas clypeata
Blue-winged Teal
Anas discors
Canvasback
Aythya valisineria
Ring-necked Duck
Aythya collaris
Hooded Merganser
Lophodytes cucullatus
Virginia Rail
Rallus limicola
Common Moorhen
Gallinula chloropus
American Coot
Fulica americana
Killdeer
Charadrius vociferus
Greater Yellowlegs
Tringa melanoleuca
Common Snipe
Gallinago gallinago
Laughing Gull
Larus atricilla
Bonaparte's Gull
Larus philadelphia
Ring-billed Gull
Larus delawarensis
Turkey Vulture
Cathartes aura
Black Vulture
Coragyps atratus
Bald Eagle
Haliaeetus leucocephalus
Northern Harrier
Circus cyaneus
Red-tailed Hawk
Buteo jamaicensis
Osprey
Pandion haliaetus
Belted Kingfisher
Ceryle alcyon
Pileated Woodpecker
Dryocopus pileatus
Tree Swallow
Tachycineta bicolor
Blue Jay
Cyanocitta cristata
American Crow
Corvus brachyrhynchos
Fish Crow
Corvus ossifragus
Tufted Titmouse
Parus bicolor
Carolina Chickadee
Parus carolinensis
Blue-gray Gnatcatcher
Polioptila caerulea
American Robin
Turdus migratorius
Loggerhead Shrike
Lanius ludovicianus
Northern Mockingbird
Mimus polyglottos
Yellow-rumped Warbler
Dendroica coronata
Pine Warbler
Dendroica pinus
Rufous-sided Towhee
Pipilo erythrophthalmus
White-throated Sparrow
Zonotrichia albicollis
Eastern Meadowlark
Sturnella magna
Red-winged Blackbird
Agelaius phoeniceus
Boat-tailed Grackle
Quiscalus major
Virginia Opossum (tracks)
Didelphis virginiana
Marsh Rabbit
Sylvilagus palustris
Raccoon (tracks) (scat)
Procyon lotor
River Otter
Lutra canadensis
Bobcat; Wildcat (tracks) (scat)
Felis rufus
White-tailed Deer
Odocoileus virginianus

(Photos by L. Greenlee)

Bald Eagle, *Haliaeetus leucocephalus.*

Donnelley Wildlife Management Area
(ACE Basin)
Colleton County
February 26, 1994

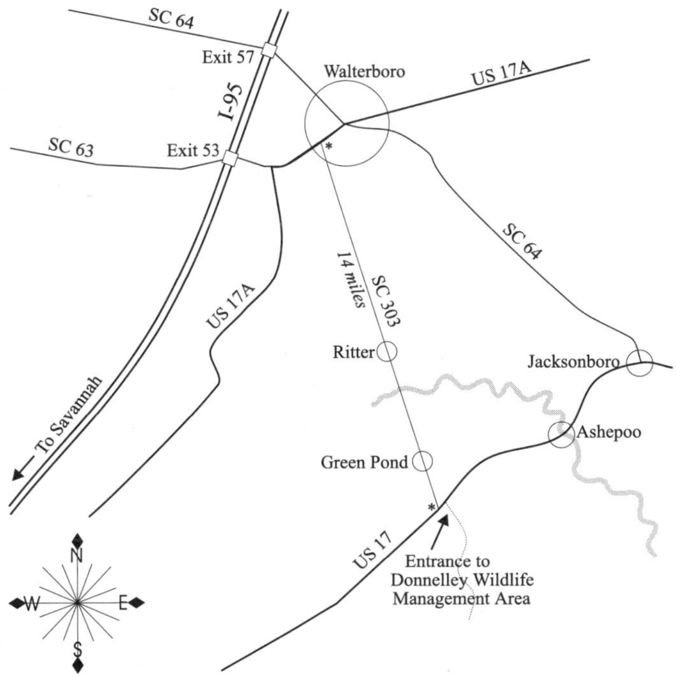

Old rice fields form the backbone of the areas of the Donnelley tract open to the public. The nature trail, which SCAN walked in a speedy five and a half hours, is about 2.5 miles long around several old rice impoundments. The trail then entered an old-growth forest with huge magnolias, live oaks, walnuts and sassafras. Many birds were seen, including residents and migrants. Many of these birds were using the old rice impoundments, as were about 20 baby alligators, who were floating on the surface soaking up the warming sun rays. Part of the trail uses old dikes, and the wildlife used the dikes, too, judging from the numerous tracks of many different kinds of animals. There is even an early-1800's small family graveyard for history enthusiasts to examine.

Directions to Donnelley Wildlife Management Area: South bound on I-95, take Exit 57,turning left onto SC64 toward Walterboro. Travel through Walterboro, staying to your right, until you come to the junction with SC 303 on your left. Take SC 303 south, passing through Green Pond, and drive 4 miles to the junction with US 17. Turn left (north toward Charleston), and then immediately turn right into the Donnelley entrance. Drive down the road about half a mile to the office on the left.

FLORA

Lichen
Leptogium cyanescens
Leptogium marginellum
Graphis scripta
Graphis atrorubens
Graphis elegans
Phaeographis dendritica
Phaeographis erumpens
Coenogonium interplexum
Coenogonium moniliforme
Bacidia incompta
Brigantiaea leucoxantha
Lecidea carneoalbens
Lecidea oreinoides
Cladonia bacillaris
Cladonia didyma
Cladonia parasitica
Cladonia subradiata
Pertusaria ostiolata
Pertusaria subpertusa
Pertusaria texana
Haematomma puniceum
Haematomma elatinum
Lecanora cupressi
Lecanora varia
Ochrolechia africana
Bulbothrix confoederata
Bulbothrix goebelii
Bulbothrix isidiza
Hypotrachyna livida
Hypotrachyna pustulifera
Parmelina aurulenta
Parmelina minarum
Parmotrema hypotropum
Parmotrema michauxianum
Parmotrema michauxianum roseum
Parmotrema perforatum
Parmotrema rampoddense
Parmotrema tinctorum
Parmotrema ultralucens
Flavoparmelia caperata
Canoparmelia caroliniana
Punctelia rudecta
Rimelia reticulata
Tuckermannopsis oakesiana
Usnea rubicunda
Usnea strigosa
Buellia punctata
Buellia dialyta
Buellia curatellae
Buellia disciformis
Dirinaria aspera
Dirinaria confusa
Dirinaria picta
Dirinaria purpurascens
Physcia americana
Physcia crispa
Physcia millegrana
Heterodermia albicans
Heterodermia obscurata
Heterodermia spaltans sp. nov.
Pyxine caesiopruinosa
Rinodina roboris
Anthracothecium caroliniana
Anthracothecium leucostomum
Pyrenula cruenta
Arthonia reniformis
Arthonia quintaria
Trypethelium virens
Trypethelium floridanum
Melanomma sp.

Virginia Chain Fern
Woodwardia virginica
Ebony Spleenwort
Asplenium platyneuron
Resurrection Fern
Polypodium polypodioides
Long-leaf Pine
Pinus palustris
Loblolly Pine
Pinus taeda
Spruce Pine
Pinus glabra
Eastern Red-cedar
Juniperus virginiana
Common Cat-tail
Typha latifolia
Cane; Switch Cane
Arundinaria gigantea
Yellow-groove Bamboo
Phyllostachys aurea
Giant Foxtail Grass
Setaria magna
Beard Grass
Erianthus sp.
Dwarf Palmetto
Sabal minor
Cabbage Palmetto
Sabal palmetto
Giant Duckweed
Spirodela punctata
Bogmat
Wolffiella floridana
Spanish-moss
Tillandsia usneoides
Soft Rush
Juncus effusus
Fringed Greenbrier
Smilax bona-nox
Snowdrops
Leucojum aestivum
Black Willow
Salix nigra
Wax-myrtle; Southern Bayberry
Myrica cerifera
Black Walnut
Juglans nigra
Hickory
Carya sp.
American Hornbeam; Ironwood
Carpinus caroliniana
Beech
Fagus grandifolia

Donnelley Wildlife Management Area (continued)
February 26, 1994

White Oak
Quercus alba
Live Oak
Quercus virginiana
Black Oak
Quercus velutina
Spanish Oak
Quercus falcata
Water Oak
Quercus nigra
American Elm
Ulmus americana
Sugarberry; Hackberry
Celtis laevigata
Mistletoe
Phoradendron serotinum
Alligator-weed
Alternanthera philoxeroides
Common Chickweed
Stellaria media
Sweetbay
Magnolia virginiana
Southern Magnolia
Magnolia grandiflora
Redbay
Persea borbonia
Sassafras
Sassafras albidum
Sweet-gum
Liquidambar styraciflua
Dewberry
Rubus trivialis
Blackberry; Brambles
Rubus sp.
Sesbania
Sesbania exaltata
Bladderpod
Glottidium vesicarium
Chinese Tallow-tree
Sapium sebiferum
Winged Sumac
Rhus copallina
American Holly
Ilex opaca
Yaupon
Ilex vomitoria
Red Maple
Acer rubrum
Muscadine; Bullace
Vitis rotundifolia
St.-Andrew's-cross
Hypericum hypericoides
Primrose-leaved Violet
Viola primulifolia
Violet
Viola sp.
Water Pennywort
Hydrocotyle ranunculoides
Intelligence-plant
Centella asiatica
Flowering Dogwood
Cornus florida
Buckthorn Bumelia
Bumelia lycioides
Persimmon
Diospyros virginiana
Horse-sugar; Sweetleaf
Symplocos tinctoria
Periwinkle
Vinca major
Dodder
Cuscuta sp.
Beautyberry
Callicarpa americana
Cross Vine
Anisostichus capreolata
Partridge-berry
Mitchella repens
Japanese Honeysuckle
Lonicera japonica
American Elderberry
Sambucus canadensis
Elephant's-foot
Elephantopus sp.
Dog-fennel
Eupatorium capillifolium
Cudweed
Gnaphalium purpureum
Sea-myrtle; Groundsel-tree
Baccharis halimifolia
Heterotheca
Heterotheca graminifolia

FAUNA

Crab Spider
Misumenops asperatus
Giant Water Bug
Belostoma lutarium
Leafy Oak Gall (wasp)
Andricus foliatus
Citrine Forktail (damselfly)
Ischnura hastata
Duckweed Firetail (damselfly)
Telebasis byersi
Colorful Foliage Ground Beetle
Apenes sinuata
Predacious Diving Beetle
Laccophilus schwartzi
Hydroporus lobatus
Agabus punctatus
Thermonetus basilaris basilaris
Water Scavenger Beetle
Tropisternus lateralis nimbatus
Cloudless Sulphur
Phoebis sennae eubule
Spring Azure (butterfly)
Celastrina ladon
Slimy Salamander
Plethodon glutinosus
Southern Cricket Frog
Acris gryllus
Green Treefrog
Hyla cinerea
Treefrog
Hyla sp.
Green Frog; Bronze Frog
Rana clamitans
Southern Leopard Frog
Rana sphenocephala
American Alligator
Alligator mississippiensis
Carolina Anole
Anolis carolinensis
Anhinga; Snakebird
Anhinga anhinga
Double-crested Cormorant
Phalacrocorax auritus
Tricolored Heron
Egretta tricolor
Little Blue Heron
Egretta caerulea
Snowy Egret
Egretta thula
Cattle Egret
Bulbicus ibis
Great Egret
Casmerodius albus
Great Blue Heron
Ardea herodias
Glossy Ibis
Plegadis falcinellus
White Ibis
Eudocimus albus
Green-winged Teal
Anas crecca
Wood Duck
Aix sponsa
Common Moorhen
Gallinula chloropus
Killdeer
Charadrius vociferus
Greater Yellowlegs
Tringa melanoleuca
Lesser Yellowlegs
Tringa flavipes
Forster's Tern
Sterna forsteri
Caspian Tern
Sterna caspia
Turkey Vulture
Cathartes aura
Black Vulture
Coragyps atratus
Bald Eagle
Haliaeetus leucocephalus
Northern Harrier
Circus cyaneus
Red-shouldered Hawk
Buteo lineatus
Red-tailed Hawk
Buteo jamaicensis
Osprey
Pandion haliaetus
Wild Turkey (tracks)
Meleagris gallopavo
Red-bellied Woodpecker
Melanerpes carolinus
Downy Woodpecker
Picoides pubescens
Yellow-bellied Sapsucker
Sphyrapicus varius
Pileated Woodpecker
Dryocopus pileatus
Eastern Phoebe
Sayornis phoebe
Tree Swallow
Tachycineta bicolor
Blue Jay
Cyanocitta cristata
American Crow
Corvus brachyrhynchos
Tufted Titmouse
Parus bicolor
Carolina Chickadee
Parus carolinensis
White-breasted Nuthatch
Sitta carolinensis
Carolina Wren
Thryothorus ludovicianus
Golden-crowned Kinglet
Regulus satrapa
Ruby-crowned Kinglet
Regulus calendula
Blue-gray Gnatcatcher
Polioptila caerulea
Eastern Bluebird
Sialia sialis
Hermit Thrush
Catharus guttatus
Solitary Vireo
Vireo solitarius
Black-and-white Warbler
Mniotilta varia
Yellow-rumped Warbler
Dendroica coronata
Yellow-throated Warbler
Dendroica dominica
Pine Warbler
Dendroica pinus
Northern Cardinal
Cardinalis cardinalis
Savannah Sparrow
Passerculus sandwichensis
Song Sparrow
Melospiza melodia
Chipping Sparrow
Spizella passerina
White-throated Sparrow
Zonotrichia albicollis
Swamp Sparrow
Melospiza georgiana
Red-winged Blackbird
Agelaius phoeniceus
Rusty Blackbird
Euphagus carolinus
Common Grackle
Quiscalus quiscula
Raccoon (tracks)
Procyon lotor
River Otter (tracks)
Lutra canadensis
Bobcat; Wildcat (tracks, scat)
Felis rufus
Wild Pig (tracks)
Sus scrofa
White-tailed Deer (tracks)
Odocoileus virginianus

Edisto River Canoe Trail
Colleton County
August 8, 1992

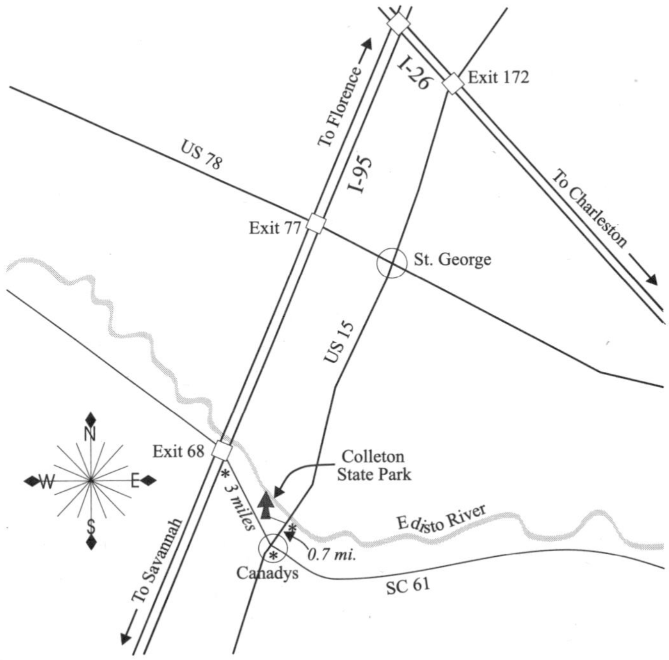

Changing one's point of view from land to water gives a better perspective of a riverine habitat. It is a very effective way to get a good look at plant material and animals that inhabit areas we humans tend to avoid -- wet and muddy. On this trip the beauty of the river was enhanced by good looks at aquatic species of trees, shrubs and birds that normally go unnoticed. Even though it was hot in August, sandbars provided places to stash the canoes while everyone got a good cooling swim.

Directions to the Edito River Canoe Trail: From I-95 between I-26 and Walterboro, take Exit 68, SC 61, east to Canadys, only 3 miles. Turn left (north) onto US 15 toward St. George. The entrance to Colleton State Park is on the left in in less than a mile, just before the Edisto River bridge.

FLORA

Netted Chain Fern
Woodwardia areolata
Resurrection Fern
Polypodium polypodioides
Loblolly Pine
Pinus taeda
Spruce Pine
Pinus glabra
Bald-cypress
Taxodium distichum
Arrowhead; Duck-potato
Sagittaria graminea graminea
Dwarf Palmetto
Sabal minor
Spanish-moss
Tillandsia usneoides
Yellow Star-grass
Hypoxis hirsuta
Black Willow
Salix nigra
Water Hickory
Carya aquatica
River Birch
Betula nigra
American Hornbeam; Ironwood
Carpinus caroliniana
Overcup Oak
Quercus lyrata
Water-elm; Planer-tree
Planera aquatica
Alligator-weed
Alternanthera philoxeroides
Climbing-hydrangea
Decumaria barbara
Sweet-gum
Liquidambar styraciflua
May Haw
Crataegus aestivalis
Hawthorn
Crataegus sp.
Water Locust
Gleditsia aquatica
Sebastian-bush
Sebastiania ligustrina
Poison-ivy
Rhus radicans
Titi
Cyrilla racemiflora
Red Maple
Acer rubrum
Pepper-vine
Ampelopsis arborea
Water Tupelo
Nyssa aquatica
Swamp Dogwood; Stiff Cornel
Cornus stricta
Water Ash; Carolina Ash
Fraxinus caroliniana
Swamp Milkweed
Asclepias perennis
Shade Mudflower
Micranthemum umbrosum
Trumpet Vine
Campsis radicans
Soldier Grainy-club
Cordyceps militaris

FAUNA

Flat Stink Bug
Brochymena sp.
Dobsonfly (egg)
Corydalis cornutus
Pedunculate Ground Beetle
Dyschirius sp.
Whirligig Beetle
Dineutus discolor
Dineutus serrulatus serrulatus
Gyrinus lugens
Green June Beetle
Cotinus nitida
Buprestid Beetle
Buprestis rufipes
Wide-necked Longhorn Beetle
Prionus laticollis
Oval Leaf Beetle
Paria thoracica
Eastern Tiger Swallowtail
Papilio glaucus
Spicebush Swallowtail
Papilio troilus
Cloudless Sulphur
Phoebis sennae eubule
Red-spotted Purple (butterfly)
Limenitis arthemis astyanax
Horace's Duskywing
Erynnis horatius
Brown Water Snake
Nerodia taxispilota
Wood Stork
Mycteria americana
White Ibis
Eudocimus albus
Spotted Sandpiper
Actitis macularia
Turkey Vulture
Cathartes aura
Yellow-billed Cuckoo
Coccyzus americanus
Ruby-throated Hummingbird
Archilochus colubris

(Photo by L. Greenlee)

Eastern tiger swallowtail, *Papilio glaucus*.

Westvaco Bluff Trail
(ACE Basin)
Colleton County
November 20, 1993

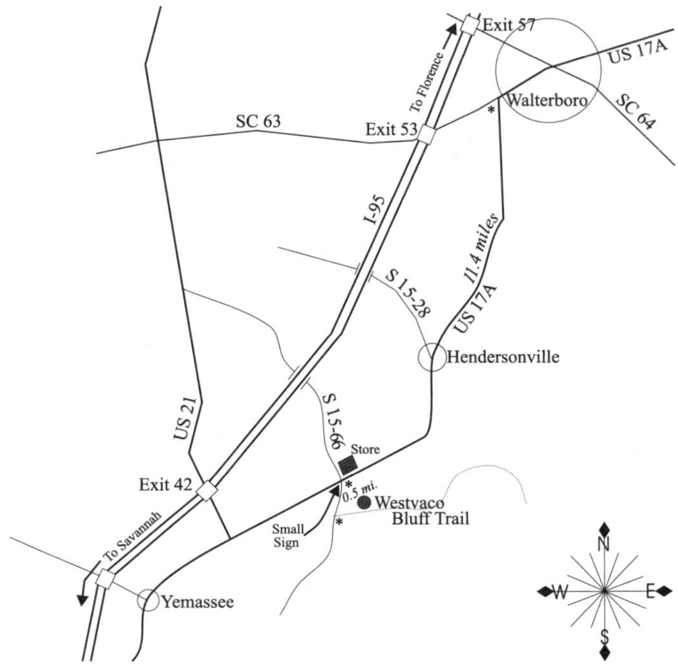

Land that is being highly cultivated for crop production can still provide a fascinating slice of the natural (i.e., non-human) world. On a pine plantation Westvaco owns in Colleton County, they have educational programs and multi-use land policies that they are willing to share on a controlled basis. The Bluff Trail goes by pine plantings of varying ages, plus old rice fields, giving an interesting look at a controlled version of old field succession. The background information we were given before we started walking helped us understand what we were seeing. Pines that mature in 15 to 18 years are impressive, as are the old water-loving species along the edges of the rice fields. The old rice fields, being disturbed less recently, provided the most unusual plant material including some uncommon ferns. In one old-growth area, a marbled orb-weaver spider had spread her web across the trail. She was centered on it, just at eye level for most of us. The Westvaco employees who accompanied us seemed to have as much fun as we did.

Privately owned by Westvaco Paper Company. Advanced permission needed to visit this site.

Directions to Westvaco Bluff Trail: From I-95 near Walterboro, take Exit 53, SC 63, east toward Walterboro for about a mile. Turn south (right) onto US 17A toward Hendersonville and Savannah. Go approximately 11.4 miles, passing through Hendersonville. Turn left (south) onto S 15-66; the turn is just past a store and is marked by a small sign. Go about one-half mile and turn left again onto the first dirt road, the entrance road. Go 0.8 miles and turn left to the picnic shelter.

FLORA

Clubmoss
Lycopodium sp.
Cinnamon Fern
Osmunda cinnamomea
Royal Fern
Osmunda regalis spectabilis
Japanese Climbing Fern
Lygodium japonicum
Bracken
Pteridium aquilinum
Southern Lady Fern
Athyrium asplenioides
Thelypteris (fern)
Thelypteris kunthii
Fern
Thelypteris sp.
Netted Chain Fern
Woodwardia areolata
Ebony Spleenwort
Asplenium platyneuron
Resurrection Fern
Polypodium polypodioides
Loblolly Pine
Pinus taeda
Short-leaf Pine
Pinus echinata
Bald-cypress
Taxodium distichum
Eastern Red-cedar
Juniperus virginiana
Cane; Switch Cane
Arundinaria gigantea
Panic Grass
Panicum sp.
Beard Grass; Plume Grass
Erianthus strictus
Beard Grass
Erianthus sp.
Dwarf Palmetto
Sabal minor
Arrowleaf Arum
Peltandra virginica
Spanish-moss
Tillandsia usneoides
Greenbrier; Bullbrier
Smilax rotundifolia
Fringed Greenbrier
Smilax bona-nox
Sawbrier; Greenbrier
Smilax glauca
Sarsaparilla-vine
Smilax pumila
Greenbrier; Catbrier
Smilax sp.
Lizard's-tail
Saururus cernuus
Black Willow
Salix nigra
Wax-myrtle; Southern Bayberry
Myrica cerifera
Mockernut Hickory
Carya tomentosa
Pignut Hickory
Carya glabra
White Oak
Quercus alba
Overcup Oak
Quercus lyrata
Swamp Chestnut Oak
Quercus michauxii
Live Oak
Quercus virginiana
Red Oak
Quercus rubra
Black Oak
Quercus velutina
Spanish Oak
Quercus falcata
Cherrybark Oak
Quercus falcata pagodaefolia
Water Oak
Quercus nigra
Willow Oak
Quercus phellos
Laurel Oak
Quercus laurifolia
American Elm
Ulmus americana
Winged Elm
Ulmus alata
Sugarberry; Hackberry
Celtis laevigata
Red Mulberry
Morus rubra

False Nettle
Boehmeria cylindrica
Mistletoe
Phoradendron serotinum
Polygonum
Polygonum sp.
Tulip-poplar
Liriodendron tulipifera
Climbing-hydrangea
Decumaria barbara
Sweet-gum
Liquidambar styraciflua
Blackberry; Brambles
Rubus sp.
Redbud; Judas-tree
Cercis canadensis
Beggar-lice
Desmodium sp.
Bush-clover
Lespedeza sp.
American Wisteria
Wisteria frutescens
Coral-bean
Erythrina herbacea
Wood-sorrel
Oxalis florida
Cranesbill
Geranium carolinianum
Trifoliate Orange
Poncirus trifoliata
Chinese Tallow-tree
Sapium sebiferum
Phyllanthus
Phyllanthus sp.
Poison-ivy
Rhus radicans
Poison-oak
Rhus toxicodendron
Winged Sumac
Rhus copallina
American Holly
Ilex opaca
Inkberry
Ilex glabra
Red Maple
Acer rubrum
Supple-jack; Rattan-vine
Berchemia scandens
Muscadine; Bullace
Vitis rotundifolia
Summer Grape
Vitis aestivalis
Pepper-vine
Ampelopsis arborea
St.-Andrew's-cross
Hypericum hypericoides
St.-Peter's-wort
Hypericum stans
Hypericum
Hypericum sp.
Violet
Viola sp.
Waxweed
Cuphea carthagensis
Ludwigia
Ludwigia sp.

Pennywort
Hydrocotyle sp.
Mock Bishop's-weed
Ptilimnium capillaceum
Black Tupelo; Blackgum
Nyssa sylvatica sylvatica
Water Tupelo
Nyssa aquatica
Flowering Dogwood
Cornus florida
Sparkleberry
Vaccinium arboreum
Highbush Blueberry
Vaccinium corymbosum
Blueberry
Vaccinium sp.
Persimmon
Diospyros virginiana
Horse-sugar; Sweetleaf
Symplocos tinctoria
Water Ash; Carolina Ash
Fraxinus caroliniana
Yellow Jessamine
Gelsemium sempervirens
Gentian
Gentiana villosa
Gentiana sp.
Climbing Dogbane
Trachelospermum difforme
Common Morning-glory
Ipomoea purpurea
Brazilian Vervain
Verbena braziliensis
Beautyberry
Callicarpa americana
Skullcap
Scutellaria sp.
Lyre-leaved Sage
Salvia lyrata
Shrubby Savory
Satureja georgiana
Purple Gerardia
Agalinis purpurea
Gerardia
Agalinis virgata
Cross Vine
Anisostichus capreolata
Trumpet Vine
Campsis radicans
Catalpa; Catawba-tree
Catalpa speciosa
Partridge-berry
Mitchella repens
Bedstraw
Galium sp.
Glade's Lobelia
Lobelia glandulosa
Pyrrhopappus
Pyrrhopappus carolinianus
Thistle
Carduus sp.
Dog-fennel
Eupatorium capillifolium
Eupatorium compositifolium
False Hoarhound
Eupatorium rotundifolium

Climbing Hempweed
Mikania scandens
Marsh-fleabane
Pluchea camphorata
Cudweed
Gnaphalium purpureum
Sea-myrtle; Groundsel-tree
Baccharis halimifolia
Frost Aster
Aster pilosus
Goldenrod
Solidago sp.
Spanish-needles
Bidens bipinnata
Emetic Russula
Russula emetica
Common Laccaria
Laccaria laccata

FAUNA

Chimney Crayfish
Cambarus diogenes
Orb-weaver
Neoscona domiciliorum
Marbled Orb-weaver
Araneus marmoreus
Short-horned Grasshopper
Metaleptea brevicornis
Wheel Bug
Arilus cristatus
Velvet-ant
Dasymutilla sp.
Fire Ant
Solenopsis geminata
Hickory Girdler Beetle
Oncideres cingulata
Cloudless Sulphur
Phoebis sennae eubule
Io Moth
Automeris io

Eastern Box Turtle
Terrapene carolina
Carolina Anole
Anolis carolinensis
White Ibis (bones)
Eudocimus albus
Turkey Vulture
Cathartes aura
Black Vulture
Coragyps atratus
Wild Turkey (tracks)
Meleagris gallopavo
Rock Dove; Pigeon
Columba livia
Red-bellied Woodpecker
Melanerpes carolinus
Yellow-bellied Sapsucker
Sphyrapicus varius
Eastern Phoebe
Sayornis phoebe
American Crow
Corvus brachyrhynchos
Fish Crow
Corvus ossifragus
Tufted Titmouse
Parus bicolor
Carolina Chickadee
Parus carolinensis
Ruby-crowned Kinglet
Regulus calendula
American Robin
Turdus migratorius
Yellow-rumped Warbler
Dendroica coronata
Northern Cardinal
Cardinalis cardinalis
Bobcat; Wildcat (scat)
Felis rufus
White-tailed Deer
Odocoileus virginianus

(Photo by L. Smith)

Westvaco Bluff Trail wends through diverse habitat including tupelo swamp.

Darlington County

Dargan's Pond, March 27, 1993 and Sara Simons' Place, December 11, 1993
Darlington County

Dargan's Pond is a fishing lake on property owned by the Pee Dee Agricultural Experiment Station, and managed by the SC Department of Natural Resources. The adjacent fields provide open habitat, and the cypress trees at the head of the pond had nesting great blue herons at the time of the SCAN visit. SCAN was only in the area for about an hour.

Directions to Dargan's Pond: From the intersection of I-20 and I-95 near Florence, go north on I-95 to Exit 170 (SC 327). Turn north on S 21-26. In about 1 mile at the Darlington County line, the road number changes to S 16-495; ignore the change and continue in the same direction for another mile. The entrance to the public boat landing at Dargan's Pond is on the left.

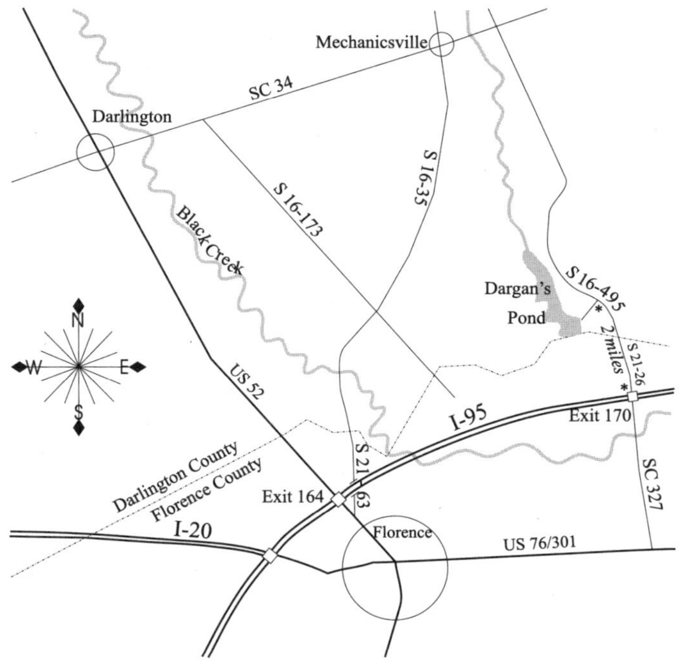

Sara Simons was gracious enough to let us spend a day exploring the pond edge, woods, and field edges near her family's cabin. It was a cold blustery day, so birds and flying insects were staying hidden in sheltered areas. Plant material was not rare, but several items were unusual in the portions that had not been farmed in many years. A nice diversity of habitats added interest, on a day when the observers were trying to hide from the wind, just like the other animals. The list is typical for coastal plain areas in December.

Sara Simons' Place is privately owned. Not open to the public. Map is not provided.

FLORA

Cinnamon Fern
Osmunda cinnamomea
Bracken
Pteridium aquilinum
Netted Chain Fern
Woodwardia areolata
Ebony Spleenwort
Asplenium platyneuron
Resurrection Fern
Polypodium polypodioides
Loblolly Pine
Pinus taeda
Short-leaf Pine
Pinus echinata
Eastern Red-cedar
Juniperus virginiana
Common Cat-tail
Typha latifolia
Cane; Switch Cane
Arundinaria gigantea
Duckmeat
Spirodela oligorrhiza
Duckweed
Lemna gibba
Spanish-moss
Tillandsia usneoides
Greenbrier; Bullbrier
Smilax rotundifolia
Fringed Greenbrier
Smilax bona-nox
Bamboo Smilax
Smilax laurifolia
Sweet-scented Smilax
Smilax smallii
Greenbrier; Catbrier
Smilax sp.
Crane-fly Orchid
Tipularia discolor
Wax-myrtle; Southern Bayberry
Myrica cerifera
Black Walnut
Juglans nigra
Mockernut Hickory
Carya tomentosa
Pignut Hickory
Carya glabra
Tag Alder
Alnus serrulata
American Hornbeam; Ironwood
Carpinus caroliniana
Beech
Fagus grandifolia
White Oak
Quercus alba
Post Oak
Quercus stellata
Red Oak
Quercus rubra
Black Oak
Quercus velutina
Spanish Oak
Quercus falcata
Water Oak
Quercus nigra
Willow Oak
Quercus phellos
Heartleaf
Hexastylis arifolia
Mexican-tea
Chenopodium ambrosioides
Pokeweed; Pokeberry
Phytolacca americana
Spatterdock; Yellow Pond-lily
Nuphar luteum
Tulip-poplar
Liriodendron tulipifera
Sweetbay
Magnolia virginiana

Redbay
Persea borbonia
Sassafras
Sassafras albidum
Virginia-willow
Itea virginica
Climbing-hydrangea
Decumaria barbara
Sweet-gum
Liquidambar styraciflua
Dewberry
Rubus trivialis
Blackberry; Brambles
Rubus sp.
Red Chokeberry
Sorbus arbutifolia
Black Cherry; Wild Cherry
Prunus serotina
Carolina Cherrylaurel
Prunus caroliniana
Redbud; Judas-tree
Cercis canadensis
Sicklepod
Cassia obtusifolia
Cranesbill
Geranium carolinianum
Chinaberry
Melia azedarach
Poison-ivy
Rhus radicans
Winged Sumac
Rhus copallina
American Holly
Ilex opaca
Strawberry-bush
Euonymus sp.
Red Maple
Acer rubrum
Muscadine; Bullace
Vitis rotundifolia
Devil's-shoestring
Sida rhombifolia
Prickly-pear
Opuntia compressa
Flowering Dogwood
Cornus florida
Spotted Wintergreen
Chimaphila maculata
Pinxter-flower; Pinxter-bloom
Rhododendron nudiflorum
Sparkleberry
Vaccinium arboreum
Highbush Blueberry
Vaccinium corymbosum
Yellow Jessamine
Gelsemium sempervirens
Dodder
Cuscuta sp.
Common Morning-glory
Ipomoea purpurea
Beautyberry
Callicarpa americana
Trumpet Vine
Campsis radicans
Partridge-berry
Mitchella repens
Japanese Honeysuckle
Lonicera japonica
Possumhaw Viburnum
Viburnum nudum
Maple-leaf Viburnum
Viburnum acerifolium
Cocklebur
Xanthium strumarium
Ragwort
Senecio sp.
Dog-fennel
Eupatorium capillifolium
Sea-myrtle; Groundsel-tree
Baccharis halimifolia
Tall Goldenrod
Solidago altissima
Heterotheca
Heterotheca graminifolia
Bladder-stalks
Physalacria inflata
Tooth Fungus
Hericium sp.
Purple-spored Puffball
Calvatia cyathiformis
Tough Puffball
Mycenastrum corium

FAUNA

Carolina Locust
Dissosteira carolina
Boll's Sand Grasshopper
Spharagemon bolli
Spur-throated Grasshopper
Melanoplus sp.
Fire Ant
Solenopsis geminata
Dogwood Club Gall (midge)
Mycodiplosis clavula
Crawling Water Beetle
Haliplus triopsus
Peltodytes oppositus
Predacious Diving Beetle
Laccophilus proximus
Hydroporus carolinus
Hydroporus clypealis
Hydroporus hybridus
Hydroporus sp.
Coptotomus lenticus
Skin Beetle
Trox tuberculatus
Trox sp.
Darkling Beetle
Diaperis maculata
Platydema erythrocera
Alobates pennsylvanica
Red-horned Grain Beetle
Platydema ruficorne
Hickory Girdler Beetle
Oncideres cingulata
Spring Peeper
Hyla crucifer
Eastern Musk Turtle
Sternotherus odoratus
Cooter
Chrysemys sp.
Anhinga; Snakebird
Anhinga anhinga
Eastern Phoebe
Sayornis phoebe
Blue Jay
Cyanocitta cristata
Tufted Titmouse
Parus bicolor
American Robin
Turdus migratorius
Brown Thrasher
Toxostoma rufum
White-throated Sparrow
Zonotrichia albicollis
Rusty Blackbird
Euphagus carolinus
Rabbit
Sylvilagus sp.
Beaver
Castor canadensis
Bobcat; Wildcat
Felis rufus
White-tailed Deer
Odocoileus virginianus

Kalmia Gardens of Coker College, November 17, 1990
Segars Heritage Preserve, May 14, 1994
Darlington County

Visiting a garden doesn't sound like a good destination for studying natural history, but Kalmia Gardens is an exception. The gardens around the Hart House attract a nice representation of local birds and insects. In addition, there are trails through woodlands and down a bluff to the edge of Black Creek, a classic blackwater creek. In November, most herbaceous plants have been rendered dormant by cold, but here along some seepage areas and under some denser canopy areas, interesting flowering plants were found. This is one of the scattered areas on the edge of the sandhills, as it changes to inner Coastal Plain, where communities of plants reminiscent of the mountains are found. Since the SCAN visit, Segars Heritage Preserve has been established across Black Creek and a bridge installed to connect the two properties, expanding the types of habitats that may be seen here. SCAN appreciated Dr. George Sawyer, Director of Kalmia Gardens, for being our host.

Directions to Kalmia Gardens of Coker College: From I-20 between Columbia and Florence, take Exit 116 (US 15) north through Bishopville, Una, and Lydia. Turn left at the 4-way intersection (north toward Darlington) and continue on US 15 for about 3 more miles to SC 151. Turn left again toward McBee, passing through suburban Hartsville, and watch carefully for a sharp right turn onto Business 151 in about 6 miles. Turn right toward downtown Hartsville. Kalmia Gardens is on the left inside a wrought-iron entrance gate.

Segars Heritage Preserve occupies the floodplain side of Black Creek on the northwest edge of Hartsville. Kalmia Gardens sits on the opposite bluff side of the creek. This preserve had just become a part of the Heritage Trust program when a group of adventurous SCAN members decided to explore the area to help inventory the flora and fauna. No trails had yet been established, and the access bridge not yet built. Because the area had been heavily damaged by Hurricane Hugo, fallen trees were covered in greenbriar and blackberry vines. The preserve also includes a seepage area creating a white-cedar swamp.

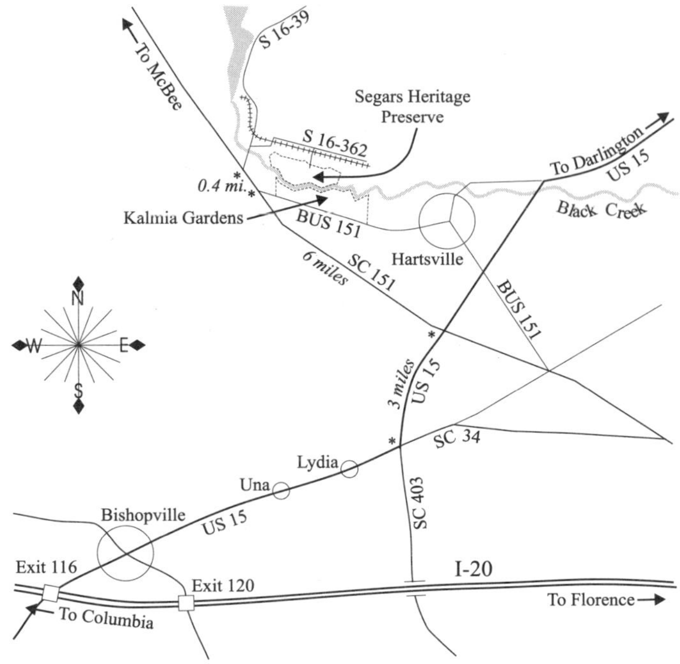

Although directions to Segars Heritage Preserve are given here, access is through the grounds of Kalmia Gardens.

Directions to Segars Heritage Preserve: From I-20 between Columbia and Florence, take Exit 116 (US 15) north through Bishopville, Una, and Lydia. Turn left at the 4-way intersection (north toward Darlington) and continue on US 15 for about 3 more miles to SC 151. Turn left again toward McBee, passing through suburban Hartsville, and watch carefully for a sharp right turn onto Business 151 in about 6 miles. Go past that intersection 0.4 miles to the next street and turn right onto S 16-39. Cross Black Creek and immediately turn right onto S 16-362 before the railroad track. That street then crosses the track and bends to the right. Continue about a mile and turn right back over the track on an obscure road marked by a Heritage Preserve sign.

Kalmia Gardens of Coker College and Segars Heritage Preserve, (continued) November 17, 1990 and May 14, 1994

FLORA

Gelatinous Lichen
Leptogium cyanescens
Graphis scripta
Bacidia schweinitzii
Bacidia incompta
Lecidea russula
Cladonia bacillaris
Cladonia coniocraea
Pertusaria leucostoma
Pertusaria ostiolata
Pertusaria paratuberculifera
Pertusaria subpertusa
Pertusaria texana
Pertusaria sinusmexicani
Pertusaria sp.
Maronea constans
Lecanora miculata
Lecanora cupressi
Lecanora pallida
Lecanora varia
Lecanora sp. #2 Brodo
Lecanora hybocarpa
Phlyctis sp.
Bulbothrix goebelii
Bulbothrix isidiza
Hypotrachyna formosana
Hypotrachyna livida
Parmelina horrescens
Parmelina horrescens asidiata
Parmelina minarum
Parmotrema hypotropum
Parmotrema michauxianum
Parmotrema perforatum
Parmotrema subtinctorium
Parmotrema xanthinum
Flavoparmelia caperata
Canoparmelia caroliniana
Punctelia rudecta
Rimelia reticulata
Ramalina americana
Usnea rubicunda
Buellia dialyta
Buellia curatellae
Buellia disciformis
Physcia americana
Physcia sp.
Heterodermia albicans
Heterodermia obscurata
Pyxine sorediata
Pyrenula cruenta
Pyrenula nitida
Trypethelium virens
Melanomma sp.

Peat Moss
Sphagnum sp.
Cinnamon Fern
Osmunda cinnamomea
Royal Fern
Osmunda regalis spectabilis
Bracken
Pteridium aquilinum
Southern Lady Fern
Athyrium asplenioides
Christmas Fern
Polystichum acrostichoides
Netted Chain Fern
Woodwardia areolata
Ebony Spleenwort
Asplenium platyneuron
Resurrection Fern
Polypodium polypodioides
Long-leaf Pine
Pinus palustris
Loblolly Pine
Pinus taeda
Pond Pine
Pinus serotina
Short-leaf Pine
Pinus echinata
Bald-cypress
Taxodium distichum
Atlantic White-cedar
Chamaecyparis thyoides
Cane; Switch Cane
Arundinaria gigantea

Bogmoss
Mayaca sp.
Spanish-moss
Tillandsia usneoides
Roseling; Spiderwort
Tradescantia rosea
Greenbrier; Bullbrier
Smilax rotundifolia
Sawbrier; Greenbrier
Smilax glauca
Bamboo Smilax
Smilax laurifolia
Greenbrier; Catbrier
Smilax sp.
Bear-grass
Yucca filamentosa
Dwarf Iris
Iris verna
Downy Rattlesnake-plantain
Goodyera pubescens
Swamp Bayberry
Myrica heterophylla
Mockernut Hickory
Carya tomentosa
Pignut Hickory
Carya glabra
American Hornbeam; Ironwood
Carpinus caroliniana
Beech
Fagus grandifolia
White Oak
Quercus alba

Post Oak
Quercus stellata
Scrubby Post Oak
Quercus margaretta
Turkey Oak
Quercus laevis
Spanish Oak
Quercus falcata
Black Jack Oak
Quercus marilandica
Water Oak
Quercus nigra
Blue-jack Oak
Quercus incana
Willow Oak
Quercus phellos
Laurel Oak
Quercus laurifolia
Red Mulberry
Morus rubra
Mistletoe
Phoradendron serotinum
Heartleaf
Hexastylis arifolia
Yellowroot
Xanthorhiza simplicissima
Tulip-poplar
Liriodendron tulipifera
Sweetbay
Magnolia virginiana
Southern Magnolia
Magnolia grandiflora
Dwarf PawPaw
Asimina parviflora
Redbay
Persea borbonia
Sassafras
Sassafras albidum
Virginia-willow
Itea virginica
Climbing-hydrangea
Decumaria barbara
Sweet-gum
Liquidambar styraciflua
Witch-hazel
Hamamelis virginiana
Witch-alder
Fothergilla gardenii
American Sycamore
Platanus occidentalis
Five-fingers
Potentilla canadensis
Blackberry; Brambles
Rubus sp.
Red Chokeberry
Sorbus arbutifolia
Shadbush; Serviceberry
Amelanchier sp.
Black Cherry; Wild Cherry
Prunus serotina
Redbud; Judas-tree
Cercis canadensis
Wild Indigo
Baptisia bracteata
Leadplant; Indigo-bush
Amorpha sp.

Beggar-lice
Desmodium sp.
Sericea
Lespedeza cuneata
Orange Milkwort
Polygala lutea
Tread-softly
Cnidoscolus stimulosus
Three-seeded Mercury
Acalypha sp.
Flowering Spurge
Euphorbia corollata
Poison Sumac
Rhus vernix
Poison-ivy
Rhus radicans
Poison-oak
Rhus toxicodendron
Winged Sumac
Rhus copallina
Titi
Cyrilla racemiflora
American Holly
Ilex opaca
Yaupon
Ilex vomitoria
Inkberry
Ilex glabra
Sweet Gallberry
Ilex coriacea
Holly
Ilex sp.
Hearts-a'burstin'-with-love
Euonymus americanus
Red Maple
Acer rubrum
Virginia-creeper
Parthenocissus quinquefolia
Muscadine; Bullace
Vitis rotundifolia
Summer Grape
Vitis aestivalis
Loblolly-bay
Gordonia lasianthus
St.-Andrew's-cross
Hypericum hypericoides
Prickly-pear
Opuntia compressa
Water-milfoil
Myriophyllum sp.
Devil's-walking-stick
Aralia spinosa
Pennywort
Hydrocotyle sibthorpioides
Golden-Alexander
Zizia sp.
Blackgum
Nyssa sylvatica
Black Tupelo; Blackgum
Nyssa sylvatica sylvatica
Swamp Tupelo; Swamp Blackgum
Nyssa sylvatica biflora
Flowering Dogwood
Cornus florida
Sweet Pepperbush
Clethra alnifolia

Spotted Wintergreen
Chimaphila maculata
Carolina-laurel
Rhododendron minus
Swamp Honeysuckle
Rhododendron viscosum
Azalea
Rhododendron sp.
Sand-myrtle
Leiophyllum buxifolium
Mountain-laurel
Kalmia latifolia
Maleberry
Lyonia ligustrina
Fetterbush
Lyonia lucida
Staggerbush
Lyonia mariana
Dog-hobble
Leucothoe axillaris
Dog-hobble
Leucothoe axillaris axillaris
Sourwood
Oxydendrum arboreum
Trailing-arbutus
Epigaea repens
Dangleberry
Gaylussacia frondosa
Sparkleberry
Vaccinium arboreum
Low-bush Blueberry
Vaccinium tenellum
Highbush Blueberry
Vaccinium corymbosum
Sheep-laurel; Lamb-kill
Kalmia angustifolia
Galax; Galaxy; Coltsfoot
Galax aphylla
Persimmon
Diospyros virginiana
Horse-sugar; Sweetleaf
Symplocos tinctoria
Yellow Jessamine
Gelsemium sempervirens
Gentian
Gentiana catesbaei
Sandhill Milkweed
Asclepias humistrata
Ground-ivy; Ponyfoot
Dichondra carolinensis
Beautyberry
Callicarpa americana
Blue-curls
Trichostema dichotomum
Shrubby Savory
Satureja georgiana
Trumpet Vine
Campsis radicans
Partridge-berry
Mitchella repens
Possumhaw Viburnum
Viburnum nudum
Wahlenbergia
Wahlenbergia marginata
Small's Ragwort
Senecio smallii

Elephant's-foot
Elephantopus tomentosus
Cudweed
Gnaphalium purpureum
Sea-myrtle; Groundsel-tree
Baccharis halimifolia
Starved Aster
Aster lateriflorus
Rosinweed
Silphium compositum
Black-knot
Apiosporina morbosa
Splash-cups
Cyathus striatus

FAUNA

Potter Wasp
Eumenes fraternus
Pied Skimmer
Libellula luctuosa
Twelve-spotted Skimmer
Libellula pulchella
Yellow-legged Meadowfly
Sympetrum vicinum
Sparkling Jewelwing Damselfly
Calopteryx dimidiata
Variable Dancer (damselfly)
Argia fumipennis
Searcher; Ground Beetle
Carabus vinctus
Pedunculate Ground Beetle
Clivina rufa
Seed-eating Ground Beetle
Amara cupreolata
Dingy Ground Beetle
Selenophorus ellipticus
Dingy Ground Beetle
Discoderus paralleleus
Passalus; Bess-beetle
Popilius disjunctus
Branch-and-leaf Buprestid
Agrilus sp.
Soldier Beetle
Cantharis sp.
Soft-winged Flower Beetle
Hypebaeus sp.
Tumbling Flower Beetle
Mordella melaena
Tumbling Flower Beetle
Mordellistena pubescens
Comb-clawed Beetle
Isomira sp.
Ant-like Leaf Beetle
Ariotus sp.
Minute Lady Beetle
Delphastus pusillus
Seven-spotted Lady Beetle
Coccinella septempunctata
Multicolored Asian Lady Beetle
Harmonia axyridis
Ant-like Longhorn Beetle
Euderces picipes
Leaf-mining Weevil
Eugnamptus collaris collaris
Acorn Weevil
Curculio sp.

Kalmia Gardens of Coker College and Segars Heritage Preserve, (continued) November 17, 1990 and May 14, 1994

Potato Bud Weevil
Anthonomus nigrinus
Carolina Anole
Anolis carolinensis
Ground Skink
Scincella lateralis
Turkey Vulture
Cathartes aura
Barred Owl
Strix varia
Chimney Swift
Chaetura pelagica
Red-bellied Woodpecker
Melanerpes carolinus
Great Crested Flycatcher
Myiarchus crinitus
Acadian Flycatcher
Empidonax virescens
American Crow
Corvus brachyrhynchos
Carolina Chickadee
Parus carolinensis
Carolina Wren
Thryothorus ludovicianus
Ruby-crowned Kinglet
Regulus calendula
Blue-gray Gnatcatcher
Polioptila caerulea
American Robin
Turdus migratorius
Brown Thrasher
Toxostoma rufum
Yellow-throated Vireo
Vireo flavifrons
Red-Eyed Vireo
Vireo olivaceus
Northern Cardinal
Cardinalis cardinalis
Rufous-sided Towhee
Pipilo erythrophthalmus
Summer Tanager
Piranga rubra

(Photo by A. Ciegler)
Carolina anole, *Anolis carolinensis*.

Williamson Park and Woods Dargan
Darlington County
March 23, 1991

Williamson Park, located within the limits of the town of Darlington, includes an extensive area along Swift Creek, a blackwater tributary of Black Creek. Highlighting the beauty of the area is a winding boardwalk through areas of water carpeted with parrot-feather. When SCAN visited, red maples were festooned with bright red "keys", as maple seeds are sometimes called. Winter birds were still feeding, but a few early spring arrivals were seen as well. This trip provided a good look at a classic blackwater creek-fed cypress swamp and all its inhabitants.

Directions to Williamson Park: From Columbia, take I-20 east toward Florence. At Exit 131, turn north on US 401 to Darlington, about 12 miles. Cross the US 52 bypass and continue to the center of Darlington. When you reach the courthouse square, continue east on SC 34 (Cashua Street) for 2 blocks beyond the square. Turn left (north) on Spring Street; the entrance to Williamson Park is about 3 blocks farther on the right side of Spring Street.

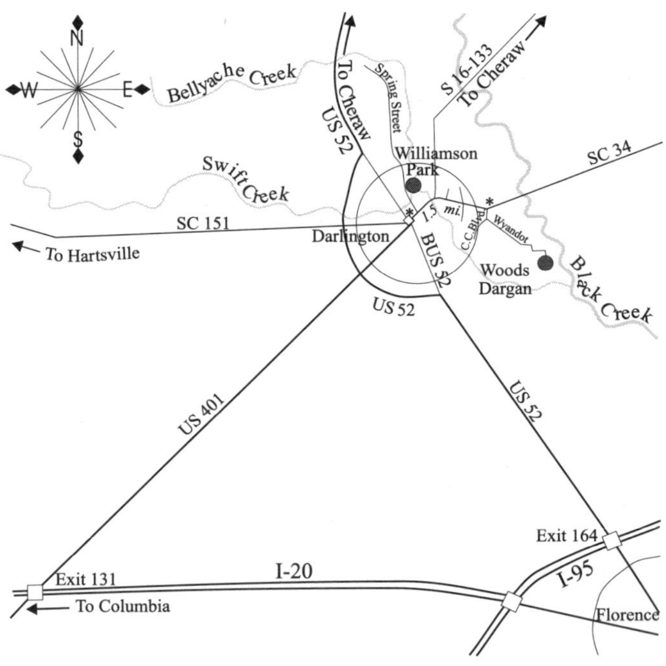

Woods Dargan protects an almost pristine riverine floodplain and adjacent high ground along Black Creek, providing a classic example of this type of ecosystem. This area is adjacent to the town of Darlington and is thereby vulnerable to the effects of development. Its protection is a wise conservation measure.

Directions to Woods Dargan: From Columbia, take I-20 east toward Florence. At Exit 131, turn north on US 401 to Darlington, about 12 miles. Cross the US 52 bypass and continue to the center of Darlington. At the intersection of SC 34 and BUS 52, continue east on SC 34 (Cashua Street) for 1.5 more miles; two blocks after crossing Swift Creek, turn right onto Country Club Boulevard. (If you come to Black Creek, turn around and go back.) Drive one block on Country Club, then fork left onto Wyandot; follow Wyandot to Woods Dargan at the end of the road near Black Creek.

Owned by the Black Creek Land Trust. Access limited. Call (843) 393-3881/2211.

FLORA

Lichen

Graphis scripta
Phaeographis dendritica
Bacidia chlorantha
Bacidia schweinitzii
Bacidia incompta
Bacidia sp.
Leptogium austroamericanum
Lecidea myriocarpoides
Lecidea sp.
Pyrrhospora varians
Cladina evansii
Cladonia capitata
Cladonia chlorophaea
Cladonia coniocraea
Cladonia cristatella scyphulifera
Cladonia floerkeana
Cladonia leporina
Cladonia macilenta
Pertusaria leucostoma
Pertusaria pustulata
Pertusaria ostiolata
Pertusaria paratuberculifera
Pertusaria subpertusa
Pertusaria texana
Pertusaria tetrathalmia
Pertusaria velata
Pertusaria macounii
Pertusaria ophthalmiza
Pertusaria sp.
Lecanora hypoptoides
Lecanora miculata
Lecanora willeyi
Lecanora cupressi
Lecanora varia
Lecanora allophana argentata
Lecanora sp.
Ochrolechia africana
Phlyctis sp.
Candelaria fibrosa
Hypotrachyna livida
Parmelina dissecta
Parmelina galbina
Parmotrema hypotropum
Parmotrema michauxianum
Parmotrema perforatum
Parmotrema reticulatum
Parmotrema robustum
Parmotrema subisidiosum
Pseudoparmelia amazonica
Pseudoparmelia caroliniana
Xanthoparmelia subramigera
Punctelia rudecta
Tuckermannopsis fendleri
Ramalina americana
Usnea strigosa
Buellia parasema
Buellia punctata
Physcia aipolia
Physcia millegrana
Physcia sorediosa
Heterodermia albicans
Heterodermia obscurata
Phaeophyscia rubropulchra
Rinodina sophodes
Physciella cloantha
Pyrenula nitida
Mycoporum pyrenocarpum
Hyperphyscia syncolla
Trypethelium virens

Southern Lady Fern
Athyrium asplenioides
Christmas Fern
Polystichum acrostichoides
Sensitive Fern
Onoclea sensibilis
Netted Chain Fern
Woodwardia areolata
Ebony Spleenwort
Asplenium platyneuron
Resurrection Fern
Polypodium polypodioides
Bald-cypress
Taxodium distichum
Brazilian Elodea
Egeria densa
Cane; Switch Cane
Arundinaria gigantea
Dwarf Palmetto
Sabal minor
Arrowleaf Arum
Peltandra virginica
Duckmeat
Spirodela oligorrhiza
Spanish-moss
Tillandsia usneoides
Woodrush
Luzula echinata
Fringed Greenbrier
Smilax bona-nox
Sawbrier; Greenbrier
Smilax glauca
Sweet-scented Smilax
Smilax smallii
Atamasco-lily
Zephyranthes atamasco
Dwarf Iris
Iris verna
Lizard's-tail
Saururus cernuus
Wax-myrtle; Southern Bayberry
Myrica cerifera
American Hornbeam; Ironwood
Carpinus caroliniana
White Oak
Quercus alba
Post Oak
Quercus stellata
Swamp Chestnut Oak
Quercus michauxii
Spanish Oak
Quercus falcata
Black Jack Oak
Quercus marilandica
Laurel Oak
Quercus laurifolia
Mistletoe
Phoradendron serotinum
Heartleaf
Hexastylis arifolia
Common Chickweed
Stellaria media
Mouse-ear Chickweed
Cerastium glomeratum
Small-flowered Crowfoot
Ranunculus abortivus
Nandina
Nandina domestica
Tulip-poplar
Liriodendron tulipifera
Sweetbay
Magnolia virginiana
Southern Magnolia
Magnolia grandiflora
Anise
Illicium sp.
Hairy Bitter Cress
Cardamine hirsuta
Sweet-gum
Liquidambar styraciflua
Indian Strawberry
Duchesnea indica
Common Apple
Malus pumila
Red Chokeberry
Sorbus arbutifolia
Shadbush
Amelanchier canadensis
Black Cherry; Wild Cherry
Prunus serotina
Carolina Cherrylaurel
Prunus caroliniana
Redbud; Judas-tree
Cercis canadensis
American Wisteria
Wisteria frutescens
Violet Wood-sorrel
Oxalis violacea
Water Starwort
Callitriche heterophylla
Water Starwort
Callitriche deflexa
Poison-ivy
Rhus radicans
Winged Sumac
Rhus copallina
Titi
Cyrilla racemiflora
American Holly
Ilex opaca
Hearts-a'burstin'-with-love
Euonymus americanus
Boxelder
Acer negundo
Red Maple
Acer rubrum
Spotted Touch-me-not
Impatiens capensis
St.-Andrew's-cross
Hypericum hypericoides
Primrose-leaved Violet
Viola primulifolia
Seed-box
Ludwigia alternifolia
Water-milfoil
Myriophyllum brasiliense
Flowering Dogwood
Cornus florida
Piedmont Azalea
Rhododendron canescens
Fetterbush
Lyonia lucida
Florida Leucothoe
Leucothoe populifolia
Coastal Honeybells
Leucothoe racemosa
Sourwood
Oxydendrum arboreum
Sparkleberry
Vaccinium arboreum
Elliott's Blueberry
Vaccinium elliottii
Privet; Ligustrum
Ligustrum japonicum
Chinese Privet
Ligustrum sinense
Yellow Jessamine
Gelsemium sempervirens
Periwinkle
Vinca major
Dodder
Cuscuta sp.
Beautyberry
Callicarpa americana
Glecoma
Glecoma hederacea
Common Henbit
Lamium amplexicaule
Red Henbit
Lamium purpureum
Stachys
Stachys sp.
Hedge-hyssop
Gratiola virginiana
Mazus
Mazus japonicus
Toadflax
Linaria canadensis
Corn Speedwell; Veronica
Veronica arvensis
Cross Vine
Anisostichus capreolata
Trumpet Vine
Campsis radicans
Partridge-berry
Mitchella repens
Cleavers
Galium aparine
Japanese Honeysuckle
Lonicera japonica
American Elderberry
Sambucus canadensis
Hawk's-beard
Crepis japonica
Climbing Hempweed
Mikania scandens
Smoky Polypore
Bjerkandera adusta

FAUNA

Common Baskettail (dragonfly)
Epitheca cynosura
Green Tiger Beetle
Cicindela sexguttata

Williamson Park and Woods Dargan (continued)
March 23, 1991

Minute Ground Beetle
Tachys ephippiatus
Tachyta nana inornata
Woodland Ground Beetle
Agonum moerens
Notch-mouthed Ground Beetle
Dicaelus dilitatus
Crawling Water Beetle
Haliplus fasciatus
Whirligig Beetle
Dineutus discolor
Gyrinus lugens
Water Scavenger Beetle
Tropisternus natator
Rove Beetle
Quedius sp.
Paederine Rove Beetle
Rugilus dentatus
Crab-like Rove Beetle
Tachinus sp.
Obscure Rove Beetle
Aleochara sp.
Carrion Beetle
Silpha inaequalis
Hister Beetle
Platysoma lecontei
Platysoma carolinus
Passalus; Bess-beetle
Popilius disjunctus
Soldier Beetle
Cantharis bilineatus
Bark-gnawing Beetle
Corticotomus sp.
Tenebroides laticollis
Checkered Beetle
Thanasimus Dubius
Red-horned Grain Beetle
Platydema ruficorne
Darkling Beetle
Uloma punctulata
Alobates pennsylvanica
Sap-feeding Beetle
Prometopia sexmaculata
Flat Bark Beetle
Laemophloeus modestus
Brontes dubius
False Skin Beetle
Diplocoelus rudis
Cylindrical Bark Beetle
Colydium lineola
Flower Longhorn Beetle
Rhagium inquisitor
Zebra Swallowtail
Eurytides marcellus
Eastern Tiger Swallowtail
Papilio glaucus
Falcate Orangetip
Anthocharis midea
Cloudless Sulphur
Phoebis sennae eubule
American Snout (butterfly)
Libytheana carinenta
Pearl Crescent
Phyciodes tharos
Silver-spotted Skipper
Epargyreus clarus
Pickerel
Esox sp.
Southern Dusky Salamander
Desmognathus auriculatus
Slimy Salamander
Plethodon glutinosus
Southern Toad
Bufo terrestris
Toad (tracks)
Bufo sp.
Northern Cricket Frog
Acris crepitans
Green Treefrog
Hyla cinerea
Cooter (tracks)
Chrysemys sp.
Carolina Anole
Anolis carolinensis
Banded Water Snake
Nerodia fasciata
Green-backed Heron
Butorides striatus
Barred Owl
Strix varia
Belted Kingfisher
Ceryle alcyon
Red-bellied Woodpecker
Melanerpes carolinus
Hairy Woodpecker
Picoides villosus
Eastern Phoebe
Sayornis phoebe
Tufted Titmouse
Parus bicolor
Carolina Chickadee
Parus carolinensis
Carolina Wren
Thryothorus ludovicianus
Ruby-crowned Kinglet
Regulus calendula
Blue-gray Gnatcatcher
Polioptila caerulea
Northern Mockingbird
Mimus polyglottos
Prothonotary Warbler
Protonotaria citrea
Yellow-rumped Warbler
Dendroica coronata
Pine Warbler
Dendroica pinus
Northern Cardinal
Cardinalis cardinalis
Common Grackle
Quiscalus quiscula
Virginia Opossum (tracks)
Didelphis virginiana
Southern Short-tailed Shrew
Blarina carolinensis
Cotton Mouse
Peromyscus gossypinus
Muskrat
Ondatra zibethicus
Raccoon (tracks)
Procyon lotor
Domestic Cat (tracks)
Felis catus
White-tailed Deer (tracks)
Odocoileus virginianus

Dillon County

Little Pee Dee State Park
Dillon County
August 26, 1989

Diverse habitats from sandhills to riverine swamp are found at Little Pee Dee State Park. The nature trail passes by most of these habitats, except river's edge. The high ground and the lake edge provide a rich area to explore in a short distance. Most of the area has been timbered, and is therefore a second-growth forest. The biodiversity can be seen in the accompanying list. Some members put in canoes for a short distance, enjoying the river's edge. This change in perspective added much to the day.

Directions to Little Dee Pee State Park: From the intersection of I-95 and I-20 near Florence, take the I-20 spur to Florence and go east on US 76 through Marion to Mullins, about 39 miles. In the center of downtown Mullins, turn left (north) onto SC 41 and drive 6 miles to Fork. In Fork, continue on SC 57 toward Floydale and Dillon. Go about 2 miles and turn right (north) onto S 17-22. After crossing the Little Pee Dee River, watch for a sign designating a right turn into Little Pee Dee State Park.

FLORA

Peat Moss
Sphagnum sp.
Sand Spikemoss
Selaginella arenicola
Cinnamon Fern
Osmunda cinnamomea
Bracken
Pteridium aquilinum
Netted Chain Fern
Woodwardia areolata
Resurrection Fern
Polypodium polypodioides
Long-leaf Pine
Pinus palustris
Loblolly Pine
Pinus taeda
Bald-cypress
Taxodium distichum
Eastern Red-cedar
Juniperus virginiana
Common Cat-tail
Typha latifolia
Bur-reed
Sparganium americanum
Cane; Switch Cane
Arundinaria gigantea
Uniola Grass
Uniola laxa
Love Grass
Eragrostis refracta
Dulichium (sedge)
Dulichium arundinaceum
Beak-rush
Rhynchospora inexpansa
Nut-rush
Scleria triglomerata
Arrowleaf Arum
Peltandra virginica
Duckmeat
Spirodela oligorrhiza
Giant Duckweed
Spirodela punctata
Water-meal
Wolffia papulifera
Yellow-eyed-grass
Xyris caroliniana
Yellow-eyed-grass
Xyris sp.

Bog-buttons
Lachnocaulon beyrichianum
Spanish-moss
Tillandsia usneoides
Dayflower
Commelina erecta
Asian Spiderwort
Aneilema keisak
Roseling; Spiderwort
Tradescantia rosea graminea
Toad Rush
Juncus bufonius
Rush
Juncus dichotomus
Juncus scirpoides
Bear-grass
Yucca filamentosa
Redroot
Lachnanthes caroliniana
Blue-eyed-grass
Sisyrinchium mucronatum
Sisyrinchium mucronatum atlanticum
Crane-fly Orchid
Tipularia discolor
Lizard's-tail
Saururus cernuus
Wax-myrtle; Southern Bayberry
Myrica cerifera
White Oak
Quercus alba
Swamp Chestnut Oak
Quercus michauxii
Turkey Oak
Quercus laevis
Spanish Oak
Quercus falcata
Black Jack Oak
Quercus marilandica
Water Oak
Quercus nigra
Blue-jack Oak
Quercus incana
Willow Oak
Quercus phellos
False Nettle
Boehmeria cylindrica
Virginia Snakeroot
Aristolochia serpentaria

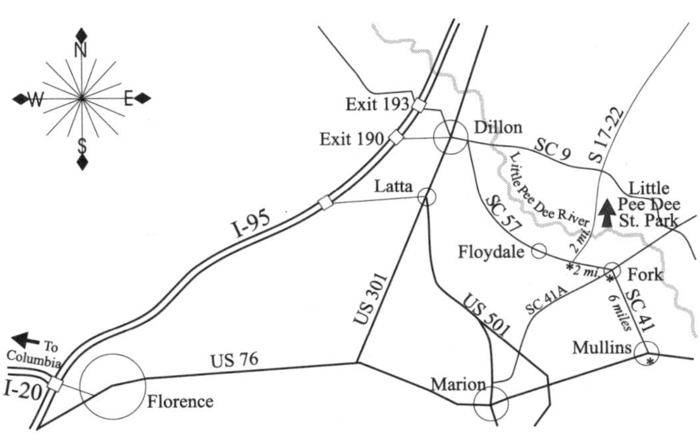

Heartleaf
Hexastylis arifolia
Smartweed
Polygonum hydropiperoides opelousanum
Jointweed
Polygonella polygama
Wire-plant
Stipulicida setacea
Water-lily
Nymphaea odorata
Water-shield
Brasenia schreberi
Sweetbay
Magnolia virginiana
Redbay
Persea borbonia
Sassafras
Sassafras albidum
Virginia-willow
Itea virginica
Sweet-gum
Liquidambar styraciflua
Red Chokeberry
Sorbus arbutifolia
Dwarf Hawthorn
Crataegus uniflora
Black Cherry; Wild Cherry
Prunus serotina
Wild Sensitive-plant
Cassia nicitans
Rattlebox
Crotalaria angulata
Zornia
Zornia bracteata
Pencil-flower
Stylosanthes biflora
Beggar-ticks
Desmodium strictum
Beggar-lice
Desmodium sp.
Bicolor; Bush-clover
Lespedeza bicolor
Sericea
Lespedeza cuneata
Bush-clover
Lespedeza hirta
Lespedeza sp.
Tephrosia
Tephrosia spicata
Climbing Butterfly-pea
Centrosema virginianum
Dollar-weed
Rhynchosia reniformis
Tread-softly
Cnidoscolus stimulosus
Tragia
Tragia urens
Spotted Spurge
Euphorbia maculata
Poison-oak
Rhus toxicodendron
Winged Sumac
Rhus copallina
Titi
Cyrilla racemiflora

American Holly
Ilex opaca
Inkberry
Ilex glabra
Red Maple
Acer rubrum
Virginia-creeper
Parthenocissus quinquefolia
Muscadine; Bullace
Vitis rotundifolia
Summer Grape
Vitis aestivalis
Loblolly-bay
Gordonia lasianthus
St.-Andrew's-cross
Hypericum hypericoides
St.-Peter's-wort
Hypericum stans
St.-John's-wort
Hypericum densiflorum
Passionflower; Maypops
Passiflora incarnata
Prickly-pear
Opuntia compressa
Water Loosestrife
Decodon verticillatus
Pale Meadow-beauty
Rhexia mariana
Ludwigia
Ludwigia leptocarpa
Ludwigia pilosa
Blackgum
Nyssa sylvatica
Water Tupelo
Nyssa aquatica
Flowering Dogwood
Cornus florida
Sweet Pepperbush
Clethra alnifolia
Spotted Wintergreen
Chimaphila maculata
Indian-pipe
Monotropa uniflora
Sheep-kill; Lamb-kill
Kalmia angustifolia caroliniana
Honey-cup
Zenobia pulverulenta
Fetterbush
Lyonia lucida
Staggerbush
Lyonia mariana
Dangleberry
Gaylussacia frondosa
Elliott's Blueberry
Vaccinium elliottii
Highbush Blueberry
Vaccinium corymbosum
Persimmon
Diospyros virginiana
Horse-sugar; Sweetleaf
Symplocos tinctoria
Yellow Jessamine
Gelsemium sempervirens
Polypremum
Polypremum procumbens
Milkweed
Asclepias sp.

Little Pee Dee State Park (continued)
August 26, 1989

Compact Dodder
Cuscuta compacta
White Morning-glory
Bonamia patens
Skullcap
Scutellaria lateriflora
Scutellaria sp.
Bugle-weed
Lycopus virginicus
Seymeria
Seymeria cassioides
Lousewort
Pedicularis canadensis
Bladderwort
Utricularia sp.
Poor-jo (or Po-jo)
Diodia teres
Richardia
Richardia brasiliensis
Oldenlandica
Oldenlandica uniflora
Possumhaw Viburnum
Viburnum nudum
Lobelia
Lobelia nuttallii
Common Ragweed
Ambrosia artemisiifolia
Thistle
Carduus repandus
Ironweed
Vernonia angustifolia angustifolia
Elephant's-foot
Elephantopus tomentosus
Elephantopus nudatus
Carphephorus
Carphephorus sp.
Joe-pye-weed
Eupatorium fistulosum
Climbing Hempweed
Mikania scandens
Marsh-fleabane
Pluchea rosea
White-topped Aster
Aster tortifolius
Goldenrod
Solidago sp.
Heterotheca
Heterotheca graminifolia
Rosinweed
Silphium compositum
Beggar-ticks
Bidens sp.

FAUNA

Filmy Dome Spider
Prolinyphia marginata
Orchard Spider
Leucauge venusta
Bird Grasshopper
Schistocerca sp.
Common Eastern Velvet-ant
Dasymutilla occidentalis occidentalis
Cypress Flower Gall (midge)
Itonida anthici
Cypress Leaf Gall (midge)
Itonida taxodii
Blue Dasher
Pachydiplax longipennis
Black-tipped Damselfly
Calopteryx aequabile
Crawling Water Beetle
Peltodytes sp.
Predacious Diving Beetle
Hydroporus vittatipennis
Coptotomus interrogatus
Whirligig Beetle
Dineutus discolor
Water Scavenger Beetle
Enochrus sp.
Dung Beetle
Phanaeus sp.
June Beetle
Phyllophaga ilicis
Marsh Beetle
Cyphon variabilis
Tick Trefoil Leaf-miner
Pachyschelus laevigatus
Darkling Beetle
Uloma punctulata
Sap-feeding Beetle
Epuraea sp.
Pallodes sp.
Minute Seed Weevil
Perigaster sp.
Eastern Tiger Swallowtail
Papilio glaucus
Spicebush Swallowtail
Papilio troilus
Palamedes Swallowtail
Papilio palamedes
Cloudless Sulphur
Phoebis sennae eubule
Sleepy Orange (butterfly)
Eurema nicippe
Pearl Crescent
Phyciodes tharos
Red Admiral
Vanessa atalanta rubria
Common Buckeye (butterfly)
Junonia coenia
Red-spotted Purple (butterfly)
Limenitis arthemis astyanax
Monarch (butterfly)
Danaus plexippus
Silver-spotted Skipper
Epargyreus clarus
Common Checkered-skipper
Pyrgus communis
Great Leopard Moth (larva)
Ecpantheria scribonia
Southern Toad
Bufo terrestris
Green Treefrog
Hyla cinerea
Green Frog; Bronze Frog
Rana clamitans
Eastern Fence Lizard
Sceloporus undulatus
Six-lined Racerunner
Cnemidophorus sexlineatus
Rough Green Snake
Opheodrys aestivus
Pileated Woodpecker
Dryocopus pileatus
American Crow
Corvus brachyrhynchos
Fish Crow
Corvus ossifragus
Carolina Chickadee
Parus carolinensis
White-eyed Vireo
Vireo griseus
Pine Warbler
Dendroica pinus

Dorchester County

Giant Portland Cement Quarry
Dorchester County
March 26, 1986

The Coastal Plain of South Carolina is underlain with thick deposits of limestone, the result of eons of sea-bed deposits. In this area, the bodies, shells, or teeth of many marine animals became fossils. As this material is mined for the making of limestone and cement products, these fossils are exposed. When arrangements can be made to collect without interfering with the operation of the quarry, it is possible to remove some very interesting specimens. Many of the best have become exhibits in the South Carolina State Museum. These offer a glimpse of the creatures you might have met in an ocean-going voyage 65 million years ago. In addition, the bodies of some land-dwelling animals were washed out to sea, and joined the marine animals in these deposits. *Archaeocete* whales are often the largest animal specimens found, but mastodon, mammoth, camel and horse are also found occasionally. Shark teeth that were shed and became fossilized are an impressive find. SCAN appreciates Giant Portland Cement granting us permission to hunt for these clues to the South Carolina past.

Privately owned. Advanced permission needed to visit this site.

Directions to Giant Portland Cement Quarry: From the intersection of I-26 and I-95, drive east toward Charleston on I-26. At Exit 177, take SC 453 north toward Holly Hill. At the very first road, turn right (east) onto County Road S 18-50. Drive approximately 0.6 mile and turn left on a dirt road just before a pond. The cement quarry is on the left.

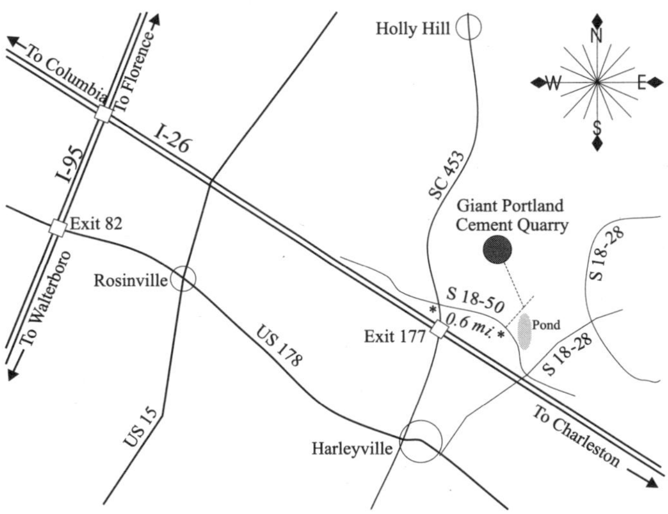

FLORA

Black Willow
Salix nigra
Poor-man's-pepper
Lepidium virginicum
Redbud; Judas-tree
Cercis canadensis
Rattlebox
Crotalaria spectabilis
Kudzu
Pueraria lobata
Red Maple
Acer rubrum
Spiny-leaved Sow-thistle
Sonchus asper
Sea-myrtle; Groundsel-tree
Baccharis halimifolia

FAUNA

Southern Mole Cricket
Scapteriscus acletus
Carolina Anole
Anolis carolinensis
Black Racer
Coluber constrictor
Great Egret
Casmerodius albus
Turkey Vulture
Cathartes aura
Red-shouldered Hawk
Buteo lineatus
Red-tailed Hawk
Buteo jamaicensis
Fish Crow
Corvus ossifragus

FOSSILS

Gastropod
Fusinus abruptus
Sand-dollar
Periarchus lyelli
Pecten
Chlamys cocoana
Pecten
Chlamys sp.
Billfish (bill fragment)
Cylindracanthus rectus
Sawfish (rostral tooth)
Pristis sp.
Stingray (barb, dental plate)
Myliobatis sp.
Sand Shark (tooth)
Odontaspis hopei
Bull Shark (tooth)
Carcharhinus leucas
Eared White Shark (tooth)
Carcharodon auriculatus
Cow Shark (tooth)
Hexanchus sp.

(Photo by L. Greenlee)
Danny Smith and Betty Gilbert study the past.

McAlhany Nature Preserve
Dorchester County
May 27, 1989

This nature preserve was still being inventoried in 1989. There was a rewarding trail alongside an oxbow lake that yielded many interesting sightings. Dr. Julian Harrison, Professor of Biology at the College of Charleston, acted as guide for the day. He showed us a drift fence with buckets at each end set up to catch insects and small reptiles for study. All the dragonflies stayed just out of reach, so none could be identified with enough certainty to be included on the flora/fauna list.

Owned by the Charleston Natural Historical Society, PO Box 504, Charleston, SC 29402. Advanced permission needed to visit this site. Phone: 843-577-7100.

Directions to McAlhany Nature Preserve: From I-95 between I-26 and Walterboro, take Exit 77, Branchville-St. George, and go west on US 78 to Reevesville, about 3 miles. Turn south (left) onto County Road S 18-16, go 0.2 miles and turn onto the first road on the right, S 18-177. In 2.9 miles, this road will merge with S 18-79; continue south another 2.4 miles. Turn right (west) onto S 18-19, cross the bridge over Cattle Creek, and watch for a locked gate on the left not too far beyond the bridge.

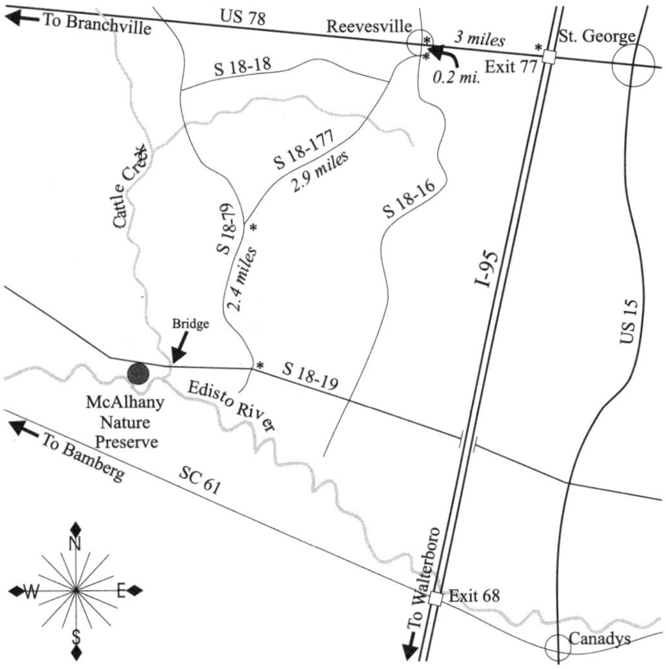

McAlhany Nature Preserve (continued)
May 27, 1989

FLORA

Lichen
Cladonia sp.
Bracken
Pteridium aquilinum
Ebony Spleenwort
Asplenium platyneuron
Mosquito Fern
Azolla caroliniana
Spruce Pine
Pinus glabra
Common Cat-tail
Typha latifolia
Oat Grass
Danthonia sp.
Nut-rush
Scleria sp.
Giant Duckweed
Spirodela punctata
Bogmat
Wolffiella floridana
Spanish-moss
Tillandsia usneoides
Roseling; Spiderwort
Tradescantia rosea
Fringed Greenbrier
Smilax bona-nox
Yellow Star-grass
Hypoxis hirsuta
Willow
Salix sp.
Wax-myrtle; Southern Bayberry
Myrica cerifera
River Birch
Betula nigra
Beech
Fagus grandifolia
Swamp Chestnut Oak
Quercus michauxii
Water Oak
Quercus nigra
Running Oak
Quercus pumila
Laurel Oak
Quercus laurifolia
Heartleaf
Hexastylis arifolia
Alligator-weed
Alternanthera philoxeroides
Spatterdock; Yellow Pond-lily
Nuphar luteum
Redbay
Persea borbonia
Sassafras
Sassafras albidum
Sweet-gum
Liquidambar styraciflua
Witch-hazel
Hamamelis virginiana
Hawthorn
Crataegus sp.
Shadbush
Amelanchier obovalis
Tread-softly
Cnidoscolus stimulosus
Sebastian-bush
Sebastiania ligustrina
Poison-ivy
Rhus radicans
Winged Sumac
Rhus copallina
American Holly
Ilex opaca
Red Buckeye
Aesculus pavia
Muscadine; Bullace
Vitis rotundifolia
Summer Grape
Vitis aestivalis
St.-Andrew's-cross
Hypericum hypericoides
Prickly-pear
Opuntia compressa
Blackgum
Nyssa sylvatica
Water Tupelo
Nyssa aquatica
Flowering Dogwood
Cornus florida
Spotted Wintergreen
Chimaphila maculata
Indian-pipe
Monotropa uniflora
Coastal Honeybells
Leucothoe racemosa
Dangleberry
Gaylussacia frondosa
Sparkleberry
Vaccinium arboreum
Elliott's Blueberry
Vaccinium elliottii
Horse-sugar; Sweetleaf
Symplocos tinctoria
Wild Olive
Osmanthus americana
Yellow Jessamine
Gelsemium sempervirens
White Milkweed
Asclepias variegata
Downy False Foxglove
Aureolaria virginica
Cross Vine
Anisostichus capreolata
Trumpet Vine
Campsis radicans
Squaw-root
Conopholis americana
Floating Bladderwort
Utricularia inflata
Partridge-berry
Mitchella repens
Small-leaf Viburnum
Viburnum obovatum
American Elderberry
Sambucus canadensis
Elephant's-foot
Elephantopus carolinianus
Heterotheca
Heterotheca graminifolia

FAUNA

Crayfish
Cambarus sp.
Carpenter Bee
Xylocopa sp.
Tiger Beetle
Cicindela unipunctata
Searcher; Ground Beetle
Carabus vinctus
Pedunculate Ground Beetle
Pasimachus sublaevis
Woodland Ground Beetle
Pterostichus sp.
Colorful Foliage Ground Beetle
Cymindis americana
Crawling Water Beetle
Peltodytes sexmaculatus
Predacious Diving Beetle
Hydroporus carolinus
Hydroporus clypealis
Hydroporus lobatus
Agabetes acuductus
Coptotomus interrogatus
Water Scavenger Beetle
Tropisternus mixtus
Tropisternus collaris striolatus
Giant Water Scavenger Beetle
Dibolocelus ovalis
Passalus; Bess-beetle
Popilius disjunctus
June Beetle
Phyllophaga sp.
Eyed Elater
Alaus oculatus
Soft-winged Flower Beetle
Ablechrus sp.
Sap-feeding Beetle
Carpophilus floralis
Flea Beetle
Dibolia sp.
Pearl Crescent
Phyciodes tharos
Mourning Cloak
Nymphalis antiopa
American Lady (butterfly)
Vanessa virginiensis
Common Buckeye (butterfly)
Junonia coenia
Green Frog; Bronze Frog
Rana clamitans
Yellow-bellied Slider
Chrysemys scripta
Ground Skink
Scincella lateralis
Six-lined Racerunner
Cnemidophorus sexlineatus
Corn Snake
Elaphe guttata
Acadian Flycatcher
Empidonax virescens
American Crow
Corvus brachyrhynchos
Tufted Titmouse
Parus bicolor
Carolina Wren
Thryothorus ludovicianus
Red-Eyed Vireo
Vireo olivaceus
Prothonotary Warbler
Protonotaria citrea
Pine Warbler
Dendroica pinus

Fairfield County

Broad River Waterfowl Area
Fairfield County
March 26, 1994

This area along the Broad River is managed for waterfowl, with shallow ponds and open grassy fields, including some planted in grains. This attracts not only waterfowl of the edible kind, but song birds, reptiles and amphibians, insects and other wildlife. It makes a very interesting place for a naturalist during nonhunting periods of the year. Much of the site is floodplain, but there is the first terrace away from the river as well. This leads to many types of habitats providing a wealth of species to be seen.

Directions to Broad River Waterfowl Area: From I-26 near Newberry, take exit 74 in Newberry County and head east on SC 34. In 12 miles, SC 34 crosses the Broad River; immediately after crossing the bridge, turn right onto Strother Road. This is a gravel road, and there is a small sign for the Broad River Waterfowl Area at the junction. At 0.8 mile, you will ford a small stream. At 1.2 miles from SC 34, turn right onto the main entrance road for the waterfowl area. If the gate is open, drive approximately 0.5 mile and park near the equipment sheds on the right; if the gate is closed, park by the gate and walk.

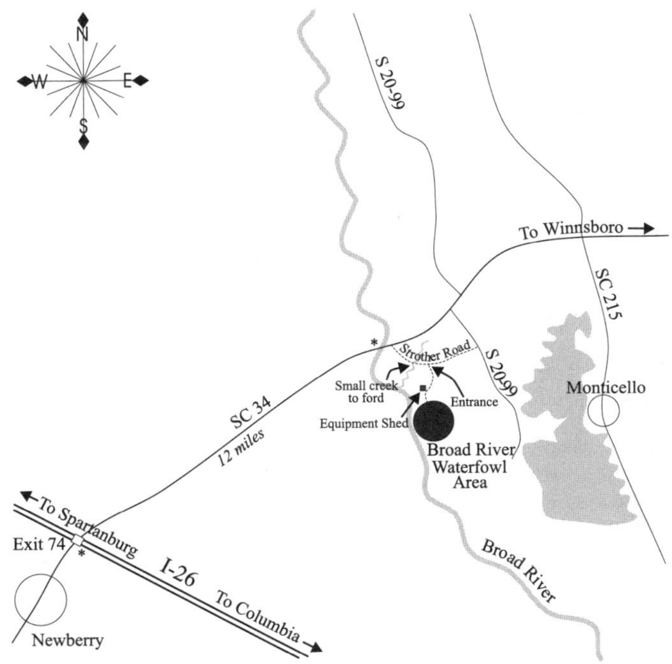

FLORA

Lichen
- *Collema furfuraceum*
- *Collema subflaccidum*
- *Leptogium cyanescens*
- *Graphis scripta*
- *Bacidia schweinitzii*
- *Lecidea carneoalbens*
- *Lecidea sp.*
- *Cladonia chlorophaea*
- *Cladonia coniocraea*
- *Cladonia cristatella*
- *Cladonia didyma*
- *Cladonia polycarpoides*
- *Cladonia peziziformis*
- *Pycnothelia papillaria*
- *Pertusaria leucostoma*
- *Pertusaria pustulata*
- *Pertusaria texana*
- *Maronea constans*
- *Lecanora allophana argentata*
- *Lecanora cupressi*
- *Lecanora pallida*
- *Lecanora varia*
- *Lecanora hybocarpa*
- *Ochrolechia africana*
- *Candelaria concolor*
- *Hypotrachyna livida*
- *Parmelina aurulenta*
- *Parmotrema dilitatum*
- *Parmotrema hypotropum*
- *Parmotrema michauxianum*
- *Parmotrema perforatum*
- *Parmotrema robustum*
- *Xanthoparmelia conspersa*
- *Flavoparmelia caperata*
- *Canoparmelia caroliniana*
- *Punctelia rudecta*
- *Imshaugia placorodia*
- *Rimelia reticulata*
- *Rimelia cetrata*
- *Rimelia cetrata f. nonreticulata*
- *Ramalina americana*
- *Usnea strigosa*
- *Buellia punctata*
- *Buellia curatellae*
- *Physcia aipolia*
- *Physcia millegrana*
- *Heterodermia obscurata*
- *Pyxine sorediata*
- *Trypethelium virens*

Meadow Spikemoss
Selaginella apoda

Loblolly Pine
Pinus taeda

Eastern Red-cedar
Juniperus virginiana

Common Cat-tail
Typha latifolia

Cane; Switch Cane
Arundinaria gigantea

Wool-grass
Scirpus cyperinus

Woodrush
Luzula sp.

Sawbrier; Greenbrier
Smilax glauca

Black Willow
Salix nigra

Eastern Cottonwood
Populus deltoides

Tag Alder
Alnus serrulata

River Birch
Betula nigra

Hop-hornbeam
Ostrya virginiana

American Hornbeam; Ironwood
Carpinus caroliniana

Beech
Fagus grandifolia

Post Oak
Quercus stellata

Cherrybark Oak
Quercus falcata pagodaefolia

Water Oak
Quercus nigra

Willow Oak
Quercus phellos

Winged Elm
Ulmus alata

Sugarberry; Hackberry
Celtis laevigata

Mouse-ear Chickweed
Cerastium glomeratum

Leather-flower
Clematis crispa

Broad River Waterfowl Area (continued)
March 26, 1994

Windflower
Thalictrum thalictroides
Small-flowered Crowfoot
Ranunculus abortivus
Mouse-ear Cress
Arabidopsis thaliana
Bitter Cress
Cardamine pensylvanica
Sweet-gum
Liquidambar styraciflua
American Sycamore
Platanus occidentalis
Wild Strawberry
Fragaria virginiana
Five-fingers
Potentilla canadensis
Rose
Rosa sp.
Hawthorn
Crataegus flabellata
Chickasaw Plum
Prunus angustifolia
Black Cherry; Wild Cherry
Prunus serotina
Redbud; Judas-tree
Cercis canadensis
Honey Locust
Gleditsia triacanthos
Vetch
Vicia sp.
Wood-sorrel
Oxalis sp.
American Holly
Ilex opaca
Holly
Ilex ambigua
Boxelder
Acer negundo
Red Maple
Acer rubrum
Painted Buckeye
Aesculus sylvatica
St.-Andrew's-cross
Hypericum hypericoides
Common Blue Violet
Viola papilionacea
Flowering Dogwood
Cornus florida
Azalea
Rhododendron sp.
Elliott's Blueberry
Vaccinium elliottii
Ash
Fraxinus sp.
Yellow Jessamine
Gelsemium sempervirens
Forget-me-not
Myosotis verna
Beautyberry
Callicarpa americana
Small Bluets
Houstonia pusilla
American Elderberry
Sambucus canadensis
Purple Jelly-drops
Ascocoryne sarcoides
Cedar Apple Rust
Gymnosporangium junipervirginianae

FAUNA

Rabid Wolf Spider
Lycosa rabida
Chinese Mantid (egg)
Tenodera aridifolia sinensis
Plant-feeding Stink Bug
Euschistus servus servus
Red Imported Fire Ant
Solenopsis invicta
Pipe-organ Wasp
Trypoxylon politum
Goldenrod Ball Gall (fly)
Eurosta solidaginis
Common Green Darner
Anax junius
Baskettail Dragonfly
Epitheca sp.
Tiger Beetle
Cicindela sp.
Searcher; Ground Beetle
Carabus vinctus
Seed-eating Ground Beetle
Amara sp.
False Bombadier Beetle
Galeritula bicolor
Feather-winged Beetle
Nossidium sp.
Click Beetle
Lacon marmoreus
Limonius sp.
Soldier Beetle
Cantharis bilineatus
Cantharis bilineatus
Darkling Beetle
Uloma punctulata
Broad-bodied Leaf Beetle
Prasocuris vittata
Zebra Swallowtail
Eurytides marcellus
Eastern Tiger Swallowtail
Papilio glaucus
Falcate Orangetip
Anthocharis midea
Spring Azure (butterfly)
Celastrina ladon
Pearl Crescent
Phyciodes tharos
Juvenal's Duskywing
Erynnis juvenalis
Evergreen Bagworm Moth
Thyridopteryx ephemeraeformis
Spring Peeper
Hyla crucifer
Upland Chorus Frog (call)
Pseudacris triseriata
Painted Turtle
Chrysemys picta
Great Blue Heron
Ardea herodias
Blue-winged Teal
Anas discors
Wood Duck
Aix sponsa
Hooded Merganser
Lophodytes cucullatus
Sora
Porzana carolina
Killdeer
Charadrius vociferus
Lesser Yellowlegs
Tringa flavipes
Wilson's Phalarope
Phalaropus tricolor
Common Snipe
Gallinago gallinago
Turkey Vulture
Cathartes aura
Black Vulture
Coragyps atratus
Red-tailed Hawk
Buteo jamaicensis
Mourning Dove
Zenaida macroura
Belted Kingfisher
Ceryle alcyon
Red-bellied Woodpecker
Melanerpes carolinus
Pileated Woodpecker
Dryocopus pileatus
Eastern Phoebe
Sayornis phoebe
Tree Swallow
Tachycineta bicolor
American Crow
Corvus brachyrhynchos
Tufted Titmouse
Parus bicolor
Carolina Chickadee
Parus carolinensis
Carolina Wren
Thryothorus ludovicianus
Ruby-crowned Kinglet
Regulus calendula
Blue-gray Gnatcatcher
Polioptila caerulea
White-eyed Vireo
Vireo griseus
Northern Parula
Parula americana
Black-and-white Warbler
Mniotilta varia
Yellow-rumped Warbler
Dendroica coronata
Yellow-throated Warbler
Dendroica dominica
Pine Warbler
Dendroica pinus
Yellowthroat
Geothlypis trichas
Northern Cardinal
Cardinalis cardinalis
Rufous-sided Towhee
Pipilo erythrophthalmus
Chipping Sparrow
Spizella passerina
White-throated Sparrow
Zonotrichia albicollis
White-crowned Sparrow
Zonotrichia leucophrys
Swamp Sparrow
Melospiza georgiana
Red-winged Blackbird
Agelaius phoeniceus
Brown-headed Cowbird
Molothrus ater
Common Grackle
Quiscalus quiscula
American Goldfinch
Carduelis tristis
Muskrat
Ondatra zibethicus
White-tailed Deer (tracks)
Odocoileus virginianus

(Photo by V. Winn)

Sam Outen, Mary Boni, Evelyn Dabbs, Mina Winn, Mary Garland Douglass, and Charlie Williams enjoy a day of fellowship.

Florence County

Burches Ferry
Florence County
November 22, 1986

Burches Ferry, a high bank exposed by the eroding action of the Greater Pee Dee River, is named for the ferry that used to cross the river from this location. It yields fossils of the Upper Cretaceous age that were deposited in a shallow sea that covered a portion of South Carolina during the later part of the dinosaur age. Although no dinosaur fossils have been found, it is possible to find the internal skeletons of small squid-like animals called belemnites and teeth of Cretaceous sharks. An oyster found here, *Exogyra costata*, serves as an index fossil and identifies the material as Upper Cretaceous in age. A fossil vertebra found at this site by Rudy Mancke several years earlier than our visit turned out to be from a marine reptile known as a Plesiosaur. This animal swam while dinosaurs were living on land nearby. Collecting is done along a small ledge over the edge of the water. It is not for the faint of heart, and not accessible during high water.

Directions to Burches Ferry: From Sumter, take US 378 east through Turbeville and Lake City to Kingsburg, about 48 miles. In Kingsburg, turn left (north) on S 21-57 toward Claussen, and go 13 miles to Mill Branch Church on the left. Burches Ferry is reached by a dirt road a little north of the church on the right.

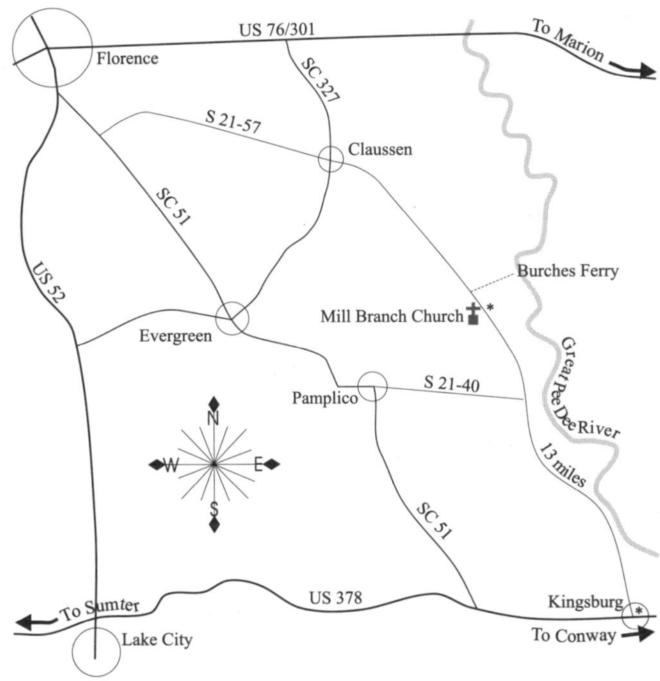

FLORA

Christmas Fern
Polystichum acrostichoides
Netted Chain Fern
Woodwardia areolata
Ebony Spleenwort
Asplenium platyneuron
Short-leaf Pine
Pinus echinata
Bald-cypress
Taxodium distichum
River Oats
Uniola latifolia
Spanish-moss
Tillandsia usneoides
Water Hickory
Carya aquatica
Tag Alder
Alnus serrulata
River Birch
Betula nigra
White Oak
Quercus alba
Overcup Oak
Quercus lyrata
Spanish Oak
Quercus falcata
Cherrybark Oak
Quercus falcata pagodaefolia
Water Oak
Quercus nigra
Sugarberry; Hackberry
Celtis laevigata
Sassafras
Sassafras albidum
Sweet-gum
Liquidambar styraciflua
American Sycamore
Platanus occidentalis
Hawthorn
Crataegus sp.
Black Cherry; Wild Cherry
Prunus serotina
Water Locust
Gleditsia aquatica
Possumhaw Holly
Ilex decidua
Pepper-vine
Ampelopsis arborea
Flowering Dogwood
Cornus florida
Sparkleberry
Vaccinium arboreum
Horse-sugar; Sweetleaf
Symplocos tinctoria
Yellow Jessamine
Gelsemium sempervirens
Beautyberry
Callicarpa americana
Barometer Earthstar
Astraeus hygrometricus

(Photo by B. Gilbert)
Lynn Greenlee busily hunts for fossils.

Burches Ferry (continued)
November 22, 1986

FAUNA

Sleepy Orange (butterfly)
Eurema nicippe
Question Mark (butterfly)
Polygonia interrogationis
Common Buckeye (butterfly)
Junonia coenia
Eastern Tent Caterpillar
Malacosoma americanum
Great Leopard Moth (larva)
Ecpantheria scribonia
Carolina Anole
Anolis carolinensis
Turkey Vulture
Cathartes aura
Red-tailed Hawk
Buteo jamaicensis
Belted Kingfisher
Ceryle alcyon
White-tailed Deer
Odocoileus virginianus

FOSSILS

Belemnite
Belemnitella americana
Marine Snail
Turritella sp.
Jingle Shell
Anomia sp.
Exogyra Snail
Exogyra costata
Fish Tooth
Enchodus ferox
Soft-shelled Turtle (fragment)
Trionyx sp.

(Photo by L. Smith)
Sleepy orange, *Eurema nicippe*.

Lynches River State Park
Florence County
November 19, 1994

Situated along the banks of the Lynches River, this state park has the classic blackwater riverine habitat, with the swamps and bluffs, plus some very fine examples of sandhills. Fall was at its peak this late in November, creating beautiful fall colors on the trees, and a lot of interesting insects attracting our attention. The diversity of species here does not include anything particularly rare, but offers a good range of flora and fauna nonetheless.

Directions to Lynches River State Park: From Florence, take US 52 south. Go through Effingham, crossing US 301 and the Lynches River (about 17-18 miles). Make the first right turn after the river onto County Road S 21-147 at the sign to Lynches River State Park. The entrance to the park is on the right.

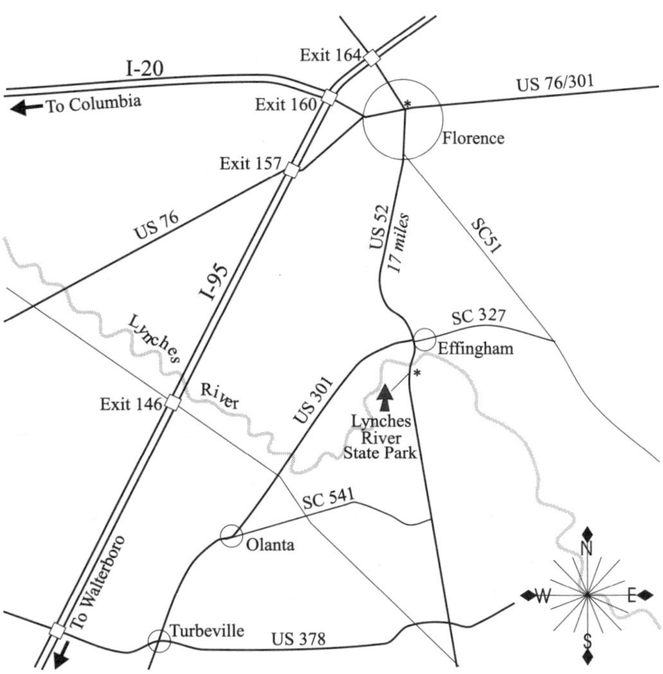

FLORA

Lichen
Collema furfuraceum
Leptogium cyanescens
Graphis scripta
Graphis atrorubens
Graphis elegans
Glyphis cicatricosa
Phaeographis dendritica
Bacidia polychroa
Bacidia schweinitzii
Bacidia incompta
Leptogium austroamericanum
Brigantiaea leucoxantha
Lecidea russula
Letrouitia vulpinum
Cladina evansii
Cladonia chlorophaea
Cladonia coniocraea
Cladonia cristatella
Cladonia didyma
Cladonia leporina
Cladonia macilenta
Cladonia polycarpoides
Cladonia peziziformis
Cladonia rappii
Pertusaria ostiolata
Pertusaria paratuberculifera
Pertusaria subpertusa
Pertusaria texana
Pertusaria multipunctoides
Maronea constans
Haematomma elatinum
Lecanora cupressi
Lecanora pulicaris
Ochrolechia africana
Phlyctis sp.
Bulbothrix confoederata
Bulbothrix goebelii

Hypotrachyna livida
Parmelina aurulenta
Parmelina minarum
Parmotrema perforatum
Parmotrema praesorediosum
Parmotrema rampoddense
Parmotrema tinctorum
Parmotrema ultralucens
Canoparmelia caroliniana
Punctelia rudecta
Rimelia reticulata
Rimelia cetrata
Ramalina americana
Usnea rubicunda
Usnea strigosa
Anaptychia palmatula
Buellia curatellae
Physcia americana
Heterodermia albicans
Heterodermia obscurata
Pyxine sorediata
Phaeophyscia imbricata
Anthracothecium pyrenuloides
Anthracothecium caroliniana
Arthonia quintaria
Trypethelium mastoideum
Trypethelium virens
Trypethelium scorites
Porina chlorotica
Porina nucula
Haircap Moss
Polytrichum commune
Cinnamon Fern
Osmunda cinnamomea
Royal Fern
Osmunda regalis spectabilis
Bracken
Pteridium aquilinum
Netted Chain Fern
Woodwardia areolata
Resurrection Fern
Polypodium polypodioides
Long-leaf Pine
Pinus palustris
Loblolly Pine
Pinus taeda
Short-leaf Pine
Pinus echinata
Bald-cypress
Taxodium distichum
Pond-cypress
Taxodium ascendens
Sandbur; Sandspurs
Cenchrus sp.
Jack-in-the-pulpit
Arisaema triphyllum
Spanish-moss
Tillandsia usneoides
Greenbrier; Bullbrier
Smilax rotundifolia
Fringed Greenbrier
Smilax bona-nox
Sawbrier; Greenbrier
Smilax glauca
Sweet-scented Smilax
Smilax smallii
Bear-grass
Yucca filamentosa
Spoonleaf Yucca
Yucca filamentosa filamentosa
Crane-fly Orchid
Tipularia discolor
Black Willow
Salix nigra
Wax-myrtle; Southern Bayberry
Myrica cerifera
Wax-myrtle
Myrica cerifera cerifera
Mockernut Hickory
Carya tomentosa
Hickory
Carya sp.
River Birch
Betula nigra
American Hornbeam; Ironwood
Carpinus caroliniana
Scrubby Post Oak
Quercus margaretta
Overcup Oak
Quercus lyrata
Swamp Chestnut Oak
Quercus michauxii
Turkey Oak
Quercus laevis
Scarlet Oak
Quercus coccinea
Spanish Oak
Quercus falcata
Water Oak
Quercus nigra
Willow Oak
Quercus phellos
Laurel Oak
Quercus laurifolia
American Elm
Ulmus americana
Sugarberry; Hackberry
Celtis laevigata
Red Mulberry
Morus rubra
Mistletoe
Phoradendron serotinum
Jointweed
Polygonella polygama
Wire-plant
Stipulicida setacea
Redbay
Persea borbonia
Sassafras
Sassafras albidum
Sweet-gum
Liquidambar styraciflua
Witch-hazel
Hamamelis virginiana
American Sycamore
Platanus occidentalis
Five-fingers
Potentilla canadensis
Blackberry; Brambles
Rubus sp.
Red Chokeberry
Sorbus arbutifolia
Swamp Haw
Crataegus viridis
Parsley-leaved Hawthorn
Crataegus marshallii
Hawthorn
Crataegus sp.
Wild Plum
Prunus americana
American Wisteria
Wisteria frutescens
Button-rosy
Polygala curtissii
Poison-ivy
Rhus radicans
False Poison Sumac
Rhus michauxii
Winged Sumac
Rhus copallina
Titi
Cyrilla racemiflora
American Holly
Ilex opaca
Possumhaw Holly
Ilex decidua
Inkberry
Ilex glabra
Hearts-a'burstin'-with-love
Euonymus americanus
Red Maple
Acer rubrum
Muscadine; Bullace
Vitis rotundifolia
Fox Grape
Vitis labrusca
St.-Andrew's-cross
Hypericum hypericoides
Prickly-pear
Opuntia compressa
Pennywort
Hydrocotyle sp.
Black Tupelo; Blackgum
Nyssa sylvatica sylvatica
Swamp Tupelo; Swamp Blackgum
Nyssa sylvatica biflora
Flowering Dogwood
Cornus florida
Swamp Dogwood; Stiff Cornel
Cornus stricta
Sweet Pepperbush
Clethra alnifolia
Spotted Wintergreen
Chimaphila maculata
Azalea
Rhododendron sp.
Fetterbush
Lyonia lucida
Sparkleberry
Vaccinium arboreum
Elliott's Blueberry
Vaccinium elliottii
Blueberry
Vaccinium sp.
Horse-sugar; Sweetleaf
Symplocos tinctoria
Fringe-tree
Chionanthus virginicus
Chinese Privet
Ligustrum sinense
Yellow Jessamine
Gelsemium sempervirens
Gentian
Gentiana sp.
Cross Vine
Anisostichus capreolata
Trumpet Vine
Campsis radicans
Partridge-berry
Mitchella repens
Japanese Honeysuckle
Lonicera japonica
Coral Honeysuckle
Lonicera sempervirens
Small-leaf Viburnum
Viburnum obovatum
Arrowwood
Viburnum dentatum
Viburnum dentatum lucidum
Viburnum
Viburnum sp.
American Elderberry
Sambucus canadensis
Ironweed
Vernonia sp.
Elephant's-foot
Elephantopus sp.
Frost Aster
Aster pilosus
Goldenrod
Solidago sp.
Heterotheca
Heterotheca graminifolia
Heterotheca mariana
Green-headed Jelly-club
Leotia viscosa
Black-knot
Apiosporina morbosa
Bearded Tooth
Hericium erinaceus
Rooting Polypore
Polyporus radicatus
Barometer Earthstar
Astraeus hygrometricus

FAUNA

American Bird Grasshopper
Schistocerca americana
Thread-legged Bug
Zelus exsanguis
Trash-bug
Hemerobiidae Family
Florida Harvester Ant
Pogonomyrmex badius
Yellowjacket
Vespula sp.
Mosquito
Psorophora ferox
Fall Woodlands Mosquito
Aedes infirmatus
Predacious Diving Beetle
Hydroporus vittatipennis
Cross-toothed Rove Beetle
Oxyporus sp.

Lynches River State Park (continued)
November 19, 1994

Crab-like Rove Beetles
Lordithon sp.
Hickory Girdler Beetle
Oncideres cingulata
Clouded Sulphur
Colias philodice
Orange Sulphur
Colias eurytheme
Cloudless Sulphur
Phoebis sennae eubule
Sleepy Orange (butterfly)
Eurema nicippe
Common Buckeye (butterfly)
Junonia coenia
Geometer Moth
Heliomata sp.
Carolina Anole
Anolis carolinensis
Eastern Fence Lizard
Sceloporus undulatus
Rat Snake
Elaphe obsoleta
Wood Duck
Aix sponsa
Red-shouldered Hawk
Buteo lineatus
Belted Kingfisher
Ceryle alcyon
Red-bellied Woodpecker
Melanerpes carolinus
Northern Flicker
Colaptes auratus
Downy Woodpecker
Picoides pubescens
Pileated Woodpecker
Dryocopus pileatus
Blue Jay
Cyanocitta cristata
American Crow
Corvus brachyrhynchos
Carolina Chickadee
Parus carolinensis
Brown-headed Nuthatch
Sitta pusilla
Eastern Bluebird
Sialia sialis
American Robin
Turdus migratorius
Northern Mockingbird
Mimus polyglottos
Cedar Waxwing
Bombycilla cedrorum
Pine Warbler
Dendroica pinus
Northern Cardinal
Cardinalis cardinalis
Rufous-sided Towhee
Pipilo erythrophthalmus
Chipping Sparrow
Spizella passerina
Dark-eyed Junco
Junco hyemalis
Gray Squirrel
Sciurus carolinensis
White-tailed Deer (tracks)
Odocoileus virginianus

Georgetown County

Huntington Beach State Park and Brookgreen Gardens

Georgetown County

September 27, 1986, and October 24, 1992

Huntington Beach State Park offers easy access to all the major ocean edge habitats. The saltwater areas attract a vast diversity of birds, and a freshwater pond only adds to the list of species to see. The marsh is accessible from a long boardwalk, or from the dike between it and the freshwater pond. There are areas of maritime forest, although better examples are found across the highway in Brookgreen Gardens. In 1992, SCAN visited both, and the list reflects that.

Directions to Huntington Beach State Park: From Georgetown, take US 17 north toward Myrtle Beach. Bypass the communities of Pawleys Island, Litchfield Beach, and North Litchfield Beach. The entrance to Huntington Beach State Park is on the right (east) side and marked with signs, about 17 miles from Georgetown; the entrance to Brookgreen Gardens is directly across the highway on the left side with the familiar, imposing statue of two fighting horses.

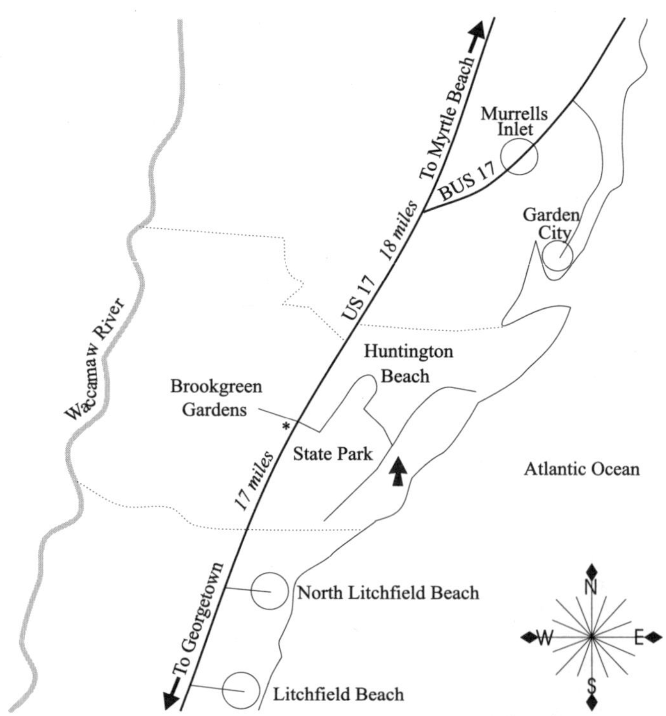

FLORA

Lichen
Leptogium cyanescens
Graphis scripta
Graphis nitida
Graphis grammatis
Sarcographa sp.
Bacidia atrogrisea
Bacidia sp.
Leptogium austroamericanum
Cladina evansii
Cladonia ochrochlora
Cladonia pityrea
Cladonia sp.
Lecanora cupressi
Lecanora pallida
Lecanora pallida cancriformis
Lecanora varia
Lecanora sp.
Ochrolechia africana
Julella sp.
Bulbothrix goebelii
Parmelina aurulenta
Parmotrema hypoleucinum
Parmotrema perforatum
Parmotrema praesorediosum
Parmotrema rigidum
Parmotrema robustum
Parmotrema tinctorum
Punctelia rudecta
Rimelia subisidiosa
Ramalina americana
Usnea strigosa
Buellia caloosensis
Buellia amphidexia
Dirinaria aspera
Dirinaria confusa
Dirinaria picta
Physcia americana
Physcia crispa
Physcia millegrana
Heterodermia obscurata
Pyxine caesiopruinosa f. aurea
Pyxine subcinerea
Pyrenula cruenta
Osmunda cinnamomea
Bracken
Pteridium aquilinum
Mosquito Fern
Azolla caroliniana
Long-leaf Pine
Pinus palustris
Loblolly Pine
Pinus taeda
Bald-cypress
Taxodium distichum
Eastern Red-cedar
Juniperus virginiana
Common Cat-tail
Typha latifolia
Sea Oats
Uniola paniculata
Cord Grass
Spartina patens
Smooth Cord Grass
Spartina alterniflora
Dune Sandbur
Cenchrus tribuloides
Seaside Panic Grass
Panicum amarum
Little Bluestem
Andropogon scoparius
Broomsedge
Andropogon virginicus
White-bracted Sedge
Dichromena colorata
Cabbage Palmetto
Sabal palmetto
Duckweed
Lemna sp.
Bogmat
Wolffiella floridana
Spanish-moss
Tillandsia usneoides
Dayflower
Commelina virginica
Fringed Greenbrier
Smilax bona-nox
Mound-lily Yucca
Yucca gloriosa
Nodding Ladies'-tresses
Spiranthes cernua
Swamp Willow
Salix caroliniana
Wax-myrtle; Southern Bayberry
Myrica cerifera
Hickory
Carya sp.
Coastal Chinquapin
Castanea pumila ashei
Live Oak
Quercus virginiana
Turkey Oak
Quercus laevis
Cherrybark Oak
Quercus falcata pagodaefolia
Water Oak
Quercus nigra
Laurel Oak
Quercus laurifolia
Red Mulberry
Morus rubra
Mexican-tea
Chenopodium ambrosioides
Seabeach Orach
Atriplex arenaria
Perennial Glasswort
Salicornia virginica
Sea-blight
Suaeda linearis
Russian-thistle
Salsola kali
Beach Amaranth
Amaranthus pumilus
Sea-purslane
Sesuvium portulacastrum
Paronychia
Paronychia riparia
Dwarf PawPaw
Asimina parviflora
Redbay
Persea borbonia
Sassafras
Sassafras albidum
Sea-rocket
Cakile harperi
Sweet-gum
Liquidambar styraciflua
Dewberry
Rubus trivialis
Redbud; Judas-tree
Cercis canadensis
Lupine
Lupinus sp.
Zornia
Zornia bracteata
Bush-clover
Lespedeza sp.
Chinese Wisteria
Wisteria sinensis
Coral-bean
Erythrina herbacea
Dollar-weed
Rhynchosia reniformis
Beach-pea; Wild Bean
Strophostyles helvola
Hercules'-club
Zanthoxylum clava-herculis
Silverleaf Croton
Croton punctatus
Spurge
Euphorbia ammannioides
Poison-oak
Rhus toxicodendron
Winged Sumac
Rhus copallina
American Holly
Ilex opaca
Myrtle-leaved Holly
Ilex cassine myrtifolia
Yaupon
Ilex vomitoria
Red Maple
Acer rubrum
Red Buckeye
Aesculus pavia
Virginia-creeper
Parthenocissus quinquefolia
Muscadine; Bullace
Vitis rotundifolia
Summer Grape
Vitis aestivalis
Pepper-vine
Ampelopsis arborea
Hypericum
Hypericum sp.
Passionflower; Maypops
Passiflora incarnata
Cutleaf Evening-primrose
Oenothera laciniata
Gaura
Gaura angustifolia
Dune Pennywort
Hydrocotyle bonariensis
Pennywort
Hydrocotyle sp.
Black Tupelo; Blackgum
Nyssa sylvatica sylvatica
Swamp Tupelo; Swamp Blackgum
Nyssa sylvatica biflora
Flowering Dogwood
Cornus florida
Spotted Wintergreen
Chimaphila maculata
Indian-pipe
Monotropa uniflora
Fetterbush
Lyonia lucida
Sparkleberry
Vaccinium arboreum
Sea-lavender
Limonium sp.
Persimmon
Diospyros virginiana
Common Marsh-pink
Sabatia stellaris
Dodder
Cuscuta sp.
Ground-ivy; Ponyfoot
Dichondra carolinensis
Fiddle-leaf Morning-glory
Ipomoea stolonifera
Brazilian Vervain
Verbena braziliensis
Fog-fruit
Lippia nodiflora
Beautyberry
Callicarpa americana
Blue-curls
Trichostema dichotomum
Bitter Mint; Hyssop
Hyptis alata
Nightshade
Solanum americanum
Smooth Water-hyssop
Bacopa monnieri
Water-hyssop
Bacopa sp.
Gerardia
Agalinis sp.
Cross Vine
Anisostichus capreolata
Trumpet Vine
Campsis radicans
Partridge-berry
Mitchella repens
Bedstraw
Galium sp.
Beach-elder
Iva imbricata
Common Ragweed
Ambrosia artemisiifolia
Elephant's-foot
Elephantopus tomentosus
Dog-fennel
Eupatorium capillifolium
Eupatorium compositifolium
Groundsel-tree
Baccharis glomeruliflora
Sea-myrtle; Groundsel-tree
Baccharis halimifolia
Horseweed; Mare's-tail
Erigeron canadensis canadensis
Aster
Aster undulatus

Huntington Beach State Park and Brookgreen Gardens (continued) September 27, 1986, and October 24, 1992

Salt-marsh Aster
Aster tenuifolius
Frost Aster
Aster pilosus
Salt-marsh Goldenrod
Solidago sempervirens
Heterotheca
Heterotheca graminifolia
Heterotheca nervosa
Camphorweed
Heterotheca subaxillaris
Sea Ox-eye
Borrichia frutescens
Silvery-violet Cort
Cortinarius alboviolaceus
Puffball
Lycoperdon sp.
Tapioca Slime
Brefeldia maxima

FAUNA

Sea-whip
Leptogorgia virgulata
Sea-pork (tunicate)
Aplidium constellatum
Marsh Periwinkle; Mud Snail
Littorina irrorata
Common Slipper (snail)
Crepidula fornicata
Eastern Mud Snail
Ilyanassa obsoleta
Knobbed Whelk
Busycon carica
Lettered Olive (snail)
Oliva sayana
White-lipped Forest Snail
Mesodon thyroidus
Squid
Loligo sp.
Turkey-wing (shell)
Arca zebra
Eastern Oyster
Crassostrea virginica
Cross-hatched Lucine
Divaricella quadrisulcata
Atlantic Razor Clam
Ensis directus
Florida Coquina
Donax variabilis
False Angel-wing (shell)
Petricola pholadiformis
Common Blue Crab
Callinectes sapidus
Blue Crab
Callinectes sp.
Mud Fiddler Crab
Uca pugnax
Sand Fiddler Crab
Uca pugilator
Fiddler Crab
Uca sp.

Wolf Spider
Lycosa sp.
Lubber Grasshopper
Brachystola magna
Field Cricket
Gryllus sp.
Antlion; Doodlebug
Myrmeleon immaculatum
Fire Ant
Solenopsis sp.
Bumblebee
Bombus sp.
Honey Bee
Apis mellifera
Golden Salt-marsh Mosquito
Aedes sollicitans
Common Green Darner
Anax junius
Eastern Pondhawk
Erythemis simplicicollis
Blue Dasher
Pachydiplax longipennis
Glider
Tramea sp.
Common Tumblebug
Canthon laevis
Rhinoceros Beetle
Strategus antaeus
Lady Beetle
Coccinella sp.
Giant Swallowtail
Papilio cresphontes
Eastern Tiger Swallowtail
Papilio glaucus
Palamedes Swallowtail
Papilio palamedes
Clouded Sulphur
Colias philodice
Cloudless Sulphur
Phoebis sennae eubule
Little Yellow (butterfly)
Eurema lisa
Gray Hairstreak
Strymon melinus
Gulf Fritillary
Agraulis vanillae
Pearl Crescent
Phyciodes tharos
Common Buckeye (butterfly)
Junonia coenia
Viceroy; Mimic
Limenitis archippus
Monarch (butterfly)
Danaus plexippus
Long-tailed Skipper
Urbanus proteus
Largemouth Bass
Micropterus salmoides
Sand Tiger Shark
Odontaspis faurus

Sting Ray
Dasyatis sp.
Sea-horse
Hippocampus reidi
Cope's Gray Treefrog
Hyla chrysoscelis
Green Treefrog
Hyla cinerea
American Alligator
Alligator mississippiensis
Yellow-bellied Slider
Chrysemys scripta
Carolina Anole
Anolis carolinensis
Six-lined Racerunner
Cnemidophorus sexlineatus
Eastern Hognose Snake
Heterodon platyrhinos
Banded Water Snake
Nerodia fasciata
Pied-billed Grebe
Podilymbus podiceps
Brown Pelican
Pelecanus occidentalis
Anhinga; Snakebird
Anhinga anhinga
Double-crested Cormorant
Phalacrocorax auritus
Tricolored Heron
Egretta tricolor
Little Blue Heron
Egretta caerulea
Snowy Egret
Egretta thula
Great Egret
Casmerodius albus
Great Blue Heron
Ardea herodias
White Ibis
Eudocimus albus
American Wigeon
Anas americana
Ring-necked Duck
Aythya collaris
Clapper Rail
Rallus longirostris
Virginia Rail
Rallus limicola
Common Moorhen
Gallinula chloropus
American Coot
Fulica americana
American Oystercatcher
Haematopus palliatus
Piping Plover
Charadrius melodus
Semipalmated Plover
Charadrius semipalmatus
Killdeer
Charadrius vociferus
Black-bellied Plover
Pluvialis squatarola
Willet
Catoptrophorus semipalmatus
Red Knot
Calidris canutus

Dunlin
Calidris alpina
Sanderling
Calidris alba
Ruddy Turnstone
Arenaria interpres
Laughing Gull
Larus atricilla
Ring-billed Gull
Larus delawarensis
Herring Gull
Larus argentatus
Great Black-backed Gull
Larus marinus
Forster's Tern
Sterna forsteri
Caspian Tern
Sterna caspia
Turkey Vulture
Cathartes aura
Black Vulture
Coragyps atratus
Bald Eagle
Haliaeetus leucocephalus
Red-shouldered Hawk
Buteo lineatus
Osprey
Pandion haliaetus
American Kestrel
Falco sparverius
Mourning Dove
Zenaida macroura
Common Ground-dove
Columbina passerina
Whip-poor-will
Caprimulgus vociferus
Belted Kingfisher
Ceryle alcyon
Red-bellied Woodpecker
Melanerpes carolinus
Red-headed Woodpecker
Melanerpes erythrocephalus
Northern Flicker
Colaptes auratus
Downy Woodpecker
Picoides pubescens
Pileated Woodpecker
Dryocopus pileatus
Great Crested Flycatcher
Myiarchus crinitus
Eastern Phoebe
Sayornis phoebe
Tree Swallow
Tachycineta bicolor
Blue Jay
Cyanocitta cristata
American Crow
Corvus brachyrhynchos
Fish Crow
Corvus ossifragus
Carolina Chickadee
Parus carolinensis
House Wren
Troglodytes aedon
Carolina Wren (sign)
Thryothorus ludovicianus

Golden-crowned Kinglet
Regulus satrapa
Loggerhead Shrike
Lanius ludovicianus
Gray Catbird
Dumetella carolinensis
Northern Mockingbird
Mimus polyglottos
European Starling
Sturnus vulgaris
Yellow-rumped Warbler
Dendroica coronata
Worm-eating Warbler
Helmitheros vermivorus
American Redstart
Setophaga ruticilla
Northern Cardinal
Cardinalis cardinalis
Rufous-sided Towhee
Pipilo erythrophthalmus
Savannah Sparrow
Passerculus sandwichensis
Song Sparrow
Melospiza melodia
Red-winged Blackbird
Agelaius phoeniceus
Rusty Blackbird
Euphagus carolinus
Common Grackle
Quiscalus quiscula
Boat-tailed Grackle
Quiscalus major
Marsh Rabbit
Sylvilagus palustris
Eastern Fox Squirrel
Sciurus niger
Gray Fox (tracks, sign)
Urocyon cinereoargenteus
Bottle-nosed Dolphin
Tursiops truncatus

Greenville County

Bunched Arrowhead Heritage Preserve
Greenville County
April 23, 1995

Bunched arrowhead, a federally endangered species, is a plant of wet seepage areas of Buncombe and Henderson counties in North Carolina and this site in South Carolina. Here it grows profusely in a very small seepage area. The emersed leaves are spatulate, about twelve inches long; white flowers are borne on a stem that rises out of the water. As the list shows, many other plants of wet soils are growing on this same site. The site is small, only 178 acres, designed to protect this small specialized habitat.

Directions to Bunched Arrowhead Heritage Preserve: From Greenville, take US 276 north for 8 miles; fork right onto US 25 just before Travelers Rest and continue heading north. Drive about 1.7 miles, turn right onto Blue Ridge Road, and go to the end of the road; turn left (north) onto Shelton Road. At the next intersection in about 0.7 miles, turn right onto McCauley Road. Drive a short distance to the paved parking area located on the right.

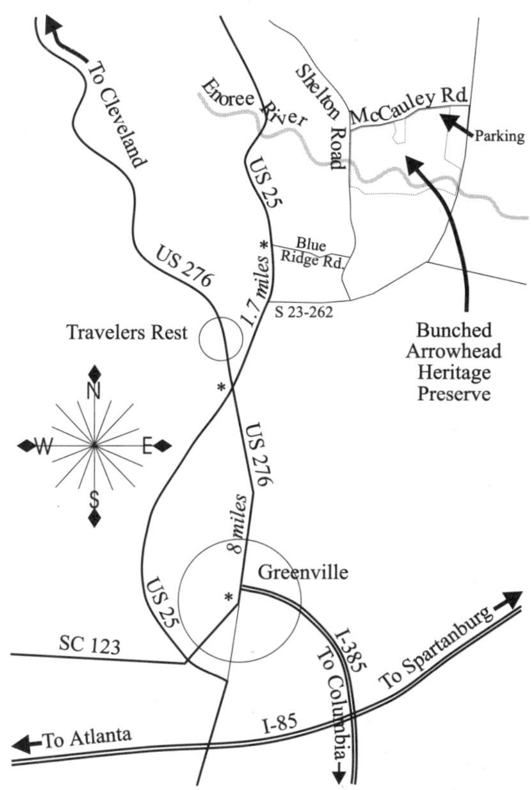

FLORA

Lichen
Lecidea mutabilis
Leptogium cyanescens
Bacidia schweinitzii
Cladina subtenuis
Cladonia coniocraea
Cladonia didyma
Cladonia parasitica
Cladonia peziziformis
Pertusaria ostiolata
Pertusaria paratuberculifera
Pertusaria macounii
Maronea constans
Haematomma puniceum
Lecanora varia
Lecanora allophana
Lecanora sp. #2 Brodo
Lecanora caesiorubella f. prolifera
Ochrolechia africana
Tephromela atra
Candelaria concolor
Hypotrachyna livida
Hypotrachyna pustulifera
Parmelina minarum
Parmotrema hypotropum
Parmotrema michauxianum
Parmotrema ultralucens
Flavoparmelia caperata
Canoparmelia caroliniana
Rimelia reticulata
Rimelia subisidiosa
Tuckermannopsis ciliaris
Tuckermannopsis viridis
Ramalina americana
Usnea strigosa
Buellia curatellae
Dirinaria confusa
Physcia aipolia
Physcia americana
Physcia millegrana
Phaeophyscia ciliata
Phaeophyscia rubropulchra
Physconia detersa
Physciella cloantha
Caloplaca flavorubescens
Trypethelium virens

Bunched Arrowhead Heritage Preserve (continued)
April 23, 1995

Haircap Moss
Polytrichum sp.
Running-pine
Lycopodium flabelliforme
Southern Lady Fern
Athyrium asplenioides
Netted Chain Fern
Woodwardia areolata
Loblolly Pine
Pinus taeda
Eastern Red-cedar
Juniperus virginiana
Bunched Arrowhead
Sagittaria fasciculata
Microstegium (grass)
Microstegium vimineum
Jack-in-the-pulpit
Arisaema triphyllum
Greenbrier; Catbrier
Smilax sp.
Solomon's-seal
Polygonatum biflorum
Wild Onion
Allium canadense
Southern Blue Flag
Iris virginica
Downy Rattlesnake-plantain
Goodyera pubescens
Crane-fly Orchid
Tipularia discolor
Tag Alder
Alnus serrulata
White Oak
Quercus alba
Spanish Oak
Quercus falcata
Willow Oak
Quercus phellos
Paper Mulberry
Broussonetia papyrifera
Sheep-sorrel; Sour-grass
Rumex acetosella
Pokeweed; Pokeberry
Phytolacca americana
Yellowroot
Xanthorhiza simplicissima
Tulip-poplar
Liriodendron tulipifera
Sassafras
Sassafras albidum
Wild Radish
Raphanus raphanistrum
Virginia-willow
Itea virginica
Climbing-hydrangea
Decumaria barbara
Sweet-gum
Liquidambar styraciflua
Five-fingers
Potentilla canadensis
Five-fingers; Cinquefoil
Potentilla recta

Blackberry; Brambles
Rubus sp.
Multiflora Rose
Rosa multiflora
Swamp Rose
Rosa palustris
Hawthorn
Crataegus sp.
Wild Plum
Prunus americana
Black Cherry; Wild Cherry
Prunus serotina
Honey Locust
Gleditsia triacanthos
Low Hop Clover
Trifolium dubium
Beggar-lice
Desmodium sp.
Bicolor; Bush-clover
Lespedeza bicolor
Vetch
Vicia sp.
Kudzu
Pueraria lobata
Cranesbill
Geranium carolinianum
Poison Sumac
Rhus vernix
Poison-ivy
Rhus radicans
Staghorn Sumac
Rhus typhina
American Holly
Ilex opaca
Hearts-a'burstin'-with-love
Euonymus americanus
Boxelder
Acer negundo
Red Maple
Acer rubrum
Virginia-creeper
Parthenocissus quinquefolia
Muscadine; Bullace
Vitis rotundifolia
Hypericum
Hypericum sp.
Primrose-leaved Violet
Viola primulifolia
Passionflower; Maypops
Passiflora incarnata
Evening-primrose
Oenothera sp.
Queen Anne's-lace; Wild Carrot
Daucus carota
Blackgum
Nyssa sylvatica
Flowering Dogwood
Cornus florida
Spotted Wintergreen
Chimaphila maculata
Deerberry; Squaw-huckleberry
Vaccinium stamineum

Blueberry
Vaccinium sp.
Ash
Fraxinus sp.
Chinese Privet
Ligustrum sinense
Yellow Jessamine
Gelsemium sempervirens
Lyre-leaved Sage
Salvia lyrata
Horse-nettle
Solanum carolinense
Toadflax
Linaria canadensis
Cross Vine
Anisostichus capreolata
Partridge-berry
Mitchella repens
Japanese Honeysuckle
Lonicera japonica
Arrowwood
Viburnum dentatum lucidum
Wild Lettuce
Lactuca sp.
Hawkweed
Hieracium sp.
Common Dandelion
Taraxacum officinale
Small's Ragwort
Senecio smallii
Batchelor's-buttons
Centaurea cyanus
Thistle
Carduus sp.
Elephant's-foot
Elephantopus sp.
Joe-pye-weed
Eupatorium sp.
Rabbit-tobacco
Gnaphalium sp.
Swamp-beacon
Mitrula paludosa

Black-knot
Apiosporina morbosa
Common Split-gill
Schizophyllum commune
Fawn Mushroom
Pluteus cervinus
Carnival-candy Slime
Arcyria denudata

FAUNA

Cherry Leaf Spindle Gall Mite
Phytopus cerasicrumena
Spittlebug
Philaenus sp.
Woolly Alder Aphid
Prociphilus tesselatus
Wool-sower Gall (wasp)
Callirhytis seminator
Eastern Tiger Swallowtail
Papilio glaucus
Eastern Box Turtle (shell)
Terrapene carolina
Killdeer
Charadrius vociferus
Mourning Dove
Zenaida macroura
Red-bellied Woodpecker
Melanerpes carolinus
American Crow
Corvus brachyrhynchos
Carolina Chickadee
Parus carolinensis
Carolina Wren
Thryothorus ludovicianus
Eastern Bluebird
Sialia sialis
Rufous-sided Towhee
Pipilo erythrophthalmus
Chipping Sparrow
Spizella passerina

(Photos by A. Ciegler and L. Smith)

Jack-in-the-pulpit, *Arisaema triphyllum*.

Science and Nature Notes: Appalachian Mountains

Mountains are certainly among the most grand geological features on Earth. They are formed by faulting, folding, volcanism, or a combination of these forces. The Earth's crust is now known to be composed of large tectonic plates which move about. These plates may move apart or crash together. This new understanding has helped explain many of the land-forms found along the edges of continents. A massive continental collision folded the Earth's crust along the eastern edge of North America creating the Appalachian Mountains. These mountains extend for approximately 2000 miles and are composed of modified marine sediments and continental crust.

The formation of the Appalachians may be divided into three phases. The first, or Taconic, phase occurred between 450 and 470 million years ago. In this phase, a small continental mass and an arc of volcanic islands were crushed against North America, forming a mountain range probably comparable to the Himalayas. These early mountains exist today as the Blue Ridge ranges. The massive granitic rocks at Caesars Head and Table Rock (Pickens County) state parks formed at this time. These rocks, initially formed miles underground, have been exposed by subsequent erosion. The Piedmont regions of North and South Carolina formed at this time as well. This period was distinguished by enormous volcanic eruptions.

The second phase of mountain building began as the continent of Africa made initial contact with northern North America, closing an ancient sea. This phase, the Acadian, took place from 360 to 400 million years ago, and is most evident in Nova Scotia and New Brunswick. The final, Alleghenian, phase occurred from 250 to 300 million years ago when Africa collided farther south, producing the Valley and Ridge ranges west of the Blue Ridge. At this point, the Appalachian formation was complete, the ancient sea was closed, and the world's continents were joined into one great land mass, Pangea.

—Danny Smith, former SCAN president

Caesars Head State Park

Greenville County

September 28, 1991, and September 23, 1995

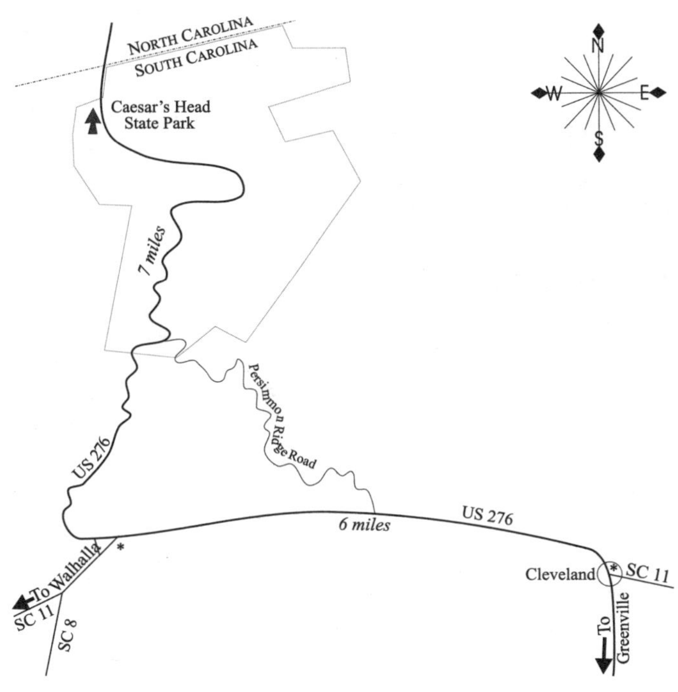

This is the most mountainous of the South Carolina state parks, offering scenery found almost nowhere else in the state. The rock for which the park is named provides an interesting habitat for some uncommon plant material. The open rock promontory also serves as a good lookout for observing the fall bird migration, with the raptors being among the most spectacular. On a good day after a cold front has passed in late September, it is possible to see thousands of hawks go by. Ravens are year-round residents here. Mountain plants add to the interest of this park, especially the spring ephemeral wildflowers, followed by flowering shrubs such as mountain-laurel and rhododendron species.

Directions to Caesars Head State Park: From Greenville, take US 276 north through Travelers Rest to Cleveland; continue 6 more miles and follow US 276 north toward Caesars Head and Brevard, North Carolina. The road winds steeply up the mountain for about 7 miles until reaching the park headquarters on the left.

Caesars Head State Park (continued)
September 28, 1991, and September 23, 1995

FLORA

Lichen
Leptogium cyanescens
Peltigera canina
Graphis scripta
Porpidia albocaerulescens
Cladina subtenuis
Cladonia chlorophaea
Cladonia coniocraea
Cladonia cristatella
Cladonia polycarpoides
Cladonia peziziformis
Pertusaria ostiolata
Pertusaria texana
Lecanora cupressi
Lecanora caesiorubella f. prolifera
Ochrolechia africana
Anzia colpodes
Bulbothrix goebelii f. minor
Candelaria concolor
Parmelina horrescens
Parmotrema tinctorum
Flavoparmelia baltimorensis
Flavoparmelia caperata
Canoparmelia caroliniana
Punctelia rudecta
Usnea rubicunda
Usnea strigosa
Trypethelium virens
Shining Clubmoss
Lycopodium lucidulum
Hay-scented Fern
Dennstaedtia punctilobula
Bracken
Pteridium aquilinum
Southern Lady Fern
Athyrium asplenioides
Christmas Fern
Polystichum acrostichoides
New York Fern
Thelypteris noveboracensis
Ebony Spleenwort
Asplenium platyneuron
White Pine
Pinus strobus
Pitch Pine
Pinus rigida
Short-leaf Pine
Pinus echinata
Virginia Pine; Scrub Pine
Pinus virginiana
Eastern Hemlock
Tsuga canadensis
Carolina Hemlock
Tsuga caroliniana
Cane; Switch Cane
Arundinaria gigantea
Microstegium (grass)
Microstegium vimineum
Jack-in-the-pulpit
Arisaema triphyllum
Dayflower
Commelina communis
Commelina sp.
Greenbrier; Bullbrier
Smilax rotundifolia
Solomon's-seal
Polygonatum biflorum
Wild Yam
Dioscorea villosa
Yellow Star-grass
Hypoxis hirsuta
Crested Iris
Iris cristata
Downy Rattlesnake-plantain
Goodyera pubescens
Crane-fly Orchid
Tipularia discolor
Pignut Hickory
Carya glabra
Cherry Birch
Betula lenta
American Chestnut
Castanea dentata
Allegheny Chinquapin
Castanea pumila pumila
White Oak
Quercus alba
Rock Chestnut Oak
Quercus prinus
Red Oak
Quercus rubra
Scarlet Oak
Quercus coccinea
Wild Ginger
Asarum canadense
Pinkweed
Polygonum pensylvanicum
Polygonum
Polygonum cespitosum longisetum
Tearthumb
Polygonum sagittatum
Virgin's-bower
Clematis virginiana
Meadow-rue
Thalictrum sp.
Tulip-poplar
Liriodendron tulipifera
Umbrella-tree
Magnolia fraseri
Sweet-shrub; Carolina-allspice
Calycanthus floridus
Sassafras
Sassafras albidum
Wild Hydrangea
Hydrangea arborescens
Hydrangea arborescens radiata
Witch-hazel
Hamamelis virginiana
Five-fingers
Potentilla canadensis
Red Clover
Trifolium pratense
Black Locust
Robinia pseudo-acacia
Wood-sorrel
Oxalis grandis
Button-rosy
Polygala curtissii
Flowering Spurge
Euphorbia corollata
Poison-ivy
Rhus radicans
American Holly
Ilex opaca
Striped Maple
Acer pensylvanicum
Red Maple
Acer rubrum
Spotted Touch-me-not
Impatiens capensis
Muscadine; Bullace
Vitis rotundifolia
Fox Grape
Vitis labrusca
Halberd-leaved Violet
Viola hastata
Common Evening-primrose
Oenothera biennis
Black Tupelo; Blackgum
Nyssa sylvatica sylvatica
Flowering Dogwood
Cornus florida
Silky Dogwood
Cornus amomum
Mountain Sweet Pepperbush
Clethra acuminata
Spotted Wintergreen
Chimaphila maculata
Indian-pipe
Monotropa uniflora
Pine-sap
Monotropa hypopithys
Rosebay; Great Laurel
Rhododendron maximum
Carolina-laurel
Rhododendron minus
Mountain-laurel
Kalmia latifolia
Dog-hobble
Leucothoe axillaris editorum
Sourwood
Oxydendrum arboreum
Trailing-arbutus
Epigaea repens
Huckleberry
Gaylussacia ursina
Galax; Galaxy; Coltsfoot
Galax aphylla
Whorled Loosestrife
Lysimachia quadrifolia
Horse-sugar; Sweetleaf
Symplocos tinctoria
Carolina Silverbell
Halesia carolina
Silverbell
Halesia sp.
American Ash; White Ash
Fraxinus americana
Common Skullcap
Scutellaria integrifolia
Carpenterweed
Prunella vulgaris
Dotted Horsemint
Monarda punctata
Horse-balm
Collinsonia canadensis
Horse-nettle
Solanum carolinense
False Foxglove
Agalinis tenuifolia
Squaw-root
Conopholis americana
Partridge-berry
Mitchella repens
Bluets
Houstonia sp.
Maple-leaf Viburnum
Viburnum acerifolium
Bluebell
Campanula divaricata
Lobelia
Lobelia puberula
Gall-of-the-earth
Prenanthes trifoliolata
Rattlesnake-root
Prenanthes altissima
Gall-of-the-earth
Prenanthes serpentaria
Prenanthes
Prenanthes sp.
Pale Indian-plantain
Cacalia atriplicifolia
Elephant's-foot
Elephantopus tomentosus
Joe-pye-weed
Eupatorium fistulosum
White Snakeroot
Eupatorium rugosum
Rabbit-tobacco
Gnaphalium obtusifolium
Heart-leaved Aster
Aster divaricatus
Heart-leaved Aster
Aster cordifolius
Aster
Aster puniceus
Aster curtisii
Aster sp.
Gray Goldenrod
Solidago nemoralis
Goldenrod
Solidago sp.
Hairy Sunflower
Helianthus tomentosus
Coreopsis
Coreopsis major
Spanish-needles
Bidens bipinnata
Beggar-ticks
Bidens pilosa
Sneeze-weed
Helenium autumnale
Yarrow; Milfoil
Achillea millefolium

Eyelash Cup
Scutellinia scutellata
Green Stain
Chlorociboria aeruginascens
Pungent Cystoderma
Cystoderma amianthinum rugusoreticulatum
Tawny Grisette
Amanita fulva
Gemmed Amanita
Amanita gemmata
Destroying-angel
Amanita virosa
Graceful Bolete
Tylopilus gracilis
Waxy-cap (mushroom)
Hygrocybe punica
Spotted-stalked Milky
Lactarius scrobiculatus
Tacky Green Russula
Russula aeruginea
Graying Yellow Russula
Russula claroflava
Spotted Mycena
Mycena maculata
Gem-studded Puffball
Lycoperdon perlatum
Earthstar
Scleroderma sp.
Stalked Puffball-in-aspic
Calostoma cinnabarina
Calostoma (mushroom)
Calostoma sp.

FAUNA

Giant Forest Snail
Mesodon normalis
White Micrathena (spider)
Micrathena mitrata
Red Velvet Mite
Trombidium sp.
June-beetle Wasp
Pelecinus polyturator
Fork-tailed Bush Katydid
Scudderia furcata
Cave Cricket
Ceuthophilus sp.
Yellowjacket
Vespula sp.
Cloudless Sulphur
Phoebis sennae eubule
Monarch (butterfly)
Danaus plexippus
Jordan's Salamander
Plethodon jordani
Eastern Garter Snake
Thamnophis sirtalis
Turkey Vulture
Cathartes aura
Mourning Dove
Zenaida macroura
Ruby-throated Hummingbird
Archilochus colubris
Downy Woodpecker
Picoides pubescens
Pileated Woodpecker
Dryocopus pileatus
Blue Jay
Cyanocitta cristata
American Crow
Corvus brachyrhynchos
Common Raven
Corvus corax
Tufted Titmouse
Parus bicolor
Carolina Chickadee
Parus carolinensis
White-breasted Nuthatch
Sitta carolinensis
Carolina Wren
Thryothorus ludovicianus
American Robin
Turdus migratorius
Brown Thrasher
Toxostoma rufum
Yellow-throated Vireo
Vireo flavifrons
Solitary Vireo
Vireo solitarius
Tennessee Warbler
Vermivora peregrina
Black-and-white Warbler
Mniotilta varia
Black-throated Blue Warbler
Dendroica caerulescens
Chestnut-sided Warbler
Dendroica pensylvanica
Magnolia Warbler
Dendroica magnolia
Black-throated Green Warbler
Dendroica virens
Yellow-throated Warbler
Dendroica dominica
Bay-breasted Warbler
Dendroica castanea
Blackpoll Warbler
Dendroica striata
Ovenbird
Seiurus aurocapillus
American Redstart
Setophaga ruticilla
Rose-breasted Grosbeak
Pheuticus ludovicianus
Rufous-sided Towhee
Pipilo erythrophthalmus
Dark-eyed Junco
Junco hyemalis
Gray Squirrel
Sciurus carolinensis

Eva Russell Chandler Heritage Preserve
Greenville County
May 8, 1993

Located on a shoulder of the Blue Ridge escarpment below Caesars Head State Park, this preserve includes habitats populated by several endangered plants. These include mountain sweet pitcher-plant, grass-of-Parnassus, divided-leaved groundsel, and Indian paintbrush. There are granite outcrops, one of which has a beautiful cataract formed by an enthusiastic stream. These areas harbor plants that are rare in the mountains. The winter of '92-'93 had been one of heavy wet snow followed by high wind, followed by an ice storm. The resulting tree fall made navigating the trails a bit hazardous, and limited the area that could be covered.

(Photo by L. Smith)

Rock outcrop.

(Photo by L. Greenlee)

Eastern box turtle, *Terrapene carolina.*

Directions to Eva Russell Chandler Heritage Preserve: From Greenville, take US 276 north through Travelers Rest to Cleveland; continue 6 more miles and follow US 276 north toward Caesars Head and Brevard, North Carolina. Watch for the Little Bald Rock community, and then go less than a mile and take a gravel road to the right, Persimmon Ridge Road. The road slopes downhill from US 276, which makes it hard to see, so watch closely. Go down Persimmon Ridge Road about 0.5 mile until you see a pull-off on the right. Trees are posted with Heritage Preserve signs. Park at the pull-off.

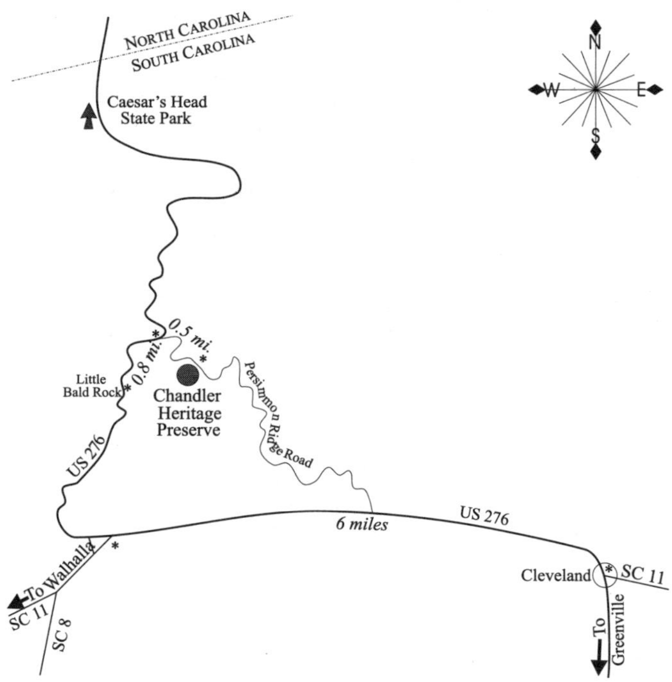

Eva Russell Chandler Heritage Preserve (continued)
May 8, 1993

FLORA

Lichen
Cladina rangiferina humilis
Cladonia polycarpoides sterilia
Polysporina simplex
Collema subflaccidum
Leptogium burnetiae
Leptogium corticola
Leptogium cyanescens
Graphis scripta
Graphis tenella
Graphis sp.
Bacidia phacodes
Bacidia abductans
Bacidia sp.
Leptogium austroamericanum
Lecidea albocaerulescens
Lecidea russula
Baeomyces fungoides
Cladina rangiferina patula
Cladina sp.
Cladonia cariosa
Cladonia caroliniana caroliniana
Cladonia chlorophaea f. intermedia
Cladonia chlorophaea costata
Cladonia chlorophaea f. pterygota
Cladonia chlorophaea lepidophora
Cladonia coniocraea
Cladonia coniocraea f. subpellucida
Cladonia coniocraea f. ceratodes
Cladonia cristatella cristatella
Cladonia cristatella pleurocarpa
Cladonia didyma squamulosa
Cladonia furcata furcata
Cladonia leporina leporina
Cladonia leporina ramosa
Cladonia leporina squamulosa
Cladonia polycarpoides squamulosa
Cladonia polycarpoides polycarpoides
Cladonia polycarpoides ramosa
Cladonia polycarpoides epiphylla
Cladonia didyma didyma
Cladonia squamosa murina
Cladonia squamosa squamosa
Cladonia squamosa phyllocoma
Cladonia squamosa levicorticata
Cladonia didyma epiphylla
Cladonia squamosa clavariella
Cladonia peziziformis
Cladonia peziziformis stenophyllodes
Pycnothelia papillaria
Pertusaria ostiolata
Pertusaria paratuberculifera
Pertusaria rubefacta
Pertusaria texana
Pertusaria sp.
Lecanora cupressi
Lecanora pallida
Phlyctis sp.
Candelaria concolor
Hypotrachyna livida
Parmelia sp.
Parmelina aurulenta
Parmelina horrescens
Parmelina minarum
Parmotrema arnoldii
Parmotrema crinitum
Parmotrema mellissii
Parmotrema michauxianum
Parmotrema xanthinum
Parmotrema chinense
Xanthoparmelia conspersa
Xanthoparmelia plittii
Xanthoparmelia subramigera
Flavoparmelia caperata
Canoparmelia caroliniana
Punctelia rudecta
Melanelia exasperata
Imshaugia aleurites
Rimelia reticulata
Tuckermannopsis oakesiana
Ramalina americana
Usnea ceratina
Usnea mutabilis
Usnea subfloridana
Usnea sp.
Buellia punctata
Buellia coccinea
Buellia curatellae
Buellia disciformis
Physcia subtilis
Heterodermia albicans
Heterodermia obscurata
Heterodermia speciosa
Pyxine sorediata
Phaeophyscia rubropulchra
Rinodina tephraspis
Physconia enteroxantha
Normandina pulchella
Trypethelium virens

Twisted-hair Spikemoss
Selaginella tortipila
Royal Fern
Osmunda regalis spectabilis
Lip Fern
Cheilanthes sp.
Bracken
Pteridium aquilinum
Southern Lady Fern
Athyrium asplenioides
Christmas Fern
Polystichum acrostichoides
New York Fern
Thelypteris noveboracensis
Resurrection Fern
Polypodium polypodioides
White Pine
Pinus strobus
Short-leaf Pine
Pinus echinata
Virginia Pine; Scrub Pine
Pinus virginiana
Eastern Hemlock
Tsuga canadensis
Eastern Red-cedar
Juniperus virginiana
Cane; Switch Cane
Arundinaria gigantea
Melic Grass
Melica mutica
Needle Grass
Stipa avenacea
Jack-in-the-pulpit
Arisaema triphyllum
Greenbrier
Smilax ecirrhata biltmoreana
Greenbrier; Catbrier
Smilax sp.
Catesby's Trillium
Trillium catesbaei
Indian Cucumber-root
Medeola virginiana
False Solomon's Seal
Smilacina racemosa
Solomon's-seal
Polygonatum biflorum
Wild Yam
Dioscorea villosa villosa
Yellow Star-grass
Hypoxis hirsuta
Crested Iris
Iris cristata
Pink Lady's-slipper
Cypripedium acaule
Downy Rattlesnake-plantain
Goodyera pubescens
Tag Alder
Alnus serrulata
Cherry Birch
Betula lenta
Beech
Fagus grandifolia
White Oak
Quercus alba
Rock Chestnut Oak
Quercus prinus
Red Oak
Quercus rubra
Buffalo-nut
Pyrularia pubera
Heartleaf
Hexastylis sp.
Sandwort
Arenaria groenlandica
Yellowroot
Xanthorhiza simplicissima
Meadow-rue
Thalictrum sp.
Tulip-poplar
Liriodendron tulipifera
Umbrella-tree
Magnolia fraseri
Dwarf PawPaw
Asimina parviflora
Sweet-shrub; Carolina-allspice
Calycanthus floridus
Sassafras
Sassafras albidum
Virginia-willow
Itea virginica
Sweet-gum
Liquidambar styraciflua
Witch-hazel
Hamamelis virginiana

Blackberry; Brambles
Rubus sp.
Indian-physic
Gillenia trifoliata
Downy Serviceberry
Amelanchier arborea
Black Cherry; Wild Cherry
Prunus serotina
Black Locust
Robinia pseudo-acacia
Bristly Locust
Robinia hispida
Hog-peanut
Amphicarpa bracteata
Flowering Spurge
Euphorbia corollata
Poison-ivy
Rhus radicans
Poison-oak
Rhus toxicodendron
Smooth Sumac
Rhus glabra
American Holly
Ilex opaca
Mountain Holly
Ilex ambigua montana
Hearts-a'burstin'-with-love
Euonymus americanus
Red Maple
Acer rubrum
Virginia-creeper
Parthenocissus quinquefolia
Muscadine; Bullace
Vitis rotundifolia
St.-Andrew's-cross
Hypericum hypericoides
Bird-foot Violet
Viola pedata
Primrose-leaved Violet
Viola primulifolia
Round-leaved Yellow Violet
Viola rotundifolia
Halberd-leaved Violet
Viola hastata
Prickly-pear
Opuntia compressa
Cow-bane
Oxypolis rigidior
Black Tupelo; Blackgum
Nyssa sylvatica sylvatica
Flowering Dogwood
Cornus florida
Silky Dogwood
Cornus amomum
Spotted Wintergreen
Chimaphila maculata
Rosebay; Great Laurel
Rhododendron maximum
Carolina-laurel
Rhododendron minus
Pinxter-flower; Pinxter-bloom
Rhododendron nudiflorum
Mountain-laurel
Kalmia latifolia
Dog-hobble
Leucothoe axillaris editorum
Sourwood
Oxydendrum arboreum
Trailing-arbutus
Epigaea repens
Sparkleberry
Vaccinium arboreum
Deerberry; Squaw-huckleberry
Vaccinium stamineum
Low-bush Blueberry
Vaccinium vacillans
Whorled Loosestrife
Lysimachia quadrifolia
Horse-sugar; Sweetleaf
Symplocos tinctoria
Carolina Silverbell
Halesia carolina
American Ash; White Ash
Fraxinus americana
Fringe-tree
Chionanthus virginicus
Angle-pod
Matelea sp.
Phlox
Phlox amoena
Beautyberry
Callicarpa americana
Lyre-leaved Sage
Salvia lyrata
False Foxglove
Aureolaria sp.
Cross Vine
Anisostichus capreolata
Broomrape
Orobanche uniflora
Partridge-berry
Mitchella repens
Bluets
Houstonia caerulea
Maple-leaf Viburnum
Viburnum acerifolium
Prenanthes
Prenanthes sp.
Rattlesnake-weed
Hieracium venosum
Ragwort
Senecio sp.
Pussy-toes
Antennaria plantaginifolia
Robin's-plantain
Erigeron pulchellus
Rosinweed
Silphium compositum

FAUNA

Rabid Wolf Spider
Lycosa rabida
Hickory Seed Gall (aphid)
Phylloxera caryoesemen
Chestnut Petiole Gall (wasp)
Neuroterus minutus
Black Oak Lamina Platform Gall
Neuroterus sp.
Blackjack Oak Apple (gall wasp)
Disholcaspis lapiei
Chestnut Oak Midrib Tumor Gall
Plagiotrochus tumificus
White Oak Club Gall (wasp)
Callirhytis clavula
Passalus; Bess-beetle
Popilius disjunctus
Chafer; Scarab Beetle
Dichelonyx fuscula
Striped Click Beetle
Ctenicera sp.
Soft-winged Flower Beetle
Collops vittatus
Leaf-mining Beetle
Anoplitis inaequalis
Evergreen Bagworm Moth
Thyridopteryx ephemeraeformis
Orange-patched Smoky Moth
Pyromorpha dimidiata
Eastern Newt
Notophthalmus viridescens
Toad
Bufo sp.
Spring Peeper
Hyla crucifer
Eastern Box Turtle
Terrapene carolina
Blue Jay
Cyanocitta cristata
American Crow
Corvus brachyrhynchos
Tufted Titmouse
Parus bicolor
Carolina Chickadee
Parus carolinensis
Carolina Wren
Thryothorus ludovicianus
Solitary Vireo
Vireo solitarius
Pine Warbler
Dendroica pinus
Hooded Warbler
Wilsonia citrina
Ovenbird
Seiurus aurocapillus
Rose-breasted Grosbeak
Pheuticus ludovicianus
Northern Cardinal
Cardinalis cardinalis
Scarlet Tanager
Piranga olivacea

Jones Gap State Park
Greenville County
August 22, 1992

Jones Gap is an easy grade approach to the top of the Blue Ridge where it forms the North Carolina - South Carolina state line. There used to be a relatively easy wagon road that followed this route mostly along the headwaters of the Middle Saluda River. Now the remnants of this road are incorporated into a hiking trail lined with flora and fauna typical of the upper Piedmont to mountain regions. Jones Gap State Park gives access to this trail from the bottom. It also includes the site of the former Cleveland Fish Hatchery, the first state-owned hatchery in South Carolina. Part of the hatchery has been restored and stocked with trout. However, Mother Nature does not always cooperate with long-planned trips. On this date, a very wet front moved in and produced a heavy steady downpour all day. This accounts for the short flora and fauna list. The day was saved by a fascinating special exhibit of Indian artifacts the park and local experts were offering on this weekend.

Directions to Jones Gap State Park: From Greenville, take US 276 north through Travelers Rest to Cleveland; continue for about 1.4 more miles on US 276 and turn right (north) onto River Falls Road, just before the F-Mart general store. Continue on winding River Falls Road (which becomes Jones Gap Road) until it reaches Jones Gap State Park at the end of the road.

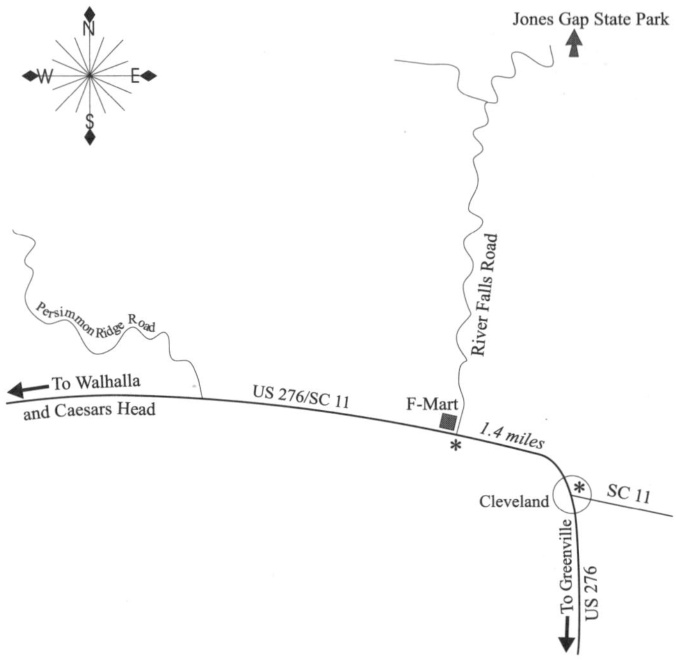

Jones Gap State Park (continued)
August 22, 1992

FLORA

Christmas Fern
Polystichum acrostichoides
New York Fern
Thelypteris noveboracensis
Resurrection Fern
Polypodium polypodioides
White Pine
Pinus strobus
Eastern Hemlock
Tsuga canadensis
Wake-robin; Red Trillium
Trillium erectum vaseyi
Indian Cucumber-root
Medeola virginiana
False Solomon's Seal
Smilacina racemosa
Solomon's-seal
Polygonatum biflorum
Crested Iris
Iris cristata
Downy Rattlesnake-plantain
Goodyera pubescens
Crane-fly Orchid
Tipularia discolor
Hickory
Carya sp.
Tag Alder
Alnus serrulata
Beech
Fagus grandifolia
White Oak
Quercus alba
Red Oak
Quercus rubra
Buffalo-nut
Pyrularia pubera
Tulip-poplar
Liriodendron tulipifera
PawPaw
Asimina triloba
Sweet-shrub; Carolina-allspice
Calycanthus floridus
Mock-orange
Philadelphus inodorus
Wild Hydrangea
Hydrangea arborescens radiata
Sweet-gum
Liquidambar styraciflua
Witch-hazel
Hamamelis virginiana
American Sycamore
Platanus occidentalis
Cocklebur
Agrimonia sp.
Naked-flower Tick-trefoil
Desmodium nudiflorum
Black Locust
Robinia pseudo-acacia
Hog-peanut
Amphicarpa bracteata
Poison-ivy
Rhus radicans
Hearts-a'burstin'-with-love
Euonymus americanus
Striped Maple
Acer pensylvanicum
Red Maple
Acer rubrum
Sugar Maple; Hard Maple
Acer saccharum
Yellow Buckeye
Aesculus octandra
Basswood; Linden
Tilia sp.
Spikenard
Aralia racemosa
Sanicle; Snakeroot
Sanicula canadensis
Flowering Dogwood
Cornus florida
Mountain Sweet Pepperbush
Clethra acuminata
Rosebay; Great Laurel
Rhododendron maximum
Carolina-laurel
Rhododendron minus
Dog-hobble
Leucothoe axillaris editorum
Sourwood
Oxydendrum arboreum
Galax; Galaxy; Coltsfoot
Galax aphylla
American Ash; White Ash
Fraxinus americana
Carpenterweed
Prunella vulgaris
Cross Vine
Anisostichus capreolata
Partridge-berry
Mitchella repens
Maple-leaf Viburnum
Viburnum acerifolium
Hawkweed
Hieracium paniculatum
Elephant's-foot
Elephantopus carolinianus

FAUNA

Orb-weaver
Neoscona arabesca
American Toad
Bufo americanus

(Photo by V. Winn)
The Middle Saluda River on a bright and sunny day. SCAN's visit was quite unlike this scene.

Roper Mountain Science Center
Greenville County
July 22, 1989

This lovely hilltop within the city limits of Greenville houses a very well-equipped center for the demonstration of scientific observation for students (of all ages). SCAN spent most of the day indoors examining, and in some cases playing with, the equipment. The telescopes allowed us to see sunspots, not something we ordinarily can do on a field trip. The plantings outside are man-made so no list was made except of a few "weeds" along the edges. Insects were active, and notes were made of those we saw. This is a facility that offers a lot of interest during the hours it is open to the public.

Operated by Greenville County School District.

Directions to Roper Mountain Science Center: From Columbia, take I-26, then take I-385 north toward Greenville. Go past the intersection with I-85 and continue on I-385 toward downtown Greenville. Exit one mile after that intersection at Exit 37, Roper Mountain Road; turn east (away from Greenville). Pass the frontage road and the next road (Greenville County Recreation Commission) and turn right at the third road into the Roper Mountain Science Center.

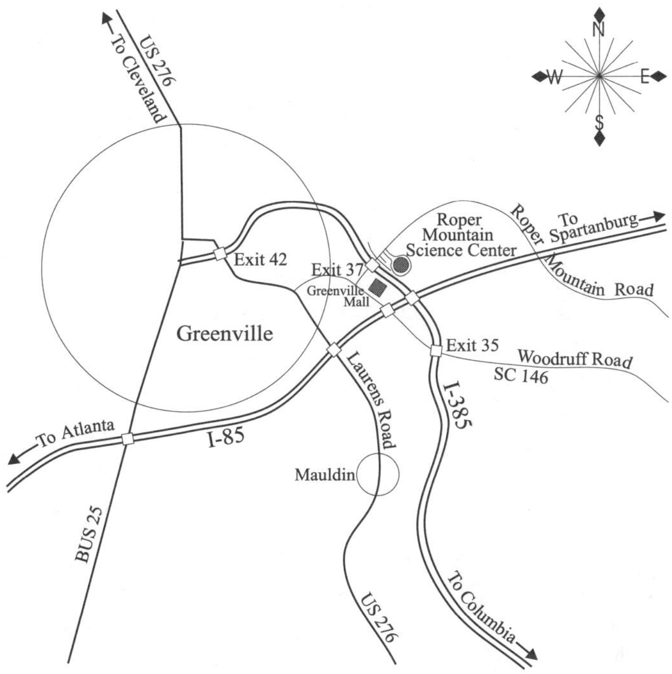

FLORA

Sensitive-brier
Schrankia microphylla
Red Clover
Trifolium pratense
Passionflower; Maypops
Passiflora incarnata
Meadow Mushroom
Agaricus campestris
Destroying-angel
Amanita virosa
Rosy Larch Bolete
Fuscoboletinus ochraceoroseus
Birch Bolete
Austroboletus betula
Emetic Russula
Russula emetica
Pinwheel Marasmius
Marasmius sp.
False Puffball
Scleroderma vulgare

FAUNA

Common Green Darner
Anax junius
Calico Pennant (dragonfly)
Celithemis elisa
Eastern Pondhawk
Erythemis simplicicollis
Eastern Amberwing (dragonfly)
Perithemis tenera
Violet-masked Glider
Tramea carolina
Water Scavenger Beetle
Berosus aculeatus
Tumbling Flower Beetle
Mordella melaena
Flea Beetle
Disonycha collata
Short-snouted Weevil
Pandeleteius sp.
Corn Pith Weevil
Centrinaspis penicellus
Eastern Tiger Swallowtail
Papilio glaucus
Cabbage White (butterfly)
Pieris rapae
Gray Hairstreak
Strymon melinus
Eastern Tailed-blue(butterfly)
Everes comyntas
Common Buckeye (butterfly)
Junonia coenia
Red-spotted Purple (butterfly)
Limenitis arthemis astyanax
Eastern Fence Lizard
Sceloporus undulatus
Six-lined Racerunner
Cnemidophorus sexlineatus

(Photo by L. Smith)

Gray hairstreak, *Strymon melinus,* on Joe-pye weed.

Wildcat Wayside
Greenville County
April 23, 1989

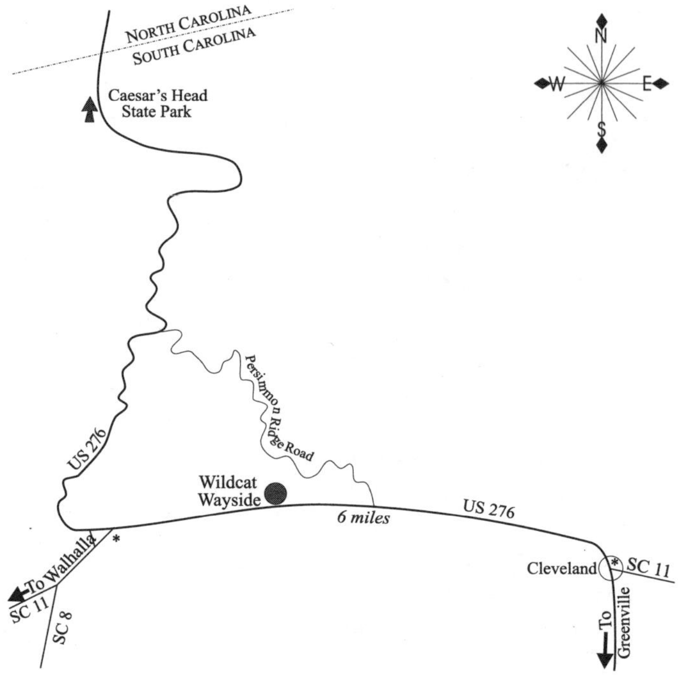

This picnic area provides access to a small valley that is rich in mountain flora and fauna. The stream that produces the waterfall visible from the highway enters the valley by sheeting down an almost vertical rock face to a pile of boulders at the base, then emerging as a recognizable stream in a conventional, albeit rocky, stream-bed. Salamanders can often be observed if one carefully turns over these rocks. The trail from the grassy picnic area to the rockface is a rhododendron tunnel with an occasional open flat area with interesting plant and animal life.

Maintained by the S.C. Department of Parks, Recreation, and Tourism.

Directions to Wildcat Wayside: From Greenville, take US 276 north through Travelers Rest to Cleveland and note the intersection where SC 11 joins US 276. Continue on US 276 and watch carefully for a sign at Persimmon Ridge Road on the right about 2-4 miles past the intersection in Cleveland. Wildcat Wayside is ahead on the right, unmarked. It has a wide parking area on the right side of the road, just past a small waterfall which is visible to slow-moving traffic.

FLORA

Rock Spikemoss
Selaginella rupestris
Bracken
Pteridium aquilinum
Southern Lady Fern
Athyrium asplenioides
Christmas Fern
Polystichum acrostichoides
New York Fern
Thelypteris noveboracensis
Ebony Spleenwort
Asplenium platyneuron
White Pine
Pinus strobus
Virginia Pine; Scrub Pine
Pinus virginiana
Eastern Hemlock
Tsuga canadensis
Catesby's Trillium
Trillium catesbaei
False Solomon's Seal
Smilacina racemosa
Solomon's-seal
Polygonatum biflorum
Blazing-star
Chamaelirium luteum
Wild Yam
Dioscorea villosa
Crested Iris
Iris cristata
Downy Rattlesnake-plantain
Goodyera pubescens
Crane-fly Orchid
Tipularia discolor
Hickory
Carya sp.
Cherry Birch
Betula lenta
Beech
Fagus grandifolia
White Oak
Quercus alba
Rock Chestnut Oak
Quercus prinus
Winged Elm
Ulmus alata
Buffalo-nut
Pyrularia pubera
Heartleaf
Hexastylis arifolia
Sandwort
Arenaria groenlandica
Yellowroot
Xanthorhiza simplicissima
Black Cohosh
Cimicifuga racemosa
Mayapple
Podophyllum peltatum
Tulip-poplar
Liriodendron tulipifera
Umbrella-tree
Magnolia fraseri
Cucumber-tree
Magnolia acuminata
Dwarf PawPaw
Asimina parviflora
Sweet-shrub; Carolina-allspice
Calycanthus floridus
Sassafras
Sassafras albidum
Wild Radish
Raphanus raphanistrum
Wild Hydrangea
Hydrangea arborescens
Sweet-gum
Liquidambar styraciflua
Witch-hazel
Hamamelis virginiana
American Sycamore
Platanus occidentalis
Five-fingers
Potentilla canadensis
Black Cherry; Wild Cherry
Prunus serotina
Redbud; Judas-tree
Cercis canadensis
Flowering Spurge
Euphorbia corollata
Poison-ivy
Rhus radicans
Poison-oak
Rhus toxicodendron
Hearts-a'burstin'-with-love
Euonymus americanus
Red Maple
Acer rubrum
Virginia-creeper
Parthenocissus quinquefolia
Muscadine; Bullace
Vitis rotundifolia
Mountain Camellia
Stewartia ovata
Sweet White Violet
Viola blanda
Halberd-leaved Violet
Viola hastata
Long-spurred Violet
Viola rostrata
Prickly-pear
Opuntia compressa
Golden-Alexander
Zizia aurea
Flowering Dogwood
Cornus florida
Alternate-leaved Dogwood
Cornus alternifolia
Spotted Wintergreen
Chimaphila maculata
Rosebay; Great Laurel
Rhododendron maximum
Mountain-laurel
Kalmia latifolia
Dog-hobble
Leucothoe axillaris
Trailing-arbutus
Epigaea repens

Sparkleberry
Vaccinium arboreum
Carolina Silverbell
Halesia carolina
Silverbell
Halesia parviflora
Fringe-tree
Chionanthus virginicus
Smooth Phlox
Phlox glaberrima
Lyre-leaved Sage
Salvia lyrata
Toadflax
Linaria canadensis
Cross Vine
Anisostichus capreolata
Partridge-berry
Mitchella repens
Bluets
Houstonia caerulea
Japanese Honeysuckle
Lonicera japonica
Maple-leaf Viburnum
Viburnum acerifolium
Prenanthes
Prenanthes sp.
Rattlesnake-weed
Hieracium venosum
Common Dandelion
Taraxacum officinale
Senecio
Senecio millefolium
Pussy-toes
Antennaria plantaginifolia
Cudweed
Gnaphalium purpureum
Robin's-plantain
Erigeron pulchellus
Tansy
Tanacetum vulgare
Common Split-gill
Schizophyllum commune

FAUNA

Ray Spider
Theridiosoma radiosa
Black & Yellow Mud-dauber
Sceliphron caementarium
June Beetle
Phyllophaga sp.
Chafer; Scarab Beetle
Dichelonyx subvittata
Zebra Swallowtail
Eurytides marcellus
Spicebush Swallowtail
Papilio troilus
Red-banded Hairstreak
Calycopis cecrops
Eastern Tailed-blue(butterfly)
Everes comyntas
Eastern Tent Caterpillar
Malacosoma americanum
Northern Dusky Salamander
Desmognathus fuscus
Carolina Anole
Anolis carolinensis
Eastern Fence Lizard
Sceloporus undulatus
Pileated Woodpecker
Dryocopus pileatus
Eastern Phoebe
Sayornis phoebe
American Crow
Corvus brachyrhynchos
Tufted Titmouse
Parus bicolor
Carolina Chickadee
Parus carolinensis
Eastern Bluebird
Sialia sialis
Yellowthroat
Geothlypis trichas
Eastern Chipmunk
Tamias striatus
Gray Squirrel
Sciurus carolinensis
Golden Mouse
Ochrotomys nuttalli

(Photo by L. Smith)
Trailing-arbutus, *Epigaea repens*.

Greenwood County

Ninety Six National Historic Site (Star Fort)
Greenwood County
July 25, 1987

This historic site features a Revolutionary War fort designed like a star, as was typical of the times. SCAN began this visit with an extensive tour of the site guided by a Park Ranger, who explained the events of the time and the role of the fort in the American Revolution. Although it was an extremely hot day even for South Carolina, members spread out over the less disturbed areas of the site and found many natural history items of interest. Oglethorpe oak grows in only a few low wooded areas in the Piedmont, including this site. Seven other species of oak were seen. It was a good day to stay in the shade and really look at the trees.

Directions to Ninety Six National Historic Site: From Greenwood, take SC 34 east 9 miles to Ninety-Six. In the center of town, turn right (south) onto SC 248. The entrance to the site is clearly marked on the left in about 1.5 miles.

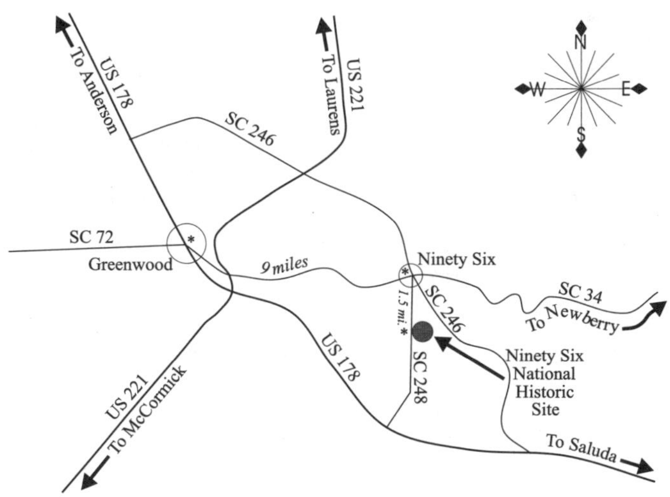

Ninety Six National Historic Site (continued)
July 25, 1987

FLORA

Christmas Fern
Polystichum acrostichoides
Loblolly Pine
Pinus taeda
Eastern Red-cedar
Juniperus virginiana
Common Cat-tail
Typha latifolia
Asparagus
Asparagus officinalis
Eastern Cottonwood
Populus deltoides
Black Walnut
Juglans nigra
Tag Alder
Alnus serrulata
White Oak
Quercus alba
Post Oak
Quercus stellata
Oglethorpe Oak
Quercus oglethorpensis
Red Oak
Quercus rubra
Black Oak
Quercus velutina
Spanish Oak
Quercus falcata
Water Oak
Quercus nigra
Willow Oak
Quercus phellos
Winged Elm
Ulmus alata
Hackberry
Celtis occidentalis georgiana
Red Mulberry
Morus rubra
Coralbeads
Cocculus carolinus
Tulip-poplar
Liriodendron tulipifera
PawPaw
Asimina triloba
Sassafras
Sassafras albidum
Sweet-gum
Liquidambar styraciflua
Black Cherry; Wild Cherry
Prunus serotina
Mimosa; Silk-tree
Albizia julibrissin
Honey Locust
Gleditsia triacanthos
Sericea
Lespedeza cuneata
Kudzu
Pueraria lobata
Poison-ivy
Rhus radicans
Winged Sumac
Rhus copallina
Hearts-a'burstin'-with-love
Euonymus americanus
Boxelder
Acer negundo
Red Maple
Acer rubrum
Spotted Touch-me-not
Impatiens capensis
Virginia-creeper
Parthenocissus quinquefolia
Muscadine; Bullace
Vitis rotundifolia
Passionflower; Maypops
Passiflora incarnata
Water-milfoil
Myriophyllum brasiliense
Queen Anne's-lace; Wild Carrot
Daucus carota
Mock Bishop's-weed
Ptilimnium capillaceum
Blackgum
Nyssa sylvatica
Flowering Dogwood
Cornus florida
Persimmon
Diospyros virginiana
American Ash; White Ash
Fraxinus americana
Butterfly-weed
Asclepias tuberosa
Brazilian Vervain
Verbena braziliensis
Trumpet Vine
Campsis radicans
Ruellia
Ruellia sp.
Black-haw; Nanny-berry
Viburnum prunifolium
Climbing Hempweed
Mikania scandens
Pussy-toes
Antennaria plantaginifolia
Black-knot
Apiosporina morbosa
Common Split-gill
Schizophyllum commune

FAUNA

Spiny-backed Spider
Micrathena gracilis
Writing Spider
Argiope aurantia
Eastern Pondhawk
Erythemis simplicicollis
Pied Skimmer
Libellula luctuosa
Eastern Amberwing (dragonfly)
Perithemis tenera
White-tail Dragonfly
Plathemis lydia
Ebony Jewelwing
Calopteryx maculata
Tiger Beetle
Cicindela sp.
Pedunculate Ground Beetle
Pasimachus sp.
Giant Stag Beetle
Lucanus elaphus
Rhinoceros Beetle
Strategus sp.
Green June Beetle
Cotinus nitida
Spicebush Swallowtail
Papilio troilus
Eastern Tailed-blue(butterfly)
Everes comyntas
Pearl Crescent
Phyciodes tharos
Common Buckeye (butterfly)
Junonia coenia
Red-spotted Purple (butterfly)
Limenitis arthemis astyanax
Hackberry Emperor (butterfly)
Asterocampa celtis
Eastern Fence Lizard
Sceloporus undulatus
Five-lined Skink
Eumeces fasciatus
Six-lined Racerunner
Cnemidophorus sexlineatus
Turkey Vulture
Cathartes aura
Northern Bobwhite
Colinus virginianus
American Robin
Turdus migratorius
Northern Mockingbird
Mimus polyglottos
Brown Thrasher
Toxostoma rufum
Northern Cardinal
Cardinalis cardinalis
Indigo Bunting
Passerina cyanea
Red-winged Blackbird
Agelaius phoeniceus
Summer Tanager
Piranga rubra
Gray Squirrel
Sciurus carolinensis

(Photo by L. Greenlee)

Writing spider, *Argiope aurantia*.

Hampton County

Lake Warren State Park
Hampton County
November 21, 1992

Lake Warren is a fairly large but shallow lake fringed by cat-tails and beautiful clumps of yellow-flowered tickseed. The short nature trail goes through a swampy bay forest with some very large loblolly-bays, plus sweet and red bay. Wild-olive is seen along the trail as well. Underneath the bays is a thicket of hollies that has been known to house a family of Swainson's warblers. Waterfowl and alligators are frequently seen in the marshy upper end of the lake where there is a long fishing pier to aid your viewing. The SCAN visit on this day produced a short list until mid-afternoon because of heavy rainfall.

Directions to Lake Warren State Park: From Columbia, take US 321 south through Fairfax to the community of Gifford, about 85 miles. Turn left (east) onto County Road S 25-41 and go 2.6 miles. Turn left again onto SC 363 and drive another 2.6 miles. Turn right on S 25-510 and drive to the park, about 1.4 miles; the entrance is on the right.

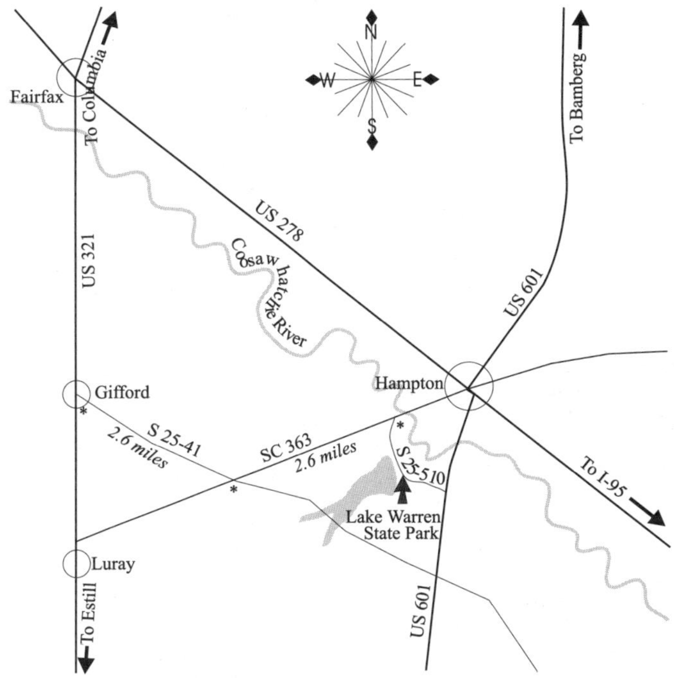

FLORA

Lichen
- *Leptogium cyanescens*
- *Graphis scripta*
- *Graphis mosquitensis*
- *Graphis elegans*
- *Graphis sp.*
- *Gyrostomum scyphuliferum*
- *Gyrostomum sp.*
- *Medusulina nitida*
- *Phaeographis dendritica*
- *Phaeographis sp.*
- *Bacidia schweinitzii*
- *Bacidia dryina*
- *Bacidia laurocerasi*
- *Leptogium austroamericanum*
- *Lecidea russula*
- *Letrouitia domingense*
- *Letrouitia sp.*
- *Cladonia calycantha*
- *Cladonia capitata*
- *Cladonia chlorophaea*
- *Cladonia sobolescens*
- *Cladonia coniocraea*
- *Cladonia cristatella cristatella*
- *Cladonia cristatella vestita*
- *Cladonia didyma*
- *Cladonia furcata*
- *Cladonia mateocyatha*
- *Cladonia ochrochlora*
- *Cladonia pityrea*
- *Cladonia polycarpoides*
- *Cladonia sp.*
- *Pertusaria leucostoma*
- *Pertusaria ostiolata*
- *Pertusaria paratuberculifera*
- *Pertusaria trachythallina*
- *Pertusaria subpertusa*
- *Pertusaria texana*
- *Pertusaria tetrathalmia*
- *Pertusaria sinusmexicani*
- *Haematomma puniceum*
- *Haematomma elatinum*
- *Lecanora miculata*
- *Lecanora allophana argentata*
- *Lecanora cupressi*
- *Lecanora pallida*
- *Lecanora pallida cancriformis*
- *Lecanora pallida f. prolifera*
- *Lecanora varia*
- *Lecanora sp.*
- *Ochrolechia pallescens*
- *Ochrolechia africana*
- *Bulbothrix goebelii*
- *Bulbothrix goebelii f. minor*
- *Hypotrachyna livida*
- *Hypotrachyna citrina sp. nov.*
- *Parmelina aurulenta*
- *Parmelina minarum*
- *Parmeliopsis subambigua*
- *Parmotrema michauxianum*
- *Parmotrema perforatum*
- *Parmotrema praesorediosum*
- *Parmotrema tinctorum*
- *Canoparmelia caroliniana*
- *Punctelia rudecta*
- *Rimelia reticulata*
- *Ramalina americana*
- *Usnea mutabilis*
- *Usnea strigosa*
- *Buellia curatellae*
- *Buellia disciformis*
- *Physcia americana*
- *Heterodermia albicans*
- *Heterodermia obscurata*
- *Heterodermia speciosa*
- *Pyxine caesiopruinosa f. aurea*
- *Rinodina roboris*
- *Pyrenula cerina*
- *Mycoporum sp.*
- *Trypethelium mastoideum*
- *Trypethelium tropicum*
- *Trypethelium floridanum*
- *Trypethelium gigantheum sp. nov.*
- *Porina sp.*
- *Schismatomma sp.*
- *Opegrapha sp.*

Cinnamon Fern
- *Osmunda cinnamomea*

Royal Fern
- *Osmunda regalis spectabilis*

Bracken
- *Pteridium aquilinum*

Southern Lady Fern
- *Athyrium asplenioides*

Virginia Chain Fern
- *Woodwardia virginica*

Netted Chain Fern
- *Woodwardia areolata*

Ebony Spleenwort
- *Asplenium platyneuron*

Mosquito Fern
- *Azolla caroliniana*

Loblolly Pine
- *Pinus taeda*

Lake Warren State Park (continued)
November 21, 1992

Pond Pine
Pinus serotina
Spruce Pine
Pinus glabra
Common Cat-tail
Typha latifolia
Southern Naiad
Najas guadalupensis
Brazilian Elodea
Egeria densa
Cane; Switch Cane
Arundinaria gigantea
Baggy-knees Grass
Sacciolepis striata
Panic Grass
Panicum verrucosum
Beard Grass; Plume Grass
Erianthus giganteus
Dulichium (sedge)
Dulichium arundinaceum
Proliferating Spike-rush
Eleocharis sp.
Wool-grass
Scirpus cyperinus
Giant Duckweed
Spirodela punctata
Bogmat
Wolffiella floridana
Spanish-moss
Tillandsia usneoides
Sawbrier; Greenbrier
Smilax glauca
Bamboo Smilax
Smilax laurifolia
Sweet-scented Smilax
Smilax smallii
Sarsaparilla-vine
Smilax pumila
Spoonleaf Yucca
Yucca filamentosa filamentosa
Water-spider Orchid
Habenaria repens
Crested Fringed Orchid
Habenaria cristata
Crane-fly Orchid
Tipularia discolor
Lizard's-tail
Saururus cernuus
Black Willow
Salix nigra
Wax-myrtle; Southern Bayberry
Myrica cerifera
Mockernut Hickory
Carya tomentosa
Pignut Hickory
Carya glabra
Tag Alder
Alnus serrulata
Beech
Fagus grandifolia
White Oak
Quercus alba

Swamp Chestnut Oak
Quercus michauxii
Spanish Oak
Quercus falcata
Water Oak
Quercus nigra
Willow Oak
Quercus phellos
Laurel Oak
Quercus laurifolia
American Elm
Ulmus americana
Red Mulberry
Morus rubra
False Nettle
Boehmeria cylindrica
Heartleaf
Hexastylis arifolia
Mexican-tea
Chenopodium ambrosioides
Alligator-weed
Alternanthera philoxeroides
Water-lily
Nymphaea odorata
Tulip-poplar
Liriodendron tulipifera
Sweetbay
Magnolia virginiana
Southern Magnolia
Magnolia grandiflora
Redbay
Persea borbonia
Sassafras
Sassafras albidum
Virginia-willow
Itea virginica
Climbing-hydrangea
Decumaria barbara
Sweet-gum
Liquidambar styraciflua
Witch-hazel
Hamamelis virginiana
Red Chokeberry
Sorbus arbutifolia arbutifolia
Black Cherry; Wild Cherry
Prunus serotina
Summer-farewell
Petalostemum pinnatum
Sesbania
Sesbania exaltata
American Wisteria
Wisteria frutescens
Coral-bean
Erythrina herbacea
Tread-softly
Cnidoscolus stimulosus
Poison Sumac
Rhus vernix
Poison-ivy
Rhus radicans
Winged Sumac
Rhus copallina

Titi
Cyrilla racemiflora
American Holly
Ilex opaca
Inkberry
Ilex glabra
Sweet Gallberry
Ilex coriacea
Red Maple
Acer rubrum
Virginia-creeper
Parthenocissus quinquefolia
Muscadine; Bullace
Vitis rotundifolia
Loblolly-bay
Gordonia lasianthus
St.-Andrew's-cross
Hypericum hypericoides
Yellow Passionflower
Passiflora lutea
Mermaid-weed
Proserpinaca pectinata
Water-milfoil
Myriophyllum pinnatum
Devil's-walking-stick
Aralia spinosa
Marsh Pennywort
Hydrocotyle umbellata
Water Pennywort
Hydrocotyle ranunculoides
Intelligence-plant
Centella asiatica
Black Tupelo; Blackgum
Nyssa sylvatica sylvatica
Swamp Tupelo; Swamp Blackgum
Nyssa sylvatica biflora
Flowering Dogwood
Cornus florida
Swamp Dogwood; Stiff Cornel
Cornus stricta
Sweet Pepperbush
Clethra alnifolia tomentosa
Spotted Wintergreen
Chimaphila maculata
Piedmont Azalea
Rhododendron canescens
Dog-hobble
Leucothoe axillaris axillaris
Blueberry
Vaccinium sp.
Water Pimpernel
Samolus parviflorus
Horse-sugar; Sweetleaf
Symplocos tinctoria
Ash
Fraxinus sp.
Wild Olive
Osmanthus americana
Yellow Jessamine
Gelsemium sempervirens
Beautyberry
Callicarpa americana
Blue-curls
Trichostema dichotomum
Shade Mudflower
Micranthemum umbrosum

Trumpet Vine
Campsis radicans
Squaw-root
Conopholis americana
Bladderwort
Utricularia sp.
Button-bush
Cephalanthus occidentalis
Partridge-berry
Mitchella repens
Possumhaw Viburnum
Viburnum nudum
American Elderberry
Sambucus canadensis
Gall-of-the-earth
Prenanthes serpentaria
Blazing-star
Liatris elegans
Dog-fennel
Eupatorium capillifolium
Eupatorium compositifolium
Climbing Hempweed
Mikania scandens
Marsh-fleabane
Pluchea camphorata
Horseweed; Mare's-tail
Erigeron canadensis ssp.
Aster
Aster dumosus
Gray Goldenrod
Solidago nemoralis
Goldenrod
Solidago fistulosa
Rough-leaved Goldenrod
Solidago rugosa
Camphorweed
Heterotheca subaxillaris
Beggar-ticks
Bidens laevis

FAUNA

Southern Toad
Bufo terrestris
Southern Cricket Frog
Acris gryllus
Pied-billed Grebe
Podilymbus podiceps
Wood Duck
Aix sponsa
Common Moorhen
Gallinula chloropus
American Coot
Fulica americana
Eastern Phoebe
Sayornis phoebe
Ruby-crowned Kinglet
Regulus calendula
Wood Thrush
Hylocichla mustelina
American Robin
Turdus migratorius

Webb Wildlife Center
Hampton County
July 28, 1990

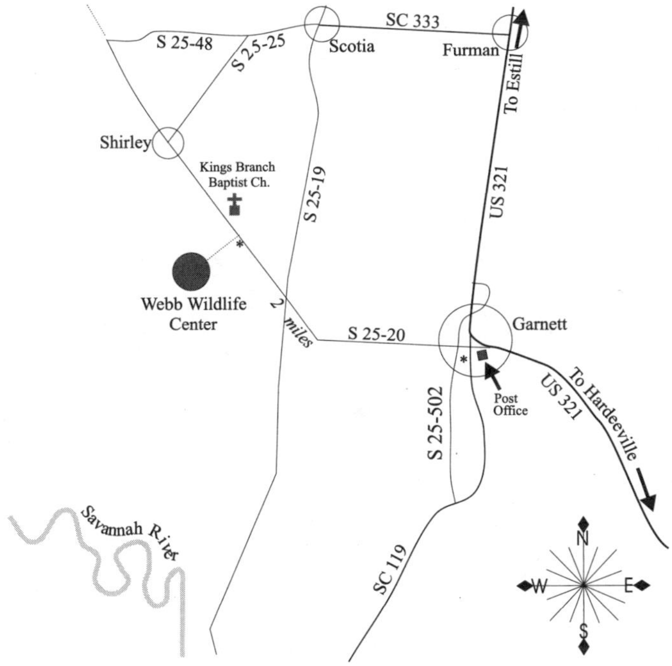

This wildlife center and game management area, operated by the Department of Natural Resources, encompasses about 6000 acres. The wildlife center itself is used as a research and demonstration facility designed to provide multi-use wildlife and timber management information that might be applied statewide. Instead of walking the nature trail through the swampy areas, SCAN, guided by Lewis Rogers of the Wildlife Center staff, walked through a pine forest, open meadow, and then a hardwood forest to observe management practices and investigate briefly the natural habitats these areas formed.

Directions to Webb Wildlife Center: From Columbia, take US 321 south to Garnett, about 100 miles. In Garnett, where US 321 curves to the left toward Hardeeville, take the right fork onto SC 119. Immediately turn right again onto County Road S 25-20 just before the post office on the left, and go about 2 miles. Turn left into the Wildlife Center before the Kings Branch Baptist Church on the right.

FLORA

Fruticose Lichen
Cladonia bacillaris
Cladonia capitata
Cladonia floerkeana
Cladonia pityrea
Cladonia squamosa
Bulbothrix goebelii
Hypotrachyna livida
Parmelia rudecta
Parmelia sp.
Parmelina galbina
Parmotrema cetratum
Parmotrema crinitum
Parmotrema michauxianum
Parmotrema perforatum
Parmotrema rampoddense
Parmotrema reticulatum
Parmotrema tinctorum
Pseudoparmelia caperata
Pseudoparmelia sphaerospora
Xanthoparmelia subramigera
Punctelia appalachensis
Punctelia bolliana
Ramalina sp.
Usnea strigosa
Physcia aipolia
Physcia americana
Heterodermia albicans
Heterodermia granulifera
Heterodermia obscurata
Pyxine caesiopruinosa

Bracken
Pteridium aquilinum
Sensitive Fern
Onoclea sensibilis
Loblolly Pine
Pinus taeda
Bald-cypress
Taxodium distichum
Cane; Switch Cane
Arundinaria gigantea
Sandbur; Sandspurs
Cenchrus sp.
Dwarf Palmetto
Sabal minor
Wax-myrtle; Southern Bayberry
Myrica cerifera
Post Oak
Quercus stellata
Spanish Oak
Quercus falcata
Laurel Oak
Quercus laurifolia
Water-elm; Planer-tree
Planera aquatica
Sweet-gum
Liquidambar styraciflua
Sicklepod
Cassia obtusifolia
Pencil-flower
Stylosanthes biflora
Milk-pea
Galactia volubilis

Polygala
Polygala grandiflora
Polygala
Polygala incarnata
Tread-softly
Cnidoscolus stimulosus
Poison-ivy
Rhus radicans
Winged Sumac
Rhus copallina
Red Maple
Acer rubrum
Piriqueta
Piriqueta caroliniana
Waxweed
Cuphea carthagensis
Button Snakeroot
Eryngium yuccifolium
Blackgum
Nyssa sylvatica
Persimmon
Diospyros virginiana
Yellow Jessamine
Gelsemium sempervirens
Angle-pod
Matelea sp.
Ground-ivy; Ponyfoot
Dichondra carolinensis
Bonamia
Bonamia humistrata
Tie-vine; Jacquemontia
Jacquemontia tamnifolia

Beautyberry
Callicarpa americana
Buchnera
Buchnera floridana
Thoroughwort
Eupatorium leucolepis
Golden Hypomyces
Hypomyces chrysopermus
Citron Amanita
Amanita citrina
Purple-bloom Russula
Russula mariae
Barometer Earthstar
Astraeus hygrometricus
Loblolly Pine Rust
Cronartium fusiforme

FAUNA

Crayfish (shell)
Cambarus sp.
Annual Cicada (shell)
Tibicen sp.
Woodland Ground Beetle
Agonum ferreus
Blister Beetle
Nemognatha piazata
Tooth-necked Longhorn Beetle
Stenodontes dasytomus
Flower Weevil
Centrinaspis sp.
Odontocorynus scutellum-album
Limnobaris denudata

Webb Wildlife Center(continued)
July 28, 1990

Black Swallowtail
Papilio polyxenes asterius
Little Yellow (butterfly)
Eurema lisa
Gulf Fritillary
Agraulis vanillae
Red-spotted Purple (butterfly)
Limenitis arthemis astyanax
Common Wood-nymph (butterfly)
Cercyonis pegala
Tent Caterpillar (tracks)
Malacasoma sp.
Eastern Box Turtle (egg)
Terrapene carolina
Black Racer
Coluber constrictor
Great Egret
Casmerodius albus
Turkey Vulture
Cathartes aura
Black Vulture
Coragyps atratus
Mississippi Kite
Ictinia mississippiensis
Northern Bobwhite
Colinus virginianus
Red-headed Woodpecker
Melanerpes erythrocephalus
Blue Jay
Cyanocitta cristata
Carolina Chickadee
Parus carolinensis
Northern Cardinal
Cardinalis cardinalis
Indigo Bunting
Passerina cyanea
Summer Tanager
Piranga rubra
Raccoon
Procyon lotor
White-tailed Deer (bones)
Odocoileus virginianus

Horry County

Cartwheel Bay Heritage Preserve
Horry County
May 24, 1986, May 16, 1987, & May 27, 1995

Cartwheel Bay Heritage Preserve is unique in that it is the only Carolina bay/pine savanna complex in South Carolina. It contains at least nine orchid species, almost every species of carnivorous plant in the state, plus numerous uncommon plants that make the area a botanist's dream. All this diversity attracts birds and insects as well. Venus'-fly-traps are usually in bloom in May, along with most of the orchids, the azalea and *Zenobia.* It will be hot, and if there has been recent rain, it may be wet. Like most bay habitats, there is also a sand ridge that remains dry.

Directions to Cartwheel Bay Heritage Preserve: From Florence, take US 76 east through Marion, Mullins, and Nichols, to the Lumber River, the boundary between Marion and Horry Counties. Continue on US 76/SC 9 for 0.7 miles, and fork off to the right onto SC 9 for 3.2 more miles. Turn right (southeast) on S 26-19 and drive about 2 miles to a dirt road on the right (opposite S 26-44 on the left). Turn right and travel a short way to Cartwheel Bay, which is marked by a sign on the left; park along the shoulder and walk in.

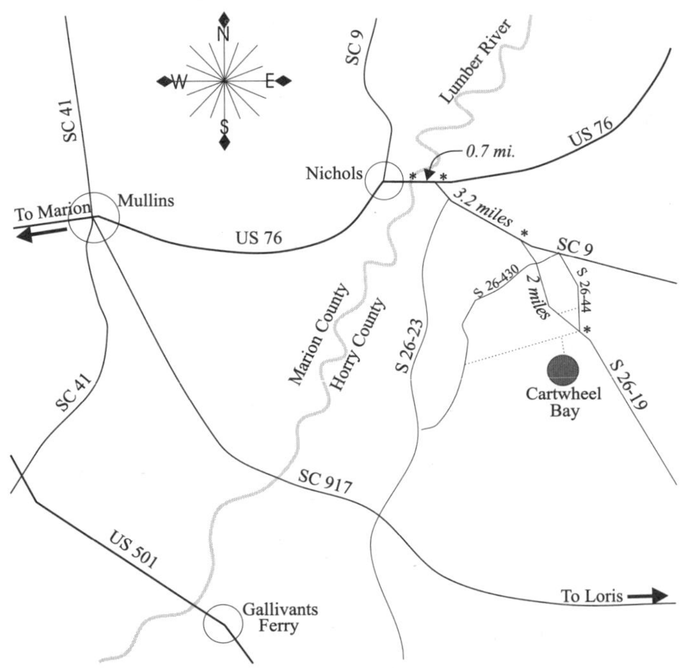

FLORA

Lichen
Graphis scripta
Cladina subtenuis
Cladonia leporina leporina
Cladonia polycarpoides
Cladonia rappii
Pertusaria copiosa
Pertusaria paratuberculifera
Pertusaria waghornei
Pertusaria sinusmexicani
Maronea constans
Haematomma elatinum
Lecanora cupressi
Lecanora varia
Ochrolechia africana
Bulbothrix confoederata
Bulbothrix goebelii
Bulbothrix isidiza
Hypotrachyna livida
Parmelina aurulenta
Parmotrema hypotropum
Parmotrema michauxianum
Parmotrema perforatum
Parmotrema tinctorum
Canoparmelia caroliniana
Punctelia rudecta
Rimelia reticulata
Rimelia subisidiosa
Usnea mutabilis
Usnea strigosa
Buellia curatellae
Heterodermia obscurata

Peat Moss
Sphagnum sp.
Southern Bog Clubmoss
Lycopodium adpressum
Clubmoss
Lycopodium sp.
Cinnamon Fern
Osmunda cinnamomea
Bracken
Pteridium aquilinum
Virginia Chain Fern
Woodwardia virginica
Long-leaf Pine
Pinus palustris
Pond Pine
Pinus serotina
Short-leaf Pine
Pinus echinata
Cane; Switch Cane
Arundinaria gigantea
Love Grass
Eragrostis sp.
Wire Grass
Aristida sp.
Panic Grass
Panicum sp.
Beak-rush
Rhynchospora debilis
Beak-rush
Rhynchospora globularis
Nut-rush
Scleria triglomerata
Nut-rush
Scleria minor
Bog-buttons
Lachnocaulon anceps
Bamboo Smilax
Smilax laurifolia
Colic-root
Aletris farinosa
Fly-poison
Amianthium muscaetoxicum
Black Snakeroot
Zigadenus densus
Yellow Star-grass
Hypoxis micrantha
Redroot
Lachnanthes caroliniana
Blue-eyed-grass
Sisyrinchium mucronatum
Rose Pogonia
Pogonia ophioglossoides
Spreading-pogonia
Cleistes divaricata
Grass-pink
Calopogon pulchellus
Pale Grass-pink
Calopogon pallidus
Grass-leaved Ladies'-tresses
Spiranthes praecox
Ladies'-tresses
Spiranthes sp.
Wax-myrtle; Southern Bayberry
Myrica cerifera
Dwarf Wax-myrtle
Myrica cerifera pumila

Scrubby Post Oak
Quercus margaretta
Black Jack Oak
Quercus marilandica
Water Oak
Quercus nigra
Blue-jack Oak
Quercus incana
Running Oak
Quercus pumila
Bastard-toadflax
Comandra umbellata
Sweetbay
Magnolia virginiana
Redbay
Persea borbonia
Sassafras
Sassafras albidum
Trumpet Pitcher-plant
Sarracenia flava
Sweet Pitcher-plant
Sarracenia rubra
Hooded Pitcher-plant
Sarracenia minor
Frog's-britches
Sarracenia purpurea
Sundew
Drosera capillaris
Venus'-fly-trap
Dionaea muscipula
Sweet-gum
Liquidambar styraciflua
Witch-alder
Fothergilla gardenii
Five-fingers
Potentilla canadensis
Potentilla simplex
Blackberry
Rubus argutus
Dewberry
Rubus trivialis
Blackberry; Brambles
Rubus sp.
Red Chokeberry
Sorbus arbutifolia
Black Cherry; Wild Cherry
Prunus serotina
Sensitive-brier
Schrankia microphylla
Wild Indigo
Baptisia cinerea
Baptisia tinctoria
Baptisia sp.
Rattlebox
Crotalaria purshii
Leadplant; Indigo-bush
Amorpha sp.
Samson Snakeroot
Psoralea psoralioides
Psoralea psoralioides psoralioides
Pencil-flower
Stylosanthes biflora
Locust
Robinia sp.
Tephrosia
Tephrosia hispidula
Tephrosia sp.

Orange Milkwort
Polygala lutea
Polygala
Polygala ramosa
Tread-softly
Cnidoscolus stimulosus
Spurge
Euphorbia curtisii
Euphorbia sp.
Poison-ivy
Rhus radicans
Poison-oak
Rhus toxicodendron
Winged Sumac
Rhus copallina
Titi
Cyrilla racemiflora
Inkberry
Ilex glabra
Sweet Gallberry
Ilex coriacea
Red Maple
Acer rubrum
Supple-jack; Rattan-vine
Berchemia scandens
Muscadine; Bullace
Vitis rotundifolia
St.-Andrew's-cross
Hypericum hypericoides
St.-Peter's-wort
Hypericum stans
St.-John's-wort
Hypericum gymnanthum
Shrubby Hypericum
Hypericum sp.
Common Blue Violet
Viola papilionacea
Bog White Violet
Viola lanceolata
Smooth Meadow-beauty
Rhexia alifanus
Pale Meadow-beauty
Rhexia mariana
Meadow-beauty
Rhexia sp.
Intelligence-plant
Centella asiatica
Eryngium
Eryngium prostratum
Button Snakeroot
Eryngium yuccifolium
Blackgum
Nyssa sylvatica
Swamp Tupelo; Swamp Blackgum
Nyssa sylvatica biflora
Sweet Pepperbush
Clethra alnifolia
Dwarf Azalea
Rhododendron atlanticum
Sheep-kill; Lamb-kill
Kalmia angustifolia caroliniana
Honey-cup
Zenobia pulverulenta
Maleberry
Lyonia ligustrina
Fetterbush
Lyonia lucida

Staggerbush
Lyonia mariana
Coastal Honeybells
Leucothoe racemosa
Dangleberry
Gaylussacia frondosa
Deerberry; Squaw-huckleberry
Vaccinium stamineum
Low-bush Blueberry
Vaccinium tenellum
Highbush Blueberry
Vaccinium corymbosum
Creeping Blueberry
Vaccinium crassifolium
Pyxie-moss
Pyxidanthera barbulata
Persimmon
Diospyros virginiana
Fringe-tree
Chionanthus virginicus
Milkweed
Asclepias michauxii
Annual Phlox
Phlox drummondii
Toadflax
Linaria canadensis
Buchnera
Buchnera floridana
Butterwort
Pinguicula pumila
Wiry Bladderwort
Utricularia subulata
Bladderwort
Utricularia sp.
Virginia Buttonweed
Diodia virginiana
Possumhaw Viburnum
Viburnum nudum
Lobelia
Lobelia nuttallii
Pyrrhopappus
Pyrrhopappus carolinianus
Sun-bonnets
Chaptalia tomentosa
Deer's-tongue; Vanilla-plant
Trilisa sp.
Eupatorium
Eupatorium sp.
Rabbit-tobacco
Gnaphalium sp.
Black-root
Pterocaulon pycnostachyum
Sea-myrtle; Groundsel-tree
Baccharis halimifolia
Swamp Fleabane
Erigeron vernus
Aster
Aster squarrosus
Heterotheca
Heterotheca graminifolia
Rosinweed
Silphium compositum
Barbara's-buttons
Marshallia obovata
Common Split-gill
Schizophyllum commune

Cartwheel Bay Heritage Preserve (continued)
May 24, 1986, May 16, 1987, and May 27, 1995

FAUNA

Writing Spider (egg)
Argiope aurantia
Goldenrod Spider
Misumena vatia
Jumping Spider
Metaphidippus sp.
Common Meadow Katydid
Orchelimum vulgare
Carolina Locust
Dissosteira carolina
Red-legged Grasshopper
Melanoplus femurrubrum
American Bird Grasshopper
Schistocerca americana
Carpenter Bee
Xylocopa virginica
Honey Bee
Apis mellifera
Calico Pennant (dragonfly)
Celithemis elisa
Halloween Pennant (dragonfly)
Celithemis eponina
Eastern Pondhawk
Erythemis simplicicollis
Blue Dasher
Pachydiplax longipennis
Violet-masked Glider
Tramea carolina
Citrine Forktail (damselfly)
Ischnura hastata
Obscure Rove Beetle
Atheta sp.
Flower Scarab Beetle
Trichiotinus piger
Flat-headed Bald Cypress Borer
Acmaeodera pulchella
Dermestid Beetle
Cryptorhopalum sp.
Tumbling Flower Beetle
Mordellistena pubescens
False Blister Beetle
Oxycopis sp.
Short-horned Leaf Beetle
Saxinus omogera
Oval Leaf Beetle
Colaspis favosa
Flea Beetle
Disonycha caroliniana
Pear-shaped Weevil
Apion sp.
Minute Seed Weevil
Ceutorhynchus sp.
Black Swallowtail
Papilio polyxenes asterius
Eastern Tiger Swallowtail
Papilio glaucus
Spicebush Swallowtail
Papilio troilus
Palamedes Swallowtail
Papilio palamedes
Gray Hairstreak
Strymon melinus
Eastern Tailed-blue(butterfly)
Everes comyntas
Pearl Crescent
Phyciodes tharos
American Lady (butterfly)
Vanessa virginiensis
Red Admiral
Vanessa atalanta rubria
Common Buckeye (butterfly)
Junonia coenia
Red-spotted Purple (butterfly)
Limenitis arthemis astyanax
Viceroy; Mimic
Limenitis archippus
Duskywing Skipper
Erynnis sp.
Hummingbird Clearwing Moth
Hemaris thysbe
Immaculate Holomelina Moth
Holomelina immaculata
Oak Toad
Bufo quercicus
Southern Toad
Bufo terrestris
Pine Woods Treefrog
Hyla femoralis
Treefrog
Hyla sp.
Broadhead Skink
Eumeces laticeps
Six-lined Racerunner
Cnemidophorus sexlineatus
Turkey Vulture
Cathartes aura
Northern Bobwhite
Colinus virginianus
Mourning Dove
Zenaida macroura
Yellow-billed Cuckoo
Coccyzus americanus
Red-headed Woodpecker
Melanerpes erythrocephalus
Northern Flicker
Colaptes auratus
Downy Woodpecker
Picoides pubescens
Pileated Woodpecker
Dryocopus pileatus
Great Crested Flycatcher
Myiarchus crinitus
Eastern Wood-pewee
Contopus virens
Willow Flycatcher (sign)
Empidonax traillii
Brown-headed Nuthatch
Sitta pusilla
Carolina Wren
Thryothorus ludovicianus
Blue-gray Gnatcatcher
Polioptila caerulea
Eastern Bluebird
Sialia sialis
Brown Thrasher
Toxostoma rufum
White-eyed Vireo
Vireo griseus
Prairie Warbler
Dendroica discolor
Pine Warbler
Dendroica pinus
Yellowthroat
Geothlypis trichas
Yellow-breasted Chat
Icteria virens
Northern Cardinal
Cardinalis cardinalis
Indigo Bunting
Passerina cyanea
Rufous-sided Towhee
Pipilo erythrophthalmus
Bachman's Sparrow
Aimophila aestivalis
Chipping Sparrow
Spizella passerina
Brown-headed Cowbird
Molothrus ater
Orchard Oriole
Icterus spurius
Summer Tanager
Piranga rubra
Bobcat; Wildcat (tracks)
Felis rufus

Ervin Dargan Heritage Preserve and Little Pee Dee River Heritage Perserve
Horry County
March 25, 1995

Ervin Dargan Heritage Preserve protects a floodplain area along the Little Pee Dee River that is relatively unspoiled. It is often affected by the water level, creating a very fertile swamp area rich in plants and animals. The insects are quite abundant as well. There are no formal trails but a road to a boat landing and a few old access roads provide easy walking.

Directions to Ervin Dargan Heritage Preserve: From Florence, take US 76/501 east toward Marion. When US 76 forks off toward Marion, continue southeast on US 501 toward Conway and Myrtle Beach about 15 miles to the Little Pee Dee River. Cross the bridge into Horry County and turn right at Gallivants Ferry (about 0.4 mile) onto S 26-99. In about a mile, fork right onto a dirt road at the sign to the public boat landing.

Little Pee Dee River Heritage Preserve is upstream from the Dargan Heritage Preserve, with similar floodplain habitat, plus xeric sandhill areas. The uncommon pyxie-moss was in bloom, and sphagnum, sundews, and loblolly-bay grew nearby. SCAN visited there in the afternoon after the morning at Dargan Heritage Preserve.

Directions to Little Pee Dee River Heritage Preserve: From Florence, take US 76 east through Marion to Mullins. Turn right (south) onto SC 917 and follow that to the Little Pee Dee River, about 5 miles. Cross the bridge into Horry County and watch for a sign to the Little Pee Dee River Heritage Preserve; turn left onto the dirt road shortly after the bridge.

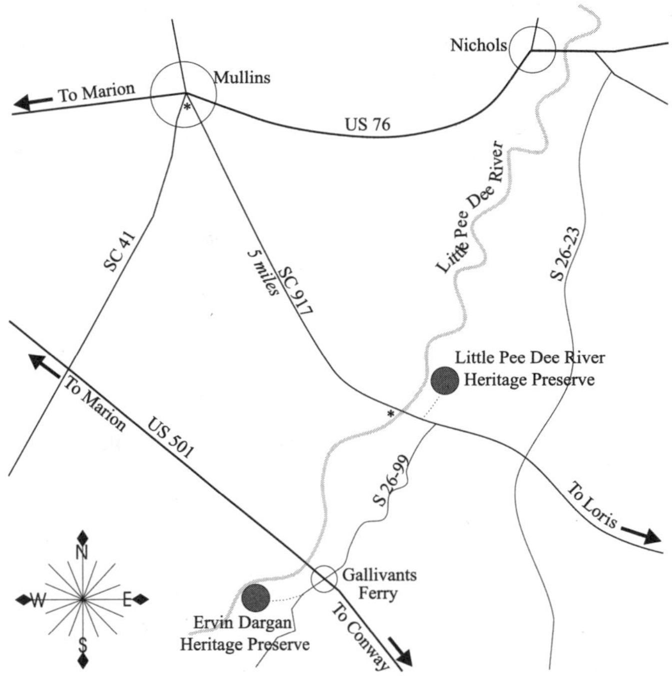

FLORA

Lichen
Graphis scripta
Graphis insidiosa
Graphis sp.
Phaeographis dendritica
Brigantiaea leucoxantha
Cladina subtenuis
Cladina subtenuis subtenuis
Cladonia coniocraea
Cladonia cristatella
Cladonia didyma
Cladonia floerkeana
Cladonia leporina
Cladonia leporina leporina
Pertusaria neoscotica
Pertusaria leucostoma
Pertusaria ostiolata
Pertusaria paratuberculifera
Pertusaria subpertusa
Pertusaria texana
Pertusaria velata
Pertusaria ophthalmiza
Pertusaria sp.
Haematomma puniceum
Lecanora cupressi
Lecanora caesiorubella f. prolifera
Ochrolechia africana
Phlyctis sp.
Bulbothrix goebelii
Bulbothrix isidiza
Hypotrachyna livida
Parmelina horrescens
Parmotrema hypotropum
Parmotrema michauxianum
Parmotrema perforatum
Parmotrema subtinctorium
Parmotrema tinctorum
Parmotrema ultralucens
Flavoparmelia caperata
Canoparmelia caroliniana
Punctelia rudecta
Rimelia reticulata
Rimelia subisidiosa
Usnea strigosa
Buellia punctata
Buellia curatellae
Buellia sp.
Physcia americana
Heterodermia obscurata
Heterodermia speciosa
Heterodermia spaltans sp. nov.
Pyxine caesiopruinosa
Trypethelium mastoideum
Trypethelium virens

Peat Moss
Sphagnum sp.
Floating Liverwort
Ricciocarpus natans
Sand Spikemoss
Selaginella arenicola
Bracken
Pteridium aquilinum
Netted Chain Fern
Woodwardia areolata
Long-leaf Pine
Pinus palustris
Loblolly Pine
Pinus taeda
Pond Pine
Pinus serotina
Bald-cypress
Taxodium distichum
Cane; Switch Cane
Arundinaria gigantea
Duckweed
Lemna sp.
Spanish-moss
Tillandsia usneoides
Bamboo Smilax
Smilax laurifolia
Sweet-scented Smilax
Smilax smallii
Willow
Salix sp.
Swamp Cottonwood
Populus heterophylla
Wax-myrtle; Southern Bayberry
Myrica cerifera
Tag Alder
Alnus serrulata
River Birch
Betula nigra
Turkey Oak
Quercus laevis
Spanish Oak
Quercus falcata
Water Oak
Quercus nigra
Willow Oak
Quercus phellos
Laurel Oak
Quercus laurifolia
Water-elm; Planer-tree
Planera aquatica
Mistletoe
Phoradendron serotinum
Alligator-weed
Alternanthera philoxeroides
Sweetbay
Magnolia virginiana
Redbay
Persea borbonia
Sassafras
Sassafras albidum
Sundew
Drosera capillaris
Sweet-gum
Liquidambar styraciflua
Dewberry
Rubus trivialis
Red Chokeberry
Sorbus arbutifolia
Shadbush
Amelanchier canadensis
Shadbush; Serviceberry
Amelanchier sp.
Black Cherry; Wild Cherry
Prunus serotina
Cranesbill
Geranium carolinianum
Poison-ivy
Rhus radicans
Titi
Cyrilla racemiflora
American Holly
Ilex opaca
Inkberry
Ilex glabra
Red Maple
Acer rubrum
Supple-jack; Rattan-vine
Berchemia scandens
Muscadine; Bullace
Vitis rotundifolia
Loblolly-bay
Gordonia lasianthus
Hypericum
Hypericum sp.
Primrose-leaved Violet
Viola primulifolia
Prickly-pear
Opuntia compressa
Water Pennywort
Hydrocotyle ranunculoides
Sweet Pepperbush
Clethra alnifolia
Fetterbush
Lyonia lucida
Sparkleberry
Vaccinium arboreum
Elliott's Blueberry
Vaccinium elliottii
Highbush Blueberry
Vaccinium corymbosum
Creeping Blueberry
Vaccinium crassifolium
Blueberry
Vaccinium sp.
Sheep-laurel; Lamb-kill
Kalmia angustifolia

Ervin Dargan Heritage Preserve and Little Pee Dee River Heritage Perserve (continued)
March 25, 1995

Pyxie-moss
Pyxidanthera barbulata barbulata
Storax; Snowbell
Styrax americana
Yellow Jessamine
Gelsemium sempervirens
Dodder
Cuscuta sp.
Toadflax
Linaria canadensis
Trumpet Vine
Campsis radicans
Japanese Honeysuckle
Lonicera japonica

FAUNA

Bowl-and-doily Spider
Frontinella pyramitela
Fishing Spider
Dolomedes sp.
Stonefly
Isoperla bilineata
American Bird Grasshopper
Schistocerca americana
Honey Bee
Apis mellifera
Bee Fly
Bombylius sp.
Springtime Darner
Basiaeschna janata
Common Baskettail (dragonfly)
Epitheca cynosura
Baskettail Dragonfly
Epitheca sp.
Painted Skimmer
Libellula semifasciata
Fragile Forktail (damselfly)
Ischnura posita
Green Tiger Beetle
Cicindela sexguttata
Crawling Water Beetle
Peltodytes sp.
Predacious Diving Beetle
Hydroporus clypealis
Hydroporus hybridus
Hydroporus cimicoides
Coptotomus interrogatus
Whirligig Beetle
Dineutus emarginatus
Gyrinus rockinghamensis
Zebra Swallowtail
Eurytides marcellus
Eastern Tiger Swallowtail
Papilio glaucus
Spring Azure (butterfly)
Celastrina ladon
Pearl Crescent
Phyciodes tharos
American Lady (butterfly)
Vanessa virginiensis
Polyphemus Moth (cocoon)
Antheraea polyphemus
Luna Moth
Actias luna
Eastern Mudminnow
Umbra pygmaea
Mummichog (fish)
Fundulus heteroclitus
Southern Toad
Bufo terrestris
Pine Woods Treefrog
Hyla femoralis
Yellow-bellied Slider
Chrysemys scripta
Carolina Anole
Anolis carolinensis
Banded Water Snake
Nerodia fasciata
Mallard
Anas platyrhynchos
Turkey Vulture
Cathartes aura
Black Vulture
Coragyps atratus
Red-shouldered Hawk
Buteo lineatus
Red-tailed Hawk
Buteo jamaicensis
American Kestrel
Falco sparverius
Barred Owl
Strix varia
Chimney Swift
Chaetura pelagica
Red-bellied Woodpecker
Melanerpes carolinus
Tree Swallow
Tachycineta bicolor
Fish Crow
Corvus ossifragus
Ruby-crowned Kinglet
Regulus calendula
Blue-gray Gnatcatcher
Polioptila caerulea
Eastern Bluebird
Sialia sialis
Hermit Thrush
Catharus guttatus
American Robin
Turdus migratorius
White-eyed Vireo (nest)
Vireo griseus
Northern Parula
Parula americana
Black-and-white Warbler
Mniotilta varia
Yellow-rumped Warbler
Dendroica coronata
Yellow-throated Warbler
Dendroica dominica
Pine Warbler
Dendroica pinus
Common Grackle
Quiscalus quiscula
Raccoon (tracks)
Procyon lotor
White-tailed Deer (tracks)
Odocoileus virginianus

FOSSILS

Scallop
Pecten sp.
Oyster
Ostrea sp.

Lewis Ocean Bay Heritage Preserve
Horry County
May 25, 1990

This Heritage Preserve of about 6000 acres actually contains more than twenty Carolina bays, allowing it to protect the largest concentration of Venus'-fly-traps in the state. It also protects a large number of other wetlands habitat plants including several orchids and various carnivorous plants. Birds and insects abound as well.

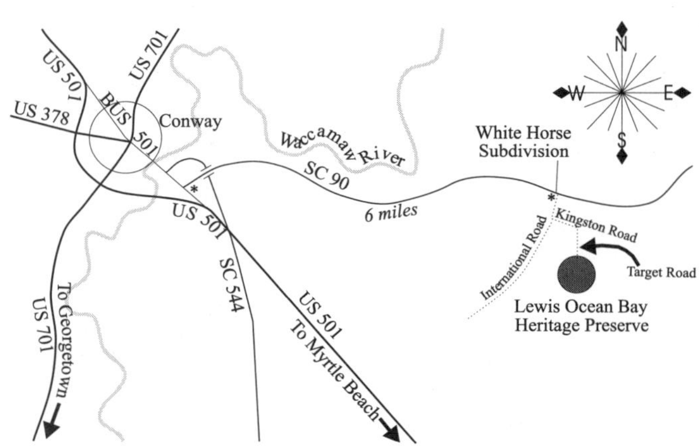

Directions to Lewis Ocean Bay Heritage Preserve: From downtown Conway, take BUS 501 southeast toward Myrtle Beach. (Do not take the US 501 Bypass, because it misses the correct intersection.) After crossing the bridge over the Waccamaw River, turn left at the traffic light onto SC 90 east. Drive about 6 miles and turn right onto International Road, a dirt road directly across from the White Horse Subdivision. Turn left at the first fork which is Kingston Road. Follow Kingston Road until it dead ends into Target Road. The entrance gate to Lewis Ocean Bay is located to the right.

FLORA

Fruticose Lichen
Cladina evansii
Cladonia coniocraea
Cladonia didyma
Cladonia furcata
Cladonia leporina
Cladonia peziziformis
Cladonia rappii
Cladonia ramulosa
Parmotrema michauxianum
Parmotrema perforatum
Cinnamon Fern
Osmunda cinnamomea
Bracken
Pteridium aquilinum
Long-leaf Pine
Pinus palustris
Loblolly Pine
Pinus taeda
Pond Pine
Pinus serotina
Swordgrass; Bulrush
Scirpus americanus
Beak-rush
Rhynchospora intermixta
Rhynchospora plumosa
Rhynchospora megalocarpa
Rhynchospora rariflora
Nut-rush
Scleria ciliata
Sedge
Carex lurida
Yellow-eyed-grass
Xyris flabelliformis
Bog-buttons
Lachnocaulon sp.
Roseling; Spiderwort
Tradescantia rosea
Toad Rush
Juncus bufonius
Fringed Greenbrier
Smilax bona-nox
Bamboo Smilax
Smilax laurifolia
Bear-grass
Yucca filamentosa
Colic-root
Aletris farinosa
Black Snakeroot
Zigadenus densus
Redroot
Lachnanthes caroliniana
Flag
Iris tridentata
Spreading-pogonia
Cleistes divaricata
Bearded Grass-pink
Calopogon barbatus
Grass-pink
Calopogon pulchellus
Grass-leaved Ladies'-tresses
Spiranthes praecox
Wax-myrtle; Southern Bayberry
Myrica cerifera
Sand Live Oak
Quercus virginiana geminata
Turkey Oak
Quercus laevis
Blue-jack Oak
Quercus incana
Sweetbay
Magnolia virginiana
Redbay
Persea borbonia
Trumpet Pitcher-plant
Sarracenia flava
Frog's-britches
Sarracenia purpurea
Sundew
Drosera intermedia
Drosera capillaris
Venus'-fly-trap
Dionaea muscipula
Wild Indigo
Baptisia cinerea
Samson Snakeroot
Psoralea psoralioides
Polygala
Polygala grandiflora
Orange Milkwort
Polygala lutea
Tread-softly
Cnidoscolus stimulosus
Inkberry
Ilex glabra
Sweet Gallberry
Ilex coriacea
Loblolly-bay
Gordonia lasianthus
Prickly-pear
Opuntia compressa
Smooth Meadow-beauty
Rhexia alifanus
Marsh Parsley
Apium leptophyllum
Sweet Pepperbush
Clethra alnifolia
Maleberry
Lyonia ligustrina
Fetterbush
Lyonia lucida
Staggerbush
Lyonia mariana
Dwarf Huckleberry
Gaylussacia dumosa
Dangleberry
Gaylussacia frondosa
Low-bush Blueberry
Vaccinium tenellum
Creeping Blueberry
Vaccinium crassifolium
Water Pimpernel
Samolus parviflorus
Fog-fruit
Lippia nodiflora
Woolly Mullein
Verbascum thapsus
Wiry Bladderwort
Utricularia subulata
Lobelia
Lobelia nuttallii
Black-root
Pterocaulon pycnostachyum
Aster
Aster squarrosus

FAUNA

Tiger Beetle
Cicindela punctulata
Palamedes Swallowtail
Papilio palamedes
Common Buckeye (butterfly)
Junonia coenia
Oak Toad
Bufo quercicus
Six-lined Racerunner
Cnemidophorus sexlineatus
Turkey Vulture
Cathartes aura
Red-cockaded Woodpecker
Picoides borealis
Eastern Phoebe
Sayornis phoebe
Northern Parula
Parula americana
Prairie Warbler
Dendroica discolor
Pine Warbler
Dendroica pinus
Yellowthroat
Geothlypis trichas
Yellow-breasted Chat
Icteria virens
Indigo Bunting
Passerina cyanea
Rufous-sided Towhee
Pipilo erythrophthalmus
White-tailed Deer (tracks)
Odocoileus virginianus

FOSSILS

Sand-dollar
Mellita carolinana
Exogyra Snail
Exogyra costata
Southern Moon Snail
Polinices duplicatus
Atlantic Slipper
Crepidula fornicata
Lettered Olive
Oliva sayana
Olive
Olivella mutica
Olivella carolinae
Ark
Anadara aequicostata
Bittersweet Clam
Glycymeris americana
Glycymeris aratus
Mussel
Brachiodontes sp.
Scallop
Carolinapecten arboreus
Kitten's-paw
Plicatula marginata
Oyster
Crassostrea virginiana
Cardita Clam
Cyclocardia granulata
Astarte Clam
Astarte glenni
Gibb's Clam
Eucrassatella speciosa
Dwarf Surf Clam
Mulinia lateralis
Wedge Rangia
Rangia cuneata
Clam
Corbicula densata
Quahog Clam
Mercenaria sp.
Broad-ribbed Chione
Chione latilirata
Cross-barred Chione
Chione cancellata
Gouldia Shell
Gouldia metastriata

(Photo by A. Ciegler)

Venus'-fly-trap, *Dionaea muscipula,* eats a wolf spider.

Playcard Environmental Education Center
Horry County
September 15, 1990, and December 10, 1994

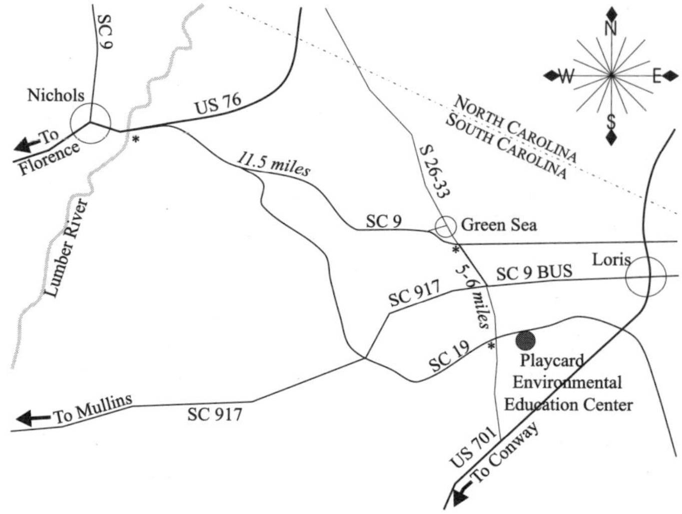

This environmental education center offers hands-on displays and demonstration areas that help school teachers and their students discover the need for a responsible approach to our environment. There are several interesting habitats accessed by short trails.

Some ponds have been set up to demonstrate the use of the Clemson Beaver Pond Leveler, a device which controls the water level in the ponds without triggering a beaver's natural instinct to stop the flow. Other demonstration areas were in the process of being developed.

Owned and operated by a private non-profit organization. Telephone: (843) 756-1277.

Directions to Playcard Environmental Education Center: From Florence, take US 76 east through Marion and Mullins to the Lumber River. Continue on US 76/SC 9 for 0.7 miles, and fork off to the right onto SC 9. About 11-12 miles after crossing the river, SC 9 bypasses the town of Green Sea. Turn right (south) onto Business SC 9/S 26-33. In about 2 miles (at Finkleas Crossroads) Business SC 9 turns to the left toward Loris; however continue straight ahead on S 26-33 for another 3 miles to Playcard Crossroads, the intersection of SC 19. Turn left onto SC 19 and go a short distance up the hill to the entrance on the right.

FLORA

Lichen
Graphis sp.
Cladonia caespiticia
Cladonia capitata
Cladonia coniocraea
Cladonia mateocyatha
Lecanora sp.
Bulbothrix goebelii
Hypogymnia krogiae
Hypotrachyna livida
Parmelia rudecta
Parmelia squarrosa
Parmelia sulcata
Parmelina dissecta
Parmelina galbina
Parmotrema cetratum
Parmotrema michauxianum
Parmotrema perforatum
Parmotrema rampoddense
Parmotrema reticulatum
Parmotrema robustum
Parmotrema subtinctorium
Pseudoparmelia caperata
Pseudoparmelia caroliniana
Usnea strigosa
Physcia aipolia
Physcia americana
Physcia millegrana
Physcia sorediosa
Physcia stellaris
Heterodermia albicans
Hetrerodermia casarettiana
Heterodermia echinata
Heterodermia obscurata
Heterodermia speciosa
Pyxine caesiopruinosa
Phaeophyscia imbricata
Anthracothecium leucostomum
Pyrenula cruenta
Pyrenula pseudobufonia
Arthopyrenia lapponina
Trypethelium mastoideum
Trypethelium virens
Trypethelium floridanum
Peat Moss
Sphagnum sp.
Southern Grape Fern
Botrychium biternatum
Cinnamon Fern
Osmunda cinnamomea
Bracken
Pteridium aquilinum
Southern Lady Fern
Athyrium asplenioides
Sensitive Fern
Onoclea sensibilis
Netted Chain Fern
Woodwardia areolata
Ebony Spleenwort
Asplenium platyneuron
Resurrection Fern
Polypodium polypodioides
Loblolly Pine
Pinus taeda
Bald-cypress
Taxodium distichum
Eastern Red-cedar
Juniperus virginiana
Bur-reed
Sparganium americanum
Cane; Switch Cane
Arundinaria gigantea
Uniola Grass
Uniola laxa
Grass
Uniola sp.
Rice Cut Grass
Leersia oryzoides
Giant Foxtail Grass
Setaria magna
Panic Grass
Panicum sp.
Broomsedge
Andropogon virginicus
Andropogon sp.
Dulichium (sedge)
Dulichium arundinaceum
Wool-grass
Scirpus cyperinus
Sedge
Carex glaucescens
Arrowleaf Arum
Peltandra virginica
Spanish-moss
Tillandsia usneoides
Asian Spiderwort
Aneilema keisak
Greenbrier; Bullbrier
Smilax rotundifolia
Fringed Greenbrier
Smilax bona-nox
Bamboo Smilax
Smilax laurifolia
Bear-grass
Yucca filamentosa
Bellwort
Uvularia pudica
Wild Yam
Dioscorea villosa
Flag; Iris
Iris sp.
Crane-fly Orchid
Tipularia discolor
Lizard's-tail
Saururus cernuus
Swamp Cottonwood
Populus heterophylla
Wax-myrtle; Southern Bayberry
Myrica cerifera
Swamp Bayberry
Myrica heterophylla
Mockernut Hickory
Carya tomentosa
Hickory
Carya sp.
Tag Alder
Alnus serrulata
White Oak
Quercus alba

Post Oak
Quercus stellata
Swamp Chestnut Oak
Quercus michauxii
Live Oak
Quercus virginiana
Red Oak
Quercus rubra
Spanish Oak
Quercus falcata
Water Oak
Quercus nigra
Willow Oak
Quercus phellos
Red Mulberry
Morus rubra
False Nettle
Boehmeria cylindrica
Mistletoe
Phoradendron serotinum
Heartleaf
Hexastylis arifolia
Lamb's-quarters; Pigweed
Chenopodium album
Pokeweed; Pokeberry
Phytolacca americana
Spatterdock; Yellow Pond-lily
Nuphar luteum
Water-lily
Nymphaea odorata
Leather-flower
Clematis crispa
Clematis
Clematis sp.
Tulip-poplar
Liriodendron tulipifera
Sweetbay
Magnolia virginiana
PawPaw
Asimina sp.
Redbay
Persea borbonia
Sassafras
Sassafras albidum
Wild Radish
Raphanus raphanistrum
Virginia-willow
Itea virginica
Sweet-gum
Liquidambar styraciflua
Witch-hazel
Hamamelis virginiana
American Sycamore
Platanus occidentalis
Blackberry; Brambles
Rubus sp.
Red Chokeberry
Sorbus arbutifolia
Black Cherry; Wild Cherry
Prunus serotina
Mimosa; Silk-tree
Albizia julibrissin
Wild Sensitive-plant
Cassia nicitans
Wild Indigo
Indigofera caroliniana
Groundnut
Apios americana
Cranesbill
Geranium carolinianum
Chinaberry
Melia azedarach
Tread-softly
Cnidoscolus stimulosus
Poison Sumac
Rhus vernix
Poison-ivy
Rhus radicans
Winged Sumac
Rhus copallina
Titi
Cyrilla racemiflora
American Holly
Ilex opaca
Inkberry
Ilex glabra
Sweet Gallberry
Ilex coriacea
Red Maple
Acer rubrum
Supple-jack; Rattan-vine
Berchemia scandens
Virginia-creeper
Parthenocissus quinquefolia
Muscadine; Bullace
Vitis rotundifolia
Summer Grape
Vitis aestivalis
Silky Camellia
Stewartia malacodendron
Loblolly-bay
Gordonia lasianthus
St.-Andrew's-cross
Hypericum hypericoides
St.-John's-wort
Hypericum walteri
Prickly-pear
Opuntia compressa
Meadow-beauty
Rhexia sp.
Ludwigia
Ludwigia leptocarpa
Ludwigia pilosa
Ludwigia sp.
Blackgum
Nyssa sylvatica
Swamp Tupelo; Swamp Blackgum
Nyssa sylvatica biflora
Flowering Dogwood
Cornus florida
Sweet Pepperbush
Clethra alnifolia
Spotted Wintergreen
Chimaphila maculata
Indian-pipe
Monotropa uniflora
Piedmont Azalea
Rhododendron canescens
Pinxter-flower; Pinxter-bloom
Rhododendron nudiflorum
Fetterbush
Lyonia lucida
Dog-hobble
Leucothoe axillaris
Coastal Honeybells
Leucothoe racemosa
Dangleberry
Gaylussacia frondosa
Sparkleberry
Vaccinium arboreum
Low-bush Blueberry
Vaccinium tenellum
Highbush Blueberry
Vaccinium corymbosum
Persimmon
Diospyros virginiana
Horse-sugar; Sweetleaf
Symplocos tinctoria
Bigleaf Snowbell
Styrax grandifolia
Chinese Privet
Ligustrum sinense
Privet; Ligustrum
Ligustrum vulgare
Yellow Jessamine
Gelsemium sempervirens
Compact Dodder
Cuscuta compacta
Dodder
Cuscuta sp.
Brazilian Vervain
Verbena braziliensis
White Vervain
Verbena urticifolia
Beautyberry
Callicarpa americana
Skullcap
Scutellaria elliptica
Common Henbit
Lamium amplexicaule
Stachys
Stachys sp.
Dotted Horsemint
Monarda punctata
Mountain Mint
Pycnanthemum muticum
Lycopus
Lycopus rubellus
False Foxglove
Aureolaria sp.
Cross Vine
Anisostichus capreolata
Trumpet Vine
Campsis radicans
Bladderwort
Utricularia biflora
Ruellia
Ruellia caroliniensis
Button-bush
Cephalanthus occidentalis
Partridge-berry
Mitchella repens
Japanese Honeysuckle
Lonicera japonica
Coral Honeysuckle
Lonicera sempervirens
Possumhaw Viburnum
Viburnum nudum
American Elderberry
Sambucus canadensis
Cardinal-flower
Lobelia cardinalis
Common Ragweed
Ambrosia artemisiifolia
Gall-of-the-earth
Prenanthes serpentaria
Ironweed
Vernonia angustifolia
Elephant's-foot
Elephantopus tomentosus
Elephantopus sp.
Thoroughwort
Eupatorium dubium
Dog-fennel
Eupatorium capillifolium
Climbing Hempweed
Mikania scandens
Marsh-fleabane
Pluchea rosea
Pluchea foetida
Sea-myrtle; Groundsel-tree
Baccharis halimifolia
Sweet Goldenrod
Solidago odora
Goldenrod
Solidago sp.
Beggar-ticks
Bidens frondosa
Spanish-needles
Bidens bipinnata
Headlike Cordyceps
Cordyceps capitata
Hypoxylon Canker
Hypoxylon sp.
Black-knot
Apiosporina morbosa
Witches'-butter
Tremella mesenterica
Spindle-shaped Yellow Coral
Clavulinopsis fusiformis
Dry-rot Fungus
Serpula sp.
Velvet Bluespread
Pulcherrichium caeruleum
Chicken Mushroom
Laetiporus sulphureus
Turkey-tail
Trametes versicolor
Crowded Parchment
Stereum complicatum
False Turkey-tail
Stereum ostrea
Silky Parchment
Stereum striatum
Red-tinged Lepiota
Lepiota rubrotincta
Cleft-foot Amanita
Amanita brunnescens
Citron Amanita
Amanita citrina
Slippery-Jack
Suillus luteus
Chanterelle Waxy-cap
Hygrophorus cantharellus

Playcard Environmental Education Center (continued) September 15, 1990, and December 10, 1994

Emetic Russula
Russula emetica
Fragile Russula
Russula fragilis
Booted Collybia
Collybia peronata
Oyster Mushroom
Pleurotus ostreatus
Mini Puffball
Lycoperdon pusillum
Columned Stinkhorn
Clathrus columnatus

FAUNA

Pregnant Gingernail Clam
Sphaerium partumeium
Crab-shaped Spiny-back
Gasteracantha cancriformis
Spiny-backed Spider
Micrathena gracilis
Orchard Spider
Leucauge venusta
Banded Garden Spider
Argiope trifasciata
Basilica Spider
Mecynogea lemniscata
Orb-weaver
Neoscona hentzi
Arrowhead Spider
Verrucosa arenata
Long-jawed Orb-weaver
Tetragnatha laboriosa
Six-spotted Fishing Spider
Dolomedes triton
Rabid Wolf Spider
Lycosa rabida
Green Lynx Spider
Peucetia viridans
Common Meadow Katydid
Orchelimum vulgare
Short-tailed Cricket
Anurogryllus muticus
Field Cricket
Gryllus sp.
Woolly Alder Aphid
Prociphilus tesselatus
Blackberry Knot Gall (wasp)
Diastrophus nebulosus
False Nettle Stem Gall (midge)
Cecidomyia boehmeriae
Cypress Leaf Gall (midge)
Itonida taxodii
Eastern Pondhawk
Erythemis simplicicollis
Blue Dasher
Pachydiplax longipennis
Searcher; Ground Beetle
Carabus vinctus

Dingy Ground Beetle
Harpalus caliginosus
Harpalus pennsylvanicus
Anisodactylus sayi
Notch-mouthed Ground Beetle
Dicaelus ovalis
Crawling Water Beetle
Peltodytes sp.
Predacious Diving Beetle
Coptotomus interrogatus
Burrowing Water Beetle
Hydrocanthus oblongus
Whirligig Beetle
Gyrinus borealis
Water Scavenger Beetle
Tropisternus quadristriatus quadristriatus
Rove Beetle
Oxytelus sp.
Omalium sp.
Crab-like Rove Beetle
Lordithon pygmaeus
Tachinus sp.
Small Carrion Beetle
Catops sp.
Hister Beetle
Margarinotus sp.
Dung Beetle
Copris carolina
Phanaeus vindex
Phanaeus nigrocyaneus
Onthophagus hecate
Onthophagus janus
Skin Beetle
Trox tuberculosus

Darkling Beetle
Hoplocephala viridipennis
Platydema excavatum
Sap-feeding Beetle
Carpophilus sp.
Minute Tree-fungus Beetle
Xestocis sp.
Flea Beetle
Altica marevagans
Eastern Tiger Swallowtail
Papilio glaucus
Palamedes Swallowtail
Papilio palamedes
Cloudless Sulphur
Phoebis sennae eubule
Silver-spotted Skipper
Epargyreus clarus
Long-tailed Skipper
Urbanus proteus
Evergreen Bagworm Moth
Thyridopteryx ephemeraeformis
Saddleback Caterpillar (larva)
Sibine stimulea
Sweetbay Silkmoth (cocoon)
Callosamia securifera
Azalea Hand-maid Moth (larva)
Datana major
Southern Toad
Bufo terrestris
Green Treefrog
Hyla cinerea
Carolina Anole
Anolis carolinensis
Ground Skink
Scincella lateralis
Great Blue Heron
Ardea herodias
Wood Duck
Aix sponsa

Turkey Vulture
Cathartes aura
Red-tailed Hawk
Buteo jamaicensis
Red-bellied Woodpecker
Melanerpes carolinus
Red-headed Woodpecker
Melanerpes erythrocephalus
Downy Woodpecker
Picoides pubescens
Pileated Woodpecker
Dryocopus pileatus
Eastern Phoebe
Sayornis phoebe
Blue Jay
Cyanocitta cristata
Tufted Titmouse
Parus bicolor
Carolina Chickadee
Parus carolinensis
White-breasted Nuthatch
Sitta carolinensis
Carolina Wren
Thryothorus ludovicianus
Ruby-crowned Kinglet
Regulus calendula
Eastern Bluebird
Sialia sialis
American Robin
Turdus migratorius
Cedar Waxwing
Bombycilla cedrorum
Yellow-rumped Warbler
Dendroica coronata
Pine Warbler
Dendroica pinus
Beaver (gnaw mark)
Castor canadensis

(Photo by L. Greenlee)
Beavers, *Castor canadensis,* are evident at Playcard.

(Photo by L. Greenlee)
Green treefrog, *Hyla cinerea.*

Southern Aggregate Coquina Mine
Horry County
December 12, 1992

Many marine fossils have been exposed here in the process of mining the earth for building materials. Because it is a very disturbed site, only a few notable observations of living fauna were recorded. The fossils were diverse and relatively plentiful in the areas visitors could safely access on this particular day. The identification assistance from Aura Baker of Myrtle Beach and Lyle Campbell of Spartanburg was much appreciated.

Privately owned and currently closed. Not accessible to collectors. No map or directions are included.

FAUNA

Black Widow
Latrodectus mactans
Writing Spider
Argiope sp.
American Bird Grasshopper
Schistocerca americana
Fire Ant
Solenopsis sp.
Cloudless Sulphur
Phoebis sennae eubule
Carolina Anole
Anolis carolinensis
Great Blue Heron
Ardea herodias
Turkey Vulture
Cathartes aura
Yellow-bellied Sapsucker
Sphyrapicus varius
Raccoon
Procyon lotor
White-tailed Deer
Odocoileus virginianus

FOSSILS

Turritella Snail
Turritella perexilis
Marine Snail
Turritella sp.
Sand-dollar
Mellita carolinana
Tagelus
Tagelus gibbus
Whelk
Busycon carica amoena
Lightning Whelk
Busycon contrarium
Whelk
Busycon maximum
Exogyra Snail
Exogyra costata
Atlantic Slipper
Crepidula fornicata
Olive
Oliva carolinensis
Cone Shell
Conus adversarius
Conus waccamawensis
Ark
Anadara lienosa
Bittersweet Clam
Glycymeris americana
Glycymeris pectinata
Scallop
Argopecten vicenarius
Carolinapecten solarioides
Pecten brouweri
Oyster
Crassostrea virginica
Ostrea lawrencei
Cardita Clam
Cyclocardia granulata
Astarte Clam
Astarte concentrica
Gibb's Clam
Eucrassatella speciosa
Dwarf Surf Clam
Mulinia lateralis
Rangia Clam
Rangia clathrodonta
Clam
Corbicula densata
Hard-shelled Clam
Mercenaria campechiensis
Cross-barred Venus
Chione cribraria
Sand Shark
Odontaspis sp.
Ground Shark (tooth)
Carcharhinus sp.
Modern Great White Shark
Carcharodon carcharius
Mackerel Shark
Lamna sp.
Horse
Equus leidyi

(Photo by B. Gilbert)
Fossil casts of gastropods are typical of South Carolina fossil deposits.

Jasper County

Savannah National Wildlife Refuge

Jasper County
December 12, 1987

Birds, huge alligators, butterflies and dragonflies all flourish at this wildlife refuge. The largest number of birds is seen on winter trips, especially waterfowl and wading birds, for which the old rice fields at the refuge are managed. Because this area has been altered extensively by planting crops and animal attractants, the flora is not unusual. However, naturalized and exotic ornamental plants survive around old home sites here, such as the Chinese parasol-tree. The refuge has a road with an auto tour or you may walk the many farm roads and dikes for better views of the birds and other animals.

Directions to Savannah National Wildlife Refuge: Take I-95 south from Walterboro toward Georgia. At Exit 5 near Hardeeville, take US 17 south toward Savannah. Drive 4.5 miles and then go straight ahead on SC 170 where US 17 diverges to the left. In about 1-2 miles you will pass a large marsh on the right and come to an intersection with the Refuge Headquarters on the right and the parking area on the left; turn in to the left. The Wildlife Drive starts just beyond the parking area.

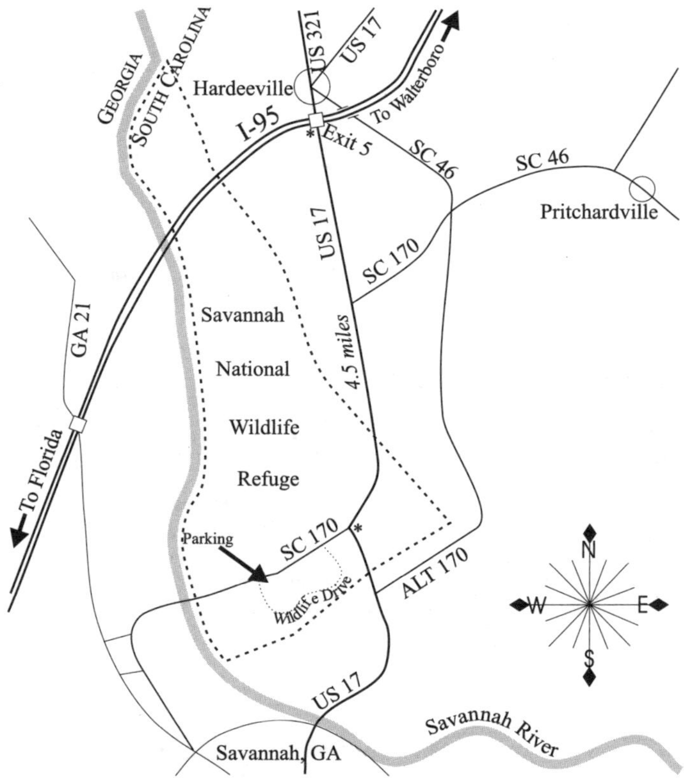

FLORA

Slash Pine
Pinus elliotti
Pond-cypress
Taxodium ascendens
Common Cat-tail
Typha latifolia
Cane; Switch Cane
Arundinaria gigantea
Southern Wild Rice
Zizaniopsis miliacea
Green Foxtail Grass
Setaria viridis
Oplismenus (grass)
Oplismenus setarius
Dwarf Palmetto
Sabal minor
Spanish-moss
Tillandsia usneoides
Sweet-scented Smilax
Smilax smallii
Swamp Willow
Salix caroliniana
Wax-myrtle; Southern Bayberry
Myrica cerifera
Live Oak
Quercus virginiana
Sugarberry; Hackberry
Celtis laevigata
Red Mulberry
Morus rubra
Sweet-gum
Liquidambar styraciflua
Dewberry
Rubus trivialis
Carolina Cherrylaurel
Prunus caroliniana
Sicklepod
Cassia obtusifolia
Bladderpod
Glottidium vesicarium
Chinaberry
Melia azedarach
Chinese Tallow-tree
Sapium sebiferum
Winged Sumac
Rhus copallina
American Holly
Ilex opaca
Yaupon
Ilex vomitoria
Red Maple
Acer rubrum
Rose Mallow; Wild Cotton
Hibiscus moscheutos
China Parasol-tree
Firmiana platanifolia
Periwinkle
Vinca major
Brazilian Vervain
Verbena braziliensis
Button-bush
Cephalanthus occidentalis
Japanese Honeysuckle
Lonicera japonica
American Elderberry
Sambucus canadensis
Elephant's-foot
Elephantopus carolinianus
Dog-fennel
Eupatorium capillifolium
False-willow
Baccharis angustifolia
Sea-myrtle; Groundsel-tree
Baccharis halimifolia
Rayless Goldenrod
Chondrophora nudata

FAUNA

Eastern Boxelder Bug
Leptocoris trivittatus
Cloudless Sulphur
Phoebis sennae eubule
Gulf Fritillary
Agraulis vanillae
American Alligator
Alligator mississippiensis
Yellow-bellied Slider
Chrysemys scripta
Cooter
Chrysemys sp.
Eastern Kingsnake
Lampropeltis getulus

Pied-billed Grebe
Podilymbus podiceps
Anhinga; Snakebird
Anhinga anhinga
Double-crested Cormorant
Phalacrocorax auritus
Green-backed Heron
Butorides striatus
Little Blue Heron
Egretta caerulea
Great Egret
Casmerodius albus
Great Blue Heron
Ardea herodias
Glossy Ibis
Plegadis falcinellus
White Ibis
Eudocimus albus
Snow Goose
Chen caerulescens
Blue Goose
Chen caerulescens
Mallard
Anas platyrhynchos
Green-winged Teal
Anas crecca
American Wigeon
Anas americana
Northern Pintail
Anas acuta
Northern Shoveler
Anas clypeata
Ruddy Duck
Oxyura jamaicensis
Wood Duck
Aix sponsa
Canvasback
Aythya valisineria
Ring-necked Duck
Aythya collaris
Lesser Scaup
Aythya affinis
Common Moorhen
Gallinula chloropus
American Coot
Fulica americana
Killdeer
Charadrius vociferus
Forster's Tern
Sterna forsteri
Turkey Vulture
Cathartes aura
Bald Eagle
Haliaeetus leucocephalus
Northern Harrier
Circus cyaneus
Red-tailed Hawk
Buteo jamaicensis
Belted Kingfisher
Ceryle alcyon
Red-bellied Woodpecker
Melanerpes carolinus
Downy Woodpecker
Picoides pubescens
Eastern Phoebe
Sayornis phoebe
Tree Swallow
Tachycineta bicolor
Blue Jay
Cyanocitta cristata
American Crow
Corvus brachyrhynchos
Carolina Chickadee
Parus carolinensis
House Wren
Troglodytes aedon
Golden-crowned Kinglet
Regulus satrapa
Ruby-crowned Kinglet
Regulus calendula
Blue-gray Gnatcatcher
Polioptila caerulea
American Robin
Turdus migratorius
Loggerhead Shrike
Lanius ludovicianus
Northern Mockingbird
Mimus polyglottos
Brown Thrasher
Toxostoma rufum
European Starling
Sturnus vulgaris
Northern Cardinal
Cardinalis cardinalis
Sharp-tailed Sparrow
Ammodramus caudacutus
Song Sparrow
Melospiza melodia
Dark-eyed Junco
Junco hyemalis
Eastern Meadowlark
Sturnella magna
Red-winged Blackbird
Agelaius phoeniceus
Common Grackle
Quiscalus quiscula
Boat-tailed Grackle
Quiscalus major
Hispid Cotton Rat (tracks) (sign)
Sigmodon hispidus
Gray Fox (scat)
Urocyon cinereoargenteus

Tillman Sand Ridge Heritage Preserve
Jasper County
July 28, 1990

This sandy area is one of the few sites in South Carolina that is home to the gopher tortoise. There are several plants that are limited to this type of habitat as well. It is a very hot, dry environment and both plants and animals have developed ways to cope. The gopher tortoise digs a large burrow several feet deep. Consequently, a variety of other animals share the burrow to escape from the heat of the sun, or escape from the regular wildfires that occur. These fires maintain this habitat, and are very important to the ecology of the region. The turkey oak, which is a dominant tree on this site, can turn its leaves vertically to escape the direct rays of the sun above and heat reflected from the white sands below. The flora-fauna list produced from this field trip is the result of a short one hour visit after a more extended visit at nearby Webb Wildlife Center in Hampton County.

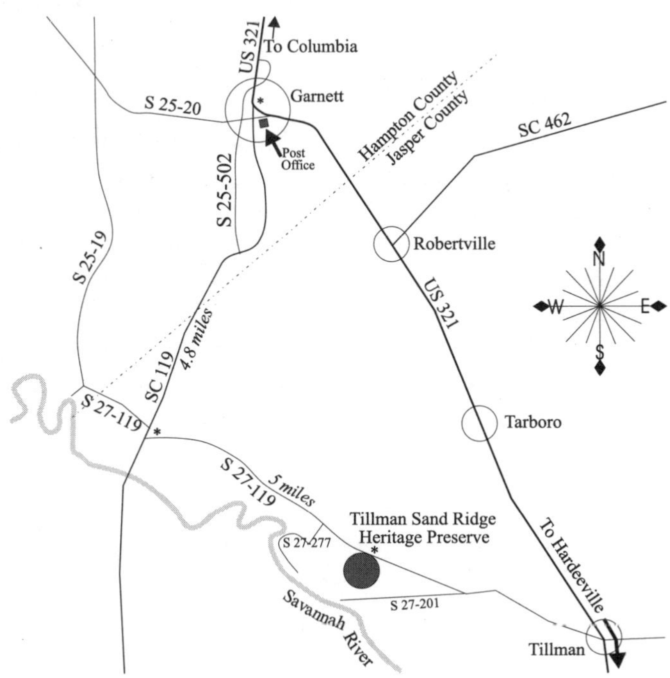

Directions to Tillman Sand Ridge Heritage Preserve: From Columbia, take US 321 south through Fairfax and Estill to Garnett, about 100 miles. In Garnett, take the right fork onto SC 119; the Post Office is on the left at the intersection. Continue south 4.8 miles farther, then turn left (east) onto County Road S 27-119 toward Tillman. The Heritage Preserve is on the right in about 5 miles, marked with an inconspicuous green sign.

Tillman Sand Ridge Heritage Preserve (continued)
July 28, 1990

FLORA

Lichen
Cladina evansii
Cladonia apodocarpa
Cladonia capitata
Cladonia coniocraea
Cladonia cylindrica
Cladonia subradiata
Bulbothrix goebelii
Hypotrachyna livida
Parmelia rudecta
Parmelia sp.
Parmotrema perforatum
Parmotrema praesorediosum
Parmotrema rampoddense
Parmotrema reticulatum
Parmotrema rigidum
Parmotrema robustum
Pseudoparmelia caperata
Pseudoparmelia caroliniana
Xanthoparmelia subramigera
Canoparmelia amazonica
Rimelia subisidiosa
Ramalina willeyi
Usnea strigosa
Heterodermia obscurata
Pyxine caesiopruinosa
Phaeophyscia ciliata

Long-leaf Pine
Pinus palustris
Slash Pine
Pinus elliotti
Wiregrass
Aristida stricta
Bulbostylis (sedge)
Bulbostylis warei
Beak-rush
Rhynchospora grayi
Saw Palmetto
Serenoa repens
White Oak
Quercus alba
Post Oak
Quercus stellata
Turkey Oak
Quercus laevis
Gopher-apple
Chrysobalanus oblongifolius
Wild Indigo
Indigofera caroliniana
Tephrosia
Tephrosia sp.
Crotonopsis
Crotonopsis linearis
Queen's-delight
Stillingia sylvatica
Prickly-pear
Opuntia compressa
Pale Meadow-beauty
Rhexia mariana
Blue-star
Amsonia ciliata
Sandhill Milkweed
Asclepias humistrata
Bonamia
Bonamia humistrata
Skullcap
Scutellaria sp.
Ironweed
Vernonia angustifolia
Blazing-star
Liatris sp.
Black-eyed-Susan
Rudbeckia hirta

FAUNA

Oval Leaf Beetle
Metachroma pallidum
Flower Weevil
Centrinaspis grisescens
Gopher Tortoise
Gopherus polyphemus
Turkey Vulture
Cathartes aura
Rufous-sided Towhee
Pipilo erythrophthalmus
Raccoon (bones)
Procyon lotor
White-tailed Deer (tracks)
Odocoileus virginianus

Kershaw County

Boykin Mill Pond
Kershaw County
June 28, 1986

One of the largest inland wading bird rookeries in South Carolina is Boykin Mill Pond. The pond, created in the 1830's, has a good stand of cypress which provides the nesting sites for the birds. As the list shows, the residents are numerous. This trip was highlighted by the use of several spotting scopes, so good views of adult birds and their nests were seen by all who attended without disturbing the birds. Flying overhead were flocks of adults heading out and returning from foraging – an impressive sight. The large number of other birds seen indicated the openness of the farm land in the immediate area of the pond. This site is no longer accessible except from public roadside. One can still watch the overhead flights of the birds, however.

Directions to Boykin Mill Pond: From Camden, take US 521 south toward Sumter; go under I-20, and after 2 miles bear right onto SC 261. Go 4 miles, crossing Swift Creek, and turn left onto Boykin Mill Road (county road S 28-2).

FLORA

Bracken
Pteridium aquilinum
Southern Lady Fern
Athyrium asplenioides
Netted Chain Fern
Woodwardia areolata
Ebony Spleenwort
Asplenium platyneuron
Resurrection Fern
Polypodium polypodioides
Loblolly Pine
Pinus taeda
Bald-cypress
Taxodium distichum
Pond-cypress
Taxodium ascendens
Common Cat-tail
Typha latifolia
Cane; Switch Cane
Arundinaria gigantea
Southern Wild Rice
Zizaniopsis miliacea
Spike-rush
Eleocharis equisetoides
Wool-grass
Scirpus cyperinus
Arrowleaf Arum
Peltandra virginica
Giant Duckweed
Spirodela punctata
Bogmat
Wolffiella floridana
Pickerelweed
Pontederia cordata
Soft Rush
Juncus effusus
Path Rush
Juncus tenuis

Sawbrier; Greenbrier
Smilax glauca
Lizard's-tail
Saururus cernuus
Black Willow
Salix nigra
Wax-myrtle; Southern Bayberry
Myrica cerifera
Tag Alder
Alnus serrulata
Post Oak
Quercus stellata
Water Oak
Quercus nigra
American Elm
Ulmus americana
Winged Elm
Ulmus alata
Sugarberry; Hackberry
Celtis laevigata
Red Mulberry
Morus rubra
False Nettle
Boehmeria cylindrica
Pokeweed; Pokeberry
Phytolacca americana
Spatterdock; Yellow Pond-lily
Nuphar luteum
Water-lily
Nymphaea odorata
Water-shield
Brasenia schreberi
Sweetbay
Magnolia virginiana
Sassafras
Sassafras albidum
Sweet-gum
Liquidambar styraciflua
Blackberry; Brambles
Rubus sp.
Cherokee Rose
Rosa laevigata
Southern Crabapple
Malus angustifolia
Black Cherry; Wild Cherry
Prunus serotina
Mimosa; Silk-tree
Albizia julibrissin
Pencil-flower
Stylosanthes biflora
Tephrosia
Tephrosia spicata
Climbing Butterfly-pea
Centrosema virginianum
Kudzu
Pueraria lobata
Poison-ivy
Rhus radicans
Winged Sumac
Rhus copallina
Titi
Cyrilla racemiflora
Boxelder
Acer negundo

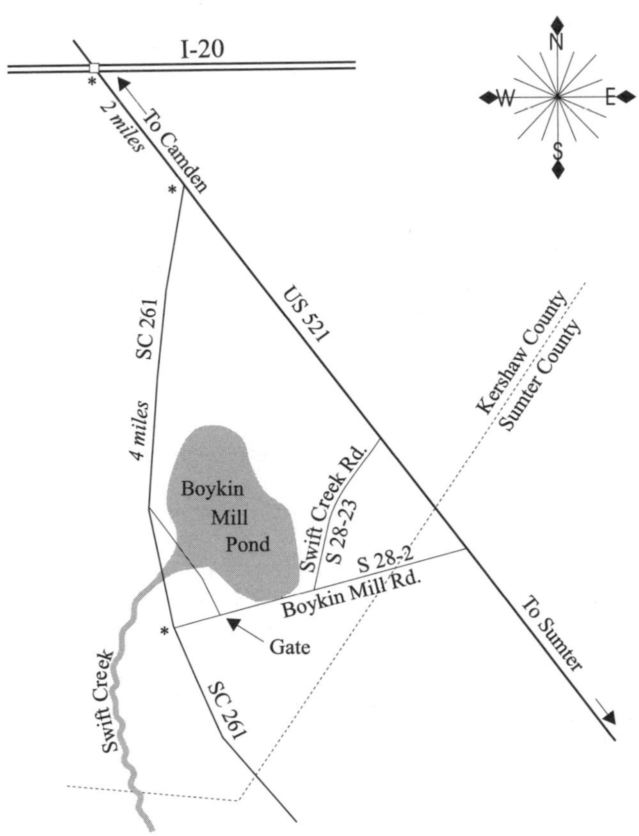

Red Maple
Acer rubrum
Virginia-creeper
Parthenocissus quinquefolia
Muscadine; Bullace
Vitis rotundifolia
Pepper-vine
Ampelopsis arborea
Devil's-shoestring
Sida rhombifolia
St.-Andrew's-cross
Hypericum hypericoides
Pin-weed
Lechea villosa
Water Loosestrife
Decodon verticillatus
Water-milfoil
Myriophyllum heterophyllum
Blackgum
Nyssa sylvatica
Swamp Tupelo; Swamp Blackgum
Nyssa sylvatica biflora
Water Tupelo
Nyssa aquatica
Fetterbush
Lyonia lucida
Coastal Honeybells
Leucothoe racemosa
Blueberry
Vaccinium sp.
Storax; Snowbell
Styrax americana
Yellow Jessamine
Gelsemium sempervirens
Polypremum
Polypremum procumbens
Small Floating-heart
Nymphoides cordata
Floating-heart
Nymphoides aquatica
Ground-ivy; Ponyfoot
Dichondra carolinensis
Brazilian Vervain
Verbena braziliensis
Beautyberry
Callicarpa americana
Carpenterweed
Prunella vulgaris
Horse-nettle
Solanum carolinense
Blue Water-hyssop
Bacopa caroliniana
Woolly Mullein
Verbascum thapsus
Trumpet Vine
Campsis radicans
Purple Bladderwort
Utricularia purpurea
Bladderwort
Utricularia biflora
English Plantain
Plantago lanceolata
Bracted Plantain
Plantago aristata
Button-bush
Cephalanthus occidentalis
Virginia Buttonweed
Diodia virginiana
American Elderberry
Sambucus canadensis
Wahlenbergia
Wahlenbergia marginata
Wild Lettuce
Lactuca canadensis
Small's Ragwort
Senecio smallii
Elephant's-foot
Elephantopus carolinianus
Sea-myrtle; Groundsel-tree
Baccharis halimifolia

FAUNA

Common Eastern Velvet-ant
Dasymutilla occidentalis occidentalis
Cypress Twig Gall (midge)
Thecodiplosis ananassi
Cypress Flower Gall (midge)
Itonida anthici
Banded Pennant (dragonfly)
Celithemis fasciata
Eastern Pondhawk
Erythemis simplicicollis
Blue Dasher
Pachydiplax longipennis
Eastern Amberwing (dragonfly)
Perithemis tenera
White-tail Dragonfly
Plathemis lydia
Violet-masked Glider
Tramea carolina
Eastern Tiger Swallowtail
Papilio glaucus
Sleepy Orange (butterfly)
Eurema nicippe
Pearl Crescent
Phyciodes tharos
Common Buckeye (butterfly)
Junonia coenia
Red-spotted Purple (butterfly)
Limenitis arthemis astyanax
Silver-spotted Skipper
Epargyreus clarus
Southern Cricket Frog
Acris gryllus
Bullfrog
Rana catesbiana
Green Frog; Bronze Frog
Rana clamitans
Carpenter Frog
Rana virgatipes
American Alligator
Alligator mississippiensis
Carolina Anole
Anolis carolinensis
Anhinga; Snakebird
Anhinga anhinga
Green-backed Heron
Butorides striatus
Little Blue Heron
Egretta caerulea

Boykin Mill Pond (continued)
June 28, 1986

Snowy Egret
Egretta thula
Cattle Egret
Bulbicus ibis
Great Egret
Casmerodius albus
Great Blue Heron
Ardea herodias
Glossy Ibis
Plegadis falcinellus
White Ibis
Eudocimus albus
Common Moorhen
Gallinula chloropus
Ruby-throated Hummingbird
Archilochus colubris
Northern Rough-winged Swallow
Stelgidopteryx serripennis
Barn Swallow
Hirundo rustica
American Crow
Corvus brachyrhynchos
Fish Crow
Corvus ossifragus
Northern Cardinal
Cardinalis cardinalis
Rufous-sided Towhee
Pipilo erythrophthalmus
Red-winged Blackbird
Agelaius phoeniceus
Summer Tanager
Piranga rubra

(Photo by A. Ciegler)
Immature little blue heron, *Egretta caerulea.*

(Photo by A. Ciegler)
White ibis, *Eudocimus albus.*

Goodale State Park
Kershaw County
October 26, 1991

Goodale is a small, underused park featuring classic sandhills habitats plus a very old mill pond with a large stand of pond cypress and other shallow pond aquatic species. There are some seepage areas that have additional carnivorous plants, and the back side of the dam has a nice population of Atlantic white-cedar. The nature trails wend through sand ridges and lower, more wet habitats, offering a wide variety of native material to be seen. This list represents about a half-day exploration.

Directions to Goodale State Park: From Columbia, take I-20 east toward Florence, and turn north onto US 521 at Exit 98 to Camden. In the center of town, turn right (north) onto US 1 toward McBee. Drive 3.2 miles on US 1 to Old Wire Road (S 28-15) and turn right. Go 2.4 miles and turn left onto S 28-331. The park entrance is on the right and marked by signs.

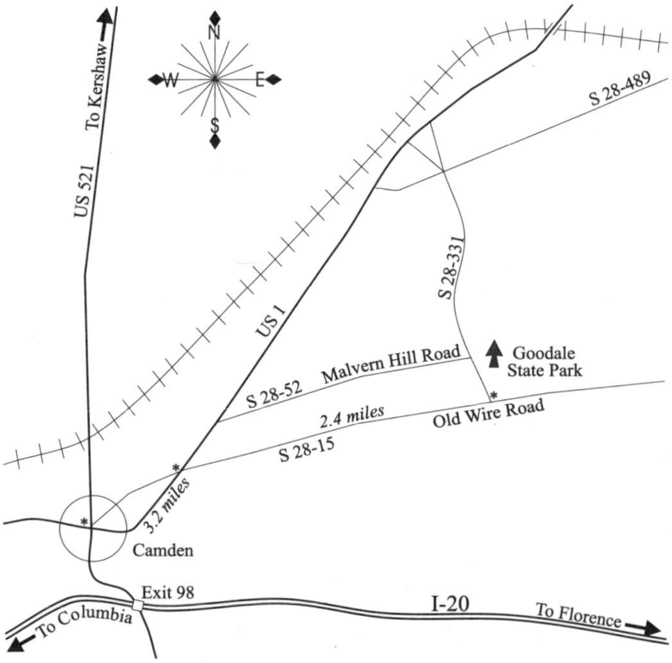

FLORA

Lichen
Graphis scripta
Lecidea russula
Cladonia calycantha
Cladonia capitata
Cladonia chlorophaea
Cladonia cristatella vestita
Cladonia didyma
Cladonia leporina
Cladonia polycarpoides
Cladonia ravenelii
Pertusaria ostiolata
Pertusaria paratuberculifera
Pertusaria texana
Pertusaria tetrathalmia
Pertusaria sp.
Haematomma puniceum
Lecanora miculata
Lecanora cupressi
Lecanora pallida
Lecanora pallida cancriformis
Lecanora varia
Lecanora allophana
Ochrolechia africana
Bulbothrix goebelii
Candelaria concolor
Cetraria sp.
Hypotrachyna livida
Parmelina aurulenta
Parmelina horrescens
Parmeliopsis subambigua
Parmotrema cetratum
Parmotrema crinitum
Parmotrema eurysacum
Parmotrema hypotropum
Parmotrema mellissii
Parmotrema michauxianum
Parmotrema perforatum
Parmotrema praesorediosum
Parmotrema subisidiosum
Parmotrema subtinctorium
Parmotrema tinctorum
Parmotrema ultralucens
Parmotrema xanthinum
Parmotrema sp.
Pseudoparmelia amazonica
Pseudoparmelia caperata
Pseudoparmelia caroliniana

Punctelia hypoleucites
Ramalina americana
Usnea ceratina
Usnea rubicunda
Usnea strigosa
Usnea trichodea
Buellia punctata
Buellia caloosensis
Buellia curtisii
Physcia aipolia
Physcia millegrana
Heterodermia albicans
Heterodermia obscurata
Heterodermia casarettiana
Pyxine caesiopruinosa
Phaeophyscia rubropulchra
Physconia detersa
Physconia enteroxantha
Xanthoria candelaria
Trypethelium virens
Slender Clubmoss
Lycopodium carolinianum
Royal Fern
Osmunda regalis spectabilis
Bracken
Pteridium aquilinum
Netted Chain Fern
Woodwardia areolata
Ebony Spleenwort
Asplenium platyneuron
Resurrection Fern
Polypodium polypodioides
Long-leaf Pine
Pinus palustris
Loblolly Pine
Pinus taeda
Bald-cypress
Taxodium distichum
Pond-cypress
Taxodium ascendens
Atlantic White-cedar
Chamaecyparis thyoides
Eastern Red-cedar
Juniperus virginiana
Common Cat-tail
Typha latifolia
Cane; Switch Cane
Arundinaria gigantea
Wire Grass
Aristida sp.
Little Bluestem
Andropogon scoparius
Golden-club
Orontium aquaticum
Bogmoss
Mayaca aubletii
Greenbrier; Bullbrier
Smilax rotundifolia
Bamboo Smilax
Smilax laurifolia
Bear-grass
Yucca filamentosa
Spoonleaf Yucca
Yucca filamentosa filamentosa
Nodding Ladies'-tresses
Spiranthes cernua cernua
Wax-myrtle; Southern Bayberry
Myrica cerifera
Mockernut Hickory
Carya tomentosa
Post Oak
Quercus stellata
Scrubby Post Oak
Quercus margaretta
Turkey Oak
Quercus laevis
Spanish Oak
Quercus falcata
Black Jack Oak
Quercus marilandica
Water Oak
Quercus nigra
Blue-jack Oak
Quercus incana
Winged Elm
Ulmus alata
Cottonweed
Froelichia gracilis
Water-lily
Nymphaea odorata
Water-shield
Brasenia schreberi
Tulip-poplar
Liriodendron tulipifera
Sweetbay
Magnolia virginiana
Southern Magnolia
Magnolia grandiflora
Redbay
Persea borbonia
Sassafras
Sassafras albidum
Sundew
Drosera leucantha
Drosera intermedia
Sweet-gum
Liquidambar styraciflua
Blackberry; Brambles
Rubus sp.
Red Chokeberry
Sorbus arbutifolia
Hawthorn
Crataegus sp.
Black Cherry; Wild Cherry
Prunus serotina
Carolina Cherrylaurel
Prunus caroliniana
Common Lupine
Lupinus diffusus
Three-seeded Mercury
Acalypha sp.
Poison-ivy
Rhus radicans
Poison-oak
Rhus toxicodendron
Winged Sumac
Rhus copallina
American Holly
Ilex opaca
Inkberry
Ilex glabra
Sweet Gallberry
Ilex coriacea
Red Maple
Acer rubrum
Virginia-creeper
Parthenocissus quinquefolia
Muscadine; Bullace
Vitis rotundifolia
St.-Andrew's-cross
Hypericum hypericoides
Pineweed
Hypericum gentianoides
Prickly-pear
Opuntia compressa
Black Tupelo; Blackgum
Nyssa sylvatica sylvatica
Swamp Tupelo; Swamp Blackgum
Nyssa sylvatica biflora
Flowering Dogwood
Cornus florida
Sweet Pepperbush
Clethra alnifolia
Spotted Wintergreen
Chimaphila maculata
Fetterbush
Lyonia lucida
Sparkleberry
Vaccinium arboreum
Highbush Blueberry
Vaccinium corymbosum
Persimmon
Diospyros virginiana
Horse-sugar; Sweetleaf
Symplocos tinctoria
Chinese Privet
Ligustrum sinense
Privet; Ligustrum
Ligustrum sp.
Yellow Jessamine
Gelsemium sempervirens
Beautyberry
Callicarpa americana
Blue-curls
Trichostema dichotomum
Dotted Horsemint
Monarda punctata
Blue Water-hyssop
Bacopa caroliniana
Gerardia
Agalinis setacea
Trumpet Vine
Campsis radicans
Purple Bladderwort
Utricularia purpurea
Bladderwort
Utricularia sp.
Poor-jo (or Po-jo)
Diodia teres
Japanese Honeysuckle
Lonicera japonica
Common Ragweed
Ambrosia artemisiifolia
Elephant's-foot
Elephantopus tomentosus
Carphephorus
Carphephorus sp.
White Snakeroot
Eupatorium rugosum
Eupatorium
Eupatorium sp.
Sea-myrtle; Groundsel-tree
Baccharis halimifolia
Horseweed; Mare's-tail
Erigeron canadensis ssp.
Aster
Aster dumosus
Sweet Goldenrod
Solidago odora
Rough-leaved Goldenrod
Solidago rugosa
Goldenrod
Solidago microcephala
Heterotheca
Heterotheca graminifolia
Camphorweed
Heterotheca subaxillaris

FAUNA

Crab-shaped Spiny-back
Gasteracantha cancriformis
Marbled Orb-weaver
Araneus marmoreus
Common Meadow Katydid
Orchelimum vulgare
Cattail Toothpick Grasshopper
Leptysma marginicollis
American Bird Grasshopper
Schistocerca americana
Antlion; Doodlebug
Myrmeleon immaculatum
Eastern Pondhawk
Erythemis simplicicollis
Blue Dragonlet
Erythrodiplax minuscula
White-tail Dragonfly
Plathemis lydia
Spreadwing Damselfly
Lestes sp.
Variable Dancer (damselfly)
Argia fumipennis
Tiger Beetle
Cicindela scutellaris unicolor
Cloudless Sulphur
Phoebis sennae eubule
Sleepy Orange (butterfly)
Eurema nicippe
Gulf Fritillary
Agraulis vanillae
Pearl Crescent
Phyciodes tharos
Common Buckeye (butterfly)
Junonia coenia
Monarch (butterfly)
Danaus plexippus
Yucca Giant-skipper (egg)
Megathymus yuccae
Woolly Bear Caterpillar
Pyrrharctia isabella
Banded Tussock Moth
Halysidota tessellaris
Dagger Moth
Acronicta sp.
Southern Toad
Bufo terrestris

Goodale State Park (continued)
October 26, 1991

Southern Cricket Frog
Acris gryllus
Bullfrog
Rana catesbiana
Yellow-bellied Slider
Chrysemys scripta
Carolina Anole
Anolis carolinensis
Ground Skink
Scincella lateralis
Six-lined Racerunner
Cnemidophorus sexlineatus
Eastern Garter Snake
Thamnophis sirtalis
Pied-billed Grebe
Podilymbus podiceps
Great Egret
Casmerodius albus
Great Blue Heron
Ardea herodias
Mallard
Anas platyrhynchos
Wood Duck
Aix sponsa
Killdeer
Charadrius vociferus
Turkey Vulture
Cathartes aura
Black Vulture
Coragyps atratus
Northern Flicker
Colaptes auratus
Pileated Woodpecker
Dryocopus pileatus
Eastern Phoebe
Sayornis phoebe
American Crow
Corvus brachyrhynchos
Fish Crow
Corvus ossifragus
Carolina Chickadee
Parus carolinensis
Brown-headed Nuthatch
Sitta pusilla
Carolina Wren
Thryothorus ludovicianus
Golden-crowned Kinglet
Regulus satrapa
Eastern Bluebird
Sialia sialis
Northern Mockingbird
Mimus polyglottos
Brown Thrasher
Toxostoma rufum
Yellow-rumped Warbler
Dendroica coronata
Common Grackle
Quiscalus quiscula
Gray Squirrel
Sciurus carolinensis
Raccoon (tracks)
Procyon lotor
White-tailed Deer (tracks)
Odocoileus virginianus

Savage Bay Heritage Preserve
Kershaw County
October 26, 1991

Savage Bay is a small bay, but deep in comparison to other Carolina bays, with standing water in its center most of the time. It has a very dense population of "bay" plants around the rim, and then the classic sand ridge that is indicative of most bays, with uncommon sandhills species such as the pine barren gentian. This bay has been timbered, but the cypress stumps have resprouted and young plants have grown to create a complete canopy over the water. The best way to see the bay is to use a canoe, or to wade. This trip is best remembered by SCAN members for the sight of many members standing around in knee-deep water while holding a variety of conversations. The sun filtering through the fall cypress leaves created a cathedral-like effect. There are no foot trails, but with old roads (be aware: some go onto private land) and by keeping sight of the vegetation along the rim of the bay, it is easy to explore the edges of this bay.

Directions to Savage Bay Heritage Preserve: From Columbia, take I-20 east toward Florence, and turn north onto US 521 at Exit 98 to Camden. In the center of town, turn right (north) onto US 1 toward McBee. Drive about 8.2 miles and turn right onto S 28-835; (if you come to the railroad underpass, you have gone too far.) Savage Bay is about half a mile on the right, marked with a sign.

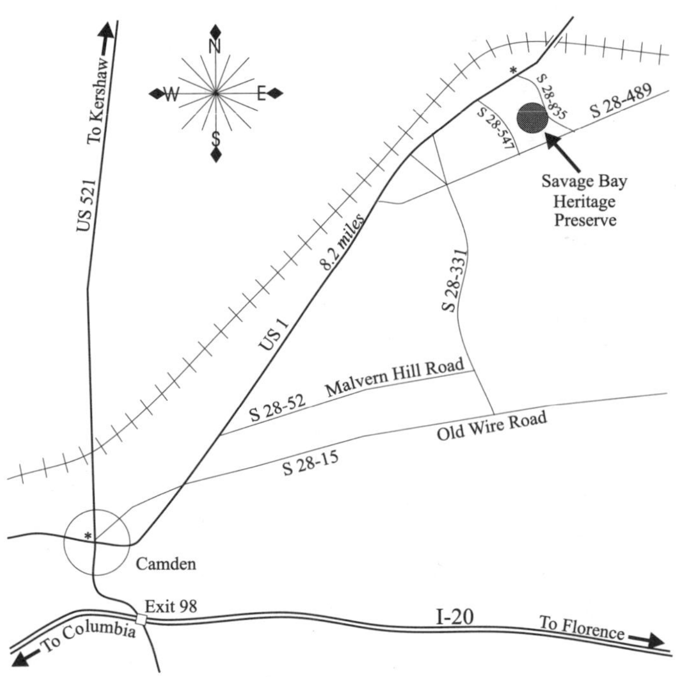

FLORA

Lichen
Graphis afzelii
Catillaria sp.
Cladonia calycantha
Cladonia chlorophaea
Cladonia cristatella
Cladonia cristatella vestita
Cladonia cristatella aurantiaca
Cladonia didyma
Cladonia leporina
Cladonia pityrea
Cladonia polycarpia
Cladonia polycarpoides
Pertusaria ostiolata
Pertusaria texana
Haematomma puniceum
Lecanora atra
Lecanora miculata
Lecanora allophana argentata
Lecanora pallida
Lecanora varia
Lecanora sp.
Ochrolechia africana
Bulbothrix confoederata
Bulbothrix goebelii
Bulbothrix isidiza
Hypotrachyna livida
Hypotrachyna pustulifera
Parmelina dissecta
Parmelina horrescens

Parmotrema cetratum
Parmotrema hypotropum
Parmotrema mellissii
Parmotrema michauxianum
Parmotrema perforatum
Parmotrema rampoddense
Parmotrema reticulatum
Parmotrema robustum
Parmotrema tinctorum
Pseudoparmelia caperata
Pseudoparmelia caroliniana
Pseudoparmelia crozalsiana
Canoparmelia texana
Ramalina willeyi
Usnea strigosa
Buellia punctata
Buellia caloosensis
Buellia leucomela
Buellia stillingiana
Physcia cloantha
Trypethelium sp.
Bracken
Pteridium aquilinum
Netted Chain Fern
Woodwardia areolata
Long-leaf Pine
Pinus palustris
Pond-cypress
Taxodium ascendens
Panic Grass
Panicum verrucosum
Beard Grass
Erianthus sp.
Little Bluestem
Andropogon scoparius
Wool-grass
Scirpus cyperinus
Pickerelweed
Pontederia cordata
Turkey Oak
Quercus laevis
Blue-jack Oak
Quercus incana
Clearweed
Pilea pumila
Wire-plant
Stipulicida setacea
Pondspice
Litsea aestivalis
Virginia-willow
Itea virginica
Sweet-gum
Liquidambar styraciflua
Red Chokeberry
Sorbus arbutifolia arbutifolia
Hog Plum
Prunus umbellata
Orange Milkwort
Polygala lutea
Poison-ivy
Rhus radicans
Winged Sumac
Rhus copallina
Titi
Cyrilla racemiflora
Inkberry
Ilex glabra
Red Maple
Acer rubrum
Muscadine; Bullace
Vitis rotundifolia
Prickly-pear
Opuntia compressa
Intelligence-plant
Centella asiatica
Swamp Tupelo; Swamp Blackgum
Nyssa sylvatica biflora
Sweet Pepperbush
Clethra alnifolia alnifolia
Honey-cup
Zenobia pulverulenta
Fetterbush
Lyonia lucida
Coastal Honeybells
Leucothoe racemosa
Persimmon
Diospyros virginiana
Pine-barren Gentian
Gentiana autumnalis
Bonamia
Bonamia sp.
Wiry Bladderwort
Utricularia subulata
Ironweed
Vernonia angustifolia angustifolia
Elephant's-foot
Elephantopus carolinianus
Blazing-star
Liatris graminifolia
Carphephorus
Carphephorus bellidifolius
Sweet Goldenrod
Solidago odora
Silphium
Silphium sp.

FAUNA

Basilica Spider
Mecynogea lemniscata
Long-jawed Orb-weaver
Tetragnatha laboriosa
Green Lynx Spider
Peucetia viridans
Jumping Spider (large)
Phidippus pulcher
Jumping Spider (small)
Metaphidippus galathea
Common Green Darner
Anax junius
Blue Dragonlet
Erythrodiplax minuscula
Spreadwing Damselfly
Lestes sp.
Tiger Beetle
Cicindela tranquebarica
Whirligig Beetle
Dineutus sp.
Large Rove Beetle
Staphylinus fossator
Darkling Beetle (larva)
Alobates pennsylvanica
Twenty-spotted Lady Beetle
Psyllobora vigintimaculata
Cloudless Sulphur
Phoebis sennae eubule
Little Yellow (butterfly)
Eurema lisa
Gulf Fritillary
Agraulis vanillae
Common Buckeye (butterfly)
Junonia coenia
Carolina Anole
Anolis carolinensis
Eastern Fence Lizard
Sceloporus undulatus
Northern Harrier
Circus cyaneus
Yellow-rumped Warbler
Dendroica coronata
Chipping Sparrow
Spizella passerina
Red-winged Blackbird
Agelaius phoeniceus
Common Grackle
Quiscalus quiscula

(Photos by L. Greenlee and A. Ciegler)

Pond-cypress, *Taxodium ascendens*.

(Photo by L. Greenlee)

The egg cases of the writing spider, *Argiope aurantia*, hang like ornaments.

Lancaster County

Flat Creek Heritage Preserve
Lancaster County
September 24, 1988, April 13, 1991, and April 16, 1994

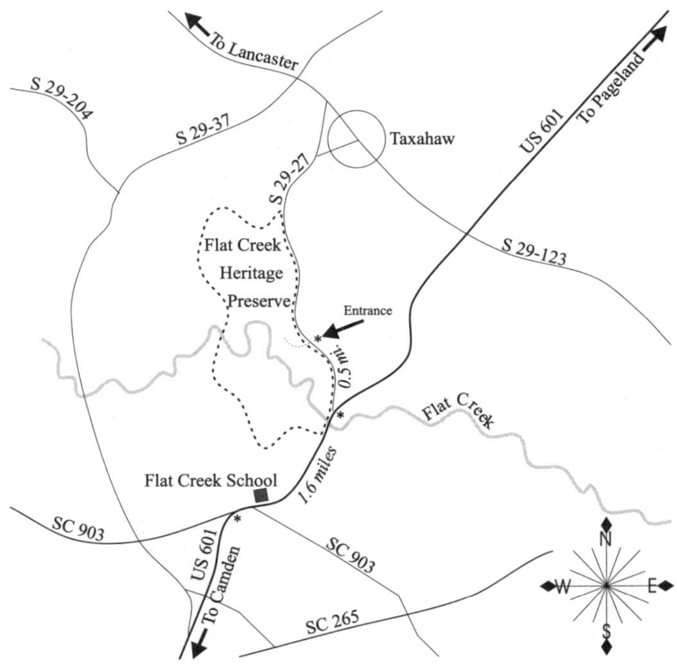

Flat Creek Heritage Preserve covers one of the most diverse small areas in the state. It includes a large exfoliating granite monadnock with classic vernal solution pools (Forty-acre Rock), a waterfall, a narrow valley that emulates mountainous cove areas, a cave, a beaver pond, a floodplain of a Piedmont creek and its bordering bluff and one of the largest basaltic dikes (Flat Creek Dike) in the eastern United States. This area harbors a very large number of unusual plants, especially in the pools on the rock and in the neutral soil over the dike. The pool sprite (*Amphianthus pusillus*) growing in the vernal pools is very rare, having been found on only one other granite outcrop. The two stonecrops are also quite rare. These plants carpet the pools in red fleshy leaves and white flowers in early spring.

Directions to Flat Creek Heritage Preserve: From Camden, drive north on US 521/601 to Kershaw (21 miles); turn right onto US 601 in downtown Kershaw and continue toward Pageland. Six miles farther, US 601 merges with SC 903. Continue on US 601/SC 903 for 1.6 miles, passing Flat Creek school on the left and crossing over the creek. Turn left onto SC 29-27. Go another half mile; the parking lot is on the left.

FLORA

Lichen

Collema subflaccidum
Leptogium burnetiae
Leptogium chloromelum
Leptogium corticola
Leptogium cyanescens
Leptogium lichenoides
Leptogium marginellum
Leptogium microstictum
Pannaria rubiginosa
Pannaria tavaresii
Parmeliella pannosa
Hydrothyria venosa
Coccocarpia erythroxyli
Nephroma helveticum
Peltigera canina
Peltigera horizontalis
Peltigera polydactyla
Lobaria pulmonaria
Lobaria quercizans
Lobaria ravenelii
Lobaria tenuis
Pseudocephalaria aurata
Pseudocephalaria crocata
Pseudocyphellaria sp.
Sticta fuliginosa
Sticta weigelii
Graphis scripta
Graphis nitida
Graphis beaumontii
Graphis afzelii
Graphis mosquitensis
Graphis atrorubens
Graphis tenella
Graphis intricata
Graphis scripta topographica
Graphis lineola
Graphis leucopepla
Graphis poitaeoides
Graphis elegans
Graphis striatula
Graphis grammatis
Glyphis cicatricosa
Gyrostomum scyphuliferum
Gyrostomum sp.
Medusulina nitida
Phaeographis erumpens
Sarcographa sp.
Xylographa disseminata
Thelotrema monosporum
Cladina arbuscula
Coenogonium interplexum
Coenogonium moniliforme
Coenogonium sp.
Dimerella lutea
Bacidia phacodes
Bacidia rosella
Bacidia chlorantha
Bacidia atrogrisea
Bilimbia lignaria
Bacidia polychroa
Bacidia schweinitzii
Bacidia incompta
Bacidia abductans
Bacidia suffusa
Bacidia dryina
Bacidia laurocerasi
Bacidia fuscus sp. nov.
Bacidia sp.
Leptogium austroamericanum
Brigantiaea leucoxantha
Catillaria glauconigrans
Lecidea varians
Lecidea nylanderi
Lecidea carneoalbens
Lecidea albocaerulescens
Lecidea congesta
Lecidea russula
Lecidea cyanea
Lecidea parasema
Lecidea parasema theioplaca
Lecidea promiscens
Lecidea myriocarpoides
Lecidea tessellata
Lecidea sp.
Porpidia macrocarpa
Rhizocarpon sp.
Letrouitia domingense
Stereocaulon dactylophyllum
Stereocaulon pileatum
Baeomyces fungoides
Cladina evansii
Cladina rangiferina
Cladina rangiferina f. crispata
Cladina stellaris
Cladonia abbreviatula
Cladonia apodocarpa
Cladonia bacillaris

Cladonia balfourii
Cladonia beaumontii
Cladonia caespiticia
Cladonia calycantha
Cladonia capitata
Cladonia cariosa
Cladonia caroliniana
Cladonia caroliniana caroliniana
Cladonia chlorophaea
Cladonia sobolescens
Cladonia coniocraea
Cladonia conista
Cladonia cristatella
Cladonia cylindrica
Cladonia didyma
Cladonia digitata
Cladonia floerkeana
Cladonia furcata
Cladonia gracilis
Cladonia incrassata
Cladonia leporina
Cladonia macilenta
Cladonia mateocyatha
Cladonia ochrochlora
Cladonia parasitica
Cladonia piedmontensis
Cladonia pityrea
Cladonia polycarpia
Cladonia polycarpoides
Cladonia pyxidata
Cladonia ravenelii
Cladonia rei
Cladonia robbinsii
Cladonia santensis
Cladonia squamosa
Cladonia strepsilis
Cladonia subsetacea
Cladonia uncialis
Cladonia verticillata
Cladonia subcariosa
Cladonia peziziformis
Pycnothelia papillaria
Umbilicaria mammulata
Melanaria macounii
Pertusaria multipunctata
Pertusaria amara
Pertusaria globularis
Pertusaria copiosa
Pertusaria neoscotica
Pertusaria leucostoma
Pertusaria pustulata
Pertusaria ostiolata
Pertusaria paratuberculifera
Pertusaria propinqua
Pertusaria plittiana
Pertusaria rubefacta
Pertusaria trachythallina
Pertusaria sinusmexicani
Pertusaria velata
Pertusaria macounii
Pertusaria xanthodes
Pertusaria tuberculifera
Pertusaria ophthalmiza
Pertusaria multipunctoides
Pertusaria valliculata
Pertusaria caroliniana sp. nov.
Maronea constans
Haematomma puniceum
Haematomma elatinum
Haematomma ochrophaeum
Lecania fuscella
Lecanora laevata
Lecanora cinerea
Lecanora deplanans
Lecanora lacustris
Lecanora atra
Lecanora hypoptoides
Lecanora miculata
Lecanora rugosa
Lecanora cenisia
Lecanora subfusca
Lecanora hageni
Lecanora willeyi
Lecanora cupressi
Lecanora pallida
Lecanora parisensis
Lecanora effusa
Lecanora varia
Candelaria concolor
Hypotrachyna livida
Parmotrema reticulatum
Xanthoparmelia conspersa
Ramalina americana
Usnea strigosa
Anaptychia palmatula
Physcia aipolia
Phaeophyscia imbricata
Trypethelium virens

Quillwort
Isoetes sp.
Rattlesnake Grape Fern
Botrychium virginianum
Common Grape Fern
Botrychium dissectum
Southern Grape Fern
Botrychium biternatum
Cinnamon Fern
Osmunda cinnamomea
Maidenhair Fern
Adiantum pedatum
Hairy Lip Fern
Cheilanthes lanosa
Woolly Lip Fern
Cheilanthes tomentosa
Lip Fern
Cheilanthes sp.
Bracken
Pteridium aquilinum
Southern Lady Fern
Athyrium asplenioides
Lady Fern
Athyrium filix-femina
Christmas Fern
Polystichum acrostichoides
Netted Chain Fern
Woodwardia areolata
Ebony Spleenwort
Asplenium platyneuron
Resurrection Fern
Polypodium polypodioides
Loblolly Pine
Pinus taeda
Short-leaf Pine
Pinus echinata
Eastern Red-cedar
Juniperus virginiana
Cane; Switch Cane
Arundinaria gigantea
Purple-top
Tridens flavus
Spike Grass
Uniola sessiliflora
River Oats
Uniola latifolia
Melic Grass
Melica mutica
Needle Grass
Stipa avenacea
Beard Grass
Gymnopogon ambiguus
Microstegium (grass)
Microstegium vimineum
Beard Grass
Erianthus sp.
Broomsedge
Andropogon ternarius
Fimbristylis (sedge)
Fimbristylis dichotoma
Jack-in-the-pulpit
Arisaema triphyllum
Giant Duckweed
Spirodela punctata
Duckweed
Lemna sp.
Water-meal
Wolffia sp.
Dayflower
Commelina erecta
Asian Spiderwort
Aneilema keisak
Spiderwort
Tradescantia sp.
Rush
Juncus georgianus
Woodrush
Luzula echinata
Carrion-flower
Smilax ecirrhata hugeri
Bamboo Smilax
Smilax laurifolia
Nodding Trillium
Trillium cernuum
Catesby's Trillium
Trillium catesbaei
False Solomon's Seal
Smilacina racemosa
Solomon's-seal
Polygonatum biflorum
Bear-grass
Yucca filamentosa
Dog's-tooth-violet
Erythronium americanum
Bellwort
Uvularia sessilifolia
False Garlic
Allium bivalve
Cinnamon-vine
Dioscorea batatas
Wild Yam
Dioscorea villosa
Blue-eyed-grass
Sisyrinchium sp.
Crested Iris
Iris cristata
Crane-fly Orchid
Tipularia discolor
Eastern Cottonwood
Populus deltoides
Black Walnut
Juglans nigra
Southern Shagbark Hickory
Carya carolinae-septentrionalis
Mockernut Hickory
Carya tomentosa
Pignut Hickory
Carya glabra
Hickory
Carya sp.
Tag Alder
Alnus serrulata
Hazel-nut
Corylus americana
Hop-hornbeam
Ostrya virginiana
American Hornbeam; Ironwood
Carpinus caroliniana
White Oak
Quercus alba
Post Oak
Quercus stellata
Rock Chestnut Oak
Quercus prinus
Turkey Oak
Quercus laevis
Black Oak
Quercus velutina
Spanish Oak
Quercus falcata
Black Jack Oak
Quercus marilandica
Willow Oak
Quercus phellos
Oak
Quercus sp.
Winged Elm
Ulmus alata
Sugarberry; Hackberry
Celtis laevigata
Hackberry
Celtis occidentalis georgiana
Red Mulberry
Morus rubra
False Nettle
Boehmeria cylindrica
Clearweed
Pilea pumila
Indian-olive
Nestronia umbellula
Wild Ginger
Asarum canadense
Heartleaf
Hexastylis arifolia
Jumpseed
Tovara virginiana
Pokeweed; Pokeberry
Phytolacca americana

Flat Creek Heritage Preserve (continued)
September 24, 1988, April 13, 1991, and April 16, 1994

Rock-portulaca
Talinum teretifolium
Purslane
Portulaca coronata
Giant Chickweed
Stellaria pubera
Rock Sandwort; Rockwort
Arenaria uniflora
Sandwort
Arenaria groenlandica
Mountain Sandwort
Arenaria groenlandica glabra
Wild Pink; Catchfly
Silene caroliniana
Water-shield
Brasenia schreberi
Yellowroot
Xanthorhiza simplicissima
Black Cohosh
Cimicifuga racemosa
Clematis
Clematis sp.
Windflower
Thalictrum thalictroides
Liverleaf
Hepatica americana
Thimbleweed
Anemone virginiana
Blue Cohosh; Papooseroot
Caulophyllum thalictroides
Mayapple
Podophyllum peltatum
Coralbeads
Cocculus carolinus
Moonseed
Menispermum canadense
Tulip-poplar
Liriodendron tulipifera
Cucumber-tree
Magnolia acuminata
PawPaw
Asimina triloba
Sweet-shrub; Carolina-allspice
Calycanthus floridus
Sassafras
Sassafras albidum
Spicebush
Lindera benzoin
Bloodroot
Sanguinaria canadensis
Draba
Draba aprica
Toothwort
Cardamine concatenata
Arabis
Arabis laevigata
Rock-cress
Arabis laevigata missouriensis
Stonecrop
Sedum pusillum
Vernal Pool Stonecrop
Sedum smallii

Stonecrop
Sedum ternatum
Mock-orange
Philadelphus inodorus
Alumroot
Heuchera americana
Foamflower
Tiarella cordifolia
Tiarella cordifolia collina
Sweet-gum
Liquidambar styraciflua
Witch-hazel
Hamamelis virginiana
American Sycamore
Platanus occidentalis
Five-fingers
Potentilla canadensis
Downy Serviceberry
Amelanchier arborea
Black Cherry; Wild Cherry
Prunus serotina
Redbud; Judas-tree
Cercis canadensis
Partridge-pea
Cassia fasciculata
Honey Locust
Gleditsia triacanthos
Sundial Lupine
Lupinus perennis
Carolina Clover
Trifolium carolinianum
Pencil-flower
Stylosanthes biflora
Groundnut
Apios americana
Hog-peanut
Amphicarpa bracteata
Violet Wood-sorrel
Oxalis violacea
Wild Geranium
Geranium maculatum
Button-rosy
Polygala curtissii
Tread-softly
Cnidoscolus stimulosus
Rushfoil
Crotonopsis elliptica
Flowering Spurge
Euphorbia corollata
Spurge
Euphorbia sp.
Poison-ivy
Rhus radicans
Fragrant Sumac
Rhus aromatica
Winged Sumac
Rhus copallina
Smooth Sumac
Rhus glabra
American Holly
Ilex opaca
Possumhaw Holly
Ilex decidua

Hearts-a'burstin'-with-love
Euonymus americanus
Bladdernut
Staphylea trifolia
Boxelder
Acer negundo
Red Maple
Acer rubrum
Southern Sugar Maple
Acer saccharum floridanum
Chalk Maple
Acer saccharum leucoderme
Painted Buckeye
Aesculus sylvatica
Red Buckeye
Aesculus pavia
Spotted Touch-me-not
Impatiens capensis
New-Jersey-tea
Ceanothus americanus
Virginia-creeper
Parthenocissus quinquefolia
Muscadine; Bullace
Vitis rotundifolia
St.-John's-wort
Hypericum lloydii
Hypericum
Hypericum sp.
Green Violet
Hybanthus concolor
Bird-foot Violet
Viola pedata
Primrose-leaved Violet
Viola primulifolia
Halberd-leaved Violet
Viola hastata
Violet
Viola sp.
Prickly-pear
Opuntia compressa
Devil-joint
Opuntia drummondii
Ludwigia
Ludwigia repens
Sanicle; Snakeroot
Sanicula canadensis
Wild Chervil
Chaerophyllum tainturieri
Meadow Parsnip
Zizia trifoliata
Blackgum
Nyssa sylvatica
Black Tupelo; Blackgum
Nyssa sylvatica sylvatica
Flowering Dogwood
Cornus florida
Sweet Pepperbush
Clethra alnifolia
Spotted Wintergreen
Chimaphila maculata
Indian-pipe
Monotropa uniflora
Pinxter-flower; Pinxter-bloom
Rhododendron nudiflorum
Azalea
Rhododendron sp.

Sourwood
Oxydendrum arboreum
Sparkleberry
Vaccinium arboreum
Buckthorn Bumelia
Bumelia lycioides
Persimmon
Diospyros virginiana
Fringe-tree
Chionanthus virginicus
Yellow Jessamine
Gelsemium sempervirens
Gentian
Gentiana villosa
Blue-star
Amsonia tabernaemontana
Angle-pod
Matelea sp.
Dodder
Cuscuta gronovii
Little Red Morning-glory
Ipomoea coccinea
Morning-glory
Ipomoea lacunosa
Phlox
Phlox nivalis
Phlox sp.
Phacelia
Phacelia dubia
Jussieu
Phacellia maculata
Wild Comfrey
Cynoglossum virginianum
Beautyberry
Callicarpa americana
Common Skullcap
Scutellaria integrifolia
Skullcap
Scutellaria sp.
Carpenterweed
Prunella vulgaris
Lyre-leaved Sage
Salvia lyrata
Sage
Salvia urticifolia
Salvia
Salvia sp.
Bugle-weed
Lycopus virginicus
Pool-sprite
Amphianthus pusillus
Hedge-hyssop
Gratiola sp.
False Pimpernel
Lindernia monticola
Woolly Mullein
Verbascum thapsus
Toadflax
Linaria canadensis
Linaria sp.
False Foxglove
Aureolaria sp.
Agalinis tenuifolia
Cross Vine
Anisostichus capreolata
Trumpet Vine
Campsis radicans

Partridge-berry
Mitchella repens
Bluets
Houstonia caerulea
Houstonia purpurea
Houstonia tenuifolia
Bedstraw
Galium sp.
Japanese Honeysuckle
Lonicera japonica
Coral Honeysuckle
Lonicera sempervirens
Black-haw; Nanny-berry
Viburnum prunifolium
American Elderberry
Sambucus canadensis
Corn-salad
Valerianella radiata
Bur Cucumber
Sicyos angulatus
Cardinal-flower
Lobelia cardinalis
Lobelia
Lobelia puberula
Common Ragweed
Ambrosia artemisiifolia
Gall-of-the-earth
Prenanthes serpentaria
Prenanthes
Prenanthes sp.
Wild Lettuce
Lactuca floridana
Rattlesnake-weed
Hieracium venosum
Hawkweed
Hieracium gronovii
Dwarf Dandelion
Krigia virginica
Pale Indian-plantain
Cacalia atriplicifolia
Small's Ragwort
Senecio smallii
Mule-ears; Bitterweed
Senecio tomentosus
Elephant's-foot
Elephantopus carolinianus
Blazing-star
Liatris secunda
Climbing Hempweed
Mikania scandens
Pussy-toes
Antennaria plantaginifolia
Rabbit-tobacco
Gnaphalium obtusifolium
Aster
Aster undulatus
Aster sp.
Green-and-gold
Chrysogonum virginianum
Coneflower
Rudbeckia laciniata
Spanish-needles
Bidens bipinnata
Bitterweed
Helenium amarum
Tree-ear
Auricularia auricula
Cinnabar-red Chanterelle
Cantharellus cinnabarinus
Fragrant Chanterelle
Craterellus odoratus
White Worm Coral
Clavaria vermicularis
Crown-tipped Coral
Clavicorona pyxidata
Violet-branched Coral
Clavulina amethystina
Yellow-tipped Coral
Ramaria formosa
Scaly Tooth
Hydnum imbricatum
Meadow Mushroom
Agaricus campestris
Parasol
Lepiota procera
American Caesar's Mushroom
Amanita caesarea
Rag-veil Amanita
Amanita cinereopannosa
Gemmed Amanita
Amanita gemmata
Fly Agaric
Amanita muscaria
Grisette
Amanita vaginata
Destroying-angel
Amanita virosa
Shaggy-stalked Bolete
Boletellus betula
Violet-gray Bolete
Tylopilus plumbeoviolaceus
Flat Crep
Crepidotus applanatus
Fading Scarlet Waxy-cap
Hygrophorus miniatus
Poison Paxillus
Paxillus involutus
Indigo Milky
Lactarius indigo
Peppery Milky
Lactarius piperatus
Common Violet-latex Milky
Lactarius uvidus
Tacky Green Russula
Russula aeruginea
Graying Yellow Russula
Russula claroflava
Fragile Russula
Russula fragilis
Blackish-red Russula
Russula kromholzii
Violet Collybia
Collybia iocephala
Purple-gilled Laccaria
Laccaria ochropurpurea
Coral Spring Mycena
Mycena acicula
Orange Moss Agaric
Rickenella fibula
Skull-shaped Puffball
Calvatia craniformis
Spiny Puffball
Lycoperdon echinatum
Gem-Studded Puffball
Lycoperdon perlatum
Elegant Stinkhorn
Mutinus elegans
Cedar Apple Rust
Gymnosporangium juniperviginianae

FAUNA

Bowl-and-doily Spider
Frontinella pyramitela
Arrow-shaped Spiny-back
Micrathena sagittata
White Micrathena (spider)
Micrathena mitrata
Writing Spider
Argiope aurantia
Orb-weaver
Neoscona sp.
Araneus bicentenarius
Marbled Orb-weaver
Araneus marmoreus
Arrowhead Spider
Verrucosa arenata
Long-jawed Orb-weaver
Tetragnatha laboriosa
Tetragnatha sp.
Six-spotted Fishing Spider
Dolomedes triton
Forest Wolf Spider
Lycosa gulosa
Jumping Spider
Menemerus bivittatus
Red Velvet Mite
Trombidium sp.
Wild Cherry Pouch Gall Mite
Eriophyes padi
Sulfur-winged Grasshopper
Arphia sulphurea
Broad-winged Katydid
Microcentrum rhombifolium
Cone-headed Katydid
Neoconcephalus sp.
Negro Bug
Corimelaena sp.
Seed Bug
Hypogeocoris sp.
Neacoryphus bicrucis
Red Bug
Largus succinctus
Flat Bug
Mezira sp.
Water-scorpion
Ranatra sp.
Hickory Leaf Gall (aphid)
Phylloxera caryoecaulis
Red Imported Fire Ant
Solenopsis invicta
Dogwood Club Gall (midge)
Lasioptera clazula
Common Green Darner
Anax junius
Eastern Pondhawk
Erythemis simplicicollis
Little Corporal Dragonfly
Ladona deplanata
Golden-winged Skimmer
Libellula auripennis
Twelve-spotted Skimmer
Libellula pulchella
White-tail Dragonfly
Plathemis lydia
Lilypad Forktail (damselfly)
Ischnura kellicotti
Tiger Beetle
Cicindela tranquebarica
Pedunculate Ground Beetle
Pasimachus depressus
Minute Ground Beetle
Mioptachys flavicauda
Predacious Diving Beetle
Laccophiilus fasciatus rufus
Water Scavenger Beetle
Tropisternus natator
Passalus; Bess-beetle
Popilius disjunctus
Dung Beetle
Ateuchus sp.
Onthophagus tuberculifrons
Flower Scarab Beetle
Euphoria sepulchralis
Yellow-marked Buprestid Beetle
Acmaeodera tubulus
Leaf Buprestid Beetle
Brachys aerosus
Soldier Beetle
Chauliognathus pennsylvanicus
Cantharis sp.
Dermestid Beetle
Cryptorhopalum sp.
Checkered Beetle
Enoclerus ichneumoneus
Soft-winged Flower Beetle
Malachius sp.
Hypebaeus oblitus
Melyrodes cribrata
Black Blister Beetle
Epicauta pennsylvanica
Darkling Beetle
Alobates pennsylvanica
False Darkling Beetle
Allopoda lutea
Lady Beetle
Brachyacantha sp.
Seven-spotted Lady Beetle
Coccinella septempunctata
Pleasing Fungus Beetle
Tritoma angulata
Flower Longhorn Beetle
Grammoptera sp.
Typocerus sp.
Cylindrical Leaf Beetle
Pachybrachis sp.
Cryptocephalus quadruplex
Warty Leaf Beetle
Chlamisus gibbosus
Oval Leaf Beetle
Nodonota sp.
Graphops curtipennis
Paria canella aterrimus
Paria canella quadrinotata
Alder Leaf Beetle (larva)
Chrysomela interrupta

Flat Creek Heritage Preserve (continued)
September 24, 1988, April 13, 1991, and April 16, 1994

Azalea Leaf Beetle
Galerucella rufosanguinea
Oak Flea Beetle
Oedionychis quercata
Flea Beetle
Oedionychis miniata
Disonycha abbreviata
Corn Flea Beetle
Chaetocnema pulicaria
Leaf-mining Beetle
Anoplitis inaequalis
Pear-shaped Weevil
Apion decoloratum
Short-snouted Weevil
Pandeleteius sp.
Flower Weevil
Aulobaris ibis
Centrinaspis perscillus
Centrinites sp.
Zebra Swallowtail
Eurytides marcellus
Eastern Tiger Swallowtail
Papilio glaucus
Spicebush Swallowtail
Papilio troilus
Falcate Orangetip
Anthocharis midea
Cloudless Sulphur
Phoebis sennae eubule
Sleepy Orange (butterfly)
Eurema nicippe
Gray Hairstreak
Strymon melinus
Eastern Tailed-blue(butterfly)
Everes comyntas
Spring Azure (butterfly)
Celastrina ladon
Variegated Fritillary
Euptoieta claudia
Pearl Crescent
Phyciodes tharos
Red-spotted Purple (butterfly)
Limenitis arthemis astyanax
Hackberry Emperor (butterfly)
Asterocampa celtis
Carolina Satyr (butterfly)
Hermeuptychia sosybius
Silver-spotted Skipper
Epargyreus clarus
Juvenal's Duskywing
Erynnis juvenalis
Yellow-washed Metarranthis
Metarranthis obfirmaria
Eastern Tent Caterpillar
Malacosoma americanum
Hummingbird Clearwing Moth
Hemaris thysbe
Large Necklace Moth
Hypsoropha monilis
Rosyside Dace
Clinostomus funduloides
Dwarf Mudpuppy
Necturus punctatus
Fowler's Toad
Bufo woodhousei
Northern Cricket Frog
Acris crepitans
Green Treefrog
Hyla cinerea
Bullfrog
Rana catesbiana
Carpenter Frog
Rana virgatipes
Eastern Narrowmouth Toad
Gastrophryne carolinensis
Yellow-bellied Slider
Chrysemys scripta
Eastern Box Turtle
Terrapene carolina
Carolina Anole
Anolis carolinensis
Eastern Fence Lizard
Sceloporus undulatus
Five-lined Skink
Eumeces fasciatus
Broadhead Skink
Eumeces laticeps
Skink
Eumeces sp.
Six-lined Racerunner
Cnemidophorus sexlineatus
Eastern Hognose Snake
Heterodon platyrhinos
Northern Water Snake
Nerodia sipedon
Anhinga; Snakebird
Anhinga anhinga
Great Blue Heron
Ardea herodias
Wood Duck
Aix sponsa
Turkey Vulture
Cathartes aura
Black Vulture
Coragyps atratus
Sharp-shinned Hawk
Accipiter striatus
Red-shouldered Hawk
Buteo lineatus
Red-tailed Hawk
Buteo jamaicensis
Osprey
Pandion haliaetus
Ruby-throated Hummingbird
Archilochus colubris
Red-bellied Woodpecker
Melanerpes carolinus
Pileated Woodpecker
Dryocopus pileatus
Eastern Phoebe
Sayornis phoebe
Blue Jay
Cyanocitta cristata
American Crow
Corvus brachyrhynchos
Tufted Titmouse
Parus bicolor
Carolina Chickadee
Parus carolinensis
Brown-headed Nuthatch
Sitta pusilla
Carolina Wren
Thryothorus ludovicianus
Ruby-crowned Kinglet
Regulus calendula
Blue-gray Gnatcatcher
Polioptila caerulea
Hermit Thrush
Catharus guttatus
Northern Mockingbird
Mimus polyglottos
Cedar Waxwing
Bombycilla cedrorum
White-eyed Vireo
Vireo griseus
Yellow-throated Vireo
Vireo flavifrons
Red-Eyed Vireo
Vireo olivaceus
Prothonotary Warbler
Protonotaria citrea
Northern Parula
Parula americana
Black-and-white Warbler
Mniotilta varia
Black-throated Blue Warbler
Dendroica caerulescens
Yellow-rumped Warbler
Dendroica coronata
Yellow-throated Warbler
Dendroica dominica
Prairie Warbler
Dendroica discolor
Pine Warbler
Dendroica pinus
Hooded Warbler
Wilsonia citrina
Ovenbird
Seiurus aurocapillus
Yellowthroat
Geothlypis trichas
Northern Cardinal
Cardinalis cardinalis
Indigo Bunting
Passerina cyanea
Rufous-sided Towhee
Pipilo erythrophthalmus
Red-winged Blackbird
Agelaius phoeniceus
Brown-headed Cowbird
Molothrus ater
Common Grackle
Quiscalus quiscula
Summer Tanager
Piranga rubra
Beaver (gnaw mark)
Castor canadensis
Raccoon (scat)
Procyon lotor
White-tailed Deer
Odocoileus virginianus

(Photo by A. Ciegler)

Pinxter-flower, *Rhododendron nudiflorum*.

Laurens County

Enoree River Site
(Sumter National Forest)
Laurens County
May 26, 1988

This site is on the south bank of the Enoree River where a low bluff creates a north-facing slope. Such slopes create conditions for very rich botanical communities. The slope and floodplain here combine to provide a combination of Piedmont and more typically mountain species. Some exceptionally large specimens of shagbark hickory and northern red oak were found along with less common plants such as fragrant sumac and ginseng. There was even an old cemetery dating from the early 1800's. There are no developed trails at this site. SCAN's experience was enhanced by the leadership of John Green, naturalist with the Spartanburg Science Center, who was already very familiar with the site and showed the members where to look.

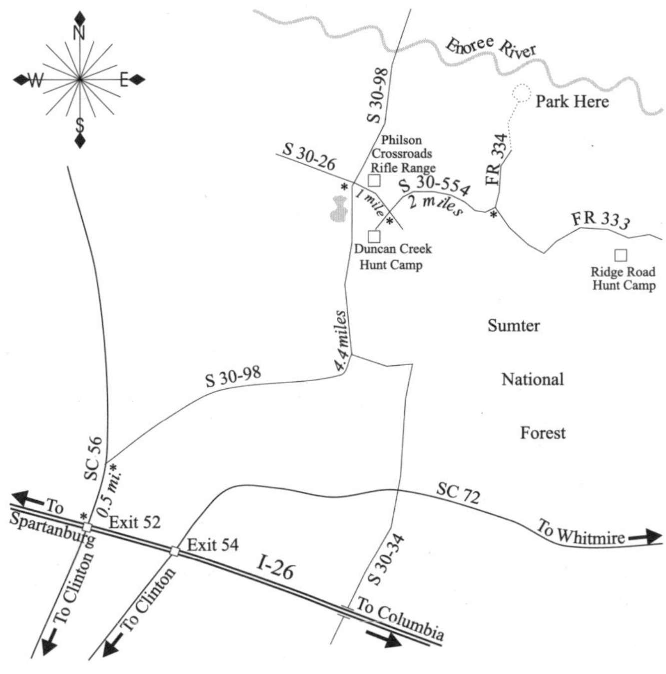

Directions to Enoree River Site: From I-26 between Columbia and Spartanburg, take Exit 52 north on SC 56. Travel 0.5 mile, then diverge to the right onto S 30-98 and drive north for 4.4 miles; (watch for the intersection with S 30-34 and be sure to take the left fork.) Pass an inconspicuous lake on the left, and turn right at the 4-way intersection posted with signs to the Rifle Range. Go about a mile, then turn left following the sign to Ridge Road Hunt Camp. Drive about 2 miles and turn left onto Forest Road 334. The pavement ends near a church on the right; continue to the parking area at the end of the road.

FLORA

Rattlesnake Grape Fern
Botrychium virginianum
Maidenhair Fern
Adiantum pedatum
Christmas Fern
Polystichum acrostichoides
Broad Beech-Fern
Thelypteris hexagonoptera
Ebony Spleenwort
Asplenium platyneuron
Eastern Red-cedar
Juniperus virginiana
Cane; Switch Cane
Arundinaria gigantea
Uniola Grass
Uniola laxa
Microstegium (grass)
Microstegium vimineum
Jack-in-the-pulpit
Arisaema triphyllum
Spiderwort
Tradescantia subaspera
Nodding Trillium
Trillium cernuum
Catesby's Trillium
Trillium catesbaei
Solomon's-seal
Polygonatum biflorum
Bellwort
Uvularia perfoliata
Yellow Star-grass
Hypoxis hirsuta
Downy Rattlesnake-plantain
Goodyera pubescens
Lizard's-tail
Saururus cernuus
Black Walnut
Juglans nigra
Bitternut Hickory
Carya cordiformis
Shagbark Hickory
Carya ovata
American Hornbeam; Ironwood
Carpinus caroliniana
Beech
Fagus grandifolia
White Oak
Quercus alba
Post Oak
Quercus stellata
Swamp Chestnut Oak
Quercus michauxii
Yellow Chestnut Oak
Quercus muehlenbergii
Spanish Oak
Quercus falcata
Cherrybark Oak
Quercus falcata pagodaefolia
Water Oak
Quercus nigra
Slippery Elm
Ulmus rubra
Sugarberry; Hackberry
Celtis laevigata
Virginia Snakeroot
Aristolochia serpentaria
Wild Ginger
Asarum canadense
Heartleaf
Hexastylis arifolia
Black Cohosh
Cimicifuga racemosa
Windflower
Thalictrum thalictroides
Liverleaf
Hepatica americana
Mayapple
Podophyllum peltatum

Enoree River Site (continued)
May 26, 1988

Tulip-poplar
Liriodendron tulipifera
Cucumber-tree
Magnolia acuminata
PawPaw
Asimina triloba
Sweet-shrub; Carolina-allspice
Calycanthus floridus
Bloodroot
Sanguinaria canadensis
Mock-orange
Philadelphus inodorus
Wild Hydrangea
Hydrangea arborescens
Sweet-gum
Liquidambar styraciflua
Cocklebur
Agrimonia sp.
Redbud; Judas-tree
Cercis canadensis
Poison-ivy
Rhus radicans
Fragrant Sumac
Rhus aromatica
American Holly
Ilex opaca
Hearts-a'burstin'-with-love
Euonymus americanus
Bladdernut
Staphylea trifolia
Boxelder
Acer negundo
Red Maple
Acer rubrum
Chalk Maple
Acer saccharum leucoderme
Painted Buckeye
Aesculus sylvatica
Spotted Touch-me-not
Impatiens capensis
Virginia-creeper
Parthenocissus quinquefolia
Muscadine; Bullace
Vitis rotundifolia
Summer Grape
Vitis aestivalis
Ginseng
Panax quinquefolium
Devil's-walking-stick
Aralia spinosa
Snakeroot
Sanicula sp.
Queen Anne's-lace; Wild Carrot
Daucus carota
Meadow Parsnip
Thaspium barbinode
Lovage
Ligusticum canadense
Blackgum
Nyssa sylvatica
Flowering Dogwood
Cornus florida
Spotted Wintergreen
Chimaphila maculata
Pinxter-flower; Pinxter-bloom
Rhododendron nudiflorum
Mountain-laurel
Kalmia latifolia
Sourwood
Oxydendrum arboreum
Creeping-Charlie
Lysimachia nummularia
Persimmon
Diospyros virginiana
Bigleaf Snowbell
Styrax grandifolia
Green Ash; Red Ash
Fraxinus pennsylvanica subintegerrima
Forget-me-not
Myosotis macrosperma
Lyre-leaved Sage
Salvia lyrata
Horse-balm
Collinsonia canadensis
Moth Mullein
Verbascum blattaria
Trumpet Vine
Campsis radicans
Beech-drops
Epifagus virginiana
Bluets
Houstonia purpurea
Cleavers
Galium aparine
Japanese Honeysuckle
Lonicera japonica
Black-haw; Nanny-berry
Viburnum prunifolium
Maple-leaf Viburnum
Viburnum acerifolium
Rattlesnake-weed
Hieracium venosum
Butterweed
Senecio glabellus
Rough Daisy Fleabane
Erigeron strigosus
Tinder Polypore
Fomes fomentarius
Violet-toothed Polypore
Trichaptum biformis
Common Split-gill
Schizophyllum commune
Crowded Parchment
Stereum complicatum
Alcohol Inky
Coprinus atramentarius
Fawn Mushroom
Pluteus cervinus
Oak-loving Collybia
Collybia dryophila

FAUNA

Negro Bug
Cormelaena sp.
Leaf-cutting Bee
Megachile sp.
Ebony Jewelwing
Calopteryx maculata
Green Tiger Beetle
Cicindela sexguttata
Woodland Ground Beetle
Pterostichus sp.
Notch-mouthed Ground Beetle
Dicaelus ambiguus
False Bombadier Beetle
Galeritula bicolor
Passalus; Bess-beetle
Popilius disjunctus
Tumblebug; Dung Beetle
Glaphyrocanthon viridis
Flower Scarab Beetle
Trichiotinus piger
Scarab Beetle
Valgus sp.
Soft-bodied Plant Beetle
Ptilodactyla sp.
Click Beetle
Agriotes sp.
Soldier Beetle
Chauliognathus marginatus
Podabrus sp.
Long-jointed Bark Beetle
Arthromacra aenea
Comb-clawed Beetle
Isomira sericea
Handsome Fungus Beetle
Stenotarsus sp.
Dusky Lady Beetle
Scymnus sp.
Pleasing Fungus Beetle
Triplax thoracica
Shining Flower Beetle
Acyclomus piceus
Flower Longhorn Beetle
Typocerus sp.
Longhorn Beetle
Hyperplatys maculatus
Shining Leaf Beetle
Lema albini
Oval Leaf Beetle
Paria canella gilvipes
Skeletonizing Leaf Beetle
Galerucella notulata
12-spotted Cucumber Beetle
Diabrotica duodecimpunctata
Flea Beetle
Altica sp.
Systena taeniata
Eastern Tailed-blue (butterfly)
Everes comyntas
American Snout (butterfly)
Libytheana carinenta
Mourning Cloak
Nymphalis antiopa
Red-spotted Purple (butterfly)
Limenitis arthemis astyanax
Hackberry Emperor (butterfly)
Asterocampa celtis
Carolina Satyr (butterfly)
Hermeuptychia sosybius
Marbled Salamander
Ambystoma opacum
Ground Skink
Scincella lateralis
Sharp-shinned Hawk
Accipiter striatus
Wild Turkey
Meleagris gallopavo
Yellow-billed Cuckoo
Coccyzus americanus
Red-bellied Woodpecker
Melanerpes carolinus
Pileated Woodpecker
Dryocopus pileatus
Acadian Flycatcher
Empidonax virescens
Tufted Titmouse
Parus bicolor
Wood Thrush
Hylocichla mustelina
Red-Eyed Vireo
Vireo olivaceus
Northern Parula
Parula americana
White-tailed Deer (scat)
Odocoileus virginianus

(Photo by L. Smith)
Horse-balm, *Collinsonia canadensis.*

(Photo by L. Smith)
Mountain-laurel, *Kalmia latifolia.*

(Photo by B. Gilbert)
Mayapple, *Podophyllum peltatum.*

Lee County

Lee State Park
Lee County
September 26, 1987, and August 24, 1991

This park offers the full range of sandhills habitat from floodplain to sand ridge turkey oak-longleaf pine areas. The nature trail skirts small ponds and then climbs to the more sandy areas. In addition there is a sandhills trail through classic turkey oak-longleaf pine habitat. To fully experience the floodplain, walking or driving the loop road provides access except during high water on the Lynches River. The list shows the diversity of plant material and associated animal life in these habitats.

Directions to Lee State Park: From Columbia, take I-20 east toward Florence, and turn north at Exit 123 onto County Road S 31-22. Drive 1.2 miles to the park entrance on the left.

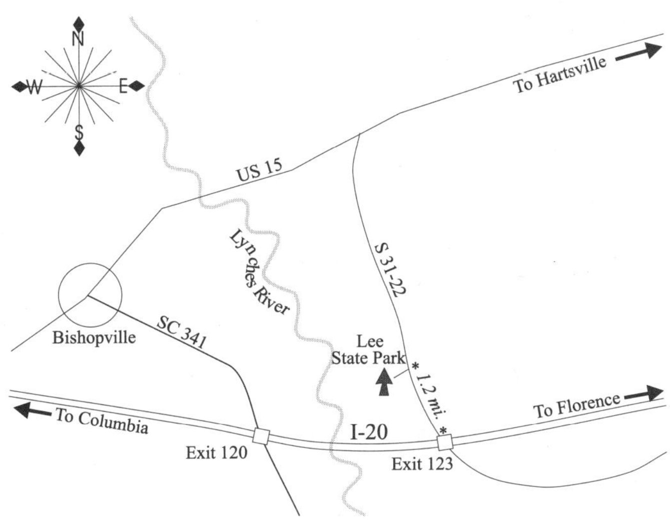

FLORA

Lichen
Graphis scripta
Graphis elegans
Bacidia laurocerasi
Letrouitia domingense
Cladina rangiferina
Cladonia bacillaris
Cladonia beaumontii
Cladonia calycantha
Cladonia caroliniana
Cladonia chlorophaea
Cladonia coniocraea
Cladonia cristatella scyphulifera
Cladonia cristatella cristatella
Cladonia didyma
Cladonia incrassata
Cladonia macilenta
Cladonia parasitica
Pertusaria ostiolata
Pertusaria paratuberculifera
Pertusaria subpertusa
Pertusaria ophthalmiza
Lecanora miculata
Lecanora pallida cancriformis
Lecanora varia
Tephromela atra
Bulbothrix goebelii
Parmelia rudecta
Parmelina dissecta
Parmelina horrescens
Parmotrema hypotropum
Parmotrema xanthinum
Pseudoparmelia caroliniana
Ramalina willeyi
Usnea strigosa
Buellia punctata
Buellia stillingiana
Physcia aipolia
Pyxine subcinerea
Physconia enteroxantha
Caloplaca camptidia
Trypethelium tropicum

Lee State Park (continued)
September 26, 1987, and August 24, 1991

Sand Spikemoss
Selaginella arenicola
Cinnamon Fern
Osmunda cinnamomea
Southern Lady Fern
Athyrium asplenioides
Netted Chain Fern
Woodwardia areolata
Resurrection Fern
Polypodium polypodioides
Long-leaf Pine
Pinus palustris
Loblolly Pine
Pinus taeda
Short-leaf Pine
Pinus echinata
Cane; Switch Cane
Arundinaria gigantea
Spike Grass
Uniola sessiliflora
River Oats
Uniola latifolia
Dulichium (sedge)
Dulichium arundinaceum
Spike-rush
Eleocharis tuberculosa
Sedge
Carex sp.
Yellow-eyed-grass
Xyris sp.
Spanish-moss
Tillandsia usneoides
Asian Spiderwort
Aneilema keisak
Bamboo Smilax
Smilax laurifolia
Bear-grass
Yucca filamentosa
Yellow Star-grass
Hypoxis hirsuta
Downy Rattlesnake-plantain
Goodyera pubescens
Crane-fly Orchid
Tipularia discolor
Lizard's-tail
Saururus cernuus
Swamp Cottonwood
Populus heterophylla
Wax-myrtle; Southern Bayberry
Myrica cerifera
Mockernut Hickory
Carya tomentosa
Tag Alder
Alnus serrulata
River Birch
Betula nigra
American Hornbeam; Ironwood
Carpinus caroliniana
White Oak
Quercus alba
Post Oak
Quercus stellata
Scrubby Post Oak
Quercus margaretta
Swamp Chestnut Oak
Quercus michauxii
Turkey Oak
Quercus laevis
Scarlet Oak
Quercus coccinea
Spanish Oak
Quercus falcata
Water Oak
Quercus nigra
Willow Oak
Quercus phellos
Laurel Oak
Quercus laurifolia
Red Mulberry
Morus rubra
False Nettle
Boehmeria cylindrica
Heartleaf
Hexastylis arifolia
Polygonum
Polygonum sp.
Jointweed
Polygonella polygama
Pokeweed; Pokeberry
Phytolacca americana
Whitlow-wort
Paronychia herniarioides
Wire-plant
Stipulicida setacea
Tulip-poplar
Liriodendron tulipifera
Sweetbay
Magnolia virginiana
Dwarf PawPaw
Asimina parviflora
Redbay
Persea borbonia
Sassafras
Sassafras albidum
Virginia-willow
Itea virginica
Sweet-gum
Liquidambar styraciflua
Witch-hazel
Hamamelis virginiana
Hawthorn
Crataegus sp.
Creeping Bush-clover
Lespedeza repens
Goat's-rue
Tephrosia virginiana
Butterfly-pea
Clitoria mariana
Galactia
Galactia sp.
Poison-ivy
Rhus radicans
Poison-oak
Rhus toxicodendron
Winged Sumac
Rhus copallina
American Holly
Ilex opaca
Hearts-a'burstin'-with-love
Euonymus americanus
Red Maple
Acer rubrum
Virginia-creeper
Parthenocissus quinquefolia
Muscadine; Bullace
Vitis rotundifolia
Pineweed
Hypericum gentianoides
Pale Meadow-beauty
Rhexia mariana
Water-willow
Ludwigia decurrens
Trailing Ludwigia
Ludwigia palustris
Blackgum
Nyssa sylvatica
Black Tupelo; Blackgum
Nyssa sylvatica sylvatica
Flowering Dogwood
Cornus florida
Sweet Pepperbush
Clethra alnifolia
Spotted Wintergreen
Chimaphila maculata
Pinxter-flower; Pinxter-bloom
Rhododendron nudiflorum
Fetterbush
Lyonia lucida
Dog-hobble
Leucothoe axillaris
Sourwood
Oxydendrum arboreum
Dwarf Huckleberry
Gaylussacia dumosa
Huckleberry
Gaylussacia sp.
Sparkleberry
Vaccinium arboreum
Elliott's Blueberry
Vaccinium elliottii
Low-bush Blueberry
Vaccinium vacillans
Persimmon
Diospyros virginiana
Horse-sugar; Sweetleaf
Symplocos tinctoria
Yellow Jessamine
Gelsemium sempervirens
Polypremum
Polypremum procumbens
Bonamia
Bonamia sp.
Beautyberry
Callicarpa americana
Glecoma
Glecoma hederacea
Lycopus
Lycopus sp.
Gerardia
Agalinis linifolia
Cross Vine
Anisostichus capreolata
Bladderwort
Utricularia sp.
Button-bush
Cephalanthus occidentalis
Virginia Buttonweed
Diodia virginiana
Poor-jo (or Po-jo)
Diodia teres
Richardia
Richardia sp.
Partridge-berry
Mitchella repens
Oldenlandica
Oldenlandica uniflora
Wahlenbergia
Wahlenbergia marginata
Cardinal-flower
Lobelia cardinalis
Lobelia
Lobelia elongata
Hawkweed
Hieracium gronovii
Elephant's-foot
Elephantopus tomentosus
Carphephorus
Carphephorus bellidifolius
Mistflower; Ageratum
Eupatorium coelestinum
Sweet Goldenrod
Solidago odora
Goldenrod
Solidago sp.
Acanthospermum
Acanthospermum australe
Eclipta
Eclipta alba
Orange Earth-tongue
Microglossum rufum
Carbon-balls (fungus)
Daldinia concentrica
Dead-man's-fingers
Xylaria polymorpha
Cauliflower Mushroom
Sparassis spathulata
Varnish Shelf
Ganoderma curtisii
Turkey-tail
Trametes versicolor
Crowded Parchment
Stereum complicatum
Yellow Blusher
Amanita flavorubescens
Bolete
Boletus curtisii
Boletus erythropus
Old-man-of-the-woods
Strobilomyces floccopus
Painted Suillus
Suillus pictus
Silvery-violet Cort
Cortinarius alboviolaceus
Chanterelle Waxy-cap
Hygrophorus cantharellus
Milky (mushroom)
Lactarius sp.
Emetic Russula
Russula emetica
Yellow Russula
Russula lutea

Pectinate Russula
Russula pectinatoides
Sulfur Tuft (mushroom)
Naematoloma fasciculare
Ringless Honey Mushroom
Armillariella tabascens
Barometer Earthstar
Astraeus hygrometricus
Earthstar
Scleroderma sp.
Chocolate-tube Slime
Stemonitis splendens

FAUNA

Crab-shaped Spiny-back
Gasteracantha cancriformis
Marbled Orb-weaver
Araneus marmoreus
Boll's Sand Grasshopper
Spharagemon bolli
Witch Hazel Cone Gall (aphid)
Hormaphis hamamelidis
Eastern Pondhawk
Erythemis simplicicollis
Great Blue Skimmer
Libellula vibrans
Blue Dasher
Pachydiplax longipennis
Meadowfly (dragonfly)
Sympetrum sp.
Glider
Tramea sp.
Dung Beetle
Phanaeus igneus
Eastern Tiger Swallowtail
Papilio glaucus
Palamedes Swallowtail
Papilio palamedes
Cloudless Sulphur
Phoebis sennae eubule
Sleepy Orange (butterfly)
Eurema nicippe
Eastern Tailed-blue(butterfly)
Everes comyntas
American Lady (butterfly)
Vanessa virginiensis
Carolina Satyr (butterfly)
Hermeuptychia sosybius
Polyphemus Moth
Antheraea polyphemus
Southern Toad
Bufo terrestris
Fowler's Toad
Bufo woodhousei
Southern Cricket Frog
Acris gryllus
Yellow-bellied Slider
Chrysemys scripta
Carolina Anole
Anolis carolinensis
Eastern Fence Lizard
Sceloporus undulatus
Red-shouldered Hawk
Buteo lineatus
Belted Kingfisher
Ceryle alcyon
Red-bellied Woodpecker
Melanerpes carolinus
Red-cockaded Woodpecker (sign)
Picoides borealis
Pileated Woodpecker
Dryocopus pileatus
Blue Jay
Cyanocitta cristata
Carolina Chickadee
Parus carolinensis
Yellow-throated Warbler
Dendroica dominica
Chipping Sparrow
Spizella passerina
Summer Tanager
Piranga rubra
Gray Squirrel
Sciurus carolinensis
White-tailed Deer
Odocoileus virginianus

(Photo by A. Ciegler)
Eastern fence lizard, *Sceloporus undulatus*.

Lynchburg Savanna Heritage Preserve
Lee County
May 22, 1993

Wet pine savannas are fascinating places. This savanna had its tree population reduced by Hurricane Hugo in 1989, and has had both natural and controlled burns since then. One natural burn occurred about a month before this SCAN visit and was followed by adequate rainfall, demonstrating by the lush new growth, the renewal value of fire. The preserve abounds in insectivorous plants, *Zenobia* in full bloom in late May, grasses, sedges, rushes, orchids, and plenty of animal life. This is a large preserve with two gates about a quarter of a mile apart which resulted, accidentally, in two different field trips for a part of the day. Fortunately, near noon, one half of the attending members spotted the other half and finished the trip together, each group having something special to show the other. It also meant there were more observations for the flora and fauna list.

Directions to Lynchburg Savanna Heritage Perserve: From Columbia, take I-20 east toward Florence, and turn south at Exit 120 onto SC 341. After 3 miles (Wisacky Crossoads) fork right onto SC 527; drive 3 miles farther to Elliott. Turn left onto US 401, then immediately right onto S 31-53 and drive 4.5 miles south. As soon as you see a fire tower, turn right onto S 31-327; do not pass the fire tower. Lynchburg Savanna is about one mile past the intersection on the right, with two parallel dirt tracks as entrances.

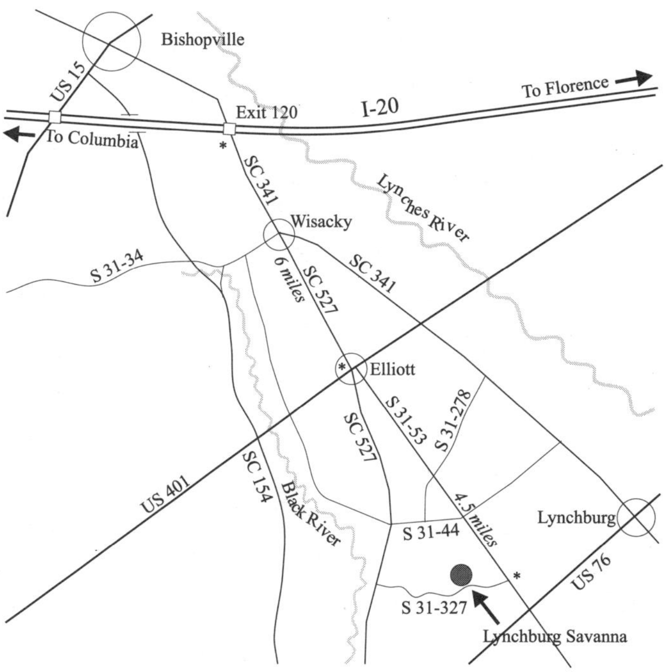

Lynchburg Savanna Heritage Preserve (continued)
May 22, 1993

FLORA

Lichen
Lecidea mutabilis
Melaspilea octomera
Melaspilea sp.
Lecidea russula
Lecidea sp.
Cladonia cristatella cristatella
Cladonia cristatella vestita
Cladonia cristatella pleurocarpa
Cladonia polycarpoides polycarpoides
Cladonia sp.
Pertusaria texana
Pertusaria sp.
Maronea constans
Haematomma puniceum
Lecania fuscella
Lecanora cupressi
Lecanora varia
Lecanora sp.
Ochrolechia africana
Bulbothrix confoederata
Bulbothrix isidiza
Hypotrachyna formosana
Hypotrachyna livida
Parmelina aurulenta
Parmotrema crinitum
Parmotrema hypotropum
Parmotrema mellissii
Parmotrema michauxianum
Parmotrema michauxianum roseum
Parmotrema perforatum
Parmotrema praesorediosum
Parmotrema robustum
Parmotrema stuppeum
Parmotrema subtinctorium
Parmotrema xanthinum
Canoparmelia caroliniana
Punctelia rudecta
Rimelia reticulata
Rimelia cetrata
Rimelia subisidiosa
Ramalina americana
Usnea strigosa
Buellia dialyta
Buellia disciformis
Physcia millegrana
Pyxine caesiopruinosa
Arthothelium sp.
Southern Bog Clubmoss
Lycopodium adpressum
Cinnamon Fern
Osmunda cinnamomea
Royal Fern
Osmunda regalis spectabilis
Bracken
Pteridium aquilinum
Virginia Chain Fern
Woodwardia virginica
Long-leaf Pine
Pinus palustris
Loblolly Pine
Pinus taeda
Pond Pine
Pinus serotina
Pond-cypress
Taxodium ascendens
Toothache Grass
Ctenium aromaticum
Panic Grass
Panicum sp.
Broomsedge
Andropogon sp.
Sedge
Carex walteriana
Arrowleaf Arum
Peltandra virginica
Yellow-eyed-grass
Xyris sp.
Bog-buttons
Lachnocaulon anceps
Greenbrier; Catbrier
Smilax sp.
Colic-root
Aletris farinosa
Black Snakeroot
Zigadenus densus
Yellow Star-grass
Hypoxis hirsuta
Hypoxis sp.
Blue-eyed-grass
Sisyrinchium sp.
Grass-pink
Calopogon pulchellus
Grass-leaved Ladies'-tresses
Spiranthes praecox
Willow
Salix humilis
Wax-myrtle; Southern Bayberry
Myrica cerifera
Swamp Bayberry
Myrica heterophylla
Black Jack Oak
Quercus marilandica
Water Oak
Quercus nigra
Running Oak
Quercus pumila
Willow Oak
Quercus phellos
Sweetbay
Magnolia virginiana
Sassafras
Sassafras albidum
Trumpet Pitcher-plant
Sarracenia flava
Hooded Pitcher-plant
Sarracenia minor
Sundew
Drosera leucantha
Drosera capillaris
Sweet-gum
Liquidambar styraciflua
Cinquefoil; Five-fingers
Potentilla sp.
Dewberry
Rubus trivialis
Blackberry; Brambles
Rubus sp.
Red Chokeberry
Sorbus arbutifolia
Black Cherry; Wild Cherry
Prunus serotina
Wild Indigo
Baptisia sp.
Samson Snakeroot
Psoralea psoralioides
Orange Milkwort
Polygala lutea
Spurge
Euphorbia curtisii
Poison-oak
Rhus toxicodendron
Winged Sumac
Rhus copallina
Titi
Cyrilla racemiflora
Inkberry
Ilex glabra
Red Maple
Acer rubrum
Hypericum
Hypericum sp.
Violet
Viola sp.
Pale Meadow-beauty
Rhexia mariana
Sundrops
Oenothera fruticosa
Mermaid-weed
Proserpinaca pectinata
Button Snakeroot
Eryngium yuccifolium
Black Tupelo; Blackgum
Nyssa sylvatica sylvatica
Sweet Pepperbush
Clethra alnifolia
Honey-cup
Zenobia pulverulenta
Maleberry
Lyonia ligustrina
Fetterbush
Lyonia lucida
Staggerbush
Lyonia mariana
Dangleberry
Gaylussacia frondosa
Low-bush Blueberry
Vaccinium tenellum
Persimmon
Diospyros virginiana
Indian-hemp
Apocynum cannabinum
Narrowleaf Milkweed
Asclepias verticillata
Milkweed
Asclepias longifolia
Common Skullcap
Scutellaria integrifolia
Mountain Mint
Pycnanthemum sp.
Lousewort
Pedicularis canadensis
Trumpet Vine
Campsis radicans
Butterwort
Pinguicula caerulea

(Photo by L. Smith)
Carolina wolf spider, *Lycosa carolinensis*.

Wiry Bladderwort
Utricularia subulata
Bluets
Houstonia caerulea
Possumhaw Viburnum
Viburnum nudum
Lobelia
Lobelia nuttallii
Leopard's-bane
Arnica acaulia
Sun-bonnets
Chaptalia tomentosa
Thistle
Carduus spinosissimus
Dog-fennel
Eupatorium capillifolium
Eupatorium compositifolium
Sea-myrtle; Groundsel-tree
Baccharis halimifolia
Swamp Fleabane
Erigeron vernus
Silphium
Silphium sp.
Coreopsis
Coreopsis sp.

FAUNA

Carolina Wolf Spider
Lycosa carolinensis
Blackgum Capsule Gall Mite
Eriophyes nyssae
Persimmon Warty Leaf Gall
Trioza diospyri
Gall (wasp)
Andricus petiolicola
Marbled Oak Gall (wasp)
Andricus pisiformis
Oak Stem Gall (wasp)
Disholcaspis spongiosa
Oak Bullet Gall (wasp)
Disholcaspis sp.
Fire Ant
Solenopsis sp.
Maple Leaf Spot Gall (midge)
Cecidomyia ocellaris
Blue Dragonlet
Erythrodiplax minuscula
Painted Skimmer
Libellula semifasciata
Violet-masked Glider
Tramea carolina
Pedunculate Ground Beetle
Ardistomis viridis
Predacious Diving Beetle
Laccophilus fasciatus
Laccophilus schwarzi
Bidessus sp.
Hydroporus lobatus
Hydroporus hybridus
Copelatus chevrolati chevrolati
Coptotomus interrogatus
Burrowing Water Beetle
Hydrocanthus iricolor
Whirligig Beetle
Dineutus carolinus
Water Scavenger Beetle
Tropisternus lateralis nimbatus
Berosus pantherinus
Paracymus subcupreus
Enochrus sublongus
Soldier Beetle
Chauliognathus marginatus
Tumbling Flower Beetle
Mordella melaena
Spotted Lady Beetle
Coleomegilla maculata
Seven-spotted Lady Beetle
Coccinella septempunctata
Flower Longhorn Beetle
Typocerus sp.
Oval Leaf Beetle
Paria canella gilvipes
Bud Weevil
Rhynchites geneus.
Babtisia-pod Weevil
Apion rostrum
Pear-shaped Weevil
Apion sp.
Eastern Tailed-blue(butterfly)
Everes comyntas
Painted Lady (butterfly)
Vanessa cardui
Common Buckeye (butterfly)
Junonia coenia
Chorus Frog
Pseudacris sp.
Spotted Turtle
Clemmys guttata
Redbelly Water Snake
Nerodia erythrogaster
Turkey Vulture
Cathartes aura
Black Vulture
Coragyps atratus
Red-tailed Hawk
Buteo jamaicensis
Northern Bobwhite
Colinus virginianus
Mourning Dove
Zenaida macroura
Eastern Screech-owl
Otus asio
Red-bellied Woodpecker
Melanerpes carolinus
Red-headed Woodpecker
Melanerpes erythrocephalus
Northern Flicker
Colaptes auratus
Downy Woodpecker
Picoides pubescens
Eastern Kingbird
Tyrannus tyrannus
Great Crested Flycatcher
Myiarchus crinitus
Blue Jay
Cyanocitta cristata
American Crow
Corvus brachyrhynchos
Fish Crow
Corvus ossifragus
Carolina Chickadee
Parus carolinensis
Brown-headed Nuthatch
Sitta pusilla
Carolina Wren
Thryothorus ludovicianus
Blue-gray Gnatcatcher
Polioptila caerulea
Eastern Bluebird
Sialia sialis
Loggerhead Shrike
Lanius ludovicianus
Gray Catbird
Dumetella carolinensis
Northern Mockingbird
Mimus polyglottos
Brown Thrasher
Toxostoma rufum
Yellowthroat
Geothlypis trichas
Yellow-breasted Chat
Icteria virens
Indigo Bunting
Passerina cyanea
Rufous-sided Towhee
Pipilo erythrophthalmus
Field Sparrow
Spizella pusilla
Chipping Sparrow
Spizella passerina
Red-winged Blackbird
Agelaius phoeniceus
Brown-headed Cowbird
Molothrus ater
Common Grackle
Quiscalus quiscula
Orchard Oriole
Icterus spurius
Summer Tanager
Piranga rubra
Raccoon (tracks)
Procyon lotor
Bobcat; Wildcat (tracks)
Felis rufus
White-tailed Deer (tracks)
Odocoileus virginianus

(Photo by L. Smith)

Spotted turtle, *Clemmys guttata*.

Mt. Pleasant Church Bay Heritage Preserve
Lee County
August 24, 1991

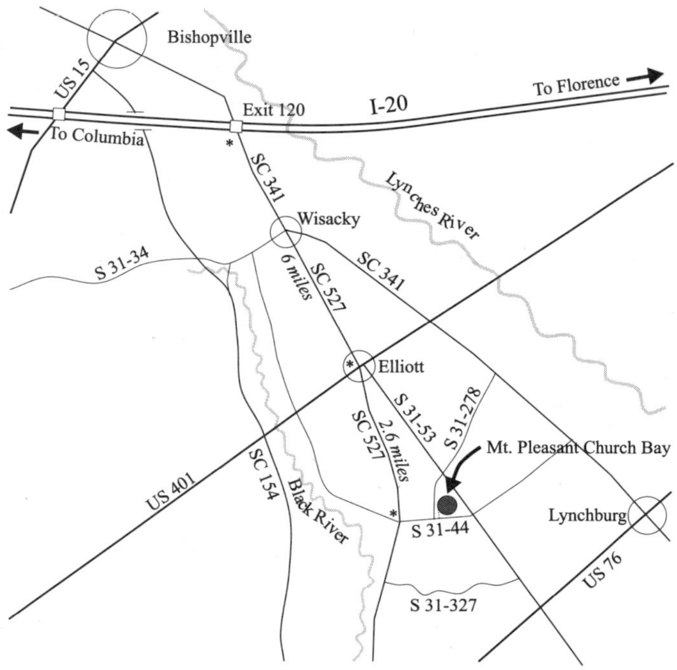

This is a bay that can be explored only by getting your feet wet, or visiting only in very dry drought conditions which would not give a normal view of this spectacular bay. The trees associated with this bay are only around the edges with a few cypress scattered across the wet areas creating an open sunny bay. The water usually ranges from about ankle to calf deep. Since most of the spectacular plants bloom in the heat of the summer, the water on your feet is a bonus! Many of the plants found here are rare and unusual, even for bays; for example, there are four species of milkwort (*Polygala*), *Oxypolis canbyi*, and hooded pitcher plant. It is possible that this bay hosts the rare pine barrens treefrog.

Directions to Mt. Pleasant Church Bay Heritage Preserve: From Columbia, take I-20 east toward Florence, and turn south at Exit 120 onto SC 341. After 3 miles (Wisacky Crossroads) fork right onto SC 527; drive 3 miles farther to Elliott. Continue on SC 527 (right fork) for 2.6 miles more, and then turn left onto S 31-44. Mt. Pleasant Church Bay is between S 31-278 and S 31-53 on the left. Park and walk north on a dirt road along a field for a quarter of a mile, and then turn right into the bay.

FLORA

Lichen
Coccocarpia erythroxyli
Cladonia calycantha
Cladonia didyma
Cladonia leporina
Pertusaria paratuberculifera
Pertusaria texana
Lecanora miculata
Lecanora varia
Ochrolechia africana
Candelaria concolor
Cetraria ciliaris
Hypotrachyna livida
Parmelia rudecta
Parmotrema hypotropum
Parmotrema perforatum
Parmotrema stuppeum
Parmotrema tinctorum
Pseudevernia consocians
Pseudoparmelia caperata
Pseudoparmelia caroliniana
Usnea strigosa
Buellia punctata
Pyxine caesiopruinosa
Rinodina sp.

Cinnamon Fern
Osmunda cinnamomea
Bracken
Pteridium aquilinum
Virginia Chain Fern
Woodwardia virginica
Loblolly Pine
Pinus taeda
Pond Pine
Pinus serotina
Pond-cypress
Taxodium ascendens
Cane; Switch Cane
Arundinaria gigantea
Smut Grass
Sporobolus poiretii
Umbrella-grass
Fuirena squarrosa
Beak-rush
Rhynchospora corniculata
Rhynchospora cephalantha
Rhynchospora filifolia
Rhynchospora torreyana
Rhynchospora perplexa
Nut-rush
Scleria triglomerata
Sedge
Carex glaucescens
Dayflower
Commelina communis

Black Willow
Salix nigra
Wax-myrtle; Southern Bayberry
Myrica cerifera
Mockernut Hickory
Carya tomentosa
Post Oak
Quercus stellata
Spanish Oak
Quercus falcata
Water Oak
Quercus nigra
Blue-jack Oak
Quercus incana
Running Oak
Quercus pumila
Willow Oak
Quercus phellos
Laurel Oak
Quercus laurifolia
False Nettle
Boehmeria cylindrica
Sweetbay
Magnolia virginiana
Redbay
Persea borbonia
Sassafras
Sassafras albidum
Hooded Pitcher-plant
Sarracenia minor

Sundew
Drosera capillaris
Drosera sp.
Sweet-gum
Liquidambar styraciflua
Red Chokeberry
Sorbus arbutifolia
Black Cherry; Wild Cherry
Prunus serotina
Mimosa; Silk-tree
Albizia julibrissin
Sicklepod
Cassia obtusifolia
Wild Sensitive-plant
Cassia nicitans
Pencil-flower
Stylosanthes biflora
Creeping Bush-clover
Lespedeza repens
Bush-clover
Lespedeza hirta
Bladderpod
Glottidium vesicarium
Climbing Butterfly-pea
Centrosema virginianum
Chinaberry
Melia azedarach
Polygala
Polygala grandiflora
Polygala mariana

Drum-heads
Polygala cruciata
Orange Milkwort
Polygala lutea
Pine-barren Milkwort
Polygala cymosa
Croton
Croton glandulosus septentrionalis
Three-seeded Mercury
Acalypha sp.
Flowering Spurge
Euphorbia corollata
Poison-oak
Rhus toxicodendron
Winged Sumac
Rhus copallina
Myrtle-leaved Holly
Ilex cassine myrtifolia
Inkberry
Ilex glabra
Red Maple
Acer rubrum
Devil's-shoestring
Sida rhombifolia
St.-Peter's-wort
Hypericum stans
Pineweed
Hypericum gentianoides
Pin-weed
Lechea sp.
Passionflower; Maypops
Passiflora incarnata
Smooth Meadow-beauty
Rhexia alifanus
Pale Meadow-beauty
Rhexia mariana
Seed-box
Ludwigia alternifolia
Narrow-leaved Ludwigia
Ludwigia linearis
Ludwigia
Ludwigia pilosa
Common Evening-primrose
Oenothera biennis
Sundrops
Oenothera fruticosa
Devil's-walking-stick
Aralia spinosa
Oxypolis
Oxypolis filiformis
Oxypolis sp.
Black Tupelo; Blackgum
Nyssa sylvatica sylvatica
Swamp Tupelo; Swamp Blackgum
Nyssa sylvatica biflora
Sweet Pepperbush
Clethra alnifolia
Honey-cup
Zenobia pulverulenta
Fetterbush
Lyonia lucida
Staggerbush
Lyonia mariana
Dangleberry
Gaylussacia frondosa
Persimmon
Diospyros virginiana

Horse-sugar; Sweetleaf
Symplocos tinctoria
Sabatia
Sabatia difformis
Sabatia campanulata
Brazilian Vervain
Verbena braziliensis
Mountain Mint
Pycnanthemum flexuosum
Horse-nettle
Solanum carolinense
Seymeria
Seymeria cassioides
Gerardia
Agalinis linifolia
Purple Gerardia
Agalinis purpurea
Trumpet Vine
Campsis radicans
Purple Bladderwort
Utricularia purpurea
English Plantain
Plantago lanceolata
Virginia Buttonweed
Diodia virginiana
Poor-jo (or Po-jo)
Diodia teres
Richardia
Richardia brasiliensis
Mexican-clover
Richardia scabra
Japanese Honeysuckle
Lonicera japonica
Creeping Cucumber
Melothria pendula
Wahlenbergia
Wahlenbergia marginata
Lobelia
Lobelia boykinii
Lobelia nuttallii
Common Ragweed
Ambrosia artemisiifolia
Ironweed
Vernonia angustifolia angustifolia
Elephant's-foot
Elephantopus sp.
Blazing-star
Liatris spicata
Liatris spicata resinosa
Eupatorium
Eupatorium recurvans
False Hoarhound
Eupatorium rotundifolium
Marsh-fleabane
Pluchea rosea
Horseweed; Mare's-tail
Erigeron canadensis canadensis
Aster
Aster squarrosus
Frost Aster
Aster pilosus pilosus
Sweet Goldenrod
Solidago odora
Tall Goldenrod
Solidago altissima
Rayless Goldenrod
Chondrophora nudata

Bitterweed
Helenium amarum

FAUNA

Green Lynx Spider
Peucetia viridans
Elegant Crab Spider
Xysticus elegans
Common Meadow Katydid
Orchelimum vulgare
Cattail Toothpick Grasshopper
Leptysma marginicollis
American Bird Grasshopper
Schistocerca americana
Bucrates Katydid
Bucrates malivolans
Assassin Bug
Pygolampis pectoralis
Common Green Darner (shell)
Anax junius
Eastern Pondhawk
Erythemis simplicicollis
Blue Dragonlet
Erythrodiplax minuscula
Blue Dasher
Pachydiplax longipennis

Violet-masked Glider
Tramea carolina
Swamp Spreadwing (damselfly)
Lestes vigilax
Blackwater Bluet (damselfly)
Enallagma weewa
Four-spotted Blister Beetle
Tetraonyx sp.
Treefrog
Hyla sp.
Eastern Mud Turtle
Kinosternon subrubrum
King Rail
Rallus elegans
Blue Jay
Cyanocitta cristata
Carolina Chickadee
Parus carolinensis
Gray Catbird
Dumetella carolinensis
Rufous-sided Towhee
Pipilo erythrophthalmus
Field Sparrow
Spizella pusilla
White-tailed Deer (tracks)
Odocoileus virginianus

(Photo by A. Ciegler)
Honey-cup, *Zenobia pulverulenta*.

(Photo by L. Smith)
Swamp spreadwing (damselfly), *Lestes vigilax*.

(Photo by L. Greenlee)
Sundew, *Drosera capillaris*.

Lexington County

Lake Murray Spillway
Lexington County
November 21, 1987

This spillway is cut into the rock that underlies this portion of Lexington County. It creates a vertical cross-section that shows the effects of geologic action, especially metamorphosis. These views are usually restricted to manmade cuts such as this spillway or a roadcut. In addition, the testing and water control have scoured a rocky riverbed and have developed a rich system of flora and fauna. Because of the steep sides of the spillway and rough nature of the rocks, it takes a moderately strenuous effort to explore this site.

Private property. Owned by SCE&G, Inc.

Directions to Lake Murray Spillway: From I-20 west of Columbia, take Exit 55 onto SC 6 north. Go about 2.9 miles past the intersection where US 378 splits off to the right. The Lake Murray Spillway can be reached (with advance permission) via SCE&G's private road between Corley Mill Road and the south end of the dam. From I-26 west of Columbia, take Exit 102 onto SC 60 and go south through Irmo about 5.7 miles.

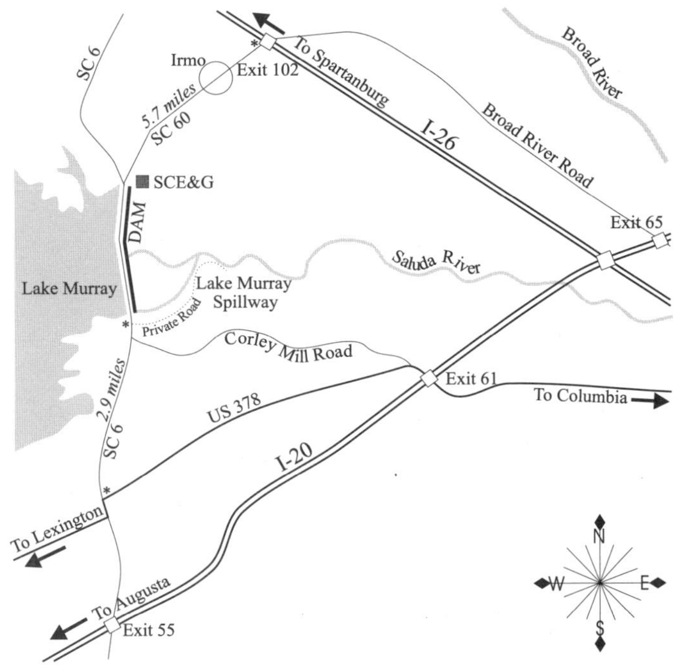

FLORA

Cinnamon Fern
Osmunda cinnamomea
Mosquito Fern
Azolla caroliniana
Loblolly Pine
Pinus taeda
Eastern Red-cedar
Juniperus virginiana
Common Cat-tail
Typha latifolia
Spike Grass
Uniola sessiliflora
Love Grass
Eragrostis sp.
Broomsedge
Andropogon virginicus
Andropogon sp.
Yellow-eyed-grass
Xyris sp.
Ladies'-tresses
Spiranthes sp.
Black Willow
Salix nigra
Eastern Cottonwood
Populus deltoides
Wax-myrtle; Southern Bayberry
Myrica cerifera
Tag Alder
Alnus serrulata
White Oak
Quercus alba
Water Oak
Quercus nigra
Alligator-weed
Alternanthera philoxeroides
Pokeweed; Pokeberry
Phytolacca americana
Sweet-gum
Liquidambar styraciflua
Mimosa; Silk-tree
Albizia julibrissin
Poison-ivy
Rhus radicans
Red Maple
Acer rubrum
Water-willow
Ludwigia decurrens
Seed-box
Ludwigia alternifolia
Evening-primrose
Oenothera sp.
Water-milfoil
Myriophyllum brasiliense
Privet; Ligustrum
Ligustrum japonicum
Yellow Jessamine
Gelsemium sempervirens
Sea-myrtle; Groundsel-tree
Baccharis halimifolia
Aster
Aster sp.
Goldenrod
Solidago sp.

FAUNA

Two-keeled Valve Snail
Valvata bicarinata
Freshwater Snail
Physella sp.
Asiatic Clam
Corbicula fluminea
Chinese Mantid
Tenodera aridifolia sinensis
Yellowjacket
Vespula sp.
Blackfly (larva)
Simulium sp.
Eastern Fence Lizard
Sceloporus undulatus
Turkey Vulture
Cathartes aura

(Photo by V. Winn)
Yelow Jessamine, *Gelsemium sempppervirens.*

Peachtree Rock Heritage Preserve
Lexington County
December 9, 1989

This preserve has a number of unusual geologic features. The peach tree-shaped rock for which it is named is soft sandstone and clay protected by a silicified layer on top which is weather resistant. There is also a waterfall caused by a creek that flows across the hard layer, cutting deeply into an area not so well protected. That the preserve was formerly part of a marine environment is evidenced by the marine fossils found at the site. (Collecting is not allowed.) Fossilized remains of marine worm and ghost shrimp burrows can be seen below the waterfall. There is also a very classic sandhills type habitat with white sand and a population of plants not widely distributed but found at the most extreme of these sandhill sites. The flora and fauna list for this trip was severely restricted by the ice storm that occurred the night before the trip.

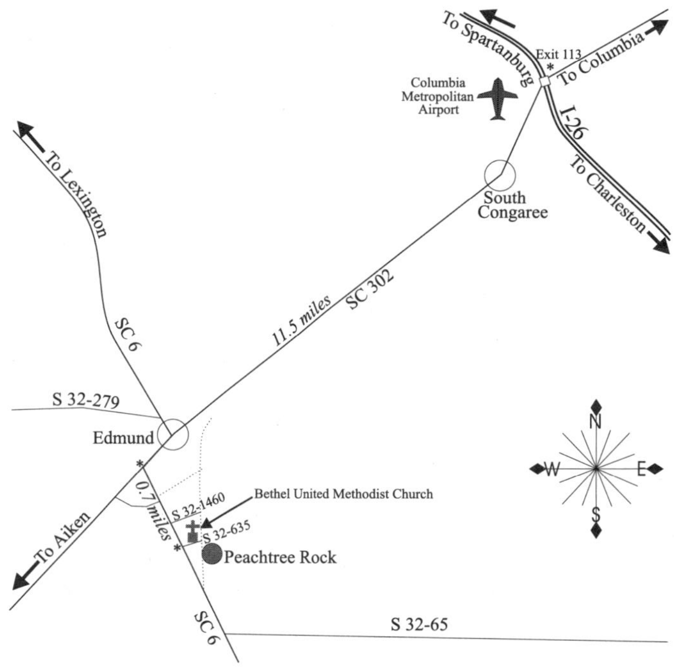

Directions to Peachtree Rock Heritage Preserve: From I-26 southwest of Columbia, take Exit 113, Airport Boulevard (SC 302), past the Columbia Airport to Edmund, where SC 6 joins from the right. Continue on SC 302 for another mile (total of 11.5 miles from the interstate exit), and follow SC 6 where it turns left. Go to the second paved road on the left, Bethel Church Road (0.7 miles) and turn left; go to the stop sign and turn right onto a dirt road. The parking lot for Peachtree Rock is on the left side of the dirt road.

FLORA

Lichen
Cladonia leporina
Parmotrema perforatum
Sand Spikemoss
Selaginella arenicola
Hairy Lip Fern
Cheilanthes lanosa
Bracken
Pteridium aquilinum
Blunt-lobed Woodsia
Woodsia obtusa
Ebony Spleenwort
Asplenium platyneuron
Resurrection Fern
Polypodium polypodioides
Long-leaf Pine
Pinus palustris
Slash Pine
Pinus elliotti
Cane; Switch Cane
Arundinaria gigantea
Bear-grass
Yucca filamentosa
Crane-fly Orchid
Tipularia discolor
Pale Hickory
Carya pallida
White Oak
Quercus alba
Turkey Oak
Quercus laevis
Black Oak
Quercus velutina
Spanish Oak
Quercus falcata
Black Jack Oak
Quercus marilandica
Water Oak
Quercus nigra
Blue-jack Oak
Quercus incana
Heartleaf
Hexastylis arifolia
Tulip-poplar
Liriodendron tulipifera
Sweetbay
Magnolia virginiana
Sweet-shrub; Carolina-allspice
Calycanthus floridus
Redbay
Persea borbonia
Sassafras
Sassafras albidum
Sweet-gum
Liquidambar styraciflua
Red Chokeberry
Sorbus arbutifolia
Hawthorn
Crataegus sp.
Black Cherry; Wild Cherry
Prunus serotina
Goat's-rue
Tephrosia virginiana
Poison-oak
Rhus toxicodendron
Titi
Cyrilla racemiflora
American Holly
Ilex opaca
Red Maple
Acer rubrum
St.-Andrew's-cross
Hypericum hypericoides
St.-John's-wort
Hypericum lloydii
Prickly-pear
Opuntia compressa
Flowering Dogwood
Cornus florida
Sweet Pepperbush
Clethra alnifolia
Rosemary
Ceratiola ericoides
Spotted Wintergreen
Chimaphila maculata
Indian-pipe
Monotropa uniflora
Pinxter-flower; Pinxter-bloom
Rhododendron nudiflorum
Sand-myrtle
Leiophyllum buxifolium
Mountain-laurel
Kalmia latifolia
Fetterbush
Lyonia lucida
Leucothoe recurva
Coastal Honeybells
Leucothoe racemosa
Sparkleberry
Vaccinium arboreum
Persimmon
Diospyros virginiana
Horse-sugar; Sweetleaf
Symplocos tinctoria
Yellow Jessamine
Gelsemium sempervirens
False Foxglove
Aureolaria pectinata

Peachtree Rock Heritage Preserve
December 9, 1989

Partridge-berry
Mitchella repens
Maple-leaf Viburnum
Viburnum acerifolium
Dog-fennel
Eupatorium capillifolium
Aster
Aster concolor
Coreopsis
Coreopsis major
Sweet Tooth
Dentinum repandum
Varnish Shelf
Ganoderma curtisii
Turkey-tail
Trametes versicolor
Puffball
Lycoperdon sp.
Barometer Earthstar
Astraeus hygrometricus

FAUNA

W.I. Ghost Shrimp (burrows)
Callianassa major
Pipe-organ Wasp
Trypoxylon politum
Soft-winged Flower Beetle
Attalus scincetus
Hickory Girdler Beetle
Oncideres cingulata
Red-spotted Purple (butterfly)
Limenitis arthemis astyanax
Evergreen Bagworm Moth
Thyridopteryx ephemeraeformis
Southern Flannel Moth
Megalopyge opercularis
Eastern Tent Caterpillar
Malacosoma americanum
Fall Webworm Moth
Hyphantria cunea
Shrew
Blarina sp.

(Photo by L. Greenlee)
Peachtree Rock, Lexington County, Sandhills region.

Marlboro County

Lake Wallace and Pledger Creek
Marlboro County
August 25, 1990

Lake Wallace is primarily oriented toward water activities, and one of the lesser-known activities here is nature watching. The north end of the lake has a waterfowl refuge with wood duck boxes and a "beauty spot" nature trail. This portion of the lakeside and lake provides a wide diversity of plants and animals. While none of the species encountered were rare, the diversity and aesthetics of the trail made the trip most enjoyable. The city of Bennettsville is to be commended for maintaining this area for a wide diversity of water-oriented activities.

Directions to Lake Wallace: Bennettsville may be reached from I-95, exit 181, north on SC 38; or from I-20, exit 116, north via Bishopville on US 15. In Bennettsville, from the intersection of Main Street (SC 385) and SC 38, take SC 385 north. Go 0.6 mile, watch for the Whitner-Evans Funeral Home, and turn left at the sign to Lake Wallace, staying on SC 385 (Gibson Highway). Drive about a mile and turn left again at another sign for the lake.

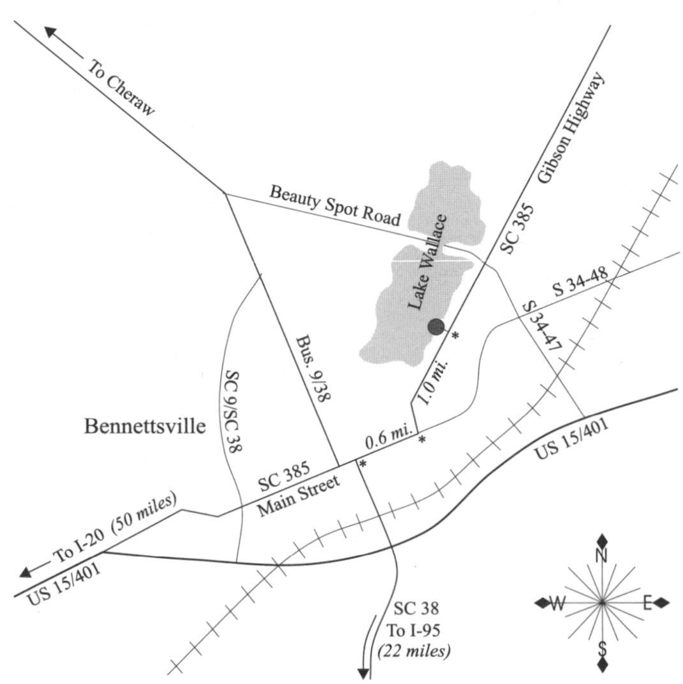

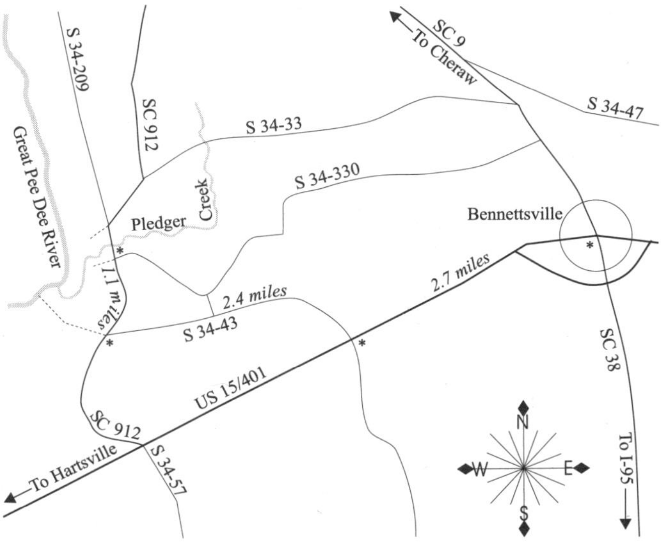

Pledger Creek offers a look at shallow-water plants and animals. Many insects were utilizing the water or visiting from drier areas to feed on the flowers of the aquatic species. The list demonstrates that a large number of plants and animals were found in the hour SCAN spent at this site.

Directions to Pledger Creek: From the center of downtown Bennettsville, take US 401/US 15 southwest toward Hartsville and Darlington. Drive 2.7 miles, then turn right (north) onto County Road 43. Go 2.4 miles to the junction with SC 912 and turn right (north). After 1.1 miles, SC 912 joins with County Road 209 and intersects S 34-330. The bridge across Pledger Creek is just north of that junction; look for a trail following the creek.

FLORA

Lichen
Lobaria quercizans
Cladonia apodocarpa
Cladonia capitata
Cladonia cristatella
Cladonia floerkeana
Cladonia sp.
Hypotrachyna livida
Parmelia rudecta
Parmelia sulcata
Parmotrema austrosinense
Parmotrema cetratum
Parmotrema dilitatum
Parmotrema hypotropum
Parmotrema perforatum
Parmotrema reticulatum
Tuckermannopsis viridis
Ramalina americana
Usnea strigosa
Physcia aipolia
Bracken
Pteridium aquilinum
Loblolly Pine
Pinus taeda
Short-leaf Pine
Pinus echinata
Bald-cypress
Taxodium distichum
Pond-cypress
Taxodium ascendens
Eastern Red-cedar
Juniperus virginiana
Common Cat-tail
Typha latifolia
Bur-reed
Sparganium americanum
Spotted Pondweed
Potamogeton pulcher
Arrowhead; Duck-potato
Sagittaria engelmanniana
Foxtail Grass
Setaria sp.
Barnyard Grass
Echinochloa crusgalli
Wool-grass
Scirpus cyperinus
Beak-rush
Rhynchospora chalarocephala
Rhynchospora glomerata
Rhynchospora inexpansa
Rhynchospora sp.
Sedge
Carex glaucescens
Carex lurida
Arrowleaf Arum
Peltandra virginica
Bogmoss
Mayaca aubletii
Yellow-eyed-grass
Xyris sp.
Dayflower
Commelina virginica
Commelina communis
Asian Spiderwort
Aneilema keisak
Greenbrier; Bullbrier
Smilax rotundifolia
Fringed Greenbrier
Smilax bona-nox
Sawbrier; Greenbrier
Smilax glauca
Sweet-scented Smilax
Smilax smallii
Yucca
Yucca sp.
Water-spider Orchid
Habenaria repens
Lizard's-tail
Saururus cernuus
Black Willow
Salix nigra
Pale Hickory
Carya pallida
Tag Alder
Alnus serrulata
River Birch
Betula nigra
White Oak
Quercus alba
Post Oak
Quercus stellata
Overcup Oak
Quercus lyrata
Red Oak
Quercus rubra
Spanish Oak
Quercus falcata
Black Jack Oak
Quercus marilandica
Water Oak
Quercus nigra
Willow Oak
Quercus phellos
Laurel Oak
Quercus laurifolia
Sugarberry; Hackberry
Celtis laevigata
Paper Mulberry
Broussonetia papyrifera
Red Mulberry
Morus rubra
Lamb's-quarters; Pigweed
Chenopodium album
Tearthumb
Polygonum sagittatum
Alligator-weed
Alternanthera philoxeroides
Pokeweed; Pokeberry
Phytolacca americana
Spatterdock; Yellow Pond-lily
Nuphar luteum
Water-lily
Nymphaea odorata
Fanwort
Cabomba caroliniana
Water-shield
Brasenia schreberi
Tulip-poplar
Liriodendron tulipifera
Sweetbay
Magnolia virginiana
Dwarf PawPaw
Asimina parviflora
Sassafras
Sassafras albidum
Poor-man's-pepper
Lepidium virginicum
Virginia-willow
Itea virginica
Sweet-gum
Liquidambar styraciflua
Himalaya-berry
Rubus bifrons
Dwarf Hawthorn
Crataegus uniflora
Black Cherry; Wild Cherry
Prunus serotina
Mimosa; Silk-tree
Albizia julibrissin
Sicklepod
Cassia obtusifolia
Wild Sensitive-plant
Cassia nicitans
Beggar-lice
Desmodium sp.
Creeping Bush-clover
Lespedeza repens
Sericea
Lespedeza cuneata
American Wisteria
Wisteria frutescens
Groundnut
Apios americana
Butterfly-pea
Clitoria mariana
Kudzu
Pueraria lobata
Tree-of-heaven
Ailanthus altissima
Chinaberry
Melia azedarach
Painted-leaf
Euphorbia heterophylla
Winged Sumac
Rhus copallina
Titi
Cyrilla racemiflora

Lake Wallace and Pledger Creek (continued) August 25, 1990

American Holly
Ilex opaca
Red Maple
Acer rubrum
Virginia-creeper
Parthenocissus quinquefolia
Muscadine; Bullace
Vitis rotundifolia
Devil's-shoestring
Sida rhombifolia
St.-Andrew's-cross
Hypericum hypericoides
St.-John's-wort
Hypericum walteri
Pineweed
Hypericum gentianoides
Passionflower; Maypops
Passiflora incarnata
Prickly-pear
Opuntia compressa
Water Loosestrife
Decodon verticillatus
Pale Meadow-beauty
Rhexia mariana
Common Meadow-beauty
Rhexia virginica virginica
Water-willow
Ludwigia decurrens
Ludwigia
Ludwigia sphaerocarpa
Trailing Ludwigia
Ludwigia palustris
Marsh Pennywort
Hydrocotyle umbellata
Blackgum
Nyssa sylvatica
Swamp Tupelo; Swamp Blackgum
Nyssa sylvatica biflora
Flowering Dogwood
Cornus florida
Sweet Pepperbush
Clethra alnifolia
Dog-hobble
Leucothoe axillaris
Sourwood
Oxydendrum arboreum
Sparkleberry
Vaccinium arboreum
Highbush Blueberry
Vaccinium corymbosum
Persimmon
Diospyros virginiana
Storax; Snowbell
Styrax americana
Chinese Privet
Ligustrum sinense
Yellow Jessamine
Gelsemium sempervirens
Small Floating-heart
Nymphoides cordata
Compact Dodder
Cuscuta compacta
Hydrolea
Hydrolea quadrivalvis
Heliotrope
Heliotropium amplexicaule
Brazilian Vervain
Verbena braziliensis
Beautyberry
Callicarpa americana
Horse-nettle
Solanum carolinense
Jimson-weed
Datura stramonium
Lindernia
Lindernia dubia
Woolly Mullein
Verbascum thapsus
Toadflax
Linaria sp.
Trumpet Vine
Campsis radicans
Bladderwort
Utricularia biflora
Button-bush
Cephalanthus occidentalis
Virginia Buttonweed
Diodia virginiana
Poor-jo (or Po-jo)
Diodia teres
Japanese Honeysuckle
Lonicera japonica
Arrowwood
Viburnum dentatum lucidum
Wahlenbergia
Wahlenbergia marginata
Cardinal-flower
Lobelia cardinalis
Common Ragweed
Ambrosia artemisiifolia
Common Dandelion
Taraxacum officinale
Elephant's-foot
Elephantopus tomentosus
Dog-fennel
Eupatorium capillifolium
Eupatorium
Eupatorium sp.
Climbing Hempweed
Mikania scandens
Rabbit-tobacco
Gnaphalium obtusifolium
Horseweed; Mare's-tail
Erigeron sp.
Sweet Goldenrod
Solidago odora
Goldenrod
Solidago sp.
Camphorweed
Heterotheca subaxillaris
Eclipta
Eclipta alba
Spanish-needles
Bidens bipinnata
Bitterweed
Helenium amarum
Citron Amanita
Amanita citrina
Red-cracked Bolete
Boletus chrysenteron
Red Gyroporus
Gyroporus purpurinus

FAUNA

Golden-silk Spider
Nephila clavipes
Writing Spider
Argiope aurantia
Long-jawed Orb-weaver
Tetragnatha sp.
Six-spotted Fishing Spider
Dolomedes triton
Carolina Mantid
Stagmomantis carolina
Eastern Pondhawk
Erythemis simplicicollis
Golden-winged Skimmer
Libellula auripennis
Slaty Skimmer
Libellula incesta
Great Blue Skimmer
Libellula vibrans
Blue Dasher
Pachydiplax longipennis
Eastern Amberwing (dragonfly)
Perithemis tenera
White-tail Dragonfly
Plathemis lydia
Violet-masked Glider
Tramea carolina
Crawling Water Beetle
Haliplus fasciatus
Peltodytes sexmaculatus
Peltodytes shermani
Peltodytes sp.
Predacious Diving Beetle
Coptotomus interrogatus
Whirligig Beetle
Dineutus discolor
Gyrinus frosti
Gyrinus analis
Water Scavenger Beetle
Berosus infuscatus
Marsh Beetle
Scirtes orbiculatus
Soldier Beetle
Chauliognathus pennsylvanicus
Fruit Weevil
Anthonomus tectus
Zebra Swallowtail
Eurytides marcellus
Eastern Tiger Swallowtail
Papilio glaucus
Spicebush Swallowtail
Papilio troilus
Sleepy Orange (butterfly)
Eurema nicippe
American Snout (butterfly)
Libytheana carinenta
Variegated Fritillary
Euptoieta claudia
Red Admiral
Vanessa atalanta rubria
Common Buckeye (butterfly)
Junonia coenia
Silver-spotted Skipper
Epargyreus clarus
Fowler's Toad
Bufo woodhousei
Green Frog; Bronze Frog
Rana clamitans
Six-lined Racerunner
Cnemidophorus sexlineatus
Great Egret
Casmerodius albus
Great Blue Heron
Ardea herodias
Mallard
Anas platyrhynchos
Lesser Yellowlegs
Tringa flavipes
Mourning Dove
Zenaida macroura
Red-bellied Woodpecker
Melanerpes carolinus
Blue Jay
Cyanocitta cristata
Fish Crow
Corvus ossifragus
Carolina Chickadee
Parus carolinensis
Northern Mockingbird
Mimus polyglottos
Northern Cardinal
Cardinalis cardinalis
Rufous-sided Towhee
Pipilo erythrophthalmus
Common Grackle
Quiscalus quiscula
Gray Squirrel
Sciurus carolinensis

(Photo by L. Greenlee)

Golden-silk spider, *Nephila clavipes*.

(Photo by L. Greenlee)

Violet-masked glider (saddlebags), *Tramea carolina*.

(Photo by A. Ciegler)

Trumpet vine, *Campsis radicans*.

McCormick County

John de la Howe School
McCormick County
April 24, 1988

This school property includes approximately 120 acres of virgin pine-oak forest. The trees are huge, with many species represented. This site has a good representation of understory plants. One unusual shrub found here is leatherwood, which produces a tough, stringy bark that American Indians used to make bow strings. The rolling terrain provides habitats of varying moisture availability, adding to the diversity of plants present.

Privately owned. Access by permission only. Phone: 864-391-2131. No map given.

FLORA

Rattlesnake Grape Fern
Botrychium virginianum
Christmas Fern
Polystichum acrostichoides
Ebony Spleenwort
Asplenium platyneuron
Loblolly Pine
Pinus taeda
Short-leaf Pine
Pinus echinata
Eastern Red-cedar
Juniperus virginiana
Jack-in-the-pulpit
Arisaema triphyllum
Bellwort
Uvularia perfoliata
Wild Yam
Dioscorea villosa
Blue-eyed-grass
Sisyrinchium angustifolium
Downy Rattlesnake-plantain
Goodyera pubescens
Crane-fly Orchid
Tipularia discolor
Black Willow
Salix nigra
Shagbark Hickory
Carya ovata
Hop-hornbeam
Ostrya virginiana
American Hornbeam; Ironwood
Carpinus caroliniana
Beech
Fagus grandifolia
White Oak
Quercus alba
Post Oak
Quercus stellata
Swamp Chestnut Oak
Quercus michauxii
Red Oak
Quercus rubra
Black Oak
Quercus velutina
Spanish Oak
Quercus falcata
Willow Oak
Quercus phellos
Heartleaf
Hexastylis arifolia
Meadow-rue
Thalictrum sp.
Tulip-poplar
Liriodendron tulipifera
Dwarf PawPaw
Asimina parviflora
Sweet-shrub; Carolina-allspice
Calycanthus floridus
Sassafras
Sassafras albidum
Yellow Cress
Rorippa islandica
Sweet-gum
Liquidambar styraciflua
American Sycamore
Platanus occidentalis
Hawthorn
Crataegus sp.
Black Cherry; Wild Cherry
Prunus serotina
Redbud; Judas-tree
Cercis canadensis
Violet Wood-sorrel
Oxalis violacea
Wild Geranium
Geranium maculatum
Poison-ivy
Rhus radicans
Hearts-a'burstin'-with-love
Euonymus americanus
Red Maple
Acer rubrum
Southern Sugar Maple
Acer saccharum floridanum
Virginia-creeper
Parthenocissus quinquefolia
Muscadine; Bullace
Vitis rotundifolia
Basswood; Linden
Tilia sp.
Modiola
Modiola caroliniana
Yellow Passionflower
Passiflora lutea

John de la Howe School (continued)
April 24, 1988

Leatherwood
Dirca palustris
Snakeroot
Sanicula sp.
Blackgum
Nyssa sylvatica
Flowering Dogwood
Cornus florida
Spotted Wintergreen
Chimaphila maculata
Deerberry; Squaw-huckleberry
Vaccinium stamineum
Elliott's Blueberry
Vaccinium elliottii
Bigleaf Snowbell
Styrax grandifolia
Ash
Fraxinus sp.
Pennywort
Obolaria virginica
Climbing Dogbane
Trachelospermum difforme
Indian-hemp
Apocynum cannabinum
Angle-pod
Matelea sp.
Lyre-leaved Sage
Salvia lyrata
Toadflax
Linaria canadensis
Cross Vine
Anisostichus capreolata
Button-bush
Cephalanthus occidentalis
Partridge-berry
Mitchella repens
Green-and-gold
Chrysogonum virginianum
Verbesina
Verbesina sp.
Yellow Morel
Morchella esculenta
Dead-man's-fingers
Xylaria polymorpha
Tree-ear
Auricularia auricula
Hophornbeam Disc (fungus)
Aleurodiscus oakesii
Violet-toothed Polypore
Trichaptum biformis
Crowded Parchment
Stereum complicatum
Fly Agaric
Amanita muscaria formosa
Grisette
Amanita vaginata
Oak-loving Collybia
Collybia dryophila
Coral Spring Mycena
Mycena acicula
Oyster Mushroom
Pleurotus ostreatus

FAUNA

Hornbeam Leaf-fold Gall(midge)
Cecidomyia pudibunda
Pedunculate Ground Beetle
Scarites sp.
Minute Ground Beetle
Bembidion confusum
Bembidion dilitatum
Dingy Ground Beetle
Agonoderus sp.
Notch-mouthed Ground Beetle
Dicaelus sp.
Crawling Water Beetle
Haliplus triopsus
Water Scavenger Beetle
Berosus sp.
June Beetle
Phyllophaga prunina
Blind Elater
Alaus myops
Net-winged Beetle
Eros humeralis
Soft-winged Flower Beetle
Malachius sp.
Hypebaeus sp.
Lagriid Bark Beetle
Statira sp.
Comb-clawed Beetle
Isomira sericea
Round-necked Longhorn Beetle
Molorchus bimaculatus
Molorchus sp.
Flower Longhorn Beetle
Toxotus cinnamopterus
Hickory Girdler Beetle (sign)
Oncideres cingulata
Leaf Beetle
Anomoea sp.
Broad-bodied Leaf Beetle
Calligrapha rhoda
Oak Flea Beetle
Oedionychis quercata
Flea Beetle
Disonycha abbreviata
Leaf-mining Beetle
Chalepus scapularis
Zebra Swallowtail
Eurytides marcellus
Spicebush Swallowtail
Papilio troilus
Fairy Moth
Adela purpurea
Northern Cricket Frog
Acris crepitans
Carolina Anole
Anolis carolinensis
Canada Goose
Branta canadensis
Red-bellied Woodpecker
Melanerpes carolinus
Downy Woodpecker
Picoides pubescens
Red-Eyed Vireo
Vireo olivaceus
Pine Warbler
Dendroica pinus
Red-winged Blackbird
Agelaius phoeniceus
Summer Tanager
Piranga rubra

(Photo by L. Greenlee)
Flowering dogwood, *Cornus florida.*

Stevens Creek Heritage Preserve
McCormick County
April 23, 1988, and March 29, 1992

This north-facing slope along the south bank of Stevens Creek is considered one of the best wildflower sites in South Carolina and possibly the entire Southeast. Imagine hillsides in full bloom with almost no bare ground visible, and you have an idea of the scope of this site at peak bloom. These flowers are mostly spring ephemerals, but a few rare shrubs share the spotlight. The dominant tree cover is old-growth hardwoods such as white oak, bitternut hickory and American ash. This plant community probably dates back to the Pleistocene Era as a glacial relict. It is unusual in that the soils are basic, but the parent material is acidic; it is also unusual because it contains so many disjunct species from both northern and southern distributions, and for its many rare species. North-facing slopes harbor a large diversity of plant life. The lack of direct sunlight results in a more constant temperature and soil moisture levels than would otherwise be found.

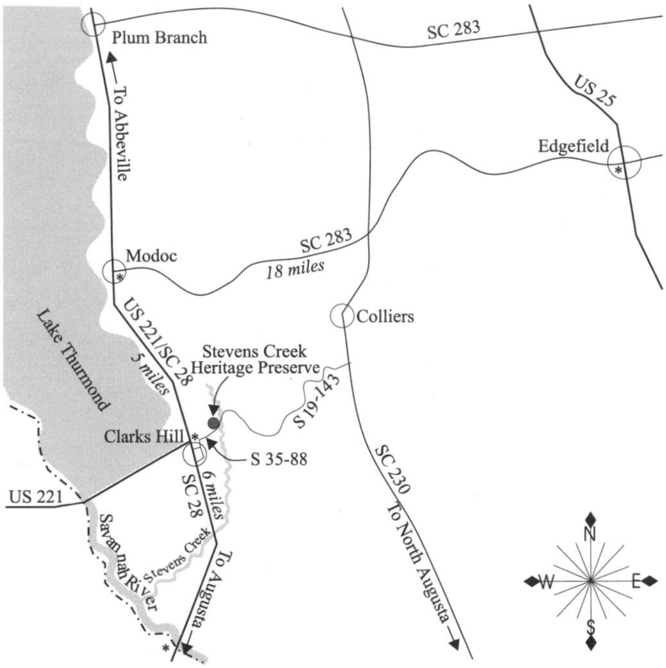

Directions to Stevens Creek Heritage Preserve: From Edgefield, drive west on SC 283 for 18 miles to Modoc. Turn left (south) onto US 221/SC 28 and travel 5 miles to the Clarks Hill Dam. Continue straight ahead on SC 28 for 0.3 miles past the intersection where US 221 crosses the dam, then turn left onto County Road 88. Stevens Creek Heritage Preserve is on the left in about a mile, marked by a small sign.

FLORA

Lichen
Collema furfuraceum
Leptogium burnetiae
Leptogium cyanescens
Pannaria rubiginosa
Peltigera canina
Graphis scripta
Graphis mosquitensis
Graphis striatula
Graphis sp.
Glyphis cicatricosa
Phaeographina sp.
Bacidia schweinitzii
Leptogium austroamericanum
Lecidea russula
Lecidea sp.
Cladonia bacillaris
Cladonia balfourii
Cladonia capitata
Cladonia chlorophaea
Cladonia sobolescens
Cladonia coniocraea
Cladonia cristatella
Cladonia cristatella cristatella
Cladonia cristatella vestita
Cladonia cristatella squamossisima
Cladonia cristatella pleurocarpa
Cladonia didyma
Cladonia macilenta
Cladonia sp.
Pertusaria neoscotica
Pertusaria leucostoma
Pertusaria ostiolata
Pertusaria paratuberculifera
Pertusaria trachythallina
Pertusaria subpertusa
Pertusaria texana
Pertusaria macounii
Pertusaria sp.
Haematomma puniceum
Lecanora lacustris
Lecanora miculata
Ochrolechia africana
Bulbothrix goebelii
Hypotrachyna livida
Parmelina aurulenta
Parmotrema cetratum
Parmotrema crinitum
Parmotrema perforatum
Parmotrema praesorediosum
Parmotrema reticulatum
Parmotrema subisidiosum
Parmotrema ultralucens
Pseudoparmelia baltimorensis
Pseudoparmelia caroliniana
Xanthoparmelia plittii
Usnea ceratina
Usnea strigosa
Buellia stigmaea
Buellia caloosensis
Buellia coccinea
Physcia americana
Physcia millegrana
Physcia subtilis
Heterodermia speciosa
Heterodermia casarettiana
Phaeophyscia ciliata
Phaeophyscia rubropulchra
Caloplaca citrina
Pyrenula laevigata
Pyrenula mamillana
Pyrenula nitida
Mycoporum compositum
Arthonia sp.
Trypethelium virens
Trypethelium sp.

Maidenhair Fern
Adiantum pedatum
Hay-scented Fern
Dennstaedtia punctilobula
Blunt-lobed Woodsia
Woodsia obtusa
Spreading Bladder Fern
Cystopteris protrusa
Marginal Shield Fern
Dryopteris marginalis
Christmas Fern
Polystichum acrostichoides
Ebony Spleenwort
Asplenium platyneuron
Loblolly Pine
Pinus taeda
Short-leaf Pine
Pinus echinata
Eastern Red-cedar
Juniperus virginiana
Cane; Switch Cane
Arundinaria gigantea
Melic Grass
Melica mutica
Jack-in-the-pulpit
Arisaema triphyllum
Spiderwort
Tradescantia hirsuticaulis
Tradescantia sp.
Woodrush
Luzula sp.
Sawbrier; Greenbrier
Smilax glauca

Stevens Creek Heritage Preserve (continued)
April 23, 1988, and March 29, 1992

Trillium
Trillium discolor
Trillium lanceolatum
Nodding Trillium
Trillium cernuum
Solomon's-seal
Polygonatum biflorum
Dog's-tooth-violet
Erythronium americanum
Bellwort
Uvularia perfoliata
Uvularia sp.
False Garlic
Allium bivalve
Blue-eyed-grass
Sisyrinchium albidum
Downy Rattlesnake-plantain
Goodyera pubescens
Bitternut Hickory
Carya cordiformis
Shagbark Hickory
Carya ovata
Mockernut Hickory
Carya tomentosa
Pignut Hickory
Carya glabra
River Birch
Betula nigra
Hop-hornbeam
Ostrya virginiana
American Hornbeam; Ironwood
Carpinus caroliniana
White Oak
Quercus alba
Swamp Chestnut Oak
Quercus michauxii
Red Oak
Quercus rubra
Black Oak
Quercus velutina
Spanish Oak
Quercus falcata
American Elm
Ulmus americana
Winged Elm
Ulmus alata
Sugarberry; Hackberry
Celtis laevigata
Wood-nettle
Laportea canadensis
Stinging Nettle
Urtica chamaedryoides
Wild Ginger
Asarum canadense
Heartleaf
Hexastylis arifolia
Spring-beauty
Claytonia virginica
Common Chickweed
Stellaria media
Giant Chickweed
Stellaria pubera
Mouse-ear Chickweed
Cerastium nutans
False Rue-anemone
Isopyrum biternatum
Windflower
Thalictrum thalictroides
Small-flowered Crowfoot
Ranunculus abortivus
Buttercup
Ranunculus sp.
Liverleaf
Hepatica americana
Windflower
Anemone lancifolia
Mayapple
Podophyllum peltatum
Tulip-poplar
Liriodendron tulipifera
Southern Magnolia
Magnolia grandiflora
Dwarf PawPaw
Asimina parviflora
PawPaw
Asimina triloba
Sweet-shrub; Carolina-allspice
Calycanthus floridus
Sassafras
Sassafras albidum
Spicebush
Lindera benzoin
Bloodroot
Sanguinaria canadensis
Dutchman's-breeches
Dicentra cucullaria
Yellow Corydalis
Corydalis flavula
Toothwort
Cardamine angustata angustata
Hairy Bitter Cress
Cardamine hirsuta
Spiny Gooseberry
Ribes echinellum
Mock-orange
Philadelphus inodorus
Alumroot
Heuchera americana
Foamflower
Tiarella cordifolia collina
Saxifrage
Saxifraga virginiensis
Sweet-gum
Liquidambar styraciflua
Witch-hazel
Hamamelis virginiana
American Sycamore
Platanus occidentalis
Five-fingers
Potentilla canadensis
Potentilla simplex
Southern Crabapple
Malus angustifolia
Littlehip Hawthorn
Crataegus spathulata
Hawthorn
Crataegus sp.
Shadbush; Serviceberry
Amelanchier sp.
Black Cherry; Wild Cherry
Prunus serotina
Redbud; Judas-tree
Cercis canadensis
Bush-clover
Lespedeza sp.
Violet Wood-sorrel
Oxalis violacea
Wild Geranium
Geranium maculatum
Poison-ivy
Rhus radicans
American Holly
Ilex opaca
Possumhaw Holly
Ilex decidua
Hearts-a'burstin'-with-love
Euonymus americanus
Bladdernut
Staphylea trifolia
Boxelder
Acer negundo
Southern Sugar Maple
Acer saccharum floridanum
Painted Buckeye
Aesculus sylvatica
Spotted Touch-me-not
Impatiens capensis
Virginia-creeper
Parthenocissus quinquefolia
Muscadine; Bullace
Vitis rotundifolia
Basswood; Linden
Tilia caroliniana
St.-Andrew's-cross
Hypericum hypericoides
Green Violet
Hybanthus concolor
Common Blue Violet
Viola papilionacea
Sweet Chervil
Osmorhiza longistylis
Meadow Parsnip
Zizia trifoliata
Blackgum
Nyssa sylvatica
Flowering Dogwood
Cornus florida
Spotted Wintergreen
Chimaphila maculata
Azalea
Rhododendron sp.
Sourwood
Oxydendrum arboreum
Sparkleberry
Vaccinium arboreum
Deerberry; Squaw-huckleberry
Vaccinium stamineum
Elliott's Blueberry
Vaccinium elliottii
Shooting-star
Dodecatheon meadia
Creeping-Charlie
Lysimachia nummularia
Persimmon
Diospyros virginiana
Carolina Silverbell
Halesia carolina
Fringe-tree
Chionanthus virginicus
Chinese Privet
Ligustrum sinense
Yellow Jessamine
Gelsemium sempervirens
Pennywort
Obolaria virginica
Baby-blue-eyes
Nemophila microcalyx
Phacelia
Phacelia dubia
Stoneseed
Lithospermum tuberosum
Lyre-leaved Sage
Salvia lyrata
Shrubby Savory
Satureja georgiana
Cross Vine
Anisostichus capreolata
Squaw-root
Conopholis americana
Partridge-berry
Mitchella repens
Cleavers
Galium aparine
Bedstraw
Galium sp.
Japanese Honeysuckle
Lonicera japonica
Coral-berry
Symphoricarpos orbiculatus
Rusty Black-haw
Viburnum rufidulum
American Elderberry
Sambucus canadensis
Pale Indian-plantain
Cacalia atriplicifolia
Butterweed
Senecio glabellus
Senecio obovatus
Thistle
Carduus sp.
Pussy-toes
Antennaria plantaginifolia
Green-and-gold
Chrysogonum virginianum australe
Yellow Morel
Morchella esculenta

FAUNA

Almond Millipede
Sigmoria aberrans
Bowl-and-doily Spider
Frontinella pyramitela
Long-jawed Orb-weaver
Tetragnatha sp.
Wolf Spider
Lycosa sp.
Jumping Spider (large)
Phidippus pulcher

Red Velvet Mite
Trombidium sp.
Oak-apple (gall wasp)
Amphibolips inanis
Crawling Water Beetle
Haliplus triopsus
Peltodytes duodecimpunctatus
Water Scavenger Beetle
Berosus pantherinus
Aphodiine Dung Beetle
Ataenius sp.
Garden Beetle
Serica sp.
June Beetle
Phyllophaga sp.
Soldier Beetle
Cantharis sp.
Malthodes sp.
Dermestid Beetle
Cryptorhopalum picicorne
Cryptorhopalum sp.
Soft-winged Flower Beetle
Collops sp.
Lagriid Bark Beetle
Statira sp.
False Ant-like Flower Beetle
Pedilus sp.
Flower Longhorn Beetle
Anoplodera sp.
Leaf Beetle
Coscinoptera dominicana
Oval Leaf Beetle
Paria canella quadrinotata
Skeletonizing Leaf Beetle
Luperodes cyanellus
Oak Flea Beetle
Oedionychis quercata
Grape Flea Beetle
Altica chalybea
Flower Weevil
Centrinaspis sp.
Zebra Swallowtail
Eurytides marcellus
Eastern Tiger Swallowtail
Papilio glaucus
Spicebush Swallowtail
Papilio troilus
Falcate Orangetip
Anthocharis midea
Spring Azure (butterfly)
Celastrina ladon
Pearl Crescent
Phyciodes tharos
Red Admiral
Vanessa atalanta rubria
Fowler's Toad
Bufo woodhousei
Northern Cricket Frog
Acris crepitans
Cope's Gray Treefrog
Hyla chrysoscelis
Upland Chorus Frog
Pseudacris triceriata
Eastern Box Turtle
Terrapene carolina
Carolina Anole
Anolis carolinensis
Ground Skink
Scincella lateralis
Turkey Vulture
Cathartes aura
Wild Turkey
Meleagris gallopavo
Northern Flicker
Colaptes auratus
Pileated Woodpecker
Dryocopus pileatus
Eastern Wood-pewee
Contopus virens
Acadian Flycatcher
Empidonax virescens
American Crow
Corvus brachyrhynchos
Tufted Titmouse
Parus bicolor
Carolina Chickadee
Parus carolinensis
Carolina Wren
Thryothorus ludovicianus
Red-Eyed Vireo
Vireo olivaceus
Prothonotary Warbler
Protonotaria citrea
Northern Parula
Parula americana
Black-and-white Warbler
Mniotilta varia
Yellow-rumped Warbler
Dendroica coronata
Pine Warbler
Dendroica pinus
Hooded Warbler
Wilsonia citrina
Summer Tanager
Piranga rubra

Newberry County

Molly's Rock Recreation Area
(Sumter National Forest)
Newberry County
June 27, 1987

Molly's Rock is like most wooded areas in the South Carolina Piedmont in its recent origins. Here, the growth is about 60 years from old field habitat to forest in succession. The resulting diversity of plants, and habitats from pond to open field, provide ample insect life and other wildlife. Dragonflies and butterflies, as you can see from the list, were plentiful.

Directions to Molly's Rock Recreation Area: From I-26 near Newberry, take Exit 74 and go north on SC 34 toward Winnsboro. After 5 miles, turn left (north) on US 176 toward Whitmire, and proceed approximately 4 miles. The Sumter National Forest-Molly's Rock Recreation Area sign will be on your right. Turn right onto the gravel road by the sign, and follow it to the picnic area where the trail begins.

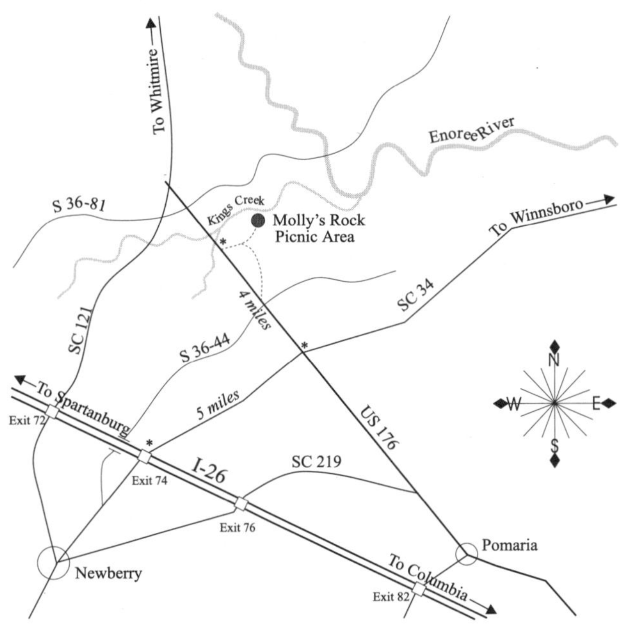

Molly's Rock Recreation Area (continued)
June 27, 1987

FLORA

Running-pine
Lycopodium flabelliforme
Rattlesnake Grape Fern
Botrychium virginianum
Southern Lady Fern
Athyrium asplenioides
Christmas Fern
Polystichum acrostichoides
Netted Chain Fern
Woodwardia areolata
Ebony Spleenwort
Asplenium platyneuron
Resurrection Fern
Polypodium polypodioides
Loblolly Pine
Pinus taeda
Eastern Red-cedar
Juniperus virginiana
Cane; Switch Cane
Arundinaria gigantea
Yellow-groove Bamboo
Phyllostachys aurea
Microstegium (grass)
Microstegium vimineum
Downy Rattlesnake-plantain
Goodyera pubescens
Black Walnut
Juglans nigra
Tag Alder
Alnus serrulata
White Oak
Quercus alba
Red Oak
Quercus rubra
Water Oak
Quercus nigra
Willow Oak
Quercus phellos
Winged Elm
Ulmus alata
Hackberry
Celtis occidentalis georgiana
Red Mulberry
Morus rubra
Heartleaf
Hexastylis sp.
Tulip-poplar
Liriodendron tulipifera
PawPaw
Asimina triloba
Sassafras
Sassafras albidum
Spicebush
Lindera benzoin
Climbing-hydrangea
Decumaria barbara
Sweet-gum
Liquidambar styraciflua
Hawthorn
Crataegus sp.
Black Cherry; Wild Cherry
Prunus serotina
Mimosa; Silk-tree
Albizia julibrissin
Redbud; Judas-tree
Cercis canadensis
Pencil-flower
Stylosanthes biflora
Naked-flower Tick-trefoil
Desmodium nudiflorum
Beggar-lice
Desmodium sp.
Tephrosia
Tephrosia spicata
Butterfly-pea
Clitoria mariana
Poison-ivy
Rhus radicans
Winged Sumac
Rhus copallina
American Holly
Ilex opaca
Possumhaw Holly
Ilex decidua
Hearts-a'burstin'-with-love
Euonymus americanus
Red Maple
Acer rubrum
Virginia-creeper
Parthenocissus quinquefolia
Muscadine; Bullace
Vitis rotundifolia
Devil's-walking-stick
Aralia spinosa
Blackgum
Nyssa sylvatica
Flowering Dogwood
Cornus florida
Spotted Wintergreen
Chimaphila maculata
Sourwood
Oxydendrum arboreum
Sparkleberry
Vaccinium arboreum
Elliott's Blueberry
Vaccinium elliottii
Persimmon
Diospyros virginiana
American Ash; White Ash
Fraxinus americana
Climbing Dogbane
Trachelospermum difforme
Beautyberry
Callicarpa americana
Carpenterweed
Prunella vulgaris
Lyre-leaved Sage
Salvia lyrata
Cross Vine
Anisostichus capreolata
Trumpet Vine
Campsis radicans
Ruellia
Ruellia caroliniensis
Virginia Buttonweed
Diodia virginiana
Partridge-berry
Mitchella repens
Japanese Honeysuckle
Lonicera japonica
Black-haw; Nanny-berry
Viburnum prunifolium
American Elderberry
Sambucus canadensis
Creeping Cucumber
Melothria pendula
Climbing Hempweed
Mikania scandens
Bitterweed
Helenium amarum

FAUNA

Writing Spider
Argiope aurantia
Orb-weaver
Araneus sp.
Long-jawed Orb-weaver
Tetragnatha sp.
Fishing Spider
Dolomedes sp.
Carolina Locust
Dissosteira carolina
Banded Pennant (dragonfly)
Celithemis fasciata
Eastern Pondhawk
Erythemis simplicicollis
Slaty Skimmer
Libellula incesta
Pied Skimmer
Libellula luctuosa
Blue Dasher
Pachydiplax longipennis
Eastern Amberwing (dragonfly)
Perithemis tenera
White-tail Dragonfly
Plathemis lydia
Ebony Jewelwing
Calopteryx maculata
Japanese Beetle
Popillia japonica
Spotless Lady Beetle
Cycloneda sanguinea
Red-banded Hairstreak
Calycopis cecrops
Eastern Tailed-blue (butterfly)
Everes comyntas
American Snout (butterfly)
Libytheana carinenta
Pearl Crescent
Phyciodes tharos
Mourning Cloak
Nymphalis antiopa
American Lady (butterfly)
Vanessa virginiensis
Common Buckeye (butterfly)
Junonia coenia
Red-spotted Purple (butterfly)
Limenitis arthemis astyanax
Hackberry Emperor (butterfly)
Asterocampa celtis
Carolina Satyr (butterfly)
Hermeuptychia sosybius
Georgia Satyr (butterfly)
Neonympha areolata
Silver-spotted Skipper
Epargyreus clarus
Io Moth
Automeris io
Green Frog; Bronze Frog
Rana clamitans
Southern Leopard Frog
Rana sphenocephala
Yellow-bellied Slider
Chrysemys scripta
Carolina Anole
Anolis carolinensis
Ground Skink
Scincella lateralis
Six-lined Racerunner
Cnemidophorus sexlineatus
Redbelly Water Snake
Nerodia erythrogaster
Red-shouldered Hawk
Buteo lineatus
Red-bellied Woodpecker
Melanerpes carolinus
Great Crested Flycatcher
Myiarchus crinitus
Eastern Wood-pewee
Contopus virens
Blue Jay
Cyanocitta cristata
American Crow
Corvus brachyrhynchos
Tufted Titmouse
Parus bicolor
Carolina Chickadee
Parus carolinensis
Carolina Wren
Thryothorus ludovicianus
Wood Thrush
Hylocichla mustelina
Northern Mockingbird
Mimus polyglottos
Red-Eyed Vireo
Vireo olivaceus
Northern Parula
Parula americana
Pine Warbler
Dendroica pinus
Hooded Warbler
Wilsonia citrina

(Photo by V. Winn)
Butterfly-pea, *Clitoria mariana*.

Oconee County

Brasstown Falls
(Sumter National Forest)
Oconee County
August 23, 1986

This waterfall is a series of short drops and cascades, but the total drop makes a very picturesque portion of Brasstown Creek. All the ledges and small outcrops seem to be ideal places for "spray habitat" plants such as filmy fern. Unfortunately SCAN was unable to find the plant on the day of our visit. The walk in from the access road travels an old logging road which goes alongside a small tributary. One hillside was covered with barren strawberry, a very uncommon plant in South Carolina. Oak-leaf hydrangea is found at this site, the only such colony in South Carolina. Although the terrain is technically still piedmont, many plants typical of mountains are found in this area.

Directions to Brasstown Falls: From the railroad bridge in Westminster, take US 76 northwest for 11 miles, and turn left at the first paved road after Brasstown Church. After a distance, the paved road will become a gravel road. Where the gravel road crosses Brasstown Creek, you will see the trail head to the falls on the right.

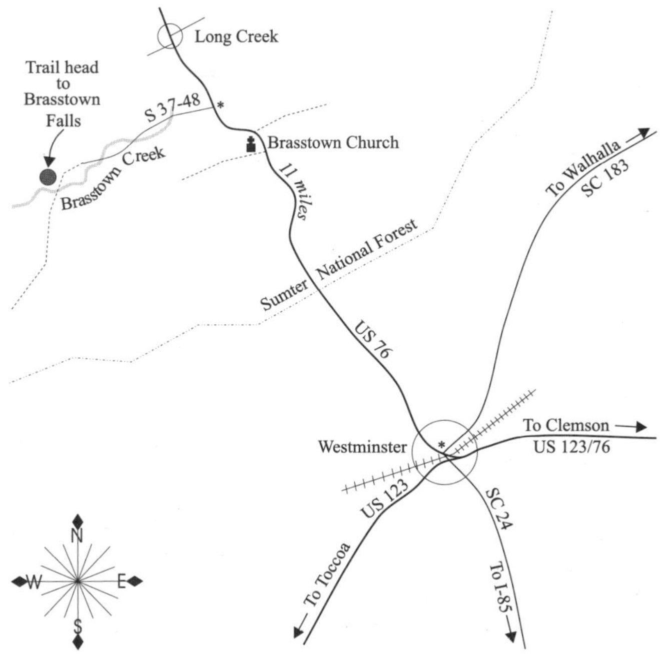

FLORA

Meadow Spikemoss
Selaginella apoda
Common Grape Fern
Botrychium dissectum
Bracken
Pteridium aquilinum
Southern Lady Fern
Athyrium asplenioides
Marginal Shield Fern
Dryopteris marginalis
Christmas Fern
Polystichum acrostichoides
New York Fern
Thelypteris noveboracensis
Ebony Spleenwort
Asplenium platyneuron
White Pine
Pinus strobus
Eastern Hemlock
Tsuga canadensis
Wapato; Duck-potato
Sagittaria latifolia pubescens
Cane; Switch Cane
Arundinaria gigantea
Microstegium (grass)
Microstegium vimineum
Jack-in-the-pulpit
Arisaema triphyllum
Dayflower
Commelina communis
Spiderwort
Tradescantia subaspera
Catesby's Trillium
Trillium catesbaei
Indian Cucumber-root
Medeola virginiana
False Solomon's Seal
Smilacina racemosa
Solomon's-seal
Polygonatum biflorum
Wild Yam
Dioscorea villosa
Downy Rattlesnake-plantain
Goodyera pubescens
Black Walnut
Juglans nigra
Tag Alder
Alnus serrulata
American Hornbeam; Ironwood
Carpinus caroliniana
Beech
Fagus grandifolia
White Oak
Quercus alba
Rock Chestnut Oak
Quercus prinus
False Nettle
Boehmeria cylindrica
Buffalo-nut
Pyrularia pubera
Tearthumb
Polygonum sagittatum
Starry Campion
Silene stellata
Yellowroot
Xanthorhiza simplicissima
Meadow-rue
Thalictrum clavatum
Tulip-poplar
Liriodendron tulipifera
Umbrella-tree
Magnolia fraseri
Sweet-shrub; Carolina-allspice
Calycanthus floridus
Sassafras
Sassafras albidum
Riverweed
Podostemum ceratophyllum
Virginia-willow
Itea virginica
Climbing-hydrangea
Decumaria barbara
Wild Hydrangea
Hydrangea arborescens
Sweet-gum
Liquidambar styraciflua
Witch-hazel
Hamamelis virginiana
Barren Strawberry
Waldsteinia fragarioides
Agrimony
Agrimonia parviflora
Downy Serviceberry
Amelanchier arborea
Sensitive-brier
Schrankia microphylla
Redbud; Judas-tree
Cercis canadensis
Groundnut
Apios americana
Butterfly-pea
Clitoria mariana

Brasstown Falls (continued)
August 23, 1986

Hog-peanut
Amphicarpa bracteata
Poison-ivy
Rhus radicans
American Holly
Ilex opaca
Hearts-a'burstin'-with-love
Euonymus americanus
Red Maple
Acer rubrum
Spotted Touch-me-not
Impatiens capensis
St.-John's-wort
Hypericum mutilum
Devil's-walking-stick
Aralia spinosa
Spikenard
Aralia racemosa
Sanicle; Snakeroot
Sanicula canadensis
Water Hemlock
Cicuta maculata
Blackgum
Nyssa sylvatica
Flowering Dogwood
Cornus florida
Alternate-leaved Dogwood
Cornus alternifolia
Silky Dogwood
Cornus amomum
Spotted Wintergreen
Chimaphila maculata
Indian-pipe
Monotropa uniflora
Rosebay; Great Laurel
Rhododendron maximum
Carolina-laurel
Rhododendron minus
Mountain-laurel
Kalmia latifolia
Dog-hobble
Leucothoe axillaris
Deerberry; Squaw-huckleberry
Vaccinium stamineum
Galax; Galaxy; Coltsfoot
Galax aphylla
Carolina Silverbell
Halesia carolina
American Ash; White Ash
Fraxinus americana
Rose-pink; Bitter-bloom
Sabatia angularis
Man-root
Ipomoea pandurata
Carpenterweed
Prunella vulgaris
Bee-balm
Monarda clinopodia
Mountain Mint
Pycnanthemum incanum
Mecardonia
Mecardonia acuminata
Hairy Gratiola
Gratiola pilosa
Lindernia
Lindernia dubia
Monkey-flower
Mimulus ringens
Woolly Mullein
Verbascum thapsus
False Foxglove
Aureolaria sp.
Lousewort
Pedicularis canadensis
Broadleaf Plantain
Plantago rugelii
Virginia Buttonweed
Diodia virginiana
Partridge-berry
Mitchella repens
Cardinal-flower
Lobelia cardinalis
Lobelia
Lobelia amoena
Lobelia puberula
Indian-tobacco
Lobelia inflata
Wild Lettuce
Lactuca biennis
Pale Indian-plantain
Cacalia atriplicifolia
Thistle
Carduus sp.
Ironweed
Vernonia glauca
Elephant's-foot
Elephantopus carolinianus
Joe-pye-weed
Eupatorium fistulosum
Pussy-toes
Antennaria solitaria
Coneflower
Rudbeckia laciniata
Wing-stem
Verbesina virginica
Coreopsis
Coreopsis major
Spanish-needles
Bidens bipinnata
Spindle-shaped Yellow Coral
Clavulinopsis fusiformis
Crested Coral
Clavulina cristata
Gelatinous-pored Polypore
Caloporus dichrous
False Caesar's Mushroom
Amanita parcivolvata
Birch Bolete
Austroboletus betula
Bolete
Austroboletus gracilis
Emetic Russula
Russula emetica
Violet Collybia
Collybia iocephala
Splash-cups
Cyathus striatus
Barometer Earthstar
Astraeus hygrometricus
Carnival Candy Slime
Arcyria denudata
Chocolate Tube Slime
Stemonitis splendens

FAUNA

Filmy Dome Spider
Prolinyphia marginata
Nursery-web Spider
Pisaurina sp.
Wheel Bug
Arilus cristatus
Dingy Ground Beetle
Anisodactylus rusticus
Common Tumblebug
Canthon laevis
Soldier Beetle
Chauliognathus pennsylvanicus
Blister Beetle
Epicauta vittata
Epicauta trichrus
Rove Beetle
Conotelus sp.
Sap-feeding Beetle
Pallodes sp.
Pleasing Fungus Beetle
Triplax sp.
Shining Flower Beetle
Olibrus sp.
12-spotted Cucumber Beetle
Diabrotica duodecimpunctata
Skeletonizing Leaf Beetle
Luperodes sp.
Pipevine Swallowtail
Battus philenor
Black Swallowtail
Papilio polyxenes asterius
Eastern Tiger Swallowtail
Papilio glaucus
Sleepy Orange (butterfly)
Eurema nicippe
Great Spangled Fritillary
Speyeria cybele
Pearl Crescent
Phyciodes tharos
Monarch (butterfly)
Danaus plexippus
Silver-spotted Skipper
Epargyreus clarus
Clymene Moth
Haploa clymene
American Toad
Bufo americanus
Eastern Fence Lizard
Sceloporus undulatus
Red-tailed Hawk
Buteo jamaicensis

(Photos by L. Greenlee and V. Winn)

Solomon's-seal, *Polygonatum biflorum* (above) and false solomon's seal, *Smilacina racemosa,* (below) in fruit.

Burrells Ford, King Creek Falls, and Cherry Hill Campground

(Sumter National Forest)

Oconee County

August 22, 1987, April 28, 1990, and May 19, 1991

This area of Sumter National Forest is on the west-facing slope of the Chattooga River watershed. The Cherry Hill campground includes a nature trail with typical mountain species. Burrells Ford and the King Creek area provide pleasant trails with creek and riverside plant and animal life. In damp weather these areas are often rich in salamanders and other amphibians. The falls are a nice drop of about 60 feet, and reached on an easy trail with lots of interesting mountain plants. The Chattooga River is a National Wild and Scenic River, and the trail alongside allows a look at relatively undisturbed mountain river edges.

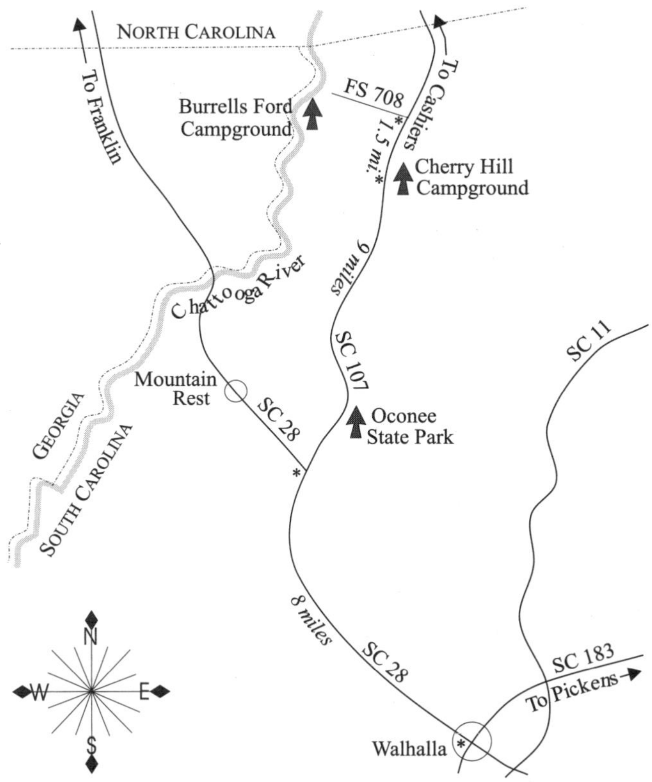

Directions to Cherry Hill Campground: From Walhalla, travel north on SC 28 for 8 miles. Take the right fork onto SC 107 and continue for about 9 more miles, bypassing Oconee State Park. Cherry Hill Campground is on the right and is marked with a sign. If you come to the Burrells Ford road, turn around and go back a mile or two.

Directions to Burrells Ford and King Creek Falls: To reach Burrells Ford Campground, continue north on SC 107 for another mile or two past Cherry Hill Campground. Turn left on the Forest Service Road 708, where there is a sign "Georgia Highway 28 — 10 miles, Chattooga River — 3 miles". Park in the parking lot just before crossing the Chattooga River and walk to the campground, about 0.25 mile. From the parking lot near the Burrells Ford campground is the trail head for the King Creek Waterfall, about 0.5 mile.

FLORA

Lichen
Collema subflaccidum
Leptogium cyanescens
Pannaria tavaresii
Parmeliella pannosa
Coccocarpia erythroxyli
Nephroma helveticum
Peltigera horizontalis
Peltigera polydactyla
Lobaria sp.
Graphis scripta
Coenogonium interplexum
Lecidea parasema
Lecidea parasema theioplaca
Lecidea tessellata
Stereocaulon pileatum
Cladonia bacillaris
Cladonia caespiticia
Cladonia coniocraea
Cladonia cristatella
Cladonia cylindrica
Cladonia didyma
Cladonia gracilis
Cladonia parasitica
Lecanora varia
Aspicilia lacustris
Bulbothrix confoederata
Cetraria oakesiana
Parmelia rudecta
Parmotrema dilitatum
Parmotrema eurysacum
Parmotrema haitiense
Parmotrema hypotropum
Parmotrema praesorediosum
Parmotrema reticulatum
Parmotrema robustum
Parmotrema subtinctorium
Parmotrema sulphuratum
Platismatia tuckermanii
Pseudevernia consocians
Pseudoparmelia caperata
Pseudoparmelia caroliniana
Xanthoparmelia subramigera
Canoparmelia caroliniana
Flavopunctelia soredica
Evernia mesomorpha
Usnea ceratina
Usnea mutabilis
Usnea strigosa
Usnea trichodea
Anaptychia palmatula
Buellia punctata
Physcia millegrana
Physcia pseudospeciosa
Physcia subtilis
Heterodermia hypoleuca
Heterodermia speciosa
Heterodermia leucomelos
Phaeophyscia adiastola
Phaeophyscia ciliata
Phaeophyscia imbricata
Rinodina tephraspis
Porina nucula
Gymnoderma lineare

Running-pine
Lycopodium flabelliforme

Meadow Spikemoss
Selaginella apoda

Rattlesnake Grape Fern
Botrychium virginianum

Cinnamon Fern
Osmunda cinnamomea

Maidenhair Fern
Adiantum pedatum

Hay-scented Fern
Dennstaedtia punctilobula

Bracken
Pteridium aquilinum

Southern Lady Fern
Athyrium asplenioides

Fancy Fern
Dryopteris intermedia

Burrells Ford, King Creek Falls and Cherry Hill Campground (continued) August 22, 1987, April 28, 1990, and May 19, 1991

Christmas Fern
Polystichum acrostichoides
Broad Beech-Fern
Thelypteris hexagonoptera
New York Fern
Thelypteris noveboracensis
Ebony Spleenwort
Asplenium platyneuron
Mountain Spleenwort
Asplenium montanum
Rock Cap Fern
Polypodium virginianum
Resurrection Fern
Polypodium polypodioides
White Pine
Pinus strobus
Loblolly Pine
Pinus taeda
Virginia Pine; Scrub Pine
Pinus virginiana
Eastern Hemlock
Tsuga canadensis
Hemlock
Tsuga sp.
Cane; Switch Cane
Arundinaria gigantea
Panic Grass
Panicum sp.
Microstegium (grass)
Microstegium vimineum
Jack-in-the-pulpit
Arisaema triphyllum
Woodrush
Luzula acuminata
Nodding Trillium
Trillium cernuum
Catesby's Trillium
Trillium catesbaei
Vasey's Trillium
Trillium vaseyi
Indian Cucumber-root
Medeola virginiana
Bluebead-lily
Clintonia borealis
Speckled Wood-lily
Clintonia umbellulata
False Solomon's Seal
Smilacina racemosa
Solomon's-seal
Polygonatum biflorum
Blazing-star
Chamaelirium luteum
Yellow Star-grass
Hypoxis hirsuta
Pink Lady's-slipper
Cypripedium acaule
Downy Rattlesnake-plantain
Goodyera pubescens
Crane-fly Orchid
Tipularia discolor
Bitternut Hickory
Carya cordiformis

Tag Alder
Alnus serrulata
Cherry Birch
Betula lenta
American Hornbeam; Ironwood
Carpinus caroliniana
Beech
Fagus grandifolia
American Chestnut
Castanea dentata
White Oak
Quercus alba
Rock Chestnut Oak
Quercus prinus
Yellow Chestnut Oak
Quercus muehlenbergii
Red Oak
Quercus rubra
Scarlet Oak
Quercus coccinea
Buffalo-nut
Pyrularia pubera
Wild Ginger
Hexastylis heterophylla
Heartleaf
Hexastylis sp.
Jumpseed
Tovara virginiana
Yellowroot
Xanthorhiza simplicissima
Baneberry
Actaea pachypoda
Virgin's-bower
Clematis virginiana
Hooked Buttercup
Ranunculus recurvatus
Liverleaf
Hepatica acutiloba
Windflower
Anemone quinquefolia
Mayapple
Podophyllum peltatum
Tulip-poplar
Liriodendron tulipifera
Umbrella-tree
Magnolia fraseri
Sweet-shrub; Carolina-allspice
Calycanthus floridus
Sassafras
Sassafras albidum
Spicebush
Lindera benzoin
Riverweed
Podostemum ceratophyllum
Wild Hydrangea
Hydrangea arborescens radiata
Alumroot
Heuchera villosa
Foamflower
Tiarella cordifolia
Sweet-gum
Liquidambar styraciflua

Witch-hazel
Hamamelis virginiana
Wild Strawberry
Fragaria virginiana
Five-fingers
Potentilla canadensis
Blackberry
Rubus argutus
Agrimony
Agrimonia parviflora
Swamp Rose
Rosa palustris
Hawthorn
Crataegus sp.
Downy Serviceberry
Amelanchier arborea
Choke Cherry
Prunus virginiana
Black Cherry; Wild Cherry
Prunus serotina
Naked-flower Tick-trefoil
Desmodium nudiflorum
Bush-clover
Lespedeza sp.
Black Locust
Robinia pseudo-acacia
Wood-sorrel
Oxalis grandis
Poison-ivy
Rhus radicans
Staghorn Sumac
Rhus typhina
Winged Sumac
Rhus copallina
Smooth Sumac
Rhus glabra
American Holly
Ilex opaca
Hearts-a'burstin'-with-love
Euonymus americanus
Red Maple
Acer rubrum
Spotted Touch-me-not
Impatiens capensis
Virginia-creeper
Parthenocissus quinquefolia
St.-Andrew's-cross
Hypericum hypericoides
Bird-foot Violet
Viola pedata
Sweet White Violet
Viola blanda
Halberd-leaved Violet
Viola hastata
Common Evening-primrose
Oenothera biennis
Angelica
Angelica sp.
Flowering Dogwood
Cornus florida
Mountain Sweet Pepperbush
Clethra acuminata
Spotted Wintergreen
Chimaphila maculata
Indian-pipe
Monotropa uniflora

Pine-sap
Monotropa hypopithys
Rosebay; Great Laurel
Rhododendron maximum
Purple Laurel
Rhododendron catawbiense
Carolina-laurel
Rhododendron minus
Pinxter-flower; Pinxter-bloom
Rhododendron nudiflorum
Swamp Honeysuckle
Rhododendron viscosum
Mountain-laurel
Kalmia latifolia
Dog-hobble
Leucothoe axillaris
Sourwood
Oxydendrum arboreum
Trailing-arbutus
Epigaea repens
Huckleberry
Gaylussacia ursina
Deerberry; Squaw-huckleberry
Vaccinium stamineum
Low-bush Blueberry
Vaccinium vacillans
Oconee-bells; Shortia
Shortia galacifolia
Galax; Galaxy; Coltsfoot
Galax aphylla
Persimmon
Diospyros virginiana
Horse-sugar; Sweetleaf
Symplocos tinctoria
Carolina Silverbell
Halesia carolina
American Ash; White Ash
Fraxinus americana
Phlox
Phlox amoena
Carpenterweed
Prunella vulgaris
Lyre-leaved Sage
Salvia lyrata
Bugle-weed
Lycopus virginicus
Horse-balm
Collinsonia canadensis
Cow-wheat
Melampyrum lineare
Squaw-root
Conopholis americana
Virginia Buttonweed
Diodia virginiana
Partridge-berry
Mitchella repens
Bluets
Houstonia serpyllifolia
Bush Honeysuckle
Diervilla sessilifolia
Japanese Honeysuckle
Lonicera japonica
Coral-berry
Symphoricarpos orbiculatus
Wild-raisin; Witherod
Viburnum cassinoides

Rusty Black-haw
Viburnum rufidulum
American Elderberry
Sambucus canadensis
Lobelia
Lobelia puberula
Prenanthes
Prenanthes sp.
Rattlesnake-weed
Hieracium venosum
Green-stemmed Joe-pye-weed
Eupatorium purpureum
Robin's-plantain
Erigeron pulchellus
Beggar-ticks
Bidens frondosa
Yarrow; Milfoil
Achillea millefolium
Ox-eye Daisy
Chrysanthemum leucanthemum
Moose-antlers
Wynnea americana
Black-club
Trichoglossum geoglossum
Jelly-club
Leotia ascomyces
Tree-ear
Auricularia auricula
Golden Chanterelle
Cantharellus cibarius
Cinnabar-red Chanterelle
Cantharellus cinnabarinus
Black-trumpet
Craterellus fallax
Fragrant Chanterelle
Craterellus odoratus
Spindle-shaped Yellow Coral
Clavulinopsis fusiformis
Coral Mushroom; Cole
Clavaria flava
Crown-tipped Coral
Clavicorona pyxidata
Crested Coral
Clavulina cristata
Cotton-base Coral
Lentaria byssiseda
Straight-branched Coral
Ramaria stricta
Ochre Spreading Tooth
Steccherinum ochraceum
Turkey-tail
Trametes versicolor
Crowded Parchment
Stereum complicatum
American Caesar's Mushroom
Amanita caesarea
King Bolete
Boletus edulis
Zeller's Bolete
Boletus zelleri
Old-man-of-the-woods
Strobilomyces floccopus
Dotted-stalk Suillus
Suillus granulatus
Bitter Bolete
Tylopilus felleus
Alcohol Inky
Coprinus atramentarius
Psathyrella (mushroom)
Psathyrella sp.
Black-nipple Fiber-head
Inocybe fuscodisca
Tacky Green Russula
Russula aeruginea
Emetic Russula
Russula emetica
Fragile Russula
Russula fragilis
Rosy Russula
Russula rosacea
Family Collybia
Clitocybula familia
Pinwheel Marasmius
Marasmius sp.
Black Jelly Oyster
Resupinatus applicatus
Skull-shaped Puffball
Calvatia craniformis
Purple-spored Puffball
Calvatia cyathiformis
Pear-shaped Puffball
Lycoperdon pyriforme
Dog Stinkhorn
Mutinus caninus
Stalked Puffball-in-aspic
Calostoma cinnabarina
Buried-stalk Puffball
Tulostoma simulans

FAUNA

Millipede
Narceus americanus
Crab Spider
Misumena sp.
Mayfly (larva)
Ephemerella sp.
Stonefly (larva)
Phasganophora capitata
Bush Katydid
Scudderia sp.
Wheel Bug
Arilus cristatus
Water-strider
Gerris sp.
Annual Cicada
Tibicen sp.
June Beetle
Phyllophaga sp.
Japanese Beetle
Popillia japonica
Marsh Beetle
Prionocyphon discoideus
Click Beetle
Ampedus areolatus
Soldier Beetle
Chauliognathus pennsylvanicus
Red-banded Hairstreak
Calycopis cecrops
Gray Hairstreak
Strymon melinus
Eastern Tailed-blue(butterfly)
Everes comyntas
Spring Azure (butterfly)
Lycaenopsis argiolus
Pearl Crescent
Phyciodes tharos
Eastern Comma; Hop Merchant
Polygonia comma
Common Buckeye (butterfly)
Junonia coenia
Silver-spotted Skipper
Epargyreus clarus
Golden Skipper
Poanes sp.
Tulip-tree Beauty Moth
Epimecia hortaria
Pistachio Emerald Moth
Helthemia pistaciaria
Eastern Newt
Notophthalmus viridescens
Mountain Dusky Salamander
Desmognathus ochrophaeus
Two-lined Salamander
Eurycea bislineata
Spring Peeper
Hyla crucifer
Common Loon
Gavia immer
Osprey
Pandion haliaetus
Chimney Swift
Chaetura pelagica
Ruby-throated Hummingbird
Archilochus colubris
Pileated Woodpecker
Dryocopus pileatus
Eastern Phoebe
Sayornis phoebe
American Crow
Corvus brachyrhynchos
Carolina Chickadee
Parus carolinensis
Prothonotary Warbler
Protonotaria citrea
Northern Parula
Parula americana
Black-throated Blue Warbler
Dendroica caerulescens
Black-throated Green Warbler
Dendroica virens
Kentucky Warbler
Oporornis formosus
Hooded Warbler
Wilsonia citrina
Worm-eating Warbler
Helmitheros vermivorus
American Goldfinch
Carduelis tristis
Gray Squirrel
Sciurus carolinensis
Beaver (gnaw mark)
Castor canadensis

(Photo L. Smith)

Turkey-tail, *Trametes versicolor.*

(Photos by A. Ciegler)

Red eft or eastern newt, *Notophthalmus viridescens.*

Buzzards Roost Heritage Preserve
Oconee County
May 18, 1991

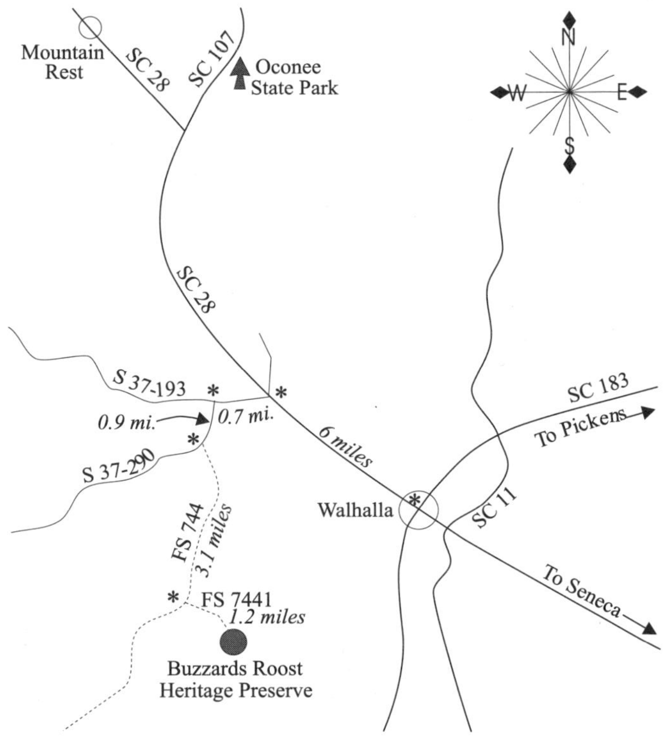

Buzzards Roost is a small monadnock with a cap of low-grade marble adjacent to Sumter National Forest. The view from the top is spectacular, although the climb is steep to reach it. The area contains specimens of several uncommon to rare plants, including the fern purple cliff-brake, table mountain pine, and August-blooming smooth coneflower. Mountain-laurel is abundant and was in full bloom on the day SCAN visited. Also along the trail that day was a molting katydid that was performing this feat in full view beside the trail. The more common plant and animal species in the preserve create a rich representation of higher elevation Piedmont flora and fauna.

Directions to Buzzards Roost Heritage Preserve: From Walhalla, take SC 28 north approximately 6 miles. Turn left onto County Road 193 (Whetstone Road) and drive 0.7 miles. Turn left onto County Road 290 (Cassidy Bridge Road) and drive 0.9 miles. Turn left again onto the gravel Forest Service road FS 744 and drive 3.1 miles. Turn left for the fourth time onto FS 7441 and drive 1.2 miles to the end of the road. Park in the turn-around.

FLORA

Lichen
Leptogium corticola
Pannaria tavaresii
Peltigera canina
Graphis scripta
Lecidea myriocarpoides
Lecidea sp.
Baeomyces fungoides
Cladonia capitata
Cladonia cariosa
Cladonia chlorophaea
Cladonia sobolescens
Cladonia coniocraea
Cladonia cristatella
Cladonia cylindrica
Cladonia incrassata
Cladonia leporina
Cladonia pityrea
Cladonia polycarpoides
Cladonia verticillata
Pycnothelia papillaria
Pertusaria leucostoma
Pertusaria paratuberculifera
Pertusaria texana
Pertusaria tetrathalmia
Pertusaria sinusmexicani
Pertusaria velata
Pertusaria tuberculifera
Pertusaria ophthalmiza
Pertusaria valliculata
Pertusaria sp.
Lecanora cupressi
Ochrolechia africana
Candelaria fibrosa
Hypotrachyna livida
Hypotrachyna ensifolia
Parmelia saxatilis
Parmelia squarrosa
Parmelina aurulenta
Parmelina horrescens
Parmotrema cetratum
Parmotrema crinitum
Parmotrema michauxianum
Parmotrema reticulatum
Pseudoparmelia caperata
Pseudoparmelia caroliniana
Imshaugia aleurites
Tuckermannopsis ciliaris
Ramalina americana
Usnea strigosa
Usnea trichodea
Anaptychia palmatula
Buellia parasema
Buellia curatellae
Buellia curtisii
Physcia stellaris
Heterodermia obscurata
Phaeophyscia pusilloides
Phaeophyscia rubropulchra
Hyperphyscia syncolla
Trypethelium virens
Trypethelium scorites

Maidenhair Fern
Adiantum pedatum
Purple Cliff-brake
Pellaea atropurpurea
Bracken
Pteridium aquilinum
Christmas Fern
Polystichum acrostichoides
Blackstem Spleenwort
Asplenium resiliens
Rock Cap Fern
Polypodium virginianum
White Pine
Pinus strobus
Table Mountain Pine
Pinus pungens
Virginia Pine; Scrub Pine
Pinus virginiana
Eastern Red-cedar
Juniperus virginiana
Cane; Switch Cane
Arundinaria gigantea
Orchard Grass
Dactylis glomerata
Melic Grass
Melica mutica
Needle Grass
Stipa avenacea
Sawbrier; Greenbrier
Smilax glauca
Indian Cucumber-root
Medeola virginiana
False Solomon's Seal
Smilacina racemosa
Solomon's-seal
Polygonatum biflorum
Bellwort
Uvularia perfoliata
Wild Yam
Dioscorea villosa
Yellow Star-grass
Hypoxis hirsuta
Dwarf Iris
Iris verna
Small Green Wood Orchid
Habenaria clavellata
Downy Rattlesnake-plantain
Goodyera pubescens
Mockernut Hickory
Carya tomentosa
Tag Alder
Alnus serrulata
White Oak
Quercus alba
Rock Chestnut Oak
Quercus prinus
Red Oak
Quercus rubra

Scarlet Oak
Quercus coccinea
Spanish Oak
Quercus falcata
Black Jack Oak
Quercus marilandica
Virginia Snakeroot
Aristolochia serpentaria
Tulip-poplar
Liriodendron tulipifera
Cucumber-tree
Magnolia acuminata
PawPaw
Asimina sp.
Sassafras
Sassafras albidum
Bloodroot
Sanguinaria canadensis
Poor-man's-pepper
Lepidium virginicum
Wild Hydrangea
Hydrangea arborescens
Sweet-gum
Liquidambar styraciflua
Five-fingers
Potentilla canadensis
Indian-physic
Gillenia trifoliata
Downy Serviceberry
Amelanchier arborea
Black Cherry; Wild Cherry
Prunus serotina
Mimosa; Silk-tree
Albizia julibrissin
Redbud; Judas-tree
Cercis canadensis
Wild Indigo
Baptisia tinctoria
White Wild Indigo
Baptisia alba
Low Hop Clover
Trifolium campestre
Leadplant; Indigo-bush
Amorpha sp.
Black Locust
Robinia pseudo-acacia
Bristly Locust
Robinia hispida
Hog-peanut
Amphicarpa bracteata
Flowering Spurge
Euphorbia corollata
Poison-oak
Rhus toxicodendron
Winged Sumac
Rhus copallina
Hearts-a'burstin'-with-love
Euonymus americanus
Red Maple
Acer rubrum
Spotted Touch-me-not
Impatiens capensis
New-Jersey-tea
Ceanothus americanus
Virginia-creeper
Parthenocissus quinquefolia

Muscadine; Bullace
Vitis rotundifolia
Summer Grape
Vitis aestivalis
Bird-foot Violet
Viola pedata
Russian-olive
Elaeagnus umbellata
Angelica
Angelica venenosa
Black Tupelo; Blackgum
Nyssa sylvatica sylvatica
Flowering Dogwood
Cornus florida
Spotted Wintergreen
Chimaphila maculata
Carolina-laurel
Rhododendron minus
Mountain-laurel
Kalmia latifolia
Maleberry
Lyonia ligustrina
Sourwood
Oxydendrum arboreum
Trailing-arbutus
Epigaea repens
Sparkleberry
Vaccinium arboreum
Deerberry; Squaw-huckleberry
Vaccinium stamineum
Low-bush Blueberry
Vaccinium vacillans
Galax; Galaxy; Coltsfoot
Galax aphylla
Persimmon
Diospyros virginiana
Horse-sugar; Sweetleaf
Symplocos tinctoria
Fringe-tree
Chionanthus virginicus
Indian-pink
Spigelia marilandica
Bindweed
Calystegia sericata
Horse-balm
Collinsonia verticillata
False Foxglove
Aureolaria pectinata
Bluets
Houstonia longifolia
Rattlesnake-weed
Hieracium venosum
Small's Ragwort
Senecio smallii
Pussy-toes
Antennaria solitaria
Antennaria plantaginifolia
Rosinweed
Silphium compositum
Green-and-gold
Chrysogonum virginianum
Wild Quinine
Parthenium integrifolium
Coreopsis
Coreopsis major
Yarrow; Milfoil
Achillea millefolium

Hairy Rubber Cup
Galiella rufa
Crown-tipped Coral
Clavicorona pyxidata
Coral Slime
Ceratiomyxa fruticulosa

FAUNA

Giant Forest Snail
Mesodon normalis
Red Velvet Mite
Trombidium sp.
Shortleg Shield-backed Katydid
Atlanticus testaceus
Searcher; Ground Beetle
Carabus sylvosus
Water Scavenger Beetle
Phaenonotum estriatum
Tumblebug; Dung Beetle
Glaphyrocanthon viridis
Aphodiine Dung Beetle
Ataenius cognatus
Earth-boring Dung Beetle
Bolboceras sp.
Garden Beetle
Serica sp.
Chafer; Scarab Beetle
Dichelonyx sp.
Buprestid Beetle
Dicerca divaricata
Soldier Beetle
Chauliognathus pennsylvanicus
Cantharis lineola
Firefly; Lightning Beetle
Phausis sp.

Flat Bark Beetle
Silvanus bidentatus
Eastern Tailed-blue(butterfly)
Everes comyntas
Red-spotted Purple (butterfly)
Limenitis arthemis astyanax
Orange-patched Smoky Moth
Pyromorpha dimidiata
Eastern Newt
Notophthalmus viridescens
Fowler's Toad
Bufo woodhousei
Cope's Gray Treefrog
Hyla chrysoscelis
Spring Peeper
Hyla crucifer
Carolina Anole
Anolis carolinensis
Eastern Fence Lizard
Sceloporus undulatus
Five-lined Skink
Eumeces fasciatus
Turkey Vulture
Cathartes aura
Carolina Wren
Thryothorus ludovicianus
Pine Warbler
Dendroica pinus
Yellow-breasted Chat
Icteria virens
Indigo Bunting
Passerina cyanea
Rufous-sided Towhee
Pipilo erythrophthalmus

(Photos by V. Winn and L. Smith)

SCAN members saw a shortleg shield-back katydid, *Atlanticus testaceus*, molt from start to finish.

Coon Branch Trail, April 22, 1995, and Lower Whitewater Falls Overlook Trail, August 28, 1993
Oconee County

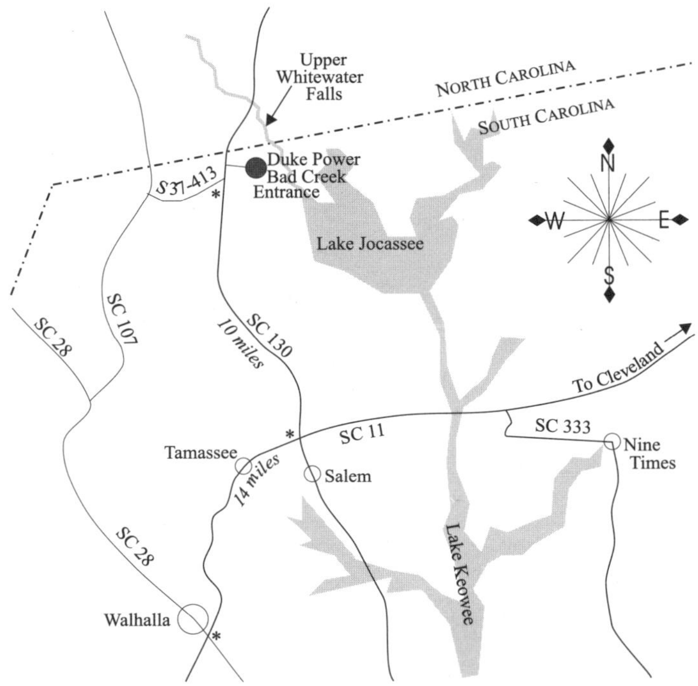

The Coon Branch Trail, accessed from the Bad Creek parking area provided by Duke Power, goes about a mile up the west side of the Whitewater River and dead-ends near a huge, unforgettable Eastern Hemlock. That tree measured twelve feet, ten inches in circumference and fifty inches in diameter. This same area was walked in 1788 by the French botanist Andre Michaux, who discovered and named many plants that were new to science at that time. As the flora and fauna list shows, late April is a good time to visit for spring flowers and migrating birds, offering looks at species of birds we see in South Carolina only during the spring or fall migration. The grade is quite easy on this trail although there is a mildly steep ridge to cross before reaching the river from the parking lot. The Coon Branch Trail is on land owned by Duke Power.

Also accessible from this point on the river is the trail to an overlook to view Lower Whitewater Falls. One crosses the river rather than turning upstream on the Coon Branch Trail. This trail follows the side of the mountain and climbs slightly, while the river drops away toward Lake Jocassee. The view across the ravine to the falls is spectacular. Along that trail are mountain plants that do well in dry habitats, as the trail quickly leaves the moisture along the river.

Directions to Coon Branch Trail and Lower Whitewater Falls Overlook Trail: From Walhalla, drive northeast on SC 11 through Tamassee to SC 130, about 14 miles. Turn left (north) on SC 130 and drive about 10 miles. Watch for the intersection of County Road 413. Continue on SC 130 about 0.4 mile past that intersection to the gated Bad Creek Entrance to the Duke Power facility on the right. (If you cross the state line into North Carolina or come to the Upper Whitewater Falls entrance, you missed the entrance; turn around and try again.) At the entrance, stop and tell the guard that you have come to hike the trails. The guard should admit you and direct you to the trail heads. Both the Lower Whitewater Falls trail and the Coon Branch trail begin within this area.

FLORA

Lichen

Collema subflaccidum
Leptogium corticola
Leptogium cyanescens
Pannaria lurida
Pannaria rubiginosa
Pannaria tavaresii
Coccocarpia cronia
Peltigera horizontalis
Sticta weigelii
Graphis scripta
Bacîdia schweinitzii
Bacidia sp.
Leptogium austroamericanum
Lecidea russula
Lecidea sp.
Baeomyces fungoides
Cladina subtenuis
Cladonia bacillaris
Cladonia chlorophaea
Cladonia chlorophaea f. intermedia
Cladonia coniocraea
Cladonia coniocraea f. phyllostrota
Cladonia coniocraea f. subpellucida
Cladonia coniocraea expansa
Cladonia cristatella
Cladonia cristatella vestita
Cladonia cristatella epiphylla
Cladonia didyma
Cladonia didyma squamulosa
Cladonia polycarpoides
Cladonia peziziformis
Cladonia peziziformis dissectula
Cladonia rappii
Pertusaria neoscotica
Pertusaria ostiolata
Pertusaria paratuberculifera
Pertusaria subpertusa
Pertusaria texana
Pertusaria sinusmexicani
Pertusaria multipunctoides
Lecanora cupressi
Lecanora varia
Lecanora sp.
Candelaria concolor
Hypotrachyna livida
Hypotrachyna revoluta
Parmelia saxatilis
Parmelia squarrosa
Parmelina aurulenta
Parmelina horrescens
Parmelina horrescens asidiata
Parmelina minarum
Parmotrema austrosinense
Parmotrema tinctorum
Platismatia tuckermanii
Pseudevernia consocians
Flavoparmelia caperata
Canoparmelia caroliniana
Punctelia rudecta
Punctelia subrudecta
Imshaugia aleurites
Rimelia reticulata
Rimelia cetrata
Ramalina americana
Usnea ceratina
Usnea rubicunda
Usnea strigosa

Usnea trichodea
Buellia punctata
Physcia aipolia
Heterodermia obscurata
Heterodermia casarettiana
Phaeophyscia rubropulchra
Physconia detersa
Normandina pulchella
Shining Clubmoss
Lycopodium lucidulum
Running-pine
Lycopodium flabelliforme
Rattlesnake Grape Fern
Botrychium virginianum
Grape Fern; Moonwort
Botrychium sp.
Maidenhair Fern
Adiantum pedatum
Hay-scented Fern
Dennstaedtia punctilobula
Bracken
Pteridium aquilinum
Southern Lady Fern
Athyrium asplenioides
Christmas Fern
Polystichum acrostichoides
Broad Beech-Fern
Thelypteris hexagonoptera
New York Fern
Thelypteris noveboracensis
Ebony Spleenwort
Asplenium platyneuron
Resurrection Fern
Polypodium polypodioides
White Pine
Pinus strobus
Short-leaf Pine
Pinus echinata
Virginia Pine; Scrub Pine
Pinus virginiana
Eastern Hemlock
Tsuga canadensis
Cane; Switch Cane
Arundinaria gigantea
Bulrush
Scirpus purshianus
Wool-grass
Scirpus cyperinus
Jack-in-the-pulpit
Arisaema triphyllum
Sawbrier; Greenbrier
Smilax glauca
Greenbrier; Catbrier
Smilax sp.
Little-sweet-Betsy
Trillium cuneatum luteum
Wake-robin
Trillium erectum
Catesby's Trillium
Trillium catesbaei
Indian Cucumber-root
Medeola virginiana
Speckled Wood-lily
Clintonia umbellulata
Solomon's-seal
Polygonatum biflorum
White Hellebore
Veratrum viride
Bellwort
Uvularia perfoliata
Uvularia sessilifolia
Uvularia sp.
Wild Yam
Dioscorea villosa villosa
Dioscorea sp.
Yellow Star-grass
Hypoxis hirsuta
Dwarf Iris
Iris verna
Yellow Lady's-slipper
Cypripedium calceolus
Showy Orchis
Orchis spectabilis
Downy Rattlesnake-plantain
Goodyera pubescens
Crane-fly Orchid
Tipularia discolor
Black Willow
Salix nigra
Mockernut Hickory
Carya tomentosa
Pignut Hickory
Carya glabra
Tag Alder
Alnus serrulata
Cherry Birch
Betula lenta
American Chestnut
Castanea dentata
White Oak
Quercus alba
Rock Chestnut Oak
Quercus prinus
Red Oak
Quercus rubra
Red Oak
Quercus rubra rubra
Black Oak
Quercus velutina
Scarlet Oak
Quercus coccinea
Southern Red Oak; Spanish Oak
Quercus falcata falcata
Buffalo-nut
Pyrularia pubera
Heartleaf
Hexastylis arifolia
Wild Ginger
Hexastylis shuttleworthii
Heartleaf
Hexastylis sp.
Yellowroot
Xanthorhiza simplicissima
Mayapple
Podophyllum peltatum
Tulip-poplar
Liriodendron tulipifera
Umbrella-tree
Magnolia fraseri
Sweet-shrub; Carolina-allspice
Calycanthus floridus
Sassafras
Sassafras albidum
Spicebush
Lindera benzoin
Bloodroot
Sanguinaria canadensis
Wild Hydrangea
Hydrangea arborescens radiata
Foamflower
Tiarella cordifolia
Sweet-gum
Liquidambar styraciflua
Witch-hazel
Hamamelis virginiana
Cinquefoil; Five-fingers
Potentilla sp.
Blackberry; Brambles
Rubus sp.
Cocklebur
Agrimonia sp.
Hawthorn
Crataegus sp.
Downy Serviceberry
Amelanchier arborea
Black Cherry; Wild Cherry
Prunus serotina
Sensitive-brier
Schrankia microphylla
Redbud; Judas-tree
Cercis canadensis
Partridge-pea
Cassia fasciculata
Naked-flower Tick-trefoil
Desmodium nudiflorum
Black Locust
Robinia pseudo-acacia
Hog-peanut
Amphicarpa bracteata
Milkwort
Polygala sp.
Poison-ivy
Rhus radicans
Poison-oak
Rhus toxicodendron
American Holly
Ilex opaca
Hearts-a'burstin'-with-love
Euonymus americanus
Red Maple
Acer rubrum
Chalk Maple
Acer saccharum leucoderme
Virginia-creeper
Parthenocissus quinquefolia
Muscadine; Bullace
Vitis rotundifolia
Basswood; Linden
Tilia sp.
St.-Andrew's-cross
Hypericum hypericoides
Hypericum
Hypericum sp.
Common Blue Violet
Viola papilionacea
Wood Violet
Viola palmata triloba
Violet
Viola septemloba
Viola sp.
Sweet White Violet
Viola blanda
Primrose-leaved Violet
Viola primulifolia
Round-leaved Yellow Violet
Viola rotundifolia
Halberd-leaved Violet
Viola hastata
Devil's-walking-stick
Aralia spinosa
Wild Sarsaparilla
Aralia nudicaulis
Black Tupelo; Blackgum
Nyssa sylvatica sylvatica
Flowering Dogwood
Cornus florida
Spotted Wintergreen
Chimaphila maculata
Pine-sap
Monotropa hypopithys
Rosebay; Great Laurel
Rhododendron maximum
Azalea
Rhododendron sp.
Mountain-laurel
Kalmia latifolia
Dog-hobble
Leucothoe axillaris editorum
Sourwood
Oxydendrum arboreum
Trailing-arbutus
Epigaea repens
Black Huckleberry
Gaylussacia baccata
Huckleberry
Gaylussacia ursina
Sparkleberry
Vaccinium arboreum
Low-bush Blueberry
Vaccinium vacillans
Galax; Galaxy; Coltsfoot
Galax aphylla
Whorled Loosestrife
Lysimachia quadrifolia
Horse-sugar; Sweetleaf
Symplocos tinctoria
Carolina Silverbell
Halesia carolina
American Ash; White Ash
Fraxinus americana
White Ash; American Ash
Fraxinus americana americana
Morning-glory
Ipomoea sp.
Skullcap
Scutellaria sp.
Squaw-root
Conopholis americana
Partridge-berry
Mitchella repens
Bluets
Houstonia serpyllifolia
Japanese Honeysuckle
Lonicera japonica
Wild-raisin; Witherod
Viburnum cassinoides

Coon Branch Trail and Lower Whitewater Falls Overlook Trail (continued) April 22, 1995, and August 28, 1993

Viburnum
Viburnum sp.
American Elderberry
Sambucus canadensis
Lobelia
Lobelia puberula
Indian-tobacco
Lobelia inflata
Elephant's-foot
Elephantopus carolinianus
Elephantopus tomentosus
Blazing-star
Liatris graminifolia
Eupatorium
Eupatorium sp.
Marsh-fleabane
Pluchea camphorata
Pussy-toes
Antennaria sp.
Robin's-plantain
Erigeron pulchellus
Heart-leaved Aster
Aster divaricatus
Sweet Goldenrod
Solidago odora
Goldenrod
Solidago sp.
Heterotheca
Heterotheca graminifolia
Rosinweed
Silphium compositum
Wild Quinine
Parthenium integrifolium
Coneflower
Rudbeckia laciniata
Sunflower
Helianthus microcephalus
Helianthus sp.
Coreopsis
Coreopsis major
Carbon-balls (fungus)
Daldinia concentrica
Black-knot
Apiosporina morbosa
Tree-ear
Auricularia auricula
Witches'-butter
Tremella mesenterica
Cauliflower Mushroom
Sparassis sp.
Mustard-yellow Polypore
Phellinus gilvus
Orange Poria
Poria spissa
Violet-toothed Polypore
Trichaptum biformis
Common Split-gill
Schizophyllum commune
Crowded Parchment
Stereum complicatum
False Turkey-tail
Stereum ostrea
Shaggy-stalked Bolete
Boletellus betula
Old-man-of-the-woods
Strobilomyces floccopus
Dotted-stalk Suillus
Suillus granulatus
Fawn Mushroom
Pluteus cervinus
Gill Fungus
Clitocybe sp.
Carnival-candy Slime
Arcyria denudata

FAUNA

Giant Forest Snail
Mesodon normalis
White-lipped Forest Snail
Mesodon thyroidus
Disk Cannibal Snail
Haplotrema concavum
Millipede
Narceus americanus
Writing Spider
Argiope aurantia
Rabid Wolf Spider
Lycosa rabida
Red Velvet Mite
Trombidium sp.
June-beetle Wasp
Pelecinus polyturator
Mayfly
Isonychia sp.
Heptageniid Mayfly
Epeorus sp.
Common Stonefly
Acroneuria internata
Common Stonefly
Paragnetina immarginata
Walkingstick
Diapheromera femorata
Carolina Locust
Dissosteira carolina
Boll's Sand Grasshopper
Spharagemon bolli
Differential Grasshopper
Melanoplus differentialis
Red-legged Grasshopper
Melanoplus femurrubrum
Spur-throated Grasshopper
Melanoplus sp.
Angular-winged Katydid
Microcentrum retinerve
True Katydid
Pterophylla camellifolia
Leaf-rolling Cricket
Camptonotus caroliniensis
Carolina Mantid (egg)
Stagmomantis carolina
Large Water-strider
Gerris remigis
Riffle Bug
Rhagovelia obesa
Dark Fishfly
Nigronia serricornis
Common Eastern Velvet-ant
Dasymutilla occidentalis occidentalis
Velvet-ant
Dasymutilla asopus bexar
Bald-faced Hornet (nest)
Vespa maculata
Yellowjacket
Vespa sp.
Eastern Yellowjacket
Vespula maculifrons
Bumblebee
Bombus impatiens
Bumblebee
Bombus sp.
Medium Evening Sedge Caddisfly
Dolophilodes sp.
Brown Checkered Summer Sedge
Polycentropus sp.
Spotted Sedge Caddisfly
Hydropsyche sp.
Caddisfly
Rhyacophila fuscula
Saddle-case Maker Caddisfly
Glossosoma sp.
Giant Brown Autumn Sedge
Pycnospsyche divergens
Pycnospsyche gentilis
Autumn Mottled Sedge Caddisfly
Neophylax mitchelli
Little Plain Brown Sedge
Lepidostoma sp.
Flower Fly
Milesia virginiensis
Shiny Bluebottle Fly
Cynomyopsis cadaverina
Common Green Darner
Anax junius
Single-striped Clubtail
Lanthus vernalis
Green Tiger Beetle
Cicindela sexguttata
Woodland Ground Beetle
Agonum extensicolle
Dingy Ground Beetle
Anisodactylus rusticus
Large Rove Beetle
Staphylinus fossator
Carrion Beetle
Nicrophorus sp.
Tumblebug; Dung Beetle
Canthon chalcites
Water-penny Beetle (larva)
Psephenus herricki
Click Beetle
Heteroderes sp.
Soldier Beetle
Chauliognathus marginatus
Chauliognathus pennsylvanicus
Soft-winged Flower Beetle
Collops quadrimaculatus
Convergent Lady Beetle
Hippodamia convergens
Seven-spotted Lady Beetle
Coccinella septempunctata
Spotless Lady Beetle
Cycloneda sanguinea
Alder Leaf Beetle
Chrysomela interrupta
E. Spotted Cucumber Beetle
Diabrotica undecimpunctata howardi
Leaf-mining Beetle
Anoplitis inaequalis
White-fringed Beetle
Graphognathus leucoloma
Broad-nosed Weevil
Phyllobius sp.
Eastern Tiger Swallowtail
Papilio glaucus
Clouded Sulphur
Colias philodice
Cloudless Sulphur
Phoebis sennae eubule
Red-banded Hairstreak
Calycopis cecrops
Eastern Tailed-blue(butterfly)
Everes comyntas
Crescent Butterfly
Phyciodes sp.
Viceroy; Mimic (cocoon)
Limenitis archippus
Silver-spotted Skipper
Epargyreus clarus
Juvenal's Duskywing
Erynnis juvenalis
Grape Leaffolder Moth
Desmia funeralis
Minor Angle (moth)
Semiothisa minorata
Angle Moth
Semiothisa sp.
Bent-line Gray (moth)
Iridopsis larvaria
Yellow Slant-line (moth)
Tetracis crocallata
Pistachio Emerald Moth
Helthemia pistaciaria
Ferguson's Scallop Shell Moth
Hydria prunivorata
Luna Moth
Actias luna
Fall Webworm Moth
Hyphantria cunea
Dusky Salamander
Desmognathus sp.
Southern Toad
Bufo terrestris
Eastern Fence Lizard
Sceloporus undulatus
Broadhead Skink
Eumeces laticeps
Turkey Vulture
Cathartes aura
Red-tailed Hawk
Buteo jamaicensis
Wild Turkey
Meleagris gallopavo
Mourning Dove
Zenaida macroura

Downy Woodpecker
Picoides pubescens
Pileated Woodpecker
Dryocopus pileatus
Eastern Kingbird
Tyrannus tyrannus
Eastern Phoebe
Sayornis phoebe
Northern Rough-winged Swallow
Stelgidopteryx serripennis
Barn Swallow
Hirundo rustica
American Crow
Corvus brachyrhynchos
Common Raven
Corvus corax
Tufted Titmouse
Parus bicolor
Carolina Chickadee
Parus carolinensis
Carolina Wren
Thryothorus ludovicianus
Ruby-crowned Kinglet
Regulus calendula
Blue-gray Gnatcatcher
Polioptila caerulea
Eastern Bluebird
Sialia sialis
Wood Thrush
Hylocichla mustelina
American Robin
Turdus migratorius
Solitary Vireo
Vireo solitarius
Red-Eyed Vireo
Vireo olivaceus
Northern Parula
Parula americana
Black-and-white Warbler
Mniotilta varia
Yellow-rumped Warbler
Dendroica coronata
Black-throated Green Warbler
Dendroica virens
Prairie Warbler
Dendroica discolor
Pine Warbler
Dendroica pinus
Hooded Warbler
Wilsonia citrina
Worm-eating Warbler
Helmitheros vermivorus
Swainson's Warbler
Limnothlypis swainsonii
American Redstart
Setophaga ruticilla
Rose-breasted Grosbeak
Pheuticus ludovicianus
Northern Cardinal
Cardinalis cardinalis
Blue Grosbeak
Guiraca caerulea
Indigo Bunting
Passerina cyanea

(Photo by L. Smith)
Lower Whitewater Falls, Oconee County, Blue Ridge region.

Rufous-sided Towhee
Pipilo erythrophthalmus
Savannah Sparrow
Passerculus sandwichensis
Chipping Sparrow
Spizella passerina
Red-winged Blackbird
Agelaius phoeniceus
Scarlet Tanager
Piranga olivacea
Summer Tanager
Piranga rubra
American Goldfinch
Carduelis tristis
Eastern Chipmunk
Tamias striatus
Wild Pig (sign)
Sus scrofa
White-tailed Deer (tracks)
Odocoileus virginianus

Devils Fork State Park
Oconee County
April 23, 1994

Devils Fork State Park is located on the west shore of Lake Jocassee, giving it a background of beautiful mountain scenery. This park occupies a portion of the limited range of Oconee-bells, harboring one of the few large stands left after the flooding of the lake. The nature trail offers an excellent opportunity to view closely this beautiful plant, as well as other species typical of the higher elevations in the state. Birds use this area on their migrations, making spring and fall good bird-observing times to visit. On this visit, SCAN members were able to see common loons in their summer plumage before they left for their summer breeding areas.

Directions to Devils Fork State Park: From Walhalla, drive northeast on SC 11 through Tamassee to SC 130, about 14 miles. Continue on SC 11 a short distance and watch for the brown sign directing you to the park. Turn left onto S 37-25; the park is about 4 miles north of SC 11. If coming from the east on SC 11, cross Lake Keowee and turn right onto S 37-127 at another brown park sign; turn right again onto S 37-25 and follow signs to the park.

(Photo by L. Smith)
Sassafras, *Sassafras albidum*.

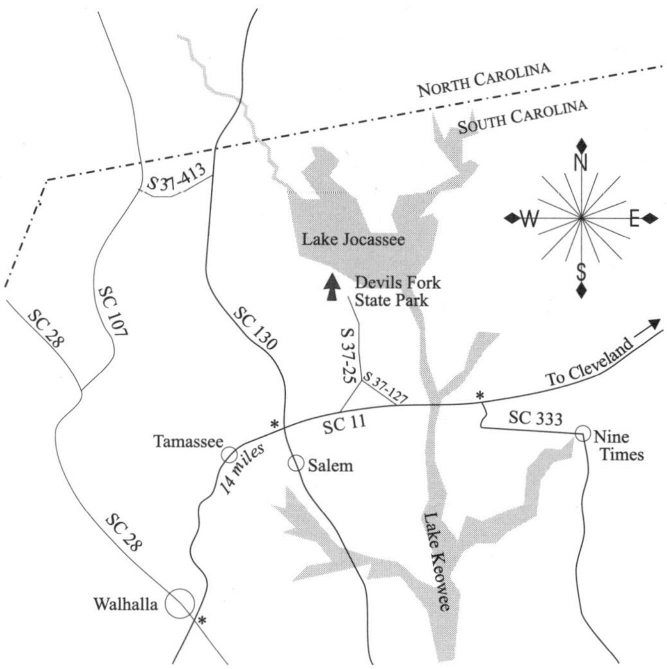

FLORA

Lichen
Collema subflaccidum
Leptogium cyanescens
Leptogium laceroides
Graphis scripta
Graphis mosquitensis
Bacidia schweinitzii
Leptogium austroamericanum
Brigantiaea leucoxantha
Lecidea russula
Baeomyces fungoides
Cladonia chlorophaea
Cladonia coniocraea
Cladonia cristatella
Cladonia cylindrica
Cladonia didyma
Cladonia peziziformis
Pertusaria amara
Pertusaria neoscotica
Pertusaria paratuberculifera
Pertusaria subpertusa
Pertusaria texana
Pertusaria multipunctoides
Lecanora minutella
Lecanora cupressi
Lecanora pallida
Lecanora varia
Lecanora allophana
Ochrolechia africana
Bulbothrix goebelii
Hypotrachyna livida
Parmelina aurulenta
Parmelina galbina
Parmelina horrescens
Parmelina minarum
Parmotrema hypotropum
Parmotrema mellissii
Parmotrema michauxianum
Parmotrema subtinctorium
Flavoparmelia caperata
Canoparmelia caroliniana
Punctelia rudecta
Melanelia exasperata
Rimelia reticulata
Tuckermannopsis oakesiana
Usnea ceratina
Usnea rubicunda
Usnea strigosa
Usnea filipendula
Anaptychia palmatula
Buellia dialyta
Buellia curatellae
Physcia aipolia
Physcia americana
Physcia stellaris
Heterodermia echinata
Heterodermia obscurata
Pyxine sorediata
Phaeophyscia imbricata
Phaeophyscia rubropulchra
Physconia enteroxantha f. detersa
Pyrenula laevigata
Trypethelium mastoideum
Trypethelium virens
Porina salicina

Southern Grape Fern
Botrychium biternatum
Cinnamon Fern
Osmunda cinnamomea
Royal Fern
Osmunda regalis spectabilis
Bracken
Pteridium aquilinum
Southern Lady Fern
Athyrium asplenioides
Christmas Fern
Polystichum acrostichoides
New York Fern
Thelypteris noveboracensis
Sensitive Fern
Onoclea sensibilis
Netted Chain Fern
Woodwardia areolata
Ebony Spleenwort
Asplenium platyneuron
White Pine
Pinus strobus
Short-leaf Pine
Pinus echinata
Virginia Pine; Scrub Pine
Pinus virginiana
Eastern Hemlock
Tsuga canadensis
Bur-reed
Sparganium americanum
Cane; Switch Cane
Arundinaria gigantea
Needle Grass
Stipa avenacea
Greenbrier; Catbrier
Smilax sp.
Indian Cucumber-root
Medeola virginiana
Blazing-star
Chamaelirium luteum
Bellwort
Uvularia sp.
Wild Yam
Dioscorea villosa
Yellow Star-grass
Hypoxis hirsuta
Dwarf Iris
Iris verna
Pink Lady's-slipper
Cypripedium acaule
Downy Rattlesnake-plantain
Goodyera pubescens
Crane-fly Orchid
Tipularia discolor
Willow
Salix sp.
Mockernut Hickory
Carya tomentosa
Tag Alder
Alnus serrulata
Beech
Fagus grandifolia
American Chestnut
Castanea dentata
White Oak
Quercus alba
Post Oak
Quercus stellata
Black Oak
Quercus velutina
Scarlet Oak
Quercus coccinea
Spanish Oak
Quercus falcata
Heartleaf
Hexastylis sp.
Yellowroot
Xanthorhiza simplicissima
Tulip-poplar
Liriodendron tulipifera
Umbrella-tree
Magnolia fraseri
Dwarf PawPaw
Asimina parviflora
Sweet-shrub; Carolina-allspice
Calycanthus floridus
Sassafras
Sassafras albidum
Virginia-willow
Itea virginica
Sweet-gum
Liquidambar styraciflua
Witch-hazel
Hamamelis virginiana
Wild Strawberry
Fragaria virginiana
Blackberry; Brambles
Rubus sp.
Red Chokeberry
Sorbus arbutifolia
Hawthorn
Crataegus sp.
Downy Serviceberry
Amelanchier arborea
Black Cherry; Wild Cherry
Prunus serotina
Black Locust
Robinia pseudo-acacia
Bristly Locust
Robinia hispida
Flowering Spurge
Euphorbia corollata
Poison-ivy
Rhus radicans
American Holly
Ilex opaca
Hearts-a'burstin'-with-love
Euonymus americanus
Red Maple
Acer rubrum
Virginia-creeper
Parthenocissus quinquefolia
Muscadine; Bullace
Vitis rotundifolia
Summer Grape
Vitis aestivalis
St.-Andrew's-cross
Hypericum stragalum
Primrose-leaved Violet
Viola primulifolia
Halberd-leaved Violet
Viola hastata
Black Tupelo; Blackgum
Nyssa sylvatica sylvatica
Flowering Dogwood
Cornus florida
Mountain Sweet Pepperbush
Clethra acuminata
Spotted Wintergreen
Chimaphila maculata
Pine-sap
Monotropa hypopithys

Devils Fork State Park (continued)
April 23, 1994

Rosebay; Great Laurel
Rhododendron maximum
Pinxter-flower; Pinxter-bloom
Rhododendron nudiflorum
Mountain-laurel
Kalmia latifolia
Dog-hobble
Leucothoe axillaris editorum
Sourwood
Oxydendrum arboreum
Trailing-arbutus
Epigaea repens
Huckleberry
Gaylussacia ursina
Sparkleberry
Vaccinium arboreum
Low-bush Blueberry
Vaccinium vacillans
Oconee-bells; Shortia
Shortia galacifolia
Galax; Galaxy; Coltsfoot
Galax aphylla
Persimmon
Diospyros virginiana
Horse-sugar; Sweetleaf
Symplocos tinctoria
Carolina Silverbell
Halesia carolina
Fringe-tree
Chionanthus virginicus
Cross Vine
Anisostichus capreolata
Partridge-berry
Mitchella repens
Possumhaw Viburnum
Viburnum nudum
Rattlesnake-weed
Hieracium venosum
Pussy-toes
Antennaria sp.
Robin's-plantain
Erigeron pulchellus
Rosinweed
Silphium compositum
Devil's-urn
Urnula craterium
Eyelash Cup
Scutellinia scutellata
Swamp-beacon
Mitrula paludosa
Carbon-cushion
Ustulina deusta
Fawn Mushroom
Pluteus cervinus
Chocolate-tube Slime
Stemonitis splendens

FAUNA

Orchard Spider
Leucauge venusta
Long-jawed Orb-weaver
Tetragnatha laboriosa
Six-spotted Fishing Spider
Dolomedes triton
Nursery-web Spider
Pisaurina mira
Scorpion
Centruroides sp.
Spur-throated Grasshopper
Melanoplus sp.
Stilt Bug
Jalysus spinosus
Toad Bug
Gelastocoris oculatus
Eastern Green Lacewing
Chrysopa ornata
Bumblebee
Bombus sp.
Common Green Darner
Anax junius
Spiketail Dragonfly
Cordulegaster sp.
Common Baskettail (dragonfly)
Epitheca cynosura
Little Corporal Dragonfly
Ladona deplanata
White-tail Dragonfly
Plathemis lydia
Violet-masked Glider
Tramea carolina
Dancer (damselfly)
Argia sp.
Green Tiger Beetle
Cicindela sexguttata
Tiger Beetle
Cicindela tortuosa
Round Sand Beetle
Omophron americanum
Minute Ground Beetle
Bembidion americanum
Bembidion sp.
Woodland Ground Beetle
Agonum errans
Agonum sp.
Crawling Water Beetle
Peltodytes sexmaculatus
Paederine Rove Beetle
Sunius sp.
Popilius disjunctus
June Beetle
Phyllophaga sp.
Click Beetle
Ampedus sp.
Soldier Beetle
Silis sp.
Flat Bark Beetle
Brontes dubius
Lady Beetle
Brachyacantha sp.
Long-horned Leaf Beetle
Donacia subtilis
Locust Leaf-mining Beetle
Chalepus dorsalis
Zebra Swallowtail
Eurytides marcellus
Eastern Tiger Swallowtail
Papilio glaucus
Eastern Tailed-blue(butterfly)
Everes comyntas
Spring Azure (butterfly)
Celastrina ladon
American Snout (butterfly)
Libytheana carinenta
American Lady (butterfly)
Vanessa virginiensis
Red Admiral
Vanessa atalanta rubria
Carolina Satyr (butterfly)
Hermeuptychia sosybius
Duskywing Skipper
Erynnis sp.
Emerald Moth
Synchlora sp.
Slimy Salamander
Plethodon glutinosus
Cope's Gray Treefrog (call)
Hyla chrysoscelis
Eastern Fence Lizard
Sceloporus undulatus
Common Loon
Gavia immer
Canada Goose
Branta canadensis
Spotted Sandpiper
Actitis macularia
Ring-billed Gull
Larus delawarensis
Turkey Vulture
Cathartes aura
Bald Eagle
Haliaeetus leucocephalus
Mourning Dove
Zenaida macroura
Downy Woodpecker
Picoides pubescens
Pileated Woodpecker
Dryocopus pileatus
Eastern Kingbird
Tyrannus tyrannus
Great Crested Flycatcher
Myiarchus crinitus
Eastern Phoebe
Sayornis phoebe
Northern Rough-winged Swallow
Stelgidopteryx serripennis
American Crow
Corvus brachyrhynchos
Tufted Titmouse
Parus bicolor
Carolina Chickadee
Parus carolinensis
White-breasted Nuthatch
Sitta carolinensis
Blue-gray Gnatcatcher
Polioptila caerulea
Wood Thrush
Hylocichla mustelina
Red-Eyed Vireo
Vireo olivaceus
Yellow-throated Warbler
Dendroica dominica
Pine Warbler
Dendroica pinus
Northern Cardinal
Cardinalis cardinalis
Common Grackle
Quiscalus quiscula
Summer Tanager
Piranga rubra
Gray Squirrel
Sciurus carolinensis
Raccoon (tracks)
Procyon lotor

(Photos by V. Winn)

Spotted wintergreen, *Chimaphila maculata.*

Station Falls
(Oconee Station State Historic Site)
Oconee County
April 29, 1990

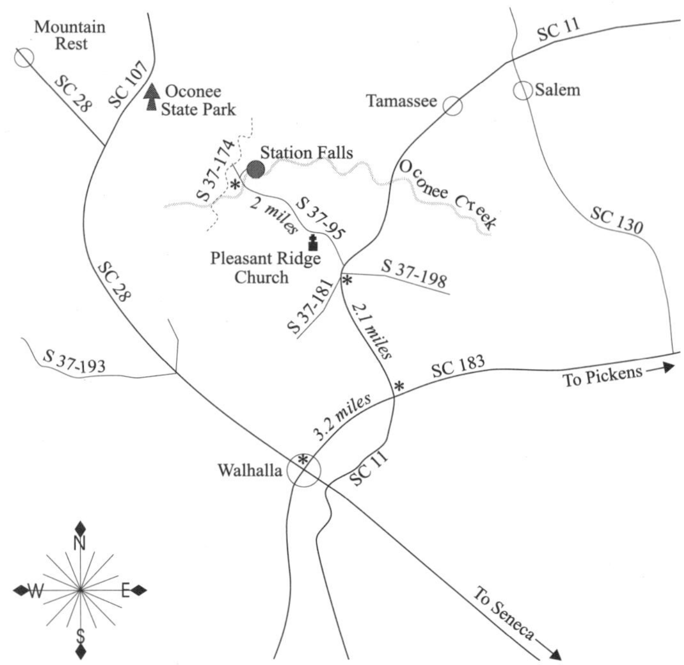

Station Falls is formed by Station Creek dropping over a rock outcrop into a rich mountain cove. In doing so, the falls create a most attractive backdrop for an unusually rich spring wildflower diversity. Hard use has damaged some of the populations of rare plants, but it is still one of the richest areas for mountain plants in the state. The beavers have dammed the creek about a thousand feet downstream from the falls, creating a nice wetlands area for plants and animals to colonize and humans to explore. For wildflower enthusiasts, this is one of the stellar sites in South Carolina.

Since SCAN's visit to Station Falls in 1990, it has become part of Oconee Station State Historic Site.

Directions to Station Falls: From Walhalla, go east on SC 183 toward Pickens for 3.2 miles. Turn left (north) onto SC 11 toward Tamassee and drive 2.1 miles. Just after County Road 181 comes in from the left and then S 37-198 joins from the right, fork left onto S 37-95. Go about 2 miles and cross Oconee Creek. Park there and follow the trail to Station Falls on the right.

FLORA

Shining Clubmoss
Lycopodium lucidulum
Rattlesnake Grape Fern
Botrychium virginianum
Maidenhair Fern
Adiantum pedatum
Southern Lady Fern
Athyrium asplenioides
Christmas Fern
Polystichum acrostichoides
Broad Beech-Fern
Thelypteris hexagonoptera
New York Fern
Thelypteris noveboracensis
Walking Fern
Asplenium rhizophyllum
Resurrection Fern
Polypodium polypodioides
Virginia Pine; Scrub Pine
Pinus virginiana
Eastern Red-cedar
Juniperus virginiana
Common Cat-tail
Typha latifolia
Cane; Switch Cane
Arundinaria gigantea
Melic Grass
Melica mutica
Nut-rush
Scleria sp.
Sedge
Carex communis
Carex laxiculmis
Jack-in-the-pulpit
Arisaema triphyllum
Sawbrier; Greenbrier
Smilax glauca
Little-sweet-Betsy
Trillium cuneatum
Wake-robin
Trillium erectum
Catesby's Trillium
Trillium catesbaei
False Solomon's Seal
Smilacina racemosa
Yellow Mandarin
Disporum lanuginosum
Solomon's-seal
Polygonatum biflorum
Blazing-star
Chamaelirium luteum
Bellwort
Uvularia perfoliata
Wild Yam
Dioscorea villosa
Yellow Star-grass
Hypoxis hirsuta
Blue-eyed-grass
Sisyrinchium angustifolium
Pink Lady's-slipper
Cypripedium acaule
Downy Rattlesnake-plantain
Goodyera pubescens
Crane-fly Orchid
Tipularia discolor
Hickory
Carya sp.
Tag Alder
Alnus serrulata
American Hornbeam; Ironwood
Carpinus caroliniana
Beech
Fagus grandifolia
Chinquapin
Castanea sp.
White Oak
Quercus alba
Spanish Oak
Quercus falcata
Slippery Elm
Ulmus rubra
Wood-nettle
Laportea canadensis
Virginia Snakeroot
Aristolochia serpentaria
Wild Ginger
Asarum canadense
Hexastylis shuttleworthii
Yellowroot
Xanthorhiza simplicissima
Monkshood
Aconitum uncinatum
Black Cohosh
Cimicifuga racemosa
Windflower
Thalictrum thalictroides
Meadow-rue
Thalictrum sp.
Bulbous Buttercup
Ranunculus bulbosus
Buttercup
Ranunculus hispidus
Liverleaf
Hepatica acutiloba
Blue Cohosh; Papooseroot
Caulophyllum thalictroides
Mayapple
Podophyllum peltatum
Tulip-poplar
Liriodendron tulipifera
Umbrella-tree
Magnolia fraseri
PawPaw
Asimina triloba
Sweet-shrub; Carolina-allspice
Calycanthus floridus
Sassafras
Sassafras albidum
Spicebush
Lindera benzoin
Bloodroot
Sanguinaria canadensis
Riverweed
Podostemum ceratophyllum
Wild Hydrangea
Hydrangea arborescens radiata
Alumroot
Heuchera americana
Foamflower
Tiarella cordifolia
Witch-hazel
Hamamelis virginiana
Five-fingers
Potentilla canadensis
Agrimony
Agrimonia parviflora

Wild Rose
Rosa carolina
Indian-physic
Gillenia trifoliata
Black Cherry; Wild Cherry
Prunus serotina
Redbud; Judas-tree
Cercis canadensis
White Wild Indigo
Baptisia alba
Naked-flower Tick-trefoil
Desmodium nudiflorum
Vetch
Vicia hugeri
Hog-peanut
Amphicarpa bracteata
Wild Geranium
Geranium maculatum
Cranesbill
Geranium carolinianum
Flowering Spurge
Euphorbia corollata
Poison-ivy
Rhus radicans
American Holly
Ilex opaca
Hearts-a'burstin'-with-love
Euonymus americanus
Red Maple
Acer rubrum
Yellow Buckeye
Aesculus octandra
Spotted Touch-me-not
Impatiens capensis
Virginia-creeper
Parthenocissus quinquefolia
Muscadine; Bullace
Vitis rotundifolia
Basswood; Linden
Tilia heterophylla
Bird-foot Violet
Viola pedata
Common Blue Violet
Viola papilionacea
Sweet White Violet
Viola blanda
Halberd-leaved Violet
Viola hastata
Canada Violet
Viola canadensis
Ginseng
Panax quinquefolium
Wild Sarsaparilla
Aralia nudicaulis
Sanicle; Snakeroot
Sanicula canadensis
Meadow Parsnip
Zizia trifoliata
Blackgum
Nyssa sylvatica
Flowering Dogwood
Cornus florida
Alternate-leaved Dogwood
Cornus alternifolia
Spotted Wintergreen
Chimaphila maculata

Indian-pipe
Monotropa uniflora
Rosebay; Great Laurel
Rhododendron maximum
Carolina-laurel
Rhododendron minus
Pinxter-flower; Pinxter-bloom
Rhododendron nudiflorum
Mountain-laurel
Kalmia latifolia
Dog-hobble
Leucothoe axillaris
Sourwood
Oxydendrum arboreum
Sparkleberry
Vaccinium arboreum
Deerberry; Squaw-huckleberry
Vaccinium stamineum
Low-bush Blueberry
Vaccinium vacillans
Horse-sugar; Sweetleaf
Symplocos tinctoria
Carolina Silverbell
Halesia carolina
American Ash; White Ash
Fraxinus americana
Pennywort
Obolaria virginica
Bindweed
Calystegia sericata
Phlox
Phlox amoena
Wild Comfrey
Cynoglossum virginianum
Lyre-leaved Sage
Salvia lyrata
Horse-balm
Collinsonia canadensis
Horse-balm
Collinsonia verticillata
Beardtongue
Penstemon sp.
False Foxglove
Aureolaria sp.
Lousewort
Pedicularis canadensis
Partridge-berry
Mitchella repens
Bluets
Houstonia longifolia
Bedstraw
Galium sp.
American Elderberry
Sambucus canadensis
Prenanthes
Prenanthes sp.
Rattlesnake-weed
Hieracium venosum
Common Dandelion
Taraxacum officinale
Small's Ragwort
Senecio smallii
Robin's-plantain
Erigeron pulchellus
Rosinweed
Silphium compositum

Green-and-gold
Chrysogonum virginianum australe
Coneflower
Rudbeckia laciniata
Coreopsis
Coreopsis major
Tree-ear
Auricularia auricula

FAUNA

Orchard Spider
Leucauge venusta
Basilica Spider
Mecynogea lemniscata
White-tail Dragonfly
Plathemis lydia
Green Tiger Beetle
Cicindela sexguttata
Marsh Beetle
Cyphon obscurus
Buprestid Beetle
Dicerca tuberculata
Firefly; Lightning Beetle
Ellychnia sp.
Twenty-spotted Lady Beetle
Psyllobora vigintimaculata
Cylindrical Bark Beetle
Philothermus glabriculus
Flower Longhorn Beetle
Toxotus cinnamopterus
Leaf-mining Beetle
Anoplitis inaequalis

Eastern Tiger Swallowtail
Papilio glaucus
Spring Azure (butterfly)
Celastrina ladon
Red-spotted Purple (butterfly)
Limenitis arthemis astyanax
Carolina Satyr (butterfly)
Hermeuptychia sosybius
Tent Caterpillar
Malacasoma sp.
Mountain Dusky Salamander
Desmognathus ochrophaeus
Shovelnose Salamander
Leurognathus marmoratus
Spring Peeper
Hyla crucifer
Eastern Box Turtle
Terrapene carolina
Turkey Vulture
Cathartes aura
Belted Kingfisher
Ceryle alcyon
Pileated Woodpecker
Dryocopus pileatus
American Crow
Corvus brachyrhynchos
Tufted Titmouse
Parus bicolor
Carolina Chickadee
Parus carolinensis
Northern Cardinal
Cardinalis cardinalis
Beaver (dam)
Castor canadensis

(Photo by L. Greenlee)

Station Falls, Oconee County, Blue Ridge region.

Orangeburg County

Santee Sinkholes and Santee State Park

Orangeburg County
June 24, 1995

This park is located on the south shore of Lake Marion providing good access to the lake with its aquatic and bird life, plus typical inner coastal plain wooded and open areas. These areas offer a good diversity of species typical of these types of habitats. The unusual feature of the park is the limestone rock near the surface, drained by underground streams. In areas where the overhead soil has collapsed, there are classic conic depressions called sinkholes. In other places the soil has been eroded so that the streams emerge from a hillside and offer caves with shallow stream bottoms as floors. It is possible to walk into these caves to areas where it is pitch black. Near these holes are calcium-loving plants, unusual in most parts of the state.

Note: The sink holes are in a restricted area with limited access to prevent damage from excessive use. Contact the park office to plan a visit to this area of the park.

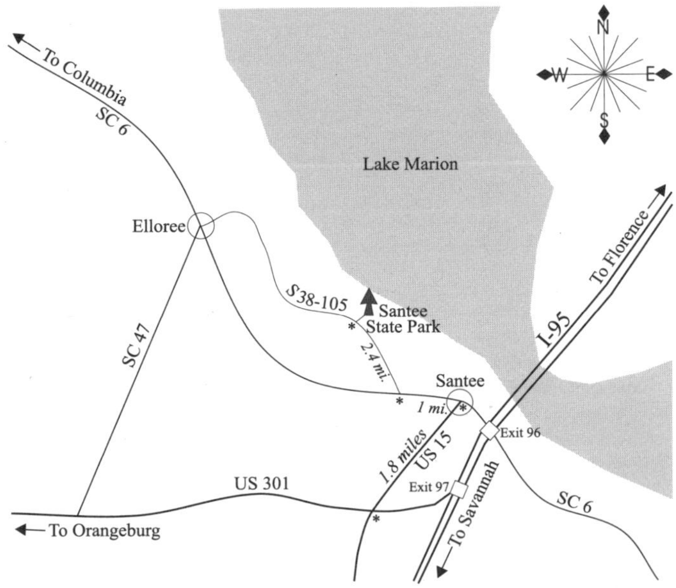

Directions to Santee State State Park: From Orangeburg, take US 301 east; cross I-26 at Exit 154 and continue for 14 miles more. At the second traffic light, turn left (north) to Santee on US 15, 1.8 miles. In the center of Santee, turn left again onto SC 6 and drive west 1 mile. Turn right at the sign to Santee State Park (S 38-105) and drive 2.4 miles to the main park entrance; the park is on the right.

FLORA

Lichen
Graphis scripta
Graphis anfractuosa
Graphis mosquitensis
Graphis insidiosa
Graphis sp.
Phaeographis dendritica
Bacidia laurocerasi
Bacidia sp.
Cladonia cristatella vestita
Cladonia polycarpoides squamulosa
Cladonia polycarpoides polycarpoides
Cladonia peziziformis
Pertusaria ostiolata
Pertusaria paratuberculifera
Pertusaria subpertusa
Pertusaria texana
Pertusaria tetrathalmia
Pertusaria sp.
Maronea constans
Haematomma puniceum
Haematomma cismonicum
Lecanora hageni
Lecanora cupressi
Lecanora varia
Lecanora sp.
Phlyctis sp.
Bulbothrix goebelii
Bulbothrix isidiza
Hypotrachyna livida
Parmelina minarum
Parmotrema michauxianum
Parmotrema perforatum
Parmotrema praesorediosum
Parmotrema subtinctorium
Canoparmelia caroliniana
Punctelia rudecta
Rimelia reticulata
Rimelia cetrata
Buellia punctata
Buellia dialyta
Buellia curatellae
Physcia americana
Physcia millegrana
Heterodermia albicans
Heterodermia obscurata
Heterodermia spaltans sp. nov.
Pyxine caesiopruinosa
Phaeophyscia rubropulchra
Anthracothecium libricola
Anthracothecium sp.
Pyrenula nitida
Arthonia sp.
Trypethelium mastoideum
Trypethelium tropicum
Trypethelium virens
Porina nucula

Cinnamon Fern
Osmunda cinnamomea
Bracken
Pteridium aquilinum
Southern Lady Fern
Athyrium asplenioides
Christmas Fern
Polystichum acrostichoides
Netted Chain Fern
Woodwardia areolata
Ebony Spleenwort
Asplenium platyneuron
Spleenwort
Asplenium sp.
Mosquito Fern
Azolla caroliniana
Loblolly Pine
Pinus taeda
Bald-cypress
Taxodium distichum
Eastern Red-cedar
Juniperus virginiana
Brazilian Elodea
Egeria densa
Oplismenus (grass)
Oplismenus setarius
Nut-rush
Scleria triglomerata
Jack-in-the-pulpit
Arisaema triphyllum
Giant Duckweed
Spirodela punctata
Spanish-moss
Tillandsia usneoides

Dayflower
Commelina communis
Water-hyacinth
Eichhornia crassipes
Carrion-flower
Smilax ecirrhata hugeri
Fringed Greenbrier
Smilax bona-nox
Sweet-scented Smilax
Smilax smallii
Wax-myrtle
Myrica cerifera cerifera
Bitternut Hickory
Carya cordiformis
Water Hickory
Carya aquatica
Mockernut Hickory
Carya tomentosa
American Hornbeam; Ironwood
Carpinus caroliniana
Beech
Fagus grandifolia
Chinquapin
Castanea sp.
White Oak
Quercus alba
Live Oak
Quercus virginiana
Black Oak
Quercus velutina
Southern Red Oak; Spanish Oak
Quercus falcata falcata
Water Oak
Quercus nigra
Laurel Oak
Quercus laurifolia
Slippery Elm
Ulmus rubra
Elm
Ulmus sp.
Sugarberry; Hackberry
Celtis laevigata
Hackberry
Celtis occidentalis georgiana
Red Mulberry
Morus rubra
False Nettle
Boehmeria cylindrica
Heartleaf
Hexastylis arifolia
Windflower
Thalictrum thalictroides
Liverleaf
Hepatica americana
Coralbeads
Cocculus carolinus
PawPaw
Asimina triloba
Sassafras
Sassafras albidum
Spicebush
Lindera benzoin
Bloodroot
Sanguinaria canadensis
Climbing-hydrangea
Decumaria barbara
Sweet-gum
Liquidambar styraciflua
Witch-hazel
Hamamelis virginiana
Redroot; Avens
Geum canadense
Cocklebur
Agrimonia pubescens microcarpa
Hawthorn
Crataegus sp.
Hog Plum
Prunus umbellata
Black Cherry; Wild Cherry
Prunus serotina
Redbud; Judas-tree
Cercis canadensis
Water Locust
Gleditsia aquatica
Beggar-lice
Desmodium sp.
Tread-softly
Cnidoscolus stimulosus
Poison-ivy
Rhus radicans
Poison-oak
Rhus toxicodendron
Winged Sumac
Rhus copallina
American Holly
Ilex opaca
Red Maple
Acer rubrum
Chalk Maple
Acer saccharum leucoderme
Red Buckeye
Aesculus pavia
Spotted Touch-me-not
Impatiens capensis
Muscadine; Bullace
Vitis rotundifolia
Summer Grape
Vitis aestivalis
Pepper-vine
Ampelopsis arborea
Basswood; Linden
Tilia sp.
Rockrose, Frostweed
Helianthemum rosmarinifolium
Passionflower; Maypops
Passiflora incarnata
Yellow Passionflower
Passiflora lutea
Pennywort
Hydrocotyle sp.
Flowering Dogwood
Cornus florida
Swamp Dogwood; Stiff Cornel
Cornus stricta
Spotted Wintergreen
Chimaphila maculata
Piedmont Azalea
Rhododendron canescens
Sparkleberry
Vaccinium arboreum
Persimmon
Diospyros virginiana
Horse-sugar; Sweetleaf
Symplocos tinctoria
American Ash; White Ash
Fraxinus americana
Fringe-tree
Chionanthus virginicus
Chinese Privet
Ligustrum sinense
Yellow Jessamine
Gelsemium sempervirens
Angle-pod
Matelea sp.
Beautyberry
Callicarpa americana
Woolly Mullein
Verbascum thapsus
Cross Vine
Anisostichus capreolata
Trumpet Vine
Campsis radicans
Squaw-root
Conopholis americana
Ruellia
Ruellia caroliniensis
Partridge-berry
Mitchella repens
Coral Honeysuckle
Lonicera sempervirens
Rusty Black-haw
Viburnum rufidulum
Wahlenbergia
Wahlenbergia marginata
Pale Indian-plantain
Cacalia atriplicifolia
Elephant's-foot
Elephantopus tomentosus
Rough Daisy Fleabane
Erigeron strigosus
Sweet Goldenrod
Solidago odora
Black-eyed-Susan
Rudbeckia sp.

FAUNA

White-lipped Forest Snail
Triodopsis albolabris
Chain River Water Snail
Goniobasis catenaria
Asiatic Clam
Corbicula sp.
Almond Millipede
Sigmoria aberrans
Spiny-backed Spider
Micrathena gracilis
Arrow-shaped Spiny-back
Micrathena sagittata
Long-jawed Orb-weaver
Tetragnatha sp.
Common Meadow Katydid
Orchelimum vulgare
Black-sided Pygmy Grasshopper
Tettigidea lateralis
Carolina Locust
Dissosteira carolina
Red-legged Grasshopper
Melanoplus femurrubrum
Thread-legged Bug
Zelus exsanguis
Froghopper
Prosapia bicincta
Common Eastern Velvet-ant
Dasymutilla occidentalis occidentalis
Red Imported Fire Ant
Solenopsis invicta
Comet Darner
Anax longipes
Southeastern Spinyleg (sign)
Dromogomphus armatus
Halloween Pennant (dragonfly)
Celithemis eponina
Eastern Pondhawk
Erythemis simplicicollis
Blue Dragonlet
Erythrodiplax minuscula
Golden-winged Skimmer
Libellula auripennis
Slaty Skimmer
Libellula incesta
Pied Skimmer
Libellula luctuosa
Blue Dasher
Pachydiplax longipennis
Eastern Amberwing (dragonfly)
Perithemis tenera
Ebony Jewelwing
Calopteryx maculata
Citrine Forktail (damselfly)
Ischnura hastata
Rose Chafer
Macrodactylus angustatus
Eyed Elater
Alaus oculatus
Net-winged Beetle
Calopteron sp.
Spotted Lady Beetle
Coleomegilla maculata
Smartweed Flea Beetle
Disonycha pensylvanica
Pipevine Swallowtail
Battus philenor
Zebra Swallowtail
Eurytides marcellus
Eastern Tiger Swallowtail
Papilio glaucus
Spicebush Swallowtail
Papilio troilus
Cloudless Sulphur
Phoebis sennae eubule
Sleepy Orange (butterfly)
Eurema nicippe
Red-spotted Purple (butterfly)
Limenitis arthemis astyanax
Viceroy; Mimic
Limenitis archippus
Carolina Satyr (butterfly)
Hermeuptychia sosybius
Common Wood-nymph (butterfly)
Cercyonis pegala
Duskywing Skipper
Erynnis sp.
Southern Skipperling
Copaeodes minimus

Santee Sinkholes and Santee State Park (continued)
June 24, 1995

Fiery Skipper
Hylephila phyleus
Banded Tussock Moth
Halysidota tessellaris
Southern Toad
Bufo terrestris
Broadhead Skink
Eumeces laticeps
Skink
Eumeces sp.
Anhinga; Snakebird
Anhinga anhinga
Double-crested Cormorant
Phalacrocorax auritus
Great Blue Heron
Ardea herodias
Mallard
Anas platyrhynchos
Laughing Gull
Larus atricilla
Turkey Vulture
Cathartes aura
Osprey
Pandion haliaetus
Northern Bobwhite
Colinus virginianus
Yellow-billed Cuckoo
Coccyzus americanus
Ruby-throated Hummingbird
Archilochus colubris
Red-bellied Woodpecker
Melanerpes carolinus
Red-headed Woodpecker
Melanerpes erythrocephalus
Downy Woodpecker
Picoides pubescens
Pileated Woodpecker
Dryocopus pileatus
Eastern Kingbird
Tyrannus tyrannus
Great Crested Flycatcher
Myiarchus crinitus
Eastern Wood-pewee
Contopus virens
Acadian Flycatcher
Empidonax virescens
Barn Swallow
Hirundo rustica
Blue Jay
Cyanocitta cristata
Fish Crow
Corvus ossifragus
Tufted Titmouse
Parus bicolor
Carolina Chickadee
Parus carolinensis
Carolina Wren
Thryothorus ludovicianus
Eastern Bluebird
Sialia sialis
Wood Thrush
Hylocichla mustelina
American Robin
Turdus migratorius
Northern Mockingbird
Mimus polyglottos
Brown Thrasher
Toxostoma rufum
Northern Parula
Parula americana
Yellow-throated Warbler
Dendroica dominica
Pine Warbler
Dendroica pinus
Northern Cardinal
Cardinalis cardinalis
Common Grackle
Quiscalus quiscula
Orchard Oriole
Icterus spurius
Summer Tanager
Piranga rubra
Gray Squirrel
Sciurus carolinensis
Raccoon (tracks)
Procyon lotor
White-tailed Deer (tracks)
Odocoileus virginianus

Pickens County

Eastatoe Creek Heritage Preserve
Pickens County
July 27, 1991

Eastatoe (pronounced East-ah-toy) is an Indian word meaning "valley of the green bird" referring to the now extinct Carolina parakeet. It is now the only known location in North America of the tunbridge fern (*Hymenophyllum tunbridgense*). Unfortunately, some of the really rare species such as this fern are in almost completely inaccessible areas of the gorge and were not encountered on this trip. The two and a half mile trail from the parking area down into the creek gorge traverses a wide spectrum of habitats, from open roadside to steep wooded hillsides with mature hemlocks to streamside floodplain at the bottom; the flora and fauna one can encounter are numerous and varied. As all persons who attended this trip can tell you, this area is prone to a regular summer phenomenon known as the late afternoon thunderstorm. They can come up very quickly and get you very wet. They also seem to sense when people are farthest from their cars! On this trip, the term "drowned rat" seemed to be used frequently as we returned to our cars. We suggest raingear even when the skies look sunny.

Directions to Eastatoe Creek Heritage Preserve: From Greenville, take US 276 north through Travelers Rest and Marietta to Cleveland. Turn left on US 276/SC 11 and go 6 miles. Where US 276 forks right toward Caesars Head, take the left fork on SC 11, and continue 11 miles farther. Turn right (north) on US 178, drive 7.4 miles to Rocky Bottom, and continue 0.9 miles farther. Make a sharp left onto a gravel road, and immediately take the right fork up the hill to the parking lot.

(Photo by A. Ciegler)
Stalked puffball-in-aspic, *Calostoma cinnabarina*.

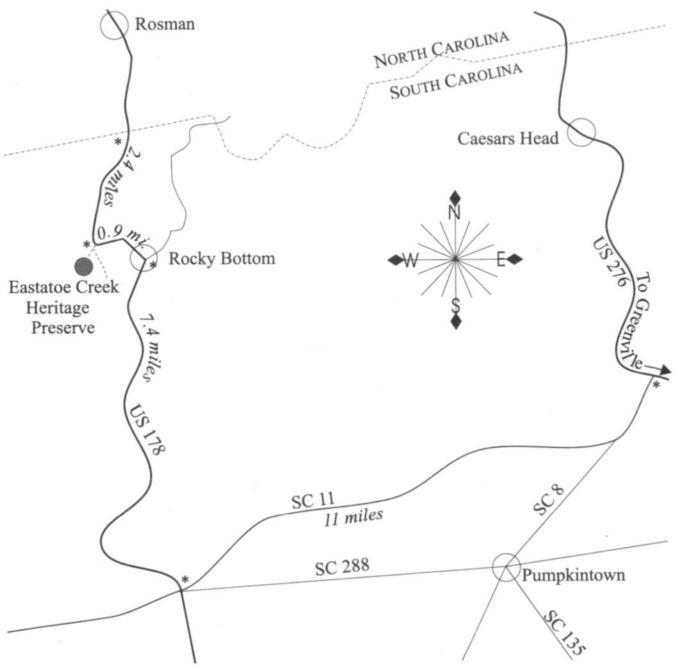

FLORA

Lichen
Collema subflaccidum
Leptogium corticola
Graphis scripta
Graphis atrorubens
Lecidea russula
Porpidia albocaerulescens
Baeomyces fungoides
Cladonia bacillaris
Cladonia capitata
Cladonia chlorophaea
Cladonia cristatella scyphulifera
Cladonia cristatella cristatella
Cladonia cristatella ramosa
Cladonia cristatella vestita
Cladonia cristatella pleurocarpa
Cladonia cristatella degenerata
Cladonia cylindrica
Cladonia leporina ramosa
Cladonia polycarpoides
Pertusaria copiosa
Pertusaria subpertusa
Pertusaria texana
Pertusaria tetrathalmia
Pertusaria ophthalmiza
Pertusaria sp.
Lecanora sp.
Ochrolechia africana
Bulbothrix confoederata
Parmelina aurulenta
Parmelina galbina
Parmotrema cetratum
Parmotrema perforatum
Parmotrema reticulatum
Pseudoparmelia caroliniana
Ramalina americana
Usnea rubicunda
Usnea strigosa
Usnea trichodea
Buellia parasema
Buellia punctata
Buellia elizae
Physcia aipolia
Physcia americana
Heterodermia obscurata
Heterodermia speciosa
Heterodermia casarettiana
Phaeophyscia imbricata
Phaeophyscia rubropulchra
Physconia enteroxantha
Pyrenula mamillana
Pyrenula nitida
Mycoporum compositum
Porina rhaphidosperma

Maidenhair Fern
Adiantum pedatum
Bracken
Pteridium aquilinum
Southern Lady Fern
Athyrium asplenioides
Christmas Fern
Polystichum acrostichoides
Broad Beech-Fern
Thelypteris hexagonoptera
New York Fern
Thelypteris noveboracensis
Ebony Spleenwort
Asplenium platyneuron
Resurrection Fern
Polypodium polypodioides
White Pine
Pinus strobus
Eastern Hemlock
Tsuga canadensis
Cane; Switch Cane
Arundinaria gigantea
Goose Grass
Eleusine indica
Dallis Grass
Paspalum dilatatum
Microstegium (grass)
Microstegium vimineum
Nut-rush
Scleria triglomerata
Jack-in-the-pulpit
Arisaema triphyllum
False Solomon's Seal
Smilacina racemosa
Solomon's-seal
Polygonatum biflorum
Carolina Lily
Lilium michauxii
Wild Yam
Dioscorea villosa
Yellow Star-grass
Hypoxis hirsuta
Crested Iris
Iris cristata
Small Green Wood Orchid
Habenaria clavellata
Downy Rattlesnake-plantain
Goodyera pubescens
Crane-fly Orchid
Tipularia discolor
Mockernut Hickory
Carya tomentosa
Cherry Birch
Betula lenta
Beech
Fagus grandifolia
American Chestnut
Castanea dentata
White Oak
Quercus alba
Rock Chestnut Oak
Quercus prinus
Red Oak
Quercus rubra
Scarlet Oak
Quercus coccinea
Wood-nettle
Laportea canadensis
Buffalo-nut
Pyrularia pubera
Dutchman's-pipe
Aristolochia macrophylla
Virginia Snakeroot
Aristolochia serpentaria
Pokeweed; Pokeberry
Phytolacca americana
Yellowroot
Xanthorhiza simplicissima
Tulip-poplar
Liriodendron tulipifera
Umbrella-tree
Magnolia fraseri
Cucumber-tree
Magnolia acuminata
Sweet-shrub; Carolina-allspice
Calycanthus floridus
Sassafras
Sassafras albidum
Climbing-hydrangea
Decumaria barbara
Mock-orange
Philadelphus inodorus
Wild Hydrangea
Hydrangea arborescens
Foamflower
Tiarella cordifolia
Sweet-gum
Liquidambar styraciflua
Indian Strawberry
Duchesnea indica
Five-fingers
Potentilla canadensis
Downy Serviceberry
Amelanchier arborea
Sensitive-brier
Schrankia microphylla
Pencil-flower
Stylosanthes biflora
Naked-flower Tick-trefoil
Desmodium nudiflorum
Sericea
Lespedeza cuneata
Black Locust
Robinia pseudo-acacia
Flowering Spurge
Euphorbia corollata
Poison-ivy
Rhus radicans
Smooth Sumac
Rhus glabra
American Holly
Ilex opaca
Hearts-a'burstin'-with-love
Euonymus americanus
Red Maple
Acer rubrum
Spotted Touch-me-not
Impatiens capensis
Virginia-creeper
Parthenocissus quinquefolia
Muscadine; Bullace
Vitis rotundifolia
Basswood; Linden
Tilia heterophylla
St.-Andrew's-cross
Hypericum stragalum
Halberd-leaved Violet
Viola hastata
Yellow Passionflower
Passiflora lutea
Spikenard
Aralia racemosa
Sanicle; Snakeroot
Sanicula canadensis
Black Tupelo; Blackgum
Nyssa sylvatica sylvatica
Flowering Dogwood
Cornus florida
Alternate-leaved Dogwood
Cornus alternifolia
Spotted Wintergreen
Chimaphila maculata
Indian-pipe
Monotropa uniflora

Eastatoe Creek Heritage Preserve (continued)
July 27, 1991

Rosebay; Great Laurel
Rhododendron maximum
Carolina-laurel
Rhododendron minus
Mountain-laurel
Kalmia latifolia
Dog-hobble
Leucothoe axillaris
Sourwood
Oxydendrum arboreum
Trailing-arbutus
Epigaea repens
Galax; Galaxy; Coltsfoot
Galax aphylla
Whorled Loosestrife
Lysimachia quadrifolia
Horse-sugar; Sweetleaf
Symplocos tinctoria
Bindweed
Calystegia sericata
Lop-seed
Phryma leptostachya
Bee-balm
Monarda clinopodia
Mountain Mint
Pycnanthemum incanum
Cross Vine
Anisostichus capreolata
Squaw-root
Conopholis americana
Partridge-berry
Mitchella repens
Bluets
Houstonia purpurea
Indian-tobacco
Lobelia inflata
Prenanthes
Prenanthes sp.
Pussy-toes
Antennaria sp.
Coreopsis
Coreopsis major
Golden Chanterelle
Cantharellus cibarius
Flame-colored Chanterelle
Cantharellus ignicolor
Chanterelle
Craterellus cornucopiodes
Cinnabar-red Polypore
Pycnoporus cinnabarinus
Indigo Milky
Lactarius indigo
Magnolia-cone Mushroom
Strobilurus conigenoides
Stalked Puffball-in-aspic
Calostoma cinnabarina

FAUNA

Giant Forest Snail
Mesodon normalis
Arrow-shaped Spiny-back
Micrathena sagittata
Writing Spider
Argiope aurantia
Orb-weaver
Neoscona sp.
June-beetle Wasp
Pelecinus polyturator
Leaf-footed Bug
Acanthocephala sp.
Oak-apple (gall wasp)
Amphibolips inanis
Green Tiger Beetle
Cicindela sexguttata
Red-bellied Tiger Beetle
Cicindela rufiventris
Purple Ground Beetle
Dicaelus purpuratus
Green Foliage Ground Beetle
Lebia viridis
Japanese Beetle
Popillia japonica
Bean Leaf Beetle
Cerotoma trifurcata
Gray-sided Oak Weevil
Pandeleteius hilaris
Eastern Tiger Swallowtail
Papilio glaucus
Banded Hairstreak
Satyrium calanus
Eastern Tailed-blue(butterfly)
Everes comyntas
Pearl Crescent
Phyciodes tharos
Silver-spotted Skipper
Epargyreus clarus
Dagger Moth (larva)
Acronicta sp.
Ashen Pinion Moth
Lithophane antennata
Cope's Gray Treefrog (call)
Hyla chrysoscelis
Gray Treefrog (call)
Hyla versicolor
Eastern Box Turtle
Terrapene carolina
Eastern Fence Lizard
Sceloporus undulatus
Skink
Eumeces sp.
Rat Snake
Elaphe obsoleta
Eastern Sreech-owl (call)
Otus asio
Chimney Swift
Chaetura pelagica
American Crow
Corvus brachyrhynchos
Northern Cardinal
Cardinalis cardinalis
Indigo Bunting
Passerina cyanea
Rufous-sided Towhee
Pipilo erythrophthalmus

Glassy Mountain Heritage Preserve
Pickens County
June 26, 1993

Glassy Mountain just outside the town of Pickens is a granite-based monadnock that has escaped most disturbances from human impact. On the top are several interesting exposed rock areas with uncommon plants and insects. On this day there was a large crop of just ripened blackberries that were happily experienced by all who attended. An extra bonus for visits to this site are the glorious views of the rolling piedmont that can be glimpsed from the rock faces. This site is rich in specimens not frequently encountered in the upstate of South Carolina.

Directions to Glassy Mountain Heritage Preserve: From the center of downtown Pickens, go northeast on SC 8 for 0.6 mile, then turn right (east toward Greenville) onto SC 183. Drive 1.3 miles and make a left fork onto South Glassy Mountain Road; (this is only a 45-degree turn instead of a full 90 degrees.) Proceed 1.1 miles. Keep left when Glassy Mountain Church Road bears right. The road will have narrowed and become rather steep. The trail head is on the right 50-75 feet past that intersection.

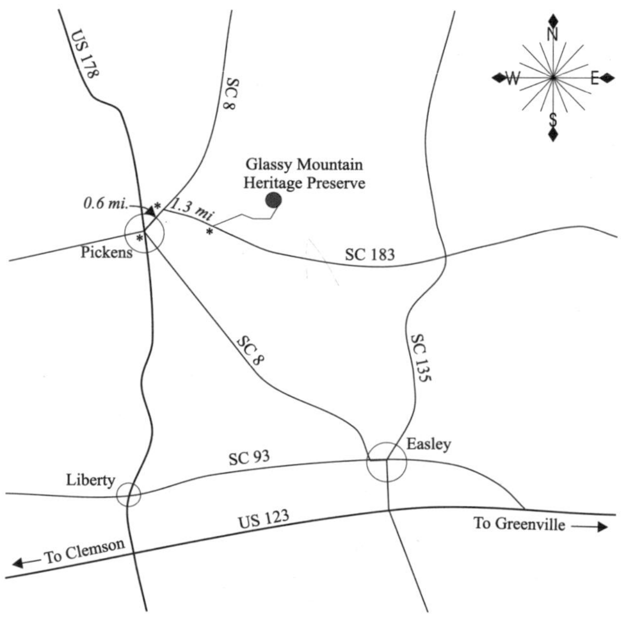

Science and Nature Notes: Monadnocks

Glassy Mountain is an excellent example of the geologic feature known as a monadnock. When the Appalachians were formed as a result of the collision of the North American tectonic plate with another set of land masses, a series of erosion events commenced affecting this homogenous granitic area. In a region of moderate rainfall, gullies began to develop on the surface. Then, as rains continued, adjacent gullies became connected on the downslopes to form valleys. Continued erosion eventually formed tributaries which dissected the mountains into short hills and spurs. Over a period of millions of years, the valleys continued to grow with erosional forces consuming much of the surrounding land mass. Steep slopes were eroded more rapidly than the valleys, so the result was a gradual decrease in relief. The valleys slowly widened until what remained are today's gently rolling Piedmont (a peneplane). A few residual remnants of the former mountains remain as isolated hills of resistant rock rising above the surrounding peneplane. These island-like masses are called monadnocks. In South Carolina, additional monadnocks include Sugarloaf Mountain in Chesterfield County and Cooks Mountain in Richland County.

FLORA

Lichen
Collema subflaccidum
Leptogium cyanescens
Leptogium phyllocarpum
Pannaria tavaresii
Coccocarpia imbricata sp. nov.
Nephroma helveticum
Peltigera canina
Lobaria pulmonaria
Sticta weigelii
Cladina arbuscula
Bacidia schweinitzii
Leptogium austroamericanum
Cladonia bacillaris
Cladonia caroliniana
Cladonia chlorophaea
Cladonia coniocraea
Cladonia didyma
Cladonia furcata subulata
Cladonia furcata f. rigidula
Cladonia furcata f. subclausa
Cladonia polycarpoides epiphylla
Cladonia squamosa
Pertusaria ostiolata
Pertusaria paratuberculifera
Pertusaria propinqua
Pertusaria subpertusa
Pertusaria texana
Pertusaria velata
Lecanora cupressi
Lecanora pallida
Lecanora varia
Lecanora sp.
Ochrolechia africana
Phlyctis sp.
Cetraria ciliaris
Cetraria sp.
Hypotrachyna livida
Parmelina horrescens
Parmelina minarum
Parmotrema austrosinense
Parmotrema mellissii
Xanthoparmelia conspersa
Xanthoparmelia plittii
Xanthoparmelia tasmanica
Flavoparmelia baltimorensis
Flavoparmelia caperata
Canoparmelia caroliniana
Punctelia rudecta
Imshaugia aleurites
Rimelia reticulata
Usnea ceratina
Usnea dasypoga
Usnea rubicunda
Usnea strigosa
Usnea subfloridana
Usnea subfusca
Buellia stellulata
Buellia punctata
Buellia disciformis
Physcia aipolia
Physcia pseudospeciosa
Physcia subtilis
Heterodermia echinata
Heterodermia speciosa
Heterodermia casarettiana
Pyxine sorediata
Phaeophyscia imbricata
Phaeophyscia rubropulchra
Dermatocarpon sp.
Normandina pulchella

Rock Spikemoss
Selaginella rupestris
Lip Fern
Cheilanthes sp.
Ebony Spleenwort
Asplenium platyneuron
Virginia Pine; Scrub Pine
Pinus virginiana
Eastern Red-cedar
Juniperus virginiana
Cane; Switch Cane
Arundinaria gigantea
Melic Grass
Melica mutica
Dallis Grass
Paspalum dilatatum
Broomsedge
Andropogon ternarius
Spiderwort
Tradescantia sp.
Greenbrier; Bullbrier
Smilax rotundifolia
Fringed Greenbrier
Smilax bona-nox
Sawbrier; Greenbrier
Smilax glauca
False Solomon's Seal
Smilacina racemosa
Solomon's-seal
Polygonatum biflorum
Bear-grass
Yucca filamentosa
Wild Yam
Dioscorea villosa
Pignut Hickory
Carya glabra
Rock Chestnut Oak
Quercus prinus
Red Oak
Quercus rubra
Black Jack Oak
Quercus marilandica
Pokeweed; Pokeberry
Phytolacca americana
Rock-portulaca
Talinum teretifolium
Common Chickweed
Stellaria media
Sandwort
Arenaria groenlandica
Deptford Pink
Dianthus armeria
Thimbleweed
Anemone virginiana
Tulip-poplar
Liriodendron tulipifera
Sassafras
Sassafras albidum
Poor-man's-pepper
Lepidium virginicum
Wild Hydrangea
Hydrangea arborescens
Alumroot
Heuchera sp.
Five-fingers
Potentilla canadensis
Wild Rose
Rosa carolina
Hawthorn
Crataegus sp.
Peach
Prunus persica
Black Cherry; Wild Cherry
Prunus serotina
Sensitive-brier
Schrankia microphylla
Wild Sensitive-plant
Cassia nicitans
Rabbit Foot Clover
Trifolium arvense
Sericea
Lespedeza cuneata
Black Locust
Robinia pseudo-acacia
Wood-sorrel
Oxalis sp.
Rushfoil
Crotonopsis elliptica
Tragia
Tragia sp.
Flowering Spurge
Euphorbia corollata
Poison-ivy
Rhus radicans
Poison-oak
Rhus toxicodendron
Winged Sumac
Rhus copallina
Smooth Sumac
Rhus glabra
Red Maple
Acer rubrum
Chalk Maple
Acer saccharum leucoderme
Painted Buckeye
Aesculus sylvatica
New-Jersey-tea
Ceanothus americanus
Virginia-creeper
Parthenocissus quinquefolia
Muscadine; Bullace
Vitis rotundifolia

Glassy Mountain Heritage Preserve (continued)
June 26, 1993

Summer Grape
Vitis aestivalis
Basswood; Linden
Tilia sp.
St.-Andrew's-cross
Hypericum hypericoides
Pineweed
Hypericum gentianoides
Yellow Passionflower
Passiflora lutea
Prickly-pear
Opuntia compressa
Meadow Parsnip
Zizia trifoliata
Black Tupelo; Blackgum
Nyssa sylvatica sylvatica
Flowering Dogwood
Cornus florida
Silky Dogwood
Cornus amomum
Spotted Wintergreen
Chimaphila maculata
Indian-pipe
Monotropa uniflora
Sourwood
Oxydendrum arboreum
Sparkleberry
Vaccinium arboreum
Deerberry; Squaw-huckleberry
Vaccinium stamineum
Low-bush Blueberry
Vaccinium vacillans
Persimmon
Diospyros virginiana
American Ash; White Ash
Fraxinus americana
Fringe-tree
Chionanthus virginicus
Chinese Privet
Ligustrum sinense
Blue-star
Amsonia tabernaemontana
Narrowleaf Milkweed
Asclepias verticillata
Butterfly-weed
Asclepias tuberosa
Milkweed
Asclepias amplexicaulis
White Milkweed
Asclepias variegata
Angle-pod
Matelea sp.
Standing-cypress
Ipomopsis rubra
Beautyberry
Callicarpa americana
Bee-balm
Monarda clinopodia
Princess-tree
Paulownia tomentosa
Woolly Mullein
Verbascum thapsus
False Foxglove
Aureolaria sp.
Trumpet Vine
Campsis radicans
Ruellia
Ruellia caroliniensis
English Plantain
Plantago lanceolata
Bluets
Houstonia purpurea
Bedstraw
Galium circaezans
Japanese Honeysuckle
Lonicera japonica
Coral Honeysuckle
Lonicera sempervirens
Black-haw; Nanny-berry
Viburnum prunifolium
Maple-leaf Viburnum
Viburnum acerifolium
Venus'-looking-glass
Specularia sp.
Common Ragweed
Ambrosia artemisiifolia
Wild Lettuce
Lactuca sp.
Sow-thistle
Sonchus sp.
Hawkweed
Hieracium sp.
Pyrrhopappus
Pyrrhopappus carolinianus
Mule-ears; Bitterweed
Senecio tomentosus
Fireweed
Erechtites hieracifolia
Pussy-toes
Antennaria plantaginifolia
Rabbit-tobacco
Gnaphalium obtusifolium
Horseweed; Mare's-tail
Erigeron sp.
White-topped Aster
Aster solidagineus
Wild Quinine
Parthenium integrifolium
Coreopsis
Coreopsis major
Coreopsis sp.

FAUNA

Walkingstick
Diapheromera carolina
Diapheromera sp.
Orange-winged Grasshopper
Pardalophora phoenicoptera
Rock-loving Grasshopper
Trimerotropis saxatilis
Spur-throated Grasshopper
Melanoplus sp.
Shield-backed Katydid
Atlanticus sp.
Red Pouch Sumac Gall (aphid)
Melaphis rhois
Hickory Vein Gall (aphid)
Phylloxera caryae-venae
Conical Hickory Leaf Gall
Phylloxera caryaefoliae
Blueberry Stem Gall (wasp)
Hemadas nubilipennis
Oak Flake Gall (wasp)
Neuroterus floccus
Chestnut Oak Vein Gall (wasp)
Plagiotrochus reticulatus
Conical Hickory Gall (midge)
Caryomyia sanguinolenta
Hickory Leaf Gall (midge)
Caryomyia consobrina
Hickory Onion Gall (midge)
Caryomyia holotricha
Hickory Tube Gall (midge)
Caryomyia tubicola
Hickory Leaf Gall (midge)
Caryomyia antennata
Hickory Midrib Gall (midge)
Caryomyia sp.
White-tail Dragonfly
Plathemis lydia
Tiger Beetle
Cicindela unipunctata
Carrion Beetle
Silpha americana
Silpha noveboracensis
Tumblebug; Dung Beetle
Glaphyrocanthon viridis
Flower Scarab Beetle
Trigonopeltastes delta
Click Beetle
Athous rufifrons
Horistonotus curiatus
Soldier Beetle
Cantharis sp.
Net-winged Beetle
Plateros sollicitus
Firefly; Lightning Beetle
Lucidota sp.
Skin Beetle
Dermestes talpinus
Tumbling Flower Beetle
Mordellistena sp.
Long-jointed Bark Beetle
Arthromacra aenea
Dusky Lady Beetle
Scymnus puncticollis

(Photo by L. Smith)
Fringe-tree, *Chionanthus virginicus,* in fruit.

Flower Longhorn Beetle
Typocerus velutinus
Typocerus sp.
Strangalina bicolor
Strangalina luteicornis
Strangalina famelica
Flea Beetle
Altica sp.
Pear-shaped Weevil
Apion crassum
Pipevine Swallowtail
Battus philenor
Sleepy Orange (butterfly)
Eurema nicippe
Red-spotted Purple (butterfly)
Limenitis arthemis astyanax
Luna Moth
Actias luna
Leconte's Haploa Moth
Haploa lecontei
Fowler's Toad
Bufo woodhousei
Eastern Box Turtle
Terrapene carolina
Eastern Fence Lizard
Sceloporus undulatus
Eastern Kingsnake
Lampropeltis getulus
Turkey Vulture
Cathartes aura
Northern Bobwhite
Colinus virginianus
Yellow-billed Cuckoo
Coccyzus americanus
Chimney Swift
Chaetura pelagica
Pileated Woodpecker
Dryocopus pileatus
Great Crested Flycatcher
Myiarchus crinitus
American Crow
Corvus brachyrhynchos
Solitary Vireo
Vireo solitarius
Pine Warbler
Dendroica pinus
Northern Cardinal
Cardinalis cardinalis
Indigo Bunting
Passerina cyanea
Rufous-sided Towhee
Pipilo erythrophthalmus
Brown-headed Cowbird (egg)
Molothrus ater

(Photo by L. Smith)
Leconte's haploa moth, *Haploa lecontei.*

Keowee-Toxaway State Park
Pickens County
April 25, 1987

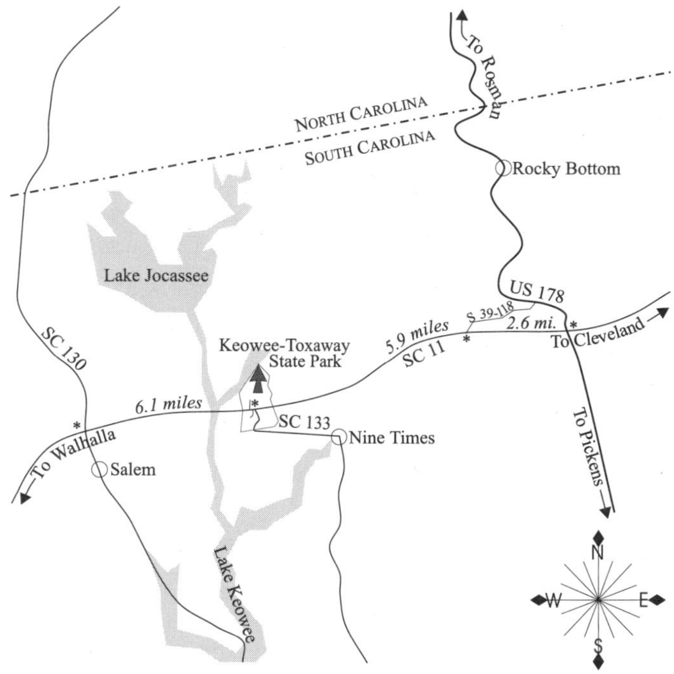

Keowee-Toxaway State Park features Native American history as well as natural history. Due to the rolling nature of the terrain, both dry and moist habitat specimens are encountered. Since SCAN members camped in the park Saturday night, a frog-hunting trip was held with great success. Spring peepers were spotted in the act of calling with their inflated vocal sacs. On the day-use trails, spring wildflowers were in bloom, as were many trees. Many of the varied, but typical upper Piedmont plants, were encountered on this trip.

Directions to Keowee-Toxaway State Park: From Greenville, travel north on US 276 through Travelers Rest and Marietta to Cleveland, about 18-20 miles. Turn left on US 276/SC 11 and go 6 miles. Where US 276 forks right toward Caesars Head State Park, take the left fork on SC 11. After 11 miles you will come to the intersection with US 178. Continue 8.5 miles farther on SC 11. Just after the junction with SC 133 from Nine Times on the left, there are two park entrance roads. The right (north) turn leads to the offices, campground, and trails; the left (south) fork leads to the nature center, picnic ground, and a short loop trail.

FLORA

Cinnamon Fern
Osmunda cinnamomea
Maidenhair Fern
Adiantum pedatum
Bracken
Pteridium aquilinum
Christmas Fern
Polystichum acrostichoides
Broad Beech Fern
Thelypteris hexagonoptera
New York Fern
Thelypteris noveboracensis
Resurrection Fern
Polypodium polypodioides
White Pine
Pinus strobus
Short-leaf Pine
Pinus echinata
Virginia Pine; Scrub Pine
Pinus virginiana
Eastern Hemlock
Tsuga canadensis
Cane; Switch Cane
Arundinaria gigantea
Melic Grass
Melica mutica
Needle Grass
Stipa avenacea
Jack-in-the-pulpit
Arisaema triphyllum
Woodrush
Luzula echinata
Trillium
Trillium discolor
Wake-robin; Red Trillium
Trillium erectum vaseyi
Catesby's Trillium
Trillium catesbaei
Indian Cucumber-root
Medeola virginiana
False Solomon's Seal
Smilacina racemosa
Solomon's-seal
Polygonatum biflorum
Blazing-star
Chamaelirium luteum
Dog's-tooth-violet
Erythronium americanum
Bellwort
Uvularia perfoliata
Bellwort
Uvularia sessilifolia
Yellow Star-grass
Hypoxis hirsuta
Dwarf Iris
Iris verna
Showy Orchis
Orchis spectabilis
Downy Rattlesnake-plantain
Goodyera pubescens
Tag Alder
Alnus serrulata
American Hornbeam; Ironwood
Carpinus caroliniana
Beech
Fagus grandifolia
White Oak
Quercus alba
Post Oak
Quercus stellata
Rock Chestnut Oak
Quercus prinus
Red Oak
Quercus rubra
Scarlet Oak
Quercus coccinea
Spanish Oak
Quercus falcata
Black Jack Oak
Quercus marilandica
Indian-olive
Nestronia umbellula
Buffalo-nut
Pyrularia pubera
Mistletoe
Phoradendron serotinum
Wild Ginger
Hexastylis shuttleworthii
Yellowroot
Xanthorhiza simplicissima
Windflower
Thalictrum thalictroides
Anemone quinquefolia
Anemone lancifolia
Tulip-poplar
Liriodendron tulipifera
Dwarf PawPaw
Asimina parviflora
Sweet-shrub; Carolina-allspice
Calycanthus floridus
Sassafras
Sassafras albidum
Bloodroot
Sanguinaria canadensis
Climbing-hydrangea
Decumaria barbara
Wild Hydrangea
Hydrangea arborescens
Foamflower
Tiarella cordifolia
Witch-hazel
Hamamelis virginiana
Five-fingers
Potentilla canadensis
Indian-physic
Gillenia trifoliata
Hawthorn
Crataegus sp.
Downy Serviceberry
Amelanchier arborea
Black Cherry; Wild Cherry
Prunus serotina
Redbud; Judas-tree
Cercis canadensis
Wild Indigo
Baptisia bracteata

Keowee-Toxaway State Park (continued)
April 25, 1987

Black Locust
Robinia pseudo-acacia
Wild Geranium
Geranium maculatum
Poison-ivy
Rhus radicans
American Holly
Ilex opaca
Hearts-a'burstin'-with-love
Euonymus americanus
Red Maple
Acer rubrum
Painted Buckeye
Aesculus sylvatica
Virginia-creeper
Parthenocissus quinquefolia
Muscadine; Bullace
Vitis rotundifolia
St.-Andrew's-cross
Hypericum hypericoides
Bird-foot Violet
Viola pedata
Common Blue Violet
Viola papilionacea
Wood Violet
Viola palmata triloba
Sweet White Violet
Viola blanda
Halberd-leaved Violet
Viola hastata
Violet
Viola walteri
Golden-Alexander
Zizia aurea
Blackgum
Nyssa sylvatica
Flowering Dogwood
Cornus florida
Alternate-leaved Dogwood
Cornus alternifolia
Spotted Wintergreen
Chimaphila maculata
Rosebay; Great Laurel
Rhododendron maximum
Carolina-laurel
Rhododendron minus
Pinxter-flower; Pinxter-bloom
Rhododendron nudiflorum
Mountain-laurel
Kalmia latifolia
Dog-hobble
Leucothoe axillaris
Sourwood
Oxydendrum arboreum
Trailing-arbutus
Epigaea repens
Sparkleberry
Vaccinium arboreum
Deerberry; Squaw-huckleberry
Vaccinium stamineum
Low-bush Blueberry
Vaccinium vacillans
Galax; Galaxy; Coltsfoot
Galax aphylla
Horse-sugar; Sweetleaf
Symplocos tinctoria
Carolina Silverbell
Halesia carolina
Bigleaf Snowbell
Styrax grandifolia
American Ash; White Ash
Fraxinus americana
Indian-hemp; Dogbane
Apocynum sp.
Creeping Phlox
Phlox stolonifera
Wild Comfrey
Cynoglossum virginianum
Cross Vine
Anisostichus capreolata
Beech-drops
Epifagus virginiana
Partridge-berry
Mitchella repens
Bluets
Houstonia caerulea
Rattlesnake-weed
Hieracium venosum
Dwarf Dandelion
Krigia virginica
Pussy-toes
Antennaria plantaginifolia
Robin's-plantain
Erigeron pulchellus
Green-and-gold
Chrysogonum virginianum australe
Coreopsis
Coreopsis major
Yellow Morel
Morchella esculenta
Wrinkled Thimble-cap
Verpa bohemica
Devil's-urn
Urnula craterium
Eyelash Cup
Scutellinia scutellata
Dead-man's-fingers
Xylaria polymorpha
Ochre Spreading Tooth
Steccherinum ochraceum
Common Split-gill
Schizophyllum commune
Fawn Mushroom
Pluteus cervinus
Fuzzy-foot
Xeromphalina campanella
Pigskin Poison Puffball
Scleroderma citrinum
Pine-oak Rust
Cronartium quercuum

FAUNA

Giant Forest Snail
Mesodon normalis
White-lipped Forest Snail
Mesodon thyroidus
Delta-spotted Spiketail
Cordulegaster diastatops
Green Tiger Beetle
Cicindela sexguttata
Zebra Swallowtail
Eurytides marcellus
Eastern Tiger Swallowtail
Papilio glaucus
Eastern Tailed-blue(butterfly)
Everes comyntas
Pearl Crescent
Phyciodes tharos
Duskywing Skipper
Erynnis sp.
Rosy Maple Moth
Dryocampa rubicunda
Polyphemus Moth (cocoon)
Antheraea polyphemus
White-dotted Prominent Moth
Nadata gibbosa
Virgin Tiger Moth
Grammia virgo
Large Necklace Moth
Hypsoropha monilis
Seal Salamander
Desmognathus monticola
Mountain Dusky Salamander
Desmognathus ochrophaeus
Spring Peeper
Hyla crucifer
Green Frog; Bronze Frog
Rana clamitans
Carolina Anole
Anolis carolinensis
Eastern Fence Lizard
Sceloporus undulatus
Ringneck Snake
Diadophis punctatus
Turkey Vulture
Cathartes aura
Red-tailed Hawk
Buteo jamaicensis
Whip-poor-will
Caprimulgus vociferus
Great Crested Flycatcher
Myiarchus crinitus
American Crow
Corvus brachyrhynchos
Tufted Titmouse
Parus bicolor
Carolina Chickadee
Parus carolinensis
White-breasted Nuthatch
Sitta carolinensis
Carolina Wren
Thryothorus ludovicianus
Blue-gray Gnatcatcher
Polioptila caerulea
Solitary Vireo
Vireo solitarius
Yellow-throated Warbler
Dendroica dominica
Hooded Warbler
Wilsonia citrina
Northern Cardinal
Cardinalis cardinalis
Gray Squirrel
Sciurus carolinensis
Beaver (gnaw marks)
Castor canadensis
Hispid Cotton Rat
Sigmodon hispidus

(Photo by L. Smith)

Maidenhair fern, Adiantum pedatum.

Table Rock State Park
Pickens County
September 29, 1991

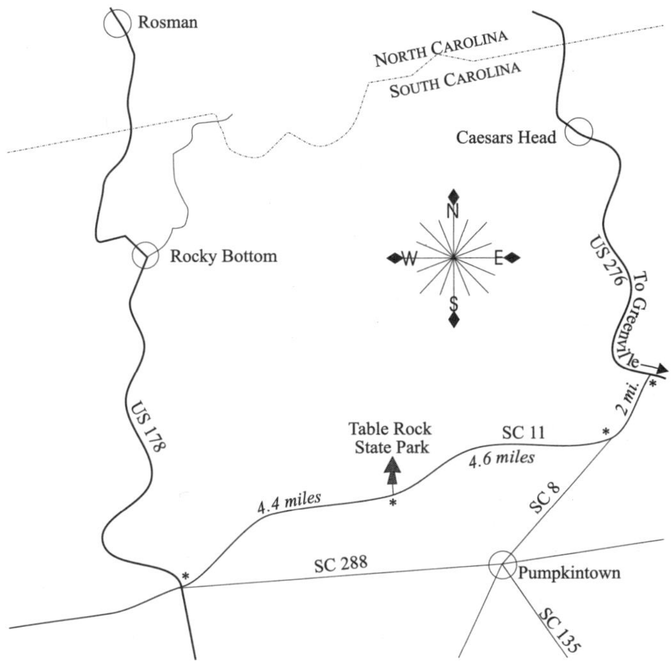

Table Rock is a flat rock-capped mountain that lends its name to the state park that includes the mountain and the area at its base. In the lower elevations is the Carrick Creek Nature Trail which follows the creek for a large portion of its almost two-mile length. The creek is one that flows picturesquely over ledges and boulders, and has several small falls and cascades. It provides stream edge and floodplain habitat for a large number of upper Piedmont and mountain species. The trail also crosses a ridge between two tributaries, offering species of drier habitats.

Directions to Table Rock State Park: From Greenville, take US 276 north through Travelers Rest and Marietta to Cleveland. Turn left on US 276/SC 11 and go 6 miles. Where US 276 forks right toward Caesars Head, take the left fork on SC 11, and continue 6.6 miles farther on SC 11. Table Rock State Park has two entrances, both on the right. The first entrance leads to the lodge; continuing on SC 11 across the lake, the second entrance leads to the campground, picnic ground, nature center, and trail heads.

FLORA

Rattlesnake Grape Fern
Botrychium virginianum
Cinnamon Fern
Osmunda cinnamomea
Maidenhair Fern
Adiantum pedatum
Southern Lady Fern
Athyrium asplenioides
Christmas Fern
Polystichum acrostichoides
Broad Beech-Fern
Thelypteris hexagonoptera
White Pine
Pinus strobus
Virginia Pine; Scrub Pine
Pinus virginiana
Eastern Hemlock
Tsuga canadensis
Cane; Switch Cane
Arundinaria gigantea
Spike Grass
Uniola sessiliflora
Jack-in-the-pulpit
Arisaema triphyllum
Greenbrier
Smilax ecirrhata biltmoreana
Sawbrier; Greenbrier
Smilax glauca
Indian Cucumber-root
Medeola virginiana
Solomon's-seal
Polygonatum biflorum
Blazing-star
Chamaelirium luteum
Bellwort
Uvularia perfoliata
Wild Yam
Dioscorea villosa
Yellow Star-grass
Hypoxis hirsuta
Yellow Star-grass
Hypoxis rigida
Crested Iris
Iris cristata
Downy Rattlesnake-plantain
Goodyera pubescens
Crane-fly Orchid
Tipularia discolor
Autumn Coral-root
Corallorhiza odontorhiza
Mockernut Hickory
Carya tomentosa
Pignut Hickory
Carya glabra
Beech
Fagus grandifolia
White Oak
Quercus alba
Rock Chestnut Oak
Quercus prinus
Red Oak
Quercus rubra
Buffalo-nut
Pyrularia pubera
Wild Ginger
Asarum canadense
Heartleaf
Hexastylis sp.
Yellowroot
Xanthorhiza simplicissima
Meadow-rue
Thalictrum sp.
Tulip-poplar
Liriodendron tulipifera
Umbrella-tree
Magnolia fraseri
Cucumber-tree
Magnolia acuminata
Sweet-shrub; Carolina-allspice
Calycanthus floridus
Bloodroot
Sanguinaria canadensis
Climbing-hydrangea
Decumaria barbara
Wild Hydrangea
Hydrangea arborescens
Hydrangea arborescens radiata
Witch-hazel
Hamamelis virginiana
Cocklebur
Agrimonia sp.
Downy Serviceberry
Amelanchier arborea arborea
Sensitive-brier
Schrankia microphylla
Naked-flower Tick-trefoil
Desmodium nudiflorum
Black Locust
Robinia pseudo-acacia
Hog-peanut
Amphicarpa bracteata
Wood-sorrel
Oxalis grandis
Flowering Spurge
Euphorbia corollata
Poison-ivy
Rhus radicans
American Holly
Ilex opaca
Hearts-a'burstin'-with-love
Euonymus americanus
Red Maple
Acer rubrum
Virginia-creeper
Parthenocissus quinquefolia
Muscadine; Bullace
Vitis rotundifolia
Summer Grape
Vitis aestivalis
St.-Andrew's-cross
Hypericum stragalum
Violet
Viola tripartita tripartita
Yellow Passionflower
Passiflora lutea
Common Evening-primrose
Oenothera biennis
Sanicle; Snakeroot
Sanicula canadensis
Cow-bane
Oxypolis rigidior
Black Tupelo; Blackgum
Nyssa sylvatica sylvatica
Flowering Dogwood
Cornus florida

Table Rock State Park (continued)
September 29, 1991

Spotted Wintergreen
Chimaphila maculata
Indian-pipe
Monotropa uniflora
Pine-sap
Monotropa hypopithys
Rosebay; Great Laurel
Rhododendron maximum
Mountain-laurel
Kalmia latifolia
Dog-hobble
Leucothoe axillaris
Sourwood
Oxydendrum arboreum
Trailing-arbutus
Epigaea repens
Deerberry; Squaw-huckleberry
Vaccinium stamineum
Elliott's Blueberry
Vaccinium elliottii
Galax; Galaxy; Coltsfoot
Galax aphylla
Horse-sugar; Sweetleaf
Symplocos tinctoria
Silverbell
Halesia sp.
White Ash; American Ash
Fraxinus americana americana
Lop-seed
Phryma leptostachya
Carpenterweed
Prunella vulgaris
Horse-balm
Collinsonia canadensis
Collinsonia tuberosa
False Foxglove
Aureolaria pedicularia
Aureolaria sp.
Cross Vine
Anisostichus capreolata
Squaw-root
Conopholis americana
Partridge-berry
Mitchella repens
Maple-leaf Viburnum
Viburnum acerifolium
Lobelia
Lobelia puberula
Prenanthes
Prenanthes sp.
Common Dandelion
Taraxacum officinale
Elephant's-foot
Elephantopus tomentosus
White Snakeroot
Eupatorium rugosum
Robin's-plantain
Erigeron pulchellus
Aster
Aster curtisii
Aster dumosus
Gray Goldenrod
Solidago nemoralis
Bearsfoot
Polymnia uvedalia
Coreopsis
Coreopsis major

FAUNA

White Micrathena (spider)
Micrathena mitrata
Orchard Spider
Leucauge venusta
Red Velvet Mite
Trombidium sp.
Round-winged Katydid
Amblycorypha oblongifolia
True Katydid
Pterophylla camellifolia
Water-strider
Gerris sp.
Yellowjacket
Vespula sp.
Cloudless Sulphur
Phoebis sennae eubule
Red-spotted Purple (butterfly)
Limenitis arthemis astyanax
Eastern Newt
Notophthalmus viridescens
Turkey Vulture
Cathartes aura
Blue Jay
Cyanocitta cristata
American Crow
Corvus brachyrhynchos
Tufted Titmouse
Parus bicolor
Red-Eyed Vireo
Vireo olivaceus

Richland County

Congaree Swamp National Monument
Richland County
May 23, 1987, and May 16, 1992

Congaree Swamp, a classic river floodplain swamp, is a site of always superlative sightings. On these two visits, the highlights were mushrooms on the 1987 trip and snakes on the 1992 trip. Trees, even after Hurricane Hugo tried to flatten the area, are the dominating first impression. Several state and national record trees are found within the monument. If you are there on a hot summer day, insects are often the next group to get your attention. Soon the birds make their presence known, as they forage in the trees or shrubs. The swamp is also renowned for the number of species and total numbers of reptiles and amphibians, and it would be a rare visit that didn't offer sightings of at least one large snake. The monument has a series of boardwalks that offer dry pathways (unless they are underwater from a current river flood) to all the major types of habitats. Also, an extensive trail system provides access to more remote areas, as does a canoe trail. Trail maps are available at the visitor's center.

Directions to Congaree Swamp National Monument: From the I-77 Bypass southeast of Columbia, take Exit 5 (Bluff Road, SC 48) and head east toward Gadsden and the Congaree Swamp National Monument. Drive 7.1 miles and note the intersection with Lower Richland Boulevard; continue on SC 48 for 0.6 mile farther. Where the road bends to the left, take the obscure right fork onto County Road 734, Old Bluff Road. Drive about 4.5 miles to the sign for the entrance to Congaree Swamp; turn right into the entrance road and go about a mile to the parking lot.

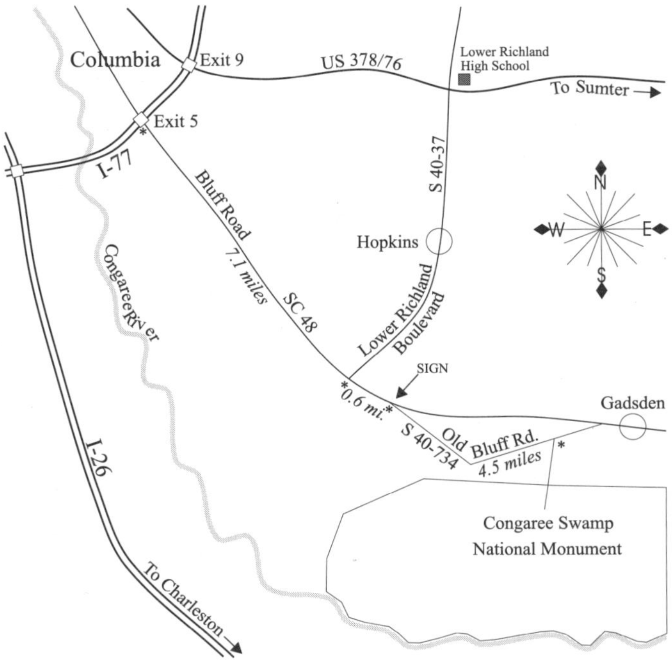

FLORA

Lichen
Collema nigrescens
Collema subflaccidum
Leptogium cyanescens
Graphis scripta
Graphis leucopepla
Graphis sp.
Gyrostomum sp.
Coenogonium interplexum
Bacidia rosella
Bacidia polychroa
Bacidia schweinitzii
Bacidia laurocerasi
Bacidia fuscus sp. nov.
Leptogium austroamericanum
Lecidea carneoalbens
Lecidea russula
Cladonia coniocraea f. phyllostrota
Cladonia coniocraea f. subpellucida
Cladonia coniocraea f. ceratodes
Cladonia coniocraea f. pycnotheliza
Pycnothelia sp.
Pertusaria leucostoma
Pertusaria ostiolata
Pertusaria multipunctoides
Pertusaria sp.
Haematomma puniceum
Lecanora cupressi
Lecanora pallida
Lecanora pallida cancriformis
Lecanora pallida f. prolifera
Lecanora horiza
Ochrolechia africana
Bulbothrix goebelii
Hypotrachyna livida f. aurea
Parmelia rudecta
Parmotrema dilitatum
Parmotrema hypotropum
Parmotrema mellissii
Parmotrema michauxianum
Parmotrema rampoddense
Parmotrema reticulatum
Parmotrema subisidiosum
Pseudoparmelia caroliniana
Usnea rubicunda
Buellia dialyta
Physcia americana
Heterodermia obscurata
Heterodermia gruoni
Heterodermia sp.
Pyxine caesiopruinosa
Phaeophyscia rubropulchra
Physconia enteroxantha
Pyrenula cruenta
Pyrenula sp.
Arthonia sp.
Trypethelium mastoideum
Trypethelium virens
Trypethelium floridanum
Trypethelium sp.
Opegrapha bonplandi

Meadow Spikemoss
Selaginella apoda
Cinnamon Fern
Osmunda cinnamomea
Royal Fern
Osmunda regalis spectabilis
Bracken
Pteridium aquilinum
Southern Lady Fern
Athyrium asplenioides
Log Fern
Dryopteris celsa
Thelypteris (fern)
Thelypteris kunthii
Sensitive Fern
Onoclea sensibilis
Netted Chain Fern
Woodwardia areolata
Ebony Spleenwort
Asplenium platyneuron
Resurrection Fern
Polypodium polypodioides
Loblolly Pine
Pinus taeda
Bald-cypress
Taxodium distichum
Arrowhead; Duck-potato
Sagittaria latifolia
Cane; Switch Cane
Arundinaria gigantea
River Oats
Uniola latifolia
Microstegium (grass)
Microstegium vimineum
Sedge
Carex sp.
Dwarf Palmetto
Sabal minor
Arrowleaf Arum
Peltandra virginica
Green-dragon
Arisaema dracontium
Jack-in-the-pulpit
Arisaema triphyllum
Spanish-moss
Tillandsia usneoides
Wild Yam
Dioscorea villosa
Yellow Star-grass
Hypoxis hirsuta
Blue-eyed-grass
Sisyrinchium albidum
Sisyrinchium sp.
Lizard's-tail
Saururus cernuus
Eastern Cottonwood
Populus deltoides
Swamp Cottonwood
Populus heterophylla
Water Hickory
Carya aquatica
Shagbark Hickory
Carya ovata
Mockernut Hickory
Carya tomentosa
American Hornbeam; Ironwood
Carpinus caroliniana
Beech
Fagus grandifolia
White Oak
Quercus alba
Overcup Oak
Quercus lyrata
Swamp Chestnut Oak
Quercus michauxii
Cherrybark Oak
Quercus falcata pagodaefolia
Water Oak
Quercus nigra
Willow Oak
Quercus phellos
Laurel Oak
Quercus laurifolia
American Elm
Ulmus americana
Winged Elm
Ulmus alata
Sugarberry; Hackberry
Celtis laevigata
Red Mulberry
Morus rubra
Wood-nettle
Laportea canadensis
False Nettle
Boehmeria cylindrica
Pokeweed; Pokeberry
Phytolacca americana
Leather-flower
Clematis crispa
Tulip-poplar
Liriodendron tulipifera
Sweetbay
Magnolia virginiana
PawPaw
Asimina triloba
Redbay
Persea borbonia
Sassafras
Sassafras albidum
Virginia-willow
Itea virginica
Climbing-hydrangea
Decumaria barbara
Sweet-gum
Liquidambar styraciflua
American Sycamore
Platanus occidentalis
Indian Strawberry
Duchesnea indica
Blackberry; Brambles
Rubus sp.
Swamp Haw
Crataegus viridis
Parsley-leaved Hawthorn
Crataegus marshallii
Hawthorn
Crataegus sp.
Black Cherry; Wild Cherry
Prunus serotina
Wood-sorrel
Oxalis sp.
Chinaberry
Melia azedarach
Poison-ivy
Rhus radicans
Winged Sumac
Rhus copallina
American Holly
Ilex opaca
Possumhaw Holly
Ilex decidua
Hearts-a'burstin'-with-love
Euonymus americanus
Boxelder
Acer negundo

Congaree Swamp National Monument (continued)
May 23, 1987, and May 16, 1992

Red Maple
Acer rubrum
Supple-jack; Rattan-vine
Berchemia scandens
Virginia-creeper
Parthenocissus quinquefolia
Muscadine; Bullace
Vitis rotundifolia
Summer Grape
Vitis aestivalis
Grape
Vitis sp.
Pepper-vine
Ampelopsis arborea
Yellow Passionflower
Passiflora lutea
Devil's-walking-stick
Aralia spinosa
Honewort
Cryptotaenia canadensis
Blackgum
Nyssa sylvatica
Black Tupelo; Blackgum
Nyssa sylvatica sylvatica
Swamp Tupelo; Swamp Blackgum
Nyssa sylvatica biflora
Water Tupelo
Nyssa aquatica
Flowering Dogwood
Cornus florida
Swamp Dogwood; Stiff Cornel
Cornus stricta
Sweet Pepperbush
Clethra alnifolia
Spotted Wintergreen
Chimaphila maculata
Dog-hobble
Leucothoe axillaris axillaris
Dangleberry
Gaylussacia frondosa
Deerberry; Squaw-huckleberry
Vaccinium stamineum
Elliott's Blueberry
Vaccinium elliottii
Water Pimpernel
Samolus parviflorus
Water Ash; Carolina Ash
Fraxinus caroliniana
American Ash; White Ash
Fraxinus americana
Green Ash; Red Ash
Fraxinus pennsylvanica subintegerrima
Swamp Privet
Forestiera acuminata
Privet; Ligustrum
Ligustrum sp.
Indian-pink
Spigelia marilandica
Angle-pod
Matelea sp.
Beautyberry
Callicarpa americana
Mazus
Mazus sp.
Cross Vine
Anisostichus capreolata
Trumpet Vine
Campsis radicans
Justicia
Justicia americana
Water-willow
Justicia ovata
Partridge-berry
Mitchella repens
Bluets
Houstonia purpurea
Bedstraw
Galium sp.
Japanese Honeysuckle
Lonicera japonica
Arrowwood
Viburnum dentatum
American Elderberry
Sambucus canadensis
Butterweed
Senecio glabellus
Climbing Hempweed
Mikania scandens
Spilanthes
Spilanthes americana repens
Ribbed-stalked Cup
Helvella acetabulum
Tree-ear
Auricularia auricula
Pale Jelly-roll
Exidia alba
Yellow-tipped Coral
Ramaria formosa
Tinder Polypore
Fomes fomentarius
Larch Polypore
Fomitopsis officinalis
Artist's Conk
Ganoderma applanatum
Turkey-tail
Trametes versicolor
Common Split-gill
Schizophyllum commune
Meadow Mushroom
Agaricus campestris
Destroying-angel
Amanita virosa
Purple-bloom Russula
Russula mariae
Pinwheel Marasmius
Marasmius rotula
Orange Pinwheel Marasmius
Marasmius siccus
Orange Mycena
Mycena leaiana
Lilac-spored Mushroom
Pleurotus sapidus

FAUNA

Orchard Spider
Leucauge venusta
Carolina Wolf Spider
Lycosa carolinensis
Wolf Spider
Lycosa sp.
Jumping Spider (large)
Phidippus pulcher
Ant
Crematogaster ashmeadi
Black Carpenter Ant
Camponotus pennsylvanicus
Phantom Cranefly
Bittacomorpha clavipes
Bumblebee Hover-fly
Volucella sp.
Swamp Darner
Epiaeschna heros
Lancet Clubtail (dragonfly)
Gomphus exilis
Royal River Cruiser
Macromia taeniolata
Great Blue Skimmer
Libellula vibrans
Ebony Jewelwing
Calopteryx maculata
Green Tiger Beetle
Cicindela sexguttata
Pedunculate Ground Beetle
Scarites subterraneus
Woodland Ground Beetle
Agonum sp.
Colorful Foliage Ground Beetle
Lebia lobulata
Lebia ornata
Whirligig Beetle
Dineutus serrulatus serrulatus
Gyrinus analis
Carrion Beetle (larva)
Silpha surinamensis
Passalus; Bess-beetle
Popilius disjunctus
Eyed Elater
Alaus oculatus
Blind Elater
Alaus myops
Firefly; Lightning Beetle
Ellychnia sp.
Darkling Beetle
Alobates pennsylvanica
Lady Beetle
Coleomegilla fuscilabris
Flower Longhorn Beetle
Typocerus sp.
Currant Tip Borer; Longhorn
Cyrtinus pygmaeus
Cylindrical Leaf Beetle
Bassareus sp.
Oval Leaf Beetle
Paria canella aterrimus
Flea Beetle
Disonycha sp.
Eggplant Flea Beetle
Epithrix fuscula
Ambrosia Beetle
Xyloborus inermis
Zebra Swallowtail
Eurytides marcellus
Eastern Tiger Swallowtail
Papilio glaucus
Cloudless Sulphur
Phoebis sennae eubule
Eastern Tailed-blue(butterfly)
Everes comyntas
Spring Azure (butterfly)
Lycaenopsis argiolus
American Snout (butterfly)
Libytheana carinenta
Question Mark (butterfly)
Polygonia interrogationis
Mourning Cloak
Nymphalis antiopa
American Lady (butterfly)
Vanessa virginiensis
Red Admiral
Vanessa atalanta rubria
Common Buckeye (butterfly)
Junonia coenia
Red-spotted Purple (butterfly)
Limenitis arthemis astyanax
Hackberry Emperor (butterfly)
Asterocampa celtis
Tawny Emperor (butterfly)
Asterocampa clyton
Carolina Satyr (butterfly)
Hermeuptychia sosybius
Monarch (butterfly)
Danaus plexippus
Evergreen Bagworm Moth
Thyridopteryx ephemeraeformis
Forest Tent Caterpillar
Malacosoma disstria
Io Moth
Automeris io
Longnose Gar (fish)
Lepisosteus osseus
Marbled Salamander
Ambystoma opacum
Cope's Gray Treefrog
Hyla chrysoscelis
Gray Treefrog
Hyla versicolor
Pickerel Frog
Rana palustris
Southern Leopard Frog
Rana sphenocephala
Snapping Turtle
Chelydra serpentina
Yellow-bellied Slider
Chrysemys scripta
Carolina Anole
Anolis carolinensis
Five-lined Skink
Eumeces fasciatus
Broadhead Skink
Eumeces laticeps
Ringneck Snake
Diadophis punctatus
Rat Snake
Elaphe obsoleta

Eastern Kingsnake
Lampropeltis getulus
Redbelly Water Snake
Nerodia erythrogaster
Banded Water Snake
Nerodia fasciata
Rough Green Snake
Opheodrys aestivus
Double-crested Cormorant
Phalacrocorax auritus
Great Blue Heron
Ardea herodias
Black Vulture
Coragyps atratus
Red-shouldered Hawk
Buteo lineatus
Mourning Dove
Zenaida macroura
Yellow-billed Cuckoo
Coccyzus americanus
Barred Owl
Strix varia
Chimney Swift
Chaetura pelagica
Ruby-throated Hummingbird
Archilochus colubris
Red-bellied Woodpecker
Melanerpes carolinus
Red-headed Woodpecker
Melanerpes erythrocephalus
Downy Woodpecker
Picoides pubescens
Yellow-bellied Sapsucker (tracks) (sign)
Sphyrapicus varius
Pileated Woodpecker
Dryocopus pileatus
Great Crested Flycatcher
Myiarchus crinitus
Eastern Wood-pewee
Contopus virens
Acadian Flycatcher
Empidonax virescens
Blue Jay
Cyanocitta cristata
American Crow
Corvus brachyrhynchos
Tufted Titmouse
Parus bicolor
Carolina Chickadee
Parus carolinensis
White-breasted Nuthatch
Sitta carolinensis
Carolina Wren
Thryothorus ludovicianus
Blue-gray Gnatcatcher
Polioptila caerulea
Wood Thrush
Hylocichla mustelina
White-eyed Vireo
Vireo griseus
Red-Eyed Vireo
Vireo olivaceus
Prothonotary Warbler
Protonotaria citrea
Northern Parula
Parula americana
Yellow-throated Warbler
Dendroica dominica
Pine Warbler
Dendroica pinus
Canada Warbler
Wilsonia canadensis
Hooded Warbler
Wilsonia citrina
Northern Cardinal
Cardinalis cardinalis
Rufous-sided Towhee
Pipilo erythrophthalmus
Common Grackle
Quiscalus quiscula
Summer Tanager
Piranga rubra
Eastern Cottontail
Sylvilagus floridanus
Gray Squirrel
Sciurus carolinensis
River Otter (tracks)
Lutra canadensis
Wild Pig (tracks) (bones)
Sus scrofa
White-tailed Deer (bones)
Odocoileus virginianus

Cooks Mountain
Richland County
March 18, 1989

Cooks Mountain, a monadnock that sits high above the right bank of the Wateree River, is owned and protected by Yancey McLeod, who has developed the property to enhance the local wildlife population. Mr. McLeod, a conservation proponent, was happy to share how he had developed the land, sited his house on the top overlooking the river, and carefully tried to keep the various species in balance. The habitats vary from near-montane to river floodplain, ponds, wooded areas, and open fields. The flora/fauna list shows the wide diversity of plants and corresponding animals found in the relatively small area. SCAN was very appreciative of Mr. McLeod sharing his fascinating property with us.

Private property. Accessible by permission only. No map provided.

FLORA

Lichen
Cladonia sp.
Bracken
Pteridium aquilinum
Christmas Fern
Polystichum acrostichoides
Resurrection Fern
Polypodium polypodioides
Long-leaf Pine
Pinus palustris
Loblolly Pine
Pinus taeda
Short-leaf Pine
Pinus echinata
Virginia Pine; Scrub Pine
Pinus virginiana
Bald-cypress
Taxodium distichum
Eel-grass
Zostera marina
Cane; Switch Cane
Arundinaria gigantea
Broomsedge
Andropogon sp.
Spanish-moss
Tillandsia usneoides
Fringed Greenbrier
Smilax bona-nox
Bamboo Smilax
Smilax laurifolia
Bear-grass
Yucca filamentosa
Wax-myrtle; Southern Bayberry
Myrica cerifera
Mockernut Hickory
Carya tomentosa
Beech
Fagus grandifolia
White Oak
Quercus alba
Post Oak
Quercus stellata
Turkey Oak
Quercus laevis
Black Oak
Quercus velutina
Black Jack Oak
Quercus marilandica
Water Oak
Quercus nigra
Mistletoe
Phoradendron serotinum
Heartleaf
Hexastylis arifolia
Sweetbay
Magnolia virginiana
Redbay
Persea borbonia
Sassafras
Sassafras albidum
Sweet-gum
Liquidambar styraciflua
American Sycamore
Platanus occidentalis
Indian Strawberry
Duchesnea indica
Southern Crabapple
Malus angustifolia
Hawthorn
Crataegus sp.
Downy Serviceberry
Amelanchier arborea
Chickasaw Plum
Prunus angustifolia
Winged Sumac
Rhus copallina
American Holly
Ilex opaca

Cooks Mountain (continued)
March 18, 1989

Water-milfoil
Myriophyllum brasiliense
Pennywort
Hydrocotyle verticillata
Water Tupelo
Nyssa aquatica
Flowering Dogwood
Cornus florida
Spotted Wintergreen
Chimaphila maculata
Mountain-laurel
Kalmia latifolia
Sourwood
Oxydendrum arboreum
Trailing-arbutus
Epigaea repens
Sparkleberry
Vaccinium arboreum
Horse-sugar; Sweetleaf
Symplocos tinctoria
Yellow Jessamine
Gelsemium sempervirens
Woolly Mullein
Verbascum thapsus
Cross Vine
Anisostichus capreolata
Beech-drops
Epifagus virginiana
Partridge-berry
Mitchella repens
Bluets
Houstonia caerulea
Small Bluets
Houstonia pusilla
Japanese Honeysuckle
Lonicera japonica
Coral Honeysuckle
Lonicera sempervirens
Hawkweed
Hieracium sp.
Dwarf Dandelion
Krigia virginica
Pussy-toes
Antennaria plantaginifolia
Multicolor Gill Polypore
Lenzites betulinae
Turkey-tail
Trametes versicolor
Earthstar Scleroderma
Scleroderma geaster

FAUNA

Oak-apple (gall wasp)
Amphibolips inanis
Baskettail Dragonfly
Epitheca sp.
Green Tiger Beetle
Cicindela sexguttata
Dingy Ground Beetle
Anisodactylus rusticus
Predacious Diving Beetle
Laccophilus fasciatus
Thermonectes basilaris
Water Scavenger Beetle
Enochrus consortus
Rove Beetle
Oxypoda sp.
Shining Fungus Beetle
Scaphisoma convexa
Passalus; Bess-beetle
Popilius disjunctus
Dung Beetle
Onthophagus tuberculifrons
Large Flat-headed Pine Borer
Chalcophora virginiensis
Darkling Beetle
Platydema erythrocera
Lady Beetle
Hyperaspis sp.
Leaf Beetle
Oedionychis concinna
Zebra Swallowtail
Eurytides marcellus
Eastern Tiger Swallowtail
Papilio glaucus
Palamedes Swallowtail
Papilio palamedes
Falcate Orangetip
Anthocharis midea
Anglewing (butterfly)
Polygonia sp.
Mourning Cloak
Nymphalis antiopa
Common Buckeye (butterfly)
Junonia coenia
Polyphemus Moth (cocoon)
Antheraea polyphemus
Fall Webworm Moth (nest)
Hyphantria cunea
Common Oak Moth
Phoberia atomaris
Killdeer
Charadrius vociferus
Turkey Vulture
Cathartes aura
Belted Kingfisher
Ceryle alcyon
Tufted Titmouse
Parus bicolor
Golden-crowned Kinglet
Regulus satrapa
Eastern Bluebird
Sialia sialis
Northern Parula
Parula americana
Pine Warbler
Dendroica pinus
White-tailed Deer (tracks)
Odocoileus virginianus

Pete's Place
Richland County
April 27, 1991

Pete's Place, a private residence in northern Richland County, lies along Cedar Creek, which has eroded a lovely shallow rocky valley in the Piedmont, just above the Fall Line. The combination of rocky outcrops and hillside adjacent to the creek's floodplain gave a wide diversity of plant material. Lush jack-in-the-pulpits graced the floodplain while a rock outcrop along the edge of an old roadbed hosted a rare native climbing fern. Box turtles, including one very tiny specimen, greeted us in each area of the property. We appreciate the Baileys' hosting us on this trip. It is nice to see such naturally rich areas still protected by local land owners.

Private property. Accessible by permission only. No map provided.

FLORA

Rattlesnake Grape Fern
Botrychium virginianum
American Climbing Fern
Lygodium palmatum
Bracken
Pteridium aquilinum
Christmas Fern
Polystichum acrostichoides
Ebony Spleenwort
Asplenium platyneuron
Resurrection Fern
Polypodium polypodioides
Loblolly Pine
Pinus taeda
Eastern Red-cedar
Juniperus virginiana
Cane; Switch Cane
Arundinaria gigantea
Melic Grass
Melica mutica
Hair Grass
Aira elegans
Microstegium (grass)
Microstegium vimineum
Green-dragon
Arisaema dracontium
Jack-in-the-pulpit
Arisaema triphyllum
Woodrush
Luzula sp.
False Garlic
Allium bivalve
Yellow Star-grass
Hypoxis hirsuta
Blue-eyed-grass
Sisyrinchium sp.
Crane-fly Orchid
Tipularia discolor
Eastern Cottonwood
Populus deltoides
Mockernut Hickory
Carya tomentosa
River Birch
Betula nigra
Hop-hornbeam
Ostrya virginiana
American Hornbeam; Ironwood
Carpinus caroliniana
White Oak
Quercus alba
Post Oak
Quercus stellata
Red Oak
Quercus rubra

Scarlet Oak
Quercus coccinea
Spanish Oak
Quercus falcata
Black Jack Oak
Quercus marilandica
Water Oak
Quercus nigra
Willow Oak
Quercus phellos
Winged Elm
Ulmus alata
Red Mulberry
Morus rubra
Virginia Snakeroot
Aristolochia serpentaria
Heartleaf
Hexastylis arifolia
Pokeweed; Pokeberry
Phytolacca americana
Buttercup
Ranunculus sardous
Tulip-poplar
Liriodendron tulipifera
PawPaw
Asimina triloba
Wild Radish
Raphanus raphanistrum
Bitter Cress
Cardamine pensylvanica
Alumroot
Heuchera americana
Foamflower
Tiarella cordifolia collina
Sweet-gum
Liquidambar styraciflua
American Sycamore
Platanus occidentalis
Five-fingers
Potentilla canadensis
Downy Serviceberry
Amelanchier arborea
Carolina Cherrylaurel
Prunus caroliniana
Redbud; Judas-tree
Cercis canadensis
Violet Wood-sorrel
Oxalis violacea
Wild Geranium
Geranium maculatum
Poison-ivy
Rhus radicans
Winged Sumac
Rhus copallina
American Holly
Ilex opaca
Hearts-a'burstin'-with-love
Euonymus americanus
Boxelder
Acer negundo
Red Maple
Acer rubrum
Chalk Maple
Acer saccharum leucoderme
Buckeye
Aesculus sp.

Spotted Touch-me-not
Impatiens capensis
Virginia-creeper
Parthenocissus quinquefolia
Muscadine; Bullace
Vitis rotundifolia
Basswood; Linden
Tilia heterophylla
Sweet Chervil
Osmorhiza longistylis
Wild Chervil
Chaerophyllum tainturieri
Golden-Alexander
Zizia aurea
Meadow Parsnip
Thaspium trifoliatum flavum
Flowering Dogwood
Cornus florida
Spotted Wintergreen
Chimaphila maculata
Mountain-laurel
Kalmia latifolia
Sourwood
Oxydendrum arboreum
Sparkleberry
Vaccinium arboreum
Elliott's Blueberry
Vaccinium elliottii
Blueberry
Vaccinium sp.
Persimmon
Diospyros virginiana
Bigleaf Snowbell
Styrax grandifolia
Ash
Fraxinus sp.

(Photo by L. Smith)

Foamflower, *Tiarella cordifolia collina.*

Fringe-tree
Chionanthus virginicus
Beautyberry
Callicarpa americana
Lyre-leaved Sage
Salvia lyrata
Toadflax
Linaria canadensis
Cross Vine
Anisostichus capreolata
Trumpet Vine
Campsis radicans
Partridge-berry
Mitchella repens
Bluets
Houstonia caerulea
Bluets
Houstonia purpurea
Venus'-looking-glass
Specularia perfoliata
Rattlesnake-weed
Hieracium venosum
Dwarf Dandelion
Krigia dandelion
Pale Indian-plantain
Cacalia atriplicifolia
Small's Ragwort
Senecio smallii
Leopard's-bane
Arnica acaulia
Thistle
Carduus lanceolatus
Green-and-gold
Chrysogonum virginianum
Dwarf-eared Coreopsis
Coreopsis auriculata

FAUNA

Bowl-and-doily Spider
Frontinella sp.
Field Cricket
Gryllus sp.
Spittlebug
Prosapia sp.
Fire Ant
Solenopsis sp.
Ebony Jewelwing
Calopteryx maculata
Twenty-spotted Lady Beetle
Psyllobora vigintimaculata
Carolina Satyr (butterfly)
Hermeuptychia sosybius
Eastern Tent Caterpillar
Malacosoma americanum
Luna Moth
Actias luna
Southern Toad
Bufo terrestris
Cope's Gray Treefrog
Hyla chrysoscelis
Eastern Box Turtle
Terrapene carolina
Carolina Anole
Anolis carolinensis
Turkey Vulture
Cathartes aura
Black Vulture
Coragyps atratus
Wild Turkey (egg)
Meleagris gallopavo
American Crow
Corvus brachyrhynchos
Tufted Titmouse
Parus bicolor
Summer Tanager
Piranga rubra
White-tailed Deer (tracks)
Odocoileus virginianus

(Photo by V. Winn)

Blue-eyed-grass, *Sisyrinchium* sp.

Spartanburg County

Croft State Park
Spartanburg County
June 24, 1989, October 23, 1993 and October 28, 1995

Croft State Park, one of the largest state parks in South Carolina, exemplifies a recovering Piedmont forest area. Approximately 95 per cent of the land in the Piedmont of South Carolina has been cleared and farmed. Erosion quickly took its toll, and the land was abandoned. Much of this park is only about sixty years into field to forest succession. The now rough terrain tells a silent tale of deep gullies from erosion. The small areas of older growth, areas along the edges of lakes and creeks, and especially along Fairforest Creek have developed a very interesting collection of plants and animals. Hardwood forests along the creek floodplains are especially attractive and interesting. On the first visit SCAN was accompanied by Dr. Robert Powell of Converse College, Dr. Gibbes Patton, retired from Wofford College, and Dick Coyle of the state park staff. They were able to explain much of the cultural history that goes along with the natural history of Croft, as it served as Camp Croft during World War II. It has only been in recent years that a thorough check of the park was made to remove practice munitions still in the soil. Their explanations and identifications greatly enhanced our visits.

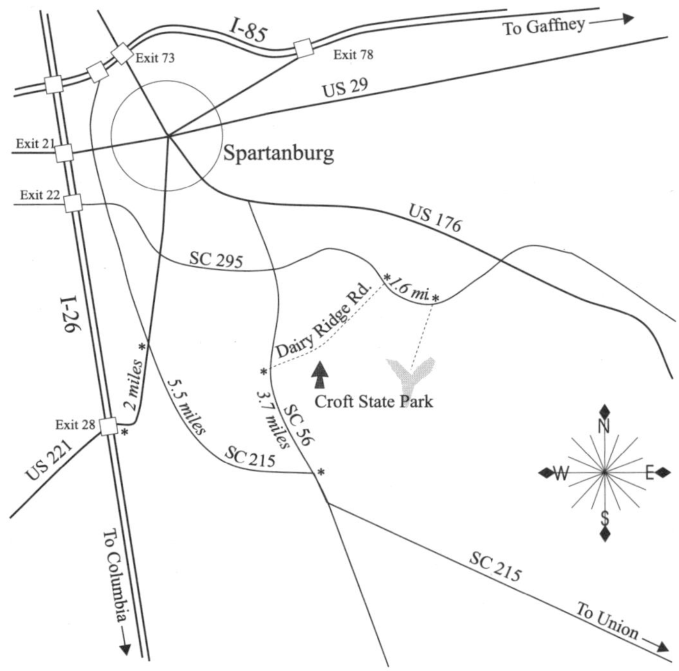

Directions to Croft State Park: From I-26 south of Spartanburg, take Exit 28 and go north on US 221 toward Spartanburg for 2 miles. Turn right onto SC 215 (south toward Union) and drive 5.5 miles. Turn left (north toward Spartanburg) and travel 3.7 miles. The main entrance to Croft State Park is on the right marked with a brown sign. To reach the Lake Johnson section of the state park, follow the main entrance road (Dairy Ridge Road) until it meets SC 295. Turn right (south) and go 1.6 miles. Turn right just past Whitestone Mall (small white building) onto Johnson Lake Road; drive to the end of the road.

FLORA

Lichen
- *Collema subflaccidum*
- *Leptogium chloromelum*
- *Leptogium cyanescens*
- *Graphis scripta*
- *Glyphis sp.*
- *Phaeographis dendritica*
- *Bacidia schweinitzii*
- *Bacidia sp.*
- *Leptogium austroamericanum*
- *Lecidea tessellata*
- *Cladina subtenuis subtenuis*
- *Cladina subtenuis cinerascens*
- *Cladonia bacillaris*
- *Cladonia chlorophaea f. intermedia*
- *Cladonia coniocraea f. subpellucida*
- *Cladonia cristatella cristatella*
- *Cladonia cristatella vestita*
- *Cladonia polycarpoides*
- *Cladonia peziziformis*
- *Cladonia peziziformis stenophyllodes*
- *Cladonia ramulosa*
- *Pertusaria paratuberculifera*
- *Pertusaria plittiana*
- *Pertusaria subpertusa*
- *Pertusaria texana*
- *Haematomma elatinum*
- *Haematomma cismonicum*
- *Lecanora cupressi*
- *Lecanora pallida*
- *Lecanora pallida cancriformis*
- *Lecanora varia*
- *Lecanora subimmergens*
- *Ochrolechia africana*
- *Hypotrachyna livida*
- *Parmelina obsessa*
- *Parmotrema crinitum*
- *Parmotrema hypotropum*
- *Parmotrema michauxianum*
- *Parmotrema perforatum*
- *Parmotrema rampoddense*
- *Flavoparmelia caperata*
- *Canoparmelia caroliniana*
- *Punctelia rudecta*
- *Rimelia reticulata*
- *Rimelia subisidiosa*
- *Ramalina americana*
- *Usnea strigosa*
- *Usnea sp.*
- *Buellia punctata*
- *Buellia curatellae*
- *Dirinaria picta*
- *Physcia aipolia*
- *Physcia americana*
- *Physcia millegrana*
- *Physcia pseudospeciosa*
- *Physcia stellaris*
- *Heterodermia obscurata*
- *Heterodermia speciosa*

Heterodermia spaltans
Pyxine caesiopruinosa
Pyxine sorediata
Phaeophyscia adiastola
Phaeophyscia imbricata
Phaeophyscia rubropulchra
Phaeophyscia rubropulchra minor
Physciella cloantha
Trypethelium mastoideum
Trypethelium virens
Running-pine
Lycopodium flabelliforme
Rattlesnake Grape Fern
Botrychium virginianum
Southern Grape Fern
Botrychium biternatum
Grape Fern; Moonwort
Botrychium sp.
Southern Lady Fern
Athyrium asplenioides
Christmas Fern
Polystichum acrostichoides
Ebony Spleenwort
Asplenium platyneuron
Resurrection Fern
Polypodium polypodioides
White Pine
Pinus strobus
Loblolly Pine
Pinus taeda
Short-leaf Pine
Pinus echinata
Virginia Pine; Scrub Pine
Pinus virginiana
Eastern Red-cedar
Juniperus virginiana
Pondweed
Potamogeton diversifolius
Southern Naiad
Najas guadalupensis
Slender Naiad
Najas minor
Cane; Switch Cane
Arundinaria gigantea
Spike Grass
Uniola sessiliflora
River Oats
Uniola latifolia
Love Grass
Eragrostis hirsuta
Dropseed
Sporobolus sp.
Foxtail Grass
Setaria sp.
Barnyard Grass
Echinochloa crusgalli
Panic Grass
Panicum boseii
Microstegium (grass)
Microstegium vimineum
Beard Grass
Erianthus sp.
Broomsedge
Andropogon virginicus
Johnson Grass
Sorghum halepense
Indian Grass
Sorghastrum elliottii
Beak-rush
Rhynchospora corniculata
Jack-in-the-pulpit
Arisaema triphyllum
Soft Rush
Juncus effusus
Greenbrier; Bullbrier
Smilax rotundifolia
Fringed Greenbrier
Smilax bona-nox
Bamboo Smilax
Smilax laurifolia
Field Garlic
Allium vineale
Crested Iris
Iris cristata
Lesser Ladies'-tresses
Spiranthes ovalis
Downy Rattlesnake-plantain
Goodyera pubescens
Crane-fly Orchid
Tipularia discolor
Black Willow
Salix nigra
Eastern Cottonwood
Populus deltoides
Black Walnut
Juglans nigra
Mockernut Hickory
Carya tomentosa
Tag Alder
Alnus serrulata
River Birch
Betula nigra
Hop-hornbeam
Ostrya virginiana
American Hornbeam; Ironwood
Carpinus caroliniana
Beech
Fagus grandifolia
American Chestnut
Castanea dentata
White Oak
Quercus alba
Post Oak
Quercus stellata
Swamp Chestnut Oak
Quercus michauxii
Red Oak
Quercus rubra
Black Oak
Quercus velutina
Scarlet Oak
Quercus coccinea
Spanish Oak
Quercus falcata
Water Oak
Quercus nigra
Willow Oak
Quercus phellos
Laurel Oak
Quercus laurifolia
Slippery Elm
Ulmus rubra
American Elm
Ulmus americana
Winged Elm
Ulmus alata
Hackberry; Sugarberry
Celtis sp.
Red Mulberry
Morus rubra
False Nettle
Boehmeria cylindrica
Virginia Snakeroot
Aristolochia serpentaria
Tearthumb
Polygonum sagittatum
Polygonum
Polygonum sp.
Black Cohosh
Cimicifuga racemosa
Liverleaf
Hepatica americana
Mayapple
Podophyllum peltatum
Coralbeads
Cocculus carolinus
Tulip-poplar
Liriodendron tulipifera
Cucumber-tree
Magnolia acuminata
Dwarf PawPaw
Asimina parviflora
PawPaw
Asimina triloba
Sassafras
Sassafras albidum
Bloodroot
Sanguinaria canadensis
Virginia-willow
Itea virginica
Climbing-hydrangea
Decumaria barbara
Sweet-gum
Liquidambar styraciflua
Witch-hazel
Hamamelis virginiana
American Sycamore
Platanus occidentalis
Cocklebur
Agrimonia sp.
Hawthorn
Crataegus sp.
Downy Serviceberry
Amelanchier arborea
Black Cherry; Wild Cherry
Prunus serotina
Mimosa; Silk-tree
Albizia julibrissin
Redbud; Judas-tree
Cercis canadensis
Honey Locust
Gleditsia triacanthos
White Clover
Trifolium repens
Beggar-lice
Desmodium sp.
Hog-peanut
Amphicarpa bracteata
Common Wood-sorrel
Oxalis acetosella
Wafer-ash; Hop-tree
Ptelea trifoliata
Three-seeded Mercury
Acalypha gracilens
Acalypha sp.
Poison-ivy
Rhus radicans
Winged Sumac
Rhus copallina
Smooth Sumac
Rhus glabra
American Holly
Ilex opaca
Possumhaw Holly
Ilex decidua
Possumhaw Holly
Ilex decidua longipes
Hearts-a'burstin'-with-love
Euonymus atropurpureus
Euonymus americanus
Boxelder
Acer negundo
Red Maple
Acer rubrum
Buckthorn
Rhamnus caroliniana
Muscadine; Bullace
Vitis rotundifolia
Fox Grape
Vitis labrusca
Pepper-vine
Ampelopsis arborea
St.-Andrew's-cross
Hypericum hypericoides
Passionflower; Maypops
Passiflora incarnata
Yellow Passionflower
Passiflora lutea
Prickly-pear
Opuntia compressa
Thorny Elaeagnus
Elaeagnus pungens
Russian-olive
Elaeagnus umbellata
Water-primrose
Ludwigia uruguayensis
Seed-box
Ludwigia alternifolia
Ludwigia
Ludwigia sp.
Snakeroot
Sanicula sp.
Blackgum
Nyssa sylvatica
Flowering Dogwood
Cornus florida
Spotted Wintergreen
Chimaphila maculata
Pine-sap
Monotropa hypopithys
Azalea
Rhododendron sp.
Mountain-laurel
Kalmia latifolia

Croft State Park (continued)
June 24, 1989, October 23, 1993 and October 28, 1995

Dog-hobble
Leucothoe axillaris
Leucothoe axillaris editorum
Sourwood
Oxydendrum arboreum
Sparkleberry
Vaccinium arboreum
Deerberry; Squaw-huckleberry
Vaccinium stamineum
Highbush Blueberry
Vaccinium corymbosum
Persimmon
Diospyros virginiana
Carolina Silverbell
Halesia carolina
American Ash; White Ash
Fraxinus americana
Red Ash; Pumpkin Ash
Fraxinus tomentosa
Ash
Fraxinus sp.
Fringe-tree
Chionanthus virginicus
Chinese Privet
Ligustrum sinense
Yellow Jessamine
Gelsemium sempervirens
Climbing-milkweed
Matelea obliqua
Angle-pod
Matelea sp.
Wild Comfrey
Cynoglossum virginianum
Verbena
Verbena sp.
Beautyberry
Callicarpa americana
Blue-curls
Trichostema dichotomum
Skullcap
Scutellaria sp.
Mountain Mint
Pycnanthemum incanum
Horse-nettle
Solanum carolinense
Cross Vine
Anisostichus capreolata
Trumpet Vine
Campsis radicans
Beech-drops
Epifagus virginiana
Ruellia
Ruellia sp.
Plantain
Plantago sp.
Button-bush
Cephalanthus occidentalis
Poor-jo (or Po-jo)
Diodia teres
Partridge-berry
Mitchella repens
Japanese Honeysuckle
Lonicera japonica
Black-haw; Nanny-berry
Viburnum prunifolium
Lobelia
Lobelia puberula
Rattlesnake-weed
Hieracium venosum
Pyrrhopappus
Pyrrhopappus carolinianus
Pale Indian-plantain
Cacalia atriplicifolia
Bull Thistle
Carduus pumilus
Ironweed
Vernonia sp.
Elephant's-foot
Elephantopus carolinianus
Climbing Hempweed
Mikania scandens
Pussy-toes
Antennaria sp.
Rabbit-tobacco
Gnaphalium obtusifolium
Daisy Fleabane
Erigeron philadelphicus
Frost Aster
Aster pilosus
Aster
Aster dumosus
Gray Goldenrod
Solidago nemoralis
Heterotheca
Heterotheca graminifolia
Eclipta
Eclipta alba
Verbesina
Verbesina sp.
Bitterweed
Helenium amarum
Hairy Rubber Cup
Galiella rufa
Brown-haired White Cup
Humaria hemisphaerica
Velvety Earth-tongue
Trichoglossum hirsutum
Purple Jelly-drops
Ascocoryne sarcoides
Yellow-green Hypomyces
Hypomyces luteovirens
Carbon-antlers
Xylaria hypoxylon
Orange Jelly
Dacrymyces palmatus
Jellied False-coral
Tremellodendron pallidum
Golden Chanterelle
Cantharellus cibarius
Cinnabar-red Chanterelle
Cantharellus cinnabarinus
Spindle-shaped Gold Coral
Clavulinopsis pulchra
White Worm Coral
Clavaria vermicularis
Crested Coral
Clavulina cristata
Clustered Coral
Ramaria botrytis
White Coral
Ramariopsis kunzei
Coral-pink Merulius
Merulius incarnatus
Bearded Tooth
Hericium erinaceus
Scaly Tooth
Hydnum imbricatum
Multicolor Gill Polypore
Lenzites betulinae
Turkey-tail
Trametes versicolor
Common Split-gill
Schizophyllum commune
Crowded Parchment
Stereum complicatum
Silky Parchment
Stereum striatum
Yellow-patches
Amanita flavoconia
Yellow Blusher
Amanita flavorubescens
Gemmed Amanita
Amanita gemmata
Strangulated Amanita
Amanita inaurata
Fly Agaric
Amanita muscaria
Salmon Unicorn Entoloma
Entoloma salmoneum
Fading Scarlet Waxy-cap
Hygrophorus miniatus
Indigo Milky
Lactarius indigo
Tacky Green Russula
Russula aeruginea
Emetic Russula
Russula emetica
Fragile Russula
Russula fragilis
Variable Russula
Russula variata
Fat-footed Clitocybe
Clitocybe clavipes
Mycena (mushroom)
Mycena inclinata
Mycena vulgaris
Gem-studded Puffball
Lycoperdon perlatum
Bird's-nest; Splash-cups
Cyathus sp.
Coral Slime
Ceratiomyxa fruticulosa
Wolf's-milk Slime
Lycogala epidendrum
Japanese Lantern Slime
Dictydium cancellatum
White-footed Slime
Diachea leucopodia

FAUNA

Forest Disk Snail
Anguispira alternata
Almond Millipede
Sigmoria aberrans
Northern Widow
Latrodectus variolus
Marbled Orb-weaver
Araneus marmoreus
Long-jawed Orb-weaver
Tetragnatha sp.
Rabid Wolf Spider
Lycosa rabida
Green Lynx Spider
Peucetia viridans
Short-winged Green Grasshopper
Dichromorpha viridis
Slant-faced Grasshopper
Syrbula admirabilis
Short-horned Grasshopper
Metaleptea brevicornis
Sulfur-winged Grasshopper
Arphia sp.
Green-striped Grasshopper
Chortophaga viridifasciata
Differential Grasshopper
Melanoplus differentialis
Red-legged Grasshopper
Melanoplus femurrubrum
Shield-backed Katydid
Atlanticus gibbosus
Field Cricket
Gryllus sp.
Rapier Meadow Katydid
Conocephalus strictus
Carolina Mantid (egg)
Stagmomantis carolina
Chinese Mantid (egg)
Tenodera aridifolia sinensis
Water Bug
Belostoma flumineum
Toad Bug
Gelastocoris oculatus
Red Imported Fire Ant
Solenopsis invicta
Paper Wasp (nest)
Polistes sp.
Eastern Pondhawk
Erythemis simplicicollis
White-tail Dragonfly
Plathemis lydia
Atlantic Bluet (damselfly)
Enallagma doubledayi
Tiger Beetle
Cicindela tranquebarica
Woodland Ground Beetle
Loxandrus minor
Pterostichus sp.
Abacidus sculptus
Dingy Ground Beetle
Harpalus pennsylvanicus
Corn-seed Beetle
Agonoderus lecontei
Notch-mouthed Ground Beetle
Dicaelus ambiguus
Dicaelus dilitatus
Predacious Diving Beetle
Laccophilus maculosus maculosus
Hydroporus carolinus
Hydroporus lobatus
Hydroporus undulatus
Ilybius biguttulus

Water Scavenger Beetle
Tropisternus collaris striolatus
Tropisternus lateralis nimbatus
Tropisternus natator
Cross-toothed Rove Beetle
Oxyporus femoralis
Shining Fungus Beetle
Scaphisoma suturalis
Passalus; Bess-beetle
Popilius disjunctus
Tumblebug; Dung Beetle
Glaphyrocanthon viridis
Hister-like Dung Beetle
Ateuchus histeroides
Earth-boring Dung Beetle
Geotrupes blackburnii blackburnii
June Beetle
Phyllophaga sp.
Rose Chafer
Macrodactylus angustatus
Large Flat-headed Pine Borer
Chalcophora virginiensis
Click Beetle
Cardiophorus sp.
Net-winged Beetle
Plateros lictor
Plateros sp.
Firefly; Lightning Beetle
Photuris sp.
Tumbling Flower Beetle
Mordellistena sp.
Darkling Beetle
Uloma imberbis
Seven-spotted Lady Beetle
Coccinella septempunctata
Hidden Snout Weevil
Conotrachelus posticatus
Eastern Tiger Swallowtail
Papilio glaucus
Eastern Tailed-blue(butterfly)
Everes comyntas
Question Mark (butterfly)
Polygonia interrogationis
Anglewing (butterfly)
Polygonia sp.
Common Buckeye (butterfly)
Junonia coenia
Viceroy; Mimic
Limenitis archippus
Monarch (butterfly)
Danaus plexippus
Duskywing Skipper
Erynnis sp.
Datana Moth Caterpillar
Datana sp.
Fall Webworm Moth
Hyphantria cunea
Mosquitofish
Gambusia sp.
Bluegill
Lepomis macrochirus
Eastern Newt
Notophthalmus viridescens
Northern Cricket Frog
Acris crepitans
Eastern Box Turtle
Terrapene carolina
Carolina Anole
Anolis carolinensis
Ground Skink
Scincella lateralis
Worm Snake
Carphophis amoenus
Pied-billed Grebe
Podilymbus podiceps
Double-crested Cormorant
Phalacrocorax auritus
Great Blue Heron
Ardea herodias
American Coot
Fulica americana
Turkey Vulture
Cathartes aura
Black Vulture
Coragyps atratus
Sharp-shinned Hawk
Accipiter striatus
Red-shouldered Hawk
Buteo lineatus
Yellow-billed Cuckoo
Coccyzus americanus
Belted Kingfisher
Ceryle alcyon
Red-bellied Woodpecker
Melanerpes carolinus
Northern Flicker
Colaptes auratus
Downy Woodpecker
Picoides pubescens
Yellow-bellied Sapsucker
Sphyrapicus varius
Eastern Phoebe
Sayornis phoebe
Acadian Flycatcher
Empidonax virescens
Blue Jay
Cyanocitta cristata
American Crow
Corvus brachyrhynchos
Tufted Titmouse
Parus bicolor
Carolina Chickadee
Parus carolinensis
Brown Creeper
Certhia americana
Red-breasted Nuthatch
Sitta canadensis
Brown-headed Nuthatch
Sitta pusilla
Carolina Wren
Thryothorus ludovicianus
Golden-crowned Kinglet
Regulus satrapa
Ruby-crowned Kinglet
Regulus calendula
Eastern Bluebird
Sialia sialis
Wood Thrush
Hylocichla mustelina
American Robin
Turdus migratorius
Northern Mockingbird
Mimus polyglottos
Brown Thrasher
Toxostoma rufum
Solitary Vireo
Vireo solitarius
Black-throated Blue Warbler
Dendroica caerulescens
Yellow-rumped Warbler
Dendroica coronata
Pine Warbler
Dendroica pinus
Ovenbird
Seiurus aurocapillus
Northern Cardinal
Cardinalis cardinalis
Chipping Sparrow
Spizella passerina
White-throated Sparrow
Zonotrichia albicollis
Red-winged Blackbird
Agelaius phoeniceus
Common Grackle
Quiscalus quiscula
House Finch
Cardodacus mexicanus
Raccoon
Procyon lotor
White-tailed Deer (tracks)
Odocoileus virginianus

Pacolet River Heritage Preserve
(Soapstone Ridge)
Spartanburg County
September 22, 1990

Pacolet River Heritage Preserve, formerly called Soapstone Ridge, was named for an outcrop of soapstone rock that lies on the bluff above the Pacolet River. Soapstone is a soft, greasy-feeling rock used for laboratory tables and many decorative uses because it is easily carved. The associated soil is high in magnesium, limiting the vegetation that will tolerate this soil composition. This site attracted humans as long ago as 6000 years, and partially carved bowls left by the Native Americans can still be seen attached to these rocks. These early workmen utilized knobs of material to minimize the amount of work they had to do to get a useful container. It is surmised that the ones still unfinished were going to be flawed, and were deliberately left behind. SCAN's enjoyment of the site was enhanced by having two members of the Wofford College faculty with us to explain what we were seeing, Dr. Terry Ferguson, an archaeologic geologist, and Dr. Doug Rayner, botanist.

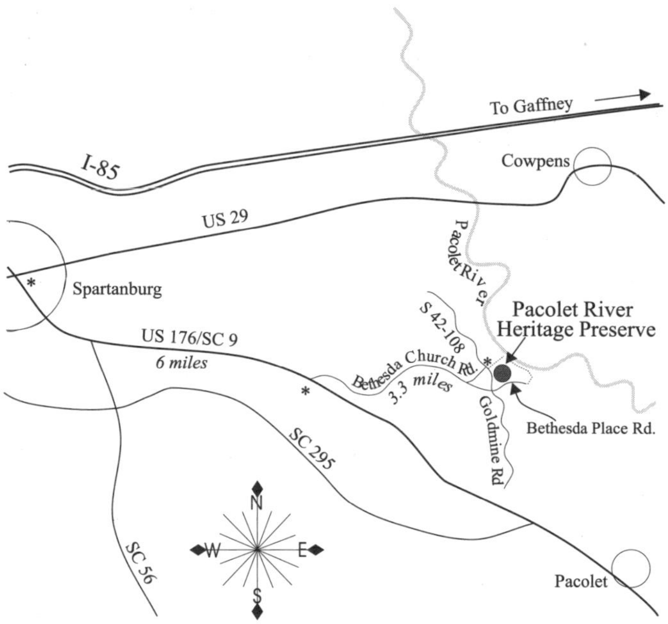

Directions to Pacolet River Heritage Preserve: From downtown Spartanburg, take US 176/SC 9 southeast toward Pacolet and Union. Drive approximately 6 miles, and turn left onto County Road 680 (Bethesda Church Road). Travel 3.3 miles until 680 ends at County Road 108, Goldmine Road. Turn right onto Goldmine, then left onto Bethesda Place Road; drive to the end and park in parking area.

Pacolet River Heritage Preserve (continued)
September 22, 1990

FLORA

Alga
Protococcus sp.
Lichen
Graphis scripta
Cladonia capitata
Cladonia caroliniana
Cladonia chlorophaea
Cladonia coniocraea
Cladonia cristatella
Cladonia cristatella scyphulifera
Cladonia macilenta
Cladonia parasitica
Cladonia polycarpoides
Cladonia robbinsii
Cladonia uncialis
Cladonia cervicornis
Pycnothelia papillaria
Cetraria oakesiana
Hypotrachyna livida
Parmelia rudecta
Parmelina dissecta
Parmotrema dilitatum
Parmotrema hypotropum
Parmotrema mellissii
Parmotrema michauxianum
Parmotrema perforatum
Parmotrema reticulatum
Parmotrema ultralucens
Parmotrema xanthinum
Parmotrema chinense
Pseudoparmelia caperata
Flavoparmelia baltimorensis
Ramalina americana
Usnea mutabilis
Usnea strigosa
Anaptychia palmatula
Physcia aipolia
Physcia americana
Heterodermia albicans
Heterodermia obscurata
Heterodermia speciosa
Pyxine caesiopruinosa
Physconia detersa
Broom Moss
Dicranum scoparium
Running-pine
Lycopodium flabelliforme
Common Grape Fern
Botrychium dissectum
Christmas Fern
Polystichum acrostichoides
Ebony Spleenwort
Asplenium platyneuron
Resurrection Fern
Polypodium polypodioides
Virginia Pine; Scrub Pine
Pinus virginiana
Eastern Red-cedar
Juniperus virginiana
Cane; Switch Cane
Arundinaria gigantea
Purple-top
Tridens flavus
Spike Grass
Uniola sessiliflora
River Oats
Uniola latifolia
Wild Rye Grass
Elymus virginicus
Paspalum (grass)
Paspalum floridanum
Fringed Greenbrier
Smilax bona-nox
Solomon's-seal
Polygonatum biflorum
False Aloe
Agave virginica
Mockernut Hickory
Carya tomentosa
Beech
Fagus grandifolia
White Oak
Quercus alba
Post Oak
Quercus stellata
Black Oak
Quercus velutina
Spanish Oak
Quercus falcata
Water Oak
Quercus nigra
Willow Oak
Quercus phellos
Winged Elm
Ulmus alata
Hackberry
Celtis occidentalis georgiana
Red Mulberry
Morus rubra
Tulip-poplar
Liriodendron tulipifera
PawPaw
Asimina sp.
Sweet-shrub; Carolina-allspice
Calycanthus floridus
Agrimony
Agrimonia parviflora
Ninebark
Physocarpus opulifolius
Red Chokeberry
Sorbus arbutifolia
Hawthorn
Crataegus sp.
Downy Serviceberry
Amelanchier arborea
Black Cherry; Wild Cherry
Prunus serotina
Mimosa; Silk-tree
Albizia julibrissin
Redbud; Judas-tree
Cercis canadensis
Partridge-pea
Cassia fasciculata
Pencil-flower
Stylosanthes biflora
Naked-flower Tick-trefoil
Desmodium nudiflorum
Flowering Spurge
Euphorbia corollata
Poison-ivy
Rhus radicans
American Holly
Ilex opaca
Hearts-a'burstin'-with-love
Euonymus americanus
Red Maple
Acer rubrum
New-Jersey-tea
Ceanothus americanus
Virginia-creeper
Parthenocissus quinquefolia
Muscadine; Bullace
Vitis rotundifolia
St.-Andrew's-cross
Hypericum hypericoides
Russian-olive
Elaeagnus umbellata
Devil's-walking-stick
Aralia spinosa
Blackgum
Nyssa sylvatica
Flowering Dogwood
Cornus florida
Spotted Wintergreen
Chimaphila maculata
Pine-sap
Monotropa hypopithys
Carolina-laurel
Rhododendron minus
Sourwood
Oxydendrum arboreum
Sparkleberry
Vaccinium arboreum
Deerberry; Squaw-huckleberry
Vaccinium stamineum
Persimmon
Diospyros virginiana
American Ash; White Ash
Fraxinus americana
Yellow Jessamine
Gelsemium sempervirens
Gentian
Gentiana villosa
Skullcap
Scutellaria sp.
Carpenterweed
Prunella vulgaris
Gerardia
Agalinis sp.
Cross Vine
Anisostichus capreolata

Trumpet Vine
Campsis radicans
Ruellia
Ruellia caroliniensis
Arrowwood
Viburnum dentatum
Lobelia
Lobelia puberula
Common Ragweed
Ambrosia artemisiifolia
Rattlesnake-root
Prenanthes altissima
Thistle
Carduus sp.
Blazing-star
Liatris graminifolia
Thoroughwort
Eupatorium album
Pussy-toes
Antennaria plantaginifolia
Goldenrod
Solidago sp.
Hairy Wood Sunflower
Helianthus atrorubens
Sunflower
Helianthus sp.
Coreopsis
Coreopsis major
Beggar-ticks
Bidens frondosa
Purple Jelly-drops
Ascocoryne sarcoides
Carbon-balls (fungus)
Daldinia concentrica
Orange Jelly
Dacrymyces deliquescens
Jellied False-coral
Tremellodendron pallidum
Cinnabar-red Chanterelle
Cantharellus cinnabarinus
Crown-tipped Coral
Clavicorona pyxidata
Ochre Spreading Tooth
Steccherinum ochraceum
Turkey-tail
Trametes versicolor
Violet-toothed Polypore
Trichaptum biformis
Rag-veil Amanita
Amanita cinereopannosa
Grisette
Amanita vaginata
Frost's Bolete
Boletus frostii
Yellow-cracked Bolete
Boletus subtomentosus
Violet-gray Bolete
Tylopilus plumbeoviolaceus
Peppery Milky
Lactarius piperatus
Orange Pinwheel Marasmius
Marasmius siccus
Rounded Earthstar
Geastrum saccatum
Dog Stinkhorn
Mutinus caninus
Pigskin Poison Puffball
Scleroderma citrinum

FAUNA

Arrow-shaped Spiny-back
Micrathena sagittata
White Micrathena (spider)
Micrathena mitrata
Marbled Orb-weaver
Araneus marmoreus
Arrowhead Spider
Verrucosa arenata
Chinese Mantid (egg)
Tenodera aridifolia sinensis
Common Green Darner
Anax junius
Dingy Ground Beetle
Harpalus pennsylvanicus
Darkling Beetle
Uloma imberbis
Hickory Girdler Beetle
Oncideres cingulata
White-fringed Beetle
Graphognathus sp.
Flower Weevil
Glyptobaris lecontei
Pseudobaris nigrina
Odontocorynus pinguescens
Pearl Crescent
Phyciodes tharos
Gemmed Satyr (butterfly)
Cyllopsis gemma
Common Wood-nymph (butterfly)
Cercyonis pegala
Woolly Bear Caterpillar (larva)
Pyrrharctia isabella
Fall Webworm Moth
Hyphantria cunea

(Photo by L. Greenlee)
Betty Gilbert takes a break.

(Photo by V. Winn)
Carbon-balls (fungus), *Daldinia concentrica.*

(Photo by L. Greenlee)
Hearts-a'burstin'-with-love, *Euonymus americanus.*

Walnut Grove Plantation
Spartanburg County
June 24, 1989

Walnut Grove is a restored plantation, based on a house built in 1765. Authentic outbuildings such as workshops and a school, and period gardens have also been restored. There is a nature trail to give a glimpse of the natural materials available in the eighteenth century. Some of the forested area is recent transition material, but some pockets of very old forest remain. SCAN was only in the area about an hour, but it proved to be an interesting site.

Directions to Walnut Grove Plantation: From I-26 south of Spartanburg, take Exit 28 and go north on US 221 toward Spartanburg. At the very first road, turn right onto County Road 196 south. After 1.0 mile it makes a right turn; continue for another half mile. Where road 196 then makes a left turn, go straight ahead onto a gravel road leading to Walnut Grove Plantation.

(Photo by V. Winn)
Running-pine, *Lycopodium flabelliforme.*

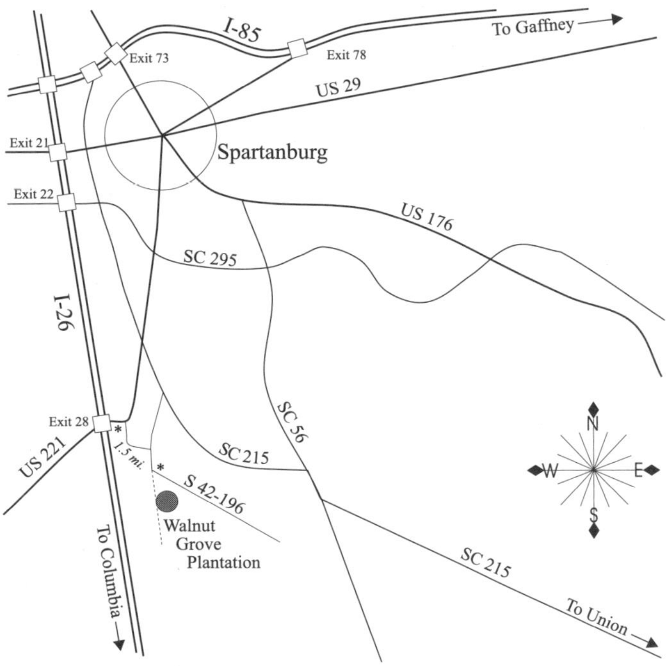

Walnut Grove Plantation (continued)
June 24, 1989

FLORA

Lichen
Cladonia sp.
Running-pine
Lycopodium flabelliforme
Rattlesnake Grape Fern
Botrychium virginianum
Christmas Fern
Polystichum acrostichoides
Ebony Spleenwort
Asplenium platyneuron
Microstegium (grass)
Microstegium vimineum
Crane-fly Orchid
Tipularia discolor
Black Walnut
Juglans nigra
Beech
Fagus grandifolia
White Oak
Quercus alba
Black Jack Oak
Quercus marilandica
Winged Elm
Ulmus alata
Red Mulberry
Morus rubra
Pokeweed; Pokeberry
Phytolacca americana
Tulip-poplar
Liriodendron tulipifera
Cucumber-tree
Magnolia acuminata
Poor-man's-pepper
Lepidium virginicum
Climbing-hydrangea
Decumaria barbara
Sweet-gum
Liquidambar styraciflua
Five-fingers
Potentilla canadensis
Chickasaw Plum; Wild Plum
Prunus angustifolia
Black Cherry; Wild Cherry
Prunus serotina
Spurge
Euphorbia sp.
American Holly
Ilex opaca
Hearts-a'burstin'-with-love
Euonymus americanus
Boxelder
Acer negundo
Red Maple
Acer rubrum
Virginia-creeper
Parthenocissus quinquefolia
Muscadine; Bullace
Vitis rotundifolia
Yellow Passionflower
Passiflora lutea
Fennel
Foeniculum vulgare
Flowering Dogwood
Cornus florida
Spotted Wintergreen
Chimaphila maculata
Sourwood
Oxydendrum arboreum
Sparkleberry
Vaccinium arboreum
Elliott's Blueberry
Vaccinium elliottii
Periwinkle
Vinca minor
Butterfly-weed
Asclepias tuberosa
Asclepias tuberosa tuberosa
Lyre-leaved Sage
Salvia lyrata
Ruellia
Ruellia purshiana
Ruellia caroliniensis
Plantain
Plantago sp.
Bluets
Houstonia purpurea
Cleavers
Galium aparine
Japanese Honeysuckle
Lonicera japonica
Prenanthes
Prenanthes sp.
Wild Lettuce
Lactuca sp.
Elephant's-foot
Elephantopus carolinianus
Daisy Fleabane
Erigeron philadelphicus

FAUNA

Bush Katydid
Scudderia sp.
Carpenter Bee
Xylocopa virginica
Ebony Jewelwing
Calopteryx maculata
Green Tiger Beetle
Cicindela sexguttata
Dingy Ground Beetle
Anisodactylus rusticus
Rose Chafer
Macrodactylus angustatus
Flower Scarab Beetle
Trigonopeltastes sp.
Soldier Beetle
Chauliognathus marginatus
Narrow-waisted Bark Beetle
Mycterus scaber
False Darkling Beetle
Canifa pallipes
False Blister Beetle
Asclera sp.
Seven-spotted Lady Beetle
Coccinella septempunctata
White-fringed Beetle
Graphognathus leucoloma
Palamedes Swallowtail
Papilio palamedes
Variegated Fritillary
Euptoieta claudia
American Lady (butterfly)
Vanessa virginiensis
Red-spotted Purple (butterfly)
Limenitis arthemis astyanax
Fowler's Toad
Bufo woodhousei
American Crow
Corvus brachyrhynchos
Pine Warbler
Dendroica pinus

(Photo by L. Smith)
Ebony jewelwing, *Calopteryx maculata.*

(Photo by L. Greenlee)
Variegated fritillary, *Euptoieta claudia.*

(Photo by L. Smith)
Chickasaw plum or wild plum, *Prunus angustifolia.*

Sumter County

Milford Plantation
Sumter County
March 28, 1987

SCAN's visit to this antebellum plantation and home of an early South Carolina governor, John L. Manning, was blessed with a lovely spring day. The plantation lies just above the hills leading down to the floodplain of the Wateree River. There were extensive trails and field edges to explore. The visit was made possible by the permission of the owner, Mr. Emory Clark, and greatly enhanced by the background information and trail guidance provided by Mr. Louis Hall.

Prvate property. Not open to the public. No map or directions provided.

FLORA

Lichen
Cladonia sp.
Rattlesnake Grape Fern
Botrychium virginianum
Southern Lady Fern
Athyrium asplenioides
Log Fern
Dryopteris celsa
Christmas Fern
Polystichum acrostichoides
Netted Chain Fern
Woodwardia areolata
Ebony Spleenwort
Asplenium platyneuron
Resurrection Fern
Polypodium polypodioides
Loblolly Pine
Pinus taeda
Deodar Cedar
Cedrus deodara
Bald-cypress
Taxodium distichum
Japanese Cryptomeria; Japanese Cedar
Cryptomeria japonica
Dwarf Palmetto
Sabal minor
Spanish-moss
Tillandsia usneoides
Spring Coral-root
Corallorhiza wisteriana
American Hornbeam; Ironwood
Carpinus caroliniana
Beech
Fagus grandifolia
Post Oak
Quercus stellata
Rock Chestnut Oak
Quercus prinus
Cherrybark Oak
Quercus falcata pagodaefolia
Water Oak
Quercus nigra
Willow Oak
Quercus phellos
Heartleaf
Hexastylis arifolia
Wild Pink; Catchfly
Silene caroliniana
Windflower
Thalictrum thalictroides
Mayapple
Podophyllum peltatum
Bloodroot
Sanguinaria canadensis
Hairy Bitter Cress
Cardamine hirsuta
Climbing-hydrangea
Decumaria barbara
Sweet-gum
Liquidambar styraciflua
Blackberry; Brambles
Rubus sp.
Hawthorn
Crataegus sp.
Hog Plum
Prunus umbellata
Redbud; Judas-tree
Cercis canadensis
Poison-ivy
Rhus radicans
American Holly
Ilex opaca
Yaupon
Ilex vomitoria
Hearts-a'burstin'-with-love
Euonymus americanus
Red Buckeye
Aesculus pavia
Violet
Viola affinis
Flowering Dogwood
Cornus florida
Dog-hobble
Leucothoe axillaris
Periwinkle
Vinca minor
Toadflax
Linaria canadensis
Cross Vine
Anisostichus capreolata
Partridge-berry
Mitchella repens
Small Bluets
Houstonia pusilla
Japanese Honeysuckle
Lonicera japonica
American Elderberry
Sambucus canadensis
Turkey-tail
Trametes versicolor
False Turkey-tail
Stereum ostrea

FAUNA

Passalus; Bess-beetle
Popilius disjunctus
Yellow-marked Buprestid Beetle
Acmaeodera tubulus
Cucumber Beetle
Diabrotica sp.
Pale-striped Flea Beetle
Systena blanda
Eastern Tiger Swallowtail
Papilio glaucus
Cabbage White (butterfly)
Pieris rapae
Falcate Orangetip
Anthocharis midea
Cloudless Sulphur
Phoebis sennae eubule
Sleepy Orange (butterfly)
Eurema nicippe
Duskywing Skipper
Erynnis sp.
Polyphemus Moth (cocoon)
Antheraea polyphemus
Southern Dusky Salamander
Desmognathus auriculatus
Eastern Mud Turtle
Kinosternon subrubrum
Carolina Anole
Anolis carolinensis
Five-lined Skink
Eumeces fasciatus
Ground Skink
Scincella lateralis
Brown Water Snake
Nerodia taxispilota
Downy Woodpecker
Picoides pubescens
Pileated Woodpecker
Dryocopus pileatus
Carolina Chickadee
Parus carolinensis
Carolina Wren
Thryothorus ludovicianus
Blue-gray Gnatcatcher
Polioptila caerulea
American Robin (nest)
Turdus migratorius
White-eyed Vireo
Vireo griseus
Red-Eyed Vireo
Vireo olivaceus
Northern Parula (and nest)
Parula americana
Yellow-throated Warbler
Dendroica dominica
Eastern Fox Squirrel
Sciurus niger
White-tailed Deer (tracks)
Odocoileus virginianus

(Photo by L. Greenlee)

SCAN members walk the trails of Milford Plantation.

Poinsett State Park
Sumter County
June 25, 1994

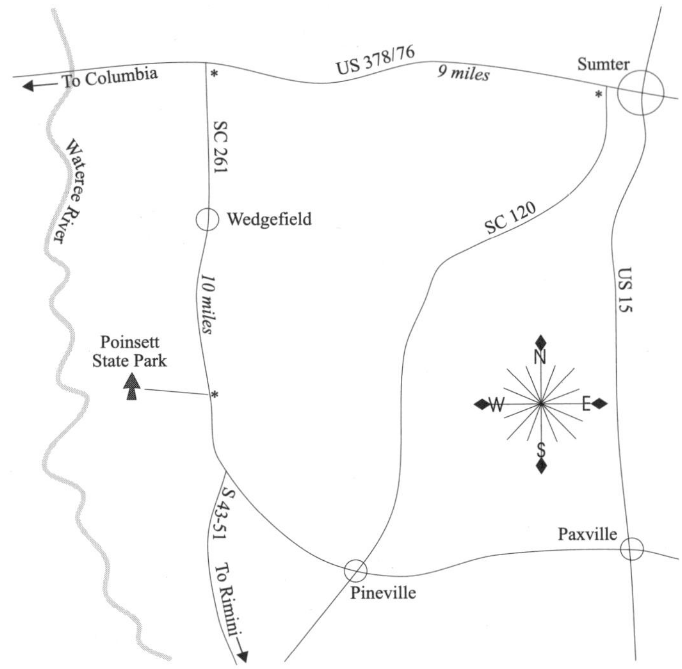

Poinsett State Park is named for the South Carolinian who was an ambassador to Mexico in the 1840's and while there discovered and brought back to the United States the plant we now call the poinsettia. This park is high on the bluff overlooking the floodplain of the Wateree River. The hilly terrain is the result of spring-fed creeks eroding deep valleys into these bluffs. This has created plant communities representative of sandhills, swamps, mountain bluffs and pine-hardwood forests. There is an old mill pond that harbors a large amount of aquatic life. Dragonflies abound, as is shown by the long list observed on this day. One uncommon sighting was the mocha emerald, a dragonfly whose adult tends to perch by hanging straight down vertically from a twig tip that is in the interior of a plant. It often takes sharp eyes to find them.

Directions to Poinsett State Park: From Sumter, go west on US 378/US 76. Note the junction with SC 120 on the left and continue travelling 9 miles to the junction with SC 261, also on the left. Turn left (south) onto SC 261 and continue through Wedgefield for 10 miles. At the sign for Poinsett State Park, turn right and follow additional signs to the park, about 2 miles.

FLORA

Lichen
Cladina arbuscula
Cladonia chlorophaea
Cladonia sobolescens
Cladonia cristatella
Cladonia leporina
Cladonia mateocyatha
Cladonia piedmontensis
Cladonia polycarpia
Cladonia polycarpoides
Cladonia uncialis
Cladonia peziziformis
Bulbothrix goebelii
Hypotrachyna livida
Parmelina horrescens
Parmelina minarum
Parmotrema cetratum
Parmotrema crinitum f. ultralucens
Parmotrema michauxianum
Parmotrema perforatum
Parmotrema rampoddense
Parmotrema ultralucens
Flavoparmelia caperata
Canoparmelia caroliniana
Punctelia bolliana
Punctelia rudecta
Rimelia reticulata
Rimelia cetrata
Rimelia subisidiosa
Usnea strigosa
Physcia aipolia
Physcia americana
Heterodermia spaltans sp. nov.
Pyxine caesiopruinosa
Physconia enteroxantha

Cinnamon Fern
Osmunda cinnamomea
Bracken
Pteridium aquilinum
Southern Lady Fern
Athyrium asplenioides
Christmas Fern
Polystichum acrostichoides
Netted Chain Fern
Woodwardia areolata
Ebony Spleenwort
Asplenium platyneuron
Resurrection Fern
Polypodium polypodioides
Loblolly Pine
Pinus taeda
Short-leaf Pine
Pinus echinata
Bald-cypress
Taxodium distichum
Bur-reed
Sparganium americanum
Cane; Switch Cane
Arundinaria gigantea
Cut Grass
Leersia sp.
Baggy-knees Grass
Sacciolepis striata
Nut-rush
Scleria triglomerata
Arrowleaf Arum
Peltandra virginica
Duckweed
Lemna sp.
Bogmoss
Mayaca fluviatilis
Spanish-moss
Tillandsia usneoides
Roseling; Spiderwort
Tradescantia rosea
Greenbrier; Bullbrier
Smilax rotundifolia
Fringed Greenbrier
Smilax bona-nox
Sawbrier; Greenbrier
Smilax glauca
Indian Cucumber-root
Medeola virginiana
Lizard's-tail
Saururus cernuus
Wax-myrtle; Southern Bayberry
Myrica cerifera
Mockernut Hickory
Carya tomentosa
Pignut Hickory
Carya glabra
Tag Alder
Alnus serrulata
Hop-hornbeam
Ostrya virginiana
American Hornbeam; Ironwood
Carpinus caroliniana
Beech
Fagus grandifolia
Chinquapin
Castanea sp.
White Oak
Quercus alba
Post Oak
Quercus stellata
Scrubby Post Oak
Quercus margaretta
Scarlet Oak
Quercus coccinea
Spanish Oak
Quercus falcata
Water Oak
Quercus nigra
Willow Oak
Quercus phellos
Laurel Oak
Quercus laurifolia
Winged Elm
Ulmus alata
Hackberry; Sugarberry
Celtis sp.
False Nettle
Boehmeria cylindrica
Heartleaf
Hexastylis arifolia
Yellowroot
Xanthorhiza simplicissima

Tulip-poplar
Liriodendron tulipifera
Sweetbay
Magnolia virginiana
Sweet-shrub; Carolina-allspice
Calycanthus floridus
Redbay
Persea borbonia
Sassafras
Sassafras albidum
Virginia-willow
Itea virginica
Climbing-hydrangea
Decumaria barbara
Sweet-gum
Liquidambar styraciflua
Witch-hazel
Hamamelis virginiana
Dwarf Hawthorn
Crataegus uniflora
Downy Serviceberry
Amelanchier arborea
Black Cherry; Wild Cherry
Prunus serotina
Carolina Cherrylaurel
Prunus caroliniana
Redbud; Judas-tree
Cercis canadensis
Zornia
Zornia bracteata
Naked-flower Tick-trefoil
Desmodium nudiflorum
Beggar-lice
Desmodium sp.
Sericea
Lespedeza cuneata
American Wisteria
Wisteria frutescens
Wild Indigo
Indigofera caroliniana
Hog-peanut
Amphicarpa bracteata
Tread-softly
Cnidoscolus stimulosus
Carolina Ipecac
Euphorbia ipecacuanhae
Flowering Spurge
Euphorbia corollata
Poison-ivy
Rhus radicans
Winged Sumac
Rhus copallina
American Holly
Ilex opaca
Sweet Gallberry
Ilex coriacea
Hearts-a'burstin'-with-love
Euonymus americanus
Red Maple
Acer rubrum
Spotted Touch-me-not
Impatiens capensis
Supple-jack; Rattan-vine
Berchemia scandens
Virginia-creeper
Parthenocissus quinquefolia
Muscadine; Bullace
Vitis rotundifolia
Summer Grape
Vitis aestivalis
Pepper-vine
Ampelopsis arborea
Silky Camellia
Stewartia malacodendron
St.-Andrew's-cross
Hypericum hypericoides
St.-John's-wort
Hypericum walteri
Hypericum
Hypericum sp.
Water Loosestrife
Decodon verticillatus
Devil's-walking-stick
Aralia spinosa
Snakeroot
Sanicula sp.
Black Tupelo; Blackgum
Nyssa sylvatica sylvatica
Swamp Tupelo; Swamp Blackgum
Nyssa sylvatica biflora
Flowering Dogwood
Cornus florida
Sweet Pepperbush
Clethra alnifolia
Swamp Honeysuckle
Rhododendron viscosum
Mountain-laurel
Kalmia latifolia
Fetterbush
Lyonia lucida
Dog-hobble
Leucothoe axillaris
Sourwood
Oxydendrum arboreum
Sparkleberry
Vaccinium arboreum
Elliott's Blueberry
Vaccinium elliottii
Low-bush Blueberry
Vaccinium tenellum
Persimmon
Diospyros virginiana
Horse-sugar; Sweetleaf
Symplocos tinctoria
American Ash; White Ash
Fraxinus americana
Yellow Jessamine
Gelsemium sempervirens
Climbing Dogbane
Trachelospermum difforme
Man-root
Ipomoea pandurata
Beautyberry
Callicarpa americana
Skullcap
Scutellaria sp.
Cross Vine
Anisostichus capreolata
Trumpet Vine
Campsis radicans
Ruellia
Ruellia caroliniensis
Button-bush
Cephalanthus occidentalis
Partridge-berry
Mitchella repens
Japanese Honeysuckle
Lonicera japonica
Possumhaw Viburnum
Viburnum nudum
Elephant's-foot
Elephantopus tomentosus
Eupatorium
Eupatorium sp.
Pussy-toes
Antennaria plantaginifolia
Sea-myrtle; Groundsel-tree
Baccharis halimifolia
Heterotheca
Heterotheca graminifolia
Variable Russula
Russula variata

FAUNA

Purse-web Spider
Sphodros bicolor
Orchard Spider
Leucauge venusta
Orb-weaver
Neoscona arabesca
Long-jawed Orb-weaver
Tetragnatha elongata
Carolina Locust
Dissosteira carolina
Wheel Bug
Arilus cristatus
Common Eastern Velvet-ant
Dasymutilla occidentalis occidentalis
Velvet-ant
Dasymutilla sp.
Mocha Emerald (dragonfly)
Somatochlora linearis
Halloween Pennant (dragonfly)
Celithemis eponina
Banded Pennant (dragonfly)
Celithemis fasciata
Eastern Pondhawk
Erythemis simplicicollis
Slaty Skimmer
Libellula incesta
Pond Skimmer
Libellula sp.
Eastern Amberwing (dragonfly)
Perithemis tenera
Black-mantled Glider
Tramea lacerata
Ebony Jewelwing
Calopteryx maculata
Pedunculate Ground Beetle
Pasimachus depressus
Coptotomus interrogatus
Whirligig Beetle
Dineutus robertsi
Gyrinus lugens
Hister-like Dung Beetle
Ateuchus histeroides
Dung Beetle
Onthophagus tuberculifrons
Flower Scarab Beetle
Trigonopeltastes delta
Tick Trefoil Leaf-miner
Pachyschelus laevigatus
Click Beetle
Limonius sp.
Net-winged Beetle
Caenia dimidiata
Tumbling Flower Beetle
Mordella atrata
Mordellistena unicolor
Spotless Lady Beetle
Cycloneda sanguinea
Flower Longhorn Beetle
Typocerus velutinus
Round-necked Longhorn Beetle
Elaphidion mucronatum
Warty Leaf Beetle
Chlamisus gibbosus
Flea Beetle
Disonycha abbreviata
Leaf-mining Beetle
Chalepus scapularis
Eastern Tiger Swallowtail
Papilio glaucus
Palamedes Swallowtail
Papilio palamedes
Fall Webworm Moth
Hyphantria cunea
Cricket Frog
Acris sp.
Green Frog; Bronze Frog
Rana clamitans
Carolina Anole
Anolis carolinensis
Eastern Fence Lizard
Sceloporus undulatus
Broadhead Skink
Eumeces laticeps
Skink
Eumeces sp.
Six-lined Racerunner
Cnemidophorus sexlineatus
Black Racer
Coluber constrictor
Little Blue Heron
Egretta caerulea
Yellow-billed Cuckoo
Coccyzus americanus
Red-bellied Woodpecker
Melanerpes carolinus
Downy Woodpecker
Picoides pubescens
Pileated Woodpecker
Dryocopus pileatus
Eastern Wood-pewee
Contopus virens
Eastern Phoebe
Sayornis phoebe
Acadian Flycatcher
Empidonax virescens
American Crow
Corvus brachyrhynchos
Fish Crow
Corvus ossifragus

Poinsett State Park (continued)
June 25, 1994

Tufted Titmouse
Parus bicolor
Carolina Chickadee
Parus carolinensis
Carolina Wren
Thryothorus ludovicianus
Wood Thrush
Hylocichla mustelina
Brown Thrasher
Toxostoma rufum
White-eyed Vireo
Vireo griseus
Yellow-throated Vireo
Vireo flavifrons
Red-Eyed Vireo
Vireo olivaceus
Northern Parula
Parula americana
Yellow-throated Warbler
Dendroica dominica
Northern Cardinal
Cardinalis cardinalis
Indigo Bunting
Passerina cyanea
Rufous-sided Towhee
Pipilo erythrophthalmus
Summer Tanager
Piranga rubra
Gray Squirrel
Sciurus carolinensis

Woods Bay State Park
Sumter, Clarendon and Florence Counties
November 18, 1995

Woods Bay is a Carolina Bay with water in it year round, shaded by an almost complete canopy of trees. Nearby, an old mill pond that is shallow and mostly open to the sun offers another body of water for aquatic species. The bay is accessible from a boardwalk that extends deep into its center. Usually canoes are available for a small fee to explore the interior more closely. The mill pond has a nature trail that goes all around it, most of it quite close to the water's edge. These two bodies of water, plus the characteristic bay sand rim, create a very diverse set of plant and animal communities. Reptiles, amphibians, and aquatic insects are readily found here. The plant material is diverse and typical of these kinds of habitats. On this day, an added treat was a glimpse of a golden eagle in a tree just outside the park boundaries.

Directions to Woods Bay State Park: From Sumter, take US 378 east about 20 miles to Turbeville. Just past the center of town, fork left (north) on US 301 toward Olanta. Watch carefully for County Road 152 in about 4.5 miles; turn left. Go about 2 miles to the park entrance on the left.

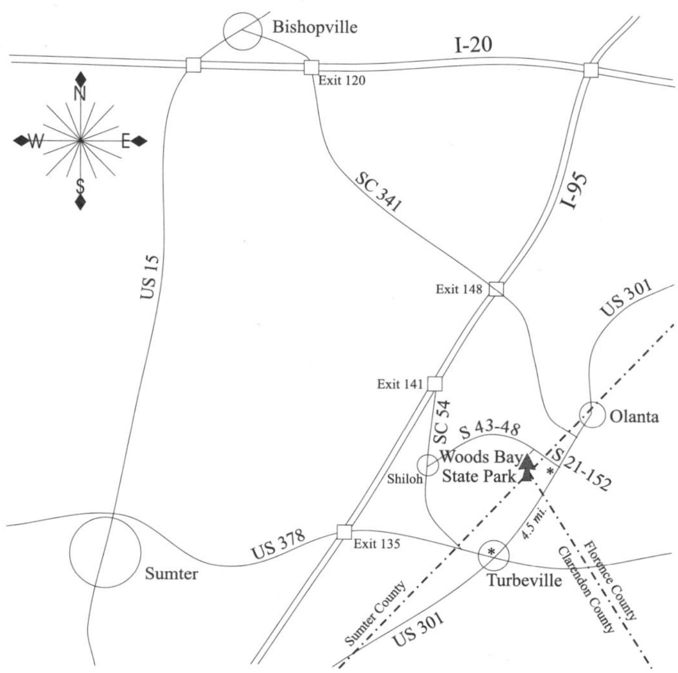

FLORA

Lichen
Leptogium cyanescens
Graphis scripta
Graphis elegans
Glyphis cicatricosa
Phaeographis dendritica
Bacidia polychroa
Bacidia schweinitzii
Bacidia laurocerasi
Lecidea russula
Cladina subtenuis
Cladonia cristatella
Cladonia didyma
Cladonia leporina
Cladonia polycarpoides
Pertusaria ostiolata
Pertusaria paratuberculifera
Pertusaria texana
Pertusaria tetrathalmia
Pertusaria sinusmexicani
Maronea constans
Haematomma puniceum
Haematomma elatinum
Lecanora cupressi
Lecanora pallida rubescens
Lecanora varia
Ochrolechia africana
Pachyospora sp
Phlyctis willeyi
Bulbothrix confoederata
Bulbothrix goebelii
Bulbothrix goebelii f. minor
Bulbothrix isidiza
Hypotrachyna livida
Parmelina horrescens
Parmelina minarum
Parmeliopsis subambigua
Parmotrema hypotropum
Parmotrema michauxianum
Parmotrema perforatum
Parmotrema tinctorum
Parmotrema xanthinum
Flavoparmelia caperata
Canoparmelia caroliniana
Punctelia rudecta
Rimelia reticulata
Rimelia subisidiosa
Usnea mutabilis
Usnea rubicunda
Usnea strigosa
Buellia punctata
Buellia curatellae
Buellia disciformis
Physcia americana
Heterodermia albicans
Heterodermia obscurata
Phaeophyscia ciliata
Phaeophyscia pusilloides
Phaeophyscia rubropulchra
Anthracothecium leucostomum
Pyrenula sp.
Trypethelium mastoideum
Trypethelium virens
Cinnamon Fern
Osmunda cinnamomea
Bracken
Pteridium aquilinum
Southern Lady Fern
Athyrium asplenioides

Virginia Chain Fern
Woodwardia virginica
Netted Chain Fern
Woodwardia areolata
Resurrection Fern
Polypodium polypodioides
Long-leaf Pine
Pinus palustris
Loblolly Pine
Pinus taeda
Pond Pine
Pinus serotina
Cane; Switch Cane
Arundinaria gigantea
Wire Grass
Aristida sp.
Smut Grass
Sporobolus poiretii
Dulichium (sedge)
Dulichium arundinaceum
Spanish-moss
Tillandsia usneoides
Red-berried Greenbrier
Smilax walteri
Bamboo Smilax
Smilax laurifolia
Sweet-scented Smilax
Smilax smallii
Bear-grass
Yucca filamentosa
Crane-fly Orchid
Tipularia discolor
Lizard's-tail
Saururus cernuus
Wax-myrtle; Southern Bayberry
Myrica cerifera
Swamp Bayberry
Myrica heterophylla
Mockernut Hickory
Carya tomentosa
Pignut Hickory
Carya glabra
White Oak
Quercus alba
Turkey Oak
Quercus laevis
Southern Red Oak; Spanish Oak
Quercus falcata falcata
Water Oak
Quercus nigra
Willow Oak
Quercus phellos
Heartleaf
Hexastylis arifolia
Jointweed
Polygonella polygama
Wire-plant
Stipulicida setacea
Sandwort; Longroot
Arenaria caroliniana
Water-lily
Nymphaea odorata
Tulip-poplar
Liriodendron tulipifera
Sweetbay
Magnolia virginiana
Redbay
Persea borbonia
Sassafras
Sassafras albidum
Virginia-willow
Itea virginica
Sweet-gum
Liquidambar styraciflua
American Sycamore
Platanus occidentalis
Shadbush; Serviceberry
Amelanchier sp.
Black Cherry; Wild Cherry
Prunus serotina
Beggar-lice
Desmodium sp.
Wood-sorrel
Oxalis sp.
Winged Sumac
Rhus copallina
Titi
Cyrilla racemiflora
American Holly
Ilex opaca
Inkberry
Ilex glabra
Hearts-a'burstin'-with-love
Euonymus americanus
Red Maple
Acer rubrum
Muscadine; Bullace
Vitis rotundifolia
Loblolly-bay
Gordonia lasianthus
St.-Andrew's-cross
Hypericum hypericoides
Prickly-pear
Opuntia compressa
Water Loosestrife
Decodon verticillatus
Devil's-walking-stick
Aralia spinosa
Swamp Tupelo; Swamp Blackgum
Nyssa sylvatica biflora
Flowering Dogwood
Cornus florida
Sweet Pepperbush
Clethra alnifolia
Spotted Wintergreen
Chimaphila maculata
Azalea
Rhododendron sp.
Fetterbush
Lyonia lucida
Coastal Honeybells
Leucothoe racemosa
Sourwood
Oxydendrum arboreum
Deerberry; Squaw-huckleberry
Vaccinium stamineum
Highbush Blueberry
Vaccinium corymbosum
Horse-sugar; Sweetleaf
Symplocos tinctoria
Yellow Jessamine
Gelsemium sempervirens
Blue-star
Amsonia ciliata
Ground-ivy; Ponyfoot
Dichondra carolinensis
Beautyberry
Callicarpa americana
Gerardia
Agalinis sp.
Cross Vine
Anisostichus capreolata
Trumpet Vine
Campsis radicans
Bladderwort
Utricularia sp.
Button-bush
Cephalanthus occidentalis
Partridge-berry
Mitchella repens
Possumhaw Viburnum
Viburnum nudum
American Elderberry
Sambucus canadensis
Common Dandelion
Taraxacum officinale
Elephant's-foot
Elephantopus sp.
Dog-fennel
Eupatorium capillifolium
Sea-myrtle; Groundsel-tree
Baccharis halimifolia
Aster
Aster sp.
Sweet Goldenrod
Solidago odora
Goldenrod
Solidago sp.
Heterotheca
Heterotheca graminifolia
Green-headed Jelly-club
Leotia viscosa
Bear's-head Tooth
Hericium coralloides
Mustard-yellow Polypore
Phellinus gilvus
Cort; Webcap
Cortinarius sp.
Peppery Milky
Lactarius piperatus
Emetic Russula
Russula emetica
Honey Mushroom
Armillariella mellea
Booted Collybia
Collybia peronata
Barometer Earthstar
Astraeus hygrometricus
Pigskin Poison Puffball
Scleroderma citrinum

FAUNA

Amphipod; Scud
Crangonyx serratus
Crab-shaped Spiny-back
Gasteracantha cancriformis
Six-spotted Fishing Spider
Dolomedes triton
Green Stink Bug
Acrosternum hilare
Ambush Bug
Phymata pennsylvanica
Water-strider
Gerris sp.
Backswimmer
Notonecta sp.
Florida Harvester Ant
Pogonomyrmex badius
Meadowfly; Meadowhawk
(dragonfly) *Sympetrum sp.*
Seven-spotted Lady Beetle
Coccinella septempunctata
Multicolored Asian Lady Beetle
Harmonia axyridis
Lady Beetle
Anatis sp.
Cloudless Sulphur
Phoebis sennae eubule
Polyphemus Moth (cocoon)
Antheraea polyphemus

(Photo by R. Emmitt)

Mating yellow-legged meadowhawks, *Sympetrum vicinum*.

Woods Bay State Park (continued)
November 18, 1995

Footpath Sallow (moth)
Metaxaglaea semitaria
Spring Peeper
Hyla crucifer
Carolina Anole
Anolis carolinensis
Anhinga; Snakebird
Anhinga anhinga
Turkey Vulture
Cathartes aura
Golden Eagle
Aquila chrysaetos
Red-tailed Hawk
Buteo jamaicensis
Wild Turkey (tracks)
Meleagris gallopavo
Red-bellied Woodpecker
Melanerpes carolinus
Northern Flicker
Colaptes auratus
Downy Woodpecker
Picoides pubescens
Yellow-bellied Sapsucker
Sphyrapicus varius
Pileated Woodpecker
Dryocopus pileatus
Eastern Phoebe
Sayornis phoebe
Tufted Titmouse
Parus bicolor
Brown Creeper
Certhia americana
Carolina Wren
Thryothorus ludovicianus
Ruby-crowned Kinglet
Regulus calendula
Gray Catbird
Dumetella carolinensis
Northern Mockingbird
Mimus polyglottos
Yellow-rumped Warbler
Dendroica coronata
Northern Cardinal
Cardinalis cardinalis
Rufous-sided Towhee
Pipilo erythrophthalmus
Common Grackle
Quiscalus quiscula
Gray Squirrel
Sciurus carolinensis
Raccoon (tracks)
Procyon lotor
Bobcat; Wildcat (tracks)
Felis rufus
White-tailed Deer (tracks)
Odocoileus virginianus

Union County

Tyger River at Sparks Creek
(Sumter National Forest)
Union County
February 23, 1991

February is always a surprising month for a natural history outing. The weather seems to have an unusually large effect on what one sees. This trip fit that description. The area visited was a wooded area and an open meadow in the mid-Piedmont. There was even a small waterfall on Sparks Creek. It was a moist, threatening day, although it never rained, creating one of the most memorable parts of the trip — the many green colors of lush mosses. Large trees, young beech trees with leaves still attached, and an old stream bed were other highlights. It proved to be a site good for practicing winter twig identification.

Directions to Tyger River at Sparks Creek: From I-26 between Columbia and Spartanburg, take Exit 44 and go east on SC 49 through Cross Anchor to Cross Keys, 8 miles. Where SC 49 makes a left turn, and Proctor's Salvage Company and a sign to Rose Hill are on the right, bear right onto Old Buncombe Road, SC 18. Go 1.9 miles to the Padgett Creek Baptist Church on the left. Turn left at the Sedalia Hunt Camp sign just past the church onto Bombing Range Road, SC 481. Go about 1.2 miles, turn right onto a dirt-gravel road, FS 320B, and drive to the end of the road. The trail follows the Tyger River upstream.

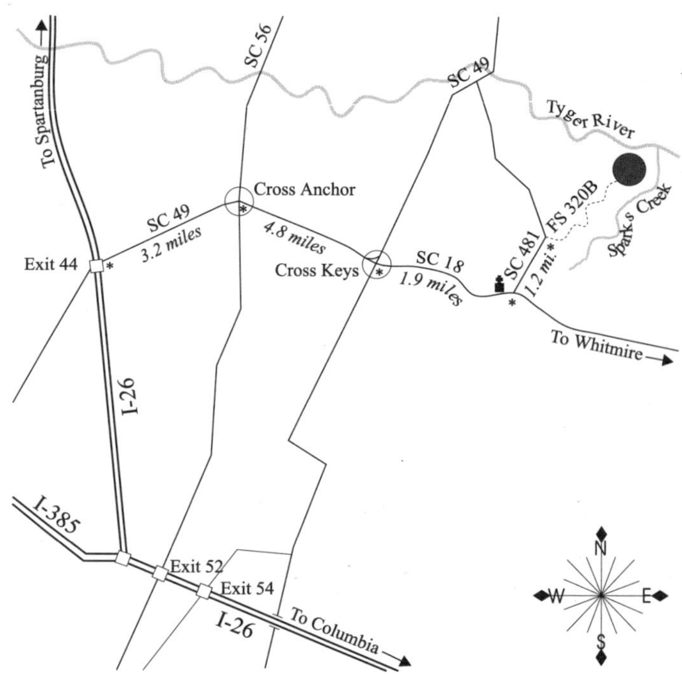

FLORA

Lichen
Leptogium microstictum
Lobaria ravenelii
Graphis scripta
Bacidia schweinitzii
Lecidea nylanderi
Cladina evansii
Cladonia capitata
Cladonia chlorophaea
Cladonia cristatella
Cladonia macilenta
Cladonia polycarpoides
Cladonia santensis
Cladonia humilis
Cladonia sp.
Pertusaria multipunctata
Pertusaria pustulata
Pertusaria paratuberculifera

Pertusaria macounii
Pertusaria xanthodes
Lecanora hageni
Lecanora cupressi
Lecanora varia
Ochrolechia africana
Candelaria concolor
Parmelia rudecta
Parmelina aurulenta
Parmelina dissecta
Parmotrema cetratum
Parmotrema hypotropum
Parmotrema perforatum
Parmotrema subtinctorium
Parmotrema tinctorum
Pseudoparmelia caperata
Pseudoparmelia crozalsiana
Ramalina americana
Usnea strigosa
Buellia disciformis
Physcia aipolia
Physcia millegrana
Physcia stellaris
Heterodermia speciosa
Phaeophyscia pusilloides
Phaeophyscia rubropulchra
Trypethelium tropicum
Trypethelium sp.
Opegrapha astraea
Lepraria membranacea
Christmas Fern
Polystichum acrostichoides
Ebony Spleenwort
Asplenium platyneuron
Resurrection Fern
Polypodium polypodioides
Loblolly Pine
Pinus taeda
Virginia Pine; Scrub Pine
Pinus virginiana
Eastern Red-cedar
Juniperus virginiana
Cane; Switch Cane
Arundinaria gigantea
River Oats
Uniola latifolia
Field Garlic
Allium vineale
Downy Rattlesnake-plantain
Goodyera pubescens
Crane-fly Orchid
Tipularia discolor
Eastern Cottonwood
Populus deltoides
Black Walnut
Juglans nigra
Bitternut Hickory
Carya cordiformis
Pignut Hickory
Carya glabra
Tag Alder
Alnus serrulata
River Birch
Betula nigra
American Hornbeam; Ironwood
Carpinus caroliniana
Beech
Fagus grandifolia
White Oak
Quercus alba
Red Oak
Quercus rubra
Spanish Oak
Quercus falcata
Water Oak
Quercus nigra
Willow Oak
Quercus phellos
Laurel Oak
Quercus laurifolia
American Elm
Ulmus americana
Winged Elm
Ulmus alata
Sugarberry; Hackberry
Celtis laevigata
Tulip-poplar
Liriodendron tulipifera
PawPaw
Asimina triloba
Sassafras
Sassafras albidum
Spicebush
Lindera benzoin
Mouse-ear Cress
Arabidopsis thaliana
Hairy Bitter Cress
Cardamine hirsuta
Sweet-gum
Liquidambar styraciflua
American Sycamore
Platanus occidentalis
Blackberry; Brambles
Rubus sp.
Avens
Geum sp.
Hawthorn
Crataegus sp.
Black Cherry; Wild Cherry
Prunus serotina
Poison-ivy
Rhus radicans
Winged Sumac
Rhus copallina
American Holly
Ilex opaca
Hearts-a'burstin'-with-love
Euonymus americanus
Boxelder
Acer negundo
Red Maple
Acer rubrum
Summer Grape
Vitis aestivalis
Violet
Viola sp.
Russian-olive
Elaeagnus umbellata
Queen Anne's-lace; Wild Carrot
Daucus carota
Flowering Dogwood
Cornus florida
Spotted Wintergreen
Chimaphila maculata
Sourwood
Oxydendrum arboreum
Sparkleberry
Vaccinium arboreum
Elliott's Blueberry
Vaccinium elliottii
Creeping-Charlie
Lysimachia nummularia
Ash
Fraxinus sp.
Chinese Privet
Ligustrum sinense
Yellow Jessamine
Gelsemium sempervirens
Woolly Mullein
Verbascum thapsus
Cross Vine
Anisostichus capreolata
Trumpet Vine
Campsis radicans
Bedstraw
Galium sp.
Japanese Honeysuckle
Lonicera japonica
Coral-berry
Symphoricarpos orbiculatus
Black-haw; Nanny-berry
Viburnum prunifolium
Viburnum
Viburnum sp.
American Elderberry
Sambucus canadensis
Black-knot
Apiosporina morbosa
Tree-ear
Auricularia auricula
Witches'-butter
Tremella mesenterica

FAUNA

White-lipped Forest Snail
Mesodon thyroidus
Carolina Mantid (egg)
Stagmomantis carolina
Chinese Mantid (egg)
Tenodera aridifolia sinensis
Blackberry Knot Gall (wasp)
Diastrophus nebulosus
Dogwood Club Gall (midge)
Lasioptera clazula
Hickory Girdler Beetle
Oncideres cingulata
Forest Tent Caterpillar (egg)
Malacosoma disstria
Polyphemus Moth (cocoon)
Antheraea polyphemus
Upland Chorus Frog (call)
Pseudacris triseriata
Turkey Vulture
Cathartes aura
American Crow
Corvus brachyrhynchos
Tufted Titmouse
Parus bicolor
Carolina Chickadee
Parus carolinensis
Pine Warbler
Dendroica pinus
White-tailed Deer
Odocoileus virginianus

(Photo by V. Winn)

White-lipped forest snail, *Mesodon thyroidus*.

York County

Broad River near Hickory Grove
York County
October 27, 1990

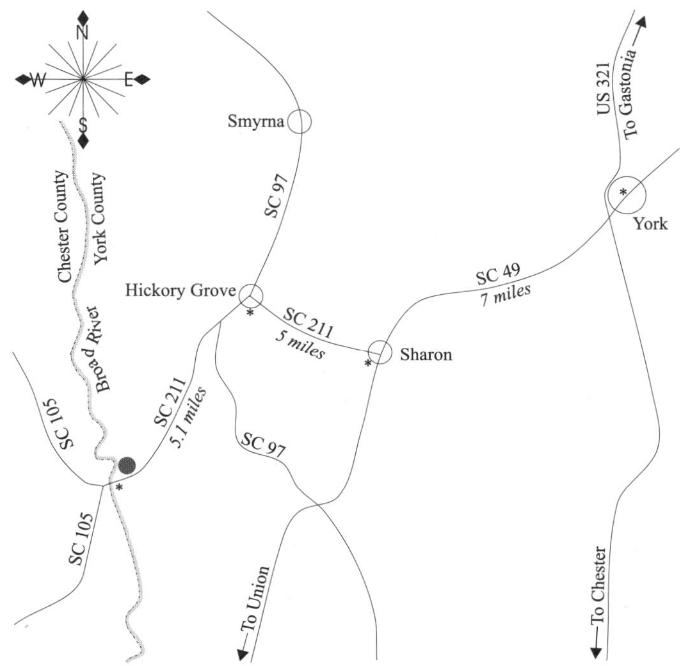

Floodplain and field edge habitats were explored on this privately-owned section of the Broad River. The species observed were not unexpected for such a Piedmont site, but certain individual trees have attained impressive size. One river birch measured nine feet, nine inches in diameter! At the time of our visit, this undammed section of the river was being considered for scenic river status; Happily it is now indeed part of the Broad River State Scenic River. Thanks to W.B. Wilkerson for sharing his property, and his perspective on this river.

Directions to Broad River near Hickory Grove: From York, drive south on SC 49 for 7 miles. Turn right (west) on SC 211 to Hickory Grove (5 miles), and then turn left in Hickory Grove, staying on SC 211. The bridge over the Broad River is in about 5.1 more miles. Park there and look for trails going north (upstream) on both sides of the river.

FLORA

Lichen
Cladonia capitata
Hypotrachyna livida
Parmotrema perforatum
Pseudoparmelia caroliniana
Ramalina americana
Usnea strigosa
Heterodermia hypoleuca
Running-pine
Lycopodium flabelliforme
Common Grape Fern
Botrychium dissectum
Christmas Fern
Polystichum acrostichoides
Ebony Spleenwort
Asplenium platyneuron
Eastern Red-cedar
Juniperus virginiana
Purple-top
Tridens flavus
River Oats
Uniola latifolia
Wild Rye Grass
Elymus virginicus
Microstegium (grass)
Microstegium vimineum
Fringed Greenbrier
Smilax bona-nox
Eastern Cottonwood
Populus deltoides
Black Walnut
Juglans nigra
Bitternut Hickory
Carya cordiformis
River Birch
Betula nigra
American Hornbeam; Ironwood
Carpinus caroliniana
Water Oak
Quercus nigra
Willow Oak
Quercus phellos
Slippery Elm
Ulmus rubra
Winged Elm
Ulmus alata
Sugarberry; Hackberry
Celtis laevigata
Hackberry
Celtis occidentalis georgiana
Red Mulberry
Morus rubra
Wood-nettle
Laportea canadensis
False Nettle
Boehmeria cylindrica
Clearweed
Pilea pumila
Smartweed; Knotweed
Polygonum setaceum
Pokeweed; Pokeberry
Phytolacca americana
Common Chickweed
Stellaria media
Coralbeads
Cocculus carolinus
Tulip-poplar
Liriodendron tulipifera
PawPaw
Asimina triloba
Spicebush
Lindera benzoin
Bitter Cress
Cardamine pensylvanica
Sweet-gum
Liquidambar styraciflua
American Sycamore
Platanus occidentalis
Cocklebur
Agrimonia sp.
Hawthorn
Crataegus sp.
Mimosa; Silk-tree
Albizia julibrissin
Honey Locust
Gleditsia triacanthos
Tree-of-heaven
Ailanthus altissima
Tarweed
Acalypha rhomboidea
Poison-ivy
Rhus radicans
Boxelder
Acer negundo
Yellow Passionflower
Passiflora lutea
Water-willow
Ludwigia decurrens
Ash
Fraxinus sp.
Chinese Privet
Ligustrum sinense
Privet; Ligustrum
Ligustrum sp.
Angle-pod
Matelea sp.
Glecoma
Glecoma hederacea
Cross Vine
Anisostichus capreolata

Trumpet Vine
Campsis radicans
Japanese Honeysuckle
Lonicera japonica
American Elderberry
Sambucus canadensis
Frost Aster
Aster pilosus
Verbesina
Verbesina alternifolia

FAUNA

Marbled Orb-weaver
Araneus marmoreus
Wolf Spider
Lycosa sp.
Carolina Locust
Dissosteira carolina
Slender Meadow Katydid
Conocephalus fasciatus
Eastern Boxelder Bug
Leptocoris trivittatus
Wheel Bug
Arilus cristatus
Toad Bug
Gelastocoris sp.
Hackberry Gall Psyllid (aphid)
Pachypsylla venusta
Poplar Petioleball Aphid
Pemphigus populitransversus
Common Eastern Velvet-ant
Dasymutilla occidentalis occidentalis
Yellowjacket
Vespula sp.
Paper Wasp
Polistes sp.
Pipe-organ Wasp
Trypoxylon politum
False Nettle Stem Gall (midge)
Cecidomyia boehmeriae
Water Scavenger Beetle
Berosus sp.
Large Rove Beetle
Staphylinus maxillosus
Flower Weevil
Pseudobaris nigrina
Pearl Crescent
Phyciodes tharos
Common Buckeye (butterfly)
Junonia coenia
Northern Cricket Frog
Acris crepitans
Cricket Frog
Acris sp.
Upland Chorus Frog (call)
Pseudacris triseriata
Great Blue Heron
Ardea herodias
Turkey Vulture
Cathartes aura
Black Vulture
Coragyps atratus
Red-tailed Hawk
Buteo jamaicensis
Northern Flicker
Colaptes auratus
Downy Woodpecker
Picoides pubescens
Yellow-bellied Sapsucker
Sphyrapicus varius
Blue Jay
Cyanocitta cristata
American Crow
Corvus brachyrhynchos
Carolina Chickadee
Parus carolinensis
Carolina Wren
Thryothorus ludovicianus
Ruby-crowned Kinglet
Regulus calendula
Northern Mockingbird
Mimus polyglottos
Cedar Waxwing
Bombycilla cedrorum
Yellow-rumped Warbler
Dendroica coronata
Northern Cardinal
Cardinalis cardinalis
Rufous-sided Towhee
Pipilo erythrophthalmus
Song Sparrow
Melospiza melodia
White-throated Sparrow
Zonotrichia albicollis
Rusty Blackbird
Euphagus carolinus
Virginia Opossum (tracks)
Didelphis virginiana
Gray Squirrel (tracks)
Sciurus carolinensis
Raccoon (tracks)
Procyon lotor
White-tailed Deer (tracks, bones)
Odocoileus virginianus
White-tailed Deer (remains)
Odocoileus virginianus

Kings Mountain State Park
York County
July 22, 1995

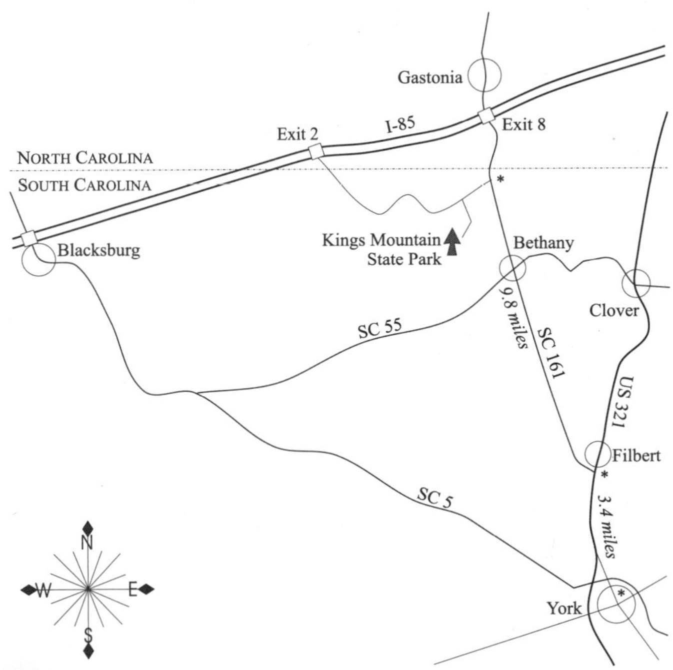

Kings Mountain is one of the several monadnocks that dot the sandhills and Piedmont of South Carolina. The park consists of over 6000 acres, plus it adjoins Kings Mountain National Military Park's 4000 acres. The trail followed on this trip started on a typical Piedmont ridge near the primitive campground, and then dropped to parallel Clark's Creek which offered a plant community more reminiscent of the mountains. One of the highlights was finding a population of big-leaved storax, *Styrax grandifolia*. It was a hot day which reduced the bird sightings but increased insect observations.

Directions to Kings Mountain State Park: From downtown York, drive north on US 321 for 3.4 miles. Fork left on SC 161 and continue north for 9.8 miles, passing through Bethany. Turn left at the brown state park sign into Kings Mountain State Park; if you come to the North Carolina state line, turn around and go back 0.2 miles.

(Photo by L. Smith)
Wild yam, *Dioscorea villosa*.

Kings Mountain State Park (continued)
July 22, 1995

FLORA

Lichen
Collema subflaccidum
Leptogium cyanescens
Sticta weigelii
Graphis scripta
Graphis elegans
Bacidia schweinitzii
Bacidia laurocerasi
Lecidea tessellata
Cladina subtenuis subtenuis
Cladonia chlorophaea
Cladonia chlorophaea f. intermedia
Cladonia sobolescens
Cladonia coniocraea
Cladonia coniocraea f. subpellucida
Cladonia cristatella cristatella
Cladonia cristatella vestita
Cladonia cylindrica
Cladonia didyma
Cladonia parasitica
Cladonia ramulosa
Pertusaria ostiolata
Pertusaria paratuberculifera
Pertusaria subpertusa
Pertusaria texana
Pertusaria macounii
Lecanora cupressi
Lecanora varia
Lecanora hybocarpa
Ochrolechia africana
Aspicilia laevata
Bulbothrix goebelii
Bulbothrix goebelii f. minor
Candelaria concolor
Hypotrachyna livida
Parmelina aurulenta
Parmelina horrescens
Parmelina horrescens asidiata
Parmelina obsessa
Parmelina minarum
Parmotrema hypotropum
Parmotrema michauxianum
Parmotrema ultralucens
Flavoparmelia baltimorensis
Flavoparmelia caperata
Canoparmelia caroliniana
Punctelia rudecta
Rimelia reticulata
Rimelia cetrata f. nonreticulata
Tuckermannopsis viridis
Ramalina americana
Usnea ceratina
Usnea mutabilis
Usnea rubicunda
Usnea strigosa
Anaptychia palmulata
Buellia stellulata
Buellia stigmaea
Buellia punctata
Buellia curatellae
Buellia sp.
Physcia aipolia
Physcia alba
Physcia americana
Physcia halei
Physcia millegrana
Physcia subtilis
Heterodermia albicans
Heterodermia obscurata
Phaeophyscia ciliata
Phaeophyscia imbricata
Phaeophyscia pusilloides
Phaeophyscia rubropulchra

Running-pine
Lycopodium flabelliforme
Cinnamon Fern
Osmunda cinnamomea
Royal Fern
Osmunda regalis spectabilis
Bracken
Pteridium aquilinum
Southern Lady Fern
Athyrium asplenioides
Christmas Fern
Polystichum acrostichoides
New York Fern
Thelypteris noveboracensis
Short-leaf Pine
Pinus echinata
Virginia Pine; Scrub Pine
Pinus virginiana
Eastern Red-cedar
Juniperus virginiana
Cane; Switch Cane
Arundinaria gigantea
Cut Grass
Leersia sp.
Microstegium (grass)
Microstegium vimineum
Greenbrier; Catbrier
Smilax sp.
False Solomon's Seal
Smilacina racemosa
Bellwort
Uvularia sessilifolia
Wild Yam
Dioscorea villosa
Downy Rattlesnake-plantain
Goodyera pubescens
Swamp Willow
Salix caroliniana
Eastern Cottonwood
Populus deltoides
Mockernut Hickory
Carya tomentosa
Pignut Hickory
Carya glabra
American Hornbeam; Ironwood
Carpinus caroliniana
Beech
Fagus grandifolia
American Chestnut
Castanea dentata
Allegheny Chinquapin
Castanea pumila pumila
White Oak
Quercus alba
Post Oak
Quercus stellata
Rock Chestnut Oak
Quercus prinus
Scarlet Oak
Quercus coccinea
Southern Red Oak; Spanish Oak
Quercus falcata falcata
Black Jack Oak
Quercus marilandica
False Nettle
Boehmeria cylindrica
Heartleaf
Hexastylis arifolia
Hexastylis sp.
Yellowroot
Xanthorhiza simplicissima
Tulip-poplar
Liriodendron tulipifera
Sassafras
Sassafras albidum
Sweet-gum
Liquidambar styraciflua
Red Chokeberry
Sorbus arbutifolia
Downy Serviceberry
Amelanchier arborea
Black Cherry; Wild Cherry
Prunus serotina
Black Locust
Robinia pseudo-acacia
Poison-ivy
Rhus radicans
American Holly
Ilex opaca
Hearts-a'burstin'-with-love
Euonymus americanus
Red Maple
Acer rubrum
Virginia-creeper
Parthenocissus quinquefolia
Muscadine; Bullace
Vitis rotundifolia
St.-Andrew's-cross
Hypericum hypericoides
Hypericum stragalum
Primrose-leaved Violet
Viola primulifolia
Black Tupelo; Blackgum
Nyssa sylvatica sylvatica
Flowering Dogwood
Cornus florida
Spotted Wintergreen
Chimaphila maculata
Azalea
Rhododendron sp.
Mountain-laurel
Kalmia latifolia
Dog-hobble
Leucothoe axillaris editorum
Sourwood
Oxydendrum arboreum
Wintergreen
Gaultheria procumbens
Huckleberry
Gaylussacia sp.
Highbush Blueberry
Vaccinium corymbosum
Persimmon
Diospyros virginiana
Horse-sugar; Sweetleaf
Symplocos tinctoria
Bigleaf Snowbell
Styrax grandifolia
White Ash; American Ash
Fraxinus americana americana
Fringe-tree
Chionanthus virginicus
False Foxglove
Aureolaria sp.
Beech-drops
Epifagus virginiana
Partridge-berry
Mitchella repens
Pussy-toes
Antennaria sp.
Sweet Goldenrod
Solidago odora
Rosinweed
Silphium compositum

FAUNA

Spiny-backed Spider
Micrathena gracilis
Orchard Spider
Leucauge venusta
Funnel-web Spider
Agelenopsis naevia
Green Lynx Spider
Peucetia viridans
Jumping Spider
Hentzia sp.
Round-headed Katydid
Amblycorypha floridana
Water-strider
Gerris sp.
Locust Treehopper
Thelia bimaculata
Velvet-ant
Dasymutilla sp.
Great Blue Skimmer
Libellula vibrans
Tiger Beetle
Cicindela unipunctata
Passalus; Bess-beetle
Popilius disjunctus
Click Beetle
Limonius sp.
Net-winged Beetle
Calopteron reticulatum
Darkling Beetle
Centronopus calcaratus
Alobates pennsylvanica
Sap-feeding Beetle
Stelidota sp.
Bean Leaf Beetle
Cerotoma trifurcata

Leaf-mining Beetle
Anoplitis inaequalis
Goldbug
Metriona bicolor
Eastern Tiger Swallowtail
Papilio glaucus
Spicebush Swallowtail
Papilio troilus
Spring Azure (butterfly)
Celastrina ladon
Red-spotted Purple (butterfly)
Limenitis arthemis astyanax
Silver-spotted Skipper
Epargyreus clarus
Tent Caterpillar
Malacasoma sp.
Eastern Box Turtle
Terrapene carolina
Five-lined or SE 5-lined Skink
Eumeces sp.

Yellow-billed Cuckoo
Coccyzus americanus
Acadian Flycatcher
Empidonax virescens
American Crow
Corvus brachyrhynchos
Carolina Chickadee
Parus carolinensis
White-breasted Nuthatch
Sitta carolinensis
Carolina Wren
Thryothorus ludovicianus
Solitary Vireo
Vireo solitarius
Red-Eyed Vireo
Vireo olivaceus
Gray Squirrel
Sciurus carolinensis
Armadillo (scute)
Chlamydotherium sp.

(Photo by R. Emmitt)

Red-spotted purple, *Limenitis arthemis astyanax*.

Georgia

Heggie's Rock
Columbia County, Georgia
March 28, 1992

Heggie's Rock, a 92-acre site, is called the best of the granitic flat rocks in eastern North America for the remarkable endemic flora restricted to these exposures. While it is similar to Flat Creek Heritage Preserve (Forty-acre Rock) in Lancaster County, SC, it is much less damaged by man. It has the best community zonation of all the granite outcrops. When the plant material in the pools are in bloom it is a lovely sight, and is known as "the rock that blushes". Our visit inspired the following poem from one member.

White clouds and blue sky
Rare plants, Sedum, Isoetes, Amphianthus, in flower
Nestled in a granitic bower.

Owned by the The Nature Conservancy of Georgia. Access by permission only. No map or directions provided.

FLORA

Lichen
Lecidea sp.
Cladina rangiferina
Cladonia caroliniana
Cladonia cristatella
Cladonia cylindrica
Cladonia piedmontensis
Cladonia robbinsii
Parmelia rudecta
Parmotrema perforatum
Parmotrema reticulatum
Parmotrema tinctorum
Pseudoparmelia baltimorensis
Pseudoparmelia caroliniana
Xanthoparmelia conspersa
Ramalina americana
Physcia aipolia
Twisted-hair Spikemoss
Selaginella tortipila
Quillwort
Isoetes melanopoda
Moonwort
Botrychium lunarioides
Hairy Lip Fern
Cheilanthes lanosa
Woolly Lip Fern
Cheilanthes tomentosa
Resurrection Fern
Polypodium polypodioides
Loblolly Pine
Pinus taeda
Short-leaf Pine
Pinus echinata
Eastern Red-cedar
Juniperus virginiana
Spiderwort
Tradescantia hirsuticaulis
Bamboo Smilax
Smilax laurifolia
Greenbrier; Catbrier
Smilax sp.
Bear-grass
Yucca filamentosa
Dog's-tooth-violet
Erythronium americanum
Bellwort
Uvularia sp.
False Garlic
Allium bivalve
Atamasco-lily
Zephyranthes atamasco
Dwarf Iris
Iris verna
Crane-fly Orchid
Tipularia discolor

Heggie's Rock (continued)
March 28, 1992

Wax-myrtle; Southern Bayberry
Myrica cerifera
Pignut Hickory
Carya glabra
Hickory
Carya sp.
Post Oak
Quercus stellata
Rock Chestnut Oak
Quercus prinus
Water Oak
Quercus nigra
Willow Oak
Quercus phellos
Winged Elm
Ulmus alata
Mistletoe
Phoradendron serotinum
Heartleaf
Hexastylis arifolia
Rock-portulaca
Talinum teretifolium
Rock Sandwort; Rockwort
Arenaria uniflora
Mountain Sandwort
Arenaria groenlandica glabra
Redbay
Persea borbonia
Draba
Draba aprica
Arabis
Arabis laevigata
Rock-cress
Arabis laevigata missouriensis
Stonecrop
Sedum pusillum
Vernal Pool Stonecrop
Sedum smallii
Lepuropetalon
Lepuropetalon spathulatum
Sweet-gum
Liquidambar styraciflua
Cinquefoil; Five-fingers
Potentilla sp.
Blackberry; Brambles
Rubus sp.
Black Cherry; Wild Cherry
Prunus serotina
Fragrant Sumac
Rhus aromatica
American Holly
Ilex opaca
Hearts-a'burstin'-with-love
Euonymus americanus
Red Maple
Acer rubrum
Supple-jack; Rattan-vine
Berchemia scandens
Buckthorn
Rhamnus caroliniana
Muscadine; Bullace
Vitis rotundifolia
St.-Andrew's-cross
Hypericum hypericoides
Hypericum
Hypericum sp.
Violet
Viola sp.
Devil-joint
Opuntia drummondii
Flowering Dogwood
Cornus florida
Spotted Wintergreen
Chimaphila maculata
Sparkleberry
Vaccinium arboreum
Elliott's Blueberry
Vaccinium elliottii
Blueberry
Vaccinium sp.
Persimmon
Diospyros virginiana
Forestiera
Forestiera ligustrina
Fringe-tree
Chionanthus virginicus
Yellow Jessamine
Gelsemium sempervirens
Phacelia
Phacelia dubia georgiana
Beautyberry
Callicarpa americana
Pool-sprite
Amphianthus pusillus
Toadflax
Linaria sp.
Cross Vine
Anisostichus capreolata
Small Bluets
Houstonia pusilla
Japanese Honeysuckle
Lonicera japonica
Dwarf Dandelion
Krigia virginica
Mule-ears; Bitterweed
Senecio tomentosus
Green-and-gold
Chrysogonum virginianum australe
Rust
Gymnosporangium sp.

FAUNA

Carolina Locust
Dissosteira carolina
American Bird Grasshopper
Schistocerca americana
Phantom Cranefly
Bittacomorpha clavipes
Predacious Diving Beetle
Hydroporus sp.
Agabus punctatus
Falcate Orangetip
Anthocharis midea
Olive Juniper Hairstreak
Callophrys gryneus gryneus
Common Buckeye (butterfly)
Junonia coenia
Gemmed Satyr (butterfly)
Cyllopsis gemma
Juvenal's Duskywing
Erynnis juvenalis
Eastern Tent Caterpillar
Malacosoma americanum
Northern Cricket Frog
Acris crepitans
Eastern Box Turtle (shell)
Terrapene carolina
Carolina Anole
Anolis carolinensis
Canada Goose
Branta canadensis
Turkey Vulture
Cathartes aura
Red-shouldered Hawk
Buteo lineatus
Barred Owl (sign)
Strix varia
Downy Woodpecker
Picoides pubescens
Carolina Chickadee
Parus carolinensis
Brown Thrasher
Toxostoma rufum
White-eyed Vireo
Vireo griseus
Yellow-rumped Warbler
Dendroica coronata
Northern Cardinal
Cardinalis cardinalis
White-tailed Deer (tracks, scat, remains)
Odocoileus virginianus

(Photo by A. Ciegler)

Mina Winn and Jan Ciegler study the vernal pool stonecrop, *Sedum smallii,* found only on monadnocks.

(Photo by L. Smith)

Atamasco-lily, *Zephyranthes atamasco.*

Black Rock Mountain State Park
Rabun County, Georgia
May 21, 1994

Perfect weather gave access to most of the best parts of this state park atop Black Rock Mountain in extreme northeast Georgia. On the north side of the mountain are some sheltered areas that have developed some uncommon plant and animal communities. The mountain species of birds nesting in the park greeted the early morning risers with a great serenade, and the nest building led to close looks at thrushes and scarlet tanagers that South Carolinians do not get to see frequently. Another unusual observation was a female giant ichneumon wasp, with her four-inch long ovipositor. She was hunting for signs of her prey, another kind of wasp known as a horntail. Wild turkeys were calmly exploring the area just off the trail. On Sunday the group explored the area around a small pond, slightly lower on the mountain than Saturday's destination. Beetles, dragonflies and butterflies were especially numerous.

Directions to Black Rock Mountain State Park: From Westminster, take US 76 northwest to Clayton, GA, about 26 miles. Turn right (north) on US 23/US 441, and go 4 miles through Clayton to Mountain City. Look for the sign to Black Rock Mountain State Park, turn left, and follow the winding mountain road about 3 miles to the top.

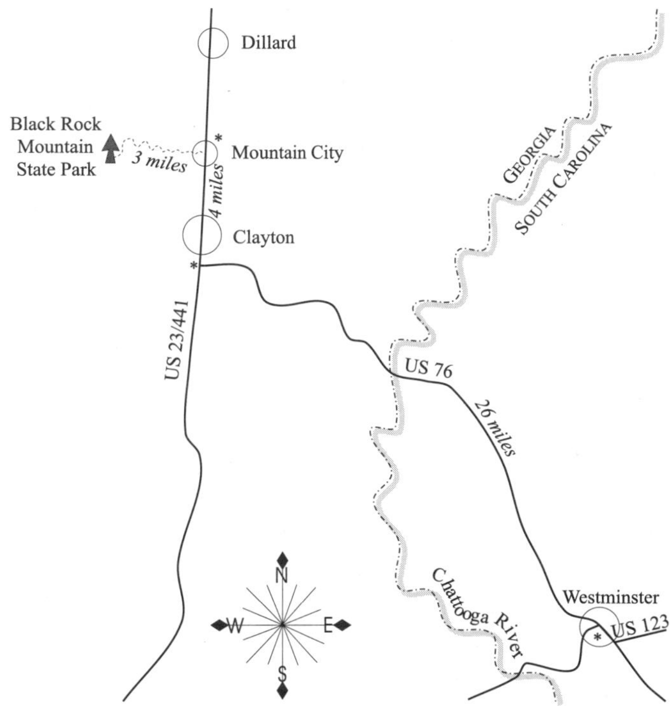

FLORA

Lichen
Collema subflaccidum
Leptogium cyanescens
Pannaria tavaresii
Pannaria sp.
Peltigera canina
Peltigera polydactyla
Sticta weigelii
Graphis scripta
Bacidia schweinitzii
Leptogium austroamericanum
Porpidia albocaerulescens
Baeomyces fungoides
Cladina rangiferina
Cladonia bacillaris
Cladonia chlorophaea
Cladonia sobolescens
Cladonia coniocraea
Cladonia cristatella
Cladonia didyma
Cladonia furcata
Cladonia polycarpoides
Umbilicaria mammulata
Pertusaria propinqua
Pertusaria texana
Lecanora cupressi
Lecanora pallida
Lecanora allophana
Ochrolechia africana
Ochrolechia mexicana
Tephromela atra
Bulbothrix goebelii
Candelaria concolor
Hypotrachyna formosana
Hypotrachyna livida
Parmelina minarum
Parmotrema austrosinense
Parmotrema crinitum
Parmotrema hypotropum
Parmotrema michauxianum
Parmotrema perforatum
Parmotrema stuppeum
Parmotrema tinctorum
Parmotrema ultralucens
Pseudevernia consocians
Flavoparmelia baltimorensis
Flavoparmelia caperata
Canoparmelia caroliniana
Punctelia rudecta
Melanelia halei
Rimelia reticulata
Tuckermannopsis ciliaris
Tuckermannopsis fendleri
Tuckermannopsis oakesiana
Ramalina americana
Usnea rubicunda
Usnea strigosa
Usnea trichodea
Buellia sp.
Physcia americana
Physcia halei
Physcia millegrana
Heterodermia albicans
Heterodermia obscurata
Heterodermia speciosa
Heterodermia leucomelos
Heterodermia casarettiana
Phaeophyscia ciliata
Phaeophyscia imbricata
Rinodina sp.
Physconia detersa
Physconia enteroxantha
Physciella cloantha
Xanthoria candelaria

Running-pine
Lycopodium flabelliforme
Rattlesnake Grape Fern
Botrychium virginianum
Cinnamon Fern
Osmunda cinnamomea
Hay-scented Fern
Dennstaedtia punctilobula
Bracken
Pteridium aquilinum
Southern Lady Fern
Athyrium asplenioides
Silvery Spleenwort
Athyrium thelypteroides
Marginal Shield Fern
Dryopteris marginalis
Christmas Fern
Polystichum acrostichoides
Broad Beech-Fern
Thelypteris hexagonoptera
New York Fern
Thelypteris noveboracensis
Rock Cap Fern
Polypodium virginianum
White Pine
Pinus strobus
Short-leaf Pine
Pinus echinata
Virginia Pine; Scrub Pine
Pinus virginiana
Eastern Hemlock
Tsuga canadensis
Eastern Red-cedar
Juniperus virginiana
Orchard Grass
Dactylis glomerata

Black Rock Mountain State Park (continued)
May 21, 1994

Jack-in-the-pulpit
Arisaema triphyllum
Spiderwort
Tradescantia virginiana
Woodrush
Luzula acuminata
Carrion-flower
Smilax herbacea
Sawbrier; Greenbrier
Smilax glauca
Little-sweet-Betsy
Trillium cuneatum
Wake-robin
Trillium erectum
Catesby's Trillium
Trillium catesbaei
Vasey's Trillium
Trillium vaseyi
Indian Cucumber-root
Medeola virginiana
Speckled Wood-lily
Clintonia umbellulata
False Solomon's Seal
Smilacina racemosa
Yellow Mandarin
Disporum lanuginosum
Solomon's-seal
Polygonatum biflorum
Fly-poison
Amianthium muscaetoxicum
False Hellebore
Veratrum parviflorum
Featherbells
Stenanthium gramineum
Bellwort
Uvularia perfoliata
Uvularia pudica
Uvularia sessilifolia
Wild Yam
Dioscorea villosa
Yellow Star-grass
Hypoxis hirsuta
Crested Iris
Iris cristata
Pink Lady's-slipper
Cypripedium acaule
Yellow Lady's-slipper
Cypripedium calceolus
Downy Rattlesnake-plantain
Goodyera pubescens
Black Walnut
Juglans nigra
Mockernut Hickory
Carya tomentosa
Pignut Hickory
Carya glabra
Hickory
Carya sp.
Hazel-nut
Corylus americana
Cherry Birch
Betula lenta

Beech
Fagus grandifolia
American Chestnut
Castanea dentata
White Oak
Quercus alba
Rock Chestnut Oak
Quercus prinus
Red Oak
Quercus rubra
Black Oak
Quercus velutina
Scarlet Oak
Quercus coccinea
Buffalo-nut
Pyrularia pubera
Giant Chickweed
Stellaria pubera
Fire Pink
Silene virginica
Yellowroot
Xanthorhiza simplicissima
Monkshood
Aconitum uncinatum
Baneberry
Actaea pachypoda
Black Cohosh
Cimicifuga racemosa
Early Meadow-rue
Thalictrum dioicum
Buttercup
Ranunculus hispidus
Blue Cohosh; Papooseroot
Caulophyllum thalictroides
Umbrella-leaf
Diphylleia cymosa
Mayapple
Podophyllum peltatum
Tulip-poplar
Liriodendron tulipifera
Umbrella-tree
Magnolia fraseri
Sweet-shrub; Carolina-allspice
Calycanthus floridus
Sassafras
Sassafras albidum
Spicebush
Lindera benzoin
Bloodroot
Sanguinaria canadensis
Wild Hydrangea
Hydrangea arborescens
False Goatsbeard
Astilbe biternata
Witch-hazel
Hamamelis virginiana
Wild Strawberry
Fragaria virginiana
Five-fingers
Potentilla canadensis
Potentilla simplex
Multiflora Rose
Rosa multiflora

Indian-physic
Gillenia trifoliata
Goat's-beard
Aruncus dioicus
Downy Serviceberry
Amelanchier arborea
Black Cherry; Wild Cherry
Prunus serotina
Naked-flower Tick-trefoil
Desmodium nudiflorum
Black Locust
Robinia pseudo-acacia
Wood Vetch
Vicia caroliniana
Hog-peanut
Amphicarpa bracteata
Wood-sorrel
Oxalis sp.
Poison-ivy
Rhus radicans
American Holly
Ilex opaca
Hearts-a'burstin'-with-love
Euonymus americanus
Striped Maple
Acer pensylvanicum
Red Maple
Acer rubrum
Yellow Buckeye
Aesculus octandra
Spotted Touch-me-not
Impatiens capensis
New-Jersey-tea
Ceanothus americanus
Virginia-creeper
Parthenocissus quinquefolia
Muscadine; Bullace
Vitis rotundifolia
Basswood; Linden
Tilia americana
Wood Violet
Viola palmata triloba
Round-leaved Yellow Violet
Viola rotundifolia
Halberd-leaved Violet
Viola hastata
Canada Violet
Viola canadensis
Spikenard
Aralia racemosa
Meadow Parsnip
Thaspium barbinode
Cow-bane
Oxypolis rigidior
Black Tupelo; Blackgum
Nyssa sylvatica sylvatica
Flowering Dogwood
Cornus florida
Alternate-leaved Dogwood
Cornus alternifolia
Mountain Sweet Pepperbush
Clethra acuminata
Spotted Wintergreen
Chimaphila maculata
Indian-pipe
Monotropa uniflora

Pine-sap
Monotropa hypopithys
Rosebay; Great Laurel
Rhododendron maximum
Purple Laurel
Rhododendron catawbiense
Mountain-laurel
Kalmia latifolia
Fetterbush
Leucothoe recurva
Sourwood
Oxydendrum arboreum
Trailing-arbutus
Epigaea repens
Huckleberry
Gaylussacia ursina
Deerberry; Squaw-huckleberry
Vaccinium stamineum
Low-bush Blueberry
Vaccinium vacillans
Highbush Blueberry
Vaccinium corymbosum
Whorled Loosestrife
Lysimachia quadrifolia
Horse-sugar; Sweetleaf
Symplocos tinctoria
Carolina Silverbell
Halesia carolina
Green Ash
Fraxinus pennsylvanica
Chinese Privet
Ligustrum sinense
Indian-hemp; Dogbane
Apocynum sp.
Poke Milkweed
Asclepias exaltata
Milkweed
Asclepias quadrifolia
Horse-balm
Collinsonia canadensis
False Foxglove
Aureolaria flava
Squaw-root
Conopholis americana
Bluets
Houstonia purpurea
Maple-leaf Viburnum
Viburnum acerifolium
Elderberry
Sambucus sp.
Rattlesnake-weed
Hieracium venosum
Pale Indian-plantain
Cacalia atriplicifolia
Small's Ragwort
Senecio smallii
Ragwort
Senecio aureus
Pussy-toes
Antennaria plantaginifolia
Robin's-plantain
Erigeron pulchellus
Rosinweed
Silphium compositum
Azalea Pinkster Apple
Exobasidium vaccinii

FAUNA

Giant Forest Snail
Mesodon normalis
Long-jawed Orb-weaver
Tetragnatha laboriosa
Yellow Crab Spider
Misumenoides formosipes
Linden Mite Gall Mite
Eriophyes abnormis
Sulfur-winged Grasshopper
Arphia sulphurea
Spur-throated Grasshopper
Melanoplus sp.
Witch Hazel Cone Gall (aphid)
Hormaphis hamamelidis
Pubescent Bract Hickory Gall
Phylloxera globosa
Reniform Midrib Carya Gall
Phylloxera caryaeren
Giant Ichneumon
Megarhyssa sp.
Spongy Oak-apple (gall wasp)
Amphibolips confluens
Nippled Oak-apple (gall wasp)
Amphibolips cookii
Chestnut Oak Midrib Tumor Gall
Plagiotrochus tumificus
Bald-faced Hornet
Vespula maculata
Maple Leaf Spot Gall (midge)
Cecidomyia ocellaris
Linden Wart Gall (midge)
Cecidomyia verrucicola
Chestnut Bud Gall (midge)
Rhopalomyia castaneae
Phantom Cranefly
Bittacomorpha clavipes
Robber Fly
Asilidae
Common Green Darner
Anax junius
Clubtail Dragonfly
Gomphus sp.
Banded Pennant (dragonfly)
Celithemis fasciata
Corporal Skimmer
Libellula exusta deplanata
Common Whitetail (dragonfly)
Libellula lydia
Blackwater Bluet (damselfly)
Enallagma weewa
Searcher; Ground Beetle
Calosoma calidum
Notch-mouthed Ground Beetle
Dicaelus dilitatus
Predacious Diving Beetle
Hydroporus niger
Water Scavenger Beetle
Tropisternus lateralis nimbatus
Enochrus ochraceus
Dung Beetle
Onthophagus orpheus orpheus
June Beetle
Phyllophaga grandis
Phyllophaga sp.
Chafer; Scarab Beetle
Dichelonyx elongata
Rhinoceros Beetle
Xyloryctes jamaicensis
Branch-and-leaf Buprestid
Agrilus politus
Metallic Wood-boring Beetle
Taphrocerus sp.
Striped Click Beetle
Ctenicera sp.
Soldier Beetle
Chauliognathus marginatus
Podabrus basilaris
Cantharis sp.
Malthodes curvatus
Checkered Beetle
Enoclerus thoracicus
Soft-winged Flower Beetle
Attalus sp.
Tumbling Flower Beetle
Mordella sp.
Darkling Beetle
Tarpela micans
Long-jointed Bark Beetle
Arthromacra aenea
False Darkling Beetle
Anaspis sp.
False Ant-like Flower Beetle
Pedilus collaris
Silken Fungus Beetle
Henoticus sp.
Lizard Beetle
Languria mozardi
Yellow-spotted Lady Beetle
Brachyacantha ursina
Lady Beetle
Coleomegilla fuscilabris
Seven-spotted Lady Beetle
Coccinella septempunctata
Black-spotted Lady Beetle
Anatis 15-punctata
Flower Longhorn Beetle
Leptura lineola
Anoplodera mutabilis
Anoplodera sp.
Gaurotes cyanipennis
Round-necked Longhorn Beetle
Tilloclytus geminata
Flat-faced Longhorn Beetle
Goes oculata
Stem Borer; Longhorn Beetle
Oberea basalis
Punctate Leaf Beetle
Orsodacne atra
Syneta sp.
Cylindrical Leaf Beetle
Pachybrachis sp.
Oval Leaf Beetle
Nodonota tristis
Striped Cucumber Beetle
Acalymna vittatum
Skeletonizing Leaf Beetle
Luperodes cyanellus
Kalmia Flea Beetle
Altica viridana
Eggplant Flea Beetle
Epithrix fuscula
Leaf-mining Beetle
Anoplitis inaequalis
Pear-shaped Weevil
Apion emaciipes
Apion sp.
Zebra Swallowtail
Eurytides marcellus
Eastern Tiger Swallowtail
Papilio glaucus
Spicebush Swallowtail
Papilio troilus
Pearl Crescent
Phyciodes tharos
Mourning Cloak
Nymphalis antiopa
Common Buckeye (butterfly)
Junonia coenia
Red-spotted Purple (butterfly)
Limenitis arthemis astyanax
Monarch (butterfly)
Danaus plexippus
Silver-spotted Skipper
Epargyreus clarus
Duskywing Skipper
Erynnis sp.
Forest Tent Caterpillar
Malacosoma disstria
Datana Moth Caterpillar
Datana sp.
Black & Yellow Lichen Moth
Lycomorpha pholus
Eight-spotted Forester Moth
Alypia octomaculata
Slimy Salamander
Plethodon glutinosus
Eastern Box Turtle
Terrapene carolina
Turkey Vulture
Cathartes aura
Wild Turkey
Meleagris gallopavo
Barred Owl
Strix varia
Northern Flicker
Colaptes auratus
Great Crested Flycatcher
Myiarchus crinitus
Eastern Wood-pewee
Contopus virens
Eastern Phoebe
Sayornis phoebe
Blue Jay
Cyanocitta cristata
American Crow
Corvus brachyrhynchos
Carolina Chickadee
Parus carolinensis
Wood Thrush
Hylocichla mustelina
Veery
Catharus fuscescens
American Robin
Turdus migratorius
Solitary Vireo
Vireo solitarius
Red-Eyed Vireo
Vireo olivaceus
Black-and-white Warbler
Mniotilta varia
Black-throated Blue Warbler
Dendroica caerulescens
Hooded Warbler
Wilsonia citrina
Ovenbird
Seiurus aurocapillus
Indigo Bunting
Passerina cyanea
Dark-eyed Junco
Junco hyemalis
Scarlet Tanager
Piranga olivacea
Eastern Cottontail
Sylvilagus floridanus

(Photo by L. Smith)

Giant ichneumon, *Megarhyssa sp.*

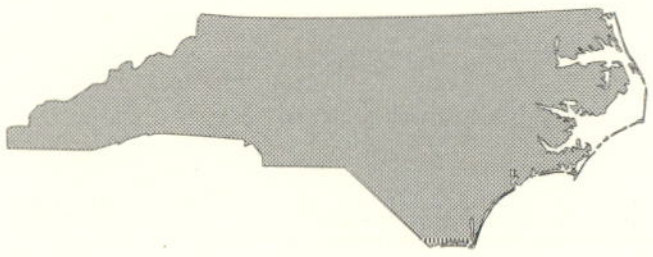

North Carolina

Texasgulf Lee Creek Phosphate Mine
Aurora, Beaufort County, North Carolina
October 15, 1988

This is a huge open pit mine operating in a large deposit of limestone, minerals and marine fossils. The array of fossils exposed is astonishing. The fossils are lying on the surface, waiting to be picked up. Some SCAN members with huge buckets and sacks tried to carry off all they saw, but were eminently unsuccessful, but very satisfied! Texasgulf controlled the access to the site, but provided the group with a bus and driver for the day! Such kind hosting was greatly appreciated.

Access by permission only

Directions to Texasgulf Lee Creek Phosphate Mine: From Florence, take I-95 north to North Carolina, passing Lumberton. At Exit 58 just past Fayetteville, turn east onto US 13 and drive 24 miles to Newton Grove. Pick up NC 55 east to Kinston, approximately 50 miles; bypass Kinston on US 70 and follow that east 31 miles to New Bern. Travel down Broad Street east and cross the Neuse River on NC 55. At the yellow flashing light in Grantsboro, 16 miles from the center of New Bern, turn left (north) onto NC 306. Follow NC 306 for 11 miles to the north, turn right for 4 miles east to Aurora, then turn left about 6 miles to the quarry entrance on the left.

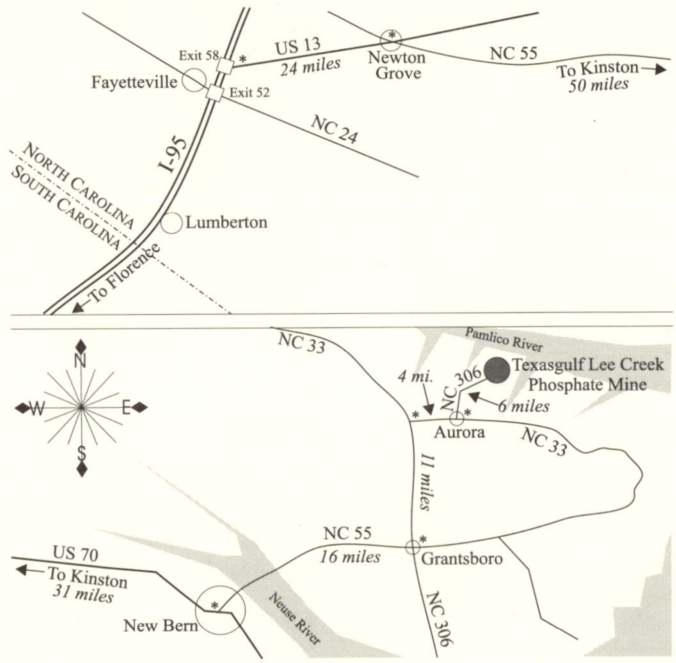

FOSSILS

Colonial Coral
Septastrea crassa
Scallop
Chesapecten jeffersonius jeffersonius
Chesapecten jeffersonius septenarius
Chesapecten madisonius
Argopecten eboreus
Placopecten clintonius
Marine Snail
Turritella sp.
Limpet
Crucibulum Lawrencei
Whelk
Busycon spiratus
Volute
Volutifusus typus
Southern Moon Snail
Polinices duplicatus
Olive
Oliva carolinensis
Cone Shell
Conus adversarius
Bittersweet Clam
Glycymeris americana
Ark
Noetia limula
Oyster
Crassostrea virginiana
Conradostrea lawrencei
Clam
Ensis directus
Mercenaria mercenaria
Chione
Chione grus
Broad-ribbed Cardita
Cardita arata
Ecphora
Ecphora quadricostata
Pterorhytis
Pterorhytis conradi
Philippi's Nutmeg Snail
Trigonostoma tenera
Lemintina
Lemintina sp.
Shark
Hemipristis serra
Shark (tooth)
Carcharodon megalodon

(Photo by L. Greenlee)

Hard hats were the order of the day.

Standing Indian Area
(Nantahala National Forest)
Macon County, North Carolina
August 27-28, 1988

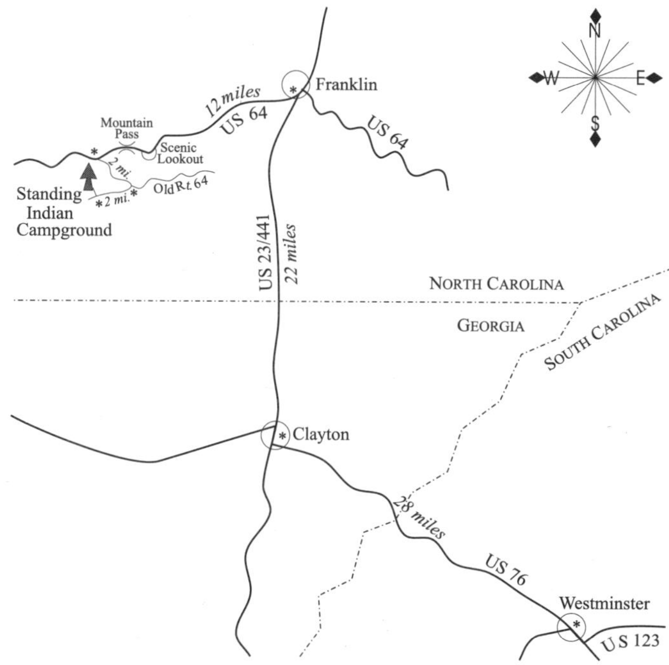

This overnight camping trip took SCAN members higher into the Appalachian Mountains than can be done within the borders of South Carolina. The Standing Indian area is beautiful, with a classic mountain flora and fauna. The area contains one of the largest tulip-poplars in the eastern U. S. as well as unique salamanders and other species. Plentiful rain gave a good crop of mushrooms. Habitats explored were wooded hillsides, an open meadow, mountain creek floodplain, and the rock bed of the creek.

Directions to Standing Indian Area: From Westminster, SC, take US 76 about 28 miles west to Clayton, Ga. Turn right (north) onto US 23/US 441, and drive approximately 22 miles to the outskirts of Franklin, N.C. Turn west onto US 64 and drive 12 miles: go past the scenic lookout, then over the mountain pass, and watch for a sign 1-2 miles past the mountain pass that says "Standing Indian Campground". Turn left onto Old Route 64, go 2 miles, turn right at another "Standing Indian" sign, and go 2 miles into the campground.

FLORA

Shining Clubmoss
Lycopodium lucidulum
Groundpine
Lycopodium obscurum
Rattlesnake Grape Fern
Botrychium virginianum
Cinnamon Fern
Osmunda cinnamomea
Maidenhair Fern
Adiantum pedatum
Southern Lady Fern
Athyrium asplenioides
Marginal Shield Fern
Dryopteris marginalis
Mountain Wood Fern
Dryopteris campyloptera
Christmas Fern
Polystichum acrostichoides
Broad Beech-Fern
Thelypteris hexagonoptera
New York Fern
Thelypteris noveboracensis
Sensitive Fern
Onoclea sensibilis
Rock Cap Fern
Polypodium virginianum
Eastern Hemlock
Tsuga canadensis
Green-dragon
Arisaema dracontium
Jack-in-the-pulpit
Arisaema triphyllum
Indian Cucumber-root
Medeola virginiana
Speckled Wood-lily
Clintonia umbellulata
False Solomon's Seal
Smilacina racemosa
False Lily-of-the-valley
Maianthemum canadense
Yellow Mandarin
Disporum lanuginosum
Solomon's-seal
Polygonatum biflorum
False Hellebore
Veratrum sp.
Turk's-cap Lily
Lilium superbum
Showy Orchis
Orchis spectabilis
Yellow Fringed Orchid
Habenaria ciliaris
Three Birds Orchid
Triphora trianthophora
Downy Rattlesnake-plantain
Goodyera pubescens
Tag Alder
Alnus serrulata
Cherry Birch
Betula lenta
Beech
Fagus grandifolia
American Chestnut
Castanea dentata
White Oak
Quercus alba
Rock Chestnut Oak
Quercus prinus
Red Oak
Quercus rubra
Clearweed
Pilea pumila
Dutchman's-pipe
Aristolochia macrophylla
Tearthumb
Polygonum sagittatum
Starry Campion
Silene stellata
Mountain Bugbane; Cohosh
Cimicifuga americana
Virgin's-bower
Clematis virginiana
Liverleaf
Hepatica acutiloba
Umbrella-leaf
Diphylleia cymosa
Tulip-poplar
Liriodendron tulipifera
Umbrella-tree
Magnolia fraseri
Sweet-shrub; Carolina-allspice
Calycanthus floridus
Sassafras
Sassafras albidum
Spicebush
Lindera benzoin
Bloodroot
Sanguinaria canadensis
Stonecrop
Sedum ternatum
Wild Hydrangea
Hydrangea arborescens
Grass-of-Parnassus
Parnassia asarifolia
False Goatsbeard
Astilbe biternata
Witch-hazel
Hamamelis virginiana
American Sycamore
Platanus occidentalis
Flowering Raspberry
Rubus odoratus
Cocklebur
Agrimonia gryposepala
Common Apple
Malus pumila
Downy Serviceberry
Amelanchier arborea
Black Cherry; Wild Cherry
Prunus serotina
Yellowwood
Cladrastis lutea
Bush-clover
Lespedeza intermedia
Black Locust
Robinia pseudo-acacia
Hog-peanut
Amphicarpa bracteata
Striped Maple
Acer pensylvanicum
Red Maple
Acer rubrum
Sugar Maple; Hard Maple
Acer saccharum

Standing Indian Area (continued)
August 27-28, 1988

Yellow Buckeye
Aesculus octandra
Jewel-weed; Touch-me-not
Impatiens pallida
Spotted Touch-me-not
Impatiens capensis
Basswood; Linden
Tilia sp.
Sweet White Violet
Viola blanda
Bronze Willow-herb
Epilobium coloratum
Wild Sarsaparilla
Aralia nudicaulis
Sweet Chervil
Osmorhiza longistylis
Angelica
Angelica sp.
Cow-bane
Oxypolis rigidior
Blackgum
Nyssa sylvatica
Flowering Dogwood
Cornus florida
Alternate-leaved Dogwood
Cornus alternifolia
Mountain Sweet Pepperbush
Clethra acuminata
Spotted Wintergreen
Chimaphila maculata
Indian-pipe
Monotropa uniflora
Rosebay; Great Laurel
Rhododendron maximum
Mountain-laurel
Kalmia latifolia
Dog-hobble
Leucothoe axillaris
Galax; Galaxy; Coltsfoot
Galax aphylla
Carolina Silverbell
Halesia carolina
Dodder
Cuscuta sp.
Phlox
Phlox sp.
Lop-seed
Phryma leptostachya
Carpenterweed
Prunella vulgaris
Bee-balm; Bergamot
Monarda didyma
Bee-balm
Monarda clinopodia
Lemon Mint
Monarda citriodora
Horse-balm
Collinsonia canadensis
Turtleheads
Chelone obliqua
Partridge-berry
Mitchella repens
Maple-leaf Viburnum
Viburnum acerifolium

American Elderberry
Sambucus canadensis
Indian-tobacco
Lobelia inflata
White Snakeroot
Eupatorium rugosum
Goldenrod
Solidago glomerata
Sneeze-weed
Helenium autumnale
Green Stain
Chlorociboria aeruginascens
Scaly Vase Chanterelle
Gomphus floccosus
White Worm Coral
Clavaria vermicularis
Violet-branched Coral
Clavulina amethystina
Flat-topped Coral
Clavariadelphus truncatus
Yellow-tipped Coral
Ramaria formosa
Sweet Tooth
Dentinum repandum
Violet-toothed Polypore
Trichaptum biformis
American Caesar's Mushroom
Amanita caesarea
Booted Amanita
Amanita cothurnata
Fly Agaric
Amanita muscaria
Destroying-angel
Amanita virosa
Shaggy-stalked Bolete
Boletellus betula
Old-man-of-the-woods
Strobilomyces floccopus
Chicken-fat Suillus
Suillus americanus
Silvery-violet Cort
Cortinarius alboviolaceus
Violet Cort
Cortinarius violaceus
Yellow Unicorn Entoloma
Entoloma murraii
Salmon Unicorn Entoloma
Entoloma salmoneum
Scarlet Waxy-cap
Hygrophorus coccineus
Buff Fishy Milky
Lactarius luteolus
Peck's Milky
Lactarius peckii
Emetic Russula
Russula emetica
Wood Clitocybe
Clitocybe ectypoides
Violet Collybia
Collybia iocephala
Zoned-cap Collybia
Crinipellis zonata
Common Laccaria
Laccaria laccata

Train-wrecker
Lentinus lepideus
Angel's-wings
Pleurocybella porrigens
Gem-studded Puffball
Lycoperdon perlatum
Splash-cups
Cyathus striatus
Pigskin Poison Puffball
Scleroderma citrinum
Stalked Puffball-in-aspic
Calostoma cinnabarina

FAUNA

Witch Hazel Cone Gall (aphid)
Hormaphis hamamelidis
Aphodiine Dung Beetle
Aphodius lividus
Asiatic Garden Beetle
Maladera castanea
Japanese Beetle
Popillia japonica
Toed-winged Beetle
Ptilodactyla serricollis
Soldier Beetle
Chauliognathus pennsylvanicus
Blister Beetle
Meloe sp.
Darkling Beetle
Tarpela micans
False Darkling Beetle
Anaspis sp.
Sap-feeding Beetle
Epuraea sp.
Pleasing Fungus Beetle
Triplax thoracica
Locust Borer; Longhorn Beetle
Megacyllene robiniae

Flea Beetle
Altica sp.
Eastern Tiger Swallowtail
Papilio glaucus
Spring Azure (butterfly)
Lycaenopsis argiolus
Great Spangled Fritillary
Speyeria cybele
Northern Dusky Salamander
Desmognathus fuscus
American Toad
Bufo americanus
Ruffed Grouse
Bonasa umbellus
Long-eared Owl
Asio otus
Barred Owl
Strix varia
Blue Jay
Cyanocitta cristata
Common Raven
Corvus corax
Tufted Titmouse
Parus bicolor
Carolina Chickadee
Parus carolinensis
Blue-gray Gnatcatcher
Polioptila caerulea
Dark-eyed Junco
Junco hyemalis
Eastern Chipmunk
Tamias striatus
Woodchuck; Groundhog
Marmota monax
Red Squirrel
Tamiasciurus hudsonicus
Red Fox
Vulpes vulpes
White-tailed Deer
Odocoileus virginianus

(Photo by L. Smith)
Yellow fringed orchid, *Habenaria ciliaris*.

Pearson's Falls and Big Bradley Falls
Polk County, North Carolina
May 12, 1990

This narrow valley leading to the lovely ninety-foot Pearson's Falls at the upper end contains a spectacular collection of spring ephemeral wildflowers. Almost all the mountain species are here, in large numbers. The access is gentle and paved. The trees are magnificent examples, too, when one remembers to take time to look up. Migrating birds are often found in high numbers at the height of the spring bloom. This property is privately owned by a local non-profit organization that collects a small fee and limits access to times when an attendant is on duty. The road into this sanctuary is breathtaking when the trillium are in bloom.

Directions to Pearson's Falls: From Spartanburg, take I-26 north into North Carolina. At the first exit after crossing the state line (Exit 36, Tryon/Columbus), turn onto NC 108 and head west into Tryon. Turn right onto US 176 toward Saluda and go a distance of 8.4 miles from the interstate. Turn left into a small street marked by a small sign for Pearson's Falls. Follow the road a mile or two to the entrance.

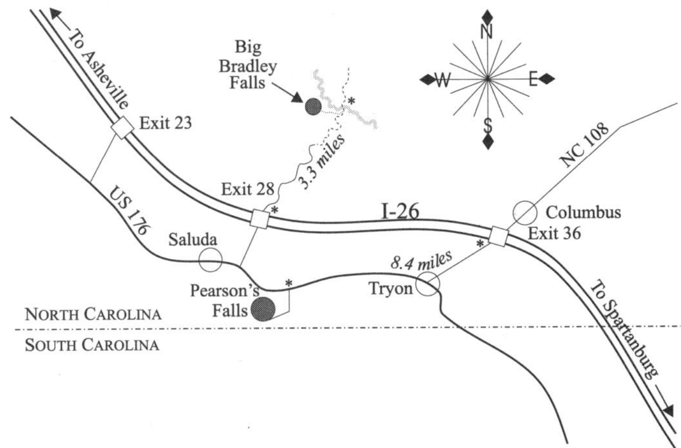

Big Bradley Falls is a high narrow falls dropping abruptly down its mountainside. The trail to the overlook is mostly old roadbed that crosses the creek at a flat but somewhat deep ford. The agile can rock-hop at low water levels. While the plant material and wildlife are typical of the mountain habitat, near the beginning of the trail one passes through an interesting high mountain meadow that looks like it ought to offer interesting birding earlier in the day that SCAN visited.

Big Bradley Falls is owned by the North Carolina Wildlife Resource Commission.

Directions to Big Bradley Falls: From Spartanburg, take I-26 north into North Carolina. At Exit 28, marked for Saluda, turn right (away from Saluda) onto Holbert Cove Road. The paved road winds past houses for 2.2 miles, then continues downhill on a narrow gravel road for an additional 1.1 mile. When the road levels out and crosses a creek, park on the right on either side of the creek. The trail leading to Big Bradley Falls begins just before the bridge at a yellow gate marked "Closed to Vehicles".

FLORA

Lichen
Coccocarpia erythroxyli
Lobaria quercizans
Cladonia polycarpoides
Hypogymnia krogiae
Hypotrachyna livida
Parmelina galbina
Parmelina minarum
Parmeliopsis subambigua
Parmotrema cetratum
Parmotrema michauxianum
Parmotrema perforatum
Parmotrema tinctorum
Pseudevernia consocians
Pseudoparmelia caperata
Pseudoparmelia caroliniana
Xanthoparmelia subramigera
Punctelia subrudecta
Imshaugia placorodia
Rimelia reticulata
Tuckermannopsis aurescens
Usnea strigosa
Physcia aipolia
Heterodermia granulifera
Heterodermia hypoleuca

Rattlesnake Grape Fern
Botrychium virginianum
Maidenhair Fern
Adiantum pedatum
Southern Lady Fern
Athyrium asplenioides
Silvery Spleenwort
Athyrium thelypteroides
Spreading Bladder Fern
Cystopteris protrusa
Christmas Fern
Polystichum acrostichoides
Broad Beech-Fern
Thelypteris hexagonoptera
New York Fern
Thelypteris noveboracensis
Sensitive Fern
Onoclea sensibilis
Walking Fern
Asplenium rhizophyllum
Ebony Spleenwort
Asplenium platyneuron
Rock Cap Fern
Polypodium virginianum
White Pine
Pinus strobus
Virginia Pine; Scrub Pine
Pinus virginiana
Eastern Hemlock
Tsuga canadensis
Eastern Red-cedar
Juniperus virginiana
Cane; Switch Cane
Arundinaria gigantea
Melic Grass
Melica mutica
Sedge
Carex retroflexa
Carex communis
Carex plantaginea
Carex digitalis
Carex laxiflora
Carex virescens
Jack-in-the-pulpit
Arisaema triphyllum
Spiderwort
Tradescantia virginiana
Carrion-flower
Smilax herbacea
Sawbrier; Greenbrier
Smilax glauca

Pearson's Falls and Big Bradley Falls
May 12, 1990

Little-sweet-Betsy
Trillium cuneatum
Wake-robin; Red Trillium
Trillium erectum vaseyi
Indian Cucumber-root
Medeola virginiana
Speckled Wood-lily
Clintonia umbellulata
False Solomon's Seal
Smilacina racemosa
Yellow Mandarin
Disporum lanuginosum
Solomon's-seal
Polygonatum biflorum
Blazing-star
Chamaelirium luteum
Wild Yam
Dioscorea villosa
Crested Iris
Iris cristata
Showy Orchis
Orchis spectabilis
Downy Rattlesnake-plantain
Goodyera pubescens
Crane-fly Orchid
Tipularia discolor
Puttyroot; Adam-and-Eve
Aplectrum hyemale
Butternut
Juglans cinerea
Shagbark Hickory
Carya ovata
Cherry Birch
Betula lenta
American Hornbeam; Ironwood
Carpinus caroliniana
Beech
Fagus grandifolia
White Oak
Quercus alba
Rock Chestnut Oak
Quercus prinus
Red Oak
Quercus rubra
Slippery Elm
Ulmus rubra
Wood-nettle
Laportea canadensis
Virginia Snakeroot
Aristolochia serpentaria
Wild Ginger
Asarum canadense
Jumpseed
Tovara virginiana
Giant Chickweed
Stellaria pubera
Yellowroot
Xanthorhiza simplicissima
Black Cohosh
Cimicifuga racemosa
Meadow-rue
Thalictrum clavatum
Hooked Buttercup
Ranunculus recurvatus
Liverleaf
Hepatica acutiloba
Blue Cohosh; Papooseroot
Caulophyllum thalictroides
Mayapple
Podophyllum peltatum
Tulip-poplar
Liriodendron tulipifera
Umbrella-tree
Magnolia fraseri
Cucumber-tree
Magnolia acuminata
Sweet-shrub; Carolina-allspice
Calycanthus floridus
Sassafras
Sassafras albidum
Spicebush
Lindera benzoin
Bloodroot
Sanguinaria canadensis
Bitter Cress
Cardamine clematitis
Mock-orange
Philadelphus inodorus
Wild Hydrangea
Hydrangea arborescens
False Goatsbeard
Astilbe biternata
Golden Saxifrage
Chrysosplenium americanum
Alumroot
Heuchera americana
Foamflower
Tiarella cordifolia
Saxifrage
Saxifraga micranthidifolia
Witch-hazel
Hamamelis virginiana
Wineberry
Rubus phoenicolasius
Indian-physic
Gillenia trifoliata
Downy Serviceberry
Amelanchier arborea
Black Cherry; Wild Cherry
Prunus serotina
Redbud; Judas-tree
Cercis canadensis
Wild Indigo
Baptisia tinctoria
Black Locust
Robinia pseudo-acacia
Hog-peanut
Amphicarpa bracteata
Poison-ivy
Rhus radicans
American Holly
Ilex opaca
Mountain Holly
Ilex ambigua montana
Hearts-a'burstin'-with-love
Euonymus americanus
Red Maple
Acer rubrum
Sugar Maple; Hard Maple
Acer saccharum
Yellow Buckeye
Aesculus octandra
Spotted Touch-me-not
Impatiens capensis
Virginia-creeper
Parthenocissus quinquefolia
Basswood; Linden
Tilia heterophylla
St.-Andrew's-cross
Hypericum hypericoides
Green Violet
Hybanthus concolor
Round-leaved Yellow Violet
Viola rotundifolia
Halberd-leaved Violet
Viola hastata
Canada Violet
Viola canadensis
Spikenard
Aralia racemosa
Sanicle; Snakeroot
Sanicula canadensis
Sweet Cicely; Anise-root
Osmorhiza claytonii
Meadow Parsnip
Zizia trifoliata
Thaspium trifoliatum
Blackgum
Nyssa sylvatica
Flowering Dogwood
Cornus florida
Alternate-leaved Dogwood
Cornus alternifolia
Spotted Wintergreen
Chimaphila maculata
Rosebay; Great Laurel
Rhododendron maximum
Carolina-laurel
Rhododendron minus
Flame Azalea
Rhododendron calendulaceum
Mountain-laurel
Kalmia latifolia

Dog-hobble
Leucothoe axillaris
Sourwood
Oxydendrum arboreum
Trailing-arbutus
Epigaea repens
Deerberry; Squaw-huckleberry
Vaccinium stamineum
Oconee-bells; Shortia
Shortia galacifolia
Persimmon
Diospyros virginiana
Carolina Silverbell
Halesia carolina
American Ash; White Ash
Fraxinus americana
Bindweed
Calystegia sericata
Waterleaf
Hydrophyllum canadense
Lyre-leaved Sage
Salvia lyrata
False Foxglove
Aureolaria sp.
Squaw-root
Conopholis americana
Partridge-berry
Mitchella repens
Bluets
Houstonia purpurea
Maple-leaf Viburnum
Viburnum acerifolium
American Elderberry
Sambucus canadensis
Venus'-looking-glass
Specularia perfoliata
Rattlesnake-root
Prenanthes altissima
Rattlesnake-weed
Hieracium venosum
Pussy-toes
Antennaria plantaginifolia
Robin's-plantain
Erigeron pulchellus
Bearsfoot
Polymnia uvedalia
Rosinweed
Silphium compositum
Coneflower
Rudbeckia laciniata
Verbesina
Verbesina sp.
Coreopsis
Coreopsis major
Ox-eye Daisy
Chrysanthemum leucanthemum

(Photo by L. Greenlee)
Red-eyed vireo nest, *Vireo olivaceus*.

FAUNA

Maple Spindle Gall Mite
Phyllocoptes aceris-crumena
Witch Hazel Cone Gall (aphid)
Hormaphis hamamelidis
Slippery Elm Pouch Gall(aphid)
Gobaishia ulmifusus
Ant-nest Scarab Beetle
Cremastocheilus castaneae
Bark-gnawing Beetle
Tenebroides dubia
Blue-black Blister Beetle
Meloe angusticollis
Cylindrical Bark Beetle
Bitoma sp.
Broad-bodied Leaf Beetle
Calligrapha rhoda
Broad-nosed Weevil
Aphrastus sp.
Eastern Tiger Swallowtail
Papilio glaucus
Pearl Crescent
Phyciodes tharos
Silver-spotted Skipper
Epargyreus clarus
Eastern Box Turtle
Terrapene carolina
Eastern Fence Lizard
Sceloporus undulatus
Ringneck Snake
Diadophis punctatus
Carolina Wren
Thryothorus ludovicianus
Wood Thrush
Hylocichla mustelina
Red-Eyed Vireo
Vireo olivaceus
Black-and-white Warbler
Mniotilta varia
Worm-eating Warbler
Helmitheros vermivorus
Northern Cardinal
Cardinalis cardinalis
Scarlet Tanager
Piranga olivacea

Charlie's Acres
Transylvania County, North Carolina
April 26, 1986, and April 25, 1992

For several years prior to his death, SCAN was blessed with the membership of Charles Moore of Brevard, N.C. He had been a wildflower enthusiast since childhood, and owned a major portion of a mountainside adjacent to Pisgah National Forest just at the outskirts of the City of Brevard. Mr. Moore was willing to share his personal paradise with groups of interested people. The area is very rich in spring wildflowers and other flora and fauna typical of the southern Appalachians. Mr. Moore also had a set of witticisms that he shared as he happily showed all the special plants his "Acres" sheltered. Some plants were specimens he had rescued from construction sites. A visit to this site was always rewarding.

No public access. No map provided. For a similar trip see the immediately adjacent area of Pisgah National Forest.

FLORA

Lichen
Leptogium burnetiae
Leptogium cyanescens
Leptogium laceroides
Pannaria tavaresii
Lobaria pulmonaria
Lobaria quercizans
Graphis scripta
Graphis scripta topographica
Graphis elegans
Xylographa disseminata
Bacidia schweinitzii
Bacidia sp.
Catillaria glauconigrans
Lecidea albocaerulescens
Lecidea congesta
Lecidea promiscens
Lecidea sp.
Rhizocarpon sp.
Stereocaulon sp.
Baeomyces fungoides
Cladonia chlorophaea
Cladonia coniocraea
Cladonia conista
Cladonia floerkeana
Cladonia furcata
Cladonia polycarpoides
Cladonia squamosa
Pertusaria ostiolata
Pertusaria plittiana
Pertusaria subpertusa
Pertusaria velata
Lecanora deplanans
Lecanora allophana argentata
Lecanora cupressi
Ochrolechia africana
Aspicilia laevata
Aspicilia cinerea
Cetraria ciliaris
Cetraria oakesiana
Cetrelia chicitae
Hypogymnia krogiae
Hypogymnia physodes
Hypogymnia sp.
Hypotrachyna revoluta
Menegrazzia terebrata
Parmelia appalachiensis
Parmelia halei
Parmelia squarrosa
Parmelia sp.
Parmelina dissecta
Parmeliopsis aleurites
Parmotrema crinitum
Parmotrema reticulatum
Parmotrema robustum
Parmotrema subisidiosum
Pseudevernia consocians
Pseudoparmelia caperata
Xanthoparmelia conspersa
Xanthoparmelia plittii
Melanelia exasperata
Ramalina americana
Usnea ceratina
Usnea strigosa
Usnea sp.
Anaptychia palmatula
Buellia parasema
Buellia punctata
Physcia millegrana
Heterodermia leucomelaena
Heterodermia speciosa
Heterodermia squamulosa
Phaeophyscia imbricata
Phaeophyscia pusilloides
Phaeophyscia rubropulchra
Rinodina milvina
Physconia enteroxantha
Shining Clubmoss
Lycopodium lucidulum
Groundpine
Lycopodium obscurum
Running-pine
Lycopodium flabelliforme
Rattlesnake Grape Fern
Botrychium virginianum
Cinnamon Fern
Osmunda cinnamomea
Bristle Fern
Trichomanes boschianum
Maidenhair Fern
Adiantum pedatum
Southern Lady Fern
Athyrium asplenioides
Christmas Fern
Polystichum acrostichoides
New York Fern
Thelypteris noveboracensis
Marsh Fern
Thelypteris palustris
Fern
Thelypteris sp.
Sensitive Fern
Onoclea sensibilis
White Pine
Pinus strobus
Pitch Pine
Pinus rigida
Norway Spruce
Picea abies
Eastern Hemlock
Tsuga canadensis
Carolina Hemlock
Tsuga caroliniana
Vanilla Grass
Anthoxanthum odoratum
Fraser's Sedge
Cymophyllus fraseri
Jack-in-the-pulpit
Arisaema triphyllum

Charlie's Acres (continued)
April 26, 1986, and April 25, 1992

Little-sweet-Betsy
Trillium cuneatum
Little-sweet-Betsy
Trillium cuneatum luteum
Trillium
Trillium discolor
Trillium lanceolatum
Nodding Trillium
Trillium cernuum
Wake-robin
Trillium erectum
Wake-robin; Red Trillium
Trillium erectum vaseyi
Catesby's Trillium
Trillium catesbaei
Great White Trillium
Trillium grandiflorum
Trillium
Trillium sp.
Indian Cucumber-root
Medeola virginiana
Solomon's-seal
Polygonatum biflorum
Polygonatum sp.
White Hellebore
Veratrum viride
False Hellebore
Veratrum sp.
Dog's-tooth-violet
Erythronium americanum
Bellwort
Uvularia perfoliata
Uvularia sessilifolia
Dwarf Iris
Iris verna
Crested Iris
Iris cristata
Showy Orchis
Orchis spectabilis
Downy Rattlesnake-plantain
Goodyera pubescens
Puttyroot; Adam-and-Eve
Aplectrum hyemale
Willow
Salix sp.
Black Walnut
Juglans nigra
Hickory
Carya sp.
Tag Alder
Alnus serrulata
Cherry Birch
Betula lenta
American Hornbeam; Ironwood
Carpinus caroliniana
Beech
Fagus grandifolia
Rock Chestnut Oak
Quercus prinus
Oak
Quercus sp.
Buffalo-nut
Pyrularia pubera
Heartleaf
Hexastylis arifolia
Common Chickweed
Stellaria media
Giant Chickweed
Stellaria pubera
Yellowroot
Xanthorhiza simplicissima
Small-flowered Crowfoot
Ranunculus abortivus
Hooked Buttercup
Ranunculus recurvatus
Buttercup
Ranunculus hispidus
Buttercup
Ranunculus sp.
Liverleaf
Hepatica acutiloba
Windflower
Anemone quinquefolia
Anemone sp.
Blue Cohosh; Papooseroot
Caulophyllum thalictroides
Umbrella-leaf
Diphylleia cymosa
Mayapple
Podophyllum peltatum
Tulip-poplar
Liriodendron tulipifera
Umbrella-tree
Magnolia fraseri
Sweet-shrub; Carolina-allspice
Calycanthus floridus
Sassafras
Sassafras albidum
Spicebush
Lindera benzoin
Bloodroot
Sanguinaria canadensis
Early Winter Cress
Barbarea verna
Toothwort
Cardamine diphylla
Toothwort; Dentaria
Cardamine sp.
Sweet Pitcher-plant
Sarracenia rubra
Frog's-britches
Sarracenia purpurea
Wild Hydrangea
Hydrangea arborescens
Hydrangea arborescens arborescens
Hydrangea arborescens discolor
Oak-leaf Hydrangea
Hydrangea quercifolia
Grass-of-Parnassus
Parnassia asarifolia
Foamflower
Tiarella cordifolia
Saxifrage
Saxifraga michauxii
Witch-hazel
Hamamelis virginiana
Witch-alder
Fothergilla major
American Sycamore
Platanus occidentalis
Cinquefoil; Five-fingers
Potentilla sp.
Five-fingers
Potentilla canadensis
Blackberry; Brambles
Rubus sp.
Rose
Rosa sp.
Shadbush; Serviceberry
Amelanchier sp.
Black Cherry; Wild Cherry
Prunus serotina
White Clover
Trifolium repens
Wild Geranium
Geranium maculatum
Fringed Polygala
Polygala paucifolia
Poison-ivy
Rhus radicans
American Holly
Ilex opaca
Hearts-a'burstin'-with-love
Euonymus americanus
Red Maple
Acer rubrum
Yellow Buckeye
Aesculus octandra
Buckeye
Aesculus sp.
Virginia-creeper
Parthenocissus quinquefolia
Basswood; Linden
Tilia sp.
Franklin-tree
Franklinia alatamaha
Hypericum
Hypericum sp.
Sweet White Violet
Viola blanda
Round-leaved Yellow Violet
Viola rotundifolia
Halberd-leaved Violet
Viola hastata
Canada Violet
Viola canadensis
Long-spurred Violet
Viola rostrata
Russian-olive
Elaeagnus umbellata
Common Evening-primrose
Oenothera biennis
Flowering Dogwood
Cornus florida
Alternate-leaved Dogwood
Cornus alternifolia
Silky Dogwood
Cornus amomum
Mountain Sweet Pepperbush
Clethra acuminata
Spotted Wintergreen
Chimaphila maculata
Indian-pipe
Monotropa uniflora
Rosebay; Great Laurel
Rhododendron maximum
Purple Laurel
Rhododendron catawbiense
Carolina-laurel
Rhododendron minus
Azalea
Rhododendron sp.
Mountain-laurel
Kalmia latifolia
Dog-hobble
Leucothoe axillaris
Leucothoe axillaris editorum
Sourwood
Oxydendrum arboreum
Huckleberry
Gaylussacia ursina
Oconee-bells; Shortia
Shortia galacifolia
Horse-sugar; Sweetleaf
Symplocos tinctoria
Carolina Silverbell
Halesia carolina
Creeping Phlox
Phlox stolonifera
Phacelia
Phacelia sp.
Veronica
Veronica officinalis
Lousewort
Pedicularis canadensis
Squaw-root
Conopholis americana
Broadleaf Plantain
Plantago rugelii
Partridge-berry
Mitchella repens
Bluets
Houstonia caerulea
Houstonia purpurea
Japanese Honeysuckle
Lonicera japonica
American Elderberry
Sambucus canadensis
Prenanthes
Prenanthes sp.
Common Dandelion
Taraxacum officinale
Small's Ragwort
Senecio smallii
Ragwort
Senecio aureus
Pussy-toes
Antennaria solitaria
Robin's-plantain
Erigeron pulchellus
Green-and-gold
Chrysogonum virginianum
Black-eyed-Susan
Rudbeckia sp.
Devil's-urn
Urnula craterium
Green Stain
Chlorociboria aeruginascens

Artist's Conk
Ganoderma applanatum
Common Split-gill
Schizophyllum commune

FAUNA

Giant Forest Snail
Mesodon normalis
White-lipped Forest Snail
Mesodon thyroidus
Ripe Forest Snail
Mesodon zaletus
White-lipped Forest Snail
Triodopsis albolabris
Almond Millipede
Sigmoria aberrans
Nursery-web Spider
Pisaurina sp.
Water-strider
Gerris sp.
Blueberry Stem Gall (wasp)
Hemadas nubilipennis
Blackberry Knot Gall (wasp)
Diastrophus nebulosus
Pipe-organ Wasp (nest)
Trypoxylon politum
Bumblebee
Bombus sp.
Green Tiger Beetle
Cicindela sexguttata
Giant Water Scavenger Beetle
Hydrophilus triangularis
June Beetle
Phyllophaga sp.
Soldier Beetle
Silis sp.
Dusky Lady Beetle
Scymnus sp.
Long-horned Leaf Beetle
Donacia sp.
Skeletonizing Leaf Beetle
Luperodes cyanellus
Flea Beetle
Oedionychis sp.
Flower Weevil
Limnobaris prolixa
Eastern Tiger Swallowtail
Papilio glaucus
Spicebush Swallowtail
Papilio troilus
White M Hairstreak
Parrhasius m-album
Spring Azure (butterfly)
Celastrina ladon
Pearl Crescent
Phyciodes tharos
American Lady (butterfly)
Vanessa virginiensis
Silver-spotted Skipper
Epargyreus clarus
Eastern Newt
Notophthalmus viridescens
Bullfrog (larva)
Rana catesbiana
Pickerel Frog
Rana palustris
Mallard
Anas platyrhynchos
Killdeer
Charadrius vociferus
Turkey Vulture
Cathartes aura
Broad-winged Hawk
Buteo platypterus
Ruffed Grouse
Bonasa umbellus
Mourning Dove
Zenaida macroura
Pileated Woodpecker
Dryocopus pileatus
Great Crested Flycatcher
Myiarchus crinitus
Eastern Wood-pewee
Contopus virens
Eastern Phoebe
Sayornis phoebe
Blue Jay
Cyanocitta cristata
American Crow
Corvus brachyrhynchos
Tufted Titmouse
Parus bicolor
Carolina Chickadee
Parus carolinensis
White-breasted Nuthatch
Sitta carolinensis
American Robin
Turdus migratorius
Red-Eyed Vireo
Vireo olivaceus
Northern Parula
Parula americana
Black-throated Blue Warbler
Dendroica caerulescens
Chestnut-sided Warbler
Dendroica pensylvanica
Hooded Warbler
Wilsonia citrina
Louisiana Waterthrush
Seiurus motacilla
Ovenbird
Seiurus aurocapillus
Rufous-sided Towhee
Pipilo erythrophthalmus
Red-winged Blackbird
Agelaius phoeniceus
American Goldfinch
Carduelis tristis
Eastern Chipmunk
Tamias striatus
Woodchuck; Groundhog
Marmota monax
Red Squirrel (scat)
Tamiasciurus hudsonicus

Horse Pasture River
(Nantahala National Forest)
Transylvania County, North Carolina
April 22, 1989

The trail followed on this trip is no longer open to the public. It crossed private lands which changed hands in 1997. The high falls, Rainbow Falls, is on National Forest Property but access is not readily available at the time of press. A trail entirely on public property which would not utilize the area of this trip will probably be available in a few years. The area has typical mountain plants and animals. The spray at the base of Rainbow Falls created habitat for unusual species that thrive on this almost constant mist. Unfortunately, the climb down to this area is very hazardous.

No map or directions provided. Check with the National Forest Service to find out status of public trail.

FLORA

Peat Moss
Sphagnum sp.
Groundpine
Lycopodium obscurum
Running-pine
Lycopodium flabelliforme
Hay-scented Fern
Dennstaedtia punctilobula
Southern Lady Fern
Athyrium asplenioides
White Pine
Pinus strobus
Pitch Pine
Pinus rigida
Eastern Hemlock
Tsuga canadensis
Jack-in-the-pulpit
Arisaema triphyllum
Rush
Juncus sp.
Catesby's Trillium
Trillium catesbaei
Painted Trillium
Trillium undulatum
Indian Cucumber-root
Medeola virginiana
False Solomon's Seal
Smilacina racemosa
Dog's-tooth-violet
Erythronium americanum
Bellwort
Uvularia sessilifolia
Yellow Star-grass
Hypoxis hirsuta
Dwarf Iris
Iris verna
Downy Rattlesnake-plantain
Goodyera pubescens
Rock Chestnut Oak
Quercus prinus
Winged Elm
Ulmus alata
Buffalo-nut
Pyrularia pubera
Heartleaf; Wild Ginger
Hexastylis contracta
Yellowroot
Xanthorhiza simplicissima
Meadow-rue
Thalictrum sp.
Small-flowered Crowfoot
Ranunculus abortivus
Windflower
Anemone quinquefolia

Horse Pasture River (continued)
April 22, 1989

Tulip-poplar
Liriodendron tulipifera
Umbrella-tree
Magnolia fraseri
Wild Hydrangea
Hydrangea arborescens
Saxifrage
Saxifraga virginiensis
Wild Strawberry
Fragaria virginiana
Five-fingers
Potentilla canadensis
Rose
Rosa sp.
Red Chokeberry
Sorbus arbutifolia
Downy Serviceberry
Amelanchier arborea
Black Cherry; Wild Cherry
Prunus serotina
American Holly
Ilex opaca
Red Maple
Acer rubrum
St.-Andrew's-cross
Hypericum hypericoides
Common Blue Violet
Viola papilionacea
Sweet White Violet
Viola blanda
Halberd-leaved Violet
Viola hastata
Devil's-walking-stick
Aralia spinosa
Golden-Alexander
Zizia aurea
Flowering Dogwood
Cornus florida
Mountain Sweet Pepperbush
Clethra acuminata
Spotted Wintergreen
Chimaphila maculata
Rosebay; Great Laurel
Rhododendron maximum
Carolina-laurel
Rhododendron minus
Mountain-laurel
Kalmia latifolia
Dog-hobble
Leucothoe axillaris
Fetterbush
Leucothoe recurva

(Photo by A. Ciegler)

Yellow star-grass, *Hypoxis hirsuta*

Trailing-arbutus
Epigaea repens
Deerberry; Squaw-huckleberry
Vaccinium stamineum
Galax; Galaxy; Coltsfoot
Galax aphylla
Horse-sugar; Sweetleaf
Symplocos tinctoria
Carolina Silverbell
Halesia carolina
Lyre-leaved Sage
Salvia lyrata
Veronica
Veronica officinalis
Partridge-berry
Mitchella repens
Diervilla
Diervilla lonicera
Black-haw; Nanny-berry
Viburnum prunifolium
Common Dandelion
Taraxacum officinale
Robin's-plantain
Erigeron pulchellus
Coneflower
Rudbeckia laciniata

FAUNA

Tiger Beetle
Cicindela repanda
Minute Ground Beetle
Bembidion versicolor
Flea Beetle
Altica sp.
Eastern Tiger Swallowtail
Papilio glaucus
Spicebush Swallowtail
Papilio troilus
Eastern Tailed-blue(butterfly)
Everes comyntas
Spring Azure (butterfly)
Lycaenopsis argiolus
Pearl Crescent
Phyciodes tharos
Silver-spotted Skipper
Epargyreus clarus
Southern Cloudywing
Thorybes bathyllus
Horace's Duskywing
Erynnis horatius
Eastern Garter Snake
Thamnophis sirtalis
Eastern Phoebe
Sayornis phoebe
Gray Catbird
Dumetella carolinensis
Chestnut-sided Warbler
Dendroica pensylvanica
Yellow Warbler
Dendroica petechia
Northern Cardinal
Cardinalis cardinalis
Rufous-sided Towhee
Pipilo erythrophthalmus
Song Sparrow
Melospiza melodia
Dark-eyed Junco
Junco hyemalis
White-throated Sparrow
Zonotrichia albicollis

(Photo by L. Smith)

Windflower, *Anemone quinquefolia.*

John Rock, Black Balsam Knob, and Cradle of Forestry
(Pisgah Ranger District, Pisgah National Forest)
Transylvania and Haywood Counties, North Carolina
April 24, 1993, and August 26-27, 1995

Both these trips were overnights, giving access to several different areas of Pisgah Ranger District. The fish hatchery is now named Davidson River Fish Hatchery and Educational Center and is operated by the state of North Carolina. Black Balsam Knob is an open bald at about 6000 feet above sea level; John Rock is the top of a granite knob with forest on one side and an open rock face on the other. The Cradle of Forestry area is about half-way up the mountain, between the two other elevations. Heavy rain from a tropical storm plagued the 1995 trip, but it also brought out certain species that are ordinarily subterranean, such as the bright red crayfish, and the many species of mushrooms. The spring trip yielded many migrating warblers. We in the Carolinas get to see them only as they pass to or from breeding grounds farther north. Gorgeous spring wildflowers were a highlight as well.

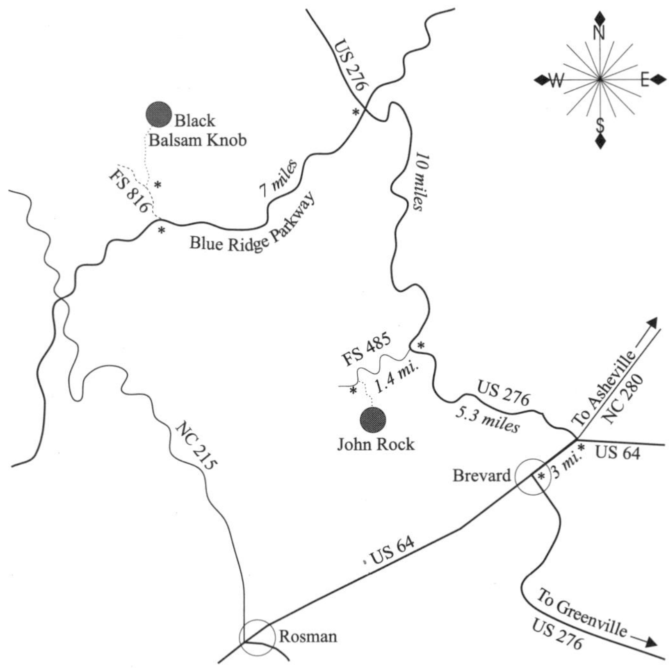

Directions to Pisgah Ranger District: From Greenville, take US 276 north to Brevard, about 50 miles. Continue north on US 276 for 3 more miles to the traffic light where US 276 turns left and US 64 turns right. To reach the Cradle of Forestry from this intersection, turn left (north) on US 276 and drive about 14 miles. The Cradle of Forestry is on the right, well-marked with signs. To reach John Rock from the same intersection north of Brevard, travel only 5.3 miles north on US 276; then turn left onto FS 485. Drive about 1.4 miles to the fish hatchery parking lot and park near the trail head. To reach Black Balsam Knob, continue past the Cradle of Forestry another 1-2 miles, then turn south on the Blue Ridge Parkway. After 7 miles, turn north on Forest Service Road 816, and park by the trailhead posted for Black Balsam Knob.

FLORA

Lichen

Collema subflaccidum
Leptogium cyanescens
Leptogium sp.
Pannaria tavaresii
Peltigera canina
Lobaria pulmonaria
Lobaria quercizans
Lobaria tenuis
Pseudocephalaria aurata
Pseudocephalaria crocata
Sticta weigelii
Graphis scripta
Cladina arbuscula
Bacidia schweinitzii
Lecidea albocaerulescens
Lecidea contigua
Lecidea tessellata
Porpidia macrocarpa
Stereocaulon sp.
Cladonia caespiticia
Cladonia chlorophaea
Cladonia coniocraea
Cladonia cristatella
Cladonia didyma
Cladonia furcata
Cladonia squamosa
Cladonia uncialis
Umbilicaria mammulata
Pertusaria neoscotica
Pertusaria rubefacta
Pertusaria waghornei
Pertusaria velata
Lecanora deplanans
Lecanora cenisia
Lecanora cupressi
Lecanora polytropa
Ochrolechia africana
Cetraria arenaria
Cetraria aurescens
Cetraria ciliaris
Cetraria sp.
Cetrelia chicitae
Cetrelia sp.
Hypogymnia krogiae
Parmelia omphalodes
Parmelia saxatilis
Parmelia squarrosa
Parmelia sulcata
Parmelina horrescens
Parmelina minarum
Pseudevernia cladonia
Pseudevernia consocians
Xanthoparmelia conspersa
Xanthoparmelia plittii
Punctelia rudecta
Punctelia subrudecta
Melanelia halei
Melanelia stygia
Melanelia subaurifera
Imshaugia aleurites
Rimelia reticulata
Tuckermannopsis oakesiana
Tuckermannopsis pinastri
Evernia mesomorpha
Usnea ceratina
Usnea mutabilis
Usnea strigosa
Usnea subfloridana
Usnea trichodea
Usnea filipendula
Buellia punctata
Buellia curatellae
Physcia millegrana
Heterodermia leucomelaena
Heterodermia obscurata
Heterodermia speciosa
Heterodermia casarettiana
Pyxine sorediata
Phaeophyscia adiastola
Phaeophyscia imbricata
Phaeophyscia rubropulchra
Dermatocarpon miniatum

John Rock, Black Balsam Knob, and Cradle of Forestry (continued)
April 24, 1993, and August 26-27, 1995

Shining Clubmoss
Lycopodium lucidulum
Running Clubmoss
Lycopodium clavatum
Groundpine
Lycopodium obscurum
Running-pine
Lycopodium flabelliforme
Rattlesnake Grape Fern
Botrychium virginianum
Hay-scented Fern
Dennstaedtia punctilobula
Bracken
Pteridium aquilinum
Southern Lady Fern
Athyrium asplenioides
Spreading Bladder Fern
Cystopteris protrusa
Shield Fern
Dryopteris sp.
Christmas Fern
Polystichum acrostichoides
New York Fern
Thelypteris noveboracensis
Mountain Spleenwort
Asplenium montanum
Rock Cap Fern
Polypodium virginianum
White Pine
Pinus strobus
Pitch Pine
Pinus rigida
Table Mountain Pine
Pinus pungens
Red Spruce; He Balsam
Picea rubens
Eastern Hemlock
Tsuga canadensis
Carolina Hemlock
Tsuga caroliniana
Fraser Fir; She Balsam
Abies fraseri
Cane; Switch Cane
Arundinaria gigantea
Jack-in-the-pulpit
Arisaema triphyllum
Woodrush
Luzula acuminata
Wake-robin
Trillium erectum
Great White Trillium
Trillium grandiflorum
Mandarin
Disporum sp.
Solomon's-seal
Polygonatum biflorum
White Hellebore
Veratrum viride
False Hellebore
Veratrum sp.
Dog's-tooth-violet
Erythronium americanum
Bellwort
Uvularia sp.
Dwarf Iris
Iris verna
Downy Rattlesnake-plantain
Goodyera pubescens
Crane-fly Orchid
Tipularia discolor
Puttyroot; Adam-and-Eve
Aplectrum hyemale
Autumn Coral-root
Corallorhiza odontorhiza
Willow
Salix sp.
Pignut Hickory
Carya glabra
Hickory
Carya sp.
Tag Alder
Alnus serrulata
Yellow Birch
Betula lutea
American Hornbeam; Ironwood
Carpinus caroliniana
Beech
Fagus grandifolia
White Oak
Quercus alba
Rock Chestnut Oak
Quercus prinus
Red Oak
Quercus rubra
Scarlet Oak
Quercus coccinea
Wood-nettle
Laportea canadensis
Heartleaf
Hexastylis sp.
Spring-beauty
Claytonia caroliniana
Giant Chickweed
Stellaria pubera
Yellowroot
Xanthorhiza simplicissima
Buttercup
Ranunculus sp.
Windflower
Anemone quinquefolia
Blue Cohosh; Papooseroot
Caulophyllum thalictroides
Mayapple
Podophyllum peltatum
Tulip-poplar
Liriodendron tulipifera
Umbrella-tree
Magnolia fraseri
Magnolia tripetala
Cucumber-tree
Magnolia acuminata
Spicebush
Lindera benzoin
Bloodroot
Sanguinaria canadensis
Toothwort
Cardamine diphylla
Cardamine concatenata
Wild Hydrangea
Hydrangea arborescens
Alumroot
Heuchera americana
Foamflower
Tiarella cordifolia
Wild Strawberry
Fragaria virginiana
Blackberry
Rubus canadensis
Mountain-ash
Sorbus americana
Downy Serviceberry
Amelanchier arborea
Fire Cherry; Pin Cherry
Prunus pensylvanica
Poison-ivy
Rhus radicans
American Holly
Ilex opaca
Mountain Holly
Ilex ambigua montana
Hearts-a'burstin'-with-love
Euonymus americanus
Strawberry-bush
Euonymus sp.
Striped Maple
Acer pensylvanicum
Red Maple
Acer rubrum
Yellow Buckeye
Aesculus octandra
Basswood; Linden
Tilia sp.
Hypericum
Hypericum sp.
Bird-foot Violet
Viola pedata

(Photo by L. Smith)

Great white trillium, *Trillium grandiflorum*.

Primrose-leaved Violet
Viola primulifolia
Round-leaved Yellow Violet
Viola rotundifolia
Halberd-leaved Violet
Viola hastata
Violet
Viola eriocarpa leiocarpa
Viola sp.
Flowering Dogwood
Cornus florida
Alternate-leaved Dogwood
Cornus alternifolia
Spotted Wintergreen
Chimaphila maculata
Rosebay; Great Laurel
Rhododendron maximum
Carolina-laurel
Rhododendron minus
Sand-myrtle
Leiophyllum buxifolium
Mountain-laurel
Kalmia latifolia
Andromeda
Pieris floribunda
Maleberry
Lyonia ligustrina
Dog-hobble
Leucothoe axillaris editorum
Leucothoe sp.
Sourwood
Oxydendrum arboreum
Trailing-arbutus
Epigaea repens
Wintergreen
Gaultheria procumbens
Huckleberry
Gaylussacia ursina
Blueberry
Vaccinium constablaei
Galax; Galaxy; Coltsfoot
Galax aphylla
Horse-sugar; Sweetleaf
Symplocos tinctoria
Carolina Silverbell
Halesia carolina
American Ash; White Ash
Fraxinus americana

Princess-tree
Paulownia tomentosa
Partridge-berry
Mitchella repens
Bluets
Houstonia serpyllifolia
Houstonia sp.
Bush Honeysuckle
Diervilla sessilifolia
Viburnum
Viburnum sp.
Common Dandelion
Taraxacum officinale
Small's Ragwort
Senecio smallii
Ragwort
Senecio aureus
Robin's-plantain
Erigeron pulchellus
Common Brown Cup
Peziza badio-confusa
Black-trumpet
Craterellus fallax
Crested Coral
Clavulina cristata
Yellow-tipped Coral
Ramaria formosa
Violet-toothed Polypore
Trichaptum biformis
Red-tinged Lepiota
Lepiota rubrotincta
Yellow-patches
Amanita flavoconia
Gemmed Amanita
Amanita gemmata
Fly Agaric
Amanita muscaria formosa
Panther Mushroom
Amanita pantherina
Destroying-angel
Amanita virosa
Frost's Bolete
Boletus frostii
Painted Suillus
Suillus pictus
Gilled Bolete
Phylloporus rhodoxanthus
Graying Yellow Russula
Russula claroflava

(Photo by V. Winn)
Fly agaric amanita, *Amanita muscaria.*

Emetic Russula
Russula emetica
Gem-studded Puffball
Lycoperdon perlatum
White-egg Bird's-nest
Crucibulum laeve
Stalked Puffball-in-aspic
Calostoma cinnabarina

FAUNA

Long-jawed Orb-weaver
Tetragnatha sp.
Red Velvet Mite
Trombidium sp.
Blueberry Stem Gall (wasp)
Hemadas nubilipennis
Post-oak Apple Gall (wasp)
Andricus unica
Oak Bullet Gall (wasp)
Disholcaspis globulus
Gouty Oak Gall (wasp)
Plagiotrochus punctatus
Bee Fly
Anthrax aralis
Common Green Darner
Anax junius
Dingy Ground Beetle
Harpalus pennsylvanicus
Harpalus spadiceus
Large Rove Beetle
Staphylinus cinnamopterus
Rove Beetle
Ontholestes sp.
Riffle Beetle
Limnius latiusculus
Soldier Beetle
Podabrus sp.
Cantharis bilineatus
Firefly; Lightning Beetle
Ellychnia sp.
Blister Beetle
Lytta aenea
Grape Flea Beetle
Altica chalybea
Flea Beetle
Altica sp.
Broad-nosed Weevil
Phyllobius sp.
Eastern Tiger Swallowtail
Papilio glaucus
Spring Azure (butterfly)
Celastrina ladon
Juvenal's Duskywing
Erynnis juvenalis
White-striped Black Moth
Trichodezia albovittata
Brook Trout
Salvelinus fontinalis
Mountain Dusky Salamander
Desmognathus ochrophaeus
Shovelnose Salamander
Leurognathus marmoratus
Eastern Box Turtle
Terrapene carolina
Timber Rattlesnake
Crotalus horridus

Killdeer
Charadrius vociferus
Turkey Vulture
Cathartes aura
Broad-winged Hawk
Buteo platypterus
Northern Flicker
Colaptes auratus
Downy Woodpecker
Picoides pubescens
Pileated Woodpecker
Dryocopus pileatus
Eastern Kingbird
Tyrannus tyrannus
Eastern Phoebe
Sayornis phoebe
Blue Jay
Cyanocitta cristata
American Crow
Corvus brachyrhynchos
Common Raven
Corvus corax
Tufted Titmouse
Parus bicolor
Carolina Chickadee
Parus carolinensis
Ruby-crowned Kinglet
Regulus calendula
Blue-gray Gnatcatcher
Polioptila caerulea
American Robin
Turdus migratorius
Gray Catbird
Dumetella carolinensis
Solitary Vireo
Vireo solitarius
Red-Eyed Vireo
Vireo olivaceus
Northern Parula
Parula americana
Black-and-white Warbler
Mniotilta varia
Black-throated Blue Warbler
Dendroica caerulescens
Blackburnian Warbler
Dendroica fusca
Cape May Warbler
Dendroica tigrina
Black-throated Green Warbler
Dendroica virens
Hooded Warbler
Wilsonia citrina
Louisiana Waterthrush
Seiurus motacilla
Ovenbird
Seiurus aurocapillus
Rufous-sided Towhee
Pipilo erythrophthalmus
Chipping Sparrow
Spizella passerina
Dark-eyed Junco
Junco hyemalis
American Goldfinch
Carduelis tristis
Woodchuck; Groundhog
Marmota monax

(Photo by L. Smith)

Black Balsam Knob, Pisgah Ranger District, Transylvania County, North Carolina, Blue Ridge region.

(Photo by L. Greenlee)

Table Rock State Park, Pickens County, Blue Ridge region.

(Photo by L. Greenlee)

Frank Hill finds the water high at Savage Bay Heritage Preserve, Kershaw County, Sandhills region.

(Photo by L. Greenlee)

Felicia Smith studies a paper wasp nest, *Polistes* sp., Old Santee Canal State Park, Berkeley County, Atlantic Coast Flatwoods region.

(Photo by L. Smith)

Yellow Prong River above second falls. Graveyard Field Loop. Blue Ridge region.

(Photos by L. Smith)

Two good close-up examples of water erosion on river rock. Below is a pot hole. Blue Ridge region.

Science and Nature Notes: Water Erosion

Water is the primary erosive force currently shaping the South Carolina landscape. Water alone can create large gullies which become valleys, while pushing the eroded sediment downstream to form flood plains and river deltas. Add small pieces of rock chipped from large rocks as they are tumbled and the stream becomes a scouring agent. Under the right conditions, large pebbles get swirled on one small area forming picturesque 'potholes' in the river bottom rock such as the one you see here. Waterfalls are created when the stream wears away a softer rock downstream and the water drops off more quickly, continuously accelerating the process. Plant material falling into the water releases acids that also can accelerate the erosion of rocks. Flood conditions erode softer soils and wash them downstream adding to the richness and depth of floodplain and delta soils.

(Photo by L. Smith)

Never a stone unturned. Rick Renfrow looks for river life.

(Photos by B. Gilbert)

Saltwater marshes and open beaches are visited by SCAN at least once a year. Pinckney Island, Beaufort County (left) and Edisto Beach, Charleston County, Tidewater region.

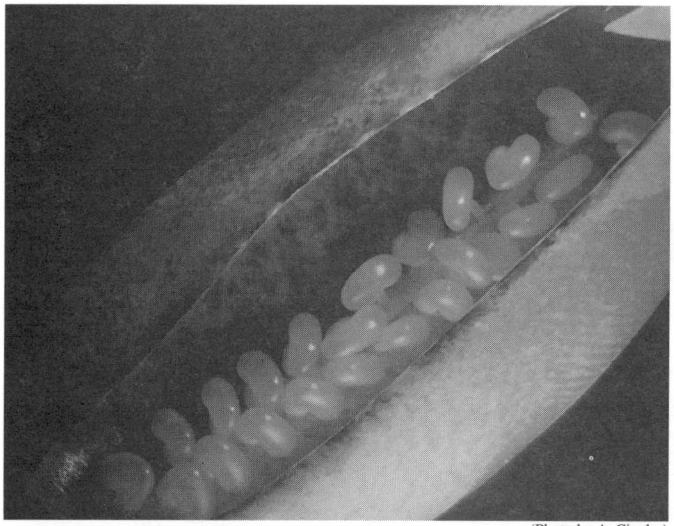

(Photo by A. Ciegler)

An artistic view of a rattlebox seed pod, *Crotalaria* sp.

(Photos by L. Smith and L. Greenlee)

Boardwalks are a favorite way to visit our swampy areas. Here we have Washo Reserve (left) in the Santee Coastal Wildlife Management Area, Charleston County, Tidewater region; and, Congaree Swamp National Monument (above), Richland County, Southern Coastal Plain region.

(Photos by R. Emmitt and A. Ciegler)

Male great blue skimmer, *Libellula vibrans*.

(Photo by L. Smith)

Blue-ringed dancer, *Argia sedula*.

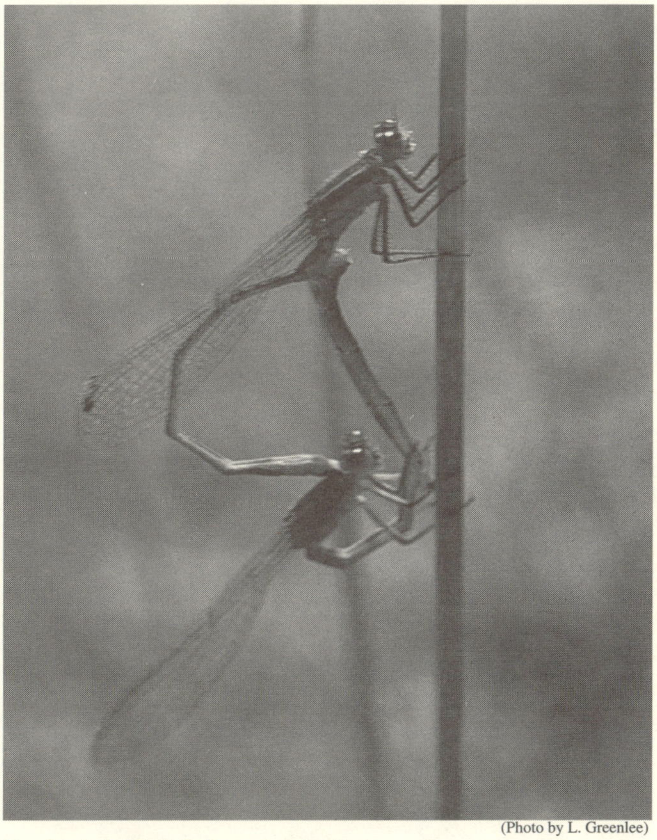

(Photo by L. Greenlee)

Rambur's forktail, *Ischnura ramburii*.

(Photo by L. Greenlee)

Black-shouldered spiny-leg, *Dromogomphus spinosus*.

(Photo by L.Greenlee)

Howard and Pam Spencer take a break at Caesars Head State Park, Greenville County, Blue Ridge region.

(Photo by V. Winn)

Gary Sowell and Lynn Greenlee take time out for a photographic opportunity, Woods Dargan, Darlington County, Sandhills region.

(Photo by V. Winn)

SCAN explores the Lynches River State Park, Florence County, Atlantic Coast Flatwoods region.

(Photo by B. Gilbert)

SCAN members head for the hills at the Martin Marietta Quarry near Cross, SC, Berkeley County, Atlantic Coastal Flatwoods region. Fossil hunting is often our December outing since flora and fauna aren't as obvious.

(Photo by Betty Gilbert)

Oleta (Beard) LaBorde (left) puts Radford to the test, Cartwheel Bay Heritage Preserve, Horry County. Atlantic Coast Flatwoods region.

(Photo by B. Gilbert)

Jo Segal (front) and Glenn Quattlebaum weather the cold at Santee National Wildlife Refuge, Clarendon County, Southern Coastal Plain region.

(Photo by A. Ciegler)

Orb-weaver web.

(Photo by L. Greenlee)

Rabid wolf spider with egg case, *Lycosa rabida*.

(Photo by L. Smith)

Egg sac of the basilica spider, *Mecynogea lemniscata*.

(Photo by L. Greenlee)

Arrowshaped spiny-back spider, *Micrathena sagittata*.

(Photo by L. Smith)

Bowl-and-doily spider web, *Frontinella pyramitela.*

(Photo by L. Smith)

Filmy-dome spider web, *Prolinyphia marginata.*

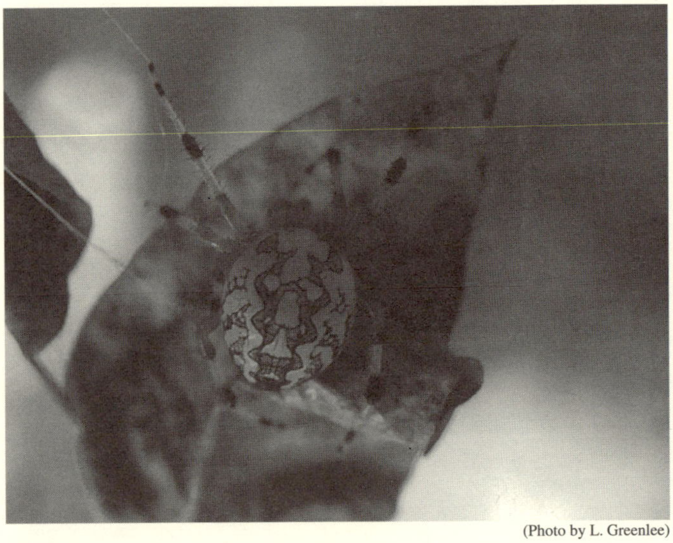

(Photo by L. Greenlee)

Marbled orb-weaver, *Araneus marmoreus*.

(Photo by L. Greenlee)

Writing spider, *Argiope aurantia*, eats a variegated fritillary, *Euptoieta claudia*.

(Photo by L. Greenlee)

Female imperial moth, *Eacles imperialis*.

(Photo by A. Ciegler)

Sam Outen prepares for his next great photograph.

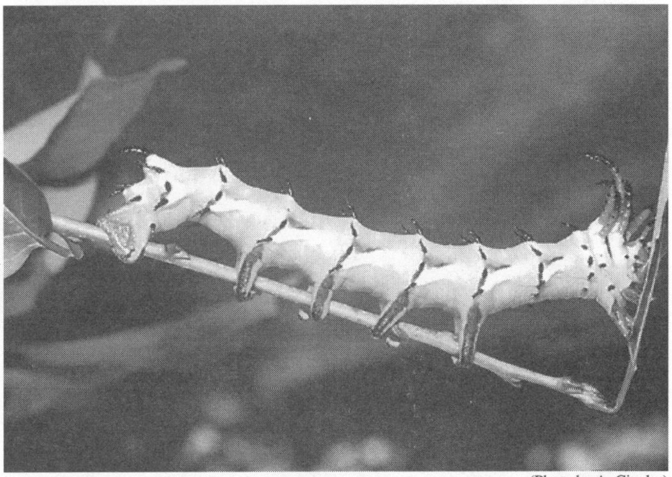

(Photo by A. Ciegler)

Hickory horned-devil larva, *Citheronia regalis*.

(Photo by L. Greenlee)

Black swallowtail, *Papilio polyxenes,* caterpillar changes into a chrysalis.

(Photo by L. Smith)

Grape leaffolder moth, *Desmia funeralis*.

(Photo by L. Greenlee)

Giant leopard moth, *Ecpantheria scribonia.*

(Photo by L. Greenlee)

Acrea moth, *Estigmene acrea.*

(Photo by L. Smith)

Female diana fritillary, female, *Speyeria diana.*

(Photo by L. Greenlee)

Male and female blinded sphinx moths, *Paonias astylus.*

(Photos by L. Greenlee)

Pipevine swallowtail, *Battus philenor*, chrysalis (left) and larva (above).

(Photo by B. Gilbert)

Luna moth, *Actias luna*.

(Photo by L. Greenlee)

Azalea caterpillars, *Datana major*.

(Photo by L. Greenlee)

Zebra longwing, *Heliconius charitonius*.

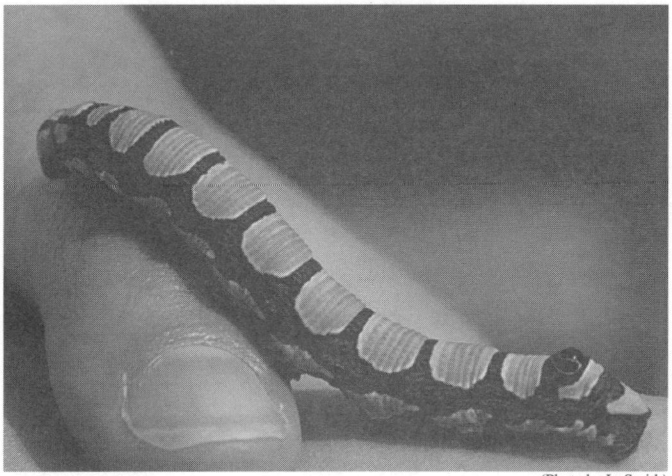

(Photo by L. Smith)

Abbot's sphinx moth larva, *Sphecodina abbottii.*

(Photo by L. Greenlee)

Cloudless sulfur, *Phoebis sennae.*

(Photo by L. Smith)

Great spangled fritillary, *Speyeria cybele*.

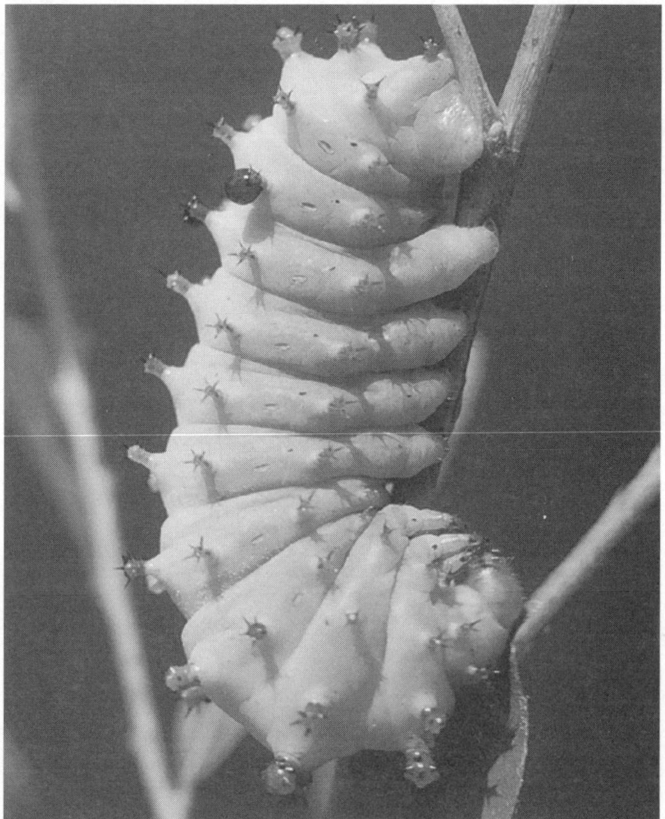

(Photos by L. Greenlee and A. Ciegler)

Cecropia moth, *Hyalophora cecropia*, larva (above) and female (left).

(Photo by A. Ciegler)

Birds-nest splashcups, *Cyathus striatus*.

(Photo by V. Winn)

Columned stinkhorn, *Clathrus columnatus*.

(Photo by V. Winn)

Jellied false coral, *Tremellodendron pallidum*.

(Photo by V. Winn)

Fluted white helvella, *Helvella crippa*.

(Photo by L. Smith)

Shaggy-stalked bolete, *Boletellus betula*.

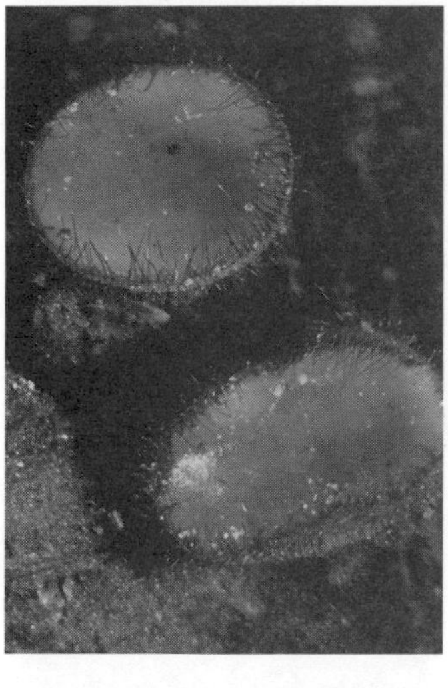

(Photo by A. Ciegler)

Eyelash splashcups, *Scutellinia scutellata*.

(Photos by L. Smith and B. Gilbert)

Views of the rare spiny gooseberry, *Ribes echinellum*.

(Photo by A. Ciegler)

Sand myrtle, *Leiophyllum buxifolium.*

(Photo by L. Smith)

Andromeda or mountain pieris, *Pieris floribunda.*

(Photo by V. Winn)

Rosebay rhododendron, *Rhododendron maximum.*

(Photo by V. Winn)

White-tailed deer, *Odocoileus virginianus*.

(Photo by L. Smith)

Cotton rat, *Sigmodon hispidus*.

Scenes of Heggie's Rock, Columbia County, Georgia, Blue Ridge region.

(Photo by L. Smith)

Vernal pool stonecrop, *Sedum smallii.*

(Photo by V. Winn)

Spiderwort, *Tradescantia hirsuticaulis.*

(Photo by L. Smith)

Alex Ciegler gets a close look at vernal pool stonecrop, *Sedum smallii.*

(Photo by L. Smith)

Pool-sprite, *Amphianthus pusillus.*

(Photo by L. Smith)

Vernal pool stonecrop, *Sedum smallii.*

(Photo by V. Winn)

Devil-joint, *Opuntia drummondii.*

(Photo by V. Winn)

Yellow Mandarin, *Disporum lanuginosum*.

(Photo by V. Winn)

Dandelion, *Taraxacum officinale*.

(Photo by L. Greenlee)

Terry Hook and Danny Smith at Pinckney Island National Wildlife Refuge, Charleston County, Tidewater region.

(Photo by L. Greenlee)

Pink lady's-slippers, *Cypripedium acaule*.

(Photo by L. Smith)

Green-and-gold, *Chrysogonum virginianum*.

(Photo by V. Winn)

White-topped sedge, *Dichranina latifolia.*

(Photo by V. Winn)

Poison-oak, Rhus toxicodendron.

(Photo by A. Ciegler)

Prickly pear, *Opuntia compressa.*

(Photo by L. Smith)

Yellow passionflower, *Passiflora lutea.*

(Photo by V. Winn)

Staggerbush, *Lyonia mariana.*

(Photo by L. Smith)

Turk's-cap lily, *Lilium superbum*.

(Photo by A. Ciegler)

Bladderwort, *Utricularia inflata*.

(Photo by L. Smith)

Wild ginger, *Asarum canadense*.

(Photo by L. Smith)

Pink-grass, *Calopogon pulchellus*.

(Photo by V. Winn)
Hooded pitcher-plant, *Sarracenia minor.*

(Photo by V. Winn)
Hooded pitcher-plant, *Sarracenia minor.*

(Photo by A. Ciegler)
Trumpets, *Sarracenia flava.*

(Photo by V. Winn)
Sundew, *Drosera leucantha.*

(Photo by A. Ciegler)
Sundew, *Drosera intermedia.*

Science & Nature Notes: Carnivorous Plants

Plants that eat animals to live sound like contradictions in terms. However, there is one group of plants that has adapted a carnivorous approach to a complete diet. These plants live in very acid, mineral poor soil most often found in bogs, swamps, freshwater marshes and savannas. They attract insects and occasionally other kinds of animals to their leaves, then trap them, absorbing the minerals in their bodies as they are dissolved on or in the leaves. Venus flytraps use a spring loaded toothed cage to entrap its meal. Butterworts (*Pinguicula*) and sundews (*Drosera*) use a sticky substance to keep the prey in place until digestion takes place. Pitcher plants have leaves in a tube shape to form a 'pitcher' that is lined with downward pointing hairs that allow an insect to climb down to the water trapped in the leaf, but do not allow it to climb out. Bladderworts, most of which grow in standing water or very wet soils, have special pouches on their roots that have an opening with sensitive trigger hairs that sense a small creature swimming by, which then open in a sucking manner to suck in the creature and some surrounding water.

(Photo by A. Ciegler)

Trumpets, *Sarracenia flava.*

(Photo by A. Ciegler)

Frog's-britches, *Sarracenia purpurea.*

(Photo by V. Winn)

Frog's-britches, *Sarracenia purpurea.*

(Photo by A. Ciegler)

Venus'-flytrap, *Dioneaea muscipula.*

(Photo by V. Winn)

Sam Pratt, Dee Hope, Pat Chaplin, Mary Boni, Loree Gandy and other SCAN members explore one of South Carolina's carolina bays which are known for their carnivorous plants.

Science & Nature Notes: Carolina Bays

Carolina bays are not bays on a shore, but an elliptical depression that tends to fill with water providing a good habitat around the edge for a type of plant commonly called 'bay'. Aerial photography gave the first look at the similarity of shape, elliptical, and orientation, northwest-southeast, of these depressions. Also unique is the shallow sand ridge around them, trapping water. Some bays have standing water year round in the center, some only seasonally, and some few seem to only have a savanna-like center. Most bays have a good stand of cypress in the permanently wet areas, then a rim of bays and other shrubs and then a sandhills-like habitat around the rim. The cause of bay formation is still a cause for much discussion among geologists. Wind and ice erosion when this was a colder climate currently seems to be the most accepted. Whatever their cause, they have created some unique habitats that are very rewarding to explore. Woods Bay State Park is a classic water-filled bay, while Lewis Ocean Bay Heritage Preserve has a huge savanna type bay.

(Photo by A. Ciegler)

Lobelia, *Lobelia* sp.

(Photo by V. Winn)

Hyssop skullcap, *Scutellia integrifolia.*

(Photo by V. Winn)

Spreading-pogonia, *Cleistes diraricata.*

(Photo by A. Ciegler)

Artistic view of grass.

(Photo by A. Ciegler)
A typical view of SCAN members — binoculars up and heads back.

(Photo by L. Greenlee)
Squawroot, *Conopholis americana.*

(Photo by V. Winn)
Mina Winn scans the tree tops for birds.

(Photo by B. Gilbert)
River-oats, *Uniola latifolia.*

(Photo by B. Gilbert)
Anna Kaye Singley takes a break from hiking to capture the beauty of nature on canvas. Anna Kaye illustrated SCAN's first book.

(Photo by B. Gilbert)

SCAN members head for the diversity of flora and fauna at one of South Carolina's unique Carolina bays. Cartwheel Bay Heritage Preserve, Horry County, Atlantic Coastal Flatwoods region.

(Photos by B. Gilbert and V. Winn)

Landsford Canal was built in the early 1800's to let boats get around the rapids on the Catawba River. Shortly after construction, the railroad system took over the transportation of goods and the locks were no longer used. Over the last 20 years, South Carolina has been gradually restoring them. Landsford Canal State Park, Chester County, Southern Piedmont Region.

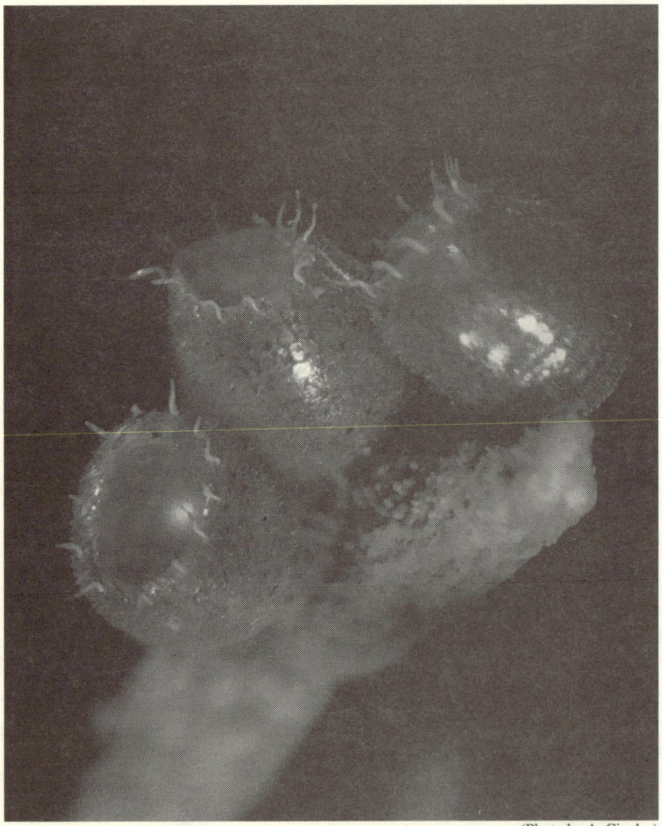

(Photo by A. Ciegler)

Insect eggs.

(Photo by L. Greenlee)

Two-spotted lady beetle, *Adalia bipunctata*.

(Photo by V. Winn)

Keeled shield-back katydid, *Neduba carinata*.

(Photo by A. Ciegler)

Centipede, *Strigamia* sp.

(Photo by V. Winn)

Locust borer, *Megacyllene robiniae*.

(Photo by V. Winn)

Eastern eyed click beetle, *Alaus oculatus*.

(Photo by A. Ciegler)

Immature katydid.

(Photo by A. Ciegler)

Spur-throated grasshopper, Melanoplus differentialis.

(Photo by L. Greenlee)

Red eft or eastern newt, *Notophthalmus viridescens*.

(Photo by L. Smith)

Southern toad, *Bufo terrestris*.

(Photo by V. Winn)

Bull frog, *Rana catesbeiana*.

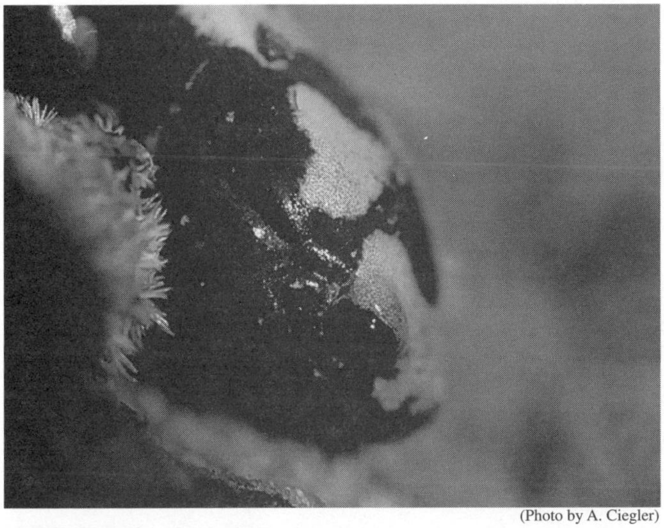

(Photo by A. Ciegler)

Marbled salamander, *Ambystoma opacum*.

(Photo by L. Greenlee)

Yellow-bellied slider, *Pseudemys scripta.*

(Photo by L. Greenlee)

Southern five-lined skink, *Eumeces inexpectatus.*

(Photo by A. Ciegler)

Timber rattlesnake, *Crotalus horridus.*

(Photo by L. Greenlee)

Rough green snake, *Opheodrys aestivus.*

(Photo by A. Ciegler)

Black rat snake, *Elaphe obsoleta.*

(Photos by V. Winn)

American Alligator, *Alligator mississippiensis*.

(Photo by L. Greenlee)

Broad-headed skink, *Eumeces laticeps*.

(Photo by L. Greenlee)

Eastern Garter Snake, Thamnophis sirtalis.

(Photo by L. Greenlee)

Mary Garland Douglass with little grass frog, *Limnaoedus ocularis*.

(Photo by A. Ciegler)

Red Smith enjoys a lunch break.

(Photo by L. Greenlee)

Ted Cottingham, David Smith, Lynn Smith, Cathe Jones, Alex Ciegler, Virginia Winn and Mina Winn break for lunch at John de la Howe School, McCormick County, Southern Piedmont region.

(Photo by L. Smith)

Dee Hope identifies flora.

(Photo by L. Greenlee)

Jan Ciegler (front) and Mina Winn look up sea life at Hunting Island State Park, Beaufort County, Tidewater region.

(Photo by A. Ciegler)

Black-necked stilt, *Himantopus mexicanus.*

(Photo by L. Greenlee)

Canada goose, *Branta canadensis,* with young.

(Photo by V. Winn)

Louisiana or tricolored heron, *Egretta tricolor*.

(Photo by A. Ciegler)

Royal terns, *Sterna maxima.*

(Photo by A. Ciegler)

Snowy egret, *Egretta thula.*

(Photo by A. Ciegler)

Barred owl, *Strix varia.*

(Photos by R.Emmitt)

Osprey, *Pandion haliaetus,* feed on live fish and usually nest near water but if suitable sites are unavailable they will nest inland. They normally nest in dead trees but are also attracted to artificial platforms.

(Photo by V. Winn)

Asiatic Claim, *Corbicula* sp.

(Photo by V. Winn)

Mallard, *Anas platyrhynchos.*

(Photo by B. Gilbert)

Neither rain, sleet, ice nor snow will keep SCAN from our appointed field trip. Landsford Canal State Park, Chester County, Piedmont region.

(Photo by L. Greenlee)

Florence High explores Playcard Environmental Center, Horry County, Atlantic Coastal Flatwoods region.

(Photo by B. Gilbert)

Patti Shrum, Gary Sowell, Jan Ciegler and Nancy Lyman look for flora and fauna at Sugarloaf Mountain, Chesterfield County, Sandhills region.

(Photos by V. Winn)

Decanie Dowling, one of SCAN's long time members, takes slides (above) at Sugarloaf Mountain in the Sandhills State Forest, Chesterfield County, Sandhills region.

(Photo by V. Winn)

Sara Lee Simons looks out from the top of Sugarloaf Mountain in the Sandhills State Forest, Chesterfield County, Sandhills region.

(Photo by V. Winn)

Mary Garland Douglass gets a closer look at aquatic fauna at Silver Bluff Audubon Preserve, Aiken County, Southern Piedmont region.

(Photo by V. Winn)

Oleta LaBorde and Decanie Dowling measuring the girth of a tree at Broad River near Hickory Grove, York County, Piedmont region.

Natural History Organizations

ACE Basin National Wildlife Refuge
Hollywood, SC 29449
Phone: 843-889-3084
**Ashipoo River Canoe Trip, page 65
**Bear Island Wildlife Management Area, page 67
**Donnelley Wildlife Management Area, page 69

Black Rock Mountain State Park
Mountain City, GA 30562
Phone: 706-746-2141
**See page 201

Carolina Sandhills National Wildlife Refuge
Route 2, Box 330
McBee, SC 29101
Phone: 843-335-8401
**See page 58

Cape Romain National Wildlife Refuge
US Department of Interior Fish and Wildlife Service
4801 Highway 17 N.
Awendaw, SC 29429
Phone: 843-928-3368
**Bull Island, page 41

Charleston Natural History Society
PO Box 504
Charleston, SC 29402
Phone: 843-577-7100
**McAlhany Nature Preserve, page 83

Congaree Swamp National Monument
200 Carolina Sims Road
Hopkins, SC 29061
Phone: 803-776-4396
**See page 176

Kalmia Gardens of Coker College
1624 West College Avenue
Hartsville, SC 29550
Phone: 843-383-8145
**Kalmia Gardens of Coker College, page 75
**Segars Heritage Preserve, page 75

Pinckney Island National Wildlife Refuge
PO Box 8487
Savannah, GA 31412
Phone: 912-652-4415
**See page 25

Pisgah Ranger District, Pisgah National Forest
1001 Pisgah Highway
Pisgah Forest, NC 28768
Phone: 828-877-3265 or 828-877-3350
**Black Balsam Knob, page 213
**Cradle of Forestry, page 213
**John Rock, page 213

Playcard Environmental Education Center
Route 3, Box 401
Loris, SC 29569
Phone: 843-756-1277
**See page 114

Roper Mountain Science Center
504 Roper Mountain Road
Greenville, SC 29615
Phone: 864-281-1188
**See page 101

Santee National Wildlife Refuge
Route 2, Box 370
Summerton, SC 29148
Phone: 803-478-2217
**See page 64

Savannah National Wildlife Refuge
PO Box 8487
Savannah, GA 31412
Phone: 912-652-4415
**See page 118

SC Department of Natural Resources
1000 Assembly Street
Columbia, SC 29201
General Information Phone: 803-734-3888
Endangered Species Phone: 803-734-3894
Wildlife Management Areas Phone: 803-734-3894
Heritage Trust Program Phone: 803-734-3894
**Bear Island WMA, page 67
**Bennetts Bay HP, page 62
**Bunched Arrowhead HP, page 93
**Buzzard's Roost HP, page 156
**Cartwheel Bay HP, page 108
**Cathedral Bay HP, page 18
**Donnelley WMA, page 69
**Eastatoe Creek HP, page 168
**Ervin Dargan HP, page 110
**Eva Russel Chandler HP, page 97
**Flat Creek HP, page 126
**Glassy Mountain HP, page 170
**Lewis Ocean Bay HP, page 112
**Little Pee Dee River HP, page 110
**Lynchburg Savanna HP, page 135
**Mt. Pleasant Church Bay HP, page 138
**Pacolet River HP, page 185
**Peachtree Rock HP, page 141
**Santee Coastal WMA, page 44
**Savage Bay HP, page 124
**Savannah River HP, page 7
**Segars HP, page 75
**Stevens Creek HP, page 147
**Tillman Sand Ridge HP, page 119
**Victoria Bluff HP, page 29
**Webb Wildlife Center, page 107

SC Department of Parks, Recreation and Tourism
1205 Pendleton Street
Edgar Brown Building
Columbia, SC 29201
State Park Office Phone: 803-734-0156
**Aiken SP, page 5
**Barnwell SP, page 20
**Caesars Head SP, page 95
**Cheraw SP, page 61
**Croft SP, page 182
**Devils Fork SP, page 161
**Edisto Beach SP, page 38
**Goodale SP, page 122
**Hunting Island SP, page 22
**Huntington Beach SP, page 90
**Jones Gap SP, page 99
**Keowee-Toxaway SP,page 173
**Kings Mountain SP, page 197
**Lake Warren SP, page 105
**Landford Canal SP, page 52
**Lee SP, page 133
**Little Pee Dee SP, page 80
**Lynches River SP, page 88
**Old Santee Canal SP, page 34
**Poinsett SP, page 190
**Rivers Bridge SP, page 19
**Santee SP, page 166
**Table Rock SP, page 175
**Woods Bay SP, page 192

SC Forestry Commission
5500 Broard River Road
PO Box 21707
Columbia, SC 29210
Phone: 803-896-8800
**Sandhills SF, page 58

Silver Bluff Audubon Preserve
Phone: 803-827-0781
**See page 12

The Nature Conservancy of Georgia
1401 Peachtree St., NE, Suite 136
Atlanta, GA 30309
Phone: 404-873-6946
**Heggies Rock, page 199

The Nature Conservancy, South Carolina Chapter
2231 Devine Street, Suite 10
PO Box 5475
Columbia, SC 29250
Phone: 803-254-9049
Note: Often has information about the newest heritage preserves.

US Forest Service
4931 Broad River Road
Columbia, SC 29212
Phone: 803-561-4000
**Burrells Ford, page 153
**Cherry Hill Campground, page 153
**Enoree River Site, page 131
**Francis Marion National Forest, page 31
**King Creek Falls, page 153
**Molly's Rock Recreation Area, page 149
**Parsons Mountain, page 3
**Tyger River at Spark Creek, page 194
**Woods Ferry Recreation Area, page 54

Westvaco Paper Company
Phone: 843-871-5000
**Westvaco Bluff Trail, page 72

Yancy McLeod Wildlife Services
1 Cooks Mountain Road
Eastover, SC 29044
Phone: 803-353-0440 or 803-353-2522
**Cooks Mountain, page 179

(Photo by B. Gilbert)

Lynn Greenlee with eastern tiger swallowtail, *Papilio glaucus*.

Bibliography

The books on which SCAN has based its flora and fauna lists are marked by two astericks. The other books listed are ones that members have found helpful. There are many more identification books on the market than are listed here. We recommend you visit your local library and book stores to find the ones most helpful to you.

** Arnett, R.H., Jr. 1960. The beetles of the United States (a manual for identification). The Catholic University of America Press, Washington. 1112 pp.

** Arnett, R.H., Jr. 1993. American insects, a handbook of the insects of America north of Mexico. The Sandhill Crane Press, Inc., Gainesville, Florida.

Baicich, Paul J. and Colin J.O. Harrison. 1997. A guide to the nests, eggs, and nestlings of North American birds. Academic Press. San Diego, Calif. 347 pp.

Baker, O.E., III, and D.B. Carmichael, Jr. 1989. South Carolina's furbearers. South Carolina Wildlife and Marine Resources Department, Columbia, South Carolina. 33 pp.

Behler, John L. and F. Wayne King. 1979. The Audubon Society field guide to North American reptiles and amphibians. Alfred A. Knopf. New York. 743 pp.

Borror, Donald J. and Richard E. White. 1970. A field guide to the insects of America north of Mexico. Houghton Mifflin Company. Boston, Mass. 404 pp.

Brockman, C. Frank. 1968. Trees of North America, a field guide to the major native and introduced species north of Mexico. Golden Press. New York. 280 pp.

Carter, R.M. 1993. Finding birds in South Carolina. University of South Carolina Press, Columbia, South Carolina. 254 pp.

** Cassie, B., J. Glassberg, P. Opler, R. Robbins, and G. Tudor.1995. North American Butterfly Association (NABA) checklist and English names of North American butterflies. North American Butterfly Association, Morristown, New Jersey. 43 pp.

** Ciegler, A. 1997. South Carolina lichens: A preliminary survey. Evansia 14(2):29-41.

Ciegler, J.C. 1997. Tiger beetles of South Carolina (Coleoptera: Cicindelidae). Coleopterists Bulletin, 51(2):177-192.

** Ciegler, J.C. 1999. Ground beetles and wrinkled bark beetles of South Carolina (Coleoptera: Geadephaga: Carabidae and Rhysodidae). Clemson, South Carolina. In press.

Claridge, Edward and Betty Ann Milligan. 1992. Animal Signatures. The Nova Scotia Museum field guide series. Nimbus Publishing. Halifax, Nova Scotia. 72 pp.

Clark, William S. 1987. A field guide to hawks. North America. Houghton Mifflin Company. Boston. 198 pp.

Cobb, Boughton. 1963. A field guide to the ferns and their related families of northeastern and central North America. Houghton Mifflin Company. Boston, Mass. 281 pp.

Conant, Roger, Robert C. Stebbins and Joseph T. Collins. 1992. Peterson first guide to reptiles and amphibians. Houghton Mifflin Company. New York, NY. 128 pp.

Conant, Roger and Joseph T. Collins. 1991. A field guide to reptiles and amphibians, eastern and central North America. Houghton Mifflin Company. Boston, Mass. 450 pp.

** Covell, C.V., Jr. 1984. A field guide to the moths of eastern North America. Houghton Mifflin Company, Boston. 496 pp.

Downie, N.M., and R.H. Arnett, Jr. 1996. The beetles of northeastern North America. Volume I and Volume II. The Sandhill Crane Press, Gainesville, Florida. 1721 pp.

Dunkle, S.W. 1989. Dragonflies of the Florida peninsula, Bermuda, and the Bahamas. Scientific Publishers, Gainesville, Florida. 154 pp.

Dunkle, S.W. 1990. Damselflies of Florida, Bermuda, and the Bahamas. Nature Guide No. 3. Scientific Publishers, Gainesville, Florida. 148 pp.

** Emerson, W.K., and M.K. Jacobson. 1976. The American Museum of Natural History guide to shells: Land, freshwater, and marine, from Nova Scotia to Florida. Alfred A. Knopf, New York. 482 pp.

Emerton, James H. 1961. The common spiders of the United States. Dover Publications, Inc. New York. 227 pp.

** Gaddy, L.L. and J.C. Morse. 1985. Common spiders of South Carolina, with an annotated checklist. Technical Bulletin 1094, South Carolina Agricultural Experiment Station, Clemson University, Clemson, South Carolina. 182 pp.

Giffen, M. 1997. South Carolina, A guide to backcountry travel and adventure. Out There Press, Asheville, North Carolina. 262 pp.

Glassberg, Jeffrey. 1999. Butterflies through binoculars, a field guide to the butterflies of eastern North America. Oxford University Press. Oxford, New York. 242 pp.

Gupton, Oscar W. and Fred C. Swope. 1989. Fall Wildflowers of the Blue Ridge and Great Smoky Mountains. University Press of Virginia. Charlottesville, Virginia. 208 pp.

** Hale, M.E. 1979. How to know the lichens. Wm. C. Brown Publishers, Dubuque, Iowa. 246 pp.

Halfpenny, James. 1986. A field guide to mammal tracking in North America. Johnson Books. Boulder. 161 pp.

Hallowell, Anne C. and Barbara G. Hallowell. 1981. Fern finder: a guide to native ferns of northeast and central North America. Nature Study Guide. Berkeley, Calif. 59 pp.

Harpootlian, P.J. 1999. The scarab beetles (Scarabaeidae) of South Carolina. Clemson University, Clemson, South Carolina. In press.

Harrison, Hal H. 1975. A field guide to birds' nests. United States east of the Mississippi River. Houghton Mifflin Company. Boston, Mass. 257 pp.

Headstrom, R. 1971. Identifying animal tracks. Mammals, birds, and other animals of the eastern United States. Dover Publications, Inc., New York. 141 pp.

Hefler, J.R. 1987. How to know the grasshoppers, crickets, cockroaches and their allies. Dover Publications, Inc., New York. 364 pp.

Holland, W.J. 1968. The moth book, a guide to the moths of North America. Dover publications, Inc. New York, NY. 479 pp.

Jerman, P.L. 1998. South Carolina nature viewing guide. South Carolina Department of Natural Resources, Columbia. 112 pp.

Johnson, B.A., ed. 1993. South Carolina Association of Naturalists: Scanning the wonders of our natural world, the first ten years. Editorial Resources Unlimited, Columbia, South Carolina. 517 pp.

** Kaplan, E.H. 1988. A field guide to southeastern and Caribbean seashores: Cape Hatteras to the Gulf Coast, Florida, and the Caribbean. Houghton Mifflin Company, Boston. 425 pp.

Kaston, B.J. 1978. How to know the spiders. The Pictue Key Nature Series. William C. Brown Company Publishers. Dubuque, Iowa. 272 pp.

Levi, H.W., and L.R. Levi. 1987. Spiders and their kin. Golden Press, New York. 160 pp.

** Lincoff, G.H. 1981. The Audubon Society field guide to North American mushrooms. Alfred A. Knopf, New York. 926 pp.

Manley, D.G. 1991. The velvet ants (Hymenoptera: Mutillidae) of South Carolina. Clemson University, Clemson, South Carolina. 55 pp.

** Martof, B.S., W.M. Palmer, J.R. Bailey, and J.R. Harrison III. 1980. Amphibians and reptiles of the Carolinas and Virginia. University of North Carolina Press, Chapel Hill, North Carolina.

McKnight, Kent H. and Vera B. McKnight. 1987. A field guide to mushrooms of North America. Houghton Mifflin Company. Boston, Mass. 429 pp.

Milne, Lorus and Margery Milne. 1995. National Audubon Society field guide to North American insects and spiders. Alfred A. Knopf. New York. 989 pp.

Mitchell, Robert T. and Herbert S. Zim. 1964. Butterflies and moths, a guide to the more common American species. Golden Press. New York. 160 pp.

Mohlenbrock, Robert H. and John W. Thieret. 1987. Trees,a quick reference guide to trees of North America. Macmillan Publishing Company. New York. 155 pp.

Moody, Richard. 1986. Fossils. How to find and identify over 300 genera. Macmillan Publishing Company. New York. 192 pp.

Murphy, Carolyn Hanna. 1995. Carolina rocks! The geology of South Carolina. Sandlapper Publishing Co., Inc. Orangeburg, South Carolina. 261 pp.

Neelands, R.W. 1974. Important trees of eastern forest. U.S. Department of Agriculture, Forest Service. 111 pp.

Niering, William A. 1995. National Audubon Society field guide to North American wildflowers. Alfred A. Knoft. New York. 887 pp.

Opler, P.A., and G.O. Krizek. 1984. Butterflies east of the Great Plains. The Johns Hopkins University Press, Baltimore. 294 pp.

Opler, Paul A. 1994. Peterson first guide to butterfies and moths. Houghton Mifflin Company. New York, NY. 128 pp.

** Otte, D. 1981. The North American grasshoppers. Volume 1. Acrididae: Gomphocerinae and Acridinae. Harvard University Press, Cambridge, Massachusetts. 275 pp.

** Otte, D. 1984. The North American grasshoppers. Volume 2. Acrididae: Oedipodinae. Harvard University Press, Cambridge, Massachusetts. 366 pp.

Peterson, Roger Tory. 1986. Peterson first guide to birds of North America. Houghton Mifflin Company. New York. 128 pp.

Petrides, George A. 1988. A field guide to eastern trees, eastern United States and Canada. Houghton Mifflin Company. Boston. 272 pp.

Porcher, Richard D. 1995. Wildflowers of the Carolina Lowcountry. University of South Carolina Press. Columbia, SC. 302 pp.

Potter, Eloise F., James F. Parnell and Robert P. Teulings. 1980. Birds of the Carolinas. The University of North Carolina Press. Chapel Hill. 408 pp.

** Purvis, O.W., B.J. Coppins, D.L. Hawksworth, P.W. James, and D.M. Moore. 1992. The lichen flora of Great Britain and Ireland. Natural History Museum Publications, London. 710 pp.

Pyle, Robert Michael. 1981. The Audubon Society field guide to North American butterflies. Alfred A. Knopf, New York. 916 pp.

** Radford, A.E., H.E. Ahles, and C.R. Bell. 1968. Manual of the vascular flora of the Carolinas. University of North Carolina Press, Chapel Hill, North Carolina. 1183 pp.

Robbins, Chandler S., Bertel Bruun and Herbert S. Zim. Birds of North America. Golden Press. New York. 340 pp.

** Ruppert, E.E., and R.S. Fox. 1988. Seashore Animals of the Southeast. University of South Carolina Press, Columbia, South Carolina. 429 pp.

** Scott, S.L., ed. 1983. Field guide to the birds of North America. National Geographic Society, Washington. 464 pp.

Shuttleworth, Floyd S. and Herbert S. Zim. 1987. Mushrooms and other non-flowering plants. Golden Press. New York. 160 pp.

Slater, J.A., and R.M. Baranowski. 1978. How to know the true bugs. Wm. C. Brown Co., Dubuque, Iowa. 256 pp.

Smith, Hobart M. and Edmond D. Brodie, Jr. 1982. Reptiles of North America. Golden Press. New York. 240 pp.

Stokes, Donald and Lillian Stokes. 1996. Stokes field guide to birds. Eastern region. Little, Brown and Company. Boston. 471 pp.

Thompson, Ida. 1982. The Audubon Society field guide to North American Fossils. Alfred A Knopf. New York. 846 pp.

** Thompson, P. 1985. Thompson's guide to freshwater fishes. How to identify the common freshwater fishes of North America. Houghton Mifflin Co., Boston. 205 pp.

Watts, May Theilgaard. 1991. Tree finder, a manual for the identification of trees by their leaves. Rochester, NY. 58 pp.

** Webster, W.D., J.F. Parnell, and W.C. Biggs, Jr. 1985. Mammals of the Carolinas, Virginia, and Maryland. University of North Carolina Press, Chapel Hill, North Carolina. 255 pp.

Westfall, M.J., Jr., and M.L. May. 1996. Damselflies of North America. Scientific Publishers, Gainesville, Florida. 650 pp.

** White, T.R., K.J. Tennessen, R.C. Fox, and P.H. Carlson. 1980. The aquatic insects of South Carolina. Part I: Anisoptera (Odonata). Station Bulletin 632, South Carolina Agricultural Experiment Station, Clemson University, Clemson, South Carolina. 153 pp.

** White, T.R., K.J. Tennessen, R.C. Fox, and P.H. Carlson. 1983. The aquatic insects of South Carolina. Part II: Zygoptera (Odonata). Station Bulletin 648, South Carolina Agricultural Experiment Station, Clemson University, Clemson, South Carolina. 72 pp.

Wilson, Jim and Anselm Atkins. 1998. Common birds of Atlanta. Hexagon Publications. Decatur, GA. 59 pp.

Wright, Amy Bartlett. 1993. Peterson's first guide to caterpillars of North America. Houghton Mifflin Company. Boston, Mass. 128 pp.

(Photo by Fran Beals)

Frances Crawford

SOUTH CAROLINA ASSOC. OF NATURALISTS

SCAN

SHARING THE ENCOUNTER OF OUR NATURAL WORLD

Common Name List

In the index which follows, flora and fauna are listed together in alphabetical order by scientific name. If you do not know the scientific name, first find the common name and make a note of the scientific name. Then refer to the Scientific Name Index to locate the county or counties in which the plant or animal has been seen.

Field trip reports are arramged by counties in alphabetical order. Species are organized under each field trip report in a taxonomic order developed by SCAN and based on standard taxonomic schemes. Flora are listed first; fauna are listed second; fossils are listed last.

Acanthospermum: *Acanthospermum australe*
Adder's-mouth, Green: *Malaxis unifolia*
Admiral, Red: *Vanessa atalanta rubria*
Agaric, Fly: *Amanita muscaria, Amanita muscaria formosa*
Agaric, Orange Moss: *Rickenella fibula*
Agrimony: *Agrimonia parviflora*
Alder, Tag: *Alnus serrulata*
Alga: *Protococcus sp.*
Alligator, American: *Alligator mississippiensis*
Alligator-weed: *Alternanthera philoxeroides*
Aloe, False: *Agave virginica*
Alumroot: *Heuchera americana, Heuchera villosa, Heuchera sp.*
Amanita, Booted: *Amanita cothurnata*
Amanita, Citron: *Amanita citrina*
Amanita, Cleft-foot: *Amanita brunnescens*
Amanita, Gemmed: *Amanita gemmata*
Amanita, Rag-veil: *Amanita cinereopannosa*
Amanita, Strangulated: *Amanita inaurata*
Amaranth, Beach: *Amaranthus pumilus*
Amberwing, Eastern (dragonfly): *Perithemis tenera*
Amphipod: *Crangonyx serratus*
Andromeda: *Pieris floribunda*
Angelica: *Angelica venenosa, Angelica sp.*
Angel's-wings: *Pleurocybella porrigens*
Angel-wing (shell): *Cyrtopleura costata*
Angel-wing, False (shell): *Petricola pholadiformis*
Angle, Minor (moth): *Semiothisa minorata*
Angle-pod: *Matelea gonocarpa, Matelea suberosa, Matelea sp.*
Anglewing (butterfly): *Polygonia sp.*
Anhinga: *Anhinga anhinga*
Anise: *Illicium sp.*
Anise-root: *Osmorhiza claytonii*
Anne's-lace, Queen: *Daucus carota*
Anoda: *Anoda cristata*
Anole, Carolina: *Anolis carolinensis*
Ant, Black Carpenter: *Camponotus pennsylvanicus*
Ant, Fire: *Solenopsis geminata, Solenopsis sp.*
Ant, Florida Carpenter: *Camponotus abdominalis*
Ant, Florida Harvester: *Pogonomyrmex badius*
Ant, Red Imported Fire: *Solenopsis invicta*
Ant: *Crematogaster ashmeadi*
Anthaenantia (grass): *Anthaenantia villosa*
Antlion: *Myrmeleon immaculatum, Myrmeleon sp.*
Aphid, Poplar Petioleball: *Pemphigus populitransversus*
Aphid, Woolly Alder: *Prociphilus tesselatus*
Apple, Azalea Pinkster: *Exobasidium vaccinii*
Apple, Blackjack Oak (gall wasp): *Disholcaspis lapiei*
Apple, Common: *Malus pumila*
Arabis: *Arabis laevigata*
Ark, Blood: *Anadara ovalis*
Ark, Incongruous: *Anadara brasiliana*
Ark, Ponderous: *Noetia ponderosa*
Ark, Transverse: *Anadara transversa*
Ark: *Anadara aequicostata, Anadara lienosa, Noetia limula*
Armadillo (scute): *Chlamydotherium sp.*
Arrowhead, Bunched: *Sagittaria fasciculata*
Arrowhead: *Sagittaria engelmanniana, Sagittaria falcata, Sagittaria graminea graminea, Sagittaria latifolia, Sagittaria sp.*
Arrowwood: *Viburnum dentatum, Viburnum dentatum lucidum*
Arum, Arrowleaf: *Peltandra virginica*
Ash, American: *Fraxinus americana*
Ash, Green: *Fraxinus pennsylvanica, Fraxinus pennsylvanica subintegerrima*

Ash, Pumpkin: *Fraxinus tomentosa*
Ash, Red: *Fraxinus tomentosa*
Ash, Water: *Fraxinus caroliniana*
Ash, White: *Fraxinus americana americana, Fraxinus americana biltmoreana*
Ash: *Fraxinus sp.*
Asparagus: *Asparagus officinalis*
Asphodel, False: *Tofieldia racemosa*
Aster, Climbing: *Aster carolinianus*
Aster, Frost: *Aster pilosus, Aster pilosus pilosus*
Aster, Heart-leaved: *Aster divaricatus, Aster cordifolius*
Aster, Salt-marsh: *Aster tenuifolius*
Aster, Starved: *Aster lateriflorus*
Aster, Stiff-leaved: *Aster linariifolius*
Aster, White-topped: *Aster paternus, Aster solidagineus, Aster tortifolius*
Aster: *Aster concolor, Aster curtisii, Aster dumosus, Aster puniceus, Aster squarrosus, Aster undulatus, Aster sp.*
Atamasco-lily: *Zephyranthes atamasco*
Avens: *Geum sp.*
Avocet, American: *Recurvirostra americana*
Azalea, Dwarf: *Rhododendron atlanticum*
Azalea, Flame: *Rhododendron calendulaceum*
Azalea, Hammocksweet: *Rhododendron viscosum serrulatum*
Azalea, Piedmont: *Rhododendron canescens*
Azalea: *Rhododendron sp.*
Azure, Spring (butterfly): *Celastrina ladon, Lycaenopsis argiolus*
Baby-blue-eyes: *Nemophila microcalyx*
Baby's-ear (shell): *Sinum perspectivum*
Backswimmer: *Notonecta sp.*
Bald-cypress: *Taxodium distichum*
Bamboo, Yellow-groove: *Phyllostachys aurea*
Band-winged Grasshopper Psinidia fenestralis
Baneberry: *Actaea pachypoda*
Barbara's-buttons: *Marshallia graminifolia, Marshallia obovata*
Barnacle, Ivory: *Balanus eburneus*
Barnacle, Rock: *Balanus sp.*
Barnacle, Sea-whip: *Conopea galeata*
Barnacle, Tropical Goose: *Lepas anatifera*
Baskettail, Common (dragonfly): *Epitheca cynosura*
Baskettail, Mantled (dragonfly): *Epitheca semiaquea*
Bass, Largemouth: *Micropterus salmoides*
Basswood: *Tilia americana, Tilia caroliniana, Tilia heterophylla, Tilia sp.*
Bastard-toadflax: *Comandra umbellata*
Batchelor's-buttons: *Centaurea cyanus*
Bayberry, Southern: *Myrica cerifera*
Bayberry, Swamp: *Myrica heterophylla*
Beach-elder: *Iva imbricata*
Beachhopper, Saltmarsh: *Orchestia grillus*
Beach-pea: *Strophostyles helvola*
Beak-rush: *Rhynchospora cephalantha, Rhynchospora chalarocephala, Rhynchospora corniculata, Rhynchospora debilis, Rhynchospora filifolia, Rhynchospora globularis, Rhynchospora glomerata, Rhynchospora grayi, Rhynchospora inexpansa, Rhynchospora intermixta, Rhynchospora macrostachya, Rhynchospora megalocarpa, Rhynchospora perplexa, Rhynchospora plumosa, Rhynchospora rariflora, Rhynchospora torreyana, Rhynchospora sp.*
Bean, Wild: *Strophostyles helvola*
Beardtongue: *Penstemon sp.*
Bear-grass: *Yucca filamentosa*
Bearsfoot: *Polymnia uvedalia*
Beautyberry: *Callicarpa americana*
Beaver (dam): *Castor canadensis*
Beaver (gnaw mark): *Castor canadensis*
Beaver: *Castor canadensis*
Bedstraw: *Galium circaezans, Galium hispidulum, Galium sp.*
Bee, Carpenter: *Xylocopa virginica, Xylocopa sp.*
Bee, Honey: *Apis mellifera*
Bee, Leaf-cutting: *Megachile sp.*
Bee-balm: *Monarda clinopodia, Monarda didyma*
Beech: *Fagus grandifolia*
Beech-drops: *Epifagus virginiana*
Beech-Fern, Broad: *Thelypteris hexagonoptera*
Beetle Short-horned Leaf: *Saxinus omogera*
Beetle, 12-spotted Cucumber: *Diabrotica duodecimpunctata*
Beetle, Alder Leaf: *Chrysomela interrupta*
Beetle, Alligatorweed Flea: *Agasicles hygrophila*
Beetle, Ambrosia: *Xyloborus inermis*
Beetle, Ant-like Leaf: *Ariotus sp.*
Beetle, Ant-like Longhorn: *Euderces picipes*
Beetle, Ant-nest Scarab: *Cremastocheilus castaneae*
Beetle, Aphodiine Dung: *Aphodius lividus, Aphodius sp., Ataenius cognatus, Ataenius sp.*
Beetle, Asiatic Garden: *Maladera castanea*
Beetle, Azalea Leaf: *Galerucella rufosanguinea*
Beetle, Bark-gnawing: *Corticotomus sp., Tenebroides dubia, Tenebroides laticollis*
Beetle, Bean Leaf: *Cerotoma trifurcata*
Beetle, Black Blister: *Epicauta pennsylvanica*
Beetle, Black-spotted Lady: *Anatis 15-punctata*
Beetle, Blister: *Epicauta trichrus, Epicauta vittata, Epicauta sp., Lytta aenea, Meloe sp., Nemognatha piazata, Pyrota lineata*
Beetle, Blue-black Blister: *Meloe angusticollis*
Beetle, Branch Twig Borer: *Dysides obscurus*
Beetle, Broad-bodied Leaf: *Calligrapha rhoda, Chrysomela scripta, Prasocuris vittata, Zymogramma suturalis*
Beetle, Buprestid: *Buprestis rufipes, Dicerca divaricata, Dicerca tuberculata*
Beetle, Burrowing Water: *Copius inflatus, Hydrocanthus iricolor, Hydrocanthus oblongus, Suphisellus sp.*

Beetle, Carrion: *Nicrophorus sp., Silpha americana, Silpha inaequalis, Silpha noveboracensis, Silpha surinamensis*
Beetle, Checkered: *Enoclerus ichneumoneus, Enoclerus thoracicus, Thanasimus Dubius*
Beetle, Click: *Agriotes sp., Ampedus areolatus, Ampedus sp., Athous rufifrons, Cardiophorus sp., Heteroderes sp., Horistonotus curiatus, Lacon marmoreus, Limonius sp.*
Beetle, Colorful Foliage Ground*: Apenes sinuata, Cymindis americana, Lebia analis, Lebia lobulata, Lebia ornata, Nemotarsus elegans*
Beetle, Comb-clawed: *Hymenorus sp., Isomira sericea, Isomira sp.*
Beetle, Contractile Scarab: *Cloeotus aphodioides*
Beetle, Convergent Lady: *Hippodamia convergens*
Beetle, Corn Flea: *Chaetocnema pulicaria*
Beetle, Corn-seed: *Agonoderus lecontei*
Beetle, Crab-like Rove: *Lordithon pygmaeus, Lordithon sp., Tachinus axillaris, Tachinus sp.*
Beetle, Crawling Water: *Haliplus fasciatus, Haliplus triopsus, Peltodytes dietrichi, Peltodytes duodecimpunctatus, Peltodytes muticus, Peltodytes oppositus, Peltodytes sexmaculatus, Peltodytes shermani, Peltodytes sp.*
Beetle, Cross-toothed Rove: *Oxyporus femoralis, Oxyporus sp.*
Beetle, Cucumber: *Diabrotica sp.*
Beetle, Cylindrical Bark: *Bitoma sp., Colydium lineola, Philothermus glabriculus*
Beetle, Dermestid: *Cryptorhopalum picicorne, Cryptorhopalum sp.*
Beetle, Dingy Ground: *Agonoderus sp., Anisodactylus rusticus, Anisodactylus sayi, Anisodactylus sp., Discoderus paralleleus, Discoderus sp., Harpalus caliginosus, Harpalus pennsylvanicus, Harpalus spadiceus, Selenophorus ellipticus, Stenolophus ochropezus*
Beetle, Dung: *Ateuchus sp., Copris carolina, Onthophagus hecate, Onthophagus janus, Onthophagus orpheus, Onthophagus orpheus orpheus, Onthophagus tuberculifrons, Onthophagus sp., Phanaeus igneus, Phanaeus nigrocyaneus, Phanaeus vindex, Phanaeus sp.*
Beetle, Dung: *Canthon chalcites, Glaphyrocanthon viridis*
Beetle, Dusky Lady: *Scymnus puncticollis, Scymnus terminatus, Scymnus sp.*
Beetle, E. Spotted Cucumber: *Diabrotica undecimpunctata howardi*
Beetle, Earth-boring Dung: B*olboceras sp., Geotrupes blackburnii blackburnii, Geotrupes egeriei*
Beetle, Eggplant Flea: *Epithrix fuscula*
Beetle, False Ant-like Flower: *Pedilus collaris, Pedilus sp.*
Beetle, False Blister: *Asclera ruficollis, Asclera sp., Oxacis sp., Oxycopis sp., Xanthochroa sp.*
Beetle, False Bombadier: *Galeritula bicolor*
Beetle, False Darkling: *Allopoda lutea, Anaspis sp., Canifa pallipes, Eustrophinus bicolor*
Beetle, False Skin: *Diplocoelus rudis*
Beetle, Feather-winged: *Nossidium sp.*
Beetle, Flat Bark: *Brontes dubius, Laemophloeus modestus, Silvanus bidentatus*
Beetle, Flat-faced Longhorn: *Goes oculata, Hippopsis lemniscata*
Beetle, Flea: *Altica marevagans, Altica sp., Crepidodera nitens, Dibolia sp., Disonycha abbreviata, Disonycha caroliniana, Disonycha collata, Disonycha sp., Oedionychis miniata, Oedionychis sp., Systena elongata, Systena frontalis, Systena taeniata*
Beetle, Flower Longhorn: *Anoplodera mutabilis, Anoplodera sp., Gaurotes cyanipennis, Grammoptera sp., Leptura lineola, Rhagium inquisitor, Strangalina bicolor, Strangalina famelica, Strangalina luteicornis, Toxotus cinnamopterus, Typocerus velutinus, Typocerus sp.*
Beetle, Flower Scarab: *Euphoria sepulchralis, Trichiotinus piger, Trigonopeltastes delta, Trigonopeltastes sp.*
Beetle, Four-spotted Blister: *Tetraonyx sp.*
Beetle, Garden: *Serica sp.*
Beetle, Giant Stag: *Lucanus elaphus*
Beetle, Giant Water Scavenger: *Dibolocelus ovalis, Hydrophilus triangularis*
Beetle, Grape Flea: *Altica chalybea*
Beetle, Grape: *Pelidnota sp.*
Beetle, Green Foliage Ground: *Lebia viridis*
Beetle, Green June: *Cotinus nitida*
Beetle, Green Tiger: *Cicindela sexguttata*
Beetle, Ground: *Calosoma scrutator*
Beetle, Handsome Fungus: *Stenotarsus sp.*
Beetle, Hickory Girdler: *Oncideres cingulata*
Beetle, Hister: *Epierus pulicarius, Hister depurator, Margarinotus sp., Platysoma carolinus, Platysoma lecontei*
Beetle, Hister-like Dung: *Ateuchus histeroides*
Beetle, Horned Fungus: *Bolitotherus cornutus*
Beetle, Japanese: *Popillia japonica*
Beetle, June: *Phyllophaga ilicis, Phyllophaga grandis, Phyllophaga prunina, Phyllophaga sp.*
Beetle, Kalmia Flea: *Altica viridana*
Beetle, Lady: *Brachyacantha sp., Coccinella sp., Coleomegilla fuscilabris, Hyperaspis sp.*
Beetle, Lagriid Bark: *Statira sp.*
Beetle, Large Rove: *Staphylinus cinnamopterus, Staphylinus fossator, Staphylinus maxillosus, Staphylinus sp.*
Beetle, Larger Elm Leaf: *Monocesta coryli*
Beetle, Leaf Buprestid: *Brachys aerosus, Brachys sp.*
Beetle, Leaf: *Anomoea sp., Coscinoptera dominicana, Oedionychis concinna*
Beetle, Leaf-mining: *Anoplitis inaequalis, Chalepus scapularis*
Beetle, Lightning: *Ellychnia sp., Lucidota atra. Lucidota sp., Phausis sp., Photinus pyralis, Photuris sp.*

Beetle, Lizard: *Languria lecontei, Languria mozardi*
Beetle, Locust Leaf-mining: *Chalepus dorsalis*
Beetle, Longhorn: *Hyperplatys maculatus, Megacyllene robiniae, Oberea basalis*
Beetle, Long-horned Leaf: *Donacia subtilis, Donacia sp.*
Beetle, Long-jointed Bark: *Arthromacra aenea*
Beetle, Marsh: *Cyphon obscurus, Cyphon variabilis, Cyphon sp., Prionocyphon discoideus, Scirtes orbiculatus, Scirtes sp.*
Beetle, Metallic Wood-boring: *Taphrocerus sp.*
Beetle, Minute Ground: *Bembidion americanum, Bembidion confusum, Bembidion dilitatum, Bembidion versicolor, Bembidion sp., Mioptachys flavicauda, Tachys ephippiatus, Tachyta nana inornata*
Beetle, Minute Lady: *Delphastus pusillus*
Beetle, Minute Tree-fungus: *Xestocis sp.*
Beetle, Multicolored Asian Lady: *Harmonia axyridis*
Beetle, Narrow-waisted Bark: *Mycterus scaber*
Beetle, Net-winged: *Caenia dimidiata, Calopteron reticulatum, Calopteron sp., Eros humeralis, Plateros lictor, Plateros sollicitus, Plateros sp.*
Beetle, Notch-mouthed Ground: *Dicaelus ambiguus, Dicaelus dilitatus, Dicaelus elongatus, Dicaelus ovalis, Dicaelus sp.*
Beetle, Oak Flea: *Oedionychis quercata*
Beetle, Obscure Rove: *Aleochara sp., Atheta flaviola, Atheta pennsylvanica, Atheta sp., Gyrophaena corruscula, Gyrophaena dissimilis*
Beetle, Oval Ground: *Oodes amaroides*
Beetle, Oval Leaf: *Colaspis favosa, Colaspis sp., Graphops curtipennis, Metachroma longicollis, Metachroma pallidum, Nodonota tristis, Nodonota sp., Paria canella aterrimus, Paria canella gilvipes, Paria canella quadrinotata, Paria sellatus, Paria thoracica, Rhabdopterus picipes, Rhabdopterus sp.*
Beetle, Paederine Rove: *Rugilus dentatus, Sunius sp.*
Beetle, Pale-striped Flea: *Systena blanda*
Beetle, Palmetto Tortoise: *Hemisphaerota cyanea*
Beetle, Pedunculate Ground: *Ardistomis viridis, Ardistomis sp., Clivina rufa, Dyschirius sp., Pasimachus depressus, Pasimachus sublaevis, Pasimachus sp., Scarites subterraneus, Scarites sp.*
Beetle, Pinhole Borer: *Platypus flavicornis*
Beetle, Pleasing Fungus: *Megalodacne fasciata, Megalodacne heros, Triplax macra, Triplax thoracica, Triplax sp., Tritoma angulata, Tritoma thoracica, Tritoma sp.*
Beetle, Predacious Diving: *Agabetes acuductus, Agabus johannis, Agabus punctatus, Bidessus sp., Copelatus chevrolati chevrolati, Coptotomus interrogatus, Coptotomus lenticus, Graphoderus liberus, Hydroporus carolinus, Hydroporus cimicoides, Hydroporus clypealis, Hydroporus hybridus, Hydroporus lobatus, Hydroporus niger, Hydroporus undulatus, Hydroporus vittatipennis, Hydroporus sp., Hydrovatus pustulatus, Ilybius biguttulus, Laccophilus fasciatus, Laccophiilus fasciatus rufus, Laccophilus maculosus maculosus, Laccophilus proximus, Laccophilus schwartzi, Thermonectes basilaris, Thermonetus basilaris basilaris*
Beetle, Punctate Leaf: *Orsodacne atra, Syneta sp.*
Beetle, Purple Ground: *Dicaelus purpuratus*
Beetle, Red Net-winged: *Dictyopterus aurora*
Beetle, Red-bellied Tiger: *Cicindela rufiventris*
Beetle, Red-horned Grain: *Platydema ruficorne*
Beetle, Rhinoceros: *Strategus antaeus, Strategus sp., Xyloryctes jamaicensis,*
Beetle, Riffle: *Limnius latiusculus*
Beetle, Round Sand: *Omophron americanum*
Beetle, Round-necked Longhorn: *Elaphidion mucronatum, Molorchus bimaculatus, Molorchus sp., Tilloclytus geminata*
Beetle, Rove: *Anthobium sp., Conotelus sp, Omalium sp., Ontholestes sp., Quedius sp., Oxypoda sp., Oxytelus sp,. Tachinus axillaris*
Beetle, Sap-feeding: *Carpophilus floralis, Carpophilus sp., Epuraea sp. Pallodes sp., Pocadius sp., Prometopia sexmaculata, Stelidota sp.*
Beetle, Scarab: *Dichelonyx elongata, Dichelonyx fuscula, Dichelonyx subvittata, Dichelonyx sp.*
Beetle, Scarab: *Valgus sp.*
Beetle, Seed-eating Ground: *Amara cupreolata, Amara sp.*
Beetle, Seven-spotted Lady: *Coccinella septempunctata*
Beetle, Shining Flower: *Acyclomus piceus, Acylomus sp., Olibrus sp.*
Beetle, Shining Fungus: *Scaphisoma convexa, Scaphisoma suturalis*
Beetle, Shining Leaf: *Lema albini, Lema sayi, Lema sp.*
Beetle, Silken Fungus: *Henoticus sp.*
Beetle, Skeletonizing Leaf: *Galerucella notulata, Galerucella nymphaeae, Luperodes cyanellus, Luperodes sp.*
Beetle, Skin: *Dermestes talpinus, Trox tuberculatus, Trox tuberculosus, Trox sp,*
Beetle, Small Carrion: *Catops sp.*
Beetle, Smartweed Flea: *Disonycha pensylvanica*
Beetle, Soft-bodied Plant: *Ptilodactyla sp.*

Beetle, Soft-winged Flower: *Ablechrus sp., Attalus circumscriptus, Attalus scincetus, Attalus sp., Collops quadrimaculatus, Collops vittatus, Collops sp., Hypebaeus oblitus, Hypebaeus sp., Melyrodes cribrata, Malachius sp.*
Beetle, Soldier: *Cantharis bilineatus, Cantharis lineola, Cantharis sp., Chauliognathus marginatus, Chauliognathus pennsylvanicus, Malthodes curvatus, Malthodes sp., Podabrus basilaris, Podabrus sp., Silis latilobus, Silis sp*
Beetle, Spotless Lady: *Cycloneda sanguinea*
Beetle, Spotted Lady: *Coleomegilla maculata*
Beetle, Stag: *Pseudolucanus placidus*
Beetle, Striped Click: *Ctenicera sp.*
Beetle, Striped Cucumber: *Acalymna vittatum*
Beetle, Tiger: *Cicindela punctulata, Cicindela repanda, Cicindela scutellaris unicolor, Cicindela tranquebarica, Cicindela tortuosa, Cicindela unipunctata, Cicindela sp.*
Beetle, Toed-winged: *Ptilodactyla serricollis*
Beetle, Tooth-necked Longhorn: *Stenodontes dasytomus*
Beetle, Tortoise: *Deloyola guttata*
Beetle, Tumbling Flower: *Mordella atrata, Mordella melaena, Mordella sp., Mordellistena ambusta, Mordellistena fuscata, Mordellistena pubescens, Mordellistena unicolor, Mordellistena sp.*
Beetle, Twenty-spotted Lady: *Psyllobora vigintimaculata*
Beetle, Twice-stabbed Lady: *Chilocorus stigma*
Beetle, Warty Leaf: *Chlamisus gibbosus*
Beetle, Water Scavenger: *Berosus aculeatus, Berosus infuscatus, Berosus pantherinus, Berosus sp., Enochrus consortus, Enochrus ochraceus, Enochrus sublongus, Enochrus sp., Helocombus bifidus, Hydrobius melaenus, Hydrochus excavatus, Hydrochus inaequalis, Hydrochara soror, Phaenonotum estriatum, Paracymus subcupreus Tropisternus blatchleyi blatchleyi, Tropisternus collaris striolatus, Tropisternus lateralis nimbatus, Tropisternus mixtus, Tropisternus natator, Tropisternus quadristriatus quadristriatus, Tropisternus striolatus*
Beetle, Water-penny: *Psephenus herricki*
Beetle, Whirligig: *Dineutus assimilis, Dineutus carolinus, Dineutus discolor, Dineutus emarginatus, Dineutus horni, Dineutus robertsi, Dineutus serrulatus serrulatus, Dineutus sp., Gyrinus analis, Gyrinus borealis, Gyrinus frosti, Gyrinus lugens, Gyrinus rockinghamensis, Gyrinus sp.*
Beetle, White Beach Tiger: *Cicindela dorsalis media*
Beetle, White-fringed: *Graphognathus leucoloma, Graphognathus sp.*
Beetle, Wide-necked Longhorn: *Prionus laticollis*
Beetle, Woodland Ground: *Abacidus sculptus, Agonum extensicolle, Agonum errans, Agonum ferreus, Agonum moerens, Agonum sp., Evarthrus sp., Loxandrus minor, Odontonyx sp., Piesmus mondelus, Pterostichus chalcites, Pterostichus sp.*
Beetle, Yellow-marked Buprestid: *Acmaeodera tubulus*
Beetle, Yellow-spotted Lady: *Brachyacantha ursina*
Beggar-lice: *Desmodium sp.*
Beggar-ticks: *Bidens frondosa, Bidens laevis, Bidens pilosa, Bidens sp., Desmodium strictum*
Belemnite: *Belemnitella americana*
Bellwort: *Uvularia perfoliata Uvularia pudica, Uvularia sessilifolia, Uvularia sp.*
Bess-beetle: *Popilius disjunctus*
Bicolor: *Lespedeza bicolor*
Billbug: *Sphenophorus sp.*
Billfish (bill fragment): *Cylindracanthus rectus*
Bindweed: *Calystegia sericata*
Bindweed: *Polygonum convolvulus*
Birch, Cherry: *Betula lenta*
Birch, River: *Betula nigra*
Birch, Yellow: *Betula lutea*
Bird's-nest, White-egg: *Crucibulum laeve*
Bird's-nest: *Cyathus sp.*
Bishop's-weed, Mock: *Ptilimnium capillaceum*
Bittern, American: *Botaurus lentiginosus*
Bitterweed: *Helenium amarum*
Blackberry: *Rubus argutus, Rubus canadensis, Rubus sp.*
Blackbird, Brewer's: *Euphagus cyanocephalus*
Blackbird, Red-winged: *Agelaius phoeniceus*
Blackbird, Rusty: *Euphagus carolinus*
Black-club: *Trichoglossum geoglossum*
Black-eyed-Susan: *Rudbeckia hirta, Rudbeckia sp.*
Blackfly: *Simulium sp.*
Blackgum, Swamp: *Nyssa sylvatica biflora*
Blackgum: *Nyssa sylvatica*
Blackgum: *Nyssa sylvatica sylvatica*
Black-haw, Rusty: *Viburnum rufidulum*
Black-haw: *Viburnum prunifolium*
Black-knot: *Apiosporina morbosa*
Black-root: *Pterocaulon pycnostachyum*
Black-trumpet: *Craterellus fallax*
Bladdernut: *Staphylea trifolia*
Bladderpod: *Glottidium vesicarium*
Bladder-stalks: *Physalacria inflata*
Bladderwort, Floating: *Utricularia inflata*
Bladderwort, Purple: *Utricularia purpurea*
Bladderwort, Wiry: *Utricularia subulata*
Bladderwort: *Utricularia biflora, Utricularia sp.*
Blazing-star: *Chamaelirium luteum, Liatris elegans, Liatris graminifolia, Liatris secunda, Liatris spicata, Liatris spicata resinosa, Liatris sp.*
Bloodroot: *Sanguinaria canadensis*
Bluebead-lily: *Clintonia borealis*

Bluebell: *Campanula divaricata*
Blueberry, Black Highbush: *Vaccinium atrococcum*
Blueberry, Creeping: *Vaccinium crassifolium*
Blueberry, Elliott's: *Vaccinium elliottii*
Blueberry, Highbush: *Vaccinium corymbosum*
Blueberry, Low-bush: *Vaccinium tenellum, Vaccinium vacillans*
Blueberry: *Vaccinium constablaei, Vaccinium myrsinites, Vaccinium sp.*
Bluebird, Eastern: *Sialia sialis*
Blue-curls: *Trichostema dichotomum*
Blue-eyed-grass: *Sisyrinchium albidum, Sisyrinchium angustifolium, Sisyrinchium mucronatum, Sisyrinchium mucronatum atlanticum, Sisyrinchium sp.*
Bluegill: *Lepomis macrochirus*
Bluespread, Velvet: *Pulcherrichium caeruleum*
Blue-star: *Amsonia tabernaemontana, Amsonia ciliata*
Bluestem, Little: *Andropogon scoparius*
Bluet (damselfly): *Enallagma sp. Houstonia caerulea, Houstonia longifolia, Houstonia purpurea, Houstonia serpyllifolia, Houstonia tenuifolia. Houstonia sp.*
Bluet, Atlantic (damselfly): *Enallagma doubledayi*
Bluet, Blackwater (damselfly): *Enallagma weewa*
Bluets, Small: *Houstonia pusilla*
Bluets, Trailing: *Houstonia procumbens*
Blusher, Yellow: *Amanita flavorubescens*
Bobcat: *Felis rufus*
Bobwhite, Northern: *Colinus virginianus*
Bog-buttons: *Lachnocaulon anceps, Lachnocaulon beyrichianum. Lachnocaulon sp.*
Bogmat: *Wolffiella floridana*
Bogmoss: *Mayaca aubletii, Mayaca fluviatilis, Mayaca sp.*
Bolete, Birch: *Austroboletus betula*
Bolete, Bitter: *Tylopilus felleus*
Bolete, Frost's: *Boletus frostii*
Bolete, Gilled: *Phylloporus rhodoxanthus*
Bolete, Graceful: *Tylopilus gracilis*
Bolete, King: *Boletus edulis*
Bolete, Red-cracked: *Boletus chrysenteron*
Bolete, Rosy Larch: *Fuscoboletinus ochraceoroseus*
Bolete, Shaggy-stalked: *Boletellus betula*
Bolete, Violet-gray: *Tylopilus plumbeoviolaceus*
Bolete, Yellow-cracked: *Boletus subtomentosus*
Bolete, Zeller's: *Boletus zelleri*
Bolete: *Austroboletus gracilis, Boletus curtisii, Boletus erythropus*
Bonamia: *Bonamia humistrata, Bonamia sp.*
Borer, Currant Tip: *Cyrtinus pygmaeus*
Borer, Flat-headed Bald Cypress: *Acmaeodera pulchella*
Borer, Large Flat-headed Pine: *Chalcophora virginiensis*
Borer, Locust: *Megacyllene robiniae*
Borer, Stem: *Oberea basalis*
Boxelder: *Acer negundo*
Brachyelytrum (grass): *Brachyelytrum erectum*
Bracken: *Pteridium aquilinum*
Bream: *Lepomis sp.*
Breath-of-spring, Sweet: *Lonicera fragrantissima*
Brittle-gill, Green: *Russula virescens*
Broomrape: *Orobanche uniflora*
Broomsedge: *Andropogon ternarius, Andropogon virginicus, Andropogon sp.*
Bryozoan, Rubbery: *Alcyonidium hauffi*
Bryozoan, Variable Crust: *Schizoporella errata*
Buchnera: *Buchnera floridana*
Buckeye, Bottlebrush: *Aesculus parviflora*
Buckeye, Common (butterfly): *Junonia coenia*
Buckeye, Painted: *Aesculus sylvatica*
Buckeye, Red: *Aesculus pavia*
Buckeye, Yellow: *Aesculus octandra*
Buckeye: *Aesculus sp.*
Buckthorn: *Rhamnus caroliniana*
Buckwheat, Climbing: *Polygonum convolvulus*
Buffalo-nut: *Pyrularia pubera*
Bufflehead: *Bucephala albeola*
Bug, Ambush: *Phymata fasciata, Phymata pennsylvanica*
Bug, Assassin: *Pygolampis pectoralis*
Bug, Broad-headed: *Alydus eurinus*
Bug, Brown Stink: *Euschistus sp.*
Bug, Eastern Boxelder: *Leptocoris trivittatus*
Bug, Flat Stink: *Brochymena sp.*
Bug, Flat: *Mezira sp.*
Bug, Giant Water: *Belostoma lutarium*
Bug, Green Stink: *Acrosternum hilare, Nezara viridula*
Bug, Leaf-footed Yucca: *Leptoglossus sp.*
Bug, Leaf-footed: *Acanthocephala femorata, Acanthocephala sp.*
Bug, Negro: *Corimelaena sp.*
Bug, Plant-feeding Stink: *Euschistus servus servus*
Bug, Red: *Largus succinctus*
Bug, Riffle: *Rhagovelia obesa*
Bug, Seed: *Hypogeocoris sp.*, *Neacoryphus bicrucis*
Bug, Stilt: *Jalysus spinosus*
Bug, Stink: *Mormidea sp.*
Bug, Thread-legged: *Zelus exsanguis*
Bug, Toad: *Gelastocoris oculatus, Gelastocoris sp.*
Bug, Water: *Belostoma flumineum*
Bug, Wheel: *Arilus cristatus*
Bugbane, Mountain: *Cimicifuga americana*
Bugle-weed: *Lycopus virginicus*
Bulbostylis (sedge): *Bulbostylis warei*
Bullfrog: *Rana catesbiana*
Bulrush: *Scirpus koilolepis, Scirpus purshianus, Scirpus robustus*
Bumblebee: *Bombus impatiens, Bombus sp.*
Bumelia, Buckthorn: *Bumelia lycioides Bumelia tenax*
Bunting, Indigo: *Passerina cyanea*
Bunting, Painted: *Passerina ciris*
Buprestid, Branch-and-leaf: *Agrilus politus, Agrilus sp.*

Burmannia: *Burmannia biflora, Burmannia capitata*
Bur-reed: *Sparganium americanum*
Bush-clover, Creeping: *Lespedeza repens*
Bush-clover: *Lespedeza hirta, Lespedeza intermedia, Lespedeza sp.*
Buttercup, Bulbous: *Ranunculus bulbosus*
Buttercup, Hooked: *Ranunculus recurvatus*
Buttercup: *Ranunculus hispidus, Ranunculus sardous, Ranunculus sp.*
Butterfly, Crescent: *Phyciodes sp.*
Butterfly-pea, Climbing: *Centrosema virginianum*
Butterfly-pea: *Clitoria mariana*
Butterfly-weed: *Asclepias tuberosa, Asclepias tuberosa tuberosa*
Butternut: *Juglans cinerea*
Butterweed: *Senecio glabellus, Senecio obovatus*
Butterwort: *Pinguicula caerulea, Pinguicula pumila*
Button-bush: *Cephalanthus occidentalis*
Button-rosy: *Polygala curtissii*
Buttonweed, Virginia: *Diodia virginiana*
Caddisfly, Autumn Mottled Sedge: *Neophylax mitchelli*
Caddisfly, Medium Evening Sedge: *Dolophilodes sp.*
Caddisfly, Saddle-case Maker: *Glossosoma sp.*
Caddisfly, Spotted Sedge: *Hydropsyche sp.*
Caddisfly: *Rhyacophila fuscula*
Calostoma (mushroom): *Calostoma sp.*
Camellia, Mountain: *Stewartia ovata*
Camellia, Silky: *Stewartia malacodendron*
Camphorweed: *Heterotheca subaxillaris, Pluchea purpurascens, Pluchea sp.*
Campion, Starry: *Silene stellata*
Cane: *Arundinaria gigantea*
Canker, Hypoxylon: *Hypoxylon sp.*
Canvasback: *Aythya valisineria*
Carbon-antlers: *Xylaria hypoxylon*
Carbon-balls (fungus): *Daldinia concentrica, Daldinia sp.*
Carbon-cushion: *Ustulina deusta*
Cardinal, Northern: *Cardinalis cardinalis*
Cardinal-flower: *Lobelia cardinalis*
Cardita, Broad-ribbed: *Cardita arata*
Carolina-allspice: *Calycanthus floridus*
Carolina-laurel: *Rhododendron minus*
Carpenterweed: *Prunella vulgaris*
Carphephorus: *Carphephorus bellidifolius, Carphephorus sp.*
Carrion-flower: *Smilax ecirrhata, Smilax ecirrhata hugeri, Smilax herbacea*
Carrot, American: *Daucus pusillus*
Carrot, Wild: *Daucus carota*
Cassena: *Ilex cassine cassine*
Cat, Domestic: *Felis catus*
Catalpa: *Catalpa speciosa, Catalpa sp.*
Catbird, Gray: *Dumetella carolinensis*
Catbrier: *Smilax sp.*
Caterpillar, Datana Moth: *Datana sp.*
Caterpillar, Eastern Tent: *Malacosoma americanum*
Caterpillar, Forest Tent: *Malacosoma disstria*
Caterpillar, Saddleback: *Sibine stimulea*
Caterpillar, Tent: *Malacasoma sp.*
Caterpillar, Woolly Bear: *Pyrrharctia isabella*
Cat's-paw: *Plicatula gibbosa*
Cat-tail, Common: *Typha latifolia*
Cat-tail, Narrow-leaved: *Typha angustifolia*
Cedar, Deodar: *Cedrus deodara*
Chafer, Rose: *Macrodactylus angustatus*
Chafer: *Dichelonyx elongata, Dichelonyx fuscula, Dichelonyx subvittata, Dichelonyx sp.*
Chanterelle, Cinnabar-red: *Cantharellus cinnabarinus*
Chanterelle, Flame-colored: *Cantharellus ignicolor*
Chanterelle, Fragrant: *Craterellus odoratus*
Chanterelle, Golden: *Cantharellus cibarius*
Chanterelle, Scaly Vase: *Gomphus floccosus*
Chanterelle: *Cantharellus sp., Craterellus cornucopiodes*
Chat, Yellow-breasted: *Icteria virens*
Checkered-skipper, Common: *Pyrgus communis*
Cherry, Black: *Prunus serotina*
Cherry, Choke: *Prunus virginiana*
Cherry, Fire: *Prunus pensylvanica*
Cherry, Wild: *Prunus serotina*
Cherrylaurel, Carolina: *Prunus caroliniana*
Chervil, Sweet: *Osmorhiza longistylis*
Chervil, Wild: *Chaerophyllum tainturieri*
Chestnut, American: *Castanea dentata*
Chickadee, Carolina: *Parus carolinensis*
Chickweed, Common: *Stellaria media*
Chickweed, Giant: *Stellaria pubera*
Chickweed, Mouse-ear: *Cerastium glomeratum, Cerastium nutans*
Chinaberry: *Melia azedarach*
Chinese-lanterns: *Physalis virginiana virginiana*
Chinquapin, Allegheny: *Castanea pumila, Castanea pumila pumila*
Chinquapin: *Castanea sp.*
Chione, Broad-ribbed: *Chione latilirata*
Chione, Cross-barred: *Chione cancellata*
Chione: *Chione grus*
Chipmunk, Eastern: *Tamias striatus*
Chokeberry, Red: *Sorbus arbutifolia, Sorbus arbutifolia arbutifolia*
Cicada, Annual: *Tibicen sp.*
Cicada, Dog-day: *Tibicen canicularis*
Cicada, Periodical: *Magicicada sp.*
Cicely, Sweet: *Osmorhiza claytonii*
Cinnamon-vine: *Dioscorea batatas*
Cinquefoil: *Potentilla sp.*
Clam, Asiatic: *Corbicula fluminea, Corbicula manilensis, Corbicula sp.*
Clam, Astarte: *Astarte glenni, Astarte concentrica*
Clam, Atlantic Razor: *Ensis directus*

Clam, Bittersweet: *Glycymeris americana Glycymeris aratus Glycymeris pectinata*
Clam, Cardita: *Cyclocardia granulata*
Clam, Dwarf Surf: *Mulinia lateralis*
Clam, Gibb's: *Eucrassatella speciosa*
Clam, Hard-shelled: *Mercenaria campechiensis*
Clam, Jack-knife: *Tagelus plebeius*
Clam, Northern Quahog: *Mercenaria mercenaria*
Clam, Pregnant Gingernail: *Sphaerium partumeium*
Clam, Quahog: *Mercenaria sp.*
Clam, Rangia: *Rangia clathrodonta*
Clam: *Corbicula densata, Ensis directus, Mercenaria mercenaria*
Clearweed: *Pilea pumila*
Cleavers: *Galium aparine*
Clematis: *Clematis sp.*
Cliff-brake, Purple: *Pellaea atropurpurea*
Climbing-hydrangea: *Decumaria barbara*
Climbing-milkweed: *Matelea obliqua*
Clitocybe, Fat-footed: *Clitocybe clavipes*
Clitocybe, Wood: *Clitocybe ectypoides*
Cloak, Mourning: *Nymphalis antiopa*
CLonghorn: *Cyrtinus pygmaeus*
Cloudywing, Southern: *Thorybes bathyllus*
Clover, Carolina: *Trifolium carolinianum*
Clover, Low Hop: *Trifolium campestre, Trifolium dubium*
Clover, Rabbit Foot: *Trifolium arvense*
Clover, Red: *Trifolium pratense*
Clover, Sour: *Melilotus indica*
Clover, White: *Trifolium repens*
Clubmoss Lycopodium sp.
Clubmoss, Running: *Lycopodium clavatum*
Clubmoss, Shining: *Lycopodium lucidulum*
Clubmoss, Slender: *Lycopodium carolinianum*
Clubmoss, Southern Bog: *Lycopodium adpressum*
Clubtail Dragonfly Gomphus sp.
Clubtail, Lancet (dragonfly): *Gomphus exilis*
Clubtail, Single-striped: *Lanthus vernalis*
Clustered Coral Ramaria botrytis
Clymene Moth Haploa clymene
Coastal Chinquapin Castanea pumila ashei
Coastal Dropseed Sporobolus virginicus
Coastal Honeybells Leucothoe racemosa
Cockle, Common: *Trachycardium auricatum*
Cockle, Giant Atlantic: *Dinocardium robustum*
Cocklebur: *Agrimonia gryposepala, Agrimonia pubescens microcarp, Agrimonia sp., Xanthium strumarium, Xanthium strumarium glabratum*
Coconut: *Cocos nucifera*
Cohosh, Black: *Cimicifuga racemosa*
Cohosh, Blue: *Caulophyllum thalictroides*
Cole: *Clavaria flava*
Colic-root: *Aletris aurea, Aletris farinosa*
Collybia (mushroom): *Collybia sp.*
Collybia, Booted: *Collybia peronata*
Collybia, Family: *Clitocybula familia*
Collybia, Little Brown: *Collybia alkalivirens*
Collybia, Oak-loving: *Collybia dryophila*
Collybia, Violet: *Collybia iocephala*
Collybia, Zoned-cap: *Crinipellis zonata*
Columbine, Wild: *Aquilegia canadensis*
Comb-jelly, Beroe's: *Beroe ovata*
Comfrey, Wild: *Cynoglossum virginianum*
Comma, Eastern: *Polygonia comma*
Conch, Florida Horse: *Pleuroploca gigantea*
Coneflower: *Rudbeckia laciniata*
Conk, Artist's: *Ganoderma applanatum*
Coot, American: *Fulica americana*
Cooter: *Chrysemys sp.*
Copperhead: *Agkistrodon contortrix*
Coquina (shell): *Donax sp.*
Coquina, Florida: *Donax variabilis*
Coral, Colonial: *Septastrea crassa*
Coral, Cotton-base: *Lentaria byssiseda*
Coral, Crested: *Clavulina cristata*
Coral, Crown-tipped: *Clavicorona pyxidata*
Coral, Flat-topped: *Clavariadelphus truncatus*
Coral, Jellied False: *Tremellodendron pallidum*
Coral, Northern Cup: *Astrangia danae*
Coral, Spindle-Shaped Gold: *Clavulinopsis pulchra*
Coral, Spindle-shaped Yellow: *Clavulinopsis fusiformis*
Coral, Star: *Favia fragum*
Coral, Straight-branched: *Ramaria stricta*
Coral, Tan Bushy Soft: *Plexaura flexuosa*
Coral, Violet-branched: *Clavulina amethystina*
Coral, White Worm: *Clavaria vermicularis*
Coral, White: *Ramariopsis kunzei*
Coral, Yellow-tipped: *Ramaria formosa*
Coralbeads: *Cocculus carolinus*
Coral-bean: *Erythrina herbacea*
Coral-berry: *Symphoricarpos orbiculatus*
Coral-root, Autumn: *Corallorhiza odontorhiza*
Coral-root, Spring: *Corallorhiza wisteriana*
Cordyceps, Headlike: *Cordyceps capitata*
Coreopsis, Dwarf-eared: *Coreopsis auriculata*
Coreopsis: *Coreopsis gladiata, Coreopsis major, Coreopsis sp.*
Cormorant, Double-crested: *Phalacrocorax auritus*
Cornel, Stiff: *Cornus stricta*
Corn-salad: *Valerianella radiata*
Cort, Silvery-violet: *Cortinarius alboviolaceus*
Cort, Violet: *Cortinarius violaceus*
Cort: *Cortinarius sp.*
Corydalis, Yellow: *Corydalis flavula*
Corydalis: *Corydalis micranthaaustralis*
Cottonmouth: *Agkistrodon piscivorus*
Cottontail, Eastern: *Sylvilagus floridanus*
Cottonweed: *Froelichia floridana, Froelichia gracilis*

Cottonwood: *Populus sp.*
Cottonwood, Eastern: *Populus deltoides*
Cottonwood, Swamp: *Populus heterophylla*
Cow-bane: *Oxypolis rigidior*
Cowbird, Brown-headed: *Molothrus ater*
Cow-wheat: *Melampyrum lineare*
Crab, Atlantic Horseshoe: *Limulus polyphemus*
Crab, Blue: *Callinectes similis, Callinectes sp.*
Crab, Common Blue: *Callinectes sapidus*
Crab, Common Mole: *Emerita talpoida*
Crab, Fiddler: *Uca sp.*
Crab, Flat-clawed Hermit: *Pagurus pollicaris*
Crab, Ghost: *Ocypode quadrata*
Crab, Green-striped Hermit: *Clibanarius vittatus*
Crab, Gulf Calico: *Hepatus epheliticus*
Crab, Hermit: *Pagurus sp., Anemone Calliactis tricolor*
Crab, Lady: *Ovalipes ocellatus*
Crab, Mottled Purse: *Persephona mediterranea*
Crab, Mud Fiddler: *Uca pugnax*
Crab, Mud: *Panopeus sp.*
Crab, Purple Marsh: *Sesarma reticulatum*
Crab, Sand Fiddler: *Uca pugilator*
Crab, Speckled: *Arenaeus cribrarius*
Crab, Spider: *Libinia sp.*
Crab, Stone: *Menippe mercenaria*
Crabapple, Southern: *Malus angustifolia*
Cranefly, Phantom: *Bittacomorpha clavipes*
Cranesbill: *Geranium carolinianum*
Crayfish, Chimney: *Cambarus diogenes*
Crayfish: *Cambarus sp.*
Creeper, Brown: *Certhia americana*
Creeping-Charlie: *Lysimachia nummularia*
Crep, Flat: *Crepidotus applanatus*
Crescent, Pearl: *Phyciodes tharos*
Crescent, Phaon: *Phyciodes phaon*
Cress, Bitter: *Cardamine clematitis, Cardamine pensylvanica*
Cress, Early Winter: *Barbarea verna*
Cress, Hairy Bitter: *Cardamine hirsuta*
Cress, Mouse-ear: *Arabidopsis thaliana*
Cress, Yellow: *Rorippa islandica*
Cricket, Cave: *Ceuthophilus sp.*
Cricket, Field: *Gryllus assimilis, Gryllus sp.*
Cricket, Leaf-rolling: *Camptonotus caroliniensis*
Cricket, Mole: *Scapteriscus sp.*
Cricket, Short-tailed: *Anurogryllus muticus*
Cricket, Snowy Tree: *Oecanthus fultoni*
Cricket, Southern Mole: *Scapteriscus acletus*
Cricket, Thick-legged Camel: *Ceuthophilus crassifemoris*
Croton, Silverleaf: *Croton punctatus*
Croton: *Croton glandulosus septentrionalis*
Crotonopsis: *Crotonopsis linearis*
Crow, American: *Corvus brachyrhynchos*
Crow, Fish: *Corvus ossifragus*
Crow: *Corvus sp.*
Crowfoot, Small-flowered: *Ranunculus abortivus*
Cruiser, River (dragonfly): *Macromia sp.*
Cruiser, Royal River: *Macromia taeniolata*
Cryptomeria, Japanese: *Cryptomeria japonica*
Cuckoo, Yellow-billed: *Coccyzus americanus*
Cucumber, Bur: *Sicyos angulatus*
Cucumber, Creeping: *Melothria pendula*
Cucumber-root, Indian: *Medeola virginiana*
Cucumber-tree: *Magnolia acuminata*
Cudweed: *Gnaphalium purpureum*
Cup Eyelash: *Scutellinia scutellata*
Cup, Brown-Haired White: *Humaria hemisphaerica*
Cup, Common Brown: *Peziza badio-confusa*
Cup, Hairy Rubber: *Galiella rufa*
Cup, Ribbed-stalked: *Helvella acetabulum*
Cylindrical Leaf Beetle: *Bassareus sp., Cryptocephalus quadruplex, Pachybrachis sp.*
Cystoderma, Pungent: *Cystoderma amianthinum rugusoreticulatum*
Dace, Rosyside: *Clinostomus funduloides*
Daddy-long-legs: *Leiobunum sp.*
Daffodil: *Narcissus pseudo-narcissus*
Dahoon: *Ilex cassine cassine*
Daisy, Ox-eye: *Chrysanthemum leucanthemum*
Damselfly, Black-tipped: *Calopteryx aequabile*
Damselfly, Blue-fronted Dancer: *Argia apicalis*
Damselfly, Carolina Spreadwing: *Lestes vidua*
Damselfly, Sparkling Jewelwing: *Calopteryx dimidiata*
Damselfly, Spreadwing: *Lestes sp.*
Dancer (damselfly): *Argia sp.*
Dancer, Blue-ringed (damselfly): *Argia sedula*
Dancer, Powdered (damselfly): *Argia moesta*
Dancer, Seepage (damselfly): *Argia bipunctulata*
Dancer, Variable (damselfly): *Argia fumipennis*
Dandelion, Common: *Taraxacum officinale*
Dandelion, Dwarf: *Krigia dandelion, Krigia virginica*
Dangleberry: *Gaylussacia frondosa*
Darkling Beetle: *Alobates pennsylvanica, Centronopus calcaratus, Diaperis maculata, Hoplocephala viridipennis, Platydema erythrocera, Platydema excavatum, Platydema sp., Strongylium sp., Tarpela micans, Uloma imberbis, Uloma punctulata*
Darner, Comet: *Anax longipes*
Darner, Common Green: *Anax junius*
Darner, Springtime: *Basiaeschna janata*
Darner, Swamp: *Epiaeschna heros*
Dasher, Blue: *Pachydiplax longipennis*
Dayflower: *Commelina communis, Commelina erecta, Commelina virginica, Commelina sp.*
Daylily: *Hemerocallis fulva*
Deadman's-fingers (sponge): *Briareum asbestinum*
Dead-man's-fingers: *Xylaria polymorpha*
Deer, White-tailed (remains): *Odocoileus virginianus*
Deer, White-tailed: *Odocoileus virginianus*

Deerberry: *Vaccinium stamineum*
Deer's-tongue: *Trilisa paniculata, Trilisa sp.*
Destroying-angel: *Amanita virosa*
Devil-joint: *Opuntia drummondii*
Devil's-shoestring: *Sida rhombifolia*
Devil's-urn: *Urnula craterium*
Devil's-walking-stick: *Aralia spinosa*
Dewberry: *Rubus trivialis*
Diervilla: *Diervilla lonicera*
Disc, Hophornbeam (fungus): *Aleurodiscus oakesii*
Disk (shell): *Dosinia discus*
Ditch-grass: *Ruppia maritima*
Dobsonfly: *Corydalis cornutus*
Dock, Curly: *Rumex crispus*
Dock, Swamp: *Rumex verticillatus*
Dodder, Compact: *Cuscuta compacta*
Dodder: *Cuscuta gronovii, Cuscuta sp.*
Dogbane, Climbing: *Trachelospermum difforme*
Dog-fennel: *Eupatorium capillifolium, Eupatorium compositifolium*
Dog-hobble: *Leucothoe axillaris, Leucothoe axillaris axillaris, Leucothoe axillaris editorum, Leucothoe sp.*
Dog's-tooth-violet: *Erythronium americanum*
Dogwood, Alternate-leaved: *Cornus alternifolia*
Dogwood, Flowering: *Cornus florida*
Dogwood, Silky: *Cornus amomum*
Dogwood, Swamp: *Cornus stricta*
Dogwood: *Cornus asperifolia*
Dollarleaf: *Desmodium rotundifolium*
Dollar-weed: *Rhynchosia reniformis*
Dolphin, Bottle-nosed: *Tursiops truncatus*
Doodlebug: *Myrmeleon immaculatum, Myrmeleon sp.*
Dove, Mourning: *Zenaida macroura*
Dove, Rock: *Columba livia*
Draba: *Draba aprica*
Dragonfly, Baskettail: *Epitheca sp.*
Dragonfly, Little Corporal: *Ladona deplanata*
Dragonfly, Spiketail: *Cordulegaster sp.*
Dragonfly, White-tail: *Plathemis lydia*
Dragonlet, Blue: *Erythrodiplax minuscula*
Dragonslayer: *Hagenius brevistylus*
Dropseed: *Sporobolus sp.*
Drum, Black (tooth): *Pogonias cromis*
Drum-heads: *Polygala cruciata*
Duck, American Black: *Anas rubripes*
Duck, Mottled: *Anas fulvigula*
Duck, Ring-necked: *Aythya collaris*
Duck, Ruddy: *Oxyura jamaicensis*
Duck, Wood: *Aix sponsa*
Duckmeat: *Spirodela oligorrhiza*
Duck-potato: *Sagittaria engelmanniana*
Duck-potato: *Sagittaria latifolia pubescens*
Duckweed, Giant: *Spirodela punctata, Spirodela sp*
Duckweed: *Lemna gibba, Lemna perpusilla, Lemna valdiviana, Lemna sp.*
Dulichium (sedge): *Dulichium arundinaceum*
Dunlin: *Calidris alpina*
Duskywing, Horace's: *Erynnis horatius*
Duskywing, Juvenal's: *Erynnis juvenalis*
Dutchman's-breeches: *Dicentra cucullaria*
Dutchman's-pipe: *Aristolochia macrophylla*
Dyschoriste: *Dyschoriste oblongifolia*
Eagle, Bald: *Haliaeetus leucocephalus*
Eagle, Golden: *Aquila chrysaetos*
Earthstar, Barometer: *Astraeus hygrometricus*
Earthstar, Beaked: *Geastrum pectinatum*
Earthstar, Rounded: *Geastrum saccatum*
Earthstar: *Geastrum sp., Scleroderma sp.*
Earth-tongue, Orange: *Microglossum rufum*
Earth-tongue, Velvety: *Trichoglossum hirsutum*
Eclipta: *Eclipta alba*
Ecphora: *Ecphora quadricostata*
Edge, Hoary (skipper): *Achalarus lyciades*
Eel-grass: *Zostera marina*
Egret, Cattle: *Bulbicus ibis*
Egret, Great: *Casmerodius albus*
Egret, Reddish: *Egretta rufescens*
Egret, Snowy: *Egretta thula*
Elaeagnus, Thorny: *Elaeagnus pungens*
Elater, Blind: *Alaus myops*
Elater, Eyed: *Alaus oculatus*
Elderberry, American: *Sambucus canadensis*
Elderberry: *Sambucus sp.*
Elephant's-foot: *Elephantopus carolinianus, Elephantopus elatus, Elephantopus nudatus, Elephantopus tomentosus, Elephantopus sp.*
Elfin, Henry's (butterfly): *Callophrys henrici*
Elm, American: *Ulmus americana*
Elm, Slippery: *Ulmus rubra*
Elm, Winged: *Ulmus alata*
Elm: *Ulmus sp.*
Elodea, Brazilian: *Egeria densa*
Elytraria: *Elytraria caroliniensis*
Emerald, Mocha (dragonfly): *Somatochlora linearis*
Emperor, Hackberry (butterfly): *Asterocampa celtis*
Emperor, Tawny (butterfly): *Asterocampa clyton*
Entoloma, Salmon Unicorn: *Entoloma salmoneum*
Entoloma, Yellow Unicorn: *Entoloma murraii*
Eryngium: *Eryngium prostratum*
Eupatorium: *Eupatorium pilosum, Eupatorium recurvans, Eupatorium sp.*
Evening-primrose, Common: *Oenothera biennis*
Evening-primrose, Cutleaf: *Oenothera laciniata*
Evening-primrose: *Oenothera sp.*
Facelis: *Facelis retusa*
Falcon, Peregrine: *Falco peregrinus*
False-willow: *Baccharis angustifolia*
Fanwort: *Cabomba caroliniana*
Featherbells: *Stenanthium gramineum*

Fennel: *Foeniculum vulgare*
Fern, American Climbing: *Lygodium palmatum*
Fern, Bristle: *Trichomanes boschianum*
Fern, Christmas: *Polystichum acrostichoides*
Fern, Cinnamon: *Osmunda cinnamomea*
Fern, Common Grape: *Botrychium dissectum*
Fern, Fancy: *Dryopteris intermedia*
Fern, Grape: *Botrychium sp.*
Fern, Hairy Lip: *Cheilanthes lanosa*
Fern, Hay-scented: *Dennstaedtia punctilobula*
Fern, Japanese Climbing: *Lygodium japonicum*
Fern, Lady: *Athyrium filix-femina*
Fern, Lip: *Cheilanthes sp.*
Fern, Log: *Dryopteris celsa*
Fern, Maidenhair: *Adiantum pedatum*
Fern, Marginal Shield: *Dryopteris marginalis*
Fern, Marsh: *Thelypteris palustris*
Fern, Mosquito: *Azolla caroliniana*
Fern, Mountain Wood: *Dryopteris campyloptera*
Fern, Netted Chain: *Woodwardia areolata*
Fern, New York: *Thelypteris noveboracensis*
Fern, Rattlesnake Grape: *Botrychium virginianum*
Fern, Resurrection: *Polypodium polypodioides*
Fern, Rock Cap: *Polypodium virginianum*
Fern, Royal: *Osmunda regalis spectabilis*
Fern, Sensitive: *Onoclea sensibilis*
Fern, Shield: *Dryopteris sp.*
Fern, Southern Grape: *Botrychium biternatum*
Fern, Southern Lady: *Athyrium asplenioides*
Fern, Spreading Bladder: *Cystopteris protrusa*
Fern, Virginia Chain: *Woodwardia virginica*
Fern, Walking: *Asplenium rhizophyllum*
Fern, Woolly Lip: *Cheilanthes tomentosa*
Fern: *Thelypteris sp.*
Fetterbush: *Leucothoe recurva, Lyonia lucida*
Fiber-head, Black-nipple: *Inocybe fuscodisca*
Fiber-head, White: *Inocybe geophylla*
Fiber-vase, Common: *Thelophora terrestris*
Fiber-vase: *Thelephora sp.*
Figwort: *Scrophularia marilandica*
Fimbristylis (sedge): *Fimbristylis dichotoma, Fimbristylis spadicea*
Finch, House: *Cardodacus mexicanus*
Finch, Purple: *Cardodacus purpureus*
Fir, Fraser: *Abies fraseri*
Firefly: *Ellychnia sp., Lucidota atra. Lucidota sp., Phausis sp., Photinus pyralis, Photuris sp.*
Firetail, Duckweed (damselfly): *Telebasis byersi*
Fireweed: *Erechtites hieracifolia*
Fish, Atlantic Silverside: *Menidia menidia*
Fish, Gar: *Lepisosteus sp.*
Fish, Puffer: *Chilomycterus sp.*
Fishfly, Dark: *Nigronia serricornis*
Five-fingers: *Potentilla canadensis, Potentilla recta, Potentilla simplex*
Flag, Southern Blue: *Iris virginica*
Flag: *Iris tridentata, Eumeces sp.*
Fleabane, Daisy: *Erigeron annuus, Erigeron philadelphicus*
Fleabane, Rough Daisy: *Erigeron strigosus*
Fleabane, Swamp: *Erigeron vernus*
Flicker, Northern: *Colaptes auratus*
Floating-heart, Small: *Nymphoides cordata*
Floating-heart: *Nymphoides aquatica*
Fly, Bee: *Anthrax aralis, Bombylius sp.*
Fly, Flower: *Milesia virginiensis*
Fly, Fruit: *Drosophila melanogaster*
Fly, House: *Musca domestica*
Fly, Pomace: *Drosophila melanogaster*
Fly, Robber: *Asilidae sp.*
Fly, Shiny Bluebottle: *Cynomyopsis cadaverina*
Flycatcher, Acadian: *Empidonax virescens*
Flycatcher, Great Crested: *Myiarchus crinitus*
Flycatcher, Willow: *Empidonax traillii*
Fly-poison: *Amianthium muscaetoxicum*
Foamflower: *Tiarella cordifolia, Tiarella cordifolia collina*
Fog-fruit: *Lippia nodiflora*
Forestiera: *Forestiera ligustrina*
Forget-me-not: *Myosotis macrosperma, Myosotis verna*
Forktail, Citrine (damselfly): *Ischnura hastata*
Forktail, Fragile (damselfly): *Ischnura posita*
Forktail, Lilypad (damselfly): *Ischnura kellicotti*
Forktail, Rambur's (damselfly): *Ischnura ramburi*
Fox, Gray: *Urocyon cinereoargenteus*
Fox, Red: *Vulpes vulpes*
Foxglove, Downy False: *Aureolaria virginica*
Foxglove, False: *Agalinis tenuifolia, Aureolaria flava, Aureolaria pectinata, Aureolaria pedicularia, Aureolaria sp.*
Franklin-tree: *Franklinia alatamaha*
Fringe-tree: *Chionanthus virginicus*
Fritillary, Great Spangled: *Speyeria cybele*
Fritillary, Gulf: *Agraulis vanillae*
Fritillary, Variegated: *Euptoieta claudia*
Frog, Brimley's Chorus: *Pseudacris brimleyi*
Frog, Carpenter: *Rana virgatipes*
Frog, Chorus: *Pseudacris sp.*
Frog, Cricket: *Acris sp.*
Frog, Green: *Rana clamitans*
Frog, Little Grass: *Limnaoedus ocularis*
Frog, Northern Cricket: *Acris crepitans*
Frog, Pickerel: *Rana palustris*
Frog, Pig: *Rana grylio*
Frog, Southern Cricket: *Acris gryllus*
Frog, Southern Leopard: *Rana sphenocephala*
Frog, Upland Chorus (call): *Pseudacris triseriata*
Frog, Upland Chorus: *Pseudacris triceriata*
Froghopper: *Prosapia bicincta*
Frog's-bit: *Limnobium spongia*
Frog's-britches: *Sarracenia purpurea*

Frostweed: *Helianthemum rosmarinifolium*
Fungus, Alder Cone: *Taphrina alni-incanae*
Fungus, Dry-rot: *Serpula sp.*
Fungus, Gill: *Clitocybe sp.*
Fungus, Tooth: *Hericium sp.*
Fuzzy-foot: *Xeromphalina campanella*
Galactia: *Galactia sp.*
Galax: *Galax aphylla*
Galaxy: *Galax aphylla*
Galerina, Deadly: *Galerina autumnalis*
Gall (wasp): *Andricus petiolicola*
Gall, Black Oak Lamina Platform: *Neuroterus sp.*
Gall, Blackberry Branch (wasp): *Diplolepis rubicola*
Gall, Blackberry Knot (wasp): *Diastrophus nebulosus*
Gall, Blueberry Stem (wasp): *Hemadas nubilipennis*
Gall, Bullet (wasp): *Cynips sp.*
Gall, Chestnut Bud (midge): *Rhopalomyia castaneae*
Gall, Chestnut Oak Midrib Tumor: *Plagiotrochus tumificus*
Gall, Chestnut Oak Vein (wasp): *Plagiotrochus reticulatus*
Gall, Chestnut Petiole (wasp): *Neuroterus minutus*
Gall, Conical Hickory (midge): *Caryomyia sanguinolenta*
Gall, Conical Hickory Leaf: *Phylloxera caryaefoliae*
Gall, Conical Oak (wasp): *Callirhytis conifera*
Gall, Conical Woolly Oak (wasp): *Andricus perplexus*
Gall, Cottonwood Petiole (aphid): *Pemphigus populicaulis*
Gall, Cypress Flower (midge): *Itonida anthici*
Gall, Cypress Leaf (midge): *Itonida taxodii*
Gall, Cypress Twig (midge): *Thecodiplosis ananassi*
Gall, Dogwood Club (midge): *Lasioptera clazula, Mycodiplosis clavula*
Gall, False Nettle Stem (midge): *Cecidomyia boehmeriae*
Gall, Goldenrod Ball (fly): *Eurosta solidaginis*
Gall, Goldenrod Bunch (midge): *Rhopalomyia solidaginis*
Gall, Gouty Oak (wasp): *Plagiotrochus punctatus*
Gall, Hackberry Blister (aphid): *Pachypsylla vesiculum*
Gall, Hickory Leaf (aphid): *Phylloxera caryoecaulis*
Gall, Hickory Leaf (midge): *Caryomyia consobrina, Caryomyia antennata*
Gall, Hickory Midrib (midge): *Caryomyia sp.*
Gall, Hickory Onion (midge): *Caryomyia holotricha*
Gall, Hickory Seed (aphid): *Phylloxera caryoesemen*
Gall, Hickory Tube (midge): *Caryomyia tubicola*
Gall, Hickory Vein (aphid): *Phylloxera caryae*-venae
Gall, Hornbeam Leaf-fold (midge): *Cecidomyia pudibunda*
Gall, Horned Oak (wasp): *Andricus clavigerus*
Gall, Huckleberry Oak (wasp): *Callirhytis lustrans*
Gall, Large Conical Hickory: *Phylloxera notabilis*
Gall, Leafy Oak (wasp): *Andricus foliatus*
Gall, Linden Wart (midge): *Cecidomyia verrucicola*
Gall, Live-oak Potato (wasp): *Plagiotrochus batatoides*
Gall, Maple Leaf Spot (midge): *Cecidomyia ocellaris*
Gall, Marbled Oak (wasp): *Andricus pisiformis*
Gall, Marginal Oak Leaf (wasp): *Dryophanta sp.*
Gall, Oak Blister (midge): *Cincticornia pustulata*
Gall, Oak Bullet (wasp): *Disholcaspis globulus, Disholcaspis sp.*
Gall, Oak Flake (wasp): *Neuroterus floccus*
Gall, Oak Leaf (wasp): *Belenocnema fossoria*
Gall, Oak Stem (wasp): *Disholcaspis spongiosa*
Gall, Persimmon Warty Leaf: *Trioza diospyri*
Gall, Pine Needle (midge): *Janetiella coloradensis*
Gall, Post-oak Apple (wasp): *Andricus unica*
Gall, Pubescent Bract Hickory: *Phylloxera globosa*
Gall, Red Pouch Sumac (aphid): *Melaphis rhois*
Gall, Reniform Midrib Carya: *Phylloxera caryaeren*
Gall, Slippery Elm Pouch (aphid): *Gobaishia ulmifusus*
Gall, Small Bullet Oak (wasp): *Andricus maxwelli*
Gall, Spiny Rose (wasp): *Diplolepis bicolor*
Gall, Spiny Witch Hazel (aphid): *Hamamelistes spinosus*
Gall, Stem (wasp): *Plagiotrochus sp.*
Gall, Verbesina Axillary Flower: *Rhopalomyia sp.*
Gall, Water Oak Midvein (wasp): *Callirhytis sp.*
Gall, White Oak Club (wasp): *Callirhytis clavula*
Gall, Witch Hazel Cone (aphid): *Hormaphis hamamelidis*
Gall, Wool-bearing (wasp): *Andricus lanigera*
Gall, Wool-sower (wasp): *Callirhytis seminator*
Gallberry, Sweet: *Ilex coriacea*
Gall-of-the-earth: *Prenanthes trifoliolata, Prenanthes serpentaria*
Gar, Longnose (fish): *Lepisosteus osseus*
Garlic, False: *Allium bivalve*
Garlic, Field: *Allium vineale*
Gastropod: *Fusinus abruptus*
Gaura: *Gaura angustifolia*
Gentian, Pine-barren: *Gentiana autumnalis*
Gentian: *Gentiana catesbaei, Gentiana villosa, Gentiana sp*
Geranium, Wild: *Geranium maculatum*
Gerardia, Purple: *Agalinis purpurea*
Gerardia: *Agalinis linifolia, Agalinis setacea, Agalinis tenella, Agalinis virgata, Agalinis sp.*
Germander: *Teucrium canadense*
Giant-skipper, Yucca: *Megathymus yuccae*
Gill, Common Split: *Schizophyllum commune*
Ginger, Wild: *Asarum canadense, Hexastylis heterophylla, Hexastylis minor, Hexastylis shuttleworthii*
Ginseng: *Panax quinquefolium*
Glasswort, Perennial: *Salicornia virginica*
Glasswort: *Salicornia bigelovii, Salicornia europaea, Salicornia sp*
Glecoma: *Glecoma hederacea*
Glider, Black-mantled: *Tramea lacerata*
Glider, Violet-masked: *Tramea carolina*
Glider, Wandering: *Pantala flavescens*
Glider: *Tramea sp.*
Gnatcatcher, Blue-gray: *Polioptila caerulea*
Goatsbeard, False: *Astilbe biternata*
Goat's-beard: *Aruncus dioicus*
Goat's-rue: *Tephrosia virginiana*

Godwit, Marbled: *Limosa fedoa*
Goldbug: *Metriona bicolor*
Golden-Alexander: *Zizia aurea, Zizia sp*
Golden-club: *Orontium aquaticum*
Goldeneye, Common: *Bucephala clangula*
Goldenrod, Gray: *Solidago nemoralis*
Goldenrod, Rayless: *Chondrophora nudata*
Goldenrod, Rough-leaved: *Solidago rugosa*
Goldenrod, Salt-marsh: *Solidago sempervirens*
Goldenrod, Slim: *Solidago stricta*
Goldenrod, Sweet: *Solidago odora*
Goldenrod, Tall: *Solidago altissima*
Goldenrod, Woody: *Solidago pauciflosculosa*
Goldenrod: *Solidago fistulosa, Solidago glomerata, Solidago microcephala, Solidago tenuifolia, Solidago sp*
Goldfinch, American: *Carduelis tristis*
Goose, Blue: *Chen caerulescens*
Goose, Canada: *Branta canadensis*
Goose, Snow: *Chen caerulescens*
Gooseberry, Spiny: *Ribes echinellum*
Gopher-apple: *Chrysobalanus oblongifolius*
Gopherweed: *Baptisia perfoliata*
Goutweed: *Aegopodium podagraria*
Grackle, Boat-tailed: *Quiscalus major*
Grackle, Common: *Quiscalus quiscula*
Grainy-club, Soldier: *Cordyceps militaris*
Grape, Fox: *Vitis labrusca*
Grape, Summer: *Vitis aestivalis*
Grape: *Vitis sp.*
Grass, Baggy-knees: *Sacciolepis striata*
Grass, Barnyard: *Echinochloa crusgalli*
Grass, Beard: *Gymnopogon ambiguus, Gymnopogon brevifolius, Erianthus contortus, Erianthus giganteus, Erianthus strictus, Erianthus sp.*
Grass, Bermuda: *Cynodon dactylon*
Grass, Cord: *Spartina cynosuroides, Spartina patens*
Grass, Crowfoot: *Dactyloctenium aegyptium*
Grass, Cut: *Leersia sp.*
Grass, Dallis: *Paspalum dilatatum*
Grass, Finger: *Chloris petraea*
Grass, Foxtail: *Alopecurus carolinianus, Setaria sp.*
Grass, Gamma: *Tripsacum dactyloides*
Grass, Giant Foxtail: *Setaria magna*
Grass, Goose: *Eleusine indica*
Grass, Green Foxtail: *Setaria viridis*
Grass, Hair: *Aira elegans*
Grass, Indian: *Sorghastrum elliottii, Sorghastrum sp.*
Grass, Johnson: *Sorghum halepense*
Grass, Love: *Eragrostis hirsuta, Eragrostis refracta, Eragrostis sp.*
Grass, Melic: *Melica mutica*
Grass, Needle: *Stipa avenacea*
Grass, Oat: *Danthonia spicata, Danthonia sp.*
Grass, Orchard: *Dactylis glomerata*
Grass, Panic: *Panicum boseii, Panicum verrucosum, Panicum sp.*
Grass, Quaking: *Briza minor*
Grass, Rice Cut: *Leersia oryzoides*
Grass, Salt: *Distichlis spicata*
Grass, Sand: *Triplasis americana, Triplasis purpurea*
Grass, Seaside Panic: *Panicum amarum*
Grass, Smooth Cord: *Spartina alterniflora*
Grass, Smut: *Sporobolus poiretii*
Grass, Spike: *Uniola sessiliflora*
Grass, Switch: *Panicum virgatum*
Grass, Three Awn: *Aristida tuberculosa*
Grass, Toothache: *Ctenium aromaticum*
Grass, Uniola: *Uniola laxa*
Grass, Vanilla: *Anthoxanthum odoratum*
Grass, Wild Rye: *Elymus virginicus*
Grass, Wire: *Aristida stricta, Aristida sp.*
Grass, Yellow Foxtail: *Setaria glauca*
Grass: *Calamovilfa brevipilis, Uniola sp.*
Grasshopper, American Bird: *Schistocerca americana*
Grasshopper, Autumn Yellow-winged: *Arphia xanthoptera*
Grasshopper, Bird: *Schistocerca sp.*
Grasshopper, Black-sided Pygmy: *Tettigidea lateralis*
Grasshopper, Boll's Sand: *Spharagemon bolli*
Grasshopper, Cattail Toothpick: *Leptysma marginicollis*
Grasshopper, Differential: *Melanoplus differentialis*
Grasshopper, Green-striped: *Chortophaga viridifasciata*
Grasshopper, Leather-color Bird: *Schistocerca alutacea*
Grasshopper, Lubber: *Brachystola magna*
Grasshopper, Meadow: *Orchelimum sp.*
Grasshopper, Mottled Sand: *Spharagemon collare*
Grasshopper, Obscure Bird: *Schistocerca obscura*
Grasshopper, Olive-green Swamp: *Paroxya clavuliger*
Grasshopper, Orange-winged: *Pardalophora phoenicoptera*
Grasshopper, Red-legged: *Melanoplus femurrubrum*
Grasshopper, Rock-loving: *Trimerotropis saxatilis*
Grasshopper, Sand: *Spharagemon cristatum*
Grasshopper, Short-horned: *Metaleptea brevicornis*
Grasshopper, Short-winged Green: *Dichromorpha viridis*
Grasshopper, Slant-faced: *Syrbula admirabilis*
Grasshopper, Spur-throated: *Melanoplus sp.*
Grasshopper, Sulfur-winged: *Arphia sulphurea, Arphia sp.*
Grasshopper: *Mermiria intertexta, Orphulella pelidna*
Grass-of-Parnassus: *Parnassia asarifolia*
Grass-pink, Pale: *Calopogon pallidus*
Grass-pink: Bearded: *Calopogon barbatus*
Grass-pink: *Calopogon pulchellus*
Gratiola, Hairy: *Gratiola pilosa*
Gray, Bent-line (moth): *Iridopsis larvaria*
Grebe, Horned: *Podiceps auritus*
Grebe, Pied-billed: *Podilymbus podiceps*
Green-and-gold: *Chrysogonum virginianum, Chrysogonum virginianum australe*
Greenbrier, Dune: *Smilax auriculata*

Greenbrier, Fringed: *Smilax bona-nox*
Greenbrier, Red-berried: *Smilax walteri*
Greenbrier: *Smilax ecirrhata biltmoreana, Smilax rotundifolia, Smilax sp.*
Greenbrier: *Smilax glauca*
Green-dragon: *Arisaema dracontium*
Grisette, Tawny: *Amanita fulva*
Grisette: *Amanita vaginata*
Grosbeak, Blue: *Guiraca caerulea*
Grosbeak, Rose-breasted: *Pheuticus ludovicianus*
Ground-cherry: *Physalis viscosa maritima, Physalis sp*
Ground-dove, Common: *Columbina passerina*
Groundhog: *Marmota monax*
Ground-ivy: *Dichondra carolinensis*
Groundnut: *Apios americana*
Groundpine: *Lycopodium obscurum*
Groundsel, Common: *Senecio vulgaris*
Groundsel-tree: *Baccharis glomeruliflora*
Groundsel-tree: *Baccharis halimifolia*
Grouse, Ruffed: *Bonasa umbellus*
Gulfweed: *Sargassum fluitans*
Gull, Bonaparte's: *Larus philadelphia*
Gull, Franklin's: *Larus pipixcan*
Gull, Great Black-backed: *Larus marinus*
Gull, Herring: *Larus argentatus*
Gull, Laughing: *Larus atricilla*
Gull, Ring-billed: *Larus delawarensis*
Gyroporus, Red: *Gyroporus purpurinus*
Hackberry: *Celtis occidentalis georgiana, Celtis sp.*
Hairstreak, Banded: *Satyrium calanus*
Hairstreak, Gray: *Strymon melinus*
Hairstreak, Olive Juniper: *Callophrys gryneus gryneus*
Hairstreak, Red-banded: *Calycopis cecrops*
Hairstreak, White M: *Parrhasius m-album*
Harrier, Northern: *Circus cyaneus*
Haw, May: *Crataegus aestivalis*
Haw, Swamp: *Crataegus viridis*
Hawk, Broad-winged: *Buteo platypterus*
Hawk, Red-shouldered: *Buteo lineatus*
Hawk, Red-tailed: *Buteo jamaicensis*
Hawk, Sharp-shinned: *Accipiter striatus*
Hawk's-beard: *Crepis japonica*
Hawkweed: *Hieracium paniculatum, Hieracium gronovii, Hieracium sp.*
Hawthorn, Cockspur: *Crataegus crus-galli*
Hawthorn, Dwarf: *Crataegus uniflora*
Hawthorn, Littlehip: *Crataegus spathulata*
Hawthorn, Parsley-leaved: *Crataegus marshallii*
Hawthorn: *Crataegus flabellata, Crataegus punctata, Crataegus sp.*
Hazel-nut: *Corylus americana*
Heartleaf: *Hexastylis arifolia, Hexastylis contracta, Hexastylis sp.*
Hearts-a'burstin'-with-love: *Euonymus americanus, Euonymus atropurpureus*
Hedge-hyssop: *Gratiola virginiana, Gratiola sp.*
Heliotrope: *Heliotropium amplexicaule*
Hellebore, False: *Veratrum parviflorum, Veratrum sp*
Hellebore, White: *Veratrum viride*
Hemlock, Carolina: *Tsuga caroliniana*
Hemlock, Eastern: *Tsuga canadensis*
Hemlock, Water: *Cicuta maculata*
Hemlock: *Tsuga sp.*
Hempweed, Climbing: *Mikania scandens*
Henbit, Common: *Lamium amplexicaule*
Henbit, Red: *Lamium purpureum*
Hercules'-club: *Zanthoxylum clava*-herculis
Heron, Great Blue: *Ardea herodias*
Heron, Green-backed: *Butorides striatus*
Heron, Little Blue: *Egretta caerulea*
Heron, Tricolored: *Egretta tricolor*
Heterotheca: *Heterotheca gossypina, Heterotheca graminifolia, Heterotheca mariana, Heterotheca nervosa, Heterotheca nervosa nervosa*
Hibiscus: *Hibiscus sp.*
Hickory, Bitternut: *Carya cordiformis*
Hickory, Mockernut: *Carya tomentosa*
Hickory, Pale: *Carya pallida*
Hickory, Pignut: *Carya glabra*
Hickory, Shagbark: *Carya ovata*
Hickory, Southern Shagbark: *Carya carolinae - septentrionalis*
Hickory, Water: *Carya aquatica*
Hickory: *Carya sp.*
Himalaya-berry: *Rubus bifrons*
Hoarhound, False: *Eupatorium rotundifolium*
Hog-peanut: *Amphicarpa bracteata*
Holly, American: *Ilex opaca*
Holly, Mountain: *Ilex ambigua montana*
Holly, Myrtle-leaved: *Ilex cassine myrtifolia*
Holly, Possumhaw: *Ilex decidua, Ilex decidua longipes*
Holly: *Ilex ambigua, Ilex sp.*
Honewort: *Cryptotaenia canadensis*
Honey-cup: *Zenobia pulverulenta*
Honeysuckle, Bush: *Diervilla sessilifolia*
Honeysuckle, Coral: *Lonicera sempervirens*
Honeysuckle, Japanese: *Lonicera japonica*
Honeysuckle, Swamp: *Rhododendron viscosum*
Hop-hornbeam: *Ostrya virginiana*
Hornbeam, American: *Carpinus caroliniana*
Hornet, Bald-faced: *Vespa maculata, Vespula maculata*
Hornwort: *Ceratophyllum demersum*
Horse: *Equus leidyi*
Horse-balm: *Collinsonia canadensis, Collinsonia tuberosa, Collinsonia verticillata*
Horsemint, Dotted: *Monarda punctata*
Horse-nettle: *Solanum carolinense*
Horse-sugar: *Symplocos tinctoria*
Horsetail, Field: *Equisetum arvense*

Horsetail: *Equisetum hyemale affine*
Horseweed: *Erigeron canadensis canadensis, Erigeron canadensis ssp., Erigeron sp.*
Hover-fly, Bumblebee: *Volucella sp.*
Huckleberry, Black: *Gaylussacia baccata*
Huckleberry, Dwarf: *Gaylussacia dumosa*
Huckleberry: *Gaylussacia ursina, Gaylussacia sp*
Hudsonia: *Hudsonia ericoides*
Hummingbird, Ruby-throated: *Archilochus colubris*
Hydrangea, Oak-leaf: *Hydrangea quercifolia*
Hydrangea, Wild: *Hydrangea arborescens, Hydrangea arborescens arborescens, Hydrangea arborescens discolor, Hydrangea arborescens radiata*
Hydrolea: *Hydrolea quadrivalvis*
Hypericum, Shrubby: *Hypericum sp.*
Hypericum: *Hypericum nudiflorum, Hypericum sp.*
Hypomyces, Golden: *Hypomyces chrysopermus*
Hypomyces, Yellow-Green: *Hypomyces luteovirens*
Ibis, Glossy: *Plegadis falcinellus*
Ibis, White: *Eudocimus albus*
Ichneumon, Giant: *Megarhyssa sp.*
Indian-hemp: *Apocynum cannabinum, Apocynum sp.*
Indian-olive: *Nestronia umbellula*
Indian-physic: *Gillenia trifoliata*
Indian-pink: *Spigelia marilandica*
Indian-pipe: *Monotropa uniflora*
Indian-plantain, Pale: *Cacalia atriplicifolia*
Indian-tobacco: *Lobelia inflata*
Indigo, White Wild: *Baptisia alba*
Indigo, Wild: *Baptisia bracteata, Baptisia cinerea, Baptisia tinctoria, Baptisia sp, Indigofera caroliniana*
Indigo-bush: *Amorpha fruticosa*
Indigo-bush: *Amorpha sp.*
Inkberry: *Ilex glabra*
Inky, Alcohol: *Coprinus atramentarius*
Intelligence-plant: *Centella asiatica*
Ipecac, Carolina: *Euphorbia ipecacuanhae*
Iresine: *Iresine rhizomatosa*
Iris, Crested: *Iris cristata*
Iris, Dwarf: *Iris verna*
Iris: *Iris sp.*
Ironweed: *Vernonia acaulis, Vernonia angustifolia, Vernonia angustifolia angustifolia, Vernonia glauca, Vernonia sp.*
Ironwood: *Carpinus caroliniana*
Ivy, English: *Hedera helix*
Jack-in-the-pulpit: *Arisaema triphyllum*
Jay, Blue: *Cyanocitta cristata*
Jelly, Orange: *Dacrymyces deliquescens, Dacrymyces palmatus*
Jelly-club, Green-headed: *Leotia viscosa*
Jelly-club: *Leotia ascomyces*
Jelly-drops, Purple: *Ascocoryne sarcoides*
Jelly-roll, Pale: *Exidia alba*
Jessamine, Yellow: *Gelsemium sempervirens*
Jewel-weed: *Impatiens pallida*
Jewelwing, Ebony: *Calopteryx maculata*
Jimson-weed: *Datura stramonium*
Joe-pye-weed, Green-stemmed: *Eupatorium purpureum*
Joe-pye-weed: *Eupatorium fistulosum, Eupatorium sp.*
Johnny-jump-up: *Viola rafinesquii*
Jointweed: *Polygonella americana, Polygonella polygama*
Jumpseed: *Tovara virginiana*
Junco, Dark-eyed: *Junco hyemalis*
Jussieu: *Phacellia maculata*
Justicia: *Justicia americana*
Katydid, Angular-winged: *Microcentrum retinerve*
Katydid, Broad-winged: *Microcentrum rhombifolium*
Katydid, Bucrates: *Bucrates malivolans*
Katydid, Bush: *Scudderia sp.*
Katydid, Common Meadow: *Orchelimum vulgare*
Katydid, Cone-headed: *Neoconocephalus retesus, Neoconcephalus sp.*
Katydid, Fork-tailed Bush: *Scudderia furcata*
Katydid, Meadow: *Orchelimum volantum*
Katydid, Montezuma: *Montezumina modesta*
Katydid, Rapier Meadow: *Conocephalus strictus*
Katydid, Round-headed: *Amblycorypha floridana*
Katydid, Round-winged: *Amblycorypha oblongifolia*
Katydid, Shield-backed: *Atlanticus gibbosus, Atlanticus sp.*
Katydid, Shortleg Shield-backed: *Atlanticus testaceus*
Katydid, Slender Meadow: *Conocephalus fasciatus*
Katydid, Texas Bush: *Scudderia texensis*
Katydid, True: *Pterophylla camellifolia*
Kestrel, American: *Falco sparverius*
Killdeer: *Charadrius vociferus*
Kingbird Eastern: *Tyrannus tyrannus*
Kingfisher, Belted: *Ceryle alcyon*
Kinglet, Golden-crowned: *Regulus satrapa*
Kinglet, Ruby-crowned: *Regulus calendula*
Kingsnake, Eastern: *Lampropeltis getulus*
Kite, American Swallow-tailed: *Elanoides forficatus*
Kite, Mississippi: *Ictinia mississippiensis*
Kitten's-paw: *Plicatula gibbosa*
Kitten's-paw: *Plicatula marginata*
Knot, Red: *Calidris canutus*
Knotweed: *Polygonum setaceum*
Kudzu: *Pueraria lobata*
Laccaria, Common: *Laccaria laccata*
Laccaria, Purple-gilled: *Laccaria ochropurpurea*
Lacewing, Eastern Green: *Chrysopa ornata*
Lacewing, Green: *Chrysopa sp.*
Ladies'-tresses, Grass-leaved: *Spiranthes praecox*
Ladies'-tresses, Lesser: *Spiranthes ovalis*
Ladies'-tresses, Nodding: *Spiranthes cernua, Spiranthes cernua cernua*
Ladies'-tresses: *Spiranthes sp.*
Lady, American (butterfly): *Vanessa virginiensis*
Lady, Painted (butterfly): *Vanessa cardui*

Lady's-slipper, Pink: *Cypripedium acaule*
Lady's-slipper, Yellow: *Cypripedium calceolus*
Lamb's-quarters: *Chenopodium album*
Laurel, Purple: *Rhododendron catawbiense*
Leadplant: *Amorpha herbacea, Amorpha sp.*
Leaf-miner, Tick Trefoil: *Pachyschelus laevigatus*
Leather-flower: *Clematis crispa*
Leatherwood: *Dirca palustris*
Lemintina: *Lemintina sp.*
Leopard's-bane: *Arnica acaulia*
Lepiota, Lemon-yellow: *Lepiota lutea*
Lepiota, Red-tinged: *Lepiota rubrotincta*
Lepuropetalon: *Lepuropetalon spathulatum*
Lettuce, Sea: *Ulva lactuca, Ulva sp.*
Lettuce, Wild: *Lactuca biennis, Lactuca canadensis, Lactuca floridana, Lactuca sp.*
Leucothoe, Florida: *Leucothoe populifolia*
Lily, Carolina: *Lilium michauxii*
Lily, Turk's-cap: *Lilium superbum*
Lily-of-the-valley, False: *Maianthemum canadense*
Limpet, Cayenne Keyhole: *Diodoro cayenensis*
Limpet: *Crucibulum Lawrencei*
Linden: *Tilia americana, Tilia caroliniana, Tilia heterophylla, Tilia sp*
Lindernia: *Lindernia dubia*
Lion's-paw (shell): *Lyropecten nodosus*
Little-sweet-Betsy: *Trillium cuneatum, Trillium cuneatum luteum*
Liverleaf: *Hepatica acutiloba, Hepatica americana*
Liverwort, Floating: *Riccia fluitans, Ricciocarpus natans*
Lizard, Eastern Fence: *Sceloporus undulatus*
Lizard, Glass: *Ophisaurus sp.*
Lizard's-tail: *Saururus cernuus*
Lobelia, Glade's: *Lobelia glandulosa*
Lobelia: *Lobelia amoena, Lobelia boykinii, Lobelia elongata, Lobelia nuttallii, Lobelia puberula, Lobelia sp.*
Loblolly-bay: *Gordonia lasianthus*
Locust, Black: *Robinia pseudo-acacia*
Locust, Bristly: *Robinia hispida*
Locust, Carolina: *Dissosteira carolina*
Locust, Honey: *Gleditsia triacanthos*
Locust, Water: *Gleditsia aquatica*
Locust: *Robinia sp.*
Loon, Common: *Gavia immer*
Loosestrife, Water: *Decodon verticillatus*
Loosestrife, Whorled: *Lysimachia quadrifolia*
Lop-seed: *Phryma leptostachya*
Lotus: *Nelumbo lutea*
Lousewort: *Pedicularis canadensis*
Lovage: *Ligusticum canadense*
Lucine, Cross-hatched: *Divaricella quadrisulcata*
Ludwigia, Narrow-leaved: *Ludwigia linearis*
Ludwigia, Trailing: *Ludwigia palustris*
Ludwigia: *Ludwigia leptocarpa, Ludwigia pilosa, Ludwigia repens, Ludwigia sphaerocarpa, Ludwigia suffruticosa, Ludwigia sp.*
Lupine, Common: *Lupinus diffusus*
Lupine, Sundial: *Lupinus perennis*
Lupine: *Lupinus sp.*
Lycopus: *Lycopus rubellus, Lycopus sp.*
Mackerel: *Scomber sp.*
Maggot, Rat-tailed: *Eristalis tenax*
Magnolia, Southern: *Magnolia grandiflora*
Maleberry: *Lyonia ligustrina*
Mallard: *Anas platyrhynchos*
Mallow, Rose: *Hibiscus moscheutos*
Mallow, Seashore: *Kosteletskya virginica*
Mandarin, Yellow: *Disporum lanuginosum*
Mandarin: *Disporum sp.*
Man-root: *Ipomoea pandurata*
Mantid, Carolina: *Stagmomantis carolina*
Mantid, Chinese: *Tenodera aridifolia sinensis*
Maple, Chalk: *Acer saccharum leucoderme*
Maple, Red: *Acer rubrum*
Maple, Silver: *Acer saccharinum*
Maple, Southern Sugar: *Acer saccharum floridanum*
Maple, Striped: *Acer pensylvanicum*
Maple, Sugar: *Acer saccharum*
Marasmius, Orange Pinwheel: *Marasmius siccus*
Marasmius, Pinwheel: *Marasmius rotula, Marasmius sp.*
Mark, Question (butterfly): *Polygonia interrogationis*
Marsh-elder: *Iva frutescens*
Marsh-fleabane: *Pluchea camphorata, Pluchea foetida, Pluchea rosea*
Marsh-pink, Common: *Sabatia stellaris*
Martin, Purple: *Progne subis*
Mayapple: *Podophyllum peltatum*
Mayfly, Heptageniid: *Epeorus sp.*
Mayfly: *Ephemerella sp., Isonychia sp.*
Mazus: *Mazus japonicus, Mazus sp.*
Meadow-beauty, Common: *Rhexia virginica, Rhexia virginica virginica*
Meadow-beauty, Pale: *Rhexia mariana*
Meadow-beauty, Smooth: *Rhexia alifanus*
Meadow-beauty, Yellow: *Rhexia lutea*
Meadow-beauty: *Rhexia sp.*
Meadowfly (dragonfly): *Sympetrum sp.*
Meadowfly, Yellow-legged: *Sympetrum vicinum*
Meadowlark, Eastern: *Sturnella magna*
Meadow-rue, Early: *Thalictrum dioicum*
Meadow-rue: *Thalictrum clavatum, Thalictrum revolutum, Thalictrum sp.*
Mecardonia: *Mecardonia acuminata*
Medic, Black: *Medicago lupulina*
Medic, Spotted: *Medicago arabica*
Melanthera: *Melanthera hastata*
Merchant, Hop: *Polygonia comma*

Mercury, Three-seeded: *Acalypha gracilens, Acalypha sp.*
Merganser, Hooded: *Lophodytes cucullatus*
Merganser, Red-breasted: *Mergus serrator*
Mermaid-weed: *Proserpinaca pectinata*
Merulius, Coral-pink: *Merulius incarnatus*
Metalmark, Little (butterfly): *Calephelis virginiensis*
Metarranthis, Yellow-washed: *Metarranthis obfirmaria*
Mexican-clover: *Richardia scabra*
Mexican-tea: *Chenopodium ambrosioides*
Micrathena, White (spider): *Micrathena mitrata*
Microstegium (grass): *Microstegium vimineum*
Milk-pea: *Galactia volubilis*
Milkweed, Narrowleaf: *Asclepias verticillata*
Milkweed, Poke: *Asclepias exaltata*
Milkweed, Sandhill: *Asclepias humistrata*
Milkweed, Swamp: *Asclepias perennis*
Milkweed, White: *Asclepias variegata*
Milkweed: *Asclepias amplexicaulis, Asclepias longifolia, Asclepias michauxii, Asclepias quadrifolia, Asclepias sp.*
Milkweed-vine: *Cynanchum palustre*
Milkwort, Orange: *Polygala lutea*
Milkwort, Pine-barren: *Polygala cymosa*
Milkwort, Racemed: *Polygala polygama*
Milkwort: *Polygala sp.*
Milky (mushroom): *Lactarius sp.*
Milky, Buff Fishy: *Lactarius luteolus*
Milky, Common Violet-latex: *Lactarius uvidus*
Milky, Corrugated-cap: *Lactarius corrugis*
Milky, Indigo: *Lactarius indigo*
Milky, Peck's: *Lactarius peckii*
Milky, Peppery: *Lactarius piperatus*
Milky, Spotted-stalked: *Lactarius scrobiculatus*
Milky-cap, Voluminous-latex: *Lactarius volemus*
Millipede, Almond: *Sigmoria aberrans*
Millipede: *Narceus americanus*
Mimic: *Limenitis archippus*
Mimosa: *Albizia julibrissin*
Mink: *Mustela vison*
Mint, Bitter: *Hyptis alata*
Mint, Lemon: *Monarda citriodora*
Mint, Mountain: *Pycnanthemum flexuosum, Pycnanthemum incanum, Pycnanthemum muticum, Pycnanthemum sp.*
Mistflower: *Eupatorium coelestinum*
Mistletoe: *Phoradendron serotinum*
Mite, Blackgum Capsule Gall: *Eriophyes nyssae*
Mite, Cherry Leaf Spindle Gall: *Phytopus cerasicrumena*
Mite, Globular Elm Leaf Gall: *Eriophyes ulmi*
Mite, Linden Mite Gall: *Eriophyes abnormis*
Mite, Maple Spindle Gall: *Phyllocoptes aceris-crumena*
Mite, Red Velvet: *Trombidium sp.*
Mite, Wild Cherry Pouch Gall: *Eriophyes padi*
Miterwort: *Cynoctotum mitreola*
Mockingbird, Northern: *Mimus polyglottos*
Mock-orange: *Philadelphus inodorus*
Modiola: *Modiola caroliniana*
Mole, Eastern: *Scalopus aquaticus*
Monarch (butterfly): *Danaus plexippus*
Monkey-flower: *Mimulus alatus, Mimulus ringens*
Monkshood: *Aconitum uncinatum*
Moonseed: *Menispermum canadense*
Moonwort: *Botrychium lunarioides*
Moonwort: *Botrychium sp.*
Moorhen, Common: *Gallinula chloropus*
Moose-antlers: *Wynnea americana*
Morel, Yellow: *Morchella esculenta*
Morning-glory, Common: *Ipomoea purpurea*
Morning-glory, Fiddle-leaf: *Ipomoea stolonifera*
Morning-glory, Ivy-leaf: *Ipomoea hederacea hederacea*
Morning-glory, Lavender Marsh: *Ipomoea sagittata*
Morning-glory, Little Red: *Ipomoea coccinea*
Morning-glory, White: *Bonamia patens*
Morning-glory: *Ipomoea lacunosa, Ipomoea sp.*
Mosquito, Black Salt-marsh: *Aedes taeniorhyncus*
Mosquito, Fall Woodlands: *Aedes infirmatus*
Mosquito, Golden Salt-marsh: *Aedes sollicitans*
Mosquito, Unmarked Slender: *Culisetia inornata*
Mosquito: *Psorophora ferox*
Mosquitofish: *Gambusia sp.*
Moss, Apple: *Bartramia pomiformis*
Moss, Broom: *Dicranum scoparium*
Moss, Haircap: *Polytrichum commune, Polytrichum sp*
Moss, Peat: *Sphagnum sp.*
Moth, Ailanthus Webworm: *Atteva punctella*
Moth, Angle: *Semiothisa sp.*
Moth, Ashen Pinion: *Lithophane antennata*
Moth, Azalea Hand-maid: *Datana major*
Moth, Banded Tussock: *Halysidota tessellaris*
Moth, Bella: *Utetheisa bella*
Moth, Black & Yellow Lichen: *Lycomorpha pholus*
Moth, Common Oak: *Phoberia atomaris*
Moth, Confused Eusarca: *Eusarca confusaria*
Moth, Dagger: *Acronicta sp.*
Moth, Eight-spotted Forester: *Alypia octomaculata*
Moth, Emerald: *Synchlora sp.*
Moth, Evergreen Bagworm: *Thyridopteryx ephemeraeformis*
Moth, Fairy: *Adela purpurea*
Moth, Fall Webworm: *Hyphantria cunea*
Moth, Ferguson's Scallop Shell: *Hydria prunivorata*
Moth, Geometer: *Heliomata sp.*
Moth, Grape Leaffolder: *Desmia funeralis*
Moth, Grapeleaf Skeletonizer: *Harrisina americana*
Moth, Great Leopard: *Ecpantheria scribonia*
Moth, Half-Yellow: *Tarachidia semiflava*
Moth, Hummingbird Clearwing: *Hemaris thysbe*
Moth, Immaculate Holomelina: *Holomelina immaculata*
Moth, Io: *Automeris io*
Moth, Large Necklace: *Hypsoropha monilis*
Moth, Leconte's Haploa: *Haploa lecontei*

Moth, Luna: *Actias luna*
Moth, Orange-patched Smoky: *Pyromorpha dimidiata*
Moth, Pink-spotted Hawk: *Agrius cingulatus*
Moth, Pistachio Emerald: *Helthemia pistaciaria*
Moth, Pitcher-plant Borer: *Papaipema appassionata*
Moth, Polyphemus: *Antheraea polyphemus*
Moth, Regal: *Citheronia regalis*
Moth, Rosy Maple: *Dryocampa rubicunda*
Moth, Salt Marsh: *Estigmene acrea*
Moth, Southern Flannel: *Megalopyge opercularis*
Moth, Tulip-tree Beauty: *Epimecia hortaria*
Moth, Virgin Tiger: *Grammia virgo*
Moth, Virginia Ctenucha: *Ctenucha virginica*
Moth, Virginian Tiger: *Spilosoma virginica*
Moth, White-dotted Prominent: *Nadata gibbosa*
Moth, White-striped Black: *Trichodezia albovittata*
Mountain-ash: *Sorbus americana*
Mountain-laurel: *Kalmia latifolia*
Mouse, Cotton: *Peromyscus gossypinus*
Mouse, Golden: *Ochrotomys nuttalli*
Mouse-tail: *Myosurus minimus*
Mud-dauber, Black & Yellow: *Sceliphron caementarium*
Mudflower, Shade: *Micranthemum umbrosum*
Mudminnow, Eastern: *Umbra pygmaea*
Mudpuppy, Dwarf: *Necturus punctatus*
Mulberry, Paper: *Broussonetia papyrifera*
Mulberry, Red: *Morus rubra*
Mule-ears: *Senecio tomentosus*
Mullein, Moth: *Verbascum blattaria*
Mullein, Woolly: *Verbascum thapsus*
Mummichog (fish): *Fundulus heteroclitus*
Murex (snail): *Chicoreus sp.*
Muscadine: *Vitis rotundifolia*
Mushroom, American Caesar's: *Amanita caesarea*
Mushroom, Cauliflower: *Sparassis spathulata, Sparassis sp.*
Mushroom, Chicken: *Laetiporus sulphureus*
Mushroom, Coral: *Clavaria flava*
Mushroom, False Caesar's: *Amanita parcivolvata*
Mushroom, Fawn: *Pluteus cervinus*
Mushroom, Honey: *Armillariella mellea*
Mushroom, Lilac-spored: *Pleurotus sapidus*
Mushroom, Magnolia-cone: *Strobilurus conigenoides*
Mushroom, Meadow: *Agaricus campestris*
Mushroom, Oyster: *Pleurotus ostreatus, Pleurotis sp.*
Mushroom, Panther: *Amanita pantherina*
Mushroom, Ringless Honey: *Armillariella tabascens*
Muskrat: *Ondatra zibethicus*
Mussel, Ribbed: *Geukensia demissa*
Mussel: *Brachiodontes sp.*
Mustard, Tansy: *Descurainia brachycarpa, Descurainia pinnata*
Mycena (mushroom): *Mycena inclinata, Mycena vulgaris*
Mycena, Common: *Mycena galericulata*
Mycena, Coral Spring: *Mycena acicula*
Mycena, Orange: *Mycena leaiana*
Mycena, Spotted: *Mycena maculata*
Naiad, Slender: *Najas minor*
Naiad, Southern: *Najas guadalupensis*
Nandina: *Nandina domestica*
Narcissus: *Narcissus tazetta x poeticus, Narcissus sp.*
Nettle, False: *Boehmeria cylindrica*
Nettle, Stinging: *Urtica chamaedryoides*
New-Jersey-tea: *Ceanothus americanus*
Newt, Eastern: *Notophthalmus viridescens*
Night-Heron, Black-crowned: *Nyctocorax nyctocorax*
Nightshade: *Solanum americanum, Solanum sp.*
Ninebark: *Physocarpus opulifolius*
Nuthatch, Brown-headed: *Sitta pusilla*
Nuthatch, Red-breasted: *Sitta canadensis*
Nuthatch, White-breasted: *Sitta carolinensis*
Nut-rush: *Scleria ciliata, Scleria minor, Scleria oligantha, Scleria triglomerata, Scleria sp.*
Oak, Black Jack: *Quercus marilandica*
Oak, Black: *Quercus velutina*
Oak, Blue-jack: *Quercus incana*
Oak, Cherrybark: *Quercus falcata pagodaefolia*
Oak, Laurel: *Quercus laurifolia*
Oak, Live: *Quercus virginiana*
Oak, Oglethorpe: *Quercus oglethorpensis*
Oak, Overcup: *Quercus lyrata*
Oak, Post: *Quercus stellata*
Oak, Red: *Quercus rubra, Quercus rubra rubra*
Oak, Rock Chestnut: *Quercus prinus*
Oak, Running: *Quercus pumila*
Oak, Sand Live: *Quercus virginiana geminata*
Oak, Sawtooth: *Quercus acutissima*
Oak, Scarlet: *Quercus coccinea*
Oak, Scrubby Post: *Quercus margaretta*
Oak, Southern Red: *Quercus falcata falcata*
Oak, Spanish: *Quercus falcata, Quercus falcata falcata*
Oak, Swamp Chestnut: *Quercus michauxii*
Oak, Swamp Red: *Quercus shumardii*
Oak, Turkey: *Quercus laevis*
Oak, Water: *Quercus nigra*
Oak, White: *Quercus alba*
Oak, Willow: *Quercus phellos*
Oak, Yellow Chestnut: *Quercus muehlenbergii*
Oak: *Quercus sp.*
Oak-apple (gall wasp): *Amphibolips inanis*
Oak-apple, Nippled (gall wasp): *Amphibolips cookii*
Oak-apple, Spongy (gall wasp): *Amphibolips confluens*
Oats, River: *Uniola latifolia*
Oats, Sea: *Uniola paniculata*
Oconee-bells: *Shortia galacifolia*
Oldenlandica: *Oldenlandica uniflora*
Old-man-of-the-woods: *Strobilomyces floccopus*
Olive (snail): *Oliva sp.*
Olive, Lettered (snail): *Oliva sayana*

Olive, Lettered: *Oliva sayana*
Olive, Wild: *Osmanthus americana*
Olive: *Olivella carolinae, Oliva carolinensis, Olivella mutica*
Onion, Wild: *Allium canadense*
Oplismenus (grass): *Oplismenus setarius*
Opossum, Virginia: *Didelphis virginiana*
Orach, Marsh: *Atriplex patula*
Orach, Seabeach: *Atriplex arenaria*
Orange, Sleepy (butterfly): *Eurema nicippe*
Orange, Trifoliate: *Poncirus trifoliata*
Orangetip, Falcate: *Anthocharis midea*
Orb-weaver, Long-jawed: *Tetragnatha elongata, Tetragnatha laboriosa, Tetragnatha sp.*
Orb-weaver, Marbled: *Araneus marmoreus*
Orb-weaver: *Araneus bicentenarius, Araneus sp., Neoscona arabesca, Neoscona domiciliorum, Neoscona hentzi, Neoscona sp.*
Orchid, Crane-fly: *Tipularia discolor*
Orchid, Crested Fringed: *Habenaria cristata*
Orchid, Small Green Wood: *Habenaria clavellata*
Orchid, Southern Rein: *Habenaria flava*
Orchid, Three Birds: *Triphora trianthophora*
Orchid, Yellow Fringed: *Habenaria ciliaris*
Orchis, Showy: *Orchis spectabilis*
Oriole, Orchard: *Icterus spurius*
Osprey: *Pandion haliaetus*
Otter, River: *Lutra canadensis*
Ovenbird: *Seiurus aurocapillus*
Owl, Barred: *Strix varia*
Owl, Long-eared: *Asio otus*
Ox-eye, Sea: *Borrichia frutescens*
Oxypolis: *Oxypolis filiformis, Oxypolis ternata, Oxypolis sp.*
Oyster, Bitter: *Panellus stipticus*
Oyster, Black Jelly: *Resupinatus applicatus*
Oyster, Eastern: *Crassostrea virginica*
Oyster, Leaflike: *Hohenbuehelia petaloides*
Oyster, Winged Pearl: *Pteria colymbus*
Oyster: *Conradostrea lawrencei, Crassostrea virginiana, Crassostrea virginica, Ostrea lawrencei, Ostrea sp.*
Oystercatcher, American: *Haematopus palliatus*
Oyster-drill, Atlantic: *Urosalpinx cinerea*
Oyster-drill, Thick-lipped: *Eupleura caudata*
Oyster-mop, Yellow: *Tricholomopsis sulfureoides*
Painted-leaf: *Euphorbia heterophylla*
Palm, Needle: *Rhapidophyllum hystrix*
Palmetto, Cabbage: *Sabal palmetto*
Palmetto, Dwarf: *Sabal minor*
Palmetto, Saw: *Serenoa repens*
Palmetto-bug: *Periplaneta fuliginosa*
Parasol: *Lepiota procera*
Parasol-tree, China: *Firmiana platanifolia*
Parchment, Crowded: *Stereum complicatum*
Parchment, Silky: *Stereum striatum*
Paronychia: *Paronychia riparia*
Parsley, Marsh: *Apium leptophyllum*
Parsnip, Meadow: *Thaspium barbinode, Thaspium trifoliatum, Thaspium trifoliatum flavum, Zizia trifoliata*
Parsnip, Water: *Sium suave*
Partridge-berry: *Mitchella repens*
Partridge-pea: *Cassia fasciculata*
Parula, Northern: *Parula americana*
Paspalum (grass): *Paspalum floridanum, Paspalum sp.*
Passalus: *Popilius disjunctus*
Passionflower, Yellow: *Passiflora lutea*
Passionflower: *Passiflora incarnata*
Paw Paw, Dwarf: *Asimina parviflora*
PawPaw: *Asimina triloba, Asimina sp.*
Paxillus, Poison: *Paxillus involutus*
Peach: *Prunus persica*
Pearlwort: *Sagina decumbens*
Pecan: *Carya illinoensis*
Pecten: *Chlamys cocoana, Chlamys sp.*
Peeper, Spring: *Hyla crucifer*
Pelican, Brown: *Pelecanus occidentalis*
Pen (shell): *Atrina sp.*
Pen, Amber (shell): *Pinna carnea*
Pen, Half-naked (shell): *Atrina seminuda*
Pen, Stiff (shell): *Atrina rigida*
Pencil-flower: *Stylosanthes biflora*
Pennant, Amanda's (dragonfly): *Celithemis amanda*
Pennant, Banded (dragonfly): *Celithemis fasciata*
Pennant, Calico (dragonfly): *Celithemis elisa*
Pennant, Four-spotted (dragonfly): *Brachymesia gravida*
Pennant, Halloween (dragonfly): *Celithemis eponina*
Pennyroyal: *Hedeoma pulegioides*
Pennywort, Dune: *Hydrocotyle bonariensis*
Pennywort, Marsh: *Hydrocotyle umbellata*
Pennywort, Water: *Hydrocotyle ranunculoides*
Pennywort: *Hydrocotyle sibthorpioides, Hydrocotyle verticillata, Hydrocotyle sp., Obolaria virginica*
Pepperbush, Mountain Sweet: *Clethra acuminata*
Pepperbush, Sweet: *Clethra alnifolia, Clethra alnifolia alnifolia, Clethra alnifolia tomentosa*
Pepper-vine: *Ampelopsis arborea, Ampelopsis cordata*
Periwinkle: *Vinca major, Vinca minor*
Persimmon: *Diospyros virginiana*
Phacelia: *Phacelia dubia, Phacelia dubia georgiana, Phacelia sp.*
Phalarope, Wilson's: *Phalaropus tricolor*
Phlox, Annual: *Phlox drummondii*
Phlox, Creeping: *Phlox stolonifera*
Phlox, Smooth: *Phlox glaberrima*
Phlox: *Phlox amoena, Phlox carolina, Phlox nivalis, Phlox sp.*
Phoebe, Eastern: *Sayornis phoebe*
Phyllanthus: *Phyllanthus sp.*
Pickerel: *Esox sp.*
Pickerelweed: *Pontederia cordata*

Pig, Wild: *Sus scrofa*
Pimpernel, False: *Lindernia monticola*
Pimpernel, Water: *Samolus parviflorus*
Pine, Loblolly: *Pinus taeda*
Pine, Long-leaf: *Pinus palustris*
Pine, Pitch: *Pinus rigida*
Pine, Pond: *Pinus serotina*
Pine, Sand: *Pinus clausa*
Pine, Scrub: *Pinus virginiana*
Pine, Short-leaf: *Pinus echinata*
Pine, Slash: *Pinus elliotti*
Pine, Spruce: *Pinus glabra*
Pine, Table Mountain: *Pinus pungens*
Pine, Virginia: *Pinus virginiana*
Pine, White: *Pinus strobus*
Pine: *Pinus sp.*
Pine-sap: *Monotropa hypopithys*
Pineweed: *Hypericum gentianoides*
Pink, Deptford: *Dianthus armeria*
Pink, Fire: *Silene virginica*
Pink, Wild: *Silene caroliniana*
Pinkweed: *Polygonum pensylvanicum*
Pintail, Northern: *Anas acuta*
Pin-weed: *Lechea villosa, Lechea sp.*
Pinxter-bloom: *Rhododendron nudiflorum*
Pinxter-flower: *Rhododendron nudiflorum*
Pipewort: *Eriocaulon decangulare, Eriocaulon sp.*
Piriqueta: *Piriqueta caroliniana*
Pitcher-plant, Hooded: *Sarracenia minor*
Pitcher-plant, Sweet: *Sarracenia rubra*
Pitcher-plant, Trumpet: *Sarracenia flava*
Plantain, Bracted: *Plantago aristata*
Plantain, Broadleaf: *Plantago rugelii*
Plantain, English: *Plantago lanceolata*
Plantain: *Plantago sp.*
Plover, Black-bellied: *Pluvialis squatarola*
Plover, Piping: *Charadrius melodus*
Plover, Semipalmated: *Charadrius semipalmatus*
Plum, Chickasaw: *Prunus angustifolia*
Plum, Hog: *Prunus umbellata*
Plum, Wild: *Prunus americana*
Pogonia, Rose: *Pogonia ophioglossoides*
Poison-ivy: *Rhus radicans*
Poison-oak: *Rhus toxicodendron*
Pokeberry: *Phytolacca americana*
Pokeweed: *Phytolacca americana*
Polycarpon: *Polycarpon tetraphyllum*
Polygala, Fringed: *Polygala paucifolia*
Polygala: *Polygala grandiflora, Polygala incarnata, Polygala mariana, Polygala ramosa, Polygala verticillata*
Polygonum: *Polygonum cespitosum longisetum, Polygonum sp*
Polypore, Cinnabar-red: *Pycnoporus cinnabarinus*
Polypore, Elegant: *Polyporus varius*
Polypore, Gelatinous-pored: *Caloporus dichrous*
Polypore, Larch: *Fomitopsis officinalis*
Polypore, Multicolor Gill: *Lenzites betulinae*
Polypore, Mustard-yellow: *Phellinus gilvus*
Polypore, Rooting: *Polyporus radicatus*
Polypore, Smoky: *Bjerkandera adusta, Bjerkandera sp.*
Polypore, Spongy Toothed : *Spongipellis pachydon*
Polypore, Tinder: *Fomes fomentarius*
Polypore, Violet Toothed: *Trichaptum biformis*
Polypremum: *Polypremum procumbens*
Pond-cypress: *Taxodium ascendens*
Pondhawk, Eastern: *Erythemis simplicicollis*
Pond-lily, Yellow: *Nuphar luteum*
Pondspice: *Litsea aestivalis*
Pondweed, Spotted: *Potamogeton pulcher*
Pondweed: *Potamogeton diversifolius, Potamogeton sp.*
Pool-sprite: *Amphianthus pusillus*
Poor-jo (or Po-jo): *Diodia teres*
Poor-man's-pepper: *Lepidium virginicum*
Poplar: *Populus sp.*
Poria, Orange: *Poria spissa*
Prenanthes: *Prenanthes sp.*
Prickly-pear: *Opuntia compressa, Opuntia sp.*
Princess-tree: *Paulownia tomentosa*
Privet, Chinese: *Ligustrum sinense*
Privet, Swamp: *Forestiera acuminata*
Privet: *Ligustrum japonicum, Ligustrum vulgare, Ligustrum sp.*
Psathyrella (mushroom): *Psathyrella sp.*
Psyllid, Hackberry Gall (aphid): *Pachypsylla venusta*
Pterorhytis: *Pterorhytis conradi*
Puffball, Buried-stalk: *Tulostoma simulans*
Puffball, False: *Scleroderma vulgare*
Puffball, Gem-Studded: *Lycoperdon perlatum*
Puffball, Granular: *Arachnion album*
Puffball, Mini: *Lycoperdon pusillum*
Puffball, Pear-shaped: *Lycoperdon pyriforme*
Puffball, Pigskin Poison: *Scleroderma citrinum*
Puffball, Purple-spored: *Calvatia cyathiformis*
Puffball, Skull-shaped: *Calvatia craniformis*
Puffball, Spiny: *Lycoperdon echinatum*
Puffball, Tough: *Mycenastrum corium*
Puffball: *Calvatia utriformis, Lycoperdon sp.*
Puffball-in-aspic, Stalked: *Calostoma cinnabarina*
Purple, Red-spotted (butterfly): *Limenitis arthemis astyanax*
Purple-bead: *Galium hispidulum*
Purple-top: *Tridens flavus*
Purslane: *Portulaca coronata, Portulaca oleracea*
Pussy-toes: *Antennaria plantaginifolia, Antennaria solitaria, Antennaria sp.*
Puttyroot: *Aplectrum hyemale*
Pyrrhopappus: *Pyrrhopappus carolinianus*
Pyxie-moss: *Pyxidanthera barbulata, Pyxidanthera barbulata barbulata, Pyxidanthera barbulata brevifolia*

Queen (butterfly): *Danaus gilippus*
Queen's-delight: *Stillingia sylvatica*
Quillwort: *Isoetes melanopoda, Isoetes sp.*
Quinine, Wild: *Parthenium integrifolium*
Rabbit, Marsh: *Sylvilagus palustris*
Rabbit: *Sylvilagus sp.*
Rabbit-tobacco: *Gnaphalium obtusifolium, Gnaphalium sp.*
Raccoon: *Procyon lotor*
Racer, Black: *Coluber constrictor*
Racerunner, Six-lined: *Cnemidophorus sexlineatus*
Radish, Wild: *Raphanus raphanistrum*
Ragweed, Common: *Ambrosia artemisiifolia*
Ragwort, Small's: *Senecio smallii*
Ragwort: *Senecio aureus, Senecio sp.*
Rail, Clapper: *Rallus longirostris*
Rail, King: *Rallus elegans*
Rail, Virginia: *Rallus limicola*
Rangia, Wedge: *Rangia cuneata*
Raspberry, Flowering: *Rubus odoratus*
Rat, Hispid Cotton: *Sigmodon hispidus*
Rattlebox: *Crotalaria angulata, Crotalaria purshii, Crotalaria spectabilis, Crotalaria sp.*,
Rattlesnake, Timber: *Crotalus horridus*
Rattlesnake-plantain, Downy: *Goodyera pubescens*
Rattlesnake-root: *Prenanthes altissima*
Rattlesnake-weed: *Hieracium venosum*
Raven, Common: *Corvus corax*
Ray, Sting: *Dasyatis sp.*
Redbay: *Persea borbonia*
Redbud: *Cercis canadensis*
Red-cedar, Eastern: *Juniperus virginiana*
Red-cedar, Southern: *Juniperus silicicola*
Redroot: *Geum canadense*
Redroot: *Lachnanthes caroliniana*
Redstart, American: *Setophaga ruticilla*
Reed, Giant: *Arundo donax*
Rhynchosia: *Rhynchosia sp.*
Rice, Southern Wild: *Zizaniopsis miliacea*
Rice, Wild: *Zizania aquatica*
Richardia: *Richardia brasiliensis, Richardia sp.*
Riverweed: *Podostemum ceratophyllum*
Robin, American: *Turdus migratorius*
Robin's-plantain: *Erigeron pulchellus*
Rock-cress: *Arabis laevigata missouriensis*
Rock-portulaca: *Talinum teretifolium*
Rockrose: *Helianthemum corymbosum, Helianthemum rosmarinifolium*
Rose, Cherokee: *Rosa laevigata*
Rose, Multiflora: *Rosa multiflora*
Rose, Swamp: *Rosa palustris*
Rose, Wild: *Rosa carolina*
Rose: *Rosa micrantha, Rosa sp.*
Rosebay: *Rhododendron maximum*
Roseling: *Tradescantia rosea graminea*
Roseling; Spiderwort: *Tradescantia rosea*
Rosemary: *Ceratiola ericoides*
Rose-pink: *Sabatia angularis*
Rosinweed: *Silphium compositum*
Rue-anemone, False: *Isopyrum biternatum*
Ruellia: *Ruellia caroliniensis, Ruellia purshiana, , Ruellia sp.*
Running-pine: *Lycopodium flabelliforme*
Rush, Black Needle: *Juncus roemerianus*
Rush, Path: *Juncus tenuis*
Rush, Toad: *Juncus bufonius*
Rush: *Juncus dichotomus, Juncus georgianus, Juncus scirpoides, Juncus sp.*
Rushfoil: *Crotonopsis elliptica*
Russian-olive: *Elaeagnus umbellata*
Russian-thistle: *Salsola kali*
Russula, Blackish-red: *Russula kromholzii*
Russula, Emetic: *Russula emetica*
Russula, Fragile: *Russula fragilis*
Russula, Graying Yellow: *Russula claroflava*
Russula, Pectinate: *Russula pectinatoides*
Russula, Purple-bloom: *Russula mariae*
Russula, Red: *Russula sp.*
Russula, Rosy: *Russula rosacea*
Russula, Tacky Green: *Russula aeruginea*
Russula, Variable: *Russula variata*
Russula, Yellow: *Russula lutea*
Rust, Cedar Apple: *Gymnosporangium junipervirginianae*
Rust, Loblolly Pine: *Cronartium fusiforme*
Rust, Pine-oak: *Cronartium quercuum*
Rust: *Gymnosporangium sp.*
Sabatia: *Sabatia calycina, Sabatia campanulata, Sabatia difformis, Sabatia sp.*
Sage, Lyre-leaved: *Salvia lyrata*
Sage, Wood: *Teucrium canadense*
Sage: *Salvia urticifolia*
Salamander, Dusky: *Desmognathus sp.*
Salamander, Jordan's: *Plethodon jordani*
Salamander, Marbled: *Ambystoma opacum*
Salamander, Mole: *Ambystoma talpoideum*
Salamander, Mountain Dusky: *Desmognathus ochrophaeus*
Salamander, Northern Dusky: *Desmognathus fuscus*
Salamander, Seal: *Desmognathus monticola*
Salamander, Shovelnose: *Leurognathus marmoratus*
Salamander, Slimy: *Plethodon glutinosus*
Salamander, Southern Dusky: *Desmognathus auriculatus*
Salamander, Two-lined: *Eurycea bislineata*
Sallow, Footpath (moth): *Metaxaglaea semitaria*
Saltwort: *Batis maritima*
Salvia: *Salvia sp.*
Sandbur, Dune: *Cenchrus tribuloides*
Sandbur: *Cenchrus sp.*
Sand-dollar: *Mellita carolinana, Periarchus lyell,*
Sanderling: *Calidris alba*
Sand-myrtle: *Leiophyllum buxifolium*

Sandpiper, Least: *Calidris minutilla*
Sandpiper, Semipalmated: *Calidris pusilla*
Sandpiper, Spotted: *Actitis macularia*
Sandpiper, Western: *Calidris mauri*
Sandwort, Mountain: *Arenaria groenlandica glabra*
Sandwort, Rock: *Arenaria uniflora*
Sandwort: *Arenaria caroliniana, Arenaria groenlandica*
Sanicle: *Sanicula canadensis*
Sapsucker, Yellow-bellied: *Sphyrapicus varius*
Sarsaparilla, Wild: *Aralia nudicaulis*
Sarsaparilla-vine: *Smilax pumila*
Sassafras: *Sassafras albidum*
Satyr, Carolina (butterfly): *Hermeuptychia sosybius*
Satyr, Gemmed (butterfly): *Cyllopsis gemma*
Satyr, Georgia (butterfly): *Neonympha areolata*
Savory, Shrubby: *Satureja georgiana*
Sawbrier: *Smilax glauca*
Sawfish (rostral tooth): *Pristis sp.*
Sawfly, Argid: *Argidae Family*
Sawgill, Tiger: *Lentinus tigrinus*
Saw-grass: *Cladium jamaicense*
Saxifrage, Golden: *Chrysosplenium americanum*
Saxifrage: *Saxifraga michauxii, Saxifraga micranthidifolia, Saxifraga virginiensis,*
Scaber-stalk, Common: *Leccinum scabrum*
Scale, Cottony Cushion: *Icerya purchasi*
Scallop: *Argopecten eboreus, Argopecten vicenarius, Carolinapecten arboreus, Carolinapecten solarioides, Chesapecten jeffersonius jeffersonius, Chesapecten jeffersonius septenarius, Chesapecten madisonius, Pecten brouweri, Pecten sp., Placopecten clintonius,*
Scaup, Greater: *Aythya marila*
Scaup, Lesser: *Aythya affinis*
Scleroderma, Earthstar: *Scleroderma geaster*
Scorpion: *Centruroides vittatus, Centruroides sp.*
Screech-owl, Eastern (call): *Otus asio*
Screech-owl, Eastern: *Otus asio*
Scud: *Crangonyx serratus*
Sea-blight: *Suaeda linearis*
Sea-cucumber, Brown: *Sclerodactyla briareus*
Sea-cucumber, Striped: *Thyonella gemmata*
Sea-horse: *Hippocampus reidi*
Seal, False Solomon's: *Smilacina racemosa*
Sea-lavender: *Limonium carolinianum, Limonium nashii, Limonium sp.*
Sea-myrtle: *Baccharis halimifolia*
Sea-pork (tunicate): *Aplidium constellatum, Aplidium sp.*
Sea-purslane: *Sesuvium portulacastrum*
Searcher, Fiery: *Calosoma scrutator*
Searcher; Ground Beetle: *Calosoma calidum, Carabus sylvosus, Carabus vinctus*
Sea-rocket: *Cakile harperi, Cakile sp.*
Seaweed, Dead Man's Fingers: *Codium sp.*
Sea-whip: *Leptogorgia virgulata*
Sebastian-bush: *Sebastiania ligustrina*
Sedge, Brown Checkered Summer: *Polycentropus sp.*
Sedge, Fraser's: *Cymophyllus fraseri*
Sedge, Giant Brown Autumn: *Pycnospsyche divergens, Pycnospsyche gentilis*
Sedge, Little Plain Brown: *Lepidostoma sp.*
Sedge, White-bracted: *Dichromena colorata, Dichromena latifolia*
Sedge: *Carex amphibola, Carex blanda, Carex caroliniana, Carex cephalophora, Carex communis, Carex complanata, Carex crinita, Carex debilis, Carex digitalis, Carex flaccosperma, Carex retroflexa, Carex glaucescens, Carex intumescens, , Carex laxiculmis, Carex laxiflora, Carex lurida, Carex oxylepis, Carex plantaginea, Carex rosea, Carex virescens, Carex vulpinoidea, , Carex walteriana, Carex willdenowii, Carex sp.*
Seed-box, Slender: *Ludwigia virgata*
Seed-box: *Ludwigia alternifolia*
Senecio: *Senecio millefolium*
Senna, Wild: *Cassia marilandica*
Sensitive-brier: *Schrankia microphylla*
Sensitive-plant, Wild: *Cassia nicitans*
Sericea: *Lespedeza cuneata*
Serviceberry, Downy: *Amelanchier arborea, Amelanchier arborea arborea*
Serviceberry: *Amelanchier sp.*
Sesbania: *Sesbania exaltata*
Seymeria: *Seymeria cassioides*
Shadbush: *Amelanchier canadensis, Amelanchier obovalis, Amelanchier sp.*
Shaggy-mane: *Coprinus comatus*
Shark (tooth): *Carcharodon megalodon*
Shark, Bull (tooth): *Carcharhinus leucas*
Shark, Cow (tooth): *Hexanchus sp.*
Shark, Eared White (tooth): *Carcharodon auriculatus*
Shark, Ground (tooth): *Carcharhinus sp.*
Shark, Lemon (tooth): *Negaprion brevirostris*
Shark, Mackerel: *Lamna sp.*
Shark, Modern Great White: *Carcharodon carcharius*
Shark, Sand (tooth): *Odontaspis hopei*
Shark, Sand Tiger (tooth): *Odontaspis taurus*
Shark, Sand Tiger: *Odontaspis faurus*
Shark, Sand: *Odontaspis sp.*
Shark: *Hemipristis serra*
Shark-eye: *Polineces duplicatus*
Sheep-kill: *Kalmia angustifolia caroliniana*
Sheep-laurel: *Kalmia angustifolia*
Sheep-sorrel: *Rumex acetosella*
Shelf, Varnish: *Ganoderma curtisii*
Shell, Cone: *Conus adversariu,s Conus waccamawensis*
Shell, Gouldia: *Gouldia metastriata*
Shell, Jingle: *Anomia simplex, Anomia sp.*
Shell, Worm: *Vermicularia knorrii*

Shellmound-shrub: *Sageretia minutiflora*
Sherardia: *Sherardia arvensis*
Shooting-star: *Dodecatheon meadia*
Shortia: *Shortia galacifolia*
Shoveler, Northern: *Anas clypeata*
Shrew, Least: *Cryptotis parva*
Shrew, Southern Short-tailed: *Blarina carolinensis*
Shrew: *Blarina sp.*
Shrike, Loggerhead: *Lanius ludovicianus*
Shrimp, Ghost: *Callianassa sp.*
Shrimp, W.I. Ghost (burrows): *Callianassa major*
Sicklepod: *Cassia obtusifolia*
Silkmoth, Sweetbay: *Callosamia securifera*
Silphium: *Silphium astericus, Silphium dentatumm, Silphium sp.*
Silverbell, Carolina: *Halesia carolina*
Silverbell: *Halesia parviflora, Halesia sp.*
Skate (egg case): *Raja sp.*
Skate, Clearnose (egg case): *Raja eglanteria*
Skimmer, Bar-winged: *Libellula axilena*
Skimmer, Black: *Rhynchops niger*
Skimmer, Corporal: *Libellula exusta deplanata*
Skimmer, Golden-winged: *Libellula auripennis*
Skimmer, Great Blue: *Libellula vibrans*
Skimmer, Painted: *Libellula semifasciata*
Skimmer, Pied: *Libellula luctuosa*
Skimmer, Pond: *Libellula sp.*
Skimmer, Roseate (dragonfly): *Orthemis ferruginea*
Skimmer, Slaty: *Libellula incesta*
Skimmer, Twelve-spotted: *Libellula pulchella*
Skink, Broadhead: *Eumeces laticeps*
Skink, Five-lined: *Eumeces fasciatus, Eumeces sp.*
Skink, Ground: *Scincella lateralis*
Skink, SE 5-lined: *Eumeces sp.*
Skink: *Eumeces sp.*
Skipper, Duskywing: *Erynnis sp.*
Skipper, Fiery: *Hylephila phyleus*
Skipper, Golden: *Poanes sp.*
Skipper, Least: *Ancyloxypha numitor*
Skipper, Long-tailed: *Urbanus proteus*
Skipper, Silver-spotted: *Epargyreus clarus*
Skipper, Zabulon: *Poanes zabulon*
Skipperling, Southern: *Copaeodes minimus*
Skullcap, Common: *Scutellaria integrifolia*
Skullcap: *Scutellaria elliptica, Scutellaria lateriflora, Scutellaria sp.*
Slant-line, Yellow (moth): *Tetracis crocallata*
Slider, Yellow-bellied: *Chrysemys scripta*
Slime, Carnival Candy: *Arcyria denudata*
Slime, Chocolate Tube: *Stemonitis splendens*
Slime, Coral: *Ceratiomyxa fruticulosa*
Slime, Japanese Lantern: *Dictydium cancellatum*
Slime, Scrambled-egg: *Fuligo septica*
Slime, Tapioca: *Brefeldia maxima*
Slime, White-Footed: *Diachea leucopodia*
Slime, Wolf's-milk: *Lycogala epidendrum*
Slipper, Atlantic: *Crepidula fornicata*
Slipper, Common (snail): *Crepidula fornicata*
Slipper, Flat (snail): *Crepidula plana*
Slippery-Jack: *Suillus luteus*
Smartweed, Dotted: *Polygonum punctatum*
Smartweed, Nodding: *Polygonum lapathifolium*
Smartweed, Water: *Polygonum densiflorum*
Smartweed: *Polygonum hydropiperoides opelousanum, Polygonum setaceum*
Smilax, Bamboo: *Smilax laurifolia*
Smilax, Sweet-scented: *Smilax smallii*
Snail, Augur: *Terebra sp.*
Snail, Chain River Water: *Goniobasis catenaria*
Snail, Disk Cannibal: *Haplotrema concavum*
Snail, Eastern Mud: *Ilyanassa obsoleta*
Snail, Elongate Cannibal: *Euglandia rosea*
Snail, Exogyra: *Exogyra costata*
Snail, Forest Disk: *Anguispira alternata*
Snail, Freshwater: *Physella sp.*
Snail, Giant Forest: *Mesodon normalis*
Snail, Juxtaposed Forest: *Triodopsis juxtidens*
Snail, Lesser Mystery: *Campeloma decisum*
Snail, Marine: *Turritella sp.*
Snail, Marsh Periwinkle; Mud: *Littorina irrorata*
Snail, Philippi's Nutmeg: *Trigonostoma tenera*
Snail, Ripe Forest: *Mesodon zaletus*
Snail, River Horn: *Elimia sp.*
Snail, Southern Moon: *Polinices duplicatus*
Snail, Turritella: *Turritella perexilis*
Snail, Two-keeled Valve: *Valvata bicarinata*
Snail, White-lipped Forest: *Mesodon thyroidus, Triodopsis albolabris*
Snake, Banded Water: *Nerodia fasciata*
Snake, Brown Water: *Nerodia taxispilota*
Snake, Corn: *Elaphe guttata*
Snake, Eastern Garter: *Thamnophis sirtalis*
Snake, Eastern Hognose: *Heterodon platyrhinos*
Snake, Northern Water: *Nerodia sipedon*
Snake, Rat: *Elaphe obsoleta*
Snake, Redbelly Water: *Nerodia erythrogaster*
Snake, Ringneck: *Diadophis punctatus*
Snake, Rough Green: *Opheodrys aestivus*
Snake, Worm: *Carphophis amoenus*
Snakebird: *Anhinga anhinga*
Snakeroot, Black: *Sanicula marilandica, Zigadenus densus*
Snakeroot, Button: *Eryngium yuccifolium*
Snakeroot, Samson: *Psoralea psoralioides, Psoralea psoralioides psoralioides*
Snakeroot, Virginia: *Aristolochia serpentaria*
Snakeroot, White: *Eupatorium rugosum*
Snakeroot, Yellow-flowered: *Sanicula gregaria*
Snakeroot: *Sanicula sp.*, *Eryngium integrifolium*

Snakeroot; Sanicle: *Sanicula smallii*
Sneeze-weed: *Helenium autumnale, Helenium flexuosum*
Snipe, Common: *Gallinago gallinago*
Snout, American (butterfly): *Libytheana carinenta*
Snowbell, Bigleaf: *Styrax grandifolia*
Snowbell: *Styrax americana*
Snowdrops: *Leucojum aestivum*
Soft Rush: *Juncus effusus*
Solomon's-seal: *Polygonatum biflorum, Polygonatum sp.*
Sora: *Porzana carolina*
Sorrel; Dock: *Rumex sp.*
Sour-grass: *Rumex acetosella*
Sourwood: *Oxydendrum arboreum*
Sow-thistle, Common: *Sonchus oleraceus*
Sow-thistle, Spiny-leaved: *Sonchus asper*
Sow-thistle: *Sonchus sp.*
Spanish-bayonet: *Yucca aloifolia*
Spanish-moss: *Tillandsia usneoides*
Spanish-needles: *Bidens bipinnata*
Sparkleberry: *Vaccinium arboreum*
Sparrow, Bachman's: *Aimophila aestivalis*
Sparrow, Chipping: *Spizella passerina*
Sparrow, Field: *Spizella pusilla*
Sparrow, Savannah: *Passerculus sandwichensis*
Sparrow, Sharp-tailed: *Ammodramus caudacutus*
Sparrow, Song: *Melospiza melodia*
Sparrow, Swamp: *Melospiza georgiana*
Sparrow, White-crowned: *Zonotrichia leucophrys*
Sparrow, White-throated: *Zonotrichia albicollis*
Spatterdock: *Nuphar luteum*
Speedwell, Corn: *Veronica arvensis*
Spicebush, Southern: *Lindera melissaefolium*
Spicebush, Swallowtail: *Papilio troilus*
Spicebush: *Lindera benzoin*
Spider, Arrowhead: *Verrucosa arenata*
Spider, Banded Garden: *Argiope trifasciata*
Spider, Basilica: *Mecynogea lemniscata*
Spider, Bowl-and-doily: *Frontinella pyramitela, Frontinella sp.*
Spider, Brownish-gray Fishing: *Dolomedes tenebrosus*
Spider, Carolina Wolf: *Lycosa carolinensis*
Spider, Crab: *Misumena sp., Misumenops asperatus, Xysticus sp.*
Spider, Elegant Crab: *Xysticus elegans*
Spider, Filmy Dome: *Prolinyphia marginata*
Spider, Fishing: *Dolomedes sp.*
Spider, Forest Wolf: *Lycosa gulosa*
Spider, Funnel-web: *Agelenopsis naevia*
Spider, Goldenrod: *Misumena vatia*
Spider, Golden-silk: *Nephila clavipes*
Spider, Grass: *Agelenopsis sp.*
Spider, Green Lynx: *Peucetia viridans*
Spider, Jumping (large): *Phidippus pulcher*
Spider, Jumping (small): *Metaphidippus galathea*
Spider, Jumping: *Hentzia sp.*
Spider, Jumping: *Menemerus bivittatus, Metaphidippus sp., Phidippus cardinalis*
Spider, Nursery-web: *Pisaurina mira, Pisaurina sp.*
Spider, Orchard: *Leucauge venusta*
Spider, Purse-web: *Sphodros bicolor*
Spider, Rabid Wolf: *Lycosa rabida*
Spider, Ray: *Theridiosoma radiosa*
Spider, Six-spotted Fishing: *Dolomedes triton*
Spider, Spiny-backed: *Micrathena gracilis*
Spider, Wolf: *Lycosa sp., Pardosa sp.*
Spider, Writing: *Argiope aurantia, Argiope sp.,*
Spider, Yellow Crab: *Misumenoides formosipes*
Spider-lily: *Hymenocallis occidentalis*
Spiderwort, Asian: *Aneilema keisak*
Spiderwort: *Tradescantia hirsuticaulis, Tradescantia subaspera, Tradescantia virginiana, Tradescantia sp.*
Spikemoss, Meadow: *Selaginella apoda*
Spikemoss, Rock: *Selaginella rupestris*
Spikemoss, Sand: *Selaginella arenicola*
Spikemoss, Twisted-hair: *Selaginella tortipila*
Spikenard: *Aralia racemosa*
Spike-rush, Proliferating: *Eleocharis sp.*
Spike-rush: *Eleocharis equisetoides, Eleocharis obtusa, Eleocharis tuberculosa*
Spiketail, Delta-spotted: *Cordulegaster diastatops*
Spilanthes: *Spilanthes americana repens*
Spiny-back, Arrow-shaped: *Micrathena sagittata*
Spiny-back, Crab-shaped: *Gasteracantha cancriformis*
Spinyleg, Black-shouldered: *Dromogomphus spinosus*
Spinyleg, Southeastern: *Dromogomphus armatus*
Spittlebug: *Prosapia sp., Philaenus sp.*
Splash-cups: *Cyathus striatus*
Spleenwort, Blackstem: *Asplenium resiliens*
Spleenwort, Bradley's: *Asplenium bradleyi*
Spleenwort, Ebony: *Asplenium platyneuron*
Spleenwort, Mountain: *Asplenium montanum*
Spleenwort, Silvery: *Athyrium thelypteroides*
Spleenwort: *Asplenium sp.*
Spreading-pogonia: *Cleistes divaricata*
Spreadwing, Swamp (damselfly): *Lestes vigilax*
Spring-beauty: *Claytonia caroliniana, Claytonia virginica*
Spruce, Norway: *Picea abies*
Spruce, Red: *Picea rubens*
Spurge, Flowering: *Euphorbia corollata*
Spurge, Prostrate: *Euphorbia supina*
Spurge, Spotted: *Euphorbia maculata*
Spurge: *Euphorbia curtisii, Euphorbia ammannioides, Euphorbia polygonifolia, Euphorbia sp.*
Squaw-huckleberry: *Vaccinium stamineum*
Squaw-huckleberry: *Vaccinium stamineum stamineum*
Squaw-root: *Conopholis americana*
Squid: *Loligo sp.*
Squirrel, Eastern Fox: *Sciurus niger*

Squirrel, Gray: *Sciurus carolinensis*
Squirrel, Red: *Tamiasciurus hudsonicus*
St.-Andrew's-cross: *Hypericum hypericoides, Hypericum stragalum,*
St.-John's-wort: *Hypericum densiflorum, Hypericum fasciculaatum, Hypericum gymnanthum, Hypericum lloydii, Hypericum mutilum, Hypericum punctatum, Hypericum virginicum, Hypericum walteri,*
St.-Peter's-wort: *Hypericum stans*
Stachys: *Stachys sp.*
Staggerbush: *Lyonia ferruginea, Lyonia mariana*
Stain, Green: *Chlorociboria aeruginascens*
Standing-cypress: *Ipomopsis rubra*
Star-grass, Yellow: *Hypoxis hirsuta, Hypoxis micrantha, Hypoxis rigida, Hypoxis sp.*
Starling, European: *Sturnus vulgaris*
Starwort, Water: *Callitriche deflexa, Callitriche heterophylla*
Stingray (barb, dental plate): *Dasyatis sp., Myliobatis sp.*
Stinkhorn, Columned: *Clathrus columnatus*
Stinkhorn, Dog: *Mutinus caninus*
Stinkhorn, Elegant: *Mutinus elegans*
Stonecrop, Vernal Pool: *Sedum smallii*
Stonecrop: *Sedum pusillum, Sedum ternatum*
Stonefly, Common: *Acroneuria internata, Paragnetina immarginata*
Stonefly: *Isoperla bilineata, Phasganophora capitata*
Stoneseed: *Lithospermum tuberosum*
Storax: *Styrax americana*
Stork, Wood: *Mycteria americana*
Strawberry, Barren: *Waldsteinia fragarioides*
Strawberry, Indian: *Duchesnea indica*
Strawberry, Wild: *Fragaria virginiana*
Strawberry-bush: *Euonymus sp.*
Styrax: *Styrax sp.*
Sugarberry: *Celtis laevigata*
Suillus, Chicken-fat: *Suillus americanus*
Suillus, Dotted-stalk: *Suillus granulatus*
Suillus, Painted: *Suillus pictus*
Sulphur, Clouded: *Colias philodice*
Sulphur, Cloudless: *Phoebis sennae* eubule
Sulphur, Orange: *Colias eurytheme*
Sumac, False Poison: *Rhus michauxii*
Sumac, Fragrant: *Rhus aromatica*
Sumac, Poison: *Rhus vernix*
Sumac, Smooth: *Rhus glabra*
Sumac, Staghorn: *Rhus typhina*
Sumac, Winged: *Rhus copallina*
Summer-farewell: *Petalostemum pinnatum*
Sun-bonnets: *Chaptalia tomentosa*
Sundew: *Drosera capillaris, Drosera intermedia, Drosera leucantha, Drosera sp.*
Sundrops: *Oenothera fruticosa*
Sunflower, Hairy Wood: *Helianthus atrorubens*
Sunflower, Hairy: *Helianthus tomentosus*
Sunflower: *Helianthus angustifolius, Helianthus microcephalus, Helianthus sp.*
Supple-jack: *Berchemia scandens*
Swallow, Bank: *Riparia riparia*
Swallow, Barn: *Hirundo rustica*
Swallow, Northern Rough-winged: *Stelgidopteryx serripennis*
Swallow, Tree: *Tachycineta bicolor*
Swallowtail, Black: *Papilio polyxenes asterius*
Swallowtail, Eastern Tiger: *Papilio glaucus*
Swallowtail, Giant: *Papilio cresphontes*
Swallowtail, Palamedes: *Papilio palamedes*
Swallowtail, Pipevine: *Battus philenor*
Swallowtail, Zebra: *Eurytides marcellus*
Swamp-beacon: *Mitrula paludosa*
Swan, Tundra: *Cygnus columbianus*
Sweetbay: *Magnolia virginiana*
Sweet-gum: *Liquidambar styraciflua*
Sweet-shrub: *Calycanthus floridus*
Swift, Chimney: *Chaetura pelagica*
Swordgrass: *Scirpus americanus*
Sycamore, American: *Platanus occidentalis*
Tagelus: *Tagelus gibbus*
Tailed-blue, Eastern (butterfly): *Everes comyntas*
Tallow-tree, Chinese: *Sapium sebiferum*
Tamarisk: *Tamarix gallica*
Tanager, Scarlet: *Piranga olivacea*
Tanager, Summer: *Piranga rubra*
Tansy: *Tanacetum vulgare*
Tarweed: *Acalypha rhomboidea*
Teal, Blue-winged: *Anas discors*
Teal, Green-winged: *Anas crecca*
Tearthumb: *Polygonum arifolium, Polygonum sagittatum*
Teesdalia: *Teesdalia nudicaulis*
Tephrosia: *Tephrosia hispidula, Tephrosia spicata, Tephrosia sp.*
Tern, Caspian: *Sterna caspia*
Tern, Forster's: *Sterna forsteri*
Tern, Royal: *Sterna maxima*
Terrapin, Diamondback: *Malaclemys terrapin*
Thelypteris (fern): *Thelypteris kunthii*
Thimble-cap, Wrinkled: *Verpa bohemica*
Thimbleweed: *Anemone virginiana*
Thistle, Bull: *Carduus pumilus*
Thistle: *Carduus lanceolatus, Carduus repandus, Carduus spinosissimus, Carduus sp.*
Thoroughwort: *Eupatorium album, Eupatorium dubium, Eupatorium leucolepis*
Thrasher, Brown: *Toxostoma rufum*
Thrush, Hermit: *Catharus guttatus*
Thrush, Wood: *Hylocichla mustelina*
Tick, Lone Star: *Amblyomma americanum*
Tick-trefoil, Naked-flower: *Desmodium nudiflorum*
Tie-vine: *Jacquemontia tamnifolia*

Titi: *Cyrilla racemiflora*
Titmouse, Tufted: *Parus bicolor*
Toad, American: *Bufo americanus*
Toad, Eastern Narrowmouth: *Gastrophryne carolinensis*
Toad, Fowler's: *Bufo woodhousei*
Toad, Oak: *Bufo quercicus*
Toad, Southern: *Bufo terrestris*
Toad: *Bufo sp.*
Toadflax: *Linaria canadensis, Linaria sp.*
Tooth, Bearded: *Hericium erinaceus*
Tooth, Bear's-head: *Hericium coralloides*
Tooth, Fish: *Enchodus ferox*
Tooth, Ochre Spreading: *Steccherinum ochraceum*
Tooth, Scaly: *Hydnum imbricatum*
Tooth, Sweet: *Dentinum repandum*
Toothwort: *Cardamine angustata angustata, Cardamine concatenata, Cardamine diphylla, Cardamine sp.*
Tortoise, Gopher: *Gopherus polyphemus*
Touch-me-not, Spotted: *Impatiens capensis*
Towhee, Rufous-sided: *Pipilo erythrophthalmus*
Tragia: *Tragia urens, Tragia urticifolia, Tragia sp.*
Trailing-arbutus: *Epigaea repens*
Train-wrecker: *Lentinus lepideus*
Trash-bug: *Hemerobiidae Family*
Tread-softly: *Cnidoscolus stimulosus*
Tree-ear: *Auricularia auricula, Auricularia sp.*
Treefrog, Cope's Gray (call): *Hyla chrysoscelis*
Treefrog, Gray (call): *Hyla versicolor*
Treefrog, Gray: *Hyla versicolor*
Treefrog, Green: *Hyla cinerea*
Treefrog, Pine Woods: *Hyla femoralis*
Treefrog, Squirrel: *Hyla squirella*
Treefrog,Cope's Gray: *Hyla chrysoscelis*
Treefrog: *Hyla sp.*
Treehopper, Locust: *Thelia bimaculata*
Tree-of-heaven: *Ailanthus altissima*
Trepocarpus: *Trepocarpus aethusae*
Trillium, Catesby's: *Trillium catesbaei*
Trillium, Great White: *Trillium grandiflorum*
Trillium, Nodding: *Trillium cernuum*
Trillium, Painted: *Trillium undulatum*
Trillium, Vasey's: *Trillium vaseyi*
Trillium: *Trillium discolor, Trillium lanceolatum, Trillium sp.*
Trout, Brook: *Salvelinus fontinalis*
Tubeworm, Shingle: *Owenia fusiformis*
Tuft, Sulfur (mushroom): *Naematoloma fasciculare*
Tulip, Banded (snail): *Fasciolaria hunteria*
Tulip-poplar: *Liriodendron tulipifera*
Tumblebug, Common: *Canthon laevis*
Tumblebug: *Canthon chalcites, Glaphyrocanthon viridis*
Tupelo, Black: *Nyssa sylvatica sylvatica*
Tupelo, Swamp: *Nyssa sylvatica biflora*
Tupelo, Water: *Nyssa aquatica*
Turkey, Wild: *Meleagris gallopavo*
Turkey-tail, False: *Stereum ostrea*
Turkey-tail: *Trametes versicolor*
Turkey-wing (shell): *Arca zebra*
Turnstone, Ruddy: *Arenaria interpres*
Turtle, Eastern Box: *Terrapene carolina*
Turtle, Eastern Mud: *Kinosternon subrubrum*
Turtle, Eastern Musk: *Sternotherus odoratus*
Turtle, Loggerhead Sea: *Caretta caretta*
Turtle, Painted: *Chrysemys picta*
Turtle, Pond (shell fragment): *Emydidae Family*
Turtle, Snapping: *Chelydra serpentina*
Turtle, Soft-shelled (fragment): *Trionyx sp.*
Turtle, Spotted: *Clemmys guttata*
Turtleheads: *Chelone obliqua*
Twayblade, Southern: *Listera australis*
Umbrella-grass: *Fuirena squarrosa*
Umbrella-leaf: *Diphylleia cymosa*
Umbrella-tree: *Magnolia fraseri, Magnolia tripetala*
Urchin, Brown Rock: *Arbacia punctulata*
Urchin, Five-holed Keyhole: *Mellita quinquiesperforata*
Vanilla-plant: *Trilisa odoratissima*
Vanilla-plant: *Trilisa sp.*
Veery: *Catharus fuscescens*
Velvet-ant, Common Eastern: *Dasymutilla occidentalis occidentalis*
Velvet-ant: *Dasymutilla asopus bexar, Dasymutilla sp.*
Venus, Cross-barred (shell): *Chione cancellata*
Venus, Cross-barred: *Chione cribraria*
Venus'-fly-trap: *Dionaea muscipula*
Venus'-looking-glass, Small: *Specularia biflora*
Venus'-looking-glass: *Specularia perfoliata, Specularia sp.*
Verbena: *Verbena scabra, Verbena sp.*
Verbesina: *Verbesina alternifolia, Verbesina occidentalis, Verbesina virginica virginica, Verbesina sp.*
Veronica: *Veronica officinalis, Veronica sp.*
Vervain, Brazilian: *Verbena braziliensis*
Vervain, White: *Verbena urticifolia*
Vetch, Wood: *Vicia caroliniana*
Vetch: *Vicia hugeri, Vicia sp.*
Viburnum, Maple-leaf: *Viburnum acerifolium*
Viburnum, Possumhaw: *Viburnum nudum*
Viburnum, Small-leaf: *Viburnum obovatum*
Viburnum: *Viburnum sp.*
Viceroy: *Limenitis archippus*
Vine, Cross: *Anisostichus capreolata*
Vine, Trumpet: *Campsis radicans*
Violet(s): *Viola affinis, Viola eriocarpa leiocarpa, Viola septemloba, Viola tripartita tripartita, Viola walteri, Viola sp., Viola spp.*
Violet, Bird-foot: *Viola pedata*
Violet, Bog White: *Viola lanceolata*
Violet, Canada: *Viola canadensis*
Violet, Common Blue: *Viola papilionacea*
Violet, Green: *Hybanthus concolor*

Violet, Halberd-leaved: *Viola hastata*
Violet, Long-spurred: *Viola rostrata*
Violet, Primrose-leaved: *Viola primulifolia*
Violet, Round-leaved Yellow: *Viola rotundifolia*
Violet, Sweet White: *Viola blanda*
Violet, Wood: *Viola palmata triloba*
Vireo, Red-Eyed: *Vireo olivaceus*
Vireo, Solitary: *Vireo solitarius*
Vireo, White-eyed: *Vireo griseus*
Vireo, Yellow-throated: *Vireo flavifrons*
Virginia-creeper: *Parthenocissus quinquefolia*
Virginia-willow: *Itea virginica*
Virgin's-bower: *Clematis virginiana*
Volute: *Volutifusus typus*
Vulture, Black: *Coragyps atratus*
Vulture, Turkey: *Cathartes aura*
Wafer-ash: *Ptelea trifoliata*
Wahlenbergia: *Wahlenbergia marginata*
Wake-robin: *Trillium erectum, Trillium erectum vaseyi*
Walkingstick: *Diapheromera carolina, Diapheromera femorata, Diapheromera sp.*
Walnut, Black: *Juglans nigra*
Wapato: *Sagittaria latifolia pubescens*
Warbler, Bay-breasted: *Dendroica castanea*
Warbler, Black-and-white: *Mniotilta varia*
Warbler, Blackburnian: *Dendroica fusca*
Warbler, Blackpoll: *Dendroica striata*
Warbler, Black-throated Blue: *Dendroica caerulescens*
Warbler, Black-throated Green: *Dendroica virens*
Warbler, Canada: *Wilsonia canadensis*
Warbler, Cape May: *Dendroica tigrina*
Warbler, Chestnut-sided: *Dendroica pensylvanica*
Warbler, Hooded: *Wilsonia citrina*
Warbler, Kentucky: *Oporornis formosus*
Warbler, Magnolia: *Dendroica magnolia*
Warbler, Palm: *Dendroica palmarum*
Warbler, Pine: *Dendroica pinus*
Warbler, Prairie: *Dendroica discolor*
Warbler, Prothonotary: *Protonotaria citrea*
Warbler, Swainson's: *Limnothlypis swainsonii*
Warbler, Tennessee: *Vermivora peregrina*
Warbler, Worm-eating: *Helmitheros vermivorus*
Warbler, Yellow: *Dendroica petechia*
Warbler, Yellow-rumped: *Dendroica coronata*
Warbler, Yellow-throated: *Dendroica dominica*
Wasp, June-beetle: *Pelecinus polyturator*
Wasp, Paper (nest): *Polistes sp.*
Wasp, Paper: *Polistes sp.*
Wasp, Pipe-organ: *Trypoxylon politum*
Wasp, Potter: *Eumenes fraternus*
Wasp, Thread-waisted: *Ammophila sp.*
Water-elm: *Planera aquatica*
Water-hyacinth: *Eichhornia crassipes*
Water-hyssop, Blue: *Bacopa caroliniana*
Water-hyssop, Smooth: *Bacopa monnieri*
Water-hyssop: *Bacopa sp.*
Waterleaf: *Hydrophyllum canadense*
Water-lily: *Nymphaea odorata*
Water-meal: *Wolffia papulifera, Wolffia sp.*
Water-milfoil: *Myriophyllum brasiliense, Myriophyllum heterophyllum, Myriophyllum pinnatum, Myriophyllum sp.*
Water-plantain, Creeping: *Echinodorus cordifolius*
Water-plantain: *Alisma subcordatum*
Water-primrose: *Ludwigia uruguayensis*
Water-scorpion, Brown: *Ranatra fusca*
Water-scorpion: *Ranatra americana, Ranatra sp.*
Water-shield: *Brasenia schreberi*
Water-spider: *Orchid Habenaria repens*
Water-strider, Large: *Gerris remigis*
Water-strider: *Gerris sp.*
Waterthrush, Louisiana: *Seiurus motacilla*
Water-weed: *Elodea sp.*
Water-willow: *Justicia ovata, Ludwigia decurrens*
Wax-myrtle, Dwarf: *Myrica cerifera* pumila
Wax-myrtle: *Myrica cerifera, Myrica cerifera cerifera*
Waxweed: *Cuphea carthagensis*
Waxwing Cedar: *Bombycilla cedrorum*
Waxy-cap (mushroom): *Hygrocybe punica*
Waxy-cap, Chanterelle: *Hygrophorus cantharellus*
Waxy-cap, Fading Scarlet: *Hygrophorus miniatus*
Waxy-cap, Salmon: *Hygrophorus pratensis*
Waxy-cap, Scarlet: *Hygrophorus coccineus*
Weevil, Acorn: *Curculio sp.*
Weevil, Babtisia-pod: *Apion rostrum*
Weevil, Broad-nosed: *Aphrastus sp., Nemocestes sp., Phyllobius sp.*
Weevil, Bud: *Rhynchites geneus*
Weevil, Cocklebur: *Rhodobaenus tredecimpunctatus*
Weevil, Corn Pith: *Centrinaspis penicellus*
Weevil, Flower: *Aulobaris ibis, Baris aerea, Centrinaspis grisescens, Centrinaspis perscillus, Centrinaspis sp., Centrinites sp., Glyptobaris lecontei, Limnobaris denudata, Limnobaris prolixa, Odontocorynus pinguescens, Odontocorynus pulverulentus, Odontocorynus scutellum-album, Pseudobaris nigrina*
Weevil, Fruit: *Anthonomus tectus, Anthonomus sp.*
Weevil, Fungus: *Euparius marmoreus*
Weevil, Grain: *Sphenophorus sp.*
Weevil, Gray-sided Oak: *Pandeleteius hilaris*
Weevil, Hibiscus Seed: *Althaeus hibisci*
Weevil, Hidden Snout: *Conotrachelus posticatus*
Weevil, Leaf-mining: *Eugnamptus collaris collaris*
Weevil, Minute Seed: *Aleutes tachygonoides, Ceutorhynchus sp., Perigaster sp.*
Weevil, Pea & Bean: *Meibomeus musculus*
Weevil, Pear-shaped: *Apion crassum, Apion decoloratum, Apion emaciipes, Apion sp.*

Weevil, Potato Bud: *Anthonomus nigrinus*
Weevil, Primitive: *Arrhenodes minutus*
Weevil, Short-snouted: *Pandeleteius sp.*
Wentletrap (snail): *Epitoneum sp.*
Whelk (egg case): *Busycon sp.*
Whelk, Channeled: *Busycon canaliculatum*
Whelk, Knobbed: *Busycon carica*
Whelk, Lightning: *Busycon contrarium*
Whelk: *Busycon carica amoena, Busycon maximum, Busycon spiratus*
Whip-poor-will: *Caprimulgus vociferus*
Whiskfern: *Psilotum nudum*
White, Cabbage (butterfly): *Pieris rapae*
White-cedar, Atlantic: *Chamaecyparis thyoides*
White-featherling: *Tofieldia glabra*
Whitetail, Common (dragonfly): *Libellula lydia*
Whitlow-wort: *Paronychia herniarioides*
Widow, Black: *Latrodectus mactans*
Widow, Northern: *Latrodectus variolus*
Wigeon, American: *Anas americana*
Wildcat: *Felis rufus*
Wild-raisin: *Viburnum cassinoides*
Willet: *Catoptrophorus semipalmatus*
Willow, Black: *Salix nigra*
Willow, Swamp: *Salix caroliniana*
Willow: *Salix humilis, Salix sp.*
Willow-herb, Bronze: *Epilobium coloratum*
Windflower: *Anemone lancifolia, Anemone quinquefolia, Anemone sp., Thalictrum thalictroides,*
Wineberry: *Rubus phoenicolasius*
Wingstem: *Leersia lenticularis*
Wing-stem: *Verbesina virginica*
Winterberry: *Ilex verticillata*
Wintergreen, Spotted: *Chimaphila maculata*
Wintergreen: *Gaultheria procumbens*
Wire-plant: *Stipulicida setacea*
Wisteria, American: *Wisteria frutescens*
Wisteria, Chinese: *Wisteria sinensis*
Witch-alder: *Fothergilla gardenii, Fothergilla major*
Witches'-butter: *Tremella mesenterica*
Witch-hazel: *Hamamelis virginiana*
Wolf, Red: *Canis rufus*
Woodchuck: *Marmota monax*
Woodcock, American: *Scolopax minor*
Wood-lily, Speckled: *Clintonia umbellulata*
Wood-nettle: *Laportea canadensis*
Wood-nymph, Common (butterfly): *Cercyonis pegala, Cercyonis pegala pegala*
Woodpecker, Downy: *Picoides pubescens*
Woodpecker, Hairy: *Picoides villosus*
Woodpecker, Pileated: *Dryocopus pileatus*
Woodpecker, Red-bellied: *Melanerpes carolinus*
Woodpecker, Red-cockaded: *Picoides borealis*
Woodpecker, Red-headed: *Melanerpes erythrocephalus*
Wood-pewee, Eastern: *Contopus virens*
Woodrush: *Luzula acuminata, Luzula echinata, Luzula sp.*
Wood-satyr, Little (butterfly): *Megisto cymela*
Woodsia, Blunt-lobed: *Woodsia obtusa*
Wood-sorrel, Common: *Oxalis acetosella*
Wood-sorrel, Violet: *Oxalis violacea*
Wood-sorrel: *Oxalis dillenii, Oxalis florida, Oxalis grandis, Oxalis sp.*
Wool-grass: *Scirpus cyperinus*
Worm, Roofing: *Owenia fusiformis*
Wren, Carolina: *Thryothorus ludovicianus*
Wren, House: *Troglodytes aedon*
Wren, Marsh: *Cistothorus palustris*
Yam, Wild: *Dioscorea villosa, Dioscorea villosa villosa, Dioscorea sp.*
Yarrow: *Achillea millefolium*
Yaupon: *Ilex vomitoria*
Yellow, Little (butterfly): *Eurema lisa*
Yellow-eyed-grass: *Xyris caroliniana, Xyris flabelliformis, Xyris sp.*
Yellowjacket, Eastern: *Vespula maculifrons*
Yellowjacket: *Vespa sp., Vespula sp.*
Yellowlegs, Greater: *Tringa melanoleuca*
Yellowlegs, Lesser: *Tringa flavipes*
Yellow-patches: *Amanita flavoconia*
Yellowroot: *Xanthorhiza simplicissima*
Yellowthroat: *Geothlypis trichas*
Yellowwood: *Cladrastis lutea*
Yucca, Mound-lily: *Yucca gloriosa*
Yucca, Spoonleaf: *Yucca filamentosa filamentosa*
Yucca: *Yucca sp.*
Zebra (butterfly): *Heliconius charitonius tuckeri*
Zigadenus: *Zigadenus glaberrimus*
Zonite, Copper Great (snail): *Mesomphix cupreus*
Zonite, Great Common (snail): *Mesomphix vulgatus*
Zornia: *Zornia bracteata*

Scientific Name Index

In this index, flora and fauna are listed together in alphabetical order by scientific (Latin) name. To find a species by common name, please consult the Common Name List that immediately precedes this index, to find the name under which the species is listed here. A character at the junction of species and county indicates that the plant, animal, or fossil may be found in at least one site located in that county's section in the main text.

Some species found in this indexare not listed in the Common Name List because no generally accepted common name is available. Chief among the species for which this is true are the lichens.

For a comprehensive list of sites and the counties under which they are listed, see the Table of Contents at the beginning of this volume.

✻	Lower Plants
❀	Vascular Plants
●	Animals
❖	Fossils

	Abbeville	Aiken	Allendale	Anderson	Bamberg	Barnwell	Beaufort	Berkeley	Charleston	Cherokee	Chester	Chesterfield	Clarendon	Colleton	Darlington	Dillon	Dorchester	Fairfield	Florence	Georgetown	Greenville	Greenwood	Hampton	Horry	Jasper	Kershaw	Lancaster	Laurens	Lee	Lexington	Marlboro	McCormick	Newberry	Oconee	Orangeburg	Pickens	Richland	Spartanburg	Sumter	Union	York	Georgia	N. Carolina
Abacidus sculptus																																						●					
Abies fraseri																																											❀
Ablechrus sp.																	●																										
Acalymna vittatum																																										●	
Acalypha gracilens																																						❀					
Acalypha rhomboidea											❀																																
Acalypha sp.		❀					❀								❀											❀			❀									❀			❀		
Acanthocephala femorata		●																																									
Acanthocephala sp.																																				●							
Acanthospermum australe																													❀														
Accipiter striatus				●					●																		●	●										●					
Acer negundo		❀			❀			❀		❀	❀				❀			❀			❀	❀				❀	❀	❀				❀					❀	❀		❀	❀		
Acer pensylvanicum																					❀																					❀	❀
Acer rubrum	❀	❀	❀	❀	❀	❀	❀	❀	❀	❀	❀	❀	❀	❀	❀	❀	❀	❀	❀	❀	❀	❀	❀	❀	❀	❀	❀	❀	❀	❀	❀	❀	❀	❀	❀	❀	❀	❀	❀	❀	❀	❀	❀
Acer saccharinum			❀																																								
Acer saccharum																					❀																						❀
Acer saccharum floridanum	❀	❀					❀	❀			❀																❀					❀											
Acer saccharum leucoderme		❀																									❀	❀						❀	❀	❀	❀						
Achalarus lyciades											●																																
Achillea millefolium																					❀													❀									
Acmaeodera pulchella																								●																			
Acmaeodera tubulus																											●												●				
Aconitum uncinatum																																		❀								❀	
Acris crepitans												●			●												●					●						●			●	●	

Species	Abbeville	Aiken	Allendale	Anderson	Bamberg	Barnwell	Beaufort	Berkeley	Charleston	Cherokee	Chester	Chesterfield	Clarendon	Colleton	Darlington	Dillon	Dorchester	Fairfield	Florence	Georgetown	Greenville	Greenwood	Hampton	Horry	Jasper	Kershaw	Lancaster	Laurens	Lee	Lexington	Marlboro	McCormick	Newberry	Oconee	Orangeburg	Pickens	Richland	Spartanburg	Sumter	Union	York	Georgia	N. Carolina
Acris gryllus		●				●		●				●		●									●			●			●														
Acris sp.								●				●																											●		●		
Acroneuria internata																																		●									
Acronicta sp.																										●										●							
Acrosternum hilare																																							●				
Actaea pachypoda																																		❀								❀	
Actias luna		●						●				●												●										●		●	●						
Actitis macularia		●												●																				●									
Acyclomus piceus																												●															
Acylomus sp.		●																																									
Adela purpurea																																●											
Adiantum pedatum																											❀	❀				❀		❀		❀							❀
Aedes infirmatus									●										●																								
Aedes sollicitans									●											●																							
Aedes taeniorhyncus									●																																		
Aegopodium podagraria								❀																																			
Aesculus octandra																					❀													❀								❀	❀
Aesculus parviflora		❀																																									
Aesculus pavia		❀	❀				❀	❀			❀						❀			❀							❀								❀				❀				
Aesculus sp.		❀									❀																										❀						❀
Aesculus sylvatica		❀									❀							❀									❀	❀				❀				❀							
Agabetes acuductus																	●																										
Agabus johannis							●																																				
Agabus punctatus							●							●																												●	
Agalinis linifolia																													❀														
Agalinis purpurea												❀		❀															❀														
Agalinis setacea																										❀																	
Agalinis sp.		❀										❀								❀																		❀	❀				
Agalinis tenella							❀																																				
Agalinis tenuifolia																					❀						❀																
Agalinis virgata														❀																													
Agaricus campestris							*														*						*										*						
Agasicles hygrophila														●																													
Agave virginica		❀																																				❀					
Agelaius phoeniceus						●	●	●	●		●			●				●		●		●			●	●	●		●			●		●				●					●
Agelenopsis naevia																																									●		
Agkistrodon contortrix		●									●																																
Agkistrodon piscivorus		●																																									
Agonoderus lecontei																																						●					
Agonoderus sp.																																●											
Agonum errans																																		●									
Agonum extensicolle																																		●									

	Abbeville	Aiken	Allendale	Anderson	Bamberg	Barnwell	Beaufort	Berkeley	Charleston	Cherokee	Chester	Chesterfield	Clarendon	Colleton	Darlington	Dillon	Dorchester	Fairfield	Florence	Georgetown	Greenville	Greenwood	Hampton	Horry	Jasper	Kershaw	Lancaster	Laurens	Lee	Lexington	Marlboro	McCormick	Newberry	Oconee	Orangeburg	Pickens	Richland	Spartanburg	Sumter	Union	York	Georgia	N. Carolina
Agonum ferreus																							●																				
Agonum moerens															●																											
Agonum sp.											●	●																						●			●							
Agraulis vanillae							●	●	●											●			●		●	●																		
Agrilus politus																																										●	
Agrilus sp.															●																											
Agrimonia gryposepala																																											❀
Agrimonia parviflora																																		❀				❀					
Agrimonia pubescens microcarpa	❀	❀																																	❀									
Agrimonia sp.						❀															❀								❀						❀		❀		❀			❀		
Agriotes sp.																												●																
Agrius cingulatus							●																																					
Ailanthus altissima											❀																				❀										❀		
Aimophila aestivalis																								●																				
Aira elegans																																					❀					
Aix sponsa		●										●		●				●	●				●	●	●	●	●																	
Alaus myops																																●					●					
Alaus oculatus		●															●																		●		●					
Albizia julibrissin				❀		❀					❀										❀		❀		❀			❀	❀	❀		❀	❀				❀			❀		
Alcyonidium hauffi							●		●																																	
Aleochara sp.	●														●																											
Aletris aurea								❀																																		
Aletris farinosa																								❀					❀													
Aleurodiscus oakesii																																*										
Aleutes tachygonoides		●																																								
Alisma subcordatum											❀																															
Alligator mississippiensis		●					●		●					●						●					●	●																		
Allium bivalve			❀						❀																		❀					❀					❀					❀	
Allium canadense																					❀																					
Allium vineale							❀																														❀		❀			
Allopoda lutea																											●															
Alnus serrulata	❀	❀		❀		❀		❀		❀	❀	❀			❀			❀	❀		❀	❀	❀	❀		❀	❀		❀	❀	❀		❀	❀		❀		❀	❀	❀			❀
Alobates pennsylvanica			●				●								●												●	●										●				●		
Alopecurus carolinianus												❀																														
Alternanthera philoxeroides								❀						❀			❀						❀	❀						❀	❀												
Althaeus hibisci											●																															
Altica chalybea																																●											●
Altica marevagans								●																●																		
Altica sp.								●																				●								●							●
Altica viridana																																										●	
Alydus eurinus		●																																								
Alypia octomaculata																																										●	

	Abbeville	Aiken	Allendale	Anderson	Bamberg	Barnwell	Beaufort	Berkeley	Charleston	Cherokee	Chester	Chesterfield	Clarendon	Colleton	Darlington	Dillon	Dorchester	Fairfield	Florence	Georgetown	Greenville	Greenwood	Hampton	Horry	Jasper	Kershaw	Lancaster	Laurens	Lee	Lexington	Marlboro	McCormick	Newberry	Oconee	Orangeburg	Pickens	Richland	Spartanburg	Sumter	Union	York	Georgia	N. Carolina
Amanita brunnescens																								*																			
Amanita caesarea																											*							*									*
Amanita cinereopannosa																											*											*					
Amanita citrina							*																*	*							*												
Amanita cothurnata																																											*
Amanita flavoconia																																						*					*
Amanita flavorubescens																													*									*					
Amanita fulva																					*																						
Amanita gemmata					*																*						*											*					*
Amanita inaurata																																						*					
Amanita muscaria									*																		*											*					*
Amanita muscaria formosa																																*											*
Amanita pantherina																																											*
Amanita parcivolvata																																		*									
Amanita vaginata																											*					*						*					
Amanita virosa									*												*						*										*						*
Amara cupreolata															●																												
Amara sp.							●											●																									
Amaranthus pumilus																				❀																							
Amblycorypha floridana	●																																								●		
Amblycorypha oblongifolia																																				●							
Amblyomma americanum											●																																
Ambrosia artemisiifolia		❀							❀							❀				❀				❀		❀	❀		❀		❀					❀		❀					
Ambystoma opacum							●																					●									●						
Ambystoma talpoideum							●																																				
Amelanchier arborea		❀								❀											❀						❀							❀		❀	❀	❀	❀		❀	❀	❀
Amelanchier arborea arborea																																				❀							
Amelanchier canadensis													❀		❀									❀																			
Amelanchier obovalis																	❀																										
Amelanchier sp.						❀									❀									❀								❀							❀				❀
Amianthium muscaetoxicum	❀					❀																		❀																		❀	
Ammodramus caudacutus																									●																		
Ammophila sp.	●																																										
Amorpha fruticosa		❀						❀						❀																													
Amorpha herbacea	❀											❀																															
Amorpha sp.											❀				❀									❀										❀									
Ampedus areolatus																																		●									
Ampedus sp.																																		●									
Ampelopsis arborea		❀			❀			❀	❀					❀					❀	❀						❀									❀		❀	❀	❀				
Ampelopsis cordata											❀																																
Amphianthus pusillus																											❀															❀	
Amphibolips confluens																																										●	

	Abbeville	Aiken	Allendale	Anderson	Bamberg	Barnwell	Beaufort	Berkeley	Charleston	Cherokee	Chester	Chesterfield	Clarendon	Colleton	Darlington	Dillon	Dorchester	Fairfield	Florence	Georgetown	Greenville	Greenwood	Hampton	Horry	Jasper	Kershaw	Lancaster	Laurens	Lee	Lexington	Marlboro	McCormick	Newberry	Oconee	Orangeburg	Pickens	Richland	Spartanburg	Sumter	Union	York	Georgia	N. Carolina
Amphibolips cookii																																										●	
Amphibolips inanis						●																										●				●	●						
Amphicarpa bracteata		❀						❀			❀										❀						❀							❀		❀		❀	❀			❀	❀
Amphiloma lanuginosum								✻																																			
Amsonia ciliata												❀													❀														❀				
Amsonia tabernaemontana																											❀									❀							
Anadara aequicostata																								❖																			
Anadara brasiliana									●																																		
Anadara lienosa																								❖																			
Anadara ovalis									●																																		
Anadara transversa									●																																		
Anaptychia palmatula	✻																		✻								✻							✻				✻					✻
Anaptychia palmulata																																									✻		
Anas acuta									●					●											●																		
Anas americana									●					●						●					●																		
Anas clypeata									●					●											●																		
Anas crecca									●					●											●																		
Anas discors									●					●				●																									
Anas fulvigula														●																													
Anas platyrhynchos									●															●	●	●					●				●								●
Anaspis sp.																																										●	●
Anatis 15-punctata																																										●	
Anatis sp.																																							●				
Anax junius		●							●									●		●	●					●	●		●					●				●				●	●
Anax longipes	●	●																																	●								
Ancyloxypha numitor		●																																									
Andricus clavigerus							●																																				
Andricus foliatus		●					●							●																													
Andricus lanigera							●																																				
Andricus maxwelli							●																																				
Andricus perplexus							●																																				
Andricus petiolicola																													●														
Andricus pisiformis																													●														
Andricus unica																																											●
Andropogon scoparius												❀								❀						❀																	
Andropogon sp.									❀			❀		❀										❀					❀	❀							❀						
Andropogon ternarius																											❀									❀							
Andropogon virginicus							❀	❀	❀					❀						❀				❀						❀								❀					
Aneilema keisak		❀						❀			❀					❀								❀			❀		❀		❀												
Anemone lancifolia																																❀				❀							
Anemone quinquefolia																																		❀		❀							❀
Anemone sp.																																											❀

	Abbeville	Aiken	Allendale	Anderson	Bamberg	Barnwell	Beaufort	Berkeley	Charleston	Cherokee	Chester	Chesterfield	Clarendon	Colleton	Darlington	Dillon	Dorchester	Fairfield	Florence	Georgetown	Greenville	Greenwood	Hampton	Horry	Jasper	Kershaw	Lancaster	Laurens	Lee	Lexington	Marlboro	McCormick	Newberry	Oconee	Orangeburg	Pickens	Richland	Spartanburg	Sumter	Union	York	Georgia	N. Carolina
Anemone virginiana								❀																			❀									❀							
Angelica sp.																																		❀									❀
Angelica venenosa											❀																							❀									
Anguispira alternata																																						●					
Anhinga anhinga							●	●	●					●	●					●					●	●	●								●				●				
Anisodactylus rusticus																																		●			●	●					
Anisodactylus sayi																								●																			
Anisodactylus sp.									●																																		
Anisostichus capreolata		❀	❀			❀	❀	❀	❀	❀	❀			❀	❀		❀		❀	❀	❀			❀			❀		❀			❀	❀	❀	❀	❀	❀	❀	❀	❀	❀	❀	
Anoda cristata				❀																																							
Anolis carolinensis	●	●	●			●	●	●	●			●		●	●		●		●	●	●			●		●	●		●			●	●	●		●	●	●	●			●	
Anomia simplex									●																																		
Anomia sp.																			❖																								
Anomoea sp.																																●											
Anoplitis inaequalis																					●						●							●							●	●	
Anoplodera mutabilis																																										●	
Anoplodera sp.																																●										●	
Antennaria plantaginifolia	❀	❀																			❀	❀					❀					❀		❀		❀	❀	❀	❀			❀	❀
Antennaria solitaria										❀																								❀									❀
Antennaria sp.																																		❀		❀		❀			❀		
Anthaenantia villosa												❀																															
Antheraea polyphemus		●	●				●		●	●														●					●							●	●		●	●			
Anthobium sp.													●																														
Anthocharis midea			●												●			●									●					●					●		●			●	
Anthonomus nigrinus															●																												
Anthonomus sp.												●																															
Anthonomus tectus		●																													●												
Anthoxanthum odoratum																																											❀
Anthracothecium caroliniana sp. nov.									*				*																														
Anthracothecium leucostomum		*												*										*															*				
Anthracothecium libricola																																			*								
Anthracothecium pyrenuloides														*					*																								
Anthracothecium sp.		*																																	*								
Anthracothecium thelomorphum								*																																			
Anthrax aralis																																											●
Anurogryllus muticus																								●																			
Anzia colpodes																					*																						
Apenes sinuata														●																													
Aphodius lividus																																											●
Aphodius sp.													●																														
Aphrastus sp.																																											●
Apion crassum																																				●							

	Abbeville	Aiken	Allendale	Anderson	Bamberg	Barnwell	Beaufort	Berkeley	Charleston	Cherokee	Chester	Chesterfield	Clarendon	Colleton	Darlington	Dillon	Dorchester	Fairfield	Florence	Georgetown	Greenville	Greenwood	Hampton	Horry	Jasper	Kershaw	Lancaster	Laurens	Lee	Lexington	Marlboro	McCormick	Newberry	Oconee	Orangeburg	Pickens	Richland	Spartanburg	Sumter	Union	York	Georgia	N. Carolina
Apion decoloratum																											●																
Apion emaciipes																																										●	
Apion rostrum																													●														
Apion sp.																								●					●													●	
Apios americana		❀						❀						❀										❀			❀				❀			❀									
Apiosporina morbosa		✻		✻		✻				✻	✻				✻				✻		✻	✻		✻										✻						✻			
Apis mellifera									●											●				●																			
Apium leptophyllum																								❀																			
Aplectrum hyemale										❀																																	❀
Aplidium constellatum																				●																							
Aplidium sp.									●																																		
Apocynum cannabinum	❀										❀																		❀			❀											
Apocynum sp.																																				❀						❀	
Aquila chrysaetos																																							●				
Aquilegia canadensis											❀																																
Arabidopsis thaliana			❀															❀																						❀			
Arabis laevigata		❀																									❀															❀	
Arabis laevigata missouriensis																											❀															❀	
Arachnion album		✻						✻																																			
Aralia nudicaulis																																		❀									❀
Aralia racemosa																					❀													❀		❀						❀	❀
Aralia spinosa		❀				❀	❀	❀	❀				❀		❀								❀					❀	❀				❀	❀			❀	❀	❀				❀
Araneus bicentenarius					●																						●																
Araneus marmoreus						●								●												●	●		●									●			❀		
Araneus sp.																																	●										
Arbacia punctulata									●																																		
Arca zebra																				●																							
Archilochus colubris					●				●					●							●					●	●							●	●		●						
Arcyria denudata																					✻													✻									
Ardea herodias	●	●	●	●		●	●	●	●		●	●		●				●		●				●	●	●	●				●				●		●	●			●		
Ardistomis sp.							●																																				
Ardistomis viridis																													●														
Arenaeus cribrarius									●																																		
Arenaria caroliniana												❀																											❀				
Arenaria groenlandica																					❀						❀									❀							
Arenaria groenlandica glabra																											❀															❀	
Arenaria interpres							●		●											●																							
Arenaria uniflora																											❀															❀	
Argia apicalis	●	●									●																																
Argia bipunctulata												●																															
Argia fumipennis		●									●				●											●																	
Argia moesta		●									●																																

	Abbeville	Aiken	Allendale	Anderson	Bamberg	Barnwell	Beaufort	Berkeley	Charleston	Cherokee	Chester	Chesterfield	Clarendon	Colleton	Darlington	Dillon	Dorchester	Fairfield	Florence	Georgetown	Greenville	Greenwood	Hampton	Horry	Jasper	Kershaw	Lancaster	Laurens	Lee	Lexington	Marlboro	McCormick	Newberry	Oconee	Orangeburg	Pickens	Richland	Spartanburg	Sumter	Union	York	Georgia	N. Carolina
Argia sedula											●																																
Argia sp.																																		●									
Argidae Family		●																																									
Argiope aurantia	●	●					●				●											●		●			●				●		●	●		●							
Argiope sp.																								●																			
Argiope trifasciata																								●																			
Argopecten eboreus																																											❖
Argopecten vicenarius																								❖																			
Arilus cristatus								●			●			●																				●					●		●		
Ariotus sp.															●																												
Arisaema dracontium																																					❀						❀
Arisaema triphyllum						❀	❀				❀								❀		❀						❀	❀				❀		❀	❀	❀	❀	❀				❀	❀
Aristida sp.												❀												❀		❀													❀				
Aristida stricta												❀													❀																		
Aristida tuberculosa		❀																																									
Aristolochia macrophylla																																				❀							❀
Aristolochia serpentaria	❀	❀														❀												❀						❀		❀	❀	❀					❀
Armillariella mellea																																							✻				
Armillariella tabascens																													✻														
Arnica acaulia																													❀								❀						
Arphia sp.																																						●					
Arphia sulphurea																											●															●	
Arphia xanthoptera		●																																									
Arrhenodes minutus								●																																			
Arthonia quintaria														✻					✻																								
Arthonia reniformis														✻																													
Arthonia rubella											✻																																
Arthonia sp.																																✻			✻		✻						
Arthopyrenia lapponina																								✻																			
Arthothelium interveniens							✻		✻																																		
Arthothelium sp.							✻																						✻														
Arthothelium taediosum							✻																																				
Arthromacra aenea																												●								●						●	
Aruncus dioicus																																										❀	
Arundinaria gigantea	❀	❀	❀	❀		❀	❀	❀	❀	❀	❀	❀	❀	❀	❀	❀		❀			❀		❀	❀	❀	❀	❀	❀	❀	❀		❀	❀	❀		❀	❀	❀	❀	❀	❀		❀
Arundo donax				❀				❀	❀																																		
Asarum canadense								❀													❀						❀	❀				❀		❀		❀							❀
Asclepias amplexicaulis																																				❀							
Asclepias exaltata																																										❀	
Asclepias humistrata		❀						❀							❀										❀																		
Asclepias longifolia																													❀														
Asclepias michauxii																								❀																			

	Abbeville	Aiken	Allendale	Anderson	Bamberg	Barnwell	Beaufort	Berkeley	Charleston	Cherokee	Chester	Chesterfield	Clarendon	Colleton	Darlington	Dillon	Dorchester	Fairfield	Florence	Georgetown	Greenville	Greenwood	Hampton	Horry	Jasper	Kershaw	Lancaster	Laurens	Lee	Lexington	Marlboro	McCormick	Newberry	Oconee	Orangeburg	Pickens	Richland	Spartanburg	Sumter	Union	York	Georgia	N. Carolina
Asclepias perennis		❀												❀																													
Asclepias quadrifolia											❀																															❀	
Asclepias sp.	❀							❀								❀																											
Asclepias tuberosa		❀																				❀														❀		❀					
Asclepias tuberosa tuberosa																																						❀					
Asclepias variegata																	❀																			❀							
Asclepias verticillata																													❀							❀							
Asclera ruficollis			●																																								
Asclera sp.																																						●					
Ascocoryne sarcoides																		✻																				✻					
Asilidae																																										●	
Asimina parviflora	❀														❀					❀	❀								❀		❀	❀		❀		❀		❀					
Asimina sp.																								❀										❀				❀					
Asimina triloba		❀						❀		❀											❀	❀					❀	❀				❀	❀	❀	❀		❀	❀		❀	❀		
Asio otus																																											●
Asparagus officinalis																						❀																					
Aspicilia cinerea																																											✻
Aspicilia lacustris																																		✻									
Aspicilia laevata																																									✻		✻
Asplenium bradleyi												❀																															
Asplenium montanum																																		❀									❀
Asplenium platyneuron	❀	❀	❀	❀	❀	❀	❀	❀	❀	❀	❀			❀	❀		❀		❀		❀		❀	❀		❀	❀	❀		❀		❀	❀	❀	❀	❀	❀	❀	❀	❀	❀		❀
Asplenium resiliens																																		❀									
Asplenium rhizophyllum																																		❀									❀
Asplenium sp.																																			❀								
Astarte concentrica																								❖																			
Astarte glenni																								❖																			
Aster carolinianus								❀						❀																													
Aster concolor																														❀													
Aster cordifolius								❀													❀																						
Aster curtisii																					❀															❀							
Aster divaricatus																					❀													❀									
Aster dumosus						❀		❀															❀			❀										❀		❀					
Aster lateriflorus															❀																												
Aster linariifolius								❀				❀																															
Aster paternus								❀																																			
Aster pilosus		❀						❀						❀					❀	❀																		❀			❀		
Aster pilosus pilosus																													❀														
Aster puniceus																					❀																						
Aster solidagineus																																				❀							
Aster sp.						❀			❀		❀	❀									❀						❀			❀									❀				
Aster squarrosus								❀				❀												❀					❀														

	Abbeville	Aiken	Allendale	Anderson	Bamberg	Barnwell	Beaufort	Berkeley	Charleston	Cherokee	Chester	Chesterfield	Clarendon	Colleton	Darlington	Dillon	Dorchester	Fairfield	Florence	Georgetown	Greenville	Greenwood	Hampton	Horry	Jasper	Kershaw	Lancaster	Laurens	Lee	Lexington	Marlboro	McCormick	Newberry	Oconee	Orangeburg	Pickens	Richland	Spartanburg	Sumter	Union	York	Georgia	N. Carolina
Aster tenuifolius							❀		❀											❀																							
Aster tortifolius																❀																											
Aster undulatus																				❀							❀																
Asterocampa celtis		●									●											●					●	●					●				●						
Asterocampa clyton																																					●						
Astilbe biternata																																										❀	❀
Astraeus hygrometricus																			*				*						*	*				*					*				
Astrangia danae									●																																		
Ataenius cognatus													●																					●									
Ataenius sp.			●																													●											
Ateuchus histeroides	●																																					●	●				
Ateuchus sp.																											●																
Atheta flaviola	●																																										
Atheta pennsylvanica	●																																										
Atheta sp.																								●																			
Athous rufifrons																																				●							
Athyrium asplenioides	❀	❀			❀	❀								❀	❀						❀		❀	❀		❀	❀		❀				❀	❀	❀	❀	❀	❀	❀		❀	❀	❀
Athyrium filix-femina																											❀																
Athyrium thelypteroides																																										❀	❀
Atlanticus gibbosus	●																																					●					
Atlanticus sp.																																				●							
Atlanticus testaceus																																		●									
Atrina rigida									●																																		
Atrina seminuda									●																																		
Atrina sp.									●																																		
Atriplex arenaria							❀		❀											❀																							
Atriplex patula									❀																																		
Attalus circumscriptus													●																														
Attalus scincetus													●																	●													
Attalus sp.																																										●	
Atteva punctella											●																																
Aulobaris ibis	●																										●																
Aureolaria flava																																										❀	
Aureolaria pectinata	❀																													❀				❀									
Aureolaria pedicularia																																				❀							
Aureolaria sp.								❀			❀										❀			❀			❀							❀		❀					❀		❀
Aureolaria virginica	❀	❀															❀																										
Auricularia auricula			*																								*					*		*			*			*			
Auricularia sp.						*																																					
Austroboletus betula																					*													*									
Austroboletus gracilis																																		*									
Automeris io														●																			●				●						

	Abbeville	Aiken	Allendale	Anderson	Bamberg	Barnwell	Beaufort	Berkeley	Charleston	Cherokee	Chester	Chesterfield	Clarendon	Colleton	Darlington	Dillon	Dorchester	Fairfield	Florence	Georgetown	Greenville	Greenwood	Hampton
Aythya affinis									●														
Aythya collaris									●					●						●			
Aythya marila									●														
Aythya valisineria														●									
Azolla caroliniana								❀	❀					❀			❀			❀			❀
Baccharis angustifolia							❀		❀														
Baccharis glomeruliflora							❀													❀			
Baccharis halimifolia						❀	❀	❀	❀			❀		❀	❀		❀			❀			
Bacidia abductans																					*		
Bacidia assulata							*																
Bacidia atrogrisea																				*			
Bacidia chlorantha															*								
Bacidia clementis									*														
Bacidia dryina							*																*
Bacidia fuscus sp. nov.																							
Bacidia incompta							*							*	*				*				
Bacidia laurocerasi											*												*
Bacidia phacodes							*														*		
Bacidia polychroa		*									*								*				
Bacidia rosella																							
Bacidia schweinitzii		*				*		*			*				*			*	*		*		*
Bacidia sp.							*				*				*					*	*		
Bacidia suffusa							*				*												
Bacopa caroliniana						❀																	
Bacopa monnieri							❀													❀			
Bacopa sp.																				❀			
Baeomyces fungoides								*			*										*		
Balanus eburneus									●														
Balanus sp.									●														
Baptisia alba		❀																					
Baptisia bracteata															❀								
Baptisia cinerea																							
Baptisia perfoliata		❀																					
Baptisia sp.	❀	❀							❀			❀											
Baptisia tinctoria																							
Barbarea verna																							
Baris aerea											●												
Bartramia pomiformis										*													
Basiaeschna janata																							
Bassareus sp.																							
Batis maritima							❀																
Battus philenor	●	●																					

	Horry	Jasper	Kershaw	Lancaster	Laurens	Lee	Lexington	Marlboro	McCormick	Newberry	Oconee	Orangeburg	Pickens	Richland	Spartanburg	Sumter	Union	York	Georgia	N. Carolina
Aythya affinis		●																		
Aythya collaris		●																		
Aythya marila																				
Aythya valisineria		●																		
Azolla caroliniana							❀					❀								
Baccharis angustifolia		❀																		
Baccharis glomeruliflora																				
Baccharis halimifolia	❀	❀	❀			❀	❀									❀				
Bacidia abductans				*																
Bacidia assulata																				
Bacidia atrogrisea				*																
Bacidia chlorantha				*																
Bacidia clementis																				
Bacidia dryina				*																
Bacidia fuscus sp. nov.				*										*						
Bacidia incompta				*																
Bacidia laurocerasi				*		*						*		*		*		*		
Bacidia phacodes				*																
Bacidia polychroa				*										*		*				
Bacidia rosella				*										*						
Bacidia schweinitzii				*					*		*		*	*	*	*	*	*	*	*
Bacidia sp.				*							*	*			*					*
Bacidia suffusa				*																
Bacopa caroliniana			❀																	
Bacopa monnieri																				
Bacopa sp.																				
Baeomyces fungoides				*							*		*						*	*
Balanus eburneus																				
Balanus sp.																				
Baptisia alba											❀									
Baptisia bracteata													❀							
Baptisia cinerea	❀																			
Baptisia perfoliata																				
Baptisia sp.	❀					❀														
Baptisia tinctoria	❀										❀									❀
Barbarea verna																				❀
Baris aerea																				
Bartramia pomiformis																				
Basiaeschna janata	●																			
Bassareus sp.														●						
Batis maritima																				
Battus philenor											●	●	●							

	Abbeville	Aiken	Allendale	Anderson	Bamberg	Barnwell	Beaufort	Berkeley	Charleston	Cherokee	Chester	Chesterfield	Clarendon	Colleton	Darlington	Dillon	Dorchester	Fairfield	Florence	Georgetown	Greenville	Greenwood	Hampton	Horry	Jasper	Kershaw	Lancaster	Laurens	Lee	Lexington	Marlboro	McCormick	Newberry	Oconee	Orangeburg	Pickens	Richland	Spartanburg	Sumter	Union	York	Georgia	N. Carolina
Belemnitella americana																			❖																								
Belenocnema fossoria							●																																				
Belostoma flumineum																																						●					
Belostoma lutarium														●																													
Bembidion americanum																																		●									
Bembidion confusum																																●											
Bembidion dilitatum																																●											
Bembidion sp.																																		●									
Bembidion versicolor																																											●
Berchemia scandens		❀	❀			❀		❀						❀										❀													❀		❀			❀	
Beroe ovata									●																																		
Berosus aculeatus																					●																						
Berosus infuscatus							●																								●												
Berosus pantherinus																													●			●											
Berosus sp.		●																														●									●		
Betula lenta																					❀													❀		❀						❀	❀
Betula lutea																																											❀
Betula nigra		❀	❀	❀						❀	❀		❀	❀			❀	❀	❀					❀					❀		❀	❀					❀	❀		❀	❀		
Bidens bipinnata		❀												❀							❀			❀			❀				❀			❀									
Bidens frondosa																								❀										❀				❀					
Bidens laevis						❀																	❀																				
Bidens pilosa																					❀																						
Bidens sp.								❀	❀							❀																											
Bidessus sp.			●																										●														
Bilimbia lignaria								*																			*																
Bitoma sp.																																											●
Bittacomorpha clavipes						●																															●					●	
Bjerkandera adusta															*																												
Bjerkandera sp.						*																																					
Blarina carolinensis															●																												
Blarina sp.																														●													
Boehmeria cylindrica		❀				❀	❀	❀	❀		❀			❀		❀							❀	❀		❀	❀		❀					❀	❀		❀	❀	❀		❀		
Bolboceras sp.																																		●									
Boletellus betula																											*							*									*
Boletus chrysenteron																															*												
Boletus curtisii																													*														
Boletus edulis																																		*									
Boletus erythropus																													*														
Boletus frostii																																						*					*
Boletus subtomentosus																																						*					
Boletus zelleri																																		*									
Bolitotherus cornutus								●																																			

	Abbeville	Aiken	Allendale	Anderson	Bamberg	Barnwell	Beaufort	Berkeley	Charleston	Cherokee	Chester	Chesterfield	Clarendon	Colleton	Darlington	Dillon	Dorchester	Fairfield	Florence	Georgetown	Greenville	Greenwood	Hampton	Horry	Jasper	Kershaw	Lancaster	Laurens	Lee	Lexington	Marlboro	McCormick	Newberry	Oconee	Orangeburg	Pickens	Richland	Spartanburg	Sumter	Union	York	Georgia	N. Carolina
Bombus impatiens																																		●									
Bombus sp.												●								●														●									●
Bombycilla cedrorum									●			●							●					●			●														●		
Bombylius sp.																								●																			
Bonamia humistrata																							❀		❀																		
Bonamia patens		❀														❀																											
Bonamia sp.																										❀			❀														
Bonasa umbellus																																											●
Borrichia frutescens							❀		❀											❀																							
Botaurus lentiginosus														●																													
Botrychium biternatum						❀																		❀			❀							❀				❀					
Botrychium dissectum		❀		❀																							❀							❀				❀			❀		
Botrychium lunarioides																																										❀	
Botrychium sp.											❀																							❀				❀					
Botrychium virginianum																											❀	❀				❀	❀	❀		❀	❀	❀	❀			❀	❀
Brachiodontes sp.																								❖																			
Brachyacantha sp.																											●							●									
Brachyacantha ursina																																										●	
Brachyelytrum erectum		❀																																									
Brachymesia gravida														●																													
Brachys aerosus																											●																
Brachys sp.												●																															
Brachystola magna														●						●																							
Branta canadensis		●										●		●																		●		●								●	
Brasenia schreberi		❀										❀				❀										❀	❀				❀												
Brefeldia maxima																				*																							
Briareum asbestinum									●																																		
Brigantiaea leucoxantha						*			*					*					*					*			*							*									
Briza minor		❀																																									
Brochymena sp.														●																													
Brontes dubius															●																			●									
Broussonetia papyrifera																					❀										❀												
Bucephala albeola							●		●																																		
Bucephala clangula									●																																		
Buchnera floridana								❀															❀	❀																			
Bucrates malivolans																													●														
Buellia amphidexia																				*																							
Buellia caloosensis								*												*						*						*											
Buellia coccinea																					*											*											
Buellia conspirans							*																																				
Buellia curatellae	*	*							*		*			*	*			*	*		*		*	*										*	*			*	*		*		*
Buellia curtisii								*																		*								*									

	Abbeville	Aiken	Allendale	Anderson	Bamberg	Barnwell	Beaufort	Berkeley	Charleston	Cherokee	Chester	Chesterfield	Clarendon	Colleton	Darlington	Dillon	Dorchester	Fairfield	Florence	Georgetown	Greenville	Greenwood	Hampton	Horry	Jasper	Kershaw	Lancaster	Laurens	Lee	Lexington	Marlboro	McCormick	Newberry	Oconee	Orangeburg	Pickens	Richland	Spartanburg	Sumter	Union	York	Georgia	N. Carolina
Buellia dialyta									*					*	*														*					*	*		*						
Buellia disciformis		*									*	*		*	*						*		*						*							*			*	*			
Buellia elizae																																				*							
Buellia leucomela																										*																	
Buellia parasema												*			*																			*		*							*
Buellia punctata	*	*				*	*	*	*			*		*	*			*			*			*		*			*					*	*	*		*	*		*		*
Buellia sp.		*					*	*																*																	*	*	
Buellia stellulata	*																																			*					*		
Buellia stigmaea																																*									*		
Buellia stillingiana								*																		*			*														
Bufo americanus																					●													●									●
Bufo quercicus																								●																			
Bufo sp.															●						●																						
Bufo terrestris												●			●	●							●	●		●			●					●	●		●						
Bufo woodhousei	●	●									●																●		●		●	●		●		●		●					
Bulbicus ibis					●			●					●	●												●																	
Bulbostylis warei																									❀																		
Bulbothrix confoederata		*					*	*						*					*					*		*			*					*		*			*				
Bulbothrix goebelii	*	*				*	*	*	*		*			*	*				*	*			*	*	*	*			*			*		*	*		*		*		*	*	
Bulbothrix goebelii f. minor		*							*		*			*							*		*																*		*		
Bulbothrix isidiza														*	*									*		*			*						*				*				
Bulbothrix laevigatula		*						*																																			
Bumelia lycioides			❀					❀	❀					❀													❀																
Bumelia tenax									❀																																		
Buprestis rufipes														●																													
Burmannia biflora						❀		❀																																			
Burmannia capitata												❀																															
Busycon canaliculatum									●																																		
Busycon carica							●		●											●																							
Busycon carica amoena																								❖																			
Busycon contrarium									●															●																			
Busycon maximum																								❖																			
Busycon sp.									●																																		
Busycon spiratus																																											❖
Buteo jamaicensis		●		●	●	●	●		●			●		●			●	●	●					●	●		●		●					●		●			●		●		
Buteo lineatus		●					●	●	●					●			●		●	●				●			●		●				●				●	●				●	
Buteo platypterus					●																																						●
Butorides striatus								●			●				●										●	●																	
Cabomba caroliniana														❀																	❀												
Cacalia atriplicifolia											❀										❀						❀					❀		❀	❀		❀	❀				❀	
Caenia dimidiata																																							●				
Cakile harperi									❀											❀																							

	Abbeville	Aiken	Allendale	Anderson	Bamberg	Barnwell	Beaufort	Berkeley	Charleston	Cherokee	Chester	Chesterfield	Clarendon	Colleton	Darlington
Cakile sp.									❀						
Calamovilfa brevipilis												❀			
Calephelis virginiensis								●							
Calicium abietinum									*						
Calidris alba							●		●						
Calidris alpina							●		●						
Calidris canutus									●						
Calidris mauri									●						
Calidris minutilla		●					●		●						
Calidris pusilla							●								
Calliactis tricolor									●						
Callianassa major															
Callianassa sp.									●						
Callicarpa americana		❀				❀	❀	❀	❀		❀			❀	❀
Calligrapha rhoda								●							
Callinectes sapidus							●		●						
Callinectes similis									●						
Callinectes sp.															
Callirhytis clavula															
Callirhytis conifera							●								
Callirhytis lustrans	●														
Callirhytis seminator															
Callirhytis sp.											●				
Callitriche deflexa												❀			❀
Callitriche heterophylla															❀
Callophrys gryneus gryneus															
Callophrys henrici												●			
Callosamia securifera									●						
Caloplaca aurantiaca												*			
Caloplaca camptidia															
Caloplaca cinnabarina											*				
Caloplaca citrina							*								
Caloplaca ferruginea											*				
Caloplaca flavorubescens											*				
Calopogon barbatus															
Calopogon pallidus															
Calopogon pulchellus								❀							
Caloporus dichrous			*												
Calopteron reticulatum															
Calopteron sp.															
Calopteryx aequabile															
Calopteryx dimidiata		●													●

	Dillon	Dorchester	Fairfield	Florence	Georgetown	Greenville	Greenwood	Hampton	Horry	Jasper	Kershaw	Lancaster	Laurens	Lee	Lexington
Cakile sp.															
Calamovilfa brevipilis															
Calephelis virginiensis															
Calicium abietinum															
Calidris alba					●										
Calidris alpina					●										
Calidris canutus					●										
Calidris mauri															
Calidris minutilla															
Calidris pusilla															
Calliactis tricolor															
Callianassa major															●
Callianassa sp.															
Callicarpa americana			❀	❀	❀	❀		❀	❀		❀	❀		❀	
Calligrapha rhoda															
Callinectes sapidus					●										
Callinectes similis															
Callinectes sp.					●										
Callirhytis clavula						●									
Callirhytis conifera															
Callirhytis lustrans															
Callirhytis seminator						●									
Callirhytis sp.															
Callitriche deflexa															
Callitriche heterophylla															
Callophrys gryneus gryneus															
Callophrys henrici															
Callosamia securifera									●						
Caloplaca aurantiaca															
Caloplaca camptidia														*	
Caloplaca cinnabarina															
Caloplaca citrina															
Caloplaca ferruginea															
Caloplaca flavorubescens						*									
Calopogon barbatus									❀						
Calopogon pallidus									❀						
Calopogon pulchellus									❀					❀	
Caloporus dichrous															
Calopteron reticulatum															
Calopteron sp.															
Calopteryx aequabile	●														
Calopteryx dimidiata															

	Marlboro	McCormick	Newberry	Oconee	Orangeburg	Pickens	Richland	Spartanburg	Sumter	Union	York	Georgia	N. Carolina
Cakile sp.													
Calamovilfa brevipilis													
Calephelis virginiensis													
Calicium abietinum													
Calidris alba													
Calidris alpina													
Calidris canutus													
Calidris mauri													
Calidris minutilla													
Calidris pusilla													
Calliactis tricolor													
Callianassa major													
Callianassa sp.													
Callicarpa americana	❀		❀		❀	❀	❀	❀	❀			❀	
Calligrapha rhoda		●											●
Callinectes sapidus													
Callinectes similis													
Callinectes sp.													
Callirhytis clavula													
Callirhytis conifera													
Callirhytis lustrans													
Callirhytis seminator													
Callirhytis sp.													
Callitriche deflexa													
Callitriche heterophylla													
Callophrys gryneus gryneus												●	
Callophrys henrici													
Callosamia securifera													
Caloplaca aurantiaca													
Caloplaca camptidia													
Caloplaca cinnabarina													
Caloplaca citrina		*											
Caloplaca ferruginea													
Caloplaca flavorubescens													
Calopogon barbatus													
Calopogon pallidus													
Calopogon pulchellus													
Caloporus dichrous				*									
Calopteron reticulatum											●		
Calopteron sp.					●								
Calopteryx aequabile													
Calopteryx dimidiata													

	Abbeville	Aiken	Allendale	Anderson	Bamberg	Barnwell	Beaufort	Berkeley	Charleston	Cherokee	Chester	Chesterfield	Clarendon	Colleton	Darlington	Dillon	Dorchester	Fairfield	Florence	Georgetown	Greenville	Greenwood	Hampton	Horry	Jasper	Kershaw	Lancaster	Laurens	Lee	Lexington	Marlboro	McCormick	Newberry	Oconee	Orangeburg	Pickens	Richland	Spartanburg	Sumter	Union	York	Georgia	N. Carolina
Calopteryx maculata	●	●			●						●											●						●					●		●		●	●	●				
Calosoma calidum																																										●	
Calosoma scrutator									●																																		
Calostoma cinnabarina																					✻													✻		✻							✻
Calostoma sp.																					✻																						
Calvatia craniformis																											✻							✻									
Calvatia cyathiformis															✻																			✻									
Calvatia utriformis		✻																																									
Calycanthus floridus	❀									❀	❀										❀						❀	❀		❀		❀		❀		❀		❀	❀			❀	❀
Calycopis cecrops								●				●									●												●	●									
Calystegia sericata																																		❀		❀							❀
Cambarus diogenes														●																													
Cambarus sp.												●					●						●																				
Campanula divaricata																					❀																						
Campeloma decisum						●																																					
Camponotus abdominalis							●																																				
Camponotus pennsylvanicus																																					●						
Campsis radicans	❀	❀		❀	❀	❀	❀	❀	❀		❀			❀	❀		❀		❀	❀		❀	❀	❀		❀	❀	❀	❀		❀		❀		❀	❀	❀	❀	❀	❀	❀		
Camptonotus caroliniensis																																		●									
Campylothelium amylosporum									✻																																		
Candelaria concolor											✻	✻						✻			✻					✻	✻		✻					✻						✻	✻	✻	
Candelaria fibrosa															✻																			✻									
Canifa pallipes																																						●					
Canis rufus									●																																		
Canoparmelia amazonica																									✻																		
Canoparmelia caroliniana	✻	✻					✻		✻		✻			✻	✻			✻	✻		✻		✻	✻					✻					✻	✻	✻		✻	✻		✻	✻	
Canoparmelia crozalsiana		✻									✻																																
Canoparmelia texana														✻												✻																	
Cantharellus cibarius																																		✻		✻		✻					
Cantharellus cinnabarinus	✻																										✻							✻				✻					
Cantharellus ignicolor																																				✻							
Cantharellus sp.									✻																																		
Cantharis bilineatus			●												●			●																									●
Cantharis lineola																																		●									
Cantharis sp.		●						●							●												●					●				●						●	
Canthon chalcites																																		●									
Canthon laevis								●												●														●									
Caprimulgus vociferus																				●																●							
Carabus sylvosus																																		●									
Carabus vinctus															●		●	●						●																			
Carcharhinus leucas																	❖																										
Carcharhinus sp.									❖															❖																			

	Abbeville	Aiken	Allendale	Anderson	Bamberg	Barnwell	Beaufort	Berkeley	Charleston	Cherokee	Chester	Chesterfield	Clarendon	Colleton	Darlington	Dillon	Dorchester	Fairfield	Florence	Georgetown	Greenville	Greenwood	Hampton	Horry	Jasper	Kershaw	Lancaster	Laurens	Lee	Lexington	Marlboro	McCormick	Newberry	Oconee	Orangeburg	Pickens	Richland	Spartanburg	Sumter	Union	York	Georgia	N. Carolina
Carcharodon auriculatus																	❖																										
Carcharodon carcharius																								❖																			
Carcharodon megalodon																																											❖
Cardamine angustata angustata																																❀											
Cardamine clematitis																																											❀
Cardamine concatenata																											❀																❀
Cardamine diphylla																																											❀
Cardamine hirsuta									❀						❀																	❀							❀	❀			
Cardamine pensylvanica									❀									❀																			❀				❀		
Cardamine sp.																																											❀
Cardinalis cardinalis	●	●	●	●	●		●	●	●		●	●	●	●	●			●	●	●	●	●	●	●	●	●	●				●			●	●	●	●	●	●		●	●	●
Cardiophorus sp.																																						●					
Cardita arata																																											❖
Cardodacus mexicanus																																						●					
Cardodacus purpureus							●																																				
Carduelis tristis							●											●																●									●
Carduus lanceolatus																																					❀						
Carduus pumilus																																						❀					
Carduus repandus																❀																											
Carduus sp.		❀					❀					❀		❀							❀											❀		❀				❀					
Carduus spinosissimus																													❀														
Caretta caretta									●																																		
Carex amphibola		❀																																									
Carex blanda		❀																																									
Carex caroliniana		❀																																									
Carex cephalophora		❀																																									
Carex communis																																		❀									❀
Carex complanata		❀																																									
Carex crinita		❀																																									
Carex debilis		❀																																									
Carex digitalis																																											❀
Carex flaccosperma		❀																																									
Carex glaucescens												❀												❀					❀		❀												
Carex intumescens		❀																																									
Carex laxiculmis																																		❀									
Carex laxiflora																																											❀
Carex lurida																								❀							❀												
Carex oxylepis		❀																																									
Carex plantaginea																																											❀
Carex retroflexa		❀																																									❀
Carex rosea		❀																																									
Carex sp.							❀				❀																		❀								❀						

	Abbeville	Aiken	Allendale	Anderson	Bamberg	Barnwell	Beaufort	Berkeley	Charleston	Cherokee	Chester	Chesterfield	Clarendon	Colleton	Darlington	Dillon	Dorchester	Fairfield	Florence	Georgetown	Greenville	Greenwood	Hampton	Horry	Jasper	Kershaw	Lancaster	Laurens	Lee	Lexington	Marlboro	McCormick	Newberry	Oconee	Orangeburg	Pickens	Richland	Spartanburg	Sumter	Union	York	Georgia	N. Carolina
Carex virescens																																											❀
Carex vulpinoidea		❀																																									
Carex walteriana																													❀														
Carex willdenowii		❀																																									
Carolinapecten arboreus																								❖																			
Carolinapecten solarioides																								❖																			
Carphephorus bellidifolius												❀														❀			❀														
Carphephorus sp.												❀				❀										❀																	
Carphophis amoenus																																						●					
Carpinus caroliniana	❀	❀	❀		❀			❀		❀	❀			❀	❀			❀	❀								❀	❀	❀			❀		❀	❀	❀	❀	❀	❀	❀	❀		❀
Carpophilus floralis																	●																										
Carpophilus sp.																								●																			
Carya aquatica								❀						❀					❀																❀		❀						
Carya carolinae-septentrionalis																											❀																
Carya cordiformis							❀																					❀				❀		❀	❀					❀	❀		
Carya glabra	❀	❀				❀	❀				❀			❀	❀						❀		❀				❀					❀		❀		❀			❀	❀	❀	❀	❀
Carya illinoensis							❀																																				
Carya ovata	❀										❀																	❀				❀					❀						❀
Carya pallida	❀																													❀	❀												
Carya sp.		❀												❀					❀	❀	❀			❀			❀							❀								❀	❀
Carya tomentosa	❀	❀				❀	❀	❀			❀	❀	❀	❀	❀				❀				❀	❀		❀	❀		❀			❀		❀	❀	❀	❀	❀	❀		❀	❀	
Caryomyia antennata																																				●							
Caryomyia consobrina																																				●							
Caryomyia holotricha																																				●							
Caryomyia sanguinolenta																																				●							
Caryomyia sp.																																				●							
Caryomyia tubicola																																				●							
Casmerodius albus	●	●					●	●	●		●	●		●			●			●			●		●	●					●												
Cassia fasciculata																											❀							❀				❀					
Cassia marilandica		❀																																									
Cassia nicitans	❀	❀									❀					❀								❀					❀		❀					❀							
Cassia obtusifolia		❀													❀								❀		❀				❀		❀												
Castanea dentata	❀																				❀													❀		❀		❀			❀	❀	❀
Castanea pumila	❀							❀					❀																														
Castanea pumila ashei									❀											❀																							
Castanea pumila pumila																					❀																				❀		
Castanea sp.																																		❀	❀				❀				
Castor canadensis	●		●	●		●			●						●									●			●							●		●							
Catalpa sp.		❀									❀																																
Catalpa speciosa											❀			❀																													
Cathartes aura	●	●	●			●	●	●	●	●	●		●	●	●		●	●	●	●	●	●	●	●	●	●	●		●	●		●		●	●	●	●	●	●	●	●	●	●
Catharus fuscescens																																										●	

	Abbeville	Aiken	Allendale	Anderson	Bamberg	Barnwell	Beaufort	Berkeley	Charleston	Cherokee	Chester	Chesterfield	Clarendon	Colleton	Darlington	Dillon	Dorchester	Fairfield	Florence	Georgetown	Greenville	Greenwood	Hampton	Horry	Jasper	Kershaw	Lancaster	Laurens	Lee	Lexington	Marlboro	McCormick	Newberry	Oconee	Orangeburg	Pickens	Richland	Spartanburg	Sumter	Union	York	Georgia	N. Carolina
Catharus guttatus							●		●					●										●			●																
Catillaria glauconigrans																											*																*
Catillaria sp.																										*																	
Catops sp.				●			●																	●																			
Catoptrophorus semipalmatus							●		●											●																							
Caulophyllum thalictroides																											❀							❀								❀	❀
Ceanothus americanus	❀							❀			❀																❀							❀		❀		❀				❀	
Cecidomyia boehmeriae																								●																	●		
Cecidomyia ocellaris																													●													●	
Cecidomyia pudibunda																																●											
Cecidomyia verrucicola																																										●	
Cedrus deodara																																							❀				
Celastrina ladon										●		●		●				●						●			●					●		●							●		●
Celithemis amanda												●																															
Celithemis elisa																					●			●																			
Celithemis eponina																								●											●				●				
Celithemis fasciata												●														●							●						●			●	
Celtis laevigata		❀					❀	❀	❀		❀			❀				❀	❀						❀	❀	❀	❀			❀	❀			❀		❀			❀	❀		
Celtis occidentalis georgiana		❀		❀					❀		❀											❀					❀						❀		❀			❀			❀		
Celtis sp.																																						❀	❀				
Cenchrus sp.									❀										❀				❀																				
Cenchrus tribuloides							❀		❀											❀																							
Centaurea cyanus																					❀																						
Centella asiatica							❀	❀				❀	❀	❀									❀	❀		❀																	
Centrinaspis grisescens		●	●																						●																		
Centrinaspis penicellus																					●																						
Centrinaspis perscillus													●														●																
Centrinaspis sp.		●																					●									●											
Centrinites sp.																											●																
Centronopus calcaratus																																									●		
Centrosema virginianum	❀															❀										❀			❀														
Centruroides sp.																																		●									
Centruroides vittatus	●																																										
Cephalanthus occidentalis	❀	❀		❀		❀	❀	❀			❀			❀									❀	❀	❀	❀			❀		❀	❀						❀	❀				
Cerastium glomeratum			❀						❀						❀			❀																									
Cerastium nutans																																❀											
Ceratiola ericoides																														❀													
Ceratiomyxa fruticulosa																																		*				*					
Ceratophyllum demersum								❀	❀																																		
Cercis canadensis	❀	❀	❀				❀	❀	❀		❀			❀	❀		❀	❀		❀	❀						❀	❀				❀	❀	❀	❀	❀	❀	❀	❀				❀
Cercyonis pegala																							●												●			●					
Cercyonis pegala pegala								●																																			

	Abbeville	Aiken	Allendale	Anderson	Bamberg	Barnwell	Beaufort	Berkeley	Charleston	Cherokee	Chester	Chesterfield	Clarendon	Colleton	Darlington	Dillon	Dorchester	Fairfield	Florence	Georgetown	Greenville	Greenwood	Hampton	Horry	Jasper	Kershaw	Lancaster	Laurens	Lee	Lexington	Marlboro	McCormick	Newberry	Oconee	Orangeburg	Pickens	Richland	Spartanburg	Sumter	Union	York	Georgia	N. Carolina
Cerotoma trifurcata																																				●					●		
Certhia americana						●																																●	●				
Ceryle alcyon		●		●	●	●	●	●	●		●	●		●	●			●	●	●					●				●					●			●	●					
Cetraria arenaria																																											*
Cetraria aurescens																																											*
Cetraria ciliaris																													*							*							*
Cetraria oakesiana																																		*				*					*
Cetraria sp.												*														*										*							*
Cetrelia chicitae																																											*
Cetrelia sp.																																											*
Ceuthophilus crassifemoris												●																															
Ceuthophilus sp.		●																			●																						
Ceutorhynchus sp.																								●																			
Chaerophyllum tainturieri		❀						❀																			❀										❀						
Chaetocnema pulicaria																											●																
Chaetura pelagica	●							●				●			●									●										●		●	●						
Chalcophora virginiensis												●																									●	●					
Chalepus dorsalis																																		●									
Chalepus scapularis								●																								●							●				
Chamaecyparis thyoides												❀			❀											❀																	
Chamaelirium luteum																					❀													❀		❀							❀
Chaptalia tomentosa																								❀					❀														
Charadrius melodus																				●																							
Charadrius semipalmatus							●		●											●																							
Charadrius vociferus		●					●	●	●					●				●		●	●				●	●											●						●
Chauliognathus marginatus		●																										●	●					●				●				●	
Chauliognathus pennsylvanicus												●															●				●			●									●
Cheilanthes lanosa		❀																									❀			❀												❀	
Cheilanthes sp.																					❀						❀									❀							
Cheilanthes tomentosa																											❀															❀	
Chelone obliqua																																											❀
Chelydra serpentina			●																																		●						
Chen caerulescens																									●																		
Chenopodium album																								❀							❀												
Chenopodium ambrosioides		❀					❀		❀		❀				❀					❀			❀																				
Chesapecten jeffersonius jeffersonius																																											❖
Chesapecten jeffersonius septenarius																																											❖
Chesapecten madisonius																																											❖
Chicoreus sp.									●																																		
Chilocorus stigma												●																															
Chilomycterus sp.									●																																		
Chimaphila maculata	❀	❀				❀	❀		❀	❀	❀	❀			❀	❀	❀		❀	❀	❀		❀	❀		❀	❀	❀	❀	❀		❀	❀	❀	❀	❀	❀	❀	❀	❀	❀	❀	❀

	Abbeville	Aiken	Allendale	Anderson	Bamberg	Barnwell	Beaufort	Berkeley	Charleston	Cherokee	Chester	Chesterfield	Clarendon	Colleton	Darlington	Dillon	Dorchester	Fairfield	Florence	Georgetown	Greenville	Greenwood	Hampton	Horry	Jasper	Kershaw	Lancaster	Laurens	Lee	Lexington	Marlboro	McCormick	Newberry	Oconee	Orangeburg	Pickens	Richland	Spartanburg	Sumter	Union	York	Georgia	N. Carolina
Chionanthus virginicus		❀				❀							❀						❀		❀			❀			❀					❀		❀	❀	❀	❀	❀			❀	❀	
Chione cancellata																								❖																			
Chione cribraria																								❖																			
Chione grus																																											❖
Chione latilirata																								❖																			
Chlamisus gibbosus											●																●												●				
Chlamydotherium sp.																																									●		
Chlamys cocoana																	❖																										
Chlamys sp.																	❖																										
Chloris petraea							❀		❀																																		
Chlorociboria aeruginascens																					*																						*
Chondrophora nudata																									❀				❀														
Chortophaga viridifasciata		●									●																											●					
Chrysanthemum leucanthemum																																		❀									❀
Chrysemys picta				●														●																									
Chrysemys scripta	●	●						●				●		●			●			●				●	●	●	●		●				●				●						
Chrysemys sp.															●										●																		
Chrysobalanus oblongifolius																									❀																		
Chrysogonum virginianum								❀																			❀					❀		❀			❀						❀
Chrysogonum virginianum australe										❀																						❀		❀		❀						❀	
Chrysomela interrupta																											●							●									
Chrysomela scripta								●																																			
Chrysopa ornata																																		●									
Chrysopa sp.												●																															
Chrysosplenium americanum																																											❀
Cicindela dorsalis media									●																																		
Cicindela punctulata								●																●																			
Cicindela repanda																																											●
Cicindela rufiventris	●										●																									●							
Cicindela scutellaris unicolor																										●																	
Cicindela sexguttata								●			●		●		●									●				●						●		●	●	●					●
Cicindela sp.									●									●				●																					
Cicindela tortuosa																																		●									
Cicindela tranquebarica												●														●	●											●					
Cicindela unipunctata	●							●									●																			●					●		
Cicuta maculata								❀																										❀									
Cimicifuga americana																																											❀
Cimicifuga racemosa	❀										❀										❀						❀	❀						❀				❀				❀	❀
Cincticornia pustulata	●																																										
Circus cyaneus				●					●		●			●											●	●																	
Cistothorus palustris							●		●																																		
Citheronia regalis						●																																					

	Abbeville	Aiken	Allendale	Anderson	Bamberg	Barnwell	Beaufort	Berkeley	Charleston	Cherokee	Chester	Chesterfield	Clarendon	Colleton	Darlington	Dillon	Dorchester	Fairfield	Florence	Georgetown	Greenville	Greenwood	Hampton	Horry	Jasper	Kershaw	Lancaster	Laurens	Lee	Lexington	Marlboro	McCormick	Newberry	Oconee	Orangeburg	Pickens	Richland	Spartanburg	Sumter	Union	York	Georgia	N. Carolina
Cladina arbuscula																											*									*			*				*
Cladina evansii		*						*				*			*				*	*				*	*		*													*			
Cladina rangiferina																											*		*													*	
Cladina rangiferina f. crispata																											*																
Cladina rangiferina humilis																					*																						
Cladina rangiferina patula																					*																						
Cladina sp.																					*																						
Cladina stellaris																											*																
Cladina subtenuis																					*			*										*					*				
Cladina subtenuis cinerascens		*																																				*					
Cladina subtenuis subtenuis		*																						*														*			*		
Cladium jamaicense														❀																													
Cladonia abbreviatula																											*																
Cladonia apodocarpa												*													*		*				*												
Cladonia bacillaris														*	*								*				*		*			*		*		*		*				*	
Cladonia balfourii								*																			*					*											
Cladonia beaumontii												*															*		*														
Cladonia caespiticia																								*			*							*									*
Cladonia calycantha							*					*											*			*	*		*														
Cladonia capitata							*	*				*			*								*	*	*	*	*				*	*		*		*		*		*	*		
Cladonia cariosa												*									*						*							*									
Cladonia caroliniana												*															*		*							*		*				*	
Cladonia caroliniana caroliniana																					*						*																
Cladonia cervicornis																																						*					
Cladonia chlorophaea		*					*		*		*	*			*			*	*		*		*			*	*		*			*		*		*		*	*	*	*	*	*
Cladonia chlorophaea costata											*										*																						
Cladonia chlorophaea f. intermedia		*																			*													*				*			*		
Cladonia chlorophaea f. pterygota		*									*										*																						
Cladonia chlorophaea lepidophora																					*																						
Cladonia coniocraea							*		*		*	*			*			*	*		*		*	*	*		*		*			*		*		*		*			*	*	*
Cladonia coniocraea expansa																																		*									
Cladonia coniocraea f. ceratodes																					*																*						
Cladonia coniocraea f. phyllostrota		*																																*			*						
Cladonia coniocraea f. pycnotheliza																																					*						
Cladonia coniocraea f. subpellucida		*																			*													*			*	*			*		
Cladonia conista																											*																*
Cladonia cristatella		*								*	*	*						*	*		*			*		*	*				*	*		*				*	*	*		*	*
Cladonia cristatella aurantiaca																										*																	
Cladonia cristatella cristatella												*									*		*						*			*				*		*			*		
Cladonia cristatella degenerata																																				*							
Cladonia cristatella epiphylla																																		*									
Cladonia cristatella pleurocarpa												*									*								*			*				*							

	Abbeville	Aiken	Allendale	Anderson	Bamberg	Barnwell	Beaufort	Berkeley	Charleston	Cherokee	Chester	Chesterfield	Clarendon	Colleton	Darlington	Dillon	Dorchester	Fairfield	Florence	Georgetown	Greenville	Greenwood	Hampton	Horry	Jasper	Kershaw	Lancaster	Laurens	Lee	Lexington	Marlboro	McCormick	Newberry	Oconee	Orangeburg	Pickens	Richland	Spartanburg	Sumter	Union	York	Georgia	N. Carolina
Cladonia cristatella ramosa																																				*							
Cladonia cristatella scyphulifera															*														*							*		*					
Cladonia cristatella squamossisima																																*											
Cladonia cristatella vestita		*																					*			*			*			*		*	*	*		*			*		
Cladonia cylindrica							*																		*		*							*		*					*	*	
Cladonia didyma							*	*	*		*	*		*				*	*		*		*	*		*	*		*			*		*		*			*		*	*	*
Cladonia didyma didyma																					*																						
Cladonia didyma epiphylla																					*																						
Cladonia didyma squamulosa																					*													*									
Cladonia digitata																											*																
Cladonia floerkeana		*													*								*	*			*				*												*
Cladonia furcata							*																*	*			*															*	*
Cladonia furcata f. rigidula																																				*							
Cladonia furcata f. subclausa																																				*							
Cladonia furcata furcata																					*																						
Cladonia furcata subulata																																				*							
Cladonia gracilis												*															*							*									
Cladonia humilis																																								*			
Cladonia incrassata												*															*		*					*									
Cladonia incrassata f. incrassata		*																																									
Cladonia incrassata f. squamulosa		*																																									
Cladonia leporina		*						*				*			*				*					*		*	*		*	*				*					*				
Cladonia leporina leporina																					*			*																			
Cladonia leporina prolifera												*																															
Cladonia leporina ramosa												*									*															*							
Cladonia leporina squamulosa												*									*																						
Cladonia macilenta															*				*								*		*			*						*		*			
Cladonia macilenta theiophila											*																																
Cladonia mateocyatha		*																					*	*			*												*				
Cladonia ochrochlora							*		*											*			*				*																
Cladonia parasitica														*							*						*		*					*				*			*		
Cladonia peziziformis		*							*		*							*	*		*			*			*							*	*			*	*				
Cladonia peziziformis dissectula																																		*									
Cladonia peziziformis stenophyllodes																					*																	*					
Cladonia piedmontensis																											*												*			*	
Cladonia pityrea							*													*			*			*	*							*									
Cladonia polycarpia											*															*	*												*				
Cladonia polycarpoides		*				*					*	*						*	*		*		*	*		*	*							*		*		*	*	*		*	*
Cladonia polycarpoides epiphylla											*										*															*							
Cladonia polycarpoides polycarpoides																					*								*						*								
Cladonia polycarpoides ramosa																					*																						
Cladonia polycarpoides squamulosa																					*														*								

	Abbeville	Aiken	Allendale	Anderson	Bamberg	Barnwell	Beaufort	Berkeley	Charleston	Cherokee	Chester	Chesterfield	Clarendon	Colleton	Darlington	Dillon	Dorchester	Fairfield	Florence	Georgetown	Greenville	Greenwood	Hampton	Horry	Jasper	Kershaw	Lancaster	Laurens	Lee	Lexington	Marlboro	McCormick	Newberry	Oconee	Orangeburg	Pickens	Richland	Spartanburg	Sumter	Union	York	Georgia	N. Carolina
Cladonia polycarpoides sterilia																					*																						
Cladonia pyxidata																											*																
Cladonia ramulosa																								*														*			*		
Cladonia rappii											*								*					*										*									
Cladonia ravenelii												*														*	*																
Cladonia rei		*																									*																
Cladonia robbinsii																											*											*				*	
Cladonia santensis																											*													*			
Cladonia sobolescens		*									*												*				*					*		*					*		*	*	
Cladonia sp.													*				*			*			*						*		*	*					*	*	*	*			
Cladonia squamosa																							*				*									*							*
Cladonia squamosa clavariella																					*																						
Cladonia squamosa levicorticata																					*																						
Cladonia squamosa murina																					*																						
Cladonia squamosa phyllocoma																					*																						
Cladonia squamosa squamosa																					*																						
Cladonia strepsilis												*															*																
Cladonia subcariosa																											*																
Cladonia subradiata									*					*											*																		
Cladonia subsetacea												*															*																
Cladonia uncialis												*															*											*	*				*
Cladonia verticillata								*																			*							*									
Cladrastis lutea		❀																																									❀
Clathroporina sp.									*																																		
Clathrus columnatus																								*																			
Clavaria flava																																		*									
Clavaria vermicularis																											*											*					*
Clavariadelphus truncatus																																											*
Clavicorona pyxidata																											*							*				*					
Clavulina amethystina																											*																*
Clavulina cristata																																		*				*					*
Clavulinopsis fusiformis																								*										*									
Clavulinopsis pulchra																																						*					
Claytonia caroliniana																																											❀
Claytonia virginica																																❀											
Cleistes divaricata								❀																❀																			
Clematis crispa																		❀						❀													❀						
Clematis sp.								❀																❀			❀																
Clematis virginiana																					❀													❀									❀
Clemmys guttata																													●														
Clethra acuminata																					❀													❀								❀	❀
Clethra alnifolia					❀	❀	❀	❀	❀			❀	❀		❀	❀			❀					❀		❀	❀		❀	❀	❀						❀		❀				

	Abbeville	Aiken	Allendale	Anderson	Bamberg	Barnwell	Beaufort	Berkeley	Charleston	Cherokee	Chester	Chesterfield	Clarendon	Colleton	Darlington
Clethra alnifolia alnifolia															
Clethra alnifolia tomentosa		❀													
Clibanarius vittatus									●						
Clinostomus funduloides															
Clintonia borealis															
Clintonia umbellulata															
Clitocybe clavipes															
Clitocybe ectypoides															
Clitocybe sp.															
Clitocybula familia									*						
Clitoria mariana		❀							❀						
Clivina rufa															●
Cloeotus aphodioides											●				
Cnemidophorus sexlineatus		●					●		●						
Cnidoscolus stimulosus		❀			❀							❀			❀
Coccinella septempunctata													●		●
Coccinella sp.															
Coccocarpia cronia															
Coccocarpia erythroxyli															
Coccocarpia imbricata sp. nov.															
Cocculus carolinus		❀									❀				
Coccyzus americanus		●			●						●			●	
Cocos nucifera									❀						
Codium sp.									*						
Coenogonium interplexum							*		*					*	
Coenogonium moniliforme							*							*	
Coenogonium sp.						*									
Colaptes auratus						●	●		●		●	●		●	
Colaspis favosa															
Colaspis sp.											●				
Coleomegilla fuscilabris												●			
Coleomegilla maculata											●			●	
Colias eurytheme															
Colias philodice							●								
Colinus virginianus		●			●	●	●						●		
Collema furfuraceum															
Collema nigrescens								*							
Collema subflaccidum											*				
Collinsonia canadensis															
Collinsonia tuberosa															
Collinsonia verticillata															
Collops quadrimaculatus															

	Dillon	Dorchester	Fairfield	Florence	Georgetown	Greenville	Greenwood	Hampton	Horry	Jasper	Kershaw	Lancaster	Laurens	Lee	Lexington
Clethra alnifolia alnifolia											❀				
Clethra alnifolia tomentosa								❀							
Clibanarius vittatus															
Clinostomus funduloides												●			
Clintonia borealis															
Clintonia umbellulata															
Clitocybe clavipes															
Clitocybe ectypoides															
Clitocybe sp.															
Clitocybula familia															
Clitoria mariana														❀	
Clivina rufa															
Cloeotus aphodioides															
Cnemidophorus sexlineatus	●	●			●	●	●		●		●	●			
Cnidoscolus stimulosus	❀	❀						❀	❀			❀			
Coccinella septempunctata												●		●	
Coccinella sp.					●										
Coccocarpia cronia															
Coccocarpia erythroxyli												*		*	
Coccocarpia imbricata sp. nov.															
Cocculus carolinus							❀					❀			
Coccyzus americanus									●				●		
Cocos nucifera															
Codium sp.															
Coenogonium interplexum												*			
Coenogonium moniliforme												*			
Coenogonium sp.												*			
Colaptes auratus				●	●				●		●			●	
Colaspis favosa									●						
Colaspis sp.															
Coleomegilla fuscilabris															
Coleomegilla maculata														●	
Colias eurytheme				●											
Colias philodice				●	●										
Colinus virginianus							●	●	●					●	
Collema furfuraceum			*	*											
Collema nigrescens															
Collema subflaccidum			*			*						*			
Collinsonia canadensis						❀							❀		
Collinsonia tuberosa															
Collinsonia verticillata															
Collops quadrimaculatus															

	Marlboro	McCormick	Newberry	Oconee	Orangeburg	Pickens	Richland	Spartanburg	Sumter	Union	York	Georgia	N. Carolina
Clethra alnifolia alnifolia													
Clethra alnifolia tomentosa													
Clibanarius vittatus													
Clinostomus funduloides													
Clintonia borealis				❀									
Clintonia umbellulata				❀								❀	❀
Clitocybe clavipes								*					
Clitocybe ectypoides													*
Clitocybe sp.				*									
Clitocybula familia				*									
Clitoria mariana	❀		❀	❀									
Clivina rufa													
Cloeotus aphodioides													
Cnemidophorus sexlineatus	●		●						●				
Cnidoscolus stimulosus					❀				❀				
Coccinella septempunctata				●				●	●			●	
Coccinella sp.													
Coccocarpia cronia				*									
Coccocarpia erythroxyli				*									*
Coccocarpia imbricata sp. nov.						*							
Cocculus carolinus					❀			❀			❀		
Coccyzus americanus					●	●	●	●	●		●		
Cocos nucifera													
Codium sp.													
Coenogonium interplexum				*			*						
Coenogonium moniliforme													
Coenogonium sp.													
Colaptes auratus		●						●	●		●	●	●
Colaspis favosa													
Colaspis sp.													
Coleomegilla fuscilabris							●					●	
Coleomegilla maculata					●								
Colias eurytheme													
Colias philodice				●									
Colinus virginianus					●	●							
Collema furfuraceum		*											
Collema nigrescens							*						
Collema subflaccidum				*		*	*	*			*	*	*
Collinsonia canadensis				❀		❀						❀	❀
Collinsonia tuberosa						❀							
Collinsonia verticillata				❀									
Collops quadrimaculatus				●									

	Abbeville	Aiken	Allendale	Anderson	Bamberg	Barnwell	Beaufort	Berkeley	Charleston	Cherokee	Chester	Chesterfield	Clarendon	Colleton	Darlington	Dillon	Dorchester	Fairfield	Florence	Georgetown	Greenville	Greenwood	Hampton	Horry	Jasper	Kershaw	Lancaster	Laurens	Lee	Lexington	Marlboro	McCormick	Newberry	Oconee	Orangeburg	Pickens	Richland	Spartanburg	Sumter	Union	York	Georgia	N. Carolina
Collops sp.																																●											
Collops vittatus																					●																						
Collybia alkalivirens									✻																																		
Collybia dryophila																												✻				✻											
Collybia iocephala																											✻							✻									✻
Collybia peronata																								✻															✻				
Collybia sp.									✻																																		
Coluber constrictor	●						●					●					●						●																●				
Columba livia		●						●	●					●																													
Columbina passerina																				●																							
Colydium lineola															●																												
Comandra umbellata																								❀																			
Commelina communis		❀						❀													❀								❀		❀			❀	❀								
Commelina erecta		❀														❀											❀																
Commelina sp.					❀						❀										❀																						
Commelina virginica		❀									❀									❀											❀												
Conocephalus fasciatus																																									●		
Conocephalus strictus																																						●					
Conopea galeata							●		●																																		
Conopholis americana		❀									❀						❀				❀		❀									❀		❀	❀	❀						❀	❀
Conotelus sp.																																		●									
Conotrachelus posticatus																																						●					
Conradostrea lawrencei																																											❖
Contopus virens		●																						●								●	●		●		●		●			●	●
Conus adversarius																								❖																			❖
Conus waccamawensis																								❖																			
Copaeodes minimus																																			●								
Copelatus chevrolati chevrolati																													●														
Copius inflatus		●												●																													
Coprinus atramentarius																												✻						✻									
Coprinus comatus							✻																																				
Copris carolina																								●																			
Coptotomus interrogatus		●				●					●					●	●							●					●		●								●				
Coptotomus lenticus															●																												
Coragyps atratus			●				●		●		●			●				●		●			●	●		●	●		●								●	●			●		
Corallorhiza odontorhiza																																				❀							❀
Corallorhiza wisteriana																																							❀				
Corbicula densata																								❖																			
Corbicula fluminea		●									●																			●													
Corbicula manilensis		●								●																																	
Corbicula sp.								●			●																								●								
Cordulegaster diastatops																																				●							

	Abbeville	Aiken	Allendale	Anderson	Bamberg	Barnwell	Beaufort	Berkeley	Charleston	Cherokee	Chester	Chesterfield	Clarendon	Colleton	Darlington	Dillon	Dorchester	Fairfield	Florence	Georgetown	Greenville	Greenwood	Hampton	Horry	Jasper	Kershaw	Lancaster	Laurens	Lee	Lexington	Marlboro	McCormick	Newberry	Oconee	Orangeburg	Pickens	Richland	Spartanburg	Sumter	Union	York	Georgia	N. Carolina
Cordulegaster sp.																																		●									
Cordyceps capitata																								✻																			
Cordyceps militaris														✻																													
Coreopsis auriculata																																					❀						
Coreopsis gladiata												❀																															
Coreopsis major	❀							❀			❀										❀									❀				❀		❀		❀					❀
Coreopsis sp.																													❀							❀							
Corimelaena sp.																											●																
Cormelaena sp.																												●															
Cornus alternifolia																					❀													❀		❀						❀	❀
Cornus amomum				❀				❀			❀										❀													❀		❀							❀
Cornus asperifolia								❀																																			
Cornus florida	❀	❀				❀	❀		❀	❀	❀	❀	❀	❀	❀	❀	❀	❀	❀	❀	❀	❀	❀	❀		❀	❀	❀	❀	❀	❀	❀	❀	❀	❀	❀	❀	❀	❀	❀	❀	❀	❀
Cornus stricta														❀					❀				❀												❀		❀						
Corticotomus sp.															●																												
Cortinarius alboviolaceus																				✻									✻														✻
Cortinarius sp.																																							✻				
Cortinarius violaceus																																											✻
Corvus brachyrhynchos		●	●	●	●		●	●	●	●	●	●	●	●	●	●	●	●	●	●	●				●	●	●		●			●	●	●		●	●	●	●	●	●	●	●
Corvus corax																					●													●									●
Corvus ossifragus								●	●		●			●		●	●			●				●		●			●		●				●				●				
Corvus sp.	●																																										
Corydalis cornutus								●						●																													
Corydalis flavula																																❀											
Corydalis micrantha australis							❀																																				
Corylus americana																											❀															❀	
Coscinoptera dominicana																																●											
Cotinus nitida											●			●								●																					
Crangonyx serratus																																							●				
Crassostrea virginiana																								❖																			❖
Crassostrea virginica							●		●											●				●																			
Crataegus aestivalis														❀																													
Crataegus crus-galli		❀																																									
Crataegus flabellata	❀																	❀																									
Crataegus marshallii			❀																❀																		❀						
Crataegus punctata		❀																																									
Crataegus sp.		❀			❀					❀				❀			❀		❀		❀					❀			❀	❀		❀	❀	❀	❀	❀	❀	❀	❀	❀	❀		
Crataegus spathulata		❀																														❀											
Crataegus uniflora																❀															❀								❀				
Crataegus viridis			❀																❀																		❀						
Craterellus cornucopiodes																																				✻							
Craterellus fallax																																		✻									✻

	Abbeville	Aiken	Allendale	Anderson	Bamberg	Barnwell	Beaufort	Berkeley	Charleston	Cherokee	Chester	Chesterfield	Clarendon	Colleton	Darlington	Dillon	Dorchester	Fairfield	Florence	Georgetown	Greenville	Greenwood	Hampton	Horry	Jasper	Kershaw	Lancaster	Laurens	Lee	Lexington	Marlboro	McCormick	Newberry	Oconee	Orangeburg	Pickens	Richland	Spartanburg	Sumter	Union	York	Georgia	N. Carolina
Craterellus odoratus																											*							*									
Cremastocheilus castaneae																																											●
Crematogaster ashmeadi																																					●						
Crepidodera nitens											●																																
Crepidotus applanatus																											*																
Crepidula fornicata									●											●				❖																			
Crepidula plana									●																																		
Crepis japonica															❀																												
Crinipellis zonata					*																																						*
Cronartium fusiforme																							*																				
Cronartium quercuum			*																																	*							
Crotalaria angulata		❀			❀			❀								❀																											
Crotalaria purshii								❀																❀																			
Crotalaria sp.									❀																																		
Crotalaria spectabilis							❀		❀								❀																										
Crotalus horridus																																											●
Croton glandulosus septentrionalis					❀																								❀														
Croton punctatus							❀		❀											❀																							
Crotonopsis elliptica																											❀									❀							
Crotonopsis linearis																									❀																		
Crucibulum Lawrencei																																											❖
Crucibulum laeve																																											*
Cryptocephalus quadruplex																											●																
Cryptomeria japonica																																							❀				
Cryptorhopalum picicorne		●																														●											
Cryptorhopalum sp.																								●			●					●											
Cryptotaenia canadensis																																					❀						
Cryptotis parva				●																																							
Ctenicera sp.																					●																					●	
Ctenium aromaticum												❀																	❀														
Ctenucha virginica							●																																				
Culisetia inornata									●																																		
Cuphea carthagensis														❀									❀																				
Curculio sp.															●																												
Cuscuta compacta																❀								❀							❀												
Cuscuta gronovii																											❀																
Cuscuta sp.						❀		❀					❀	❀	❀					❀				❀																			❀
Cyanocitta cristata		●		●			●	●	●			●		●	●				●	●	●		●	●	●		●		●		●		●		●	●	●	●			●	●	●
Cyathus sp.																																						*					
Cyathus striatus								*							*																			*									*
Cyclocardia granulata																								❖																			
Cycloneda sanguinea		●						●	●			●																					●	●					●				

	Abbeville	Aiken	Allendale	Anderson	Bamberg	Barnwell	Beaufort	Berkeley	Charleston	Cherokee	Chester	Chesterfield	Clarendon	Colleton	Darlington	Dillon	Dorchester	Fairfield	Florence	Georgetown	Greenville	Greenwood	Hampton	Horry	Jasper	Kershaw	Lancaster	Laurens	Lee	Lexington	Marlboro	McCormick	Newberry	Oconee	Orangeburg	Pickens	Richland	Spartanburg	Sumter	Union	York	Georgia	N. Carolina
Cylindracanthus rectus																	❖																										
Cyllopsis gemma									●																													●				●	
Cymindis americana																	●																										
Cymophyllus fraseri																																											❀
Cynanchum palustre							❀																																				
Cynips sp.							●																																				
Cynoctotum mitreola		❀				❀																																					
Cynodon dactylon									❀																																		
Cynoglossum virginianum																											❀							❀		❀		❀					
Cynomyopsis cadaverina																																		●									
Cyphon obscurus																																		●									
Cyphon sp.								●																																			
Cyphon variabilis																●																											
Cypripedium acaule																					❀													❀								❀	
Cypripedium calceolus																																		❀								❀	
Cyrilla racemiflora		❀				❀		❀				❀		❀	❀	❀			❀				❀	❀		❀			❀	❀	❀								❀				
Cyrtinus pygmaeus																																					●						
Cyrtopleura costata									●																																		
Cystoderma amianthinum rugusoreticulatum																					*																						
Cystopteris protrusa		❀																														❀											❀
Dacrymyces deliquescens																																						*					
Dacrymyces palmatus																																						*					
Dactylis glomerata												❀																						❀								❀	
Dactyloctenium aegyptium							❀																																				
Daldinia concentrica							*		*																				*					*				*					
Daldinia sp.							*																																				
Danaus gilippus							●		●																																		
Danaus plexippus		●					●		●			●				●				●	●					●								●			●	●				●	
Danthonia sp.																	❀																										
Danthonia spicata		❀																																									
Dasyatis sp.									●											●																							
Dasymutilla asopus bexar																																		●									
Dasymutilla occidentalis occidentalis	●	●									●					●										●								●	●				●		●		
Dasymutilla sp.									●					●																									●		●		
Datana major																								●																			
Datana sp.																																						●				●	
Datura stramonium																															❀												
Daucus carota	❀										❀										❀	❀						❀												❀			
Daucus pusillus	❀																																										
Decodon verticillatus		❀				❀										❀										❀					❀								❀				
Decumaria barbara		❀				❀		❀			❀			❀	❀						❀		❀										❀	❀	❀	❀	❀	❀	❀				
Deloyola guttata											●																																

	Abbeville	Aiken	Allendale	Anderson	Bamberg	Barnwell	Beaufort	Berkeley	Charleston	Cherokee	Chester	Chesterfield	Clarendon	Colleton	Darlington	Dillon	Dorchester	Fairfield	Florence	Georgetown	Greenville	Greenwood	Hampton	Horry	Jasper	Kershaw	Lancaster	Laurens	Lee	Lexington	Marlboro	McCormick	Newberry	Oconee	Orangeburg	Pickens	Richland	Spartanburg	Sumter	Union	York	Georgia	N. Carolina
Delphastus pusillus															●																												
Dendroica caerulescens									●												●						●							●				●				●	●
Dendroica castanea																					●																						
Dendroica coronata				●			●		●					●	●			●		●				●		●	●					●		●				●	●		●	●	
Dendroica discolor											●													●			●							●									
Dendroica dominica			●				●		●					●				●			●			●			●		●					●	●	●	●		●				
Dendroica fusca																																											●
Dendroica magnolia																					●																						
Dendroica palmarum							●																																				
Dendroica pensylvanica																					●																						●
Dendroica petechia																																											●
Dendroica pinus		●	●			●	●		●			●	●	●	●	●	●	●	●		●			●			●					●	●	●	●	●	●	●		●			
Dendroica striata																					●																						
Dendroica tigrina																																											●
Dendroica virens																					●													●									●
Dennstaedtia punctilobula																					❀											❀		❀								❀	❀
Dentinum repandum																														*													*
Dermatocarpon miniatum																																											*
Dermatocarpon sp.																																				*							
Dermestes talpinus																																				●							
Descurainia brachycarpa																																											
Desmia funeralis																																		●									
Desmodium nudiflorum	❀	❀																			❀												❀	❀		❀		❀	❀			❀	
Desmodium rotundifolium	❀																																										
Desmodium sp.					❀				❀		❀			❀	❀	❀					❀										❀		❀		❀			❀	❀				
Desmodium strictum																❀																											
Desmognathus auriculatus															●																								●				
Desmognathus fuscus																					●																						●
Desmognathus monticola																																				●							
Desmognathus ochrophaeus																																		●		●							●
Desmognathus sp.																																		●									
Diabrotica duodecimpunctata		●										●																●						●									
Diabrotica sp.																																							●				
Diabrotica undecimpunctata howardi											●																							●									
Diachea leucopodia																																						*					
Diadophis punctatus	●																																			●	●						●
Dianthus armeria																																				❀							
Diaperis maculata								●							●																												
Diapheromera carolina																																				●							
Diapheromera femorata	●	●																																●									
Diapheromera sp.																																				●							
Diastrophus nebulosus										●	●													●																●			●

	Abbeville	Aiken	Allendale	Anderson	Bamberg	Barnwell	Beaufort	Berkeley	Charleston	Cherokee	Chester	Chesterfield	Clarendon	Colleton	Darlington	Dillon	Dorchester	Fairfield	Florence	Georgetown	Greenville	Greenwood	Hampton	Horry	Jasper	Kershaw	Lancaster	Laurens	Lee	Lexington	Marlboro	McCormick	Newberry	Oconee	Orangeburg	Pickens	Richland	Spartanburg	Sumter	Union	York	Georgia	N. Carolina
Dibolia sp.																	●																										
Dibolocelus ovalis																	●																										
Dicaelus ambiguus																												●										●					
Dicaelus dilitatus	●										●				●																							●				●	
Dicaelus elongatus							●																																				
Dicaelus ovalis																								●																			
Dicaelus purpuratus																																				●							
Dicaelus sp.																																●											
Dicentra cucullaria																																❀											
Dicerca divaricata																																		●									
Dicerca tuberculata																																		●									
Dichelonyx elongata																																										●	
Dichelonyx fuscula																					●																						
Dichelonyx sp.																																		●									
Dichelonyx subvittata																					●																						
Dichondra carolinensis		❀	❀		❀				❀						❀					❀			❀			❀													❀				
Dichromena colorata									❀											❀																							
Dichromena latifolia								❀																																			
Dichromorpha viridis	●																																					●					
Dicranum scoparium																																						*					
Dictydium cancellatum																																						*					
Dictyopterus aurora						●																																					
Didelphis virginiana		●					●							●	●																										●		
Diervilla lonicera																																											❀
Diervilla sessilifolia																																		❀									❀
Dimerella lutea																											*																
Dineutus assimilis						●																																					
Dineutus carolinus								●				●																	●														
Dineutus discolor														●	●	●															●												
Dineutus emarginatus	●					●		●				●												●																			
Dineutus horni												●																															
Dineutus robertsi											●																												●				
Dineutus serrulatus serrulatus		●												●																							●						
Dineutus sp.		●						●																		●																	
Dinocardium robustum									●																																		
Diodia teres	❀	❀										❀				❀										❀			❀		❀							❀					
Diodia virginiana																								❀		❀			❀		❀		❀	❀									
Diodoro cayenensis									●																																		
Dionaea muscipula																								❀																			
Dioscorea batatas								❀																			❀																
Dioscorea sp.											❀																							❀									
Dioscorea villosa		❀	❀			❀															❀			❀			❀					❀		❀		❀	❀				❀	❀	❀

	Abbeville	Aiken	Allendale	Anderson	Bamberg	Barnwell	Beaufort	Berkeley	Charleston	Cherokee	Chester	Chesterfield	Clarendon	Colleton	Darlington	Dillon	Dorchester	Fairfield	Florence	Georgetown	Greenville	Greenwood	Hampton
Dioscorea villosa villosa																					❀		
Diospyros virginiana	❀	❀		❀			❀	❀	❀		❀	❀		❀	❀	❀				❀		❀	❀
Diphylleia cymosa																							
Diplocoelus rudis															●								
Diplolepis bicolor				●																			
Diplolepis rubicola											●												
Dirca palustris																							
Dirinaria aspera									✻					✻						✻			
Dirinaria confusa									✻					✻						✻	✻		
Dirinaria picta							✻		✻					✻						✻			
Dirinaria purpurascens							✻		✻					✻									
Dirinaria sp.							✻																
Discoderus paralleleus															●								
Discoderus sp.					●																		
Disholcaspis globulus											●												
Disholcaspis lapiei																					●		
Disholcaspis sp.																							
Disholcaspis spongiosa																							
Disonycha abbreviata																							
Disonycha caroliniana								●															
Disonycha collata																					●		
Disonycha pensylvanica									●														
Disonycha sp.		●						●															
Disporum lanuginosum																							
Disporum sp.																							
Dissosteira carolina															●								
Distichlis spicata									❀														
Divaricella quadrisulcata									●											●			
Dodecatheon meadia		❀																					
Dolomedes sp.																							
Dolomedes tenebrosus		●					●																
Dolomedes triton		●	●																				
Dolophilodes sp.																							
Donacia sp.																							
Donacia subtilis																							
Donax sp.									●														
Donax variabilis									●											●			
Dosinia discus									●														
Draba aprica																							
Dromogomphus armatus																							
Dromogomphus spinosus											●												
Drosera capillaris								❀				❀											

	Horry	Jasper	Kershaw	Lancaster	Laurens	Lee	Lexington	Marlboro	McCormick	Newberry	Oconee	Orangeburg	Pickens	Richland	Spartanburg	Sumter	Union	York	Georgia	N. Carolina
Dioscorea villosa villosa											❀									
Diospyros virginiana	❀		❀	❀	❀	❀	❀	❀	❀	❀	❀	❀	❀	❀	❀	❀		❀	❀	❀
Diphylleia cymosa																			❀	❀
Diplocoelus rudis																				
Diplolepis bicolor																				
Diplolepis rubicola																				
Dirca palustris									❀											
Dirinaria aspera																				
Dirinaria confusa																				
Dirinaria picta															✻					
Dirinaria purpurascens																				
Dirinaria sp.																				
Discoderus paralleleus																				
Discoderus sp.																				
Disholcaspis globulus																				●
Disholcaspis lapiei																				
Disholcaspis sp.						●														
Disholcaspis spongiosa						●														
Disonycha abbreviata				●					●							●				
Disonycha caroliniana	●																			
Disonycha collata																				
Disonycha pensylvanica												●								
Disonycha sp.														●						
Disporum lanuginosum											❀								❀	❀
Disporum sp.																				❀
Dissosteira carolina	●									●	●	●				●		●	●	
Distichlis spicata																				
Divaricella quadrisulcata																				
Dodecatheon meadia									❀											
Dolomedes sp.	●									●										
Dolomedes tenebrosus																				
Dolomedes triton	●			●				●			●					●				
Dolophilodes sp.											●									
Donacia sp.																				●
Donacia subtilis											●									
Donax sp.																				
Donax variabilis																				
Dosinia discus																				
Draba aprica				❀															❀	
Dromogomphus armatus												●								
Dromogomphus spinosus																				
Drosera capillaris	❀					❀														

	Abbeville	Aiken	Allendale	Anderson	Bamberg	Barnwell	Beaufort	Berkeley	Charleston	Cherokee	Chester	Chesterfield	Clarendon	Colleton	Darlington	Dillon	Dorchester	Fairfield	Florence	Georgetown	Greenville	Greenwood	Hampton	Horry	Jasper	Kershaw	Lancaster	Laurens	Lee	Lexington	Marlboro	McCormick	Newberry	Oconee	Orangeburg	Pickens	Richland	Spartanburg	Sumter	Union	York	Georgia	N. Carolina
Drosera intermedia								❀				❀												❀		❀																	
Drosera leucantha													❀													❀			❀														
Drosera sp.																													❀														
Drosophila melanogaster	●																																										
Dryocampa rubicunda																																				●							
Dryocopus pileatus	●	●	●			●	●	●	●		●	●		●		●		●	●	●	●			●		●	●	●	●			●		●	●	●	●		●				●
Dryophanta sp.							●																																				
Dryopteris campyloptera																																											❀
Dryopteris celsa																																					❀		❀				
Dryopteris intermedia																																		❀									
Dryopteris marginalis																																❀		❀								❀	❀
Dryopteris sp.																																											❀
Duchesnea indica		❀						❀			❀				❀																					❀	❀						
Dulichium arundinaceum		❀				❀						❀				❀							❀	❀					❀										❀				
Dumetella carolinensis		●					●	●	●			●								●									●										●				●
Dyschirius sp.														●																													
Dyschoriste oblongifolia					❀																																						
Dysides obscurus								●																																			
Echinochloa crusgalli																															❀							❀					
Echinodorus cordifolius		❀						❀																																			
Eclipta alba								❀																					❀		❀							❀					
Ecpantheria scribonia																●			●																								
Ecphora quadricostata																																											❖
Egeria densa		❀						❀							❀								❀												❀								
Egretta caerulea		●					●	●	●					●						●					●	●													●				
Egretta rufescens		●																																									
Egretta thula		●					●		●					●						●						●																	
Egretta tricolor							●		●					●						●																							
Eichhornia crassipes														❀																					❀								
Elaeagnus pungens		❀				❀																																❀					
Elaeagnus umbellata	❀	❀									❀																							❀				❀		❀			❀
Elanoides forficatus			●																																								
Elaphe guttata							●										●																										
Elaphe obsoleta								●											●																	●	●						
Elaphidion mucronatum																																							●				
Eleocharis equisetoides																										❀																	
Eleocharis obtusa		❀																																									
Eleocharis sp.																							❀																				
Eleocharis tuberculosa																													❀														
Elephantopus carolinianus											❀						❀				❀				❀	❀	❀							❀				❀					
Elephantopus elatus									❀																																		
Elephantopus nudatus																❀																											

	Abbeville	Aiken	Allendale	Anderson	Bamberg	Barnwell	Beaufort	Berkeley	Charleston	Cherokee	Chester	Chesterfield	Clarendon	Colleton	Darlington	Dillon	Dorchester	Fairfield	Florence	Georgetown	Greenville	Greenwood	Hampton	Horry	Jasper	Kershaw	Lancaster	Laurens	Lee	Lexington	Marlboro	McCormick	Newberry	Oconee	Orangeburg	Pickens	Richland	Spartanburg	Sumter	Union	York	Georgia	N. Carolina
Elephantopus sp.		❀				❀	❀				❀	❀		❀					❀		❀			❀					❀										❀				
Elephantopus tomentosus	❀	❀							❀		❀	❀			❀	❀				❀	❀			❀		❀			❀		❀			❀	❀	❀			❀				
Eleusine indica		❀						❀																												❀							
Elimia sp.		●																																									
Ellychnia sp.												●	●																					●			●						●
Elodea sp.								❀																																			
Elymus virginicus	❀	❀						❀	❀																													❀			❀		
Elytraria caroliniensis								❀																																			
Emerita talpoida									●																																		
Empidonax traillii																								●																			
Empidonax virescens	●	●			●			●						●	●		●											●				●			●		●	●	●		●		
Emydidae Family									❖																																		
Enallagma doubledayi												●																										●					
Enallagma sp.								●			●																																
Enallagma weewa																													●													●	
Enchodus ferox																			❖																								
Enochrus consortus																																					●						
Enochrus ochraceus		●					●																																			●	
Enochrus sp.																●																											
Enochrus sublongus																													●														
Enoclerus ichneumoneus																											●																
Enoclerus thoracicus																																										●	
Ensis directus									●											●																							❖
Entoloma murraii																																											✻
Entoloma salmoneum																																						✻					✻
Eopyrenula leucoplaca									✻																																		
Epargyreus clarus	●	●	●								●				●	●								●		●	●				●		●	●		●					●	●	●
Epeorus sp.																																		●									
Ephemerella sp.																																		●									
Epiaeschna heros									●																												●						
Epicauta pennsylvanica		●										●															●																
Epicauta sp.					●																																						
Epicauta trichrus																																		●									
Epicauta vittata																																		●									
Epierus pulicarius							●																																				
Epifagus virginiana	❀																											❀								❀	❀	❀			❀		
Epigaea repens	❀											❀			❀						❀													❀		❀	❀					❀	❀
Epilobium coloratum																																											❀
Epimecia hortaria																																		●									
Epitheca cynosura															●									●										●									
Epitheca semiaquea												●																															
Epitheca sp.																		●						●													●						

	Abbeville	Aiken	Allendale	Anderson	Bamberg	Barnwell	Beaufort	Berkeley	Charleston	Cherokee	Chester	Chesterfield	Clarendon	Colleton	Darlington	Dillon	Dorchester	Fairfield	Florence	Georgetown	Greenville	Greenwood	Hampton	Horry	Jasper	Kershaw	Lancaster	Laurens	Lee	Lexington	Marlboro	McCormick	Newberry	Oconee	Orangeburg	Pickens	Richland	Spartanburg	Sumter	Union	York	Georgia	N. Carolina
Epithrix fuscula																																					●					●	
Epitoneum sp.									●																																		
Epuraea sp.																●																											●
Equisetum arvense													❀																														
Equisetum hyemale affine											❀																																
Equus leidyi																								❖																			
Eragrostis hirsuta																																						❀					
Eragrostis refracta																❀																											
Eragrostis sp.												❀												❀						❀													
Erechtites hieracifolia		❀					❀																													❀							
Erianthus contortus				❀																																							
Erianthus giganteus																							❀																				
Erianthus sp.														❀												❀	❀											❀					
Erianthus strictus														❀																													
Erigeron annuus		❀																																									
Erigeron canadensis canadensis		❀																		❀									❀														
Erigeron canadensis ssp.																							❀			❀																	
Erigeron philadelphicus																																						❀					
Erigeron pulchellus																					❀													❀		❀						❀	❀
Erigeron sp.																															❀					❀							
Erigeron strigosus																												❀							❀								
Erigeron vernus																								❀					❀														
Eriocaulon decangulare								❀																																			
Eriocaulon sp.												❀																															
Eriophyes abnormis																																										●	
Eriophyes nyssae											●																		●														
Eriophyes padi											●																●																
Eriophyes ulmi											●																																
Eristalis tenax										●																																	
Eros humeralis																																●											
Eryngium integrifolium												❀																															
Eryngium prostratum																								❀																			
Eryngium yuccifolium								❀				❀											❀	❀					❀														
Erynnis horatius														●																													●
Erynnis juvenalis			●															●									●							●								●	●
Erynnis sp.																								●										●	●	●		●	●			●	
Erythemis simplicicollis	●	●						●	●			●		●						●	●	●		●		●	●		●		●		●		●			●	●				
Erythrina herbacea			❀				❀	❀	❀					❀						❀			❀																				
Erythrodiplax minuscula		●										●														●			●						●								
Erythronium americanum										❀																	❀					❀				❀						❀	❀
Esox sp.															●																												
Estigmene acrea						●																																					

	Abbeville	Aiken	Allendale	Anderson	Bamberg	Barnwell	Beaufort	Berkeley	Charleston	Cherokee	Chester	Chesterfield	Clarendon	Colleton	Darlington	Dillon	Dorchester	Fairfield	Florence	Georgetown	Greenville	Greenwood
Eucrassatella speciosa																						
Euderces picipes															●							
Eudocimus albus		●						●	●					●						●		
Euglandia rosea							●															
Eugnamptus collaris collaris								●							●							
Eumeces fasciatus											●											●
Eumeces laticeps		●																				
Eumeces sp.	●							●			●											
Eumenes fraternus							●							●								
Euonymus americanus	❀	❀	❀							❀	❀				❀				❀		❀	❀
Euonymus atropurpureus																						
Euonymus sp.															❀							
Euparius marmoreus		●																				
Eupatorium album																						
Eupatorium capillifolium		❀					❀		❀		❀	❀		❀	❀					❀		
Eupatorium coelestinum		❀						❀	❀													
Eupatorium compositifolium		❀										❀		❀						❀		
Eupatorium dubium																						
Eupatorium fistulosum												❀				❀					❀	
Eupatorium leucolepis																						
Eupatorium pilosum												❀										
Eupatorium purpureum																						
Eupatorium recurvans																						
Eupatorium rotundifolium												❀		❀								
Eupatorium rugosum												❀									❀	
Eupatorium sp.	❀										❀										❀	
Euphagus carolinus														●	●					●		
Euphagus cyanocephalus									●													
Euphorbia ammannioides							❀		❀											❀		
Euphorbia corollata	❀				❀			❀							❀						❀	
Euphorbia curtisii												❀										
Euphorbia heterophylla		❀																				
Euphorbia ipecacuanhae												❀										
Euphorbia maculata																❀						
Euphorbia polygonifolia							❀		❀													
Euphorbia sp.												❀										
Euphorbia supina	❀										❀											
Euphoria sepulchralis																						
Eupleura caudata									●													
Euptoieta claudia		●							●													
Eurema lisa		●			●		●	●	●			●								●		
Eurema nicippe		●				●	●		●					●		●			●			

	Hampton	Horry	Jasper	Kershaw	Lancaster	Laurens	Lee	Lexington	Marlboro	McCormick	Newberry	Oconee	Orangeburg	Pickens	Richland	Spartanburg	Sumter	Union	York	Georgia	N. Carolina
Eucrassatella speciosa		❖																			
Euderces picipes																					
Eudocimus albus			●	●																	
Euglandia rosea																					
Eugnamptus collaris collaris																					
Eumeces fasciatus					●							●			●		●				
Eumeces laticeps		●			●							●	●		●		●				
Eumeces sp.					●								●	●			●		●		
Eumenes fraternus																					
Euonymus americanus					❀	❀	❀			❀	❀	❀		❀	❀	❀	❀	❀	❀	❀	❀
Euonymus atropurpureus																❀					
Euonymus sp.																					❀
Euparius marmoreus																					
Eupatorium album																❀					
Eupatorium capillifolium	❀	❀	❀				❀	❀	❀								❀				
Eupatorium coelestinum							❀														
Eupatorium compositifolium	❀						❀														
Eupatorium dubium		❀																			
Eupatorium fistulosum												❀									
Eupatorium leucolepis	❀																				
Eupatorium pilosum																					
Eupatorium purpureum												❀									
Eupatorium recurvans							❀														
Eupatorium rotundifolium							❀														
Eupatorium rugosum				❀										❀							❀
Eupatorium sp.		❀		❀					❀			❀					❀				
Euphagus carolinus																			●		
Euphagus cyanocephalus																					
Euphorbia ammannioides																					
Euphorbia corollata					❀		❀					❀		❀		❀	❀				
Euphorbia curtisii		❀					❀														
Euphorbia heterophylla									❀												
Euphorbia ipecacuanhae																	❀				
Euphorbia maculata																					
Euphorbia polygonifolia																					
Euphorbia sp.		❀			❀											❀					
Euphorbia supina																					
Euphoria sepulchralis					●																
Eupleura caudata																					
Euptoieta claudia					●				●							●					
Eurema lisa	●			●																	
Eurema nicippe				●	●		●		●			●	●	●			●				

	Abbeville	Aiken	Allendale	Anderson	Bamberg	Barnwell	Beaufort	Berkeley	Charleston	Cherokee	Chester	Chesterfield	Clarendon	Colleton	Darlington	Dillon	Dorchester	Fairfield	Florence	Georgetown	Greenville	Greenwood	Hampton	Horry	Jasper	Kershaw	Lancaster	Laurens	Lee	Lexington	Marlboro	McCormick	Newberry	Oconee	Orangeburg	Pickens	Richland	Spartanburg	Sumter	Union	York	Georgia	N. Carolina
Eurosta solidaginis		●					●											●																									
Eurycea bislineata																																		●									
Eurytides marcellus	●											●			●			●			●			●			●				●	●		●	●	●	●					●	
Eusarca confusaria						●																																					
Euschistus servus servus																		●																									
Euschistus sp.		●																																									
Eustrophinus bicolor			●																																								
Evarthrus sp.	●	●																																									
Everes comyntas		●	●						●			●	●								●	●		●			●	●	●				●	●		●	●	●					●
Evernia mesomorpha																																		*									*
Exidia alba																																					*						
Exobasidium vaccinii																																										*	
Exogyra costata																			❖					❖																			
Facelis retusa									❀			❀																															
Fagus grandifolia	❀									❀	❀			❀	❀		❀	❀			❀		❀					❀				❀		❀	❀	❀	❀	❀	❀	❀	❀	❀	❀
Falco peregrinus									●																																		
Falco sparverius				●			●		●	●		●		●						●				●																			
Fasciolaria hunteria									●																																		
Favia fragum									●																																		
Felis catus															●																												
Felis rufus							●		●					●	●									●					●										●				
Fimbristylis dichotoma																											❀																
Fimbristylis spadicea							❀		❀																																		
Firmiana platanifolia																									❀																		
Flavoparmelia baltimorensis	*										*										*															*		*			*	*	
Flavoparmelia caperata	*								*		*	*		*	*			*			*			*										*		*		*	*		*	*	
Flavopunctelia soredica																																		*									
Foeniculum vulgare																																						❀					
Fomes fomentarius																												*									*						
Fomitopsis officinalis																																					*						
Forestiera acuminata																																					❀						
Forestiera ligustrina		❀																																								❀	
Fothergilla gardenii		❀										❀			❀									❀																			
Fothergilla major																																											❀
Fragaria virginiana																		❀																❀								❀	❀
Franklinia alatamaha																																											❀
Fraxinus americana		❀					❀				❀										❀	❀											❀	❀	❀	❀	❀	❀	❀				❀
Fraxinus americana americana																																		❀		❀					❀		
Fraxinus americana biltmoreana		❀																																									
Fraxinus caroliniana		❀												❀																							❀						
Fraxinus pennsylvanica																																										❀	
Fraxinus pennsylvanica subintegerrima																												❀									❀						

	Abbeville	Aiken	Allendale	Anderson	Bamberg	Barnwell	Beaufort	Berkeley	Charleston	Cherokee	Chester	Chesterfield	Clarendon	Colleton	Darlington	Dillon	Dorchester	Fairfield	Florence	Georgetown	Greenville	Greenwood
Fraxinus sp.		❀				❀		❀			❀							❀			❀	
Fraxinus tomentosa																						
Froelichia floridana												❀										
Froelichia gracilis																						
Frontinella pyramitela										●												
Frontinella sp.																						
Fuirena squarrosa		❀						❀				❀										
Fulica americana							●		●					●						●		
Fuligo septica	✻																					
Fundulus heteroclitus																						
Fuscoboletinus ochraceoroseus																					✻	
Fusinus abruptus																	❖					
Galactia sp.																						
Galactia volubilis								❀														
Galax aphylla										❀					❀						❀	
Galerina autumnalis									✻													
Galeritula bicolor						●												●				
Galerucella notulata														●								
Galerucella nymphaeae												●										
Galerucella rufosanguinea																						
Galiella rufa																						
Galium aparine		❀													❀							
Galium circaezans	❀																					
Galium hispidulum							❀	❀	❀													
Galium sp.									❀		❀			❀						❀		
Gallinago gallinago														●				●				
Gallinula chloropus							●	●	●					●						●		
Gambusia sp.						●																
Ganoderma applanatum							✻															
Ganoderma curtisii	✻																					
Gasteracantha cancriformis					●	●		●	●			●										
Gastrophryne carolinensis																						
Gaultheria procumbens																						
Gaura angustifolia																				❀		
Gaurotes cyanipennis																						
Gavia immer							●		●													
Gaylussacia baccata																						
Gaylussacia dumosa																						
Gaylussacia frondosa						❀	❀	❀				❀	❀		❀	❀	❀					
Gaylussacia sp.		❀																				
Gaylussacia ursina																					❀	
Geastrum pectinatum							✻															

	Hampton	Horry	Jasper	Kershaw	Lancaster	Laurens	Lee	Lexington	Marlboro	McCormick	Newberry	Oconee	Orangeburg	Pickens	Richland	Spartanburg	Sumter	Union	York	Georgia	N. Carolina
Fraxinus sp.	❀									❀					❀	❀		❀	❀		
Fraxinus tomentosa																❀					
Froelichia floridana																					
Froelichia gracilis				❀																	
Frontinella pyramitela		●			●					●											
Frontinella sp.															●						
Fuirena squarrosa							❀														
Fulica americana	●		●													●					
Fuligo septica																					
Fundulus heteroclitus		●																			
Fuscoboletinus ochraceoroseus																					
Fusinus abruptus																					
Galactia sp.							❀														
Galactia volubilis	❀																				
Galax aphylla												❀		❀							❀
Galerina autumnalis																					
Galeritula bicolor						●															
Galerucella notulata						●															
Galerucella nymphaeae																					
Galerucella rufosanguinea					●																
Galiella rufa												✻				✻					
Galium aparine						❀				❀						❀					
Galium circaezans														❀							
Galium hispidulum																					
Galium sp.					❀					❀		❀			❀			❀			
Gallinago gallinago																					
Gallinula chloropus	●		●	●																	
Gambusia sp.																●					
Ganoderma applanatum															✻						✻
Ganoderma curtisii							✻	✻													
Gasteracantha cancriformis		●		●			●										●				
Gastrophryne carolinensis					●																
Gaultheria procumbens																			❀		❀
Gaura angustifolia																					
Gaurotes cyanipennis																				●	
Gavia immer												●									
Gaylussacia baccata												❀									
Gaylussacia dumosa		❀					❀														
Gaylussacia frondosa		❀					❀								❀						
Gaylussacia sp.							❀												❀		
Gaylussacia ursina												❀								❀	❀
Geastrum pectinatum																					

	Abbeville	Aiken	Allendale	Anderson	Bamberg	Barnwell	Beaufort	Berkeley	Charleston	Cherokee	Chester	Chesterfield	Clarendon	Colleton	Darlington	Dillon	Dorchester	Fairfield	Florence	Georgetown	Greenville	Greenwood	Hampton	Horry	Jasper	Kershaw	Lancaster	Laurens	Lee	Lexington	Marlboro	McCormick	Newberry	Oconee	Orangeburg	Pickens	Richland	Spartanburg	Sumter	Union	York	Georgia	N. Carolina
Geastrum saccatum	✻								✻																													✻					
Geastrum sp.									✻																																		
Gelastocoris oculatus																																		●				●					
Gelastocoris sp.																																									●		
Gelsemium sempervirens	❀	❀	❀		❀	❀	❀		❀	❀	❀	❀		❀	❀	❀	❀	❀	❀		❀		❀	❀		❀	❀		❀	❀	❀	❀			❀		❀	❀	❀	❀		❀	
Gentiana autumnalis												❀														❀																	
Gentiana catesbaei															❀																												
Gentiana sp.														❀					❀																								
Gentiana villosa														❀													❀											❀					
Geothlypis trichas		●						●	●									●			●			●			●		●														
Geotrupes blackburnii blackburnii												●																										●					
Geotrupes egeriei							●																																				
Geranium carolinianum				❀			❀		❀		❀			❀	❀						❀			❀										❀									
Geranium maculatum		❀						❀																			❀					❀		❀		❀	❀						❀
Gerris remigis																																		●									
Gerris sp.																																		●		●			●		●		●
Geukensia demissa							●		●																																		
Geum canadense		❀																																	❀								
Geum sp.																																								❀			
Gillenia trifoliata																					❀													❀		❀						❀	❀
Glaphyrocanthon viridis									●																			●						●		●		●					
Glecoma hederacea											❀				❀														❀												❀		
Gleditsia aquatica														❀					❀																❀								
Gleditsia triacanthos				❀					❀		❀							❀			❀	❀					❀											❀			❀		
Glossosoma sp.																																		●									
Glottidium vesicarium							❀		❀					❀											❀				❀														
Glycymeris americana																								❖																			❖
Glycymeris aratus																								❖																			
Glycymeris pectinata																								❖																			
Glyphis cicatricosa		✻						✻	✻										✻								✻					✻							✻				
Glyphis sp.																																						✻					
Glyptobaris lecontei																																						●					
Gnaphalium obtusifolium		❀		❀	❀				❀		❀										❀						❀				❀					❀		❀					
Gnaphalium purpureum														❀	❀						❀																						
Gnaphalium sp.		❀																			❀			❀																			
Gobaishia ulmifusus																																											●
Goes oculata																																										●	
Gomphus exilis																																					●						
Gomphus floccosus																																											✻
Gomphus sp.																																										●	
Goniobasis catenaria																																			●								
Goniobasis sp.		●																																									

	Abbeville	Aiken	Allendale	Anderson	Bamberg	Barnwell	Beaufort	Berkeley	Charleston	Cherokee	Chester	Chesterfield	Clarendon	Colleton	Darlington	Dillon	Dorchester	Fairfield	Florence	Georgetown	Greenville	Greenwood	Hampton	Horry	Jasper	Kershaw	Lancaster	Laurens	Lee	Lexington	Marlboro	McCormick	Newberry	Oconee	Orangeburg	Pickens	Richland	Spartanburg	Sumter	Union	York	Georgia	N. Carolina
Goodyera pubescens	❀										❀				❀						❀							❀	❀			❀	❀	❀		❀		❀		❀	❀	❀	❀
Gopherus polyphemus																									●																		
Gordonia lasianthus		❀					❀						❀		❀	❀							❀	❀															❀				
Gouldia metastriata																								❖																			
Grammia virgo																																				●							
Grammoptera sp.																											●																
Graphina sophisticascens											*																																
Graphis afzelii							*																			*	*																
Graphis anfractuosa																																			*								
Graphis atrorubens														*					*								*									*							
Graphis beaumontii							*																				*																
Graphis elegans							*	*	*					*					*				*				*		*										*		*		*
Graphis grammatis																				*							*																
Graphis intricata																											*																
Graphis leucopepla									*																		*										*						
Graphis lineola		*																									*																
Graphis mosquitensis		*						*			*												*				*					*		*	*								
Graphis nitida								*												*							*																
Graphis poitaeoides		*																									*																
Graphis scripta		*				*	*	*	*		*			*	*			*	*	*	*		*	*		*	*		*			*		*	*	*	*	*	*	*	*	*	*
Graphis scripta topographica																											*																*
Graphis sp.		*				*	*	*													*		*	*								*			*		*						
Graphis striatula																											*					*											
Graphis tenella																					*						*																
Graphoderus liberus								●																																			
Graphognathus leucoloma		●																																●				●					
Graphognathus sp.		●																																				●					
Graphops curtipennis																											●																
Gratiola pilosa																																		❀									
Gratiola sp.																											❀																
Gratiola virginiana															❀																												
Gryllus assimilis		●																																									
Gryllus sp.	●								●											●				●													●	●					
Guiraca caerulea																																		●									
Gymnoderma lineare																																		*									
Gymnopogon ambiguus																											❀																
Gymnopogon brevifolius												❀																															
Gymnosporangium junipervirginianae																		*									*																
Gymnosporangium sp.		*																																								*	
Gyrinus analis								●																							●						●						
Gyrinus borealis		●																						●																			
Gyrinus frosti																															●												

	Abbeville	Aiken	Allendale	Anderson	Bamberg	Barnwell	Beaufort	Berkeley	Charleston	Cherokee	Chester	Chesterfield	Clarendon	Colleton	Darlington	Dillon	Dorchester	Fairfield	Florence	Georgetown	Greenville	Greenwood	Hampton	Horry	Jasper	Kershaw	Lancaster	Laurens	Lee	Lexington	Marlboro	McCormick	Newberry	Oconee	Orangeburg	Pickens	Richland	Spartanburg	Sumter	Union	York	Georgia	N. Carolina
Gyrinus lugens								●						●	●																								●				
Gyrinus rockinghamensis																								●																			
Gyrinus sp.						●																																					
Gyrophaena corruscula	●																																										
Gyrophaena dissimilis	●																																										
Gyroporus purpurinus																															*												
Gyrostomum scyphuliferum						*			*														*				*																
Gyrostomum sp.																							*				*										*						
Habenaria ciliaris																																											❀
Habenaria clavellata						❀																												❀		❀							
Habenaria cristata																							❀																				
Habenaria flava					❀																																						
Habenaria repens																							❀								❀												
Haematomma cismonicum									*					*																					*			*					
Haematomma elatinum							*		*					*					*				*	*			*											*	*				
Haematomma ochrophaeum								*																			*																
Haematomma puniceum		*				*	*	*	*					*							*		*	*		*	*		*			*			*		*		*				
Haematopus palliatus									●											●																							
Hagenius brevistylus		●																																									
Halesia carolina	❀										❀										❀											❀		❀		❀		❀				❀	❀
Halesia parviflora																					❀																						
Halesia sp.																					❀															❀							
Haliaeetus leucocephalus		●							●					●						●					●									●									
Haliplus fasciatus											●				●																●												
Haliplus triopsus		●										●			●																	●											
Halysidota tessellaris		●																								●									●								
Hamamelis virginiana	❀	❀				❀	❀	❀		❀					❀		❀		❀		❀		❀	❀			❀		❀			❀		❀	❀	❀		❀	❀			❀	❀
Hamamelistes spinosus						●																																					
Haploa clymene											●																							●									
Haploa lecontei																																				●							
Haplotrema concavum																																		●									
Harmonia axyridis		●													●																								●				
Harpalus caliginosus																								●																			
Harpalus pennsylvanicus																								●														●					●
Harpalus spadiceus																																											●
Harrisina americana								●																																			
Hedeoma pulegioides		❀																																									
Hedera helix	❀					❀																																					
Helenium amarum		❀																									❀		❀		❀		❀					❀					
Helenium autumnale																					❀																						❀
Helenium flexuosum								❀																																			
Helianthemum corymbosum							❀		❀																																		

	Abbeville	Aiken	Allendale	Anderson	Bamberg	Barnwell	Beaufort	Berkeley	Charleston	Cherokee	Chester	Chesterfield	Clarendon	Colleton	Darlington	Dillon	Dorchester	Fairfield	Florence	Georgetown	Greenville	Greenwood	Hampton	Horry	Jasper	Kershaw	Lancaster	Laurens	Lee	Lexington	Marlboro	McCormick	Newberry	Oconee	Orangeburg	Pickens	Richland	Spartanburg	Sumter	Union	York	Georgia	N. Carolina
Helianthemum rosmarinifolium		❀						❀																											❀								
Helianthus angustifolius												❀																															
Helianthus atrorubens																																						❀					
Helianthus microcephalus																																		❀									
Helianthus sp.																																		❀				❀					
Helianthus tomentosus																					❀																						
Heliconius charitonius tuckeri		●				●	●		●																																		
Heliomata sp.																			●																								
Heliotropium amplexicaule		❀																													❀												
Helmitheros vermivorus																				●														●									●
Helocombus bifidus													●																														
Helthemia pistaciaria																																		●									
Helvella acetabulum																																					✻						
Hemadas nubilipennis																																				●							●
Hemaris thysbe																								●			●																
Hemerobiidae Family																			●																								
Hemerocallis fulva							❀																																				
Hemipristis serra																																											❖
Hemisphaerota cyanea							●							●																													
Henoticus sp.																																										●	
Hentzia sp.																																									●		
Hepatica acutiloba																																		❀									❀
Hepatica americana	❀	❀									❀																❀	❀				❀			❀			❀					
Hepatus epheliticus									●																																		
Hericium coralloides							✻																																✻				
Hericium erinaceus																			✻																			✻					
Hericium sp.															✻																												
Hermeuptychia sosybius	●	●	●								●		●														●	●	●				●	●	●		●						
Heteroderes sp.																																		●									
Heterodermia albicans		✻				✻	✻	✻	✻		✻			✻	✻				✻		✻		✻	✻		✻									✻			✻	✻		✻	✻	
Heterodermia casarettiana								✻				✻														✻						✻		✻		✻						✻	✻
Heterodermia echinata																								✻										✻		✻							
Heterodermia granulifera								✻															✻																				✻
Heterodermia gruoni																																					✻						
Heterodermia hypoleuca						✻		✻																										✻							✻		✻
Heterodermia leucomelaena																																											✻
Heterodermia leucomelos																																		✻								✻	
Heterodermia obscurata		✻				✻	✻	✻			✻			✻	✻			✻	✻	✻	✻		✻	✻	✻	✻								✻	✻	✻	✻	✻	✻		✻	✻	✻
Heterodermia sp.																																					✻						
Heterodermia spaltans sp. nov.									✻					✻										✻									✻	✻				✻					
Heterodermia speciosa									✻		✻	✻									✻		✻	✻								✻		✻		✻		✻		✻		✻	✻
Heterodermia squamulosa																																											✻

	Abbeville	Aiken	Allendale	Anderson	Bamberg	Barnwell	Beaufort	Berkeley	Charleston	Cherokee	Chester	Chesterfield	Clarendon	Colleton	Darlington	Dillon	Dorchester	Fairfield	Florence	Georgetown	Greenville	Greenwood	Hampton	Horry	Jasper	Kershaw	Lancaster	Laurens	Lee	Lexington	Marlboro	McCormick	Newberry	Oconee	Orangeburg	Pickens	Richland	Spartanburg	Sumter	Union	York	Georgia	N. Carolina
Heterodon platyrhinos								●												●							●																
Heterotheca gossypina												❀																															
Heterotheca graminifolia												❀		❀	❀	❀	❀		❀	❀				❀		❀								❀				❀	❀				
Heterotheca mariana												❀							❀																								
Heterotheca nervosa								❀												❀																							
Heterotheca nervosa nervosa							❀																																				
Heterotheca subaxillaris							❀		❀											❀			❀			❀					❀												
Hetrerodermia casarettiana									✻															✻																			
Heuchera americana	❀	❀																									❀					❀		❀			❀						❀
Heuchera sp.																																				❀							
Heuchera villosa																																		❀									
Hexanchus sp.																	❖																										
Hexastylis arifolia	❀	❀	❀		❀	❀					❀				❀	❀	❀				❀		❀	❀			❀	❀	❀	❀		❀		❀	❀		❀		❀		❀	❀	❀
Hexastylis contracta																																											❀
Hexastylis heterophylla																																		❀									
Hexastylis minor										❀																																	
Hexastylis shuttleworthii																																		❀		❀							
Hexastylis sp.												❀									❀												❀	❀		❀					❀		❀
Hibiscus moscheutos								❀			❀														❀																		
Hibiscus sp.				❀																																							
Hieracium gronovii																											❀		❀														
Hieracium paniculatum																					❀																						
Hieracium sp.																					❀															❀	❀						
Hieracium venosum	❀	❀																			❀						❀	❀						❀		❀	❀	❀				❀	❀
Hippocampus reidi																				●																							
Hippodamia convergens					●		●	●																										●									
Hippopsis lemniscata								●																																			
Hirundo rustica		●							●																	●								●	●								
Hister depurator							●																																				
Hohenbuehelia petaloides	✻																																										
Holomelina immaculata																								●																			
Hoplocephala viridipennis								●																●																			
Horistonotus curiatus																																				●							
Hormaphis hamamelidis	●																												●													●	●
Houstonia caerulea																					❀						❀		❀							❀	❀						❀
Houstonia longifolia		❀																																❀									
Houstonia procumbens							❀		❀																																		
Houstonia purpurea					❀																						❀	❀								❀	❀	❀				❀	❀
Houstonia pusilla			❀						❀			❀						❀																			❀		❀			❀	
Houstonia serpyllifolia																																		❀									❀
Houstonia sp.																					❀																						❀
Houstonia tenuifolia																											❀																

	Abbeville	Aiken	Allendale	Anderson	Bamberg	Barnwell	Beaufort	Berkeley	Charleston	Cherokee	Chester	Chesterfield	Clarendon	Colleton	Darlington	Dillon	Dorchester	Fairfield	Florence	Georgetown	Greenville	Greenwood	Hampton	Horry	Jasper	Kershaw	Lancaster	Laurens	Lee	Lexington	Marlboro	McCormick	Newberry	Oconee	Orangeburg	Pickens	Richland	Spartanburg	Sumter	Union	York	Georgia	N. Carolina
Hudsonia ericoides												❀																															
Humaria hemisphaerica																																						*					
Hybanthus concolor											❀																❀					❀											❀
Hydnum imbricatum																											*											*					
Hydrangea arborescens																					❀							❀						❀		❀						❀	❀
Hydrangea arborescens arborescens																																											❀
Hydrangea arborescens discolor																																											❀
Hydrangea arborescens radiata																					❀													❀		❀							
Hydrangea quercifolia																																											❀
Hydria prunivorata																																		●									
Hydrobius melaenus							●																																				
Hydrocanthus iricolor												●																	●														
Hydrocanthus oblongus																								●																			
Hydrochara soror							●																																				
Hydrochus excavatus											●																																
Hydrochus inaequalis			●																																								
Hydrocotyle bonariensis							❀		❀											❀																							
Hydrocotyle ranunculoides								❀	❀					❀									❀	❀																			
Hydrocotyle sibthorpioides															❀																												
Hydrocotyle sp.			❀		❀	❀	❀		❀					❀					❀	❀															❀								
Hydrocotyle umbellata		❀																					❀								❀												
Hydrocotyle verticillata		❀																																			❀						
Hydrolea quadrivalvis						❀																									❀												
Hydrophilus triangularis																																											●
Hydrophyllum canadense																																											❀
Hydroporus carolinus															●		●																					●					
Hydroporus cimicoides																								●																			
Hydroporus clypealis						●									●		●							●																			
Hydroporus hybridus															●									●					●														
Hydroporus lobatus														●			●												●									●					
Hydroporus niger											●																															●	
Hydroporus sp.						●						●			●																											●	
Hydroporus undulatus																																						●					
Hydroporus vittatipennis																●			●																								
Hydropsyche sp.																																		●									
Hydrothyria venosa																											*																
Hydrovatus pustulatus								●																																			
Hygrocybe punica																					*																						
Hygrophorus cantharellus																								*					*														
Hygrophorus coccineus																																											*
Hygrophorus miniatus																											*											*					
Hygrophorus pratensis	*																																										

	Abbeville	Aiken	Allendale	Anderson	Bamberg	Barnwell	Beaufort	Berkeley	Charleston	Cherokee	Chester	Chesterfield	Clarendon	Colleton	Darlington	Dillon	Dorchester	Fairfield	Florence	Georgetown	Greenville	Greenwood	Hampton	Horry	Jasper	Kershaw	Lancaster	Laurens	Lee	Lexington	Marlboro	McCormick	Newberry	Oconee	Orangeburg	Pickens	Richland	Spartanburg	Sumter	Union	York	Georgia	N. Carolina
Hyla chrysoscelis		●																		●												●		●		●	●						
Hyla cinerea						●	●		●					●	●	●				●				●			●																
Hyla crucifer						●	●								●			●			●													●		●			●				
Hyla femoralis		●					●	●			●													●																			
Hyla sp.														●										●					●														
Hyla squirella						●																																					
Hyla versicolor																																				●	●						
Hylephila phyleus																																			●								
Hylocichla mustelina		●																					●					●					●	●	●		●	●	●			●	●
Hymenocallis occidentalis		❀									❀																																
Hymenorus sp.									●																																		
Hypebaeus oblitus																											●																
Hypebaeus sp.															●																	●											
Hyperaspis sp.																																					●						
Hypericum densiflorum																❀																											
Hypericum fasciculaatum								❀																																			
Hypericum gentianoides		❀										❀														❀			❀		❀					❀							
Hypericum gymnanthum																								❀																			
Hypericum hypericoides	❀	❀				❀	❀	❀	❀		❀	❀		❀	❀	❀	❀	❀	❀		❀		❀	❀		❀				❀	❀	❀		❀		❀		❀	❀		❀	❀	❀
Hypericum lloydii																											❀			❀													
Hypericum mutilum	❀	❀																																❀									
Hypericum nudiflorum											❀																																
Hypericum punctatum		❀																																									
Hypericum sp.		❀				❀			❀		❀	❀		❀						❀	❀			❀			❀		❀					❀					❀			❀	❀
Hypericum stans								❀				❀		❀		❀								❀					❀														
Hypericum stragalum																																		❀		❀					❀		
Hypericum virginicum						❀																																					
Hypericum walteri						❀																		❀							❀								❀				
Hyperphyscia syncolla								*							*																			*									
Hyperplatys maculatus																												●															
Hyphantria cunea								●																						●				●			●	●	●				
Hypogeocoris sp.																											●																
Hypogymnia krogiae																								*																			*
Hypogymnia physodes																																											*
Hypogymnia sp.																																											*
Hypomyces chrysopermus																							*																				
Hypomyces luteovirens																																						*					
Hypotrachyna citrina sp. nov.																							*																				
Hypotrachyna ensifolia																																		*									
Hypotrachyna formosana						*									*														*													*	
Hypotrachyna livida	*	*				*	*	*	*	*	*	*		*	*			*	*		*		*	*	*	*	*		*		*	*		*	*	*		*	*		*	*	*
Hypotrachyna livida f. aurea																																					*						

	Abbeville	Aiken	Allendale	Anderson	Bamberg	Barnwell	Beaufort	Berkeley	Charleston	Cherokee	Chester	Chesterfield	Clarendon	Colleton	Darlington	Dillon	Dorchester	Fairfield	Florence	Georgetown	Greenville	Greenwood	Hampton	Horry	Jasper	Kershaw	Lancaster	Laurens	Lee	Lexington	Marlboro	McCormick	Newberry	Oconee	Orangeburg	Pickens	Richland	Spartanburg	Sumter	Union	York	Georgia	N. Carolina
Hypotrachyna pustulifera		*							*					*							*					*																	
Hypotrachyna revoluta																																		*									*
Hypoxis hirsuta	❀	❀												❀			❀				❀							❀	❀					❀		❀	❀					❀	❀
Hypoxis micrantha																								❀																			
Hypoxis rigida																																				❀							
Hypoxis sp.																													❀														
Hypoxylon sp.																								*																			
Hypsoropha monilis																											●									●							
Hyptis alata																				❀																							
Icerya purchasi								●																																			
Icteria virens																								●					●					●									
Icterus spurius																								●					●						●								
Ictinia mississippiensis																							●																				
Ilex ambigua																		❀																									
Ilex ambigua montana				❀																	❀																						❀
Ilex cassine cassine							❀	❀																																			
Ilex cassine myrtifolia					❀			❀												❀									❀														
Ilex coriacea		❀				❀	❀					❀	❀		❀								❀	❀		❀													❀				
Ilex decidua		❀						❀			❀								❀								❀					❀	❀				❀	❀					
Ilex decidua longipes																																						❀					
Ilex glabra		❀				❀	❀	❀				❀	❀	❀	❀	❀			❀				❀	❀		❀			❀										❀				
Ilex opaca	❀	❀	❀			❀	❀	❀	❀	❀	❀	❀	❀	❀	❀	❀	❀	❀	❀	❀	❀		❀	❀	❀	❀	❀	❀	❀	❀	❀	❀	❀	❀	❀	❀	❀	❀	❀	❀	❀	❀	❀
Ilex sp.		❀													❀																												
Ilex verticillata														❀																													
Ilex vomitoria							❀	❀	❀					❀	❀					❀					❀														❀				
Illicium sp.															❀																												
Ilyanassa obsoleta																				●																							
Ilybius biguttulus																																						●					
Impatiens capensis		❀	❀								❀				❀						❀	❀					❀	❀				❀		❀	❀	❀	❀		❀			❀	❀
Impatiens pallida																																											❀
Imshaugia aleurites																					*													*		*							*
Imshaugia placorodia																		*																									*
Indigofera caroliniana							❀																	❀	❀														❀				
Inocybe fuscodisca																																		*									
Inocybe geophylla							*																																				
Ipomoea coccinea																											❀																
Ipomoea hederacea hederacea									❀																																		
Ipomoea lacunosa																											❀																
Ipomoea pandurata																																		❀					❀				
Ipomoea purpurea														❀	❀																												
Ipomoea sagittata														❀																													
Ipomoea sp.									❀		❀																							❀									

	Abbeville	Aiken	Allendale	Anderson	Bamberg	Barnwell	Beaufort	Berkeley	Charleston	Cherokee	Chester	Chesterfield	Clarendon	Colleton	Darlington	Dillon	Dorchester	Fairfield	Florence	Georgetown	Greenville	Greenwood	Hampton	Horry	Jasper	Kershaw	Lancaster	Laurens	Lee	Lexington	Marlboro	McCormick	Newberry	Oconee	Orangeburg	Pickens	Richland	Spartanburg	Sumter	Union	York	Georgia	N. Carolina
Ipomoea stolonifera									❀											❀																							
Ipomopsis rubra																																				❀							
Iresine rhizomatosa							❀																																				
Iridopsis larvaria																																		●									
Iris cristata																					❀						❀									❀		❀				❀	❀
Iris sp.																								❀																			
Iris tridentata								❀																❀																			
Iris verna												❀	❀		❀																			❀		❀						❀	❀
Iris virginica		❀																			❀																						
Ischnura hastata												●		●										●											●								
Ischnura kellicotti		●																									●																
Ischnura posita		●																						●																			
Ischnura ramburi											●																																
Isoetes melanopoda																																										❀	
Isoetes sp.																											❀																
Isomira sericea																												●				●											
Isomira sp.															●																												
Isonychia sp.																																		●									
Isoperla bilineata																								●																			
Isopyrum biternatum																																❀											
Itea virginica		❀				❀		❀	❀			❀	❀	❀	❀	❀					❀		❀	❀		❀			❀		❀			❀			❀	❀	❀				
Itonida anthici		●														●										●																	
Itonida taxodii																●								●																			
Iva frutescens							❀		❀					❀																													
Iva imbricata							❀		❀											❀																							
Jacquemontia tamnifolia		❀																					❀																				
Jalysus spinosus																																		●									
Janetiella coloradensis										●																																	
Juglans cinerea																																											❀
Juglans nigra		❀					❀	❀			❀			❀	❀							❀					❀	❀					❀	❀				❀		❀	❀	❀	❀
Julella sp.																				*																							
Junco hyemalis							●		●										●		●				●																	●	●
Juncus bufonius																❀								❀																			
Juncus dichotomus																❀																											
Juncus effusus		❀				❀	❀							❀												❀												❀					
Juncus georgianus																											❀																
Juncus roemerianus							❀		❀					❀																													
Juncus scirpoides																❀																											
Juncus sp.									❀																																		❀
Juncus tenuis		❀																								❀																	
Juniperus silicicola							❀		❀																																		
Juniperus virginiana	❀	❀		❀		❀	❀	❀	❀	❀	❀	❀		❀	❀	❀		❀		❀	❀	❀		❀		❀	❀	❀		❀	❀	❀	❀	❀	❀	❀	❀	❀		❀	❀	❀	❀

Scientific Name Index

	Abbeville	Aiken	Allendale	Anderson	Bamberg	Barnwell	Beaufort	Berkeley	Charleston	Cherokee	Chester	Chesterfield	Clarendon	Colleton	Darlington	Dillon	Dorchester	Fairfield	Florence	Georgetown	Greenville	Greenwood	Hampton	Horry	Jasper	Kershaw	Lancaster	Laurens	Lee	Lexington	Marlboro	McCormick	Newberry	Oconee	Orangeburg	Pickens	Richland	Spartanburg	Sumter	Union	York	Georgia	N. Carolina
Junonia coenia		●				●	●	●	●				●			●	●		●	●	●	●		●		●			●		●		●	●			●	●			●	●	
Justicia americana																																					❀						
Justicia ovata					❀																																❀						
Kalmia angustifolia													❀		❀									❀																			
Kalmia angustifolia caroliniana																❀								❀																			
Kalmia latifolia		❀								❀		❀			❀						❀							❀		❀				❀		❀	❀	❀	❀		❀	❀	❀
Kinosternon subrubrum		●																											●										●				
Kosteletskya virginica														❀																													
Krigia dandelion																																					❀						
Krigia virginica		❀	❀										❀														❀									❀	❀					❀	
Laccaria laccata														✻																													✻
Laccaria ochropurpurea																											✻																
Laccophiilus fasciatus rufus																											●																
Laccophilus fasciatus			●								●																		●								●						
Laccophilus maculosus maculosus																																						●					
Laccophilus proximus															●																												
Laccophilus schwartzi														●																													
Laccophilus schwarzi																													●														
Lachnanthes caroliniana					❀			❀				❀				❀								❀																			
Lachnocaulon anceps								❀				❀												❀					❀														
Lachnocaulon beyrichianum																❀																											
Lachnocaulon sp.																								❀																			
Lacon marmoreus																		●																									
Lactarius corrugis									✻																																		
Lactarius indigo																											✻									✻		✻					
Lactarius luteolus																																											✻
Lactarius peckii																																											✻
Lactarius piperatus	✻																										✻											✻	✻				
Lactarius scrobiculatus																					✻																						
Lactarius sp.																													✻														
Lactarius uvidus																											✻																
Lactarius volemus	✻																																										
Lactuca biennis																																		❀									
Lactuca canadensis																										❀																	
Lactuca floridana																											❀																
Lactuca sp.					❀			❀			❀										❀															❀		❀					
Ladona deplanata											●																●							●									
Laemophloeus modestus															●																												
Laetiporus sulphureus							✻		✻															✻																			
Lamium amplexicaule			❀	❀			❀		❀						❀									❀																			
Lamium purpureum															❀																												
Lamna sp.																								❖																			

Scientific Name Index

	Abbeville	Aiken	Allendale	Anderson	Bamberg	Barnwell	Beaufort	Berkeley	Charleston	Cherokee	Chester	Chesterfield	Clarendon	Colleton	Darlington	Dillon	Dorchester	Fairfield	Florence	Georgetown	Greenville	Greenwood	Hampton	Horry	Jasper	Kershaw	Lancaster	Laurens	Lee	Lexington	Marlboro	McCormick	Newberry	Oconee	Orangeburg	Pickens	Richland	Spartanburg	Sumter	Union	York	Georgia	N. Carolina
Lampropeltis getulus																									●											●	●						
Languria lecontei		●																																									
Languria mozardi																																										●	
Lanius ludovicianus								●	●					●						●					●				●														
Lanthus vernalis																																		●									
Laportea canadensis		❀																														❀		❀		❀	❀				❀		❀
Largus succinctus		●																									●																
Larus argentatus							●		●											●																							
Larus atricilla							●	●	●					●						●															●								
Larus delawarensis							●		●					●						●														●									
Larus marinus							●													●																							
Larus philadelphia							●							●																													
Larus pipixcan									●																																		
Lasioptera clazula																											●													●			
Latrodectus mactans																								●																			
Latrodectus variolus																																						●					
Lebia analis		●																																									
Lebia lobulata																																					●						
Lebia ornata																																					●						
Lebia viridis																																				●							
Lecanactis sp.								*																																			
Lecania fuscella								*																			*		*														
Lecanora allophana											*										*					*								*								*	
Lecanora allophana argentata								*							*			*					*			*																	*
Lecanora atra																										*	*																
Lecanora caesiorubella f. prolifera																					*			*																			
Lecanora caesiorubella subs. caesiorubella									*																																		
Lecanora cenisia																											*																*
Lecanora cinerea																											*																
Lecanora cupressi		*				*	*	*	*		*	*		*	*			*	*	*	*		*	*		*	*		*					*	*	*	*	*	*	*	*	*	*
Lecanora deplanans																											*																*
Lecanora effusa																											*																
Lecanora hageni								*																			*								*					*			
Lecanora horiza																																					*						
Lecanora hybocarpa		*							*						*			*																							*		
Lecanora hypoptoides									*			*			*												*																
Lecanora lacustris																											*					*											
Lecanora laevata																											*																
Lecanora miculata								*				*			*								*			*	*		*			*											
Lecanora minutella																																		*									
Lecanora pallida		*					*	*			*				*			*		*	*		*			*	*							*		*	*	*				*	
Lecanora pallida cancriformis																				*			*			*			*								*	*					

Scientific Name Index

	Abbeville	Aiken	Allendale	Anderson	Bamberg	Barnwell	Beaufort	Berkeley	Charleston	Cherokee	Chester	Chesterfield	Clarendon	Colleton	Darlington	Dillon	Dorchester	Fairfield	Florence	Georgetown	Greenville	Greenwood	Hampton	Horry	Jasper	Kershaw	Lancaster	Laurens	Lee	Lexington	Marlboro	McCormick	Newberry	Oconee	Orangeburg	Pickens	Richland	Spartanburg	Sumter	Union	York	Georgia	N. Carolina
Lecanora pallida f. prolifera																							*														*						
Lecanora pallida rubescens									*																														*				
Lecanora parisensis																											*																
Lecanora polytropa																																											*
Lecanora pulicaris																			*																								
Lecanora rugosa												*															*																
Lecanora saligna												*																															
Lecanora sp.									*						*					*			*	*		*			*					*	*	*							
Lecanora sp. #2 Brodo									*						*						*																						
Lecanora subfusca		*																									*																
Lecanora subimmergens																																						*					
Lecanora varia	*	*				*	*	*	*		*	*		*	*			*		*	*		*	*		*	*		*					*	*	*		*	*	*	*		
Lecanora willeyi															*												*																
Leccinum scabrum	*																																										
Lechea sp.																													❀														
Lechea villosa																										❀																	
Lecidea albocaerulescens																					*						*																*
Lecidea carneoalbens														*				*									*										*						
Lecidea christophera									*																																		
Lecidea congesta																											*																*
Lecidea contigua																																											*
Lecidea cyanea																											*																
Lecidea mutabilis																					*								*														
Lecidea myriocarpoides															*												*							*									
Lecidea nylanderi												*															*													*			
Lecidea oreinoides														*																													
Lecidea parasema																											*							*									
Lecidea parasema theioplaca																											*							*									
Lecidea promiscens																											*																*
Lecidea russula		*				*		*							*				*		*		*			*	*		*			*		*		*	*		*				
Lecidea sp.												*		*	*			*									*		*			*		*								*	*
Lecidea tessellata																											*							*				*			*		*
Lecidea varians																											*																
Leersia lenticularis		❀																																									
Leersia oryzoides						❀																		❀																			
Leersia sp.																																							❀		❀		
Leiobunum sp.									●		●																																
Leiophyllum buxifolium												❀			❀															❀													❀
Lema albini																												●															
Lema sayi	●	●									●																																
Lema sp.										●		●																															
Lemintina sp.																																											❖

Scientific Name Index

	Abbeville	Aiken	Allendale	Anderson	Bamberg	Barnwell	Beaufort	Berkeley	Charleston	Cherokee	Chester	Chesterfield	Clarendon	Colleton	Darlington	Dillon	Dorchester	Fairfield	Florence	Georgetown	Greenville	Greenwood	Hampton	Horry	Jasper	Kershaw	Lancaster	Laurens	Lee	Lexington	Marlboro	McCormick	Newberry	Oconee	Orangeburg	Pickens	Richland	Spartanburg	Sumter	Union	York	Georgia	N. Carolina
Lemna gibba															❀																												
Lemna sp.						❀	❀	❀	❀					❀						❀				❀			❀												❀				
Lentaria byssiseda																																		*									
Lentinus lepideus																																											*
Lentinus tigrinus								*																																			
Lenzites betulinae																																					*	*					
Leotia ascomyces																																		*									
Leotia viscosa																			*																				*				
Lepas anatifera									●																																		
Lepidium virginicum				❀													❀														❀			❀		❀		❀					
Lepidostoma sp.																																		●									
Lepiota lutea		*																																									
Lepiota procera																											*																
Lepiota rubrotincta																								*																			*
Lepisosteus osseus																																					●						
Lepisosteus sp.									●																																		
Lepomis macrochirus																																						●					
Lepomis sp.	●																																										
Lepraria finkii											*																																
Lepraria membranacea											*																													*			
Lepraria sp.									*	*		*																															
Leptocoris trivittatus																									●																●		
Leptogium austroamericanum		*					*	*			*				*				*	*	*		*				*					*		*		*	*	*				*	
Leptogium burnetiae						*															*						*					*											*
Leptogium chloromelum																											*											*					
Leptogium corticola																					*						*							*		*							
Leptogium cyanescens		*					*	*	*		*			*	*			*	*	*	*		*				*					*		*		*	*	*	*		*	*	*
Leptogium laceroides																																		*									*
Leptogium lichenoides																											*																
Leptogium marginellum														*													*																
Leptogium microstictum																											*													*			
Leptogium phyllocarpum																																				*							
Leptogium sp.																																											*
Leptoglossus sp.									●																																		
Leptogorgia virgulata							●		●											●																							
Leptura lineola																																										●	
Leptysma marginicollis						●																				●			●														
Lepuropetalon spathulatum																																										❀	
Lespedeza bicolor																❀					❀																						
Lespedeza cuneata		❀									❀				❀	❀						❀									❀					❀			❀				
Lespedeza hirta																❀													❀														
Lespedeza intermedia																																											❀

Scientific Name Index

	Abbeville	Aiken	Allendale	Anderson	Bamberg	Barnwell	Beaufort	Berkeley	Charleston	Cherokee	Chester	Chesterfield	Clarendon	Colleton	Darlington	Dillon	Dorchester	Fairfield	Florence	Georgetown	Greenville	Greenwood
Lespedeza repens												❀										
Lespedeza sp.														❀		❀				❀		
Lestes sp.																						
Lestes vidua												●										
Lestes vigilax																						
Letrouitia domingense								*	*													
Letrouitia sp.																						
Letrouitia vulpinum																			*			
Leucauge venusta	●	●							●		●					●						
Leucojum aestivum														❀								
Leucothoe axillaris		❀								❀			❀		❀						❀	
Leucothoe axillaris axillaris						❀									❀							
Leucothoe axillaris editorum																					❀	
Leucothoe populifolia															❀							
Leucothoe racemosa		❀				❀						❀			❀		❀					
Leucothoe recurva																						
Leucothoe sp.																						
Leurognathus marmoratus																						
Liatris elegans																						
Liatris graminifolia												❀										
Liatris secunda		❀																				
Liatris sp.																						
Liatris spicata																						
Liatris spicata resinosa																						
Libellula auripennis		●						●						●								
Libellula axilena	●																					
Libellula exusta deplanata																						
Libellula incesta		●									●											
Libellula luctuosa		●													●							●
Libellula lydia		●																				
Libellula pulchella															●							
Libellula semifasciata																						
Libellula sp.																						
Libellula vibrans	●	●							●		●											
Libinia sp.									●													
Libytheana carinenta							●				●				●							
Ligusticum canadense																						
Ligustrum japonicum							❀	❀							❀							
Ligustrum sinense		❀				❀	❀	❀			❀				❀				❀		❀	
Ligustrum sp.																						
Ligustrum vulgare																						
Lilium michauxii																						

	Hampton	Horry	Jasper	Kershaw	Lancaster	Laurens	Lee	Lexington	Marlboro	McCormick	Newberry	Oconee	Orangeburg	Pickens	Richland	Spartanburg	Sumter	Union	York	Georgia	N. Carolina
Lespedeza repens							❀		❀												
Lespedeza sp.										❀		❀									
Lestes sp.				●																	
Lestes vidua																					
Lestes vigilax							●														
Letrouitia domingense	*				*		*														
Letrouitia sp.	*																				
Letrouitia vulpinum																					
Leucauge venusta		●										●		●	●		●		●		
Leucojum aestivum																					
Leucothoe axillaris		❀					❀		❀			❀		❀		❀	❀				❀
Leucothoe axillaris axillaris	❀														❀						
Leucothoe axillaris editorum												❀				❀			❀		❀
Leucothoe populifolia																					
Leucothoe racemosa		❀		❀				❀									❀				
Leucothoe recurva								❀												❀	❀
Leucothoe sp.																					❀
Leurognathus marmoratus												●									●
Liatris elegans	❀																				
Liatris graminifolia				❀								❀				❀					
Liatris secunda					❀																
Liatris sp.			❀																		
Liatris spicata							❀														
Liatris spicata resinosa							❀														
Libellula auripennis					●				●				●								
Libellula axilena																					
Libellula exusta deplanata																				●	
Libellula incesta									●		●		●				●				
Libellula luctuosa											●		●								
Libellula lydia																				●	
Libellula pulchella					●																
Libellula semifasciata		●					●														
Libellula sp.																	●				
Libellula vibrans							●		●						●				●		
Libinia sp.																					
Libytheana carinenta						●			●		●	●			●						
Ligusticum canadense						❀															
Ligustrum japonicum								❀													
Ligustrum sinense		❀		❀					❀	❀			❀	❀		❀		❀	❀	❀	
Ligustrum sp.				❀											❀				❀		
Ligustrum vulgare		❀																			
Lilium michauxii														❀							

Scientific Name Index

	Abbeville	Aiken	Allendale	Anderson	Bamberg	Barnwell	Beaufort	Berkeley	Charleston	Cherokee	Chester	Chesterfield	Clarendon	Colleton	Darlington	Dillon	Dorchester	Fairfield	Florence	Georgetown	Greenville	Greenwood	Hampton	Horry	Jasper	Kershaw	Lancaster	Laurens	Lee	Lexington	Marlboro	McCormick	Newberry	Oconee	Orangeburg	Pickens	Richland	Spartanburg	Sumter	Union	York	Georgia	N. Carolina
Lilium superbum																																											❀
Limenitis archippus								●												●				●										●	●			●					
Limenitis arthemis astyanax	●	●	●					●			●			●		●					●	●	●	●		●	●	●		●			●	●	●	●	●	●			●	●	
Limnaoedus ocularis							●	●																																			
Limnius latiusculus																																											●
Limnobaris denudata																							●																				
Limnobaris prolixa																																											●
Limnobium spongia														❀																													
Limnothlypis swainsonii																																		●									
Limonium carolinianum							❀		❀																																		
Limonium nashii							❀		❀																																		
Limonium sp.							❀													❀																							
Limonius sp.																		●																					●		●		
Limosa fedoa									●																																		
Limulus polyphemus									●																																		
Linaria canadensis			❀						❀				❀		❀						❀			❀			❀					❀					❀		❀				
Linaria sp.																											❀				❀											❀	
Lindera benzoin		❀						❀			❀																❀					❀	❀	❀	❀					❀	❀	❀	❀
Lindera melissaefolium								❀																																			
Lindernia dubia																															❀			❀									
Lindernia monticola																											❀																
Lippia nodiflora							❀		❀											❀				❀																			
Liquidambar styraciflua	❀	❀	❀	❀	❀	❀	❀	❀	❀	❀	❀	❀	❀	❀	❀	❀	❀	❀	❀	❀	❀	❀	❀	❀	❀	❀	❀	❀	❀	❀	❀	❀	❀	❀	❀	❀	❀	❀	❀	❀	❀	❀	
Liriodendron tulipifera	❀	❀		❀		❀	❀	❀		❀	❀	❀		❀	❀						❀	❀	❀	❀		❀	❀	❀	❀	❀	❀	❀	❀	❀		❀	❀	❀	❀	❀	❀	❀	❀
Listera australis							❀																																				
Lithophane antennata																																				●							
Lithospermum tuberosum																																❀											
Litsea aestivalis							❀	❀																		❀																	
Littorina irrorata							●		●											●																							
Lobaria pulmonaria																											*									*							*
Lobaria quercizans																											*				*												*
Lobaria ravenelii																											*													*			
Lobaria sp.								*																										*									
Lobaria tenuis																											*																*
Lobelia amoena																																		❀									
Lobelia boykinii								❀																					❀														
Lobelia cardinalis		❀			❀	❀		❀			❀			❀										❀			❀		❀		❀			❀									
Lobelia elongata																													❀														
Lobelia glandulosa														❀																													
Lobelia inflata																																		❀		❀							❀
Lobelia nuttallii								❀				❀				❀								❀					❀														
Lobelia puberula																					❀						❀							❀		❀		❀					

Scientific Name Index

	Abbeville	Aiken	Allendale	Anderson	Bamberg	Barnwell	Beaufort	Berkeley	Charleston	Cherokee	Chester	Chesterfield	Clarendon	Colleton	Darlington	Dillon	Dorchester	Fairfield	Florence	Georgetown	Greenville	Greenwood	Hampton	Horry	Jasper	Kershaw	Lancaster	Laurens	Lee	Lexington	Marlboro	McCormick	Newberry	Oconee	Orangeburg	Pickens	Richland	Spartanburg	Sumter	Union	York	Georgia	N. Carolina
Lobelia sp.								❀				❀																															
Loligo sp.																				●																							
Lonicera fragrantissima		❀																																									
Lonicera japonica	❀	❀					❀	❀	❀		❀			❀	❀				❀		❀			❀	❀	❀	❀	❀	❀		❀	❀	❀	❀		❀	❀	❀	❀	❀	❀	❀	❀
Lonicera sempervirens								❀	❀										❀					❀			❀								❀	❀	❀						
Lophodytes cucullatus							●		●					●				●																									
Lordithon pygmaeus																								●																			
Lordithon sp.																			●																								
Loxandrus minor																																						●					
Lucanus elaphus																						●																					
Lucidota atra		●						●																																			
Lucidota sp.																																				●							
Ludwigia alternifolia							❀		❀						❀														❀	❀								❀					
Ludwigia decurrens						❀					❀																		❀	❀	❀										❀		
Ludwigia leptocarpa																❀								❀																			
Ludwigia linearis																													❀														
Ludwigia palustris																													❀		❀												
Ludwigia pilosa																❀								❀					❀														
Ludwigia repens																											❀																
Ludwigia sp.		❀				❀					❀			❀										❀														❀					
Ludwigia sphaerocarpa																															❀												
Ludwigia suffruticosa		❀																																									
Ludwigia uruguayensis								❀																														❀					
Ludwigia virgata								❀																																			
Luperodes cyanellus																																●										●	●
Luperodes sp.												●																						●									
Lupinus diffusus												❀														❀																	
Lupinus perennis																											❀																
Lupinus sp.																				❀																							
Lutra canadensis							●		●					●																							●						
Luzula acuminata																																		❀								❀	❀
Luzula echinata		❀								❀					❀												❀									❀							
Luzula sp.																		❀														❀					❀						
Lycaenopsis argiolus																																		●			●						●
Lycogala epidendrum								*																														*					
Lycomorpha pholus																																										●	
Lycoperdon echinatum																											*																
Lycoperdon perlatum																					*						*											*					*
Lycoperdon pusillum																								*																			
Lycoperdon pyriforme																																		*									
Lycoperdon sp.																				*										*													
Lycopodium adpressum								❀				❀												❀					❀														

Scientific Name Index

	Abbeville	Aiken	Allendale	Anderson	Bamberg	Barnwell	Beaufort	Berkeley	Charleston	Cherokee	Chester	Chesterfield	Clarendon	Colleton	Darlington	Dillon	Dorchester	Fairfield	Florence	Georgetown	Greenville	Greenwood	Hampton	Horry	Jasper	Kershaw	Lancaster	Laurens	Lee	Lexington	Marlboro	McCormick	Newberry	Oconee	Orangeburg	Pickens	Richland	Spartanburg	Sumter	Union	York	Georgia	N. Carolina
Lycopodium carolinianum																										❀																	
Lycopodium clavatum																																											❀
Lycopodium flabelliforme										❀	❀										❀												❀	❀				❀			❀	❀	❀
Lycopodium lucidulum																					❀													❀									❀
Lycopodium obscurum																																											❀
Lycopodium sp.														❀										❀																			
Lycopus rubellus																								❀																			
Lycopus sp.																													❀														
Lycopus virginicus																❀											❀							❀									
Lycosa carolinensis			●									●																	●								●						
Lycosa gulosa																											●																
Lycosa rabida												●						●			●			●										●				●					
Lycosa sp.							●				●									●												●					●				●		
Lygodium japonicum														❀																													
Lygodium palmatum																																					❀						
Lyonia ferruginea							❀																																				
Lyonia ligustrina												❀			❀									❀					❀					❀									❀
Lyonia lucida		❀				❀	❀	❀	❀			❀	❀		❀	❀			❀	❀				❀		❀			❀	❀									❀				
Lyonia mariana								❀				❀			❀	❀								❀					❀														
Lyropecten nodosus									●																																		
Lysimachia nummularia																												❀				❀								❀			
Lysimachia quadrifolia	❀																				❀													❀		❀						❀	
Lytta aenea																																											●
Macrodactylus angustatus																																			●			●					
Macromia sp.														●																													
Macromia taeniolata																																					●						
Magicicada sp.						●																																					
Magnolia acuminata	❀																				❀						❀	❀						❀		❀		❀					❀
Magnolia fraseri																					❀													❀		❀						❀	❀
Magnolia grandiflora		❀				❀	❀	❀	❀					❀	❀								❀			❀						❀											
Magnolia tripetala	❀																																										❀
Magnolia virginiana		❀				❀	❀	❀	❀			❀	❀	❀	❀	❀							❀	❀		❀			❀	❀	❀						❀		❀				
Maianthemum canadense																																											❀
Malacasoma sp.																							●											●							●		
Malachius sp.																											●					●											
Malaclemys terrapin									●																																		
Malacosoma americanum				●															●		●						●			●							●					●	
Malacosoma disstria																																					●			●		●	
Maladera castanea																																											●
Malaxis unifolia								❀																																			
Malthodes curvatus																																										●	
Malthodes sp.																																●											

Scientific Name Index

	Abbeville	Aiken	Allendale	Anderson	Bamberg	Barnwell	Beaufort	Berkeley	Charleston	Cherokee	Chester	Chesterfield	Clarendon	Colleton	Darlington	Dillon	Dorchester	Fairfield	Florence	Georgetown	Greenville	Greenwood	Hampton	Horry	Jasper	Kershaw	Lancaster	Laurens	Lee	Lexington	Marlboro	McCormick	Newberry	Oconee	Orangeburg	Pickens	Richland	Spartanburg	Sumter	Union	York	Georgia	N. Carolina
Malus angustifolia																										❀						❀					❀						
Malus pumila															❀																												❀
Marasmius rotula																																					*						
Marasmius siccus																																					*	*					
Marasmius sp.																					*													*									
Margarinotus sp.																								●																			
Marmota monax																																											●
Maronea constans		*							*		*	*		*	*			*	*		*			*			*		*						*				*				
Marshallia graminifolia												❀																															
Marshallia obovata																								❀																			
Matelea gonocarpa		❀									❀																																
Matelea obliqua																																						❀					
Matelea sp.	❀	❀									❀										❀		❀				❀					❀			❀	❀	❀	❀			❀		
Matelea suberosa									❀																																		
Mayaca aubletii												❀														❀					❀												
Mayaca fluviatilis																																							❀				
Mayaca sp.		❀													❀																												
Mazus japonicus															❀																												
Mazus sp.																																					❀						
Mecardonia acuminata																																		❀									
Mecynogea lemniscata						●	●					●		●										●		●								●									
Medeola virginiana						❀					❀										❀													❀		❀			❀			❀	❀
Medicago arabica									❀																																		
Medicago lupulina									❀																																		
Medusulina nitida								*	*														*				*																
Megachile sp.																												●															
Megacyllene robiniae																																											●
Megalodacne fasciata											●																																
Megalodacne heros	●																																										
Megalopyge opercularis							●	●				●																		●													
Megarhyssa sp.																																										●	
Megathymus yuccae																										●																	
Megisto cymela		●																																									
Meibomeus musculus												●																															
Melampyrum lineare																																		❀									
Melanaria macounii																											*																
Melanelia exasperata																					*													*									*
Melanelia halei											*																															*	*
Melanelia stygia																																											*
Melanelia subaurifera																																											*
Melanerpes carolinus	●	●	●		●	●	●	●	●		●	●		●	●			●	●	●	●			●	●		●	●	●		●	●	●		●		●	●	●				
Melanerpes erythrocephalus		●						●	●		●									●			●	●					●						●		●						

Scientific Name Index

	Abbeville	Aiken	Allendale	Anderson	Bamberg	Barnwell	Beaufort	Berkeley	Charleston	Cherokee	Chester	Chesterfield	Clarendon	Colleton	Darlington	Dillon	Dorchester	Fairfield	Florence	Georgetown	Greenville	Greenwood	Hampton	Horry	Jasper	Kershaw	Lancaster	Laurens	Lee	Lexington	Marlboro	McCormick	Newberry	Oconee	Orangeburg	Pickens	Richland	Spartanburg	Sumter	Union	York	Georgia	N. Carolina
Melanomma sp.														*	*																												
Melanoplus differentialis		●																																●				●					
Melanoplus femurrubrum									●															●										●	●			●					
Melanoplus sp.	●	●			●						●	●			●																			●		●						●	
Melanotheca cruenta								*																																			
Melanthera hastata								❀																																			
Melaphis rhois																																				●							
Melaspilea octomera																													*														
Melaspilea sp.																													*														
Meleagris gallopavo		●					●		●		●			●														●				●		●			●		●			●	
Melia azedarach							❀		❀						❀									❀	❀				❀		❀						❀						
Melica mutica		❀																			❀						❀					❀		❀		❀	❀						❀
Melilotus indica								❀																																			
Mellita carolinana																								❖																			
Mellita quinquiesperforata									●																																		
Meloe angusticollis																																											●
Meloe sp.																																											●
Melospiza georgiana			●						●					●				●																									
Melospiza melodia				●			●		●					●						●					●																●		●
Melothria pendula							❀																						❀				❀										
Melyrodes cribrata																											●																
Menegrazzia terebrata																																											*
Menemerus bivittatus																											●																
Menidia menidia									●																																		
Menippe mercenaria									●																																		
Menispermum canadense																											❀																
Mercenaria campechiensis																								❖																			
Mercenaria mercenaria									●																																		❖
Mercenaria sp.																								❖																			
Mergus serrator							●		●																																		
Mermiria intertexta												●																															
Merulius incarnatus																																						*					
Mesodon normalis																					●													●		●						●	●
Mesodon thyroidus		●						●												●														●		●				●			●
Mesodon zaletus																																											●
Mesomphix cupreus		●																																									
Mesomphix vulgatus	●																																										
Metachroma longicollis		●																																									
Metachroma pallidum																									●																		
Metaleptea brevicornis		●												●																								●					
Metaphidippus galathea																										●																	
Metaphidippus sp.						●																		●																			

Scientific Name Index

	Abbeville	Aiken	Allendale	Anderson	Bamberg	Barnwell	Beaufort	Berkeley	Charleston	Cherokee	Chester	Chesterfield	Clarendon	Colleton	Darlington	Dillon	Dorchester	Fairfield	Florence	Georgetown	Greenville	Greenwood	Hampton	Horry	Jasper	Kershaw	Lancaster	Laurens	Lee	Lexington	Marlboro	McCormick	Newberry	Oconee	Orangeburg	Pickens	Richland	Spartanburg	Sumter	Union	York	Georgia	N. Carolina
Metarranthis obfirmaria																											●																
Metaxaglaea semitaria																																							●				
Metriona bicolor		●																																							●		
Mezira sp.																											●																
Micarea chlorosticta							*																																				
Micranthemum umbrosum						❀		❀						❀									❀																				
Micrathena gracilis											●											●		●											●						●		
Micrathena mitrata	●																				●						●									●		●					
Micrathena sagittata		●									●	●															●								●	●		●					
Microcentrum retinerve																																		●									
Microcentrum rhombifolium																											●																
Microglossum rufum																													*														
Micropterus salmoides								●			●									●																							
Microstegium vimineum		❀									❀										❀						❀	❀					❀	❀		❀	❀	❀			❀		
Mikania scandens		❀				❀	❀	❀	❀					❀	❀	❀						❀	❀	❀			❀				❀		❀				❀	❀					
Milesia virginiensis																																		●									
Mimulus alatus		❀																																									
Mimulus ringens																																		❀									
Mimus polyglottos							●	●	●			●		●	●				●	●		●			●	●	●		●		●		●		●			●	●		●		
Mioptachys flavicauda																											●																
Misumena sp.																																		●									
Misumena vatia																								●																			
Misumenoides formosipes		●																																								●	
Misumenops asperatus														●																													
Mitchella repens	❀	❀	❀		❀	❀	❀		❀	❀	❀			❀	❀		❀		❀	❀	❀		❀	❀			❀		❀	❀		❀	❀	❀	❀	❀	❀	❀	❀		❀		❀
Mitrula paludosa																					*													*									
Mniotilta varia		●					●	●						●				●			●			●			●					●		●								●	●
Modiola caroliniana																																❀											
Molorchus bimaculatus																																●											
Molorchus sp.													●																			●											
Molothrus ater																		●						●			●		●							●							
Monarda citriodora																																											❀
Monarda clinopodia																																		❀		❀							❀
Monarda didyma																																											❀
Monarda punctata		❀							❀												❀			❀		❀																	
Monocesta coryli											●																																
Monotropa hypopithys	❀																				❀													❀		❀		❀				❀	
Monotropa uniflora							❀									❀	❀			❀	❀			❀			❀			❀				❀		❀						❀	❀
Montezumina modesta											●																																
Morchella esculenta																																*				*							
Mordella atrata									●																														●				
Mordella melaena	●							●	●						●						●								●														

	Abbeville	Aiken	Allendale	Anderson	Bamberg	Barnwell	Beaufort	Berkeley	Charleston	Cherokee	Chester	Chesterfield	Clarendon	Colleton	Darlington	Dillon	Dorchester	Fairfield	Florence	Georgetown	Greenville	Greenwood	Hampton	Horry	Jasper	Kershaw	Lancaster	Laurens	Lee	Lexington	Marlboro	McCormick	Newberry	Oconee	Orangeburg	Pickens	Richland	Spartanburg	Sumter	Union	York	Georgia	N. Carolina
Mordella sp.																																										●	
Mordellistena ambusta									●																																		
Mordellistena fuscata		●																																									
Mordellistena pubescens															●									●																			
Mordellistena sp.																																				●		●					
Mordellistena unicolor																																							●				
Mormidea sp.								●																																			
Morus rubra	❀	❀				❀		❀	❀		❀			❀	❀				❀	❀		❀	❀	❀	❀	❀	❀		❀		❀		❀		❀		❀	❀			❀		
Mulinia lateralis									●															❖																			
Musca domestica	●											●																															
Mustela vison									●																																		
Mutinus caninus																																		✻				✻					
Mutinus elegans	✻																										✻																
Mycena acicula																											✻					✻											
Mycena galericulata									✻																																		
Mycena inclinata																																						✻					
Mycena leaiana																																					✻						
Mycena maculata																					✻																						
Mycena vulgaris																																						✻					
Mycenastrum corium															✻																												
Mycodiplosis clavula											●				●																												
Mycoporum compositum																																✻				✻							
Mycoporum pyrenocarpum							✻								✻																												
Mycoporum sp.																							✻																				
Mycteria americana		●												●																													
Mycterus scaber																																						●					
Myiarchus crinitus		●						●	●						●					●				●					●				●	●	●	●	●					●	●
Myliobatis sp.									❖								❖																										
Myosotis macrosperma																												❀															
Myosotis verna																		❀																									
Myosurus minimus												❀																															
Myrica cerifera		❀	❀			❀	❀	❀	❀				❀	❀	❀	❀	❀		❀	❀			❀	❀	❀	❀			❀	❀							❀		❀			❀	
Myrica cerifera cerifera																			❀																❀								
Myrica cerifera pumila																								❀																			
Myrica heterophylla		❀				❀						❀			❀									❀					❀										❀				
Myriophyllum brasiliense		❀													❀							❀								❀							❀						
Myriophyllum heterophyllum																										❀																	
Myriophyllum pinnatum														❀									❀																				
Myriophyllum sp.												❀			❀																												
Myrmeleon immaculatum																				●						●																	
Myrmeleon sp.									●																																		
Nadata gibbosa																																				●							

Scientific Name Index

	Abbeville	Aiken	Allendale	Anderson	Bamberg	Barnwell	Beaufort	Berkeley	Charleston	Cherokee	Chester	Chesterfield	Clarendon	Colleton	Darlington	Dillon	Dorchester	Fairfield	Florence	Georgetown	Greenville	Greenwood	Hampton	Horry	Jasper	Kershaw	Lancaster	Laurens	Lee	Lexington	Marlboro	McCormick	Newberry	Oconee	Orangeburg	Pickens	Richland	Spartanburg	Sumter	Union	York	Georgia	N. Carolina
Naematoloma fasciculare																													*														
Najas guadalupensis																							❀															❀					
Najas minor																																						❀					
Nandina domestica															❀																												
Narceus americanus																																		●									
Narcissus pseudo-narcissus							❀																																				
Narcissus sp.							❀		❀																																		
Narcissus tazetta x poeticus							❀																																				
Neacoryphus bicrucis																											●																
Necturus punctatus																											●																
Negaprion brevirostris									❖																																		
Nelumbo lutea								❀																																			
Nemocestes sp.											●																																
Nemognatha piazata																							●																				
Nemophila microcalyx		❀																														❀											
Nemotarsus elegans		●																																									
Neoconcephalus sp.																											●																
Neoconocephalus retesus		●																																									
Neonympha areolata																																	●										
Neophylax mitchelli																																		●									
Neoscona arabesca		●																			●																		●				
Neoscona domiciliorum		●									●	●		●																													
Neoscona hentzi						●						●												●																			
Neoscona sp.	●																										●									●							
Nephila clavipes									●																						●												
Nephroma helveticum																											*							*		*							
Nerodia erythrogaster																													●				●				●						
Nerodia fasciata						●						●		●	●					●				●													●						
Nerodia sipedon																											●																
Nerodia taxispilota														●																									●				
Nestronia umbellula																											❀									❀							
Neuroterus floccus																																				●							
Neuroterus minutus																					●																						
Neuroterus sp.																					●																						
Nezara viridula						●		●																																			
Nicrophorus sp.																																		●									
Nigronia serricornis																																		●									
Nodonota sp.																											●																
Nodonota tristis																																										●	
Noetia limula																																											❖
Noetia ponderosa									●																																		
Normandina pulchella		*																			*													*		*							

Scientific Name Index

	Abbeville	Aiken	Allendale	Anderson	Bamberg	Barnwell	Beaufort	Berkeley	Charleston	Cherokee	Chester	Chesterfield	Clarendon	Colleton	Darlington	Dillon	Dorchester	Fairfield	Florence	Georgetown	Greenville	Greenwood	Hampton	Horry	Jasper	Kershaw	Lancaster	Laurens	Lee	Lexington	Marlboro	McCormick	Newberry	Oconee	Orangeburg	Pickens	Richland	Spartanburg	Sumter	Union	York	Georgia	N. Carolina
Nossidium sp.																		●																									
Notonecta sp.		●																																					●				
Notophthalmus viridescens																					●													●		●		●					●
Nuphar luteum															❀		❀							❀		❀					❀												
Nyctocorax nyctocorax														●																													
Nymphaea odorata		❀				❀			❀			❀		❀		❀							❀	❀		❀					❀								❀				
Nymphalis antiopa										●							●											●					●				●					●	
Nymphoides aquatica																										❀																	
Nymphoides cordata		❀																								❀					❀												
Nyssa aquatica		❀						❀	❀					❀		❀	❀									❀											❀						
Nyssa sylvatica					❀			❀	❀	❀	❀				❀	❀	❀				❀	❀	❀	❀		❀	❀	❀	❀		❀	❀	❀	❀		❀	❀	❀					❀
Nyssa sylvatica biflora		❀				❀	❀	❀			❀	❀		❀	❀				❀	❀			❀	❀		❀			❀		❀						❀		❀				
Nyssa sylvatica sylvatica	❀	❀				❀		❀			❀	❀		❀	❀				❀	❀	❀		❀			❀	❀		❀					❀		❀	❀		❀		❀	❀	
Oberea basalis																																										●	
Obolaria virginica																																❀		❀									
Ochrolechia africana		*				*	*	*	*		*	*		*	*			*	*	*	*		*	*		*			*			*		*		*	*	*	*	*	*	*	*
Ochrolechia farinacea								*																																			
Ochrolechia mexicana																																										*	
Ochrolechia pallescens																							*																				
Ochrotomys nuttalli										●											●																						
Ocypode quadrata							●		●																																		
Odocoileus virginianus	●	●	●	●			●	●	●	●	●	●		●	●			●	●				●	●	●	●	●	●	●					●	●		●	●	●	●	●	●	●
Odontaspis faurus																				●																							
Odontaspis hopei									❖								❖																										
Odontaspis sp.																								❖																			
Odontaspis taurus									❖																																		
Odontocorynus pinguescens																																						●					
Odontocorynus pulverulentus		●																																									
Odontocorynus scutellum-album		●																					●																				
Odontonyx sp.							●																																				
Oecanthus fultoni		●																																									
Oedionychis concinna																																					●						
Oedionychis miniata																											●																
Oedionychis quercata																											●					●											
Oedionychis sp.											●																																●
Oenothera biennis																					❀								❀					❀		❀							❀
Oenothera fruticosa																													❀														
Oenothera laciniata																				❀																							
Oenothera sp.																					❀									❀													
Oldenlandica uniflora																❀													❀														
Olibrus sp.		●																																●									
Oliva carolinensis																								❖																			❖

Scientific Name Index

	Abbeville	Aiken	Allendale	Anderson	Bamberg	Barnwell	Beaufort	Berkeley	Charleston	Cherokee	Chester	Chesterfield	Clarendon	Colleton	Darlington	Dillon	Dorchester	Fairfield	Florence	Georgetown	Greenville	Greenwood	Hampton	Horry	Jasper	Kershaw	Lancaster	Laurens	Lee	Lexington	Marlboro	McCormick	Newberry	Oconee	Orangeburg	Pickens	Richland	Spartanburg	Sumter	Union	York	Georgia	N. Carolina
Oliva sayana									●											●				❖																			
Oliva sp.									●																																		
Olivella carolinae																								❖																			
Olivella mutica																								❖																			
Omalium sp.																								●																			
Omophron americanum																																		●									
Oncideres cingulata							●		●			●		●	●				●											●		●						●		●			
Ondatra zibethicus															●			●																									
Onoclea sensibilis		❀	❀		❀			❀	❀						❀								❀	❀										❀			❀						❀
Ontholestes sp.																																											●
Onthophagus hecate									●				●											●																			
Onthophagus janus																								●																			
Onthophagus orpheus			●																																								
Onthophagus orpheus orpheus																																										●	
Onthophagus sp.												●																															
Onthophagus tuberculifrons																											●										●		●				
Oodes amaroides												●																															
Opegrapha astraea											*																													*			
Opegrapha bonplandi																																					*						
Opegrapha sp.																							*																				
Opheodrys aestivus													●			●																					●						
Ophisaurus sp.							●																																				
Oplismenus setarius		❀					❀		❀																❀										❀								
Oporornis formosus																																		●									
Opuntia compressa		❀			❀		❀		❀			❀	❀		❀	❀	❀		❀		❀			❀	❀	❀	❀			❀	❀					❀		❀	❀				
Opuntia drummondii							❀																				❀															❀	
Opuntia sp.									❀																																		
Orchelimum sp.		●										●																															
Orchelimum volantum		●																																									
Orchelimum vulgare						●																		●		●			●						●								
Orchestia grillus									●																																		
Orchis spectabilis																																		❀		❀							❀
Orobanche uniflora																					❀																						
Orontium aquaticum		❀			❀			❀				❀		❀												❀																	
Orphulella pelidna												●																															
Orsodacne atra																																										●	
Orthemis ferruginea									●																																		
Osmanthus americana						❀	❀		❀								❀						❀																				
Osmorhiza claytonii																																											❀
Osmorhiza longistylis		❀																														❀					❀						❀
Osmunda cinnamomea		❀			❀	❀	❀	❀	❀			❀	❀	❀	❀	❀			❀	❀			❀	❀			❀		❀	❀				❀	❀	❀	❀		❀		❀	❀	❀
Osmunda regalis spectabilis		❀				❀	❀	❀	❀			❀		❀	❀				❀		❀		❀			❀			❀					❀			❀				❀		

Scientific Name Index

	Abbeville	Aiken	Allendale	Anderson	Bamberg	Barnwell	Beaufort	Berkeley	Charleston	Cherokee	Chester	Chesterfield	Clarendon	Colleton	Darlington	Dillon	Dorchester	Fairfield	Florence	Georgetown	Greenville	Greenwood
Ostrea lawrencei																						
Ostrea sp.																						
Ostrya virginiana	❀		❀				❀											❀				
Otus asio									●													
Ovalipes ocellatus									●													
Owenia fusiformis									●													
Oxacis sp.								●														
Oxalis acetosella																						
Oxalis dillenii							❀															
Oxalis florida		❀												❀								
Oxalis grandis																					❀	
Oxalis sp.																		❀				
Oxalis violacea															❀							
Oxycopis sp.																						
Oxydendrum arboreum	❀	❀								❀	❀	❀			❀						❀	
Oxypoda sp.													●									
Oxypolis filiformis																						
Oxypolis rigidior																					❀	
Oxypolis sp.																						
Oxypolis ternata												❀										
Oxyporus femoralis																						
Oxyporus sp.																			●			
Oxytelus sp.																						
Oxyura jamaicensis																						
Pachybrachis sp.																						
Pachydiplax longipennis		●						●	●		●			●		●				●		
Pachyospora sp																						
Pachypsylla venusta				●																		
Pachypsylla vesiculum											●											
Pachyschelus laevigatus		●			●				●		●					●						
Pagurus pollicaris									●													
Pagurus sp.									●													
Pallodes sp.	●	●							●							●						
Panax quinquefolium																						
Pandeleteius hilaris					●				●													
Pandeleteius sp.																					●	
Pandion haliaetus							●	●	●		●	●		●						●		
Panellus stipticus								✻														
Panicum amarum							❀		❀											❀		
Panicum boseii																						
Panicum sp.											❀			❀								
Panicum verrucosum												❀										

	Hampton	Horry	Jasper	Kershaw	Lancaster	Laurens	Lee	Lexington	Marlboro	McCormick	Newberry	Oconee	Orangeburg	Pickens	Richland	Spartanburg	Sumter	Union	York	Georgia	N. Carolina
Ostrea lawrencei		❖																			
Ostrea sp.		❖																			
Ostrya virginiana					❀					❀					❀	❀	❀				
Otus asio							●							●							
Ovalipes ocellatus																					
Owenia fusiformis																					
Oxacis sp.																					
Oxalis acetosella																❀					
Oxalis dillenii																					
Oxalis florida																					
Oxalis grandis												❀		❀							
Oxalis sp.														❀	❀		❀			❀	
Oxalis violacea					❀					❀					❀						
Oxycopis sp.		●																			
Oxydendrum arboreum					❀	❀	❀		❀	❀	❀	❀		❀	❀	❀	❀	❀	❀	❀	❀
Oxypoda sp.															●						
Oxypolis filiformis							❀														
Oxypolis rigidior														❀						❀	❀
Oxypolis sp.							❀														
Oxypolis ternata																					
Oxyporus femoralis																●					
Oxyporus sp.																					
Oxytelus sp.		●																			
Oxyura jamaicensis			●																		
Pachybrachis sp.					●															●	
Pachydiplax longipennis		●		●			●		●		●		●								
Pachyospora sp																	✻				
Pachypsylla venusta																			●		
Pachypsylla vesiculum																					
Pachyschelus laevigatus																	●				
Pagurus pollicaris																					
Pagurus sp.																					
Pallodes sp.												●									
Panax quinquefolium						❀						❀									
Pandeleteius hilaris														●							
Pandeleteius sp.					●																
Pandion haliaetus					●							●	●								
Panellus stipticus																					
Panicum amarum																					
Panicum boseii																❀					
Panicum sp.		❀					❀					❀									
Panicum verrucosum	❀			❀																	

	Abbeville	Aiken	Allendale	Anderson	Bamberg	Barnwell	Beaufort	Berkeley	Charleston	Cherokee	Chester	Chesterfield	Clarendon	Colleton	Darlington	Dillon	Dorchester	Fairfield	Florence	Georgetown	Greenville	Greenwood	Hampton	Horry	Jasper	Kershaw	Lancaster	Laurens	Lee	Lexington	Marlboro	McCormick	Newberry	Oconee	Orangeburg	Pickens	Richland	Spartanburg	Sumter	Union	York	Georgia	N. Carolina
Panicum virgatum							❀																																				
Pannaria lurida																																		*									
Pannaria rubiginosa																											*					*		*									
Pannaria sp.																																										*	
Pannaria tavaresii																											*							*		*						*	*
Panopeus sp.							●																																				
Pantala flavescens		●							●																																		
Papaipema appassionata								●																																			
Papilio cresphontes																				●																							
Papilio glaucus	●		●		●						●	●	●	●	●	●		●		●	●			●		●	●		●		●	●		●	●	●	●	●	●		●	●	●
Papilio palamedes								●	●							●				●				●					●								●	●	●				
Papilio polyxenes asterius		●			●				●					●									●	●										●									
Papilio troilus		●									●	●	●	●		●					●	●		●			●				●	●			●						●	●	●
Paracymus subcupreus																													●														
Paragnetina immarginata																																		●									
Pardalophora phoenicoptera																																				●							
Pardosa sp.			●																																								
Paria canella aterrimus																											●										●						
Paria canella gilvipes																												●	●														
Paria canella quadrinotata																											●					●											
Paria sellatus								●																																			
Paria thoracica														●																													
Parmelia appalachiensis																																											*
Parmelia halei																																											*
Parmelia omphalodes																																											*
Parmelia rudecta																							*	*	*				*		*			*			*	*		*		*	
Parmelia saxatilis																																		*									*
Parmelia sp.																					*		*		*																		*
Parmelia squarrosa																								*										*									*
Parmelia sulcata										*														*							*												*
Parmeliella pannosa																											*							*									
Parmelina aurulenta	*	*						*	*		*	*		*				*	*	*	*		*	*		*			*			*		*		*				*	*		
Parmelina dissecta		*						*							*									*		*			*									*		*			*
Parmelina galbina	*						*		*		*	*			*								*	*										*		*							*
Parmelina horrescens	*					*	*		*		*	*		*	*						*			*		*			*					*		*			*		*		*
Parmelina horrescens asidiata															*																			*							*		
Parmelina minarum	*	*					*		*		*	*		*	*				*		*		*											*	*	*			*		*	*	*
Parmelina obsessa																																						*			*		
Parmeliopsis aleurites																																											*
Parmeliopsis ambigua		*																																									
Parmeliopsis subambigua							*	*	*														*			*													*				*
Parmotrema arnoldii																					*																						

Scientific Name Index

	Abbeville	Aiken	Allendale	Anderson	Bamberg	Barnwell	Beaufort	Berkeley	Charleston	Cherokee	Chester	Chesterfield	Clarendon	Colleton	Darlington	Dillon	Dorchester	Fairfield	Florence	Georgetown	Greenville	Greenwood	Hampton	Horry	Jasper	Kershaw	Lancaster	Laurens	Lee	Lexington	Marlboro	McCormick	Newberry	Oconee	Orangeburg	Pickens	Richland	Spartanburg	Sumter	Union	York	Georgia	N. Carolina
Parmotrema austrosinense																															*			*		*						*	
Parmotrema cetratum								*															*	*		*					*	*		*		*			*	*			*
Parmotrema chinense																					*																	*					
Parmotrema crinitum						*		*	*												*		*			*			*			*		*				*				*	*
Parmotrema crinitum f. ultralucens														*																									*				
Parmotrema cristiferum								*																																			
Parmotrema dilitatum		*						*	*									*													*			*			*	*					
Parmotrema eurysacum																										*								*									
Parmotrema haitiense												*																						*									
Parmotrema hypoleucinum																				*																							
Parmotrema hypotropum	*	*					*	*	*	*	*	*		*	*			*			*			*		*			*		*			*			*	*	*	*	*	*	
Parmotrema mellissii						*						*		*							*					*			*					*		*	*	*					
Parmotrema michauxianum	*	*					*	*	*		*	*		*	*			*			*		*	*		*			*					*	*		*	*	*		*	*	*
Parmotrema michauxianum roseum														*															*														
Parmotrema perforatum	*	*			*		*	*	*	*	*	*		*	*			*	*	*			*	*	*	*			*	*	*	*			*	*		*	*	*	*	*	*
Parmotrema praesorediosum		*					*	*	*					*					*	*			*		*	*			*			*		*	*								
Parmotrema rampoddense		*				*	*	*	*					*					*				*	*	*	*											*	*	*				
Parmotrema reticulatum						*	*	*							*								*	*	*	*	*				*	*		*		*	*	*				*	*
Parmotrema rigidum																				*					*																		
Parmotrema robustum		*					*							*	*			*		*				*	*	*			*					*									*
Parmotrema sp.																										*																	
Parmotrema stuppeum							*																						*													*	
Parmotrema subisidiosum		*				*	*	*							*											*						*					*						*
Parmotrema subtinctorium		*					*	*				*			*									*		*			*					*	*					*			
Parmotrema sulphuratum																																		*									
Parmotrema tinctorum		*					*	*	*					*					*	*	*		*	*		*			*					*					*	*		*	*
Parmotrema ultralucens						*			*					*					*		*			*		*						*						*	*		*	*	
Parmotrema xanthinum		*						*			*				*						*					*			*									*	*				
Parnassia asarifolia																																											❀
Paronychia herniarioides																													❀														
Paronychia riparia																				❀																							
Paroxya clavuliger		●																																									
Parrhasius m-album																																											●
Parthenium integrifolium																																		❀		❀							
Parthenocissus quinquefolia	❀	❀	❀			❀	❀		❀		❀	❀			❀	❀				❀	❀	❀	❀	❀		❀	❀	❀	❀		❀	❀	❀	❀		❀	❀	❀	❀		❀	❀	❀
Parula americana		●	●					●	●					●				●						●			●	●				●	●	●	●		●		●				●
Parus bicolor		●	●		●		●	●	●		●		●	●	●		●	●			●			●			●	●				●	●	●	●	●	●	●	●	●			●
Parus carolinensis	●	●	●			●	●		●		●		●	●	●	●		●	●	●	●		●	●	●	●	●		●		●	●	●	●	●	●	●	●	●	●	●	●	●
Pasimachus depressus	●																										●												●				
Pasimachus sp.												●										●																					
Pasimachus sublaevis																	●																										
Paspalum dilatatum																																				❀							

Scientific Name Index

	Abbeville	Aiken	Allendale	Anderson	Bamberg	Barnwell	Beaufort	Berkeley	Charleston	Cherokee	Chester	Chesterfield	Clarendon	Colleton	Darlington	Dillon	Dorchester	Fairfield	Florence	Georgetown	Greenville	Greenwood	Hampton	Horry	Jasper	Kershaw	Lancaster	Laurens	Lee	Lexington	Marlboro	McCormick	Newberry	Oconee	Orangeburg	Pickens	Richland	Spartanburg	Sumter	Union	York	Georgia	N. Carolina
Paspalum floridanum																																						❀					
Paspalum sp.											❀																																
Passerculus sandwichensis							●							●						●														●									
Passerina ciris									●																																		
Passerina cyanea																						●	●	●			●		●					●		●			●			●	
Passiflora incarnata	❀							❀			❀	❀				❀				❀	❀	❀							❀		❀				❀			❀					
Passiflora lutea		❀						❀			❀												❀									❀			❀	❀	❀	❀			❀		
Paulownia tomentosa																																				❀							❀
Paxillus involutus																											*																
Pecten brouweri																								❖																			
Pecten sp.									❖															❖																			
Pedicularis canadensis																❀													❀					❀									❀
Pedilus collaris																																										●	
Pedilus sp.																																●											
Pelecanus occidentalis							●		●											●																							
Pelecinus polyturator																					●													●		●							
Pelidnota sp.											●																																
Pellaea atropurpurea																																		❀									
Peltandra virginica		❀				❀								❀	❀	❀								❀		❀			❀		❀						❀		❀				
Peltigera canina											*										*						*					*		*		*						*	*
Peltigera horizontalis																											*							*									
Peltigera polydactyla											*																*							*								*	
Peltodytes dietrichi									●																																		
Peltodytes duodecimpunctatus																																●											
Peltodytes muticus		●																																									
Peltodytes oppositus															●																												
Peltodytes sexmaculatus						●		●			●						●														●			●									
Peltodytes shermani		●				●					●																				●												
Peltodytes sp.		●									●					●								●							●												
Pemphigus populicaulis											●																																
Pemphigus populitransversus											●																														●		
Penstemon sp.																																		❀									
Periarchus lyelli																	❖																										
Perigaster sp.																●																											
Periplaneta fuliginosa									●																																		
Perithemis tenera	●	●						●	●					●							●	●				●					●		●		●				●				
Peromyscus gossypinus															●																												
Persea borbonia		❀	❀			❀	❀	❀	❀			❀	❀	❀	❀	❀	❀		❀	❀			❀	❀		❀			❀	❀							❀		❀			❀	
Persephona mediterranea									●																																		
Pertusaria amara											*	*															*							*									
Pertusaria caroliniana sp. nov.											*																*																
Pertusaria copiosa		*						*	*															*			*									*							

Scientific Name Index

	Abbeville	Aiken	Allendale	Anderson	Bamberg	Barnwell	Beaufort	Berkeley	Charleston	Cherokee	Chester	Chesterfield	Clarendon	Colleton	Darlington	Dillon	Dorchester	Fairfield	Florence	Georgetown	Greenville	Greenwood	Hampton	Horry	Jasper	Kershaw	Lancaster	Laurens	Lee	Lexington	Marlboro	McCormick	Newberry	Oconee	Orangeburg	Pickens	Richland	Spartanburg	Sumter	Union	York	Georgia	N. Carolina
Pertusaria globularis												*															*																
Pertusaria leucostoma							*								*			*					*	*			*					*		*			*						
Pertusaria macounii															*						*						*					*								*	*		
Pertusaria multipunctata																											*													*			
Pertusaria multipunctoides	*																		*								*							*			*						
Pertusaria neoscotica									*															*			*					*		*									*
Pertusaria ophthalmiza															*									*			*		*					*		*							
Pertusaria ostiolata		*				*	*	*	*		*			*	*				*		*		*	*		*	*		*			*		*	*	*	*		*		*		*
Pertusaria paratuberculifera		*				*	*	*	*		*				*				*		*		*	*		*	*		*			*		*	*	*		*	*	*	*		
Pertusaria plittiana								*																			*											*					*
Pertusaria propinqua							*		*			*															*									*						*	
Pertusaria pustulata												*			*			*									*													*			
Pertusaria rubefacta																					*						*																*
Pertusaria sinusmexicani		*					*		*						*								*	*			*							*					*				
Pertusaria sp.							*					*			*						*			*		*			*			*		*	*	*	*						
Pertusaria subpertusa		*				*	*		*		*			*	*				*				*	*					*			*		*	*	*		*			*		*
Pertusaria tetrathalmia		*							*						*								*			*								*	*	*			*				
Pertusaria texana	*	*				*	*	*	*		*			*	*			*	*		*		*	*		*			*			*		*	*	*		*	*		*	*	
Pertusaria trachythallina		*						*						*									*				*					*											
Pertusaria tuberculifera																											*							*									
Pertusaria valliculata								*																			*							*									
Pertusaria velata															*									*			*							*		*							*
Pertusaria waghornei																								*																			*
Pertusaria xanthodes																											*													*			
Petalostemum pinnatum																							❀																				
Petricola pholadiformis																				●																							
Peucetia viridans		●	●	●																				●		●			●									●			●		
Peziza badio-confusa																																											*
Phacelia dubia			❀																								❀					❀											
Phacelia dubia georgiana																																										❀	
Phacelia sp.																																											❀
Phacellia maculata																											❀																
Phaenonotum estriatum																																		●									
Phaeographina sp.																																*											
Phaeographis dendritica		*					*	*	*		*			*	*				*				*	*											*			*	*				
Phaeographis erumpens														*													*																
Phaeographis punctiformis							*																																				
Phaeographis sp.																							*																				
Phaeophyscia adiastola								*																										*				*					*
Phaeophyscia ciliata								*			*										*				*							*		*					*		*	*	
Phaeophyscia imbricata								*											*					*			*							*		*		*			*	*	*
Phaeophyscia pusilloides							*																											*					*	*	*		*

Scientific Name Index

	Abbeville	Aiken	Allendale	Anderson	Bamberg	Barnwell	Beaufort	Berkeley	Charleston	Cherokee	Chester	Chesterfield	Clarendon	Colleton	Darlington	Dillon	Dorchester	Fairfield	Florence	Georgetown	Greenville	Greenwood	Hampton	Horry	Jasper	Kershaw	Lancaster	Laurens	Lee	Lexington	Marlboro	McCormick	Newberry	Oconee	Orangeburg	Pickens	Richland	Spartanburg	Sumter	Union	York	Georgia	N. Carolina
Phaeophyscia rubropulchra											*				*						*					*						*		*	*	*	*	*	*	*	*		*
Phaeophyscia rubropulchra minor																																						*					
Phaeophyscia sciastra											*																																
Phalacrocorax auritus		●					●	●	●					●						●					●										●		●	●					
Phalaropus tricolor																		●																									
Phanaeus igneus																													●														
Phanaeus nigrocyaneus																								●																			
Phanaeus sp.																●																											
Phanaeus vindex																								●																			
Phasganophora capitata																																		●									
Phausis sp.																																		●									
Phellinus gilvus							*																											*					*				
Pheuticus ludovicianus																					●													●									
Phidippus cardinalis		●																																									
Phidippus pulcher																										●						●					●						
Philadelphus inodorus		❀									❀										❀						❀	❀				❀				❀							❀
Philaenus sp.		●																			●																						
Philothermus glabriculus																																		●									
Phlox amoena																					❀													❀									
Phlox carolina	❀																																										
Phlox drummondii																								❀																			
Phlox glaberrima																					❀																						
Phlox nivalis													❀														❀																
Phlox sp.		❀																									❀																❀
Phlox stolonifera																																				❀							❀
Phlyctidia ludoviciensis	*																																										
Phlyctis sp.		*						*							*				*		*			*											*	*							
Phlyctis willeyi	*																																						*				
Phoberia atomaris																																					●						
Phoebis sennae eubule		●				●	●	●	●			●		●	●	●			●	●	●			●	●	●	●		●					●	●	●	●		●				
Phoradendron serotinum				❀		❀	❀		❀				❀	❀	❀				❀					❀												❀	❀					❀	
Photinus pyralis	●	●																																									
Photuris sp.																																						●					
Phryma leptostachya		❀						❀			❀																									❀							❀
Phyciodes phaon							●																																				
Phyciodes sp.																																		●									
Phyciodes tharos		●	●					●			●	●			●	●	●	●		●		●		●		●	●					●	●	●		●		●			●	●	●
Phyllanthus sp.														❀																													
Phyllobius sp.																																		●									●
Phyllocoptes aceris-crumena																																											●
Phyllophaga grandis																																										●	
Phyllophaga ilicis																●																											

Scientific Name Index

	Abbeville	Aiken	Allendale	Anderson	Bamberg	Barnwell	Beaufort	Berkeley	Charleston	Cherokee	Chester	Chesterfield	Clarendon	Colleton	Darlington	Dillon	Dorchester	Fairfield	Florence	Georgetown	Greenville	Greenwood	Hampton	Horry	Jasper	Kershaw	Lancaster	Laurens	Lee	Lexington	Marlboro	McCormick	Newberry	Oconee	Orangeburg	Pickens	Richland	Spartanburg	Sumter	Union	York	Georgia	N. Carolina
Phyllophaga prunina																																●											
Phyllophaga sp.		●									●						●				●											●		●				●				●	●
Phylloporus rhodoxanthus																																											*
Phyllostachys aurea														❀																			❀										
Phylloxera caryae-venae																																				●							
Phylloxera caryaefoliae																																				●							
Phylloxera caryaeren																																										●	
Phylloxera caryoecaulis																											●																
Phylloxera caryoesemen																					●																						
Phylloxera globosa																																										●	
Phylloxera notabilis											●																																
Phymata fasciata		●																																									
Phymata pennsylvanica																																							●				
Physalacria inflata															*																												
Physalis sp.								❀																																			
Physalis virginiana virginiana		❀																																									
Physalis viscosa maritima							❀																																				
Physcia aipolia		*					*	*				*			*			*			*		*	*		*	*		*		*			*		*		*	*	*	*	*	*
Physcia alba		*																																							*		
Physcia americana	*						*		*		*			*	*				*	*	*		*	*								*		*	*	*	*	*	*		*	*	
Physcia cloantha						*																				*																	
Physcia crispa								*	*					*						*																							
Physcia halei																																									*	*	
Physcia millegrana		*					*	*	*		*			*	*			*		*	*			*		*			*			*		*	*			*		*	*	*	*
Physcia pseudospeciosa											*																							*		*		*					
Physcia sorediosa		*							*						*									*																			
Physcia sp.															*																												
Physcia stellaris											*													*										*				*		*			
Physcia subtilis																					*											*		*		*					*		
Physciella cloantha															*						*																	*				*	
Physconia detersa																					*					*								*				*				*	
Physconia enteroxantha							*					*		*							*					*			*							*	*		*			*	*
Physconia enteroxantha f. detersa																																		*									
Physella sp.																														●													
Physocarpus opulifolius																																						❀					
Phytolacca americana		❀		❀		❀		❀	❀		❀				❀						❀			❀		❀	❀		❀	❀	❀					❀	❀	❀			❀		
Phytopus cerasicrumena																					●																						
Picea abies																																											❀
Picea rubens																																											❀
Picoides borealis												●												●					●														
Picoides pubescens		●			●		●	●	●					●					●	●	●			●	●				●			●		●	●		●	●	●		●	●	●
Picoides villosus									●						●																												

Scientific Name Index

	Abbeville	Aiken	Allendale	Anderson	Bamberg	Barnwell	Beaufort	Berkeley	Charleston	Cherokee	Chester	Chesterfield	Clarendon	Colleton	Darlington	Dillon	Dorchester	Fairfield	Florence	Georgetown	Greenville	Greenwood	Hampton	Horry	Jasper	Kershaw	Lancaster	Laurens	Lee	Lexington	Marlboro	McCormick	Newberry	Oconee	Orangeburg	Pickens	Richland	Spartanburg	Sumter	Union	York	Georgia	N. Carolina
Pieris floribunda																																											❀
Pieris rapae																					●																		●				
Piesmus mondelus										●																																	
Pilea pumila		❀						❀	❀		❀															❀	❀														❀		❀
Pinguicula caerulea																													❀														
Pinguicula pumila																								❀																			
Pinna carnea									●																																		
Pinus clausa												❀																															
Pinus echinata	❀										❀	❀		❀	❀				❀		❀			❀			❀		❀		❀	❀		❀		❀	❀	❀	❀		❀	❀	
Pinus elliotti							❀					❀													❀					❀													
Pinus glabra							❀		❀					❀			❀						❀																				
Pinus palustris		❀					❀	❀	❀			❀		❀	❀	❀			❀	❀				❀	❀	❀			❀	❀							❀		❀				
Pinus pungens																																		❀									❀
Pinus rigida																					❀																						❀
Pinus serotina		❀				❀	❀					❀			❀								❀	❀					❀										❀				
Pinus sp.											❀																																
Pinus strobus																					❀													❀		❀		❀				❀	❀
Pinus taeda	❀	❀	❀	❀	❀	❀	❀	❀	❀	❀	❀		❀	❀	❀	❀		❀	❀	❀	❀	❀	❀	❀		❀	❀		❀	❀	❀	❀	❀	❀	❀		❀	❀	❀	❀		❀	
Pinus virginiana										❀		❀									❀													❀		❀	❀	❀		❀	❀	❀	❀
Pipilo erythrophthalmus		●					●		●		●	●	●	●	●			●	●	●	●			●	●	●	●		●		●			●		●	●		●		●		●
Piranga olivacea																					●													●								●	●
Piranga rubra	●	●			●			●	●		●				●							●	●	●		●	●		●			●		●	●		●		●				
Piriqueta caroliniana								❀															❀																				
Pisaurina mira						●	●																											●									
Pisaurina sp.																																		●									●
Placopecten clintonius																																											❖
Plagiotrochus batatoides									●																																		
Plagiotrochus punctatus																																											●
Plagiotrochus reticulatus																																				●							
Plagiotrochus sp.							●																																				
Plagiotrochus tumificus																					●																					●	
Planera aquatica		❀						❀						❀									❀	❀																			
Plantago aristata		❀																								❀																	
Plantago lanceolata	❀										❀															❀			❀							❀							
Plantago rugelii									❀																									❀									❀
Plantago sp.																																						❀					
Platanus occidentalis		❀	❀					❀		❀	❀				❀			❀	❀		❀			❀			❀					❀					❀	❀	❀	❀	❀		❀
Plateros lictor																																						●					
Plateros sollicitus																																				●							
Plateros sp.																																						●					
Plathemis lydia											●											●				●	●				●		●	●		●		●					
Platismatia tuckermanii																																		*									

Scientific Name Index

	Abbeville	Aiken	Allendale	Anderson	Bamberg	Barnwell	Beaufort	Berkeley	Charleston	Cherokee	Chester	Chesterfield	Clarendon	Colleton	Darlington	Dillon	Dorchester	Fairfield	Florence	Georgetown	Greenville	Greenwood	Hampton	Horry	Jasper	Kershaw	Lancaster	Laurens	Lee	Lexington	Marlboro	McCormick	Newberry	Oconee	Orangeburg	Pickens	Richland	Spartanburg	Sumter	Union	York	Georgia	N. Carolina
Platydema erythrocera								●							●																						●						
Platydema excavatum																								●																			
Platydema ruficorne															●																												
Platydema sp.								●																																			
Platypus flavicornis								●																																			
Platysoma carolinus							●								●																												
Platysoma lecontei															●																												
Plegadis falcinellus														●											●	●																	
Plethodon glutinosus														●	●																			●								●	
Plethodon jordani																					●																						
Pleurocybella porrigens																																											✻
Pleuroploca gigantea									●																																		
Pleurotis sp.			✻																																								
Pleurotus ostreatus																								✻								✻											
Pleurotus sapidus																																					✻						
Plexaura flexuosa									●																																		
Plicatula gibbosa									●																																		
Plicatula marginata																								❖																			
Pluchea camphorata		❀				❀								❀									❀											❀									
Pluchea foetida																								❀																			
Pluchea purpurascens							❀		❀																																		
Pluchea rosea								❀	❀							❀								❀					❀														
Pluchea sp.						❀																																					
Pluteus cervinus																					✻							✻						✻		✻							
Pluvialis squatarola							●		●											●																							
Poanes sp.																																		●									
Poanes zabulon		●																																									
Pocadius sp.		●																																									
Podabrus basilaris																																										●	
Podabrus sp.																												●															●
Podiceps auritus									●																																		
Podilymbus podiceps							●		●			●		●						●			●		●	●												●					
Podophyllum peltatum								❀													❀						❀	❀				❀		❀				❀	❀			❀	❀
Podostemum ceratophyllum																																		❀									
Pogonia ophioglossoides								❀																❀																			
Pogonias cromis									❖																																		
Pogonomyrmex badius							●												●																				●				
Polineces duplicatus									●																																		
Polinices duplicatus																								❖																			❖
Polioptila caerulea		●	●					●	●		●			●	●			●						●	●		●		●					●		●	●		●				●
Polistes sp.						●								●																								●			●		
Polycarpon tetraphyllum								❀																																			

Scientific Name Index

	Abbeville	Aiken	Allendale	Anderson	Bamberg	Barnwell	Beaufort	Berkeley	Charleston	Cherokee	Chester	Chesterfield	Clarendon	Colleton	Darlington	Dillon	Dorchester	Fairfield	Florence	Georgetown	Greenville	Greenwood	Hampton	Horry	Jasper	Kershaw	Lancaster	Laurens	Lee	Lexington	Marlboro	McCormick	Newberry	Oconee	Orangeburg	Pickens	Richland	Spartanburg	Sumter	Union	York	Georgia	N. Carolina
Polycentropus sp.																																		●									
Polygala cruciata												❀																	❀														
Polygala curtissii																			❀		❀						❀																
Polygala cymosa								❀																					❀														
Polygala grandiflora																							❀	❀					❀														
Polygala incarnata								❀				❀											❀																				
Polygala lutea						❀		❀				❀			❀									❀		❀			❀														
Polygala mariana																													❀														
Polygala paucifolia																																											❀
Polygala polygama		❀																																									
Polygala ramosa								❀				❀												❀																			
Polygala sp.																																		❀									
Polygala verticillata		❀																																									
Polygonatum biflorum											❀										❀						❀	❀				❀		❀		❀		❀				❀	❀
Polygonatum sp.																																											❀
Polygonella americana		❀																																									
Polygonella polygama												❀				❀			❀										❀										❀				
Polygonia comma																																		●									
Polygonia interrogationis																			●																		●	●					
Polygonia sp.						●			●																												●	●					
Polygonum arifolium														❀																													
Polygonum cespitosum longisetum																					❀																						
Polygonum convolvulus									❀																																		
Polygonum densiflorum									❀																																		
Polygonum hydropiperoides opelousanum																	❀																										
Polygonum lapathifolium											❀																																
Polygonum pensylvanicum											❀										❀																						
Polygonum punctatum									❀																																		
Polygonum sagittatum																					❀										❀			❀				❀					❀
Polygonum setaceum																																									❀		
Polygonum sp.		❀				❀		❀	❀					❀															❀									❀					
Polymnia uvedalia		❀					❀	❀																												❀							❀
Polypodium polypodioides	❀	❀			❀	❀	❀	❀	❀			❀		❀	❀	❀			❀		❀			❀		❀	❀		❀	❀			❀	❀		❀	❀	❀	❀	❀		❀	
Polypodium virginianum																																		❀								❀	❀
Polyporus radicatus																			*																								
Polyporus varius									*																																		
Polypremum procumbens	❀	❀						❀	❀		❀	❀				❀										❀			❀														
Polysporina simplex																					*																						
Polystichum acrostichoides	❀	❀						❀		❀	❀				❀				❀		❀	❀					❀	❀				❀	❀	❀	❀	❀	❀	❀	❀	❀	❀	❀	❀
Polytrichum commune																			*																								
Polytrichum sp.																					*																						
Poncirus trifoliata														❀																													

Scientific Name Index

	Abbeville	Aiken	Allendale	Anderson	Bamberg	Barnwell	Beaufort	Berkeley	Charleston	Cherokee	Chester	Chesterfield	Clarendon	Colleton	Darlington	Dillon	Dorchester	Fairfield	Florence	Georgetown	Greenville	Greenwood	Hampton	Horry	Jasper	Kershaw	Lancaster	Laurens	Lee	Lexington	Marlboro	McCormick	Newberry	Oconee	Orangeburg	Pickens	Richland	Spartanburg	Sumter	Union	York	Georgia	N. Carolina
Pontederia cordata								❀			❀			❀												❀																	
Popilius disjunctus	●	●	●				●		●		●				●		●				●						●	●						●			●	●	●		●		
Popillia japonica	●	●									●																						●	●		●							●
Populus deltoides			❀					❀			❀							❀				❀					❀			❀							❀	❀		❀	❀		
Populus heterophylla								❀			❀													❀					❀								❀						
Populus sp.																																											
Poria spissa																																		✻									
Porina chlorotica																			✻																								
Porina nucula								✻											✻															✻	✻								
Porina rhaphidosperma						✻																														✻							
Porina salicina																																		✻									
Porina sp.									✻														✻																				
Porpidia albocaerulescens																					✻															✻						✻	
Porpidia macrocarpa																											✻																✻
Portulaca coronata																											❀																
Portulaca oleracea									❀																																		
Porzana carolina							●		●									●																									
Potamogeton diversifolius																																						❀					
Potamogeton pulcher						❀			❀																						❀												
Potamogeton sp.		❀						❀																																			
Potentilla canadensis								❀							❀			❀	❀		❀			❀			❀					❀		❀		❀	❀	❀				❀	❀
Potentilla recta																					❀																						
Potentilla simplex																								❀								❀										❀	
Potentilla sp.																													❀					❀								❀	❀
Prasocuris vittata																		●																									
Prenanthes altissima																					❀																	❀					❀
Prenanthes serpentaria																					❀		❀	❀			❀																
Prenanthes sp.	❀																				❀						❀							❀		❀		❀					❀
Prenanthes trifoliolata																					❀																						
Prionocyphon discoideus																																		●									
Prionus laticollis														●																													
Pristis sp.																	❖																										
Prociphilus tesselatus				●				●													●			●																			
Procyon lotor		●					●	●	●	●				●	●								●	●	●	●	●		●					●	●			●	●		●		
Progne subis								●	●																																		
Prolinyphia marginata																●																		●									
Prometopia sexmaculata															●																												
Prosapia bicincta		●										●																							●								
Prosapia sp.																																					●						
Proserpinaca pectinata								❀															❀						❀														
Protococcus sp.																																						✻					
Protonotaria citrea		●						●							●		●										●					●		●			●						

Scientific Name Index

	Abbeville	Aiken	Allendale	Anderson	Bamberg	Barnwell	Beaufort	Berkeley	Charleston	Cherokee	Chester	Chesterfield	Clarendon	Colleton	Darlington	Dillon	Dorchester	Fairfield	Florence	Georgetown	Greenville	Greenwood	Hampton	Horry	Jasper	Kershaw	Lancaster	Laurens	Lee	Lexington	Marlboro	McCormick	Newberry	Oconee	Orangeburg	Pickens	Richland	Spartanburg	Sumter	Union	York	Georgia	N. Carolina
Prunella vulgaris	❀	❀									❀										❀					❀	❀						❀	❀		❀		❀					❀
Prunus americana						❀													❀		❀																						
Prunus angustifolia		❀						❀										❀																			❀	❀					
Prunus caroliniana		❀				❀	❀		❀						❀										❀	❀											❀		❀				
Prunus pensylvanica																																											❀
Prunus persica																																				❀							
Prunus serotina	❀	❀		❀		❀	❀		❀	❀	❀	❀	❀		❀	❀		❀	❀		❀	❀	❀	❀		❀	❀		❀	❀	❀	❀	❀	❀	❀	❀	❀	❀	❀	❀	❀	❀	❀
Prunus umbellata																										❀									❀				❀				
Prunus virginiana																																		❀									
Psathyrella sp.																																		*									
Psephenus herricki																																		●									
Pseudacris brimleyi							●																																				
Pseudacris sp.																													●														
Pseudacris triceriata																																●											
Pseudacris triseriata				●						●								●																						●	●		
Pseudevernia cladonia																																											*
Pseudevernia consocians																													*					*								*	*
Pseudobaris nigrina																																						●			●		
Pseudocephalaria aurata																											*																*
Pseudocephalaria crocata																											*																*
Pseudocyphellaria sp.																											*																
Pseudolucanus placidus											●																																
Pseudoparmelia amazonica															*											*																	
Pseudoparmelia baltimorensis																																*										*	
Pseudoparmelia caperata																							*	*	*	*			*					*				*		*			*
Pseudoparmelia caroliniana		*				*	*	*							*									*	*	*			*			*		*		*	*				*	*	*
Pseudoparmelia crozalsiana																										*														*			
Pseudoparmelia sp.										*																																	
Pseudoparmelia sphaerospora																							*																				
Psilotum nudum							❀																																				
Psinidia fenestralis									●																																		
Psoralea psoralioides																								❀					❀														
Psoralea psoralioides psoralioides																								❀																			
Psorophora ferox																			●																								
Psyllobora vigintimaculata																										●								●			●						
Ptelea trifoliata																																						❀					
Pteria colymbus									●																																		
Pteridium aquilinum	❀	❀			❀	❀	❀	❀	❀		❀	❀	❀	❀	❀	❀	❀		❀	❀	❀		❀	❀		❀	❀		❀	❀	❀			❀	❀	❀	❀		❀		❀	❀	❀
Pterocaulon pycnostachyum								❀																❀																			
Pterophylla camellifolia																																		●		●							
Pterorhytis conradi																																											❖
Pterostichus chalcites											●																																

Scientific Name Index

	Abbeville	Aiken	Allendale	Anderson	Bamberg	Barnwell	Beaufort	Berkeley	Charleston	Cherokee	Chester	Chesterfield	Clarendon	Colleton	Darlington	Dillon	Dorchester	Fairfield	Florence	Georgetown	Greenville	Greenwood	Hampton	Horry	Jasper	Kershaw	Lancaster	Laurens	Lee	Lexington	Marlboro	McCormick	Newberry	Oconee	Orangeburg	Pickens	Richland	Spartanburg	Sumter	Union	York	Georgia	N. Carolina
Pterostichus sp.		●				●											●											●										●					
Ptilimnium capillaceum		❀												❀								❀																					
Ptilodactyla serricollis																																											●
Ptilodactyla sp.																												●															
Pueraria lobata																	❀				❀	❀				❀					❀												
Pulcherrichium caeruleum		*																						*																			
Punctelia appalachensis												*											*																				
Punctelia bolliana																							*																*				
Punctelia hypoleucites																										*																	
Punctelia rudecta	*	*					*		*		*	*		*	*			*	*	*	*		*	*					*					*	*	*		*	*		*	*	*
Punctelia subrudecta																																		*									*
Pycnanthemum flexuosum																													❀														
Pycnanthemum incanum																																		❀		❀		❀					
Pycnanthemum muticum																								❀																			
Pycnanthemum sp.											❀																		❀														
Pycnoporus cinnabarinus																																				*							
Pycnospsyche divergens																																		●									
Pycnospsyche gentilis																																		●									
Pycnothelia papillaria											*	*						*			*						*							*				*					
Pycnothelia sp.																																					*						
Pygolampis pectoralis																													●														
Pyrenula cerina																							*																				
Pyrenula cruenta						*	*	*	*					*	*					*				*													*						
Pyrenula laevigata														*																		*		*									
Pyrenula mamillana							*	*																								*				*							
Pyrenula nitida															*																	*			*	*							
Pyrenula pseudobufonia																								*																			
Pyrenula sp.						*																															*		*				
Pyrgus communis			●													●																											
Pyromorpha dimidiata																					●													●									
Pyrota lineata		●																																									
Pyrrharctia isabella																										●												●					
Pyrrhopappus carolinianus									❀					❀										❀												❀		❀					
Pyrrhospora varians															*																												
Pyrularia pubera																					❀													❀		❀						❀	❀
Pyxidanthera barbulata												❀												❀																			
Pyxidanthera barbulata barbulata		❀										❀												❀																			
Pyxidanthera barbulata brevifolia												❀																															
Pyxine caesiopruinosa		*					*		*		*			*									*	*	*	*			*						*		*	*	*				
Pyxine caesiopruinosa f. aurea		*																		*			*																				
Pyxine eschweileri							*																																				
Pyxine sorediata									*		*			*	*			*	*		*													*		*		*					*

Scientific Name Index

	Abbeville	Aiken	Allendale	Anderson	Bamberg	Barnwell	Beaufort	Berkeley	Charleston	Cherokee	Chester	Chesterfield	Clarendon	Colleton	Darlington	Dillon	Dorchester	Fairfield	Florence	Georgetown	Greenville	Greenwood	Hampton	Horry	Jasper	Kershaw	Lancaster	Laurens	Lee	Lexington	Marlboro	McCormick	Newberry	Oconee	Orangeburg	Pickens	Richland	Spartanburg	Sumter	Union	York	Georgia	N. Carolina
Pyxine subcinerea																				*									*														
Quedius sp.													●		●																												
Quercus acutissima														❀																													
Quercus alba	❀	❀			❀	❀	❀	❀	❀		❀			❀	❀	❀			❀		❀	❀	❀	❀	❀		❀	❀	❀	❀	❀	❀	❀	❀	❀	❀	❀	❀	❀	❀	❀	❀	❀
Quercus coccinea	❀										❀								❀		❀								❀					❀		❀	❀	❀	❀		❀	❀	❀
Quercus falcata		❀				❀		❀			❀		❀	❀	❀	❀			❀		❀	❀	❀	❀		❀	❀	❀	❀	❀	❀	❀		❀		❀	❀	❀	❀	❀			
Quercus falcata falcata	❀		❀				❀				❀	❀																						❀	❀				❀		❀		
Quercus falcata pagodaefolia														❀				❀	❀	❀								❀									❀		❀				
Quercus incana								❀				❀			❀	❀								❀		❀			❀	❀													
Quercus laevis		❀						❀				❀	❀		❀	❀			❀	❀				❀	❀	❀	❀		❀	❀							❀		❀				
Quercus laurifolia		❀					❀	❀	❀			❀		❀	❀		❀		❀	❀			❀	❀					❀		❀				❀		❀	❀	❀	❀			
Quercus lyrata		❀				❀		❀				❀		❀					❀												❀						❀						
Quercus margaretta		❀										❀			❀				❀					❀		❀			❀										❀				
Quercus marilandica	❀					❀		❀	❀			❀	❀	❀	❀	❀								❀		❀	❀		❀	❀	❀			❀		❀	❀	❀			❀		
Quercus michauxii			❀		❀			❀						❀	❀	❀	❀		❀				❀	❀				❀	❀			❀					❀	❀					
Quercus muehlenbergii																												❀						❀									
Quercus nigra		❀	❀		❀	❀	❀	❀	❀		❀		❀	❀	❀	❀	❀	❀	❀	❀		❀	❀	❀		❀		❀	❀	❀	❀		❀		❀		❀	❀	❀	❀	❀	❀	
Quercus oglethorpensis																						❀																					
Quercus phellos	❀	❀	❀				❀		❀		❀			❀	❀	❀		❀	❀		❀	❀	❀	❀			❀		❀		❀	❀	❀				❀	❀	❀	❀	❀	❀	
Quercus prinus																					❀						❀							❀		❀			❀		❀	❀	❀
Quercus pumila								❀					❀				❀							❀					❀														
Quercus rubra	❀	❀									❀			❀	❀						❀	❀		❀							❀	❀	❀	❀		❀	❀	❀		❀		❀	❀
Quercus rubra rubra																																		●									
Quercus shumardii		❀						❀			❀																																
Quercus sp.																											❀																❀
Quercus stellata	❀	❀				❀	❀	❀	❀		❀	❀	❀	❀	❀			❀				❀	❀	❀	❀	❀	❀	❀	❀		❀	❀		❀		❀	❀	❀	❀		❀	❀	
Quercus velutina	❀	❀				❀								❀	❀							❀					❀			❀		❀		❀	❀		❀	❀				❀	
Quercus virginiana		❀				❀	❀	❀	❀					❀						❀				❀	❀										❀								
Quercus virginiana geminata		❀																					❀	❀																			
Quiscalus major							●	●	●					●						●					●																		
Quiscalus quiscula		●		●			●	●	●					●	●			●		●				●	●	●	●		●		●			●	●		●	●	●				
Raja eglanteria									●																																		
Raja sp.									●																																		
Rallus elegans																													●														
Rallus limicola														●						●																							
Rallus longirostris							●		●											●																							
Ramalina americana		*						*	*	*	*	*		*	*			*	*	*	*		*			*	*		*		*			*		*		*		*	*	*	*
Ramalina sp.																							*																				
Ramalina willeyi							*																		*	*			*														
Ramaria botrytis																																						*					
Ramaria formosa																											*										*						*
Ramaria stricta																																		*									

Scientific Name Index

	Abbeville	Aiken	Allendale	Anderson	Bamberg	Barnwell	Beaufort	Berkeley	Charleston	Cherokee	Chester	Chesterfield	Clarendon	Colleton	Darlington	Dillon	Dorchester	Fairfield	Florence	Georgetown	Greenville	Greenwood	Hampton	Horry	Jasper	Kershaw	Lancaster	Laurens	Lee	Lexington	Marlboro	McCormick	Newberry	Oconee	Orangeburg	Pickens	Richland	Spartanburg	Sumter	Union	York	Georgia	N. Carolina
Ramariopsis kunzei																																							✻					
Rana catesbiana							●	●						●												●	●																●
Rana clamitans	●	●									●			●		●	●									●					●		●			●			●				
Rana grylio														●																													
Rana palustris																																					●						●
Rana sphenocephala						●	●	●						●																		●				●					
Rana virgatipes																										●	●																
Ranatra americana		●																																								
Ranatra fusca								●																																		
Ranatra sp.									●																	●																
Rangia clathrodonta																								❖																		
Rangia cuneata																								❖																		
Ranunculus abortivus															❀			❀															❀											❀
Ranunculus bulbosus																																		❀									
Ranunculus hispidus																																		❀								❀	❀
Ranunculus recurvatus																																		❀									❀
Ranunculus sardous																																					❀						
Ranunculus sp.																																❀											❀
Raphanus raphanistrum							❀														❀			❀													❀						
Recurvirostra americana									●																																	
Regulus calendula			●	●		●	●	●	●					●	●			●					●	●	●		●							●				●	●		●		●
Regulus satrapa						●			●					●						●					●	●											●	●					
Resupinatus applicatus																																		✻									
Rhabdopterus picipes											●																															
Rhabdopterus sp.								●																																		
Rhagium inquisitor												●			●																											
Rhagovelia obesa																																		●									
Rhamnus caroliniana		❀						❀																													❀				❀	
Rhapidophyllum hystrix							❀																																			
Rhexia alifanus								❀				❀													❀					❀														
Rhexia lutea								❀																																		
Rhexia mariana		❀				❀		❀				❀				❀								❀	❀				❀		❀												
Rhexia sp.				❀					❀															❀																		
Rhexia virginica		❀										❀																														
Rhexia virginica virginica												❀																			❀											
Rhizocarpon sp.																											✻																✻
Rhodobaenus tredecimpunctatus											●																															
Rhododendron atlanticum								❀																❀																		
Rhododendron calendulaceum																																											❀
Rhododendron canescens		❀				❀								❀								❀	❀											❀								
Rhododendron catawbiense																																		❀								❀	❀
Rhododendron maximum																					❀													❀		❀						❀	❀

Scientific Name Index

	Abbeville	Aiken	Allendale	Anderson	Bamberg	Barnwell	Beaufort	Berkeley	Charleston	Cherokee	Chester	Chesterfield	Clarendon	Colleton	Darlington	Dillon	Dorchester	Fairfield	Florence	Georgetown	Greenville	Greenwood	Hampton	Horry	Jasper	Kershaw	Lancaster	Laurens	Lee	Lexington	Marlboro	McCormick	Newberry	Oconee	Orangeburg	Pickens	Richland	Spartanburg	Sumter	Union	York	Georgia	N. Carolina
Rhododendron minus															❀						❀													❀		❀		❀					❀
Rhododendron nudiflorum													❀		❀						❀			❀			❀	❀	❀	❀				❀		❀							
Rhododendron sp.	❀					❀									❀			❀	❀								❀					❀		❀				❀	❀		❀		❀
Rhododendron viscosum															❀																			❀					❀				
Rhododendron viscosum serrulatum		❀																																									
Rhopalomyia castaneae																																										●	
Rhopalomyia solidaginis							●																																				
Rhopalomyia sp.		●																																									
Rhus aromatica																											❀	❀														❀	
Rhus copallina	❀	❀	❀		❀	❀	❀	❀	❀	❀	❀	❀		❀	❀	❀	❀		❀	❀		❀	❀	❀	❀	❀	❀		❀		❀		❀	❀	❀	❀	❀	❀	❀	❀			
Rhus glabra	❀			❀							❀										❀						❀							❀		❀		❀					
Rhus michauxii																			❀																								
Rhus radicans	❀	❀	❀	❀	❀	❀	❀	❀	❀	❀	❀		❀	❀	❀		❀		❀		❀	❀	❀	❀		❀	❀	❀	❀	❀		❀	❀	❀	❀	❀	❀	❀	❀	❀	❀	❀	❀
Rhus toxicodendron	❀	❀							❀		❀	❀	❀	❀	❀	❀				❀	❀			❀		❀			❀	❀				❀	❀	❀							
Rhus typhina																					❀													❀									
Rhus vernix		❀				❀		❀				❀			❀						❀		❀	❀																			
Rhyacophila fuscula																																		●									
Rhynchites geneus																													●														
Rhynchops niger							●		●																																		
Rhynchosia reniformis								❀								❀				❀																							
Rhynchosia sp.											❀																																
Rhynchospora cephalantha																													❀														
Rhynchospora chalarocephala												❀																			❀												
Rhynchospora corniculata																													❀									❀					
Rhynchospora debilis																								❀																			
Rhynchospora filifolia																													❀														
Rhynchospora globularis																								❀																			
Rhynchospora glomerata																															❀												
Rhynchospora grayi																									❀																		
Rhynchospora inexpansa																❀															❀												
Rhynchospora intermixta																								❀																			
Rhynchospora macrostachya								❀																																			
Rhynchospora megalocarpa																								❀																			
Rhynchospora perplexa																													❀														
Rhynchospora plumosa																								❀																			
Rhynchospora rariflora																								❀																			
Rhynchospora sp.					❀																										❀												
Rhynchospora torreyana																													❀														
Ribes echinellum																																❀											
Riccia fluitans							*																																				
Ricciocarpus natans							*		*															*																			
Richardia brasiliensis		❀				❀	❀									❀													❀														

Scientific Name Index

	Abbeville	Aiken	Allendale	Anderson	Bamberg	Barnwell	Beaufort	Berkeley	Charleston	Cherokee	Chester	Chesterfield	Clarendon	Colleton	Darlington	Dillon	Dorchester	Fairfield	Florence	Georgetown	Greenville	Greenwood	Hampton	Horry	Jasper	Kershaw	Lancaster	Laurens	Lee	Lexington	Marlboro	McCormick	Newberry	Oconee	Orangeburg	Pickens	Richland	Spartanburg	Sumter	Union	York	Georgia	N. Carolina
Richardia scabra		❀										❀																	❀														
Richardia sp.																													❀														
Rickenella fibula																											*																
Rimelia cetrata											*							*	*										*					*	*				*				
Rimelia cetrata f. nonreticulata																		*																							*		
Rimelia reticulata	*	*					*		*		*	*		*	*			*	*		*		*	*					*					*	*	*		*	*		*	*	*
Rimelia subisidiosa									*		*			*						*	*			*	*				*									*	*				
Rinodina milvina																																											*
Rinodina roboris									*					*									*																				
Rinodina sophodes							*								*																												
Rinodina sp.																													*													*	
Rinodina tephraspis																					*													*									
Riparia riparia		●																																									
Robinia hispida	❀																				❀													❀									
Robinia pseudo-acacia		❀									❀										❀													❀		❀					❀	❀	❀
Robinia sp.																								❀																			
Rorippa islandica																																❀											
Rosa carolina																																		❀		❀							
Rosa laevigata									❀																	❀																	
Rosa micrantha				❀																																							
Rosa multiflora																					❀																					❀	
Rosa palustris				❀				❀						❀							❀													❀									
Rosa sp.																		❀																									❀
Rubus argutus		❀									❀													❀										❀									
Rubus bifrons																															❀												
Rubus canadensis																																											❀
Rubus odoratus																																											❀
Rubus phoenicolasius																																											❀
Rubus sp.	❀	❀							❀		❀	❀		❀	❀				❀		❀			❀		❀			❀					❀			❀		❀	❀		❀	❀
Rubus trivialis							❀		❀					❀	❀					❀				❀	❀				❀														
Rudbeckia hirta	❀																								❀																		
Rudbeckia laciniata																											❀							❀									❀
Rudbeckia sp.																																			❀								❀
Ruellia caroliniensis		❀									❀													❀									❀		❀	❀		❀	❀				
Ruellia purshiana																																						❀					
Ruellia sp.																						❀																❀					
Rugilus dentatus															●																												
Rumex acetosella																					❀																						
Rumex crispus				❀																																							
Rumex sp.									❀																																		
Rumex verticillatus								❀																																			
Ruppia maritima														❀																													

Scientific Name Index

	Abbeville	Aiken	Allendale	Anderson	Bamberg	Barnwell	Beaufort	Berkeley	Charleston	Cherokee	Chester	Chesterfield	Clarendon	Colleton	Darlington	Dillon	Dorchester	Fairfield	Florence	Georgetown	Greenville	Greenwood	Hampton	Horry	Jasper	Kershaw	Lancaster	Laurens	Lee	Lexington	Marlboro	McCormick	Newberry	Oconee	Orangeburg	Pickens	Richland	Spartanburg	Sumter	Union	York	Georgia	N. Carolina
Russula aeruginea									*												*						*							*				*					
Russula claroflava																					*						*																*
Russula emetica	*	*					*							*							*			*					*					*				*	*				*
Russula fragilis	*																							*			*							*				*					
Russula kromholzii																											*																
Russula lutea																													*														
Russula mariae																							*														*						
Russula pectinatoides																													*														
Russula rosacea																																		*									
Russula sp.							*		*																																		
Russula variata	*																																					*	*				
Russula virescens		*																																									
Sabal minor		❀	❀				❀	❀	❀					❀	❀								❀		❀												❀		❀				
Sabal palmetto							❀		❀					❀						❀																							
Sabatia angularis																																		❀									
Sabatia calycina					❀																																						
Sabatia campanulata																													❀														
Sabatia difformis																													❀														
Sabatia sp.											❀																																
Sabatia stellaris																				❀																							
Sacciolepis striata		❀					❀																❀																❀				
Sageretia minutiflora									❀																																		
Sagina decumbens			❀																																								
Sagittaria engelmanniana																															❀												
Sagittaria falcata														❀																													
Sagittaria fasciculata																					❀																						
Sagittaria graminea graminea								❀						❀																													
Sagittaria latifolia											❀																										❀						
Sagittaria latifolia pubescens																																		❀									
Sagittaria sp.									❀			❀																															
Salicornia bigelovii							❀		❀																																		
Salicornia europaea									❀																																		
Salicornia sp.									❀																																		
Salicornia virginica							❀		❀											❀																							
Salix caroliniana									❀											❀					❀																❀		
Salix humilis																													❀														
Salix nigra		❀		❀		❀		❀	❀		❀			❀			❀	❀	❀				❀			❀			❀	❀	❀	❀		❀				❀					
Salix sp.							❀	❀	❀								❀							❀										❀									❀
Salsola kali									❀											❀																							
Salvelinus fontinalis																																											●
Salvia lyrata											❀		❀	❀							❀						❀	❀				❀	❀	❀			❀	❀					❀
Salvia sp.																											❀																

Scientific Name Index

	Abbeville	Aiken	Allendale	Anderson	Bamberg	Barnwell	Beaufort	Berkeley	Charleston	Cherokee	Chester	Chesterfield	Clarendon	Colleton	Darlington	Dillon	Dorchester	Fairfield	Florence	Georgetown	Greenville	Greenwood	Hampton	Horry	Jasper	Kershaw	Lancaster	Laurens	Lee	Lexington	Marlboro	McCormick	Newberry	Oconee	Orangeburg	Pickens	Richland	Spartanburg	Sumter	Union	York	Georgia	N. Carolina
Salvia urticifolia																											❀																
Sambucus canadensis	❀	❀				❀	❀	❀	❀		❀			❀	❀		❀	❀	❀				❀	❀	❀	❀	❀					❀	❀	❀			❀		❀	❀	❀		❀
Sambucus sp.																																										❀	
Samolus parviflorus								❀			❀												❀	❀													❀						
Sanguinaria canadensis	❀					❀	❀	❀		❀																	❀	❀				❀		❀	❀	❀		❀	❀			❀	❀
Sanicula canadensis						❀															❀						❀							❀		❀							❀
Sanicula gregaria								❀																																			
Sanicula marilandica	❀																																										
Sanicula smallii	❀																																										
Sanicula sp.									❀		❀																	❀				❀						❀	❀				
Sapium sebiferum								❀	❀					❀											❀																		
Sarcographa sp.								*												*							*																
Sargassum fluitans									*																																		
Sarracenia flava								❀				❀	❀											❀					❀														
Sarracenia minor								❀				❀												❀					❀														
Sarracenia purpurea												❀												❀																			❀
Sarracenia rubra												❀												❀																			❀
Sassafras albidum	❀	❀			❀	❀	❀	❀	❀		❀	❀	❀	❀	❀	❀	❀		❀	❀	❀	❀	❀	❀		❀	❀		❀	❀	❀	❀	❀	❀	❀	❀	❀	❀	❀	❀	❀	❀	❀
Satureja georgiana		❀												❀	❀																	❀											
Satyrium calanus																																				●							
Saururus cernuus		❀	❀		❀		❀	❀			❀			❀	❀	❀							❀	❀		❀		❀	❀		❀						❀		❀				
Saxifraga michauxii																																											❀
Saxifraga micranthidifolia																																											❀
Saxifraga virginiensis																																❀											❀
Saxinus omogera																								●																			
Sayornis phoebe				●		●	●	●	●		●			●	●			●		●	●		●	●	●	●	●							●				●	●			●	●
Scalopus aquaticus							●																																				
Scaphisoma convexa																																					●						
Scaphisoma suturalis																																						●					
Scapteriscus acletus																	●																										
Scapteriscus sp.									●																																		
Scarites sp.																																●											
Scarites subterraneus																																					●						
Sceliphron caementarium																					●																						
Sceloporus undulatus	●										●	●				●			●		●	●				●	●		●	●				●		●			●				●
Schismatomma rediunta									*																																		
Schismatomma sp.																							*																				
Schistocerca alutacea		●							●			●																															
Schistocerca americana						●													●					●		●			●													●	
Schistocerca obscura									●																																		
Schistocerca sp.																●																											
Schizophyllum commune		*						*					*								*	*		*				*						*		*	*	*					*

Scientific Name Index

	Abbeville	Aiken	Allendale	Anderson	Bamberg	Barnwell	Beaufort	Berkeley	Charleston	Cherokee	Chester	Chesterfield	Clarendon	Colleton	Darlington	Dillon	Dorchester	Fairfield	Florence	Georgetown	Greenville	Greenwood	Hampton	Horry	Jasper	Kershaw	Lancaster	Laurens	Lee	Lexington	Marlboro	McCormick	Newberry	Oconee	Orangeburg	Pickens	Richland	Spartanburg	Sumter	Union	York	Georgia	N. Carolina
Schizoporella errata									●																																		
Schrankia microphylla		❀							❀			❀									❀			❀										❀		❀							
Scincella lateralis	●	●	●			●	●	●	●			●			●		●							●		●		●				●	●					●	●				
Scirpus americanus																								❀																			
Scirpus cyperinus		❀						❀										❀					❀	❀		❀					❀			❀									
Scirpus koilolepis		❀																																									
Scirpus purshianus																																		❀									
Scirpus robustus														❀																													
Scirtes orbiculatus																															●												
Scirtes sp.		●																																									
Sciurus carolinensis		●				●	●		●										●		●	●				●			●		●			●	●	●	●		●		●		
Sciurus niger		●					●		●											●																			●				
Scleria ciliata																								❀																			
Scleria minor																								❀																			
Scleria oligantha	❀	❀																																									
Scleria sp.							❀										❀																	❀									
Scleria triglomerata							❀		❀							❀								❀					❀						❀	❀			❀				
Sclerodactyla briareus									●																																		
Scleroderma citrinum																																				*		*	*				*
Scleroderma geaster			*																																		*						
Scleroderma sp.																					*								*														
Scleroderma vulgare																					*																						
Scoliciosporum umbrinum							*																																				
Scolopax minor						●																																					
Scomber sp.									●																																		
Scrophularia marilandica								❀																																			
Scudderia furcata	●																				●																						
Scudderia sp.																																		●				●					
Scudderia texensis												●																															
Scutellaria elliptica	❀	❀																						❀																			
Scutellaria integrifolia		❀																			❀						❀		❀														
Scutellaria lateriflora																❀																											
Scutellaria sp.						❀								❀		❀									❀		❀							❀				❀	❀				
Scutellinia scutellata																					*													*		*							
Scymnus puncticollis																																				●							
Scymnus sp.		●																										●															●
Scymnus terminatus								●																																			
Sebastiania ligustrina														❀			❀																										
Sedum pusillum																											❀															❀	
Sedum smallii																											❀															❀	
Sedum ternatum										❀	❀																❀																❀
Seiurus aurocapillus																					●						●											●				●	●

Scientific Name Index

	Abbeville	Aiken	Allendale	Anderson	Bamberg	Barnwell	Beaufort	Berkeley	Charleston	Cherokee	Chester	Chesterfield	Clarendon	Colleton	Darlington	Dillon	Dorchester	Fairfield	Florence	Georgetown	Greenville	Greenwood	Hampton	Horry	Jasper	Kershaw	Lancaster	Laurens	Lee	Lexington	Marlboro	McCormick	Newberry	Oconee	Orangeburg	Pickens	Richland	Spartanburg	Sumter	Union	York	Georgia	N. Carolina
Seiurus motacilla																																											●
Selaginella apoda						❀												❀																❀			❀						
Selaginella arenicola												❀				❀								❀					❀	❀													
Selaginella rupestris																					❀															❀							
Selaginella tortipila																					❀																					❀	
Selenophorus ellipticus															●																												
Semiothisa minorata																																		●									
Semiothisa sp.																																		●									
Senecio aureus																																										❀	❀
Senecio glabellus								❀																				❀				❀					❀						
Senecio millefolium																					❀																						
Senecio obovatus																																❀											
Senecio smallii		❀													❀						❀					❀	❀							❀			❀					❀	❀
Senecio sp.															❀						❀																						
Senecio tomentosus																											❀									❀						❀	
Senecio vulgaris				❀																																							
Septastrea crassa																																											❖
Serenoa repens							❀																		❀																		
Serica sp.												●																				●		●									
Serpula sp.																								✻																			
Sesarma reticulatum									●																																		
Sesbania exaltata									❀					❀									❀																				
Sesuvium portulacastrum																				❀																							
Setaria glauca				❀																																							
Setaria magna									❀					❀										❀																			
Setaria sp.									❀																						❀							❀					
Setaria viridis																									❀																		
Setophaga ruticilla																				●	●													●									
Seymeria cassioides												❀				❀													❀														
Sherardia arvensis												❀																															
Shortia galacifolia																																		❀									❀
Sialia sialis		●		●			●		●			●	●	●					●		●			●		●			●					●	●		●	●					
Sibine stimulea																								●																			
Sicyos angulatus		❀																									❀																
Sida rhombifolia						❀									❀											❀			❀		❀												
Sigmodon hispidus																									●											●							
Sigmoria aberrans	●										●																					●			●			●					●
Silene caroliniana																											❀												❀				
Silene stellata																																		❀									❀
Silene virginica								❀																																		❀	
Silis latilobus												●																															
Silis sp.																																		●									●

Scientific Name Index

	Abbeville	Aiken	Allendale	Anderson	Bamberg	Barnwell	Beaufort	Berkeley	Charleston	Cherokee	Chester	Chesterfield	Clarendon	Colleton	Darlington	Dillon	Dorchester	Fairfield	Florence	Georgetown	Greenville	Greenwood	Hampton	Horry	Jasper	Kershaw	Lancaster	Laurens	Lee	Lexington	Marlboro	McCormick	Newberry	Oconee	Orangeburg	Pickens	Richland	Spartanburg	Sumter	Union	York	Georgia	N. Carolina
Silpha americana																																					●						
Silpha inaequalis															●																												
Silpha noveboracensis																																					●						
Silpha surinamensis																																						●					
Silphium astericus					❀																																					
Silphium compositum													❀			❀	❀					❀			❀											❀							❀	❀	❀
Silphium dentatum	❀	❀																																								
Silphium sp.											❀															❀			❀														
Silvanus bidentatus																																		●								
Simulium sp.																														●												
Sinum perspectivum									●																																	
Sisyrinchium albidum																																❀					❀					
Sisyrinchium angustifolium																																❀		❀								
Sisyrinchium mucronatum		❀														❀							❀																				
Sisyrinchium mucronatum atlanticum									❀							❀																											
Sisyrinchium sp.			❀																							❀		❀									❀					
Sitta canadensis																																							●				
Sitta carolinensis							●	●							●								●			●											●		●	●				●		●
Sitta pusilla							●		●			●	●						●				●		●	●		●										●					
Sium suave								❀								❀																													
Smilacina racemosa		❀										❀										❀						❀								❀		❀						❀	❀	❀
Smilax auriculata									❀																																	
Smilax bona-nox	❀	❀		❀	❀	❀	❀	❀	❀		❀			❀	❀		❀		❀	❀				❀								❀				❀	❀	❀	❀	❀		❀		
Smilax ecirrhata						❀																																					
Smilax ecirrhata biltmoreana																					❀															❀								
Smilax ecirrhata hugeri																											❀									❀									
Smilax glauca	❀	❀		❀	❀	❀					❀			❀	❀			❀	❀				❀			❀					❀	❀		❀		❀			❀			❀	❀
Smilax herbacea											❀																															❀	❀
Smilax laurifolia		❀			❀	❀	❀	❀				❀	❀		❀								❀	❀		❀	❀		❀								❀	❀	❀			❀	
Smilax pumila		❀					❀		❀					❀									❀																					
Smilax rotundifolia		❀					❀	❀			❀			❀	❀				❀		❀			❀		❀					❀					❀		❀	❀				
Smilax smallii		❀						❀						❀				❀				❀	❀	❀						❀				❀				❀				
Smilax sp.											❀			❀	❀						❀								❀						❀							❀	❀	
Smilax walteri						❀																																	❀				
Solanum americanum																				❀																						
Solanum carolinense									❀												❀				❀			❀		❀						❀					
Solanum sp.									❀																																	
Solenopsis geminata														●	●																											
Solenopsis invicta						●			●					●				●									●							●			●					
Solenopsis sp.									●					●						●				●					●								●						
Solidago altissima															❀															❀													
Solidago fistulosa																							❀																			

Scientific Name Index

	Abbeville	Aiken	Allendale	Anderson	Bamberg	Barnwell	Beaufort	Berkeley	Charleston	Cherokee	Chester	Chesterfield	Clarendon	Colleton	Darlington	Dillon	Dorchester	Fairfield	Florence	Georgetown	Greenville	Greenwood	Hampton	Horry	Jasper	Kershaw	Lancaster	Laurens	Lee	Lexington	Marlboro	McCormick	Newberry	Oconee	Orangeburg	Pickens	Richland	Spartanburg	Sumter	Union	York	Georgia	N. Carolina
Solidago glomerata																																											❀
Solidago microcephala																										❀																	
Solidago nemoralis												❀									❀		❀													❀		❀					
Solidago odora												❀												❀		❀			❀		❀			❀	❀				❀		❀		
Solidago pauciflosculosa												❀																															
Solidago rugosa																							❀			❀																	
Solidago sempervirens									❀											❀																							
Solidago sp.	❀	❀							❀					❀		❀			❀		❀			❀					❀	❀	❀			❀				❀	❀				
Solidago stricta												❀																															
Solidago tenuifolia									❀																																		
Somatochlora linearis																																							●				
Sonchus asper																	❀																										
Sonchus oleraceus								❀																																			
Sonchus sp.																																				❀							
Sorbus americana																																											❀
Sorbus arbutifolia								❀				❀	❀		❀	❀			❀					❀		❀			❀	❀				❀				❀			❀		❀
Sorbus arbutifolia arbutifolia		❀				❀																	❀			❀																	
Sorghastrum elliottii																																						❀					
Sorghastrum sp.							❀					❀																															
Sorghum halepense	❀			❀																																		❀					
Sparassis sp.																																		*									
Sparassis spathulata																													*														
Sparganium americanum		❀							❀							❀								❀							❀			❀					❀				
Spartina alterniflora							❀		❀					❀						❀																							
Spartina cynosuroides									❀					❀																													
Spartina patens							❀		❀					❀						❀																							
Specularia biflora		❀																																									
Specularia perfoliata		❀																																			❀						❀
Specularia sp.																																				❀							
Speyeria cybele																																		●									●
Sphaerium partumeium																								●																			
Sphagnum sp.									*				*		*	*								*																			*
Spharagemon bolli	●	●										●			●														●					●									
Spharagemon collare					●																																						
Spharagemon cristatum	●																																										
Sphenophorus sp.								●																																			
Sphodros bicolor																																							●				
Sphyrapicus varius			●				●		●	●				●										●													●	●	●		●		
Spigelia marilandica		❀						❀																										❀			❀						
Spilanthes americana repens		❀						❀																													❀						
Spilosoma virginica												●																															
Spiranthes cernua						❀						❀							❀																								

Scientific Name Index

	Abbeville	Aiken	Allendale	Anderson	Bamberg	Barnwell	Beaufort	Berkeley	Charleston	Cherokee	Chester	Chesterfield	Clarendon	Colleton	Darlington	Dillon	Dorchester	Fairfield	Florence	Georgetown	Greenville	Greenwood	Hampton	Horry	Jasper	Kershaw	Lancaster	Laurens	Lee	Lexington	Marlboro	McCormick	Newberry	Oconee	Orangeburg	Pickens	Richland	Spartanburg	Sumter	Union	York	Georgia	N. Carolina
Spiranthes cernua cernua																										❀																	
Spiranthes ovalis																																						❀					
Spiranthes praecox																								❀					❀														
Spiranthes sp.		❀																						❀						❀													
Spirodela oligorrhiza															❀	❀																											
Spirodela punctata														❀		❀	❀						❀			❀	❀								❀								
Spirodela sp.								❀																																			
Spizella passerina							●		●		●	●		●				●	●		●			●		●			●					●				●					●
Spizella pusilla											●																		●														
Spongipellis pachydon								*																																			
Sporobolus poiretii		❀							❀																				❀										❀				
Sporobolus sp.							❀					❀																										❀					
Sporobolus virginicus							❀		❀																																		
Stachys sp.															❀									❀																			
Stagmomantis carolina				●		●	●			●	●																				●			●				●		●			
Staphylea trifolia		❀									❀																❀	❀				❀											
Staphylinus cinnamopterus																																											●
Staphylinus fossator																										●								●									
Staphylinus maxillosus							●																																		●		
Staphylinus sp.		●																																									
Statira sp.																																●											
Staurothele diffractella							*																																				
Steccherinum ochraceum																																		*		*		*					
Stelgidopteryx serripennis		●																								●								●									
Stelidota sp.																																									●		
Stellaria media		❀	❀				❀		❀					❀	❀																	❀				❀					❀		❀
Stellaria pubera																											❀					❀										❀	❀
Stemonitis splendens							*																						*					*									
Stenanthium gramineum																																										❀	
Stenodontes dasytomus																							●																				
Stenolophus ochropezus												●																															
Stenotarsus sp.																												●															
Stereocaulon dactylophyllum																											*																
Stereocaulon pileatum																											*							*									
Stereocaulon sp.																																											*
Stereum complicatum		*	*				*		*															*				*	*			*		*				*					
Stereum ostrea			*																					*										*					*				
Stereum striatum					*																			*														*					
Sterna caspia							●		●					●						●																							
Sterna forsteri							●		●					●						●					●																		
Sterna maxima							●		●																																		
Sternotherus odoratus					●										●																												

Scientific Name Index

	Abbeville	Aiken	Allendale	Anderson	Bamberg	Barnwell	Beaufort	Berkeley	Charleston	Cherokee	Chester	Chesterfield	Clarendon	Colleton	Darlington	Dillon	Dorchester	Fairfield	Florence	Georgetown	Greenville	Greenwood	Hampton	Horry	Jasper	Kershaw	Lancaster	Laurens	Lee	Lexington	Marlboro	McCormick	Newberry	Oconee	Orangeburg	Pickens	Richland	Spartanburg	Sumter	Union	York	Georgia	N. Carolina
Stewartia malacodendron																								❀															❀				
Stewartia ovata																					❀																						
Sticta fuliginosa																											*																
Sticta weigelii																											*							*		*					*	*	*
Stillingia sylvatica																									❀																		
Stipa avenacea		❀						❀	❀												❀						❀							❀		❀							
Stipulicida setacea		❀										❀				❀			❀							❀			❀										❀				
Strangalina bicolor																																				●							
Strangalina famelica																																				●							
Strangalina luteicornis																																				●							
Strategus antaeus																				●																							
Strategus sp.												●										●																					
Strix varia		●													●									●													●					●	●
Strobilomyces floccopus	*																												*					*									*
Strobilurus conigenoides									*																											*							
Strongylium sp.		●																																									
Strophostyles helvola									❀											❀																							
Strymon melinus							●													●	●			●			●							●									
Sturnella magna	●			●					●					●											●																		
Sturnus vulgaris							●		●											●					●																		
Stylosanthes biflora	❀				❀						❀					❀							❀	❀		❀	❀		❀				❀			❀		❀					
Styrax americana								❀																❀		❀					❀												
Styrax grandifolia	❀							❀			❀													❀				❀				❀				❀	❀				❀		
Styrax sp.						❀																																					
Suaeda linearis							❀		❀											❀																							
Suillus americanus																																											*
Suillus granulatus																																		*									
Suillus luteus																								*																			
Suillus pictus																													*														*
Sunius sp.																																		●									
Suphisellus sp.								●																																			
Sus scrofa		●	●						●					●																				●			●						
Sylvilagus floridanus		●																																			●					●	
Sylvilagus palustris														●						●																							
Sylvilagus sp.		●					●								●																												
Sympetrum sp.																													●										●				
Sympetrum vicinum															●																												
Symphoricarpos orbiculatus																																❀		❀						❀			
Symplocos tinctoria		❀				❀	❀		❀		❀	❀	❀	❀	❀	❀	❀		❀		❀		❀	❀		❀			❀	❀				❀	❀	❀	❀		❀		❀	❀	❀
Synchlora sp.																																		●									
Syneta sp.																																										●	
Syrbula admirabilis												●																										●					

Scientific Name Index

	Abbeville	Aiken	Allendale	Anderson	Bamberg	Barnwell	Beaufort	Berkeley	Charleston	Cherokee	Chester	Chesterfield	Clarendon	Colleton	Darlington	Dillon	Dorchester	Fairfield	Florence	Georgetown	Greenville	Greenwood	Hampton	Horry	Jasper	Kershaw	Lancaster	Laurens	Lee	Lexington	Marlboro	McCormick	Newberry	Oconee	Orangeburg	Pickens	Richland	Spartanburg	Sumter	Union	York	Georgia	N. Carolina
Systena blanda																																							●				
Systena elongata											●			●																													
Systena frontalis		●																																									
Systena taeniata																												●															
Tachinus axillaris																																											
Tachinus pallipes				●									●																														
Tachinus sp.															●									●																			
Tachycineta bicolor							●		●					●				●		●				●	●																		
Tachys ephippiatus															●																												
Tachyta nana inornata															●																												
Tagelus gibbus																								❖																			
Tagelus plebeius									●																																		
Talinum teretifolium																											❀									❀						❀	
Tamarix gallica									❀																																		
Tamias striatus																					●													●									●
Tamiasciurus hudsonicus																																											●
Tanacetum vulgare																					❀																						
Taphrina alni-incanae				*																																							
Taphrocerus sp.								●																																		●	
Tarachidia semiflava								●																																			
Taraxacum officinale									❀												❀										❀			❀		❀			❀				❀
Tarpela micans																																										●	●
Taxodium ascendens					❀			❀											❀						❀	❀			❀		❀												
Taxodium distichum		❀	❀		❀			❀	❀					❀	❀	❀			❀	❀			❀	❀		❀					❀				❀		❀		❀				
Teesdalia nudicaulis												❀																															
Telebasis byersi														●																													
Tenebroides dubia																																											●
Tenebroides laticollis															●																												
Tenodera aridifolia sinensis				●														●												●								●		●			
Tephromela atra														*							*								*													*	
Tephrosia hispidula								❀																❀																			
Tephrosia sp.					❀																			❀	❀																		
Tephrosia spicata	❀	❀														❀										❀							❀										
Tephrosia virginiana								❀				❀																	❀	❀													
Terebra sp.									●																																		
Terebratulina sp.																																											
Terrapene carolina						●					●			●							●		●				●					●		●		●	●	●			●	●	●
Tetracis crocallata																																		●									
Tetragnatha elongata						●																																	●				
Tetragnatha laboriosa												●												●		●	●							●								●	
Tetragnatha sp.																											●				●	●	●		●			●					●
Tetraonyx sp.																													●														

	Abbeville	Aiken	Allendale	Anderson	Bamberg	Barnwell	Beaufort	Berkeley	Charleston	Cherokee	Chester	Chesterfield	Clarendon	Colleton	Darlington	Dillon	Dorchester	Fairfield	Florence	Georgetown	Greenville	Greenwood	Hampton	Horry	Jasper	Kershaw	Lancaster	Laurens	Lee	Lexington	Marlboro	McCormick	Newberry	Oconee	Orangeburg	Pickens	Richland	Spartanburg	Sumter	Union	York	Georgia	N. Carolina
Tettigidea lateralis																																			●								
Teucrium canadense							❀		❀																																		
Thalictrum clavatum																																		❀									❀
Thalictrum dioicum																																										❀	
Thalictrum revolutum											❀																																
Thalictrum sp.																					❀											❀		❀		❀							❀
Thalictrum thalictroides		❀																❀									❀	❀				❀		❀	❀	❀			❀				
Thamnophis sirtalis																					●					●																	●
Thanasimus Dubius															●																												
Thaspium barbinode	❀							❀																				❀														❀	
Thaspium trifoliatum																																											❀
Thaspium trifoliatum flavum																																					❀						
Thecodiplosis ananassi																										●																	
Thelephora sp.	*																																										
Thelia bimaculata																																									●		
Thelophora terrestris												*																															
Thelotrema monosporum							*																				*																
Thelotrema sp.							*																																				
Thelypteris hexagonoptera	❀																											❀						❀		❀						❀	❀
Thelypteris kunthii								❀						❀																							❀						
Thelypteris noveboracensis																					❀													❀		❀					❀	❀	❀
Thelypteris palustris																																											❀
Thelypteris sp.														❀																													❀
Theridiosoma radiosa																					●																						
Thermonectes basilaris																																					●						
Thermonetus basilaris basilaris																																											
Thorybes bathyllus														●																													●
Thryothorus ludovicianus	●	●	●		●		●	●		●	●	●	●	●		●	●		●	●			●		●	●		●			●	●	●	●	●	●	●	●		●		●	
Thyonella gemmata									●																																		
Thyridopteryx ephemeraeformis									●									●			●			●						●							●						
Tiarella cordifolia																											❀							❀		❀							❀
Tiarella cordifolia collina																											❀					❀					❀						
Tibicen canicularis		●																																									
Tibicen sp.																							●											●									
Tilia americana	❀																																									❀	
Tilia caroliniana		❀																														❀											
Tilia heterophylla								❀																										❀		❀	❀						❀
Tilia sp.																					❀											❀		❀	❀	❀							❀
Tillandsia usneoides		❀	❀			❀	❀	❀	❀					❀	❀	❀	❀		❀	❀			❀	❀	❀				❀						❀		❀		❀				
Tilloclytus geminata																																										●	
Tipularia discolor	❀					❀	❀			❀	❀				❀	❀			❀		❀		❀	❀			❀		❀	❀		❀		❀		❀	❀	❀	❀	❀		❀	❀
Tofieldia glabra												❀																															

Scientific Name Index

	Abbeville	Aiken	Allendale	Anderson	Bamberg	Barnwell	Beaufort	Berkeley	Charleston	Cherokee	Chester	Chesterfield	Clarendon	Colleton	Darlington	Dillon	Dorchester	Fairfield	Florence	Georgetown	Greenville	Greenwood	Hampton	Horry	Jasper	Kershaw	Lancaster	Laurens	Lee	Lexington	Marlboro	McCormick	Newberry	Oconee	Orangeburg	Pickens	Richland	Spartanburg	Sumter	Union	York	Georgia	N. Carolina
Tofieldia racemosa								❀				❀																															
Tovara virginiana		❀																									❀							❀									❀
Toxostoma rufum				●			●		●						●						●	●		●	●	●			●						●			●	●			●	
Toxotus cinnamopterus																																●		●									
Trachelospermum difforme														❀																		❀	❀						❀				
Trachycardium auricatum									●																																		
Tradescantia hirsuticaulis		❀																														❀										❀	
Tradescantia rosea		❀													❀		❀							❀															❀				
Tradescantia rosea graminea					❀											❀																											
Tradescantia sp.																											❀					❀				❀							
Tradescantia subaspera																												❀						❀									
Tradescantia virginiana																																										❀	❀
Tragia sp.											❀																									❀							
Tragia urens																❀																											
Tragia urticifolia		❀																																									
Tramea carolina	●	●						●	●												●			●		●			●		●			●									
Tramea lacerata		●																																					●				
Tramea sp.								●	●											●									●														
Trametes versicolor	✻	✻																						✻					✻	✻				✻			✻	✻	✻				
Tremella mesenterica	✻																							✻										✻						✻			
Tremellodendron pallidum																																						✻					
Trepocarpus aethusae		❀																																									
Trichaptum biformis	✻													✻														✻				✻		✻				✻					✻
Trichiotinus piger								●																●				●															
Trichodezia albovittata																																											●
Trichoglossum geoglossum																																		✻									
Trichoglossum hirsutum																																						✻					
Trichomanes boschianum																																											❀
Trichostema dichotomum							❀								❀					❀			❀			❀												❀					
Tridens flavus							❀																				❀											❀			❀		
Trifolium arvense																																				❀							
Trifolium campestre																																		❀									
Trifolium carolinianum																											❀																
Trifolium dubium																					❀																						
Trifolium pratense																					❀																						
Trifolium repens																																						❀					❀
Trigonopeltastes delta								●	●																											●			●				
Trigonopeltastes sp.																																						●					
Trigonostoma tenera																																											❖
Trilisa odoratissima							❀																																				
Trilisa paniculata												❀																															
Trilisa sp.																								❀																			

Scientific Name Index

	Abbeville	Aiken	Allendale	Anderson	Bamberg	Barnwell	Beaufort	Berkeley	Charleston	Cherokee	Chester	Chesterfield	Clarendon	Colleton	Darlington	Dillon	Dorchester	Fairfield	Florence	Georgetown	Greenville	Greenwood	Hampton	Horry	Jasper	Kershaw	Lancaster	Laurens	Lee	Lexington	Marlboro	McCormick	Newberry	Oconee	Orangeburg	Pickens	Richland	Spartanburg	Sumter	Union	York	Georgia	N. Carolina
Trillium catesbaei																					❀						❀	❀						❀		❀						❀	❀
Trillium cernuum																											❀	❀				❀		❀									❀
Trillium cuneatum							❀	❀																										❀								❀	❀
Trillium cuneatum luteum																																		❀									❀
Trillium discolor																																❀				❀							❀
Trillium erectum																																		❀								❀	❀
Trillium erectum vaseyi																					❀															❀							❀
Trillium grandiflorum																																											❀
Trillium lanceolatum																																❀											❀
Trillium sp.																																											❀
Trillium undulatum																																											❀
Trillium vaseyi																																		❀								❀	
Trimerotropis saxatilis																																				●							
Tringa flavipes									●					●				●													●												
Tringa melanoleuca							●		●					●																													
Triodopsis albolabris						●																													●								●
Triodopsis juxtidens	●	●																																									
Trionyx sp.																			❖																								
Trioza diospyri																													●														
Triphora trianthophora							❀																																				❀
Triplasis americana												❀																															
Triplasis purpurea							❀																																				
Triplax macra	●																																										
Triplax sp.																																		●									
Triplax thoracica																												●															●
Tripsacum dactyloides									❀																																		
Tritoma angulata																											●																
Tritoma sp.						●																																					
Tritoma thoracica			●																																								
Trocholomopsis sulfureoides							✻																																				
Troglodytes aedon									●											●					●																		
Trombidium sp.										●											●						●					●		●		●							●
Tropisternus blatchleyi blatchleyi												●																															
Tropisternus collaris striolatus		●										●					●																					●					
Tropisternus lateralis nimbatus							●							●															●									●				●	
Tropisternus mixtus								●									●																										
Tropisternus natator															●												●											●					
Tropisternus quadristriatus quadristriatus									●		●													●																			
Tropisternus striolatus		●																																									
Trox sp.															●																												
Trox tuberculatus															●																												
Trox tuberculosus																								●																			

Scientific Name Index

	Abbeville	Aiken	Allendale	Anderson	Bamberg	Barnwell	Beaufort	Berkeley	Charleston	Cherokee	Chester	Chesterfield	Clarendon	Colleton	Darlington	Dillon	Dorchester	Fairfield	Florence	Georgetown	Greenville	Greenwood	Hampton	Horry	Jasper	Kershaw	Lancaster	Laurens	Lee	Lexington	Marlboro	McCormick	Newberry	Oconee	Orangeburg	Pickens	Richland	Spartanburg	Sumter	Union	York	Georgia	N. Carolina
Trypethelium floridanum								*						*									*	*													*						
Trypethelium gigantheum sp. nov.																							*																				
Trypethelium mastoideum		*					*	*	*										*				*	*										*	*		*	*	*				
Trypethelium scorites								*				*							*															*									
Trypethelium sp.												*														*						*					*			*			
Trypethelium tropicum						*		*				*											*						*						*					*			
Trypethelium virens						*	*	*	*					*	*			*	*		*			*		*	*					*		*	*		*	*	*				
Trypoxylon politum								●										●												●											●		●
Tsuga canadensis																					❀													❀		❀						❀	❀
Tsuga caroliniana																					❀																						❀
Tsuga sp.																																		❀									
Tuckermannopsis aurescens																																											*
Tuckermannopsis ciliaris																					*													*								*	
Tuckermannopsis fendleri															*																											*	
Tuckermannopsis oakesiana		*												*							*													*								*	*
Tuckermannopsis pinastri																																											*
Tuckermannopsis viridis												*									*										*										*		
Tulostoma simulans																																		*									
Turdus migratorius						●	●	●	●		●	●		●	●				●		●	●	●	●	●									●	●			●	●			●	●
Turritella perexilis																								❖																			
Turritella sp.																			❖					❖																			❖
Tursiops truncatus									●											●																							
Tylopilus felleus																																		*									
Tylopilus gracilis																					*																						
Tylopilus plumbeoviolaceus																											*											*					
Typha angustifolia									❀																																		
Typha latifolia		❀	❀			❀	❀	❀	❀		❀			❀	❀	❀	❀	❀		❀		❀	❀		❀	❀				❀	❀			❀									
Typocerus sp.												●															●	●	●							●	●						
Typocerus velutinus																																				●			●				
Tyrannus tyrannus								●																					●					●	●								●
Uca pugilator									●											●																							
Uca pugnax							●		●											●																							
Uca sp.							●		●											●																							
Ulmus alata	❀	❀	❀					❀		❀	❀			❀				❀			❀	❀				❀	❀					❀	❀				❀	❀	❀	❀	❀	❀	❀
Ulmus americana		❀	❀					❀			❀			❀					❀				❀			❀						❀					❀	❀		❀			
Ulmus rubra		❀						❀			❀																	❀						❀	❀			❀			❀		❀
Ulmus sp.																																			❀								
Uloma imberbis							●																															●					
Uloma punctulata						●									●	●		●																									
Ulva lactuca							*																																				
Ulva sp.									*																																		
Umbilicaria mammulata																											*															*	*

Scientific Name Index

	Abbeville	Aiken	Allendale	Anderson	Bamberg	Barnwell	Beaufort	Berkeley	Charleston	Cherokee	Chester	Chesterfield	Clarendon	Colleton	Darlington	Dillon	Dorchester	Fairfield	Florence	Georgetown	Greenville	Greenwood	Hampton	Horry	Jasper	Kershaw	Lancaster	Laurens	Lee	Lexington	Marlboro	McCormick	Newberry	Oconee	Orangeburg	Pickens	Richland	Spartanburg	Sumter	Union	York	Georgia	N. Carolina
Umbra pygmaea																								●																			
Uniola latifolia	❀	❀	❀					❀			❀								❀								❀		❀								❀	❀		❀	❀		
Uniola laxa			❀			❀	❀									❀								❀				❀															
Uniola paniculata							❀		❀											❀																							
Uniola sessiliflora	❀	❀					❀		❀																		❀		❀	❀						❀		❀					
Uniola sp.																								❀																			
Vaccinium tenellum								❀							❀									❀					❀										❀				
Vaccinium vacillans	❀																				❀								❀					❀		❀						❀	
Valerianella radiata																											❀																
Valgus sp.																												●															
Valvata bicarinata																														●													
Vanessa atalanta rubria	●										●					●								●							●	●		●			●						
Vanessa cardui																													●														
Vanessa virginiensis						●			●		●						●							●					●				●	●			●	●					●
Veratrum parviflorum																																										❀	
Veratrum sp.																																											❀
Veratrum viride																																		❀									❀
Verbascum blattaria		❀																										❀															
Verbascum thapsus			❀				❀	❀			❀													❀		❀	❀				❀			❀	❀	❀	❀			❀			
Verbena braziliensis		❀						❀	❀					❀						❀		❀		❀	❀	❀			❀		❀												
Verbena scabra							❀																																				
Verbena sp.		❀																																				❀					
Verbena urticifolia																								❀																			
Verbesina alternifolia																																									❀		
Verbesina occidentalis								❀																																			
Verbesina sp.		❀									❀																					❀						❀					❀
Verbesina virginica									❀																									❀									
Verbesina virginica virginica								❀																																			
Vermicularia knorrii									●																																		
Vermivora peregrina																					●																						
Vernonia acaulis												❀																															
Vernonia angustifolia																								❀	❀																		
Vernonia angustifolia angustifolia																❀										❀			❀														
Vernonia glauca																																		❀									
Vernonia sp.		❀									❀								❀																			❀					
Veronica arvensis									❀						❀																												
Veronica officinalis																																											❀
Veronica sp.																																											
Verpa bohemica																																				*							
Verrucosa arenata	●					●					●													●			●											●					
Vespa maculata														●																				●									
Vespa sp.																																		●									

Scientific Name Index

	Abbeville	Aiken	Allendale	Anderson	Bamberg	Barnwell	Beaufort	Berkeley	Charleston	Cherokee	Chester	Chesterfield	Clarendon	Colleton	Darlington	Dillon	Dorchester	Fairfield	Florence	Georgetown	Greenville	Greenwood	Hampton	Horry	Jasper	Kershaw	Lancaster	Laurens	Lee	Lexington	Marlboro	McCormick	Newberry	Oconee	Orangeburg	Pickens	Richland	Spartanburg	Sumter	Union	York	Georgia	N. Carolina
Vespula maculata						●	●	●																																		●	
Vespula maculifrons																																		●									
Vespula sp.						●													●		●									●						●					●		
Viburnum acerifolium															❀						❀							❀		❀						❀						❀	❀
Viburnum cassinoides																																		❀									
Viburnum dentatum								❀						❀					❀																		❀	❀					
Viburnum dentatum lucidum																			❀		❀										❀												
Viburnum nudum	❀	❀				❀						❀	❀	❀	❀	❀							❀	❀					❀					❀					❀				
Viburnum obovatum								❀									❀		❀																								
Viburnum prunifolium	❀	❀						❀			❀											❀					❀	❀					❀			❀		❀		❀			❀
Viburnum rufidulum																																❀		❀	❀								
Viburnum sp.										❀									❀															❀						❀			❀
Vicia caroliniana																																										❀	
Vicia hugeri																																		❀									
Vicia sp.																		❀			❀																						
Vinca major									❀					❀	❀										❀																		
Vinca minor																																						❀	❀				
Viola affinis																																							❀				
Viola blanda																					❀													❀		❀							❀
Viola canadensis																																		❀								❀	❀
Viola eriocarpa leiocarpa																																											❀
Viola hastata										❀											❀						❀							❀		❀						❀	❀
Viola lanceolata							❀						❀											❀																			
Viola palmata triloba																																		❀		❀						❀	
Viola papilionacea																		❀						❀								❀		❀		❀							❀
Viola pedata	❀											❀									❀						❀							❀		❀							❀
Viola primulifolia	❀								❀			❀		❀	❀						❀			❀			❀							❀							❀		❀
Viola rafinesquii														❀																													
Viola rostrata																					❀																						❀
Viola rotundifolia																					❀													❀								❀	❀
Viola septemloba																																		❀									
Viola sp.		❀			❀			❀	❀					❀													❀		❀					❀						❀		❀	❀
Viola spp.																																											
Viola tripartita tripartita																																				❀							
Viola walteri																																				❀							
Vireo flavifrons															●						●						●												●				
Vireo griseus		●	●				●	●	●					●		●		●						●			●										●		●			●	
Vireo olivaceus	●	●						●	●		●				●		●										●	●				●	●	●		●	●		●		●	●	●
Vireo solitarius							●		●					●							●													●		●		●			●	●	●
Vitis aestivalis										❀				❀	❀	❀	❀			❀				❀				❀						❀	❀	❀	❀		❀	❀			
Vitis labrusca									❀										❀		❀																	❀					
Vitis rotundifolia	❀	❀	❀			❀	❀	❀	❀		❀	❀	❀	❀	❀	❀	❀		❀	❀	❀	❀	❀	❀		❀	❀	❀	❀		❀	❀	❀	❀	❀	❀	❀	❀	❀		❀	❀	

Scientific Name Index

	Abbeville	Aiken	Allendale	Anderson	Bamberg	Barnwell	Beaufort	Berkeley	Charleston	Cherokee	Chester	Chesterfield	Clarendon	Colleton	Darlington	Dillon	Dorchester	Fairfield	Florence	Georgetown	Greenville	Greenwood	Hampton	Horry	Jasper	Kershaw	Lancaster	Laurens	Lee	Lexington	Marlboro	McCormick	Newberry	Oconee	Orangeburg	Pickens	Richland	Spartanburg	Sumter	Union	York	Georgia	N. Carolina
Vitis sp.		❀									❀			❀																							❀						
Volucella sp.																																					●						
Volutifusus typus																																											❖
Vulpes vulpes																																											●
Wahlenbergia marginata		❀										❀			❀											❀			❀		❀				❀								
Waldsteinia fragarioides																																		❀									
Wilsonia canadensis																																					●						
Wilsonia citrina		●																			●						●					●	●	●		●	●					●	●
Wisteria frutescens		❀				❀		❀	❀				❀	❀	❀				❀				❀								❀								❀				
Wisteria sinensis		❀				❀	❀		❀											❀																							
Wolffia papulifera						❀										❀																											
Wolffia sp.									❀																		❀																
Wolffiella floridana						❀		❀	❀					❀			❀			❀			❀			❀																	
Woodsia obtusa																														❀		❀											
Woodwardia areolata		❀			❀	❀	❀	❀	❀			❀		❀	❀	❀			❀		❀		❀	❀		❀	❀		❀				❀	❀	❀		❀		❀				
Woodwardia virginica					❀	❀	❀	❀	❀			❀		❀									❀	❀					❀										❀				
Wynnea americana																																		✻									
Xanthium strumarium				❀											❀																												
Xanthium strumarium glabratum									❀																																		
Xanthochroa sp.								●																																			
Xanthoparmelia conspersa																		✻			✻						✻									✻						✻	✻
Xanthoparmelia plittii		✻																			✻											✻				✻							✻
Xanthoparmelia sp.												✻																															
Xanthoparmelia subramigera		✻													✻						✻		✻		✻									✻									✻
Xanthoparmelia tasmanica	✻																																			✻							
Xanthorhiza simplicissima															❀						❀						❀							❀		❀			❀		❀	❀	❀
Xanthoria candelaria																										✻																✻	
Xeromphalina campanella																																				✻							
Xestocis sp.																								●																			
Xylaria hypoxylon																																						✻					
Xylaria polymorpha																													✻			✻				✻							
Xyloborus inermis																																					●						
Xylocopa sp.																	●																										
Xylocopa virginica																								●														●					
Xylographa disseminata																											✻																✻
Xyloryctes jamaicensis																																										●	
Xyris caroliniana																❀																											
Xyris flabelliformis																								❀																			
Xyris sp.		❀				❀		❀				❀	❀			❀													❀	❀	❀												
Xysticus elegans																													●														
Xysticus sp.											●																																
Yucca aloifolia							❀		❀																																		

Scientific Name Index

	Abbeville	Aiken	Allendale	Anderson	Bamberg	Barnwell	Beaufort	Berkeley	Charleston	Cherokee	Chester	Chesterfield	Clarendon	Colleton	Darlington	Dillon	Dorchester	Fairfield	Florence	Georgetown	Greenville	Greenwood	Hampton	Horry	Jasper	Kershaw	Lancaster	Laurens	Lee	Lexington	Marlboro	McCormick	Newberry	Oconee	Orangeburg	Pickens	Richland	Spartanburg	Sumter	Union	York	Georgia	N. Carolina
Yucca filamentosa									❀		❀		❀		❀	❀			❀					❀		❀	❀		❀	❀						❀	❀		❀			❀	
Yucca filamentosa filamentosa		❀					❀												❀				❀			❀																	
Yucca gloriosa								❀	❀											❀																							
Yucca sp.																															❀												
Zanthoxylum clava-herculis							❀	❀	❀											❀																							
Zelus exsanguis																			●																●								
Zenaida macroura		●					●	●	●			●						●		●	●			●					●		●			●			●						●
Zenobia pulverulenta																❀								❀		❀			❀														
Zephyranthes atamasco							❀								❀																											❀	
Zigadenus densus																								❀					❀														
Zigadenus glaberrimus												❀																															
Zizania aquatica														❀																													
Zizaniopsis miliacea								❀																	❀	❀																	
Zizia aurea																					❀															❀	❀						❀
Zizia sp.															❀																												
Zizia trifoliata																											❀					❀		❀		❀							❀
Zonotrichia albicollis			●				●		●					●	●			●																				●			●		●
Zonotrichia leucophrys				●														●																									
Zornia bracteata								❀								❀				❀																			❀				
Zostera marina																																					❀						
Zymogramma suturalis		●																																									

SC Association of Naturalists (SCAN)
PO Box 5651, Columbia, SC 29250-5651

MINI MEMBERSHIP APPLICATION

SCAN is composed of a wide and diverse membership with many varied interests, but most of all, we are people who have an amatur or professional interest in the natural history of South Carolina. Members must be 18 years of age or older.

Dues are $12.00 for individuals, with an extra $5.00 for each additional member sharing the same newsletter. New members, who join after July 1st, only pay for half a year ($6.00 for individuals/$2.50 for each additional) since they will be rejoining in January. Membership runs on a calendar year.

Date: ______________ County: ____________________

Amount Enclosed: ______________ Ck #: ____________

Name #1: ______________________________________

Name #2: ______________________________________

Name #3: ______________________________________

Address: ______________________________________

PO Box: _______________________________________

City: ________________ State: ____ Zip: ___________

Phone: _(________)_____________________________

E-Mail: _______________________________________

SC Association of Naturalists (SCAN)
PO Box 5651, Columbia, SC 29250-5651

MINI MEMBERSHIP APPLICATION

SCAN is composed of a wide and diverse membership with many varied interests, but most of all, we are people who have an amatur or professional interest in the natural history of South Carolina. Members must be 18 years of age or older.

Dues are $12.00 for individuals, with an extra $5.00 for each additional member sharing the same newsletter. New members, who join after July 1st, only pay for half a year ($6.00 for individuals/$2.50 for each additional) since they will be rejoining in January. Membership runs on a calendar year.

Date: ______________ County: ____________________

Amount Enclosed: ______________ Ck #: ____________

Name #1: ______________________________________

Name #2: ______________________________________

Name #3: ______________________________________

Address: ______________________________________

PO Box: _______________________________________

City: ________________ State: ____ Zip: ___________

Phone: _(________)_____________________________

E-Mail: _______________________________________

MAIL IN AN APPLICATION AND JOIN THE FUN. LEARN ABOUT SOUTH CAROLINA'S NATURAL HISTORY.

SC Association of Naturalists (SCAN)
PO Box 5651, Columbia, SC 29250-5651

MINI MEMBERSHIP APPLICATION

SCAN is composed of a wide and diverse membership with many varied interests, but most of all, we are people who have an amatur or professional interest in the natural history of South Carolina. Members must be 18 years of age or older.

Dues are $12.00 for individuals, with an extra $5.00 for each additional member sharing the same newsletter. New members, who join after July 1st, only pay for half a year ($6.00 for individuals/$2.50 for each additional) since they will be rejoining in January. Membership runs on a calendar year.

Date: ______________ County: ____________________

Amount Enclosed: ______________ Ck #: ____________

Name #1: ______________________________________

Name #2: ______________________________________

Name #3: ______________________________________

Address: ______________________________________

PO Box: _______________________________________

City: ________________ State: ____ Zip: ___________

Phone: _(________)_____________________________

E-Mail: _______________________________________

SC Association of Naturalists (SCAN)
PO Box 5651, Columbia, SC 29250-5651

MINI MEMBERSHIP APPLICATION

SCAN is composed of a wide and diverse membership with many varied interests, but most of all, we are people who have an amatur or professional interest in the natural history of South Carolina. Members must be 18 years of age or older.

Dues are $12.00 for individuals, with an extra $5.00 for each additional member sharing the same newsletter. New members, who join after July 1st, only pay for half a year ($6.00 for individuals/$2.50 for each additional) since they will be rejoining in January. Membership runs on a calendar year.

Date: ______________ County: ____________________

Amount Enclosed: ______________ Ck #: ____________

Name #1: ______________________________________

Name #2: ______________________________________

Name #3: ______________________________________

Address: ______________________________________

PO Box: _______________________________________

City: ________________ State: ____ Zip: ___________

Phone: _(________)_____________________________

E-Mail: _______________________________________

Errata

Please make note of the following corrections to the preceeding text:

Page vii: Should be "Edisto Beach State Park" (not Edisto State Park).

Page xii: Should be "Edisto Beach State Park" (not BeachPark)

Page 37: Caption to bottom photos should be "Hairy flower bees" (not Hairy bees).

Page 97: Footer should be "Caesars Head State Park" (not Eva Russell Caesars Head).

Page 104: The writing spider photograph is upside down.

Page 140: Photo caption should be *"Gelsemium sempervirens"* (not *semppervirens*).

Page 144: Golden-silk spider photograph is upside down.

Page 155: Footer should be "Burrells Ford" (not Burrels).

Pages 201 & 202: Black Rock Mountain State Park, GA. was an overnight trip. Dates should be May 21-22, 1994 (not May 21, 1994).

Page 211: Horse Pasture River, first paragraph, second line should be "property (not Property).

Pages 213 & 214: John Rock, Black Balsam Knob, and Cradle of Forestry. Both trips were overnights. First date should be "April 24-25, 1993" (not April 24, 1993).

Page 224: Bottom, left photo caption. Delete the second "female".

Page 233: Bottom, right photo cation should be "Grass-pink" (not Pink-grass).

Page 235: Bottom, left photo should be *"Dionaea muscipula"* (not *Dioneaea*)

Page 248: SC Department of Natural Resources (half way down the list) should be "Eva Russell Chandler" (not Russel).

Page 249: Address for SC Forestry Commission should be "5500 Broad River Road (not Broard).

Page 251: Kaston, should be "The Picture Key" (not Pictue).

Page 252 Niering, should be "Alfred Knopf" (not Knoft).

Page 283: Paragraph 2, line 1 should be "index are" (not indexare).